COMPETITIVE STRATEGIES IN
EUROPEAN BANKING

COMPETITIVE STRATEGIES IN EUROPEAN BANKING

Jordi Canals

CLARENDON PRESS · OXFORD
1993

Oxford University Press, Walton Street, Oxford OX2 6DP
Oxford New York Toronto
Delhi Bombay Calcutta Madras Karachi
Kuala Lumpur Singapore Hong Kong Tokyo
Nairobi Dar es Salaam Cape Town
Melbourne Auckland Madrid
and associated companies in
Berlin Ibadan

Oxford is a trade mark of Oxford University Press

Published in the United States
by Oxford University Press Inc., New York

British Library Cataloguing in Publication Data
Data available

Library of Congress Cataloging in Publication Data
Canals, Jordi.
[Estrategias del sector bancario en Europa. English]
Competitive strategies in European banking / Jordi Canals.
Includes bibliographical references and index.
1. Banks and banking—Europe. 2. Finance—Europe. I. Title.
HG2974.C3613 1992 332.1'094—dc20 92-28440
ISBN 0–19–877349–8

Typeset by Best-set Typesetter Ltd., Hong Kong
Printed in Great Britain
on acid-free paper by
Bookcraft (Bath) Ltd.,
Midsomer Norton, Avon

To My Parents

Foreword to the English Edition

A number of words about the banking business have been used so often that they have become clichés. Words such as deregulation, technology, and internationalization appear in almost everything written about the industry. However, their frequent use does not reflect a casual choice by authors. It reflects fundamental changes underway in the business.

Professor Canals provides a useful service in this book by analysing the changes taking place and their implications for the strategies of European banks. In past years banking strategy was something of an oxymoron. Because of geographic and product constraints, the choices available to senior bank managers were often more tactical than strategic in nature. This has changed dramatically.

European and other banks now face major strategic challenges and choices. As Professor Canals describes, senior management now must make important decisions about size or scale in their various businesses, about the market segments they wish to serve, and whether to focus on particular products or to provide a wide product range to capture economies of scope in their delivery system. To take an important example, changes in technology have greatly lowered the cost of certain kinds of transactions and, at the same time, made possible large economies of scale. Thus, banks must decide whether to grow those businesses to achieve the economies of scale or be driven out by competitors who do build to take advantage of the technology.

In an important sense, Professor Canals's book is kind to bankers. His excellent discussion of strategy is written in the context of banks becoming better, more efficient competitors. But, as he discusses, new competitors have emerged with different cost structures and different approaches to the market. These new competitors, together with the changing technology and other factors, will have a major impact on the nature and structure of financial institutions in the future. For some banks, it will not be enough to develop a more coherent strategy and become more efficient. This makes the lessons of Professor Canals's book all the more important.

Dwight B. Crane

George Gund Professor of Finance and Banking
Harvard Business School
May 1992

Foreword to the Spanish Edition

Few sectors of the European economy have undergone such profound changes in recent years as the banking industry. These changes, however, are in line with those that have taken place in other industrialized countries. Nevertheless, in the years to come further important structural alterations will no doubt take place in the role of banking in developed countries. In this context, the existence of works such as this book by Professor Jordi Canals is of incalculable value, not only for understanding what is happening now, but also in order to prepare the strategy of change for the immediate future.

The 1970s marked the start of a transformation process in the banking industry that owed much of its initial thrust both to the deregulation and liberalization of banking activities in the individual countries and to its growing internationalization. In addition, central banks have taken on a growing role in supervising the activities of financial institutions in order to ensure their solvency. All these forces for change have brought about greater competition between institutions, more effective performance, a better service to customers, and major changes in the banking business as such.

The internationalization of banking, which had previously been limited by geography, has been one of the main features of this change. This internationalization has been particularly marked in the last few years within the EC countries, in preparation for what will be from 1993 the single market for financial services. The transformations that have taken place in the European banking sector to come to terms with the new situation are highly expressive of this internationalization process, as Professor Canals's book shows. The sector is moving towards an unprecedented redrawing of frontiers, and the final outcome is still not clear in all its aspects.

An understanding of the evolution of the European banking systems is essential if we are to take steps in the right direction in this new situation. Part II of the book offers valuable information on the current trends in the main EC countries. This is one of the most valuable contributions that Professor Canals, a well-known authority in the banking industry, makes through this book.

I believe that this book will not only be a useful teaching tool (a profession of which the author is one of its most renowned exponents) but will also guide and advise bankers on the coming changes. Professor Canals must be thanked for his effort and his useful conclusions.

The Spanish banking industry, which has acquired an undeniable strength in recent years, is at this moment involved in a process of change and preparation for the European challenge of 1993. The authorities are trying to lift the obstacles that could prevent effective competition in the Single Market, mainly through the abolition of cash ratios. Banking institutions are finding themselves operating in an increasingly difficult market at the same time as their financial costs have increased considerably, partly as a result of greater competition, and they are justifiably concerned at the imminent disappearance of all barriers within the EC in 1993. In such a situation, better organization, reduced operating costs, and improved quality of service are the arms that will be vital to effectively fight the battle in the new European market. Professor Canals's book will be of major help to bankers in making their job easier, as well as providing a significant contribution to the body of recently published research work on the banking industry.

José Ramón Álvarez Rendueles
Former Governor of the Bank of Spain

Madrid
1990

Acknowledgements

This work could not have been written without the help of a number of people, not least my colleagues at IESE (International Graduate School of Management), including Eduard Ballarín, Jordi Gual, and Joan E. Ricart.

The first chapter of this book is inspired by a work completed while I was at the Brookings Institution (Washington, DC) as Guest Scholar. I owe thanks to my colleagues at Brookings, Ralph Bryant, Robert Lawrence, and Robert Litan, for their comments on a number of the principal ideas contained in this section.

The majority of the study was carried out at Harvard Business School during 1989–90, while I was there as Post-Doctoral Fellow. The excellent intellectual environment characteristic of that institution provided the perfect stimulus to complete this book. I owe thanks to John McArthur, Dean of the Business School, for the opportunity to conduct this study, as well as for his continuous intellectual support. I have had the opportunity to discuss various chapters with Professors Chris Bartlett, Dick Caves, David Collis, Dwight Crane, Herman Daems, Pankaj Ghemawat, John Goodman, Sam Hayes, David Meershawm, Mike Porter, Mike Rukstad, and Mike Yoshino, all of whom provided very valuable comments. Professors Dick Dooley, Ron Fox, Jack Gabarro, Jay Light, and Tom Piper were responsible for helping me to maintain a fast pace in my work as well as providing me with much assistance and support.

Professors Arnoldo Hax, Don Lessard, and Franco Modigliani, of MIT, and John Zysman of the University of California, Berkeley, offered excellent suggestions to improve the final draft.

I am very grateful to IESE and its Board of Directors, especially to IESE's Dean, Carlos Cavallé, for their constant encouragement to undertake this initiative.

The Spanish Ministry of Education and Science and the Fulbright Foundation deserve a special place in this section of thanks, for the financial help which enabled me to complete this project.

A previous version of this book was published in Spain by Editorial Ariel. Eric Svenson helped me in preparing the translation from the Spanish version. Lourdes Bosch, Eulália Escolá, and Anna Ticó helped to prepare various drafts of the manuscript, overcoming with their excellent work the numerous difficulties which appeared during its editing.

The Desk-Editing Department at Oxford University Press was most helpful in improving the quality of the manuscript.

As readers can see, considerable help has been received in all aspects of the research and writing of this book, but errors and omissions are the responsibility of the author alone.

Boston   J. C.
September 1990

Contents

List of Figures

List of Tables

Introduction

The European banking industry is experiencing profound change. This change is the result of several factors, including disintermediation, more economic instability in international markets, deregulation in industrialized countries, and the creation of the European market in 1992. While deregulation and economic instability are also present in other financial systems, such as in the USA and Japan, the creation of a unified financial market after 1992 is especially important for European banks, as well as those non-European banks that have extensive operations throughout the Continent. The gradual integration of Eastern Europe into the economic system of the European Community adds another factor of considerable interest.

As a result of these forces, the structure of the banking sector can be expected to look very different by the mid-1990s. Many large financial institutions have opted to position themselves to compete more effectively against the large American and Japanese banks. For example, the Deutsche Bank—one of the three large German banks—has proceeded in recent years with the acquisition of other banks in a number of European countries: the branch of Bank of America in Italy, the Sociedad de Investimentos in Portugal, and Morgan Grenfell in the United Kingdom; at the same time, it has also taken majority positions in other non-German banks throughout the Community.

Mergers and share exchanges have been utilized by various banks to broaden the scope of their activities and expand across a wider geographic area. Examples of mergers include the new Banco Bilbao Vizcaya in Spain, and the merger between Bank of Rome, Cassa di Risparmio di Roma, and Bank of Santo Spirito in Italy.

Meanwhile, other banks have preferred mutual share exchanges, as is the case of Commerzbank and Banco Hispano Americano, and of San Paolo and Compagnie Financiére de Suez. Such activities affect not only commercial banks, but also the savings banks. In countries such as Spain, restrictions have been lifted so that savings banks can compete more effectively with banks in the financial services industry, resulting in an increase in the number of offices on the part of the savings banks aspiring to serve a wide market. At the same time, the small size of some institutions has unleashed a wave of mergers and acquisitions with the aim of expansion into a wider geographic market.

This trend has also affected other banks of countries outside the EC. For instance, Kansallis-Osake-Pankki, the first bank of Finland, has recently reached an accord with Gotabanken, the sixth bank of Sweden in terms of equity, to form a Scandinavian group capable of competing against other large EC banks. Legislation in these countries is very restrictive with regard to such transactions, and the two banks have not been able to merge fully. Nevertheless, the agreement will allow them to offer financial services as though they were a single entity. Meanwhile, the North American banks' desire to get the maximum benefit from their extensive presence in Europe, and the Japanese banks' interest in this market, especially in light of 1992, make the European banking sector even more interesting.

The future trajectory of the European banking sector remains very uncertain. Compared to other sectors in the EC it is highly fragmented, and the separation between commercial, investment, and savings banks will also have an important bearing on the industry's structure and future outlook. The formation of 'supernational' banks (as in the case of the Nordic banks mentioned above), for example, will be difficult not simply because of legislation regarding mergers or the participation of foreign corporations, but also because in certain countries, such as France and Italy, government ownership in financial institutions is considerable. Another difficulty is the differences in legislation between countries with respect to the protection of banks' assets. Some German banks limit ownership to 5 per cent for each shareholder. In Switzerland, restrictions are even greater: although there are various types of shares, the voting rights are limited to Swiss citizens.

It has been frequently pointed out that size will not be banks' only competitive weapon in the unified market after 1992. The larger such institutions grow, the larger the number of niches that will be left open. Therefore, smaller banks will have greater opportunities to specialize. All of this suggests the configuration of a very different European financial map by the mid-1990s.

The objective of this book is to describe the principal forces of change in the European banking industry, as well as to analyse the strategies of the more influential enterprises operating in this sector. With this goal in mind, the behaviour of 25 commercial banks in five European countries has been studied. The countries are West Germany, Spain, France, Great Britain, and Italy; the selection is based as much on their relative economic weight inside the Community as on the size of their banking sectors. The analysis is based mainly on information obtained from inter-views with bank directors and documentation made public by the banks themselves.

This study consists of three parts. The first represents an analysis of the principal environmental forces affecting the financial sector: deregulation, economic instability, automation, and financial innovation. Without an understanding of the direction in which these forces are moving, it would be difficult to conduct an accurate analysis of the changes taking place within the sector, and to anticipate firms' strategies. As we shall see, each of these factors has important implications with regard to the structure of the European banking industry, and, in turn, the design and execution of companies' strategic plans.

The second part of the work presents a comparative study of the recent evolution of the banking sector in the same five European countries. The analysis of each country includes a study of the recent evolution of its financial sector and, in turn, each sector's inherent economic outlook in terms of profitability, costs, productivity, and competition (including analyses of firms' specific strategies), as well as the changes in the regulatory environment.

The third part of the book is dedicated to an analysis of some of the strategic choices of European banks, such as scale, diversification, and internationalization. For this analysis, some classical tools of strategic management have been employed, such as industry and competitor analyses—which integrate environmental changes affecting the industry—as well as the value-chain. Finally, we will evaluate the pace of change within the European banking industry, as well as how some of the banks are adapting to the new environment.

PART I

1
The Forces of Change in the European Banking Industry

1. INTRODUCTION

Much of the responsibility for the major changes which have affected the European banking industry is not attributable directly to the forces indigenous to the sector itself—in other words, to the specific actions of the firms that comprise it—but rather to changes in the legal and economic environment in which the industry operates. For this reason, it is necessary to understand this environment in detail, as well as how it has affected, and can be expected to affect, the future structure of the European banking industry. Changes in the legal and economic environment are certainly not indigenous to Europe, however. In the United States, such changes have also given way to a very profound transformation of the American banking industry.[1]

In this chapter, we will analyse a number of environmental forces which have had, and will continue to have, a considerable impact upon the European banking industry: deregulation, globalization, worldwide economic instability, the introduction of new information technologies, and the process of financial innovation which derives from the combination of these factors.

2. DEREGULATION OF FINANCIAL INSTITUTIONS AND MARKETS

To understand the process of financial deregulation, we must remember that the supply of financial services appears in response to decisions regarding savings and investments by firms or individuals. Those entities with a surplus of funds lend this surplus to financial intermediaries, directly in the capital markets, or indirectly through banks. Figure 1.1 demonstrates these alternative types of financial activity. The first corresponds to financial intermediaries comprised of banks (banks, savings banks, credit co-operatives, investment banks, etc.). In essence, the activities of these entities entails the assumption of risk: the acquisition

---

[1] Crane and Hayes (1982, 1983), Ballarín (1985), and Cooper and Fraser (1986) provide extensive analyses of the changes.

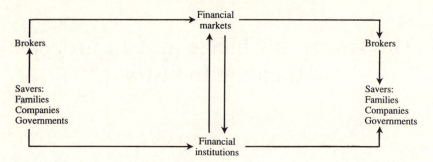

FIG. 1.1   *Types of financial intermediation*

of funds for their later sales in the form of credit or other types of instruments for the borrower of funds.

Two types of financial intermediaries—banks and savings banks—act as depositors of an important volume of capital. The accumulation of this capital permits banks to diversify the risks of their investments, as well as to take advantage of economies of scale in their different operations by which they seek to achieve the highest level of profitability. The law of large numbers permits them to keep liquid only a reduced part of their deposits, enabling them to invest the rest in assets with less liquidity but greater profitability. In this way, financial intermediaries transform the deposits they have received into profitable assets, but with less liquidity and different maturities than the initial deposits. Through this process it is possible to achieve saving incentives—since the process increases the possibilities of fund placement—and to make available funds for investment, thus promoting economic growth.[2]

The second type of institution, the so called 'brokers', put savers and investors in contact with the various kinds of financial markets. Their principal function consists of helping those seeking funds (or investors, such as firms or governments) to obtain money from the capital markets. The intermediation process of this type of asset is free of the risk that threatens banks, from both interest rate fluctuation and the possible insolvency of those seeking funds.

However, leaving aside this question, which is certainly important with regard to one or the other type of risk, historically the radical difference between the commercial and non-commercial banks emerges in the former's capacity to create money as a result of being in a position to keep only a portion of their total deposits liquid.

The capacity to create money has been, without doubt, the first and most important motivation for the monetary authorities to regulate the

---

[2] A more detailed description of this process can be seen in Campbell (1982).

operations of commercial banks with a view to assuring their solvency. By definition, the major part of the banks' deposits are subject to withdrawal at any moment, while their assets have a period of maturity and, consequently, lack liquidity in the short term. If a large number of depositors were to withdraw their deposits, the corresponding bank, and the entire banking system in general, would have serious problems which, in turn, would lead to financial panic such as that experienced by the USA at the beginning of the 1930s. The reaction of the financial authorities to this aspect of the system has been to superimpose two main types of regulatory measure: the reserve requirement (which assures a minimum of liquid capital relative to total assets), and the solvency ratio (a percentage of liquid assets relative to a bank's total number of outstanding shares).

Another regulatory instrument is the securing of deposits arranged through a government agency and with the funds coming from the banks themselves. This mechanism has functioned very efficiently since its creation in the 1930s, even in times of great financial instability. However, it has not eliminated the need for the regulation of the banking system's solvency: the system of deposit insurance is really no more than a kind of protective net to prevent serious damage in the event of bankruptcy of a financial institution. Nevertheless, regulations aimed at preventing the risk of insolvency continue to be necessary.[3]

The second reason which historically has justified the regulation of the banking system has been the need to ensure that credit is offered honestly and efficiently. The aim of these regulatory measures is to prevent financial institutions from mishandling funds (for example, directing them to activities lying outside the interest of their depositors), engaging in fraudulent activities, embezzling, or concentrating excessive risk in certain sectors or companies.

Finally, the third reason invoked for banking regulation has been the supposed need to prevent banking institutions from becoming too large and accumulating excessive power, as much from a social point of view as from the perspective of the preservation of free competition within the sector.

These reasons have justified the financial regulatory system which, applied one way or another, was in place in the USA and in Western Europe until a few years ago.[4] The system grew out of the measures approved by Franklin Roosevelt between 1933 and 1934 as a means of dealing with the financial crisis that dragged the US economy down into the bleak years of the Great Depression.

---

[3] A study of these and other regulatory measures of the banking system can be found in Baltensperger and Dermine (1987).

[4] See Kennedy (1973).

The regulatory plan which followed the Great Depression consisted of four groups of measures, each of which allowed financial entities a different degree of latitude. The first group created a federal insurance system providing coverage of up to $2,500, as well as the Federal Deposit Insurance Corporation (FDIC) which was to administer this system under the supervision of the Federal Government.

The second group of measures established the Banking Act of 1933, which limited the capacity of banks in certain areas to compete with one another. The reason for this was embedded in the general opinion that an excess of competition in the financial sector was responsible for the economic crisis of the 1930s. Of particular concern was price competition (interest rates), which seemed to have caused many firms to take out high-interest loans, generating greater bankruptcy risks. In accordance with this interpretation, Congress decided to include in the 1933 Banking Act a clause prohibiting banks from paying interest on current accounts. At the same time Congress forced the Federal Reserve to regulate the interest rates that could be paid on savings account deposits.

Regulation of the geographic expansion of credit entities, included in the Banking Act of 1933, constituted the third group of measures adopted under President Roosevelt. With this, the McFadden Act of 1927 was modified in two ways: the first liberalized certain restrictions on the ability of national banks to operate in certain States; the second put limits on the formation of banking groups through the establishment of subsidiaries, which some companies had used as a means of expanding across State boundaries.

Lastly, the separation of activities of commercial banks from those of investment banks, promulgated by the Glass–Steagall Act, constituted the fourth group of regulatory measures. This prevented State banks from acting as intermediaries in mergers and acquisitions and the issuing of corporate debt. Some exceptions were made for certain small banks acting on behalf of the local public sector.

These four measures—deposit insurance, interest-rate regulation, geographic limitation, and the separation of commercial banks from other agents of financial intermediation—were increasingly applied in Western Europe, until the framework for a regulatory apparatus of the financial system began to emerge in the 1970s.

In the same way that the regulatory process in banking was generally universal, so was the later trend toward deregulation which occurred at the beginning of the 1980s. This process actually started in the 1960s when European and American banks began to discover ways of evading legislation, especially with regard to two specific types of regulation: those related to interest rates and those connected with geographic restrictions.

This process was consolidated in the 1970s as a result of three distinct factors. The first had to do with the high inflation rate. With interest rates reaching double figures, nominal rates reached comparable levels, such that regulation prohibiting the payment of interest above a certain ceiling (established during periods of low inflation) lost meaning. As we shall see later, many firms embarked upon a process of financial innovation offering their clients products which were not recognized as deposits and which offered interest rates more reflective of the situation in the financial markets.

The second factor was related to technological innovation which, as will be discussed later in more detail, has caused an erosion of the geographic barriers: to obtain funds or credit it is no longer necessary to have a complete banking organization in a specific location. Technological innovation applied to the processing of financial information has made this a factor of prime importance in the modern management of financial institutions, and has accelerated the penetration of non-banking enterprises into those activities which were once exclusive to commercial banks.[5]

The third area of change lay in the transformation of the international financial system. The initiation of a system with floating exchange rates and the globalization of financial markets, with the subsequent increase of capital flows between countries, unleashed a greater degree of rivalry within the different national markets, thus increasing deregulatory pressures. All three of these issues will be analysed in greater detail in the following section.

The progressive liberalization of interest rates in the United States began in 1972, when (1) the savings banks in Massachusetts were authorized to establish new financial instruments without restrictions on interest rates, and (2) money-market funds were approved. The latter, after a slow awakening, eventually exceeded $3 billion in 1977 and $233 billion in 1986.[6]

Two things helped these funds to expand. The first was the growth of inflation, which became a disincentive to savings despite ridiculously high interest rates. The second factor was the introduction of the Cash Management Account by Merrill Lynch in 1977. With this, investors had access to various financial instruments at the same time: a current account with access to a credit card and a money fund which permitted immediate access to a loan up to a maximum of 50 per cent of the account-holder's balance.

The rapid growth of the Cash Management Account of Merrill Lynch

---

[5] Concerning the role of technology in the banking sector, see US Congress (1984). Ballarín (1985) deals extensively with the entry of new competitors in the American banking industry as a result of their competitive advantage in the area of information and its use.

[6] See Gilbert (1986).

had a strong impact both on the money funds and on the liberalization of interest rates. After various partial measures, regulation of interest rates in the USA disappeared in 1986, and the same has occurred in the major European countries in recent years.

A similar process has taken place with regard to the capacity of the US State banks to expand beyond State boundaries. In accordance with McFadden and the Bank Holding Company Acts, banks were permitted to establish operations in other States, but only to give credit, accept deposits, and offer credit to international commercial operations. Throughout recent years, these restrictions have been relaxed, permitting banks to operate more freely in different States. In fact, by the end of 1990 the US Secretary of the Treasury announced the overhaul of the other regulatory measures, that is to say, a reform of deposit insurance and a proposal that would allow banks to enter new business and spread across the country. In other countries such as Spain or France, deregulation in these areas has functioned in a different way, permitting certain entities that were previously restricted—such as savings institutions—to expand geographically.

The process of geographic deregulation in Europe is significant because it affects a central issue: the possibility to achieve economies of scale. We will come back to this question later.

The entry of non-banking companies into traditional banking areas— namely lending—is another recent and significant trend which has forced governments to permit banks to develop operations beyond those which, as a result of legal restrictions, they had conducted until only a few years ago.

From the perspective of lending, banks have lost ground on various fronts.[7] The first is that of consumer credit, as a result of a greater presence of savings banks in this sector. A second area is that of commercial credit. Traditionally, banks have dominated this market in a majority of countries; however, companies have increasingly come to rely on the issuing of commercial paper to raise capital. The same has occurred with long-term loans, which corporations have substituted in large part for medium-term debt issued in the capital markets. Something similar has also occurred with deposits, where a wide range of products, initially introduced into the market through Merrill Lynch's Cash Management Account, have diminished banks' traditional capacity to attract savings deposits.

These changes have led to the emergence of competition within the banking sector based on the differentiation of prices for financial services. This key factor has come to complement that of the relationships between

---

[7] See Pavel and Rosenblum (1985).

banks and customers, which was almost exclusive up until the 1970s. The advantages of the latter stem from the fact that information is an important asset in the activity of financial management, and that the demand for financial products on the part of a single client can generate an important knowledge base of potential utility to the bank in the future. Nevertheless, the importance of customer—bank relationships, innovation, and the price of financial products will all play a decisive role in years to come.

The impact of these forces on the banks' operating results was significant. Between 1979 and 1984, the net rate of return on US banks' assets fell from 0.8 to 0.65.[8] Part of the reduction can be explained by the rise in the number of bad loans, many of which were made in an effort to cope with the growing intensity of competition although many of these loans were granted with full knowledge of the relatively higher risks they carried. Yet, a good part of the explanation can also be found in the deterioration of the sector's profitability, resulting primarily from structural changes.

For this reason, commercial banks have sought to gain regulatory approval to expand beyond the realm of traditional business into other fields of financial activity. Some of the new operations accessible to them relate to the placement and exchange of share titles, and the commissions gained from such transactions have had a beneficial effect on the banks' operating results. In reality, the spirit of the Glass—Steagall Act has been superseded in order to give the banking sector the operating capacity it needs to remain a profitable industry.

The events which took place in the United Kingdom between 1986 and 1987 were particularly important; two are especially noteworthy. The first was the opening of the London capital markets to foreign firms, and the second was the abolition of fixed commissions on stock market transactions—and thus of the old British distinction between traders and dealers. This led many banks to establish offices in London in order to take advantage of the perceived opportunities emerging from a less-regulated British market. Simultaneously, these actions encouraged other countries, such as France, Italy, and Spain, to initiate a similar process of market reform in order to face the growing competition of the British market.

As can be seen, the deregulatory process has moved significantly forward. The relevant question, therefore, is not how to stop it, but rather how to find concrete ways of establishing regulatory measures, which, without diminishing the efficiency of the system, still protect the legitimate interest of other stakeholders.

---

[8] See Danker and McLaughlin (1985).

3. THE FINANCIAL LIBERALIZATION PROCESS IN THE EUROPEAN COMMUNITY

Among those industrialized countries that have pursued a policy of financial deregulation, considerable progress has been made within the European Community. The reason for this is that the deregulatory process in this case is tied to another process: the creation of a single European market by 1992. With regard to financial deregulation within the scheme of a common market, two components of change are especially significant. The first has to do with geographic liberalization with regard to banks' cross-border movements and the deregulation of the supply of financial services within the Community. The second has to do with the liberalization of capital flows between EC countries. The former action corresponds to the Second Banking Directive, the latter corresponds to the legislation approved by the EC Council of Ministers in 1988.

There is a third component worth mentioning: the creation of a European currency, a project studied in the Delors Report, made public in 1989. This is important, although its role is less crucial for the creation of a unified financial market than the other two components.

Financial unification in the European Community was initiated in 1977 with the approval of the First Banking Directive, which applied to all banking institutions. The directive established the minimum requirements for the authorization and supervision of these institutions and represented a first step toward the principle of supervision in the country of origin. The directive required member countries to have a system for authorization of new banking entities based on two principal criteria: a minimum amount of capital, and an honest, experienced management team. In this way the Directive formulated the principles for the later harmonization of the banks' liquidity and solvency ratios.

In practice, member countries possess stricter regulations than those laid down by the First Directive. The result is that significant differences still exist between countries. For this reason, although banks within the EC possess a basic right to extend their activities to other countries of the Community, in practice the disparity between national regulations makes it enormously difficult for them to do so.

A unified financial market, however, requires more than just geographic integration. The ultimate objective of the European Community is to remove all barriers impeding the free trade of financial services. The harmonization of these regulations constitutes the basis for achieving a truly unified market.

According to studies on this subject, the different legal barriers that exist in the banking sector (barriers to the establishment of banks and barriers to the trade of financial services) threaten to cause price differences for similar financial services offered throughout countries of the

TABLE 1.1. *Potential and estimated reduction in prices of financial services*

Country	Potential reduction (%)	Estimated reduction (%)
France	24	12
Germany	25	10
Italy	28	14
Spain	34	21
United Kingdom	13	7

Source: Cecchini (1988).

European Community.[9] We can thus hypothesize that the price reduction of financial services due to deregulation and greater competition within the sector will be more significant in countries such as Spain, Italy, France, and Germany, and less pronounced in the UK, as suggested by the data presented in Table 1.1.

As demonstrated, the potential benefits of these reductions are quite substantial, and could be even greater with further integration of the EC capital markets, leading to greater price competition and lower prices. Consequently, the process of integration will take place in some segments which are growing increasingly competitive. This will force banks to re-evaluate their strategic imperatives in order to survive in this rapidly changing and much more aggressive environment.

In its 1985 White Paper, the European Commission adopted new criteria for the unification of the financial markets based on the principle of 'mutual recognition'. Accordingly, it will not be necessary to adapt legislation at the national level of each member country. This principle will be complemented with control by the home-country. This means that supervision of the activity of a bank from its acceptance in a particular country is undertaken by the competent authority in that country.[10]

This principle is taken up in the Second Directive, which the European Commission began to study in 1987. It was approved by the Council of Ministers in June 1989 in the form of the Single Banking Licence, which is issued by the authorities of the country of origin. This permits any bank to establish and offer a broad range of financial services in other member countries on the basis of only one licence. This is a very important step because it permits further financial integration without excessively expanding the body of existing regulations.

[9] See Price Waterhouse (1988*a*).
[10] See Clarotti (1988) and Madroñero (1988).

The Second Directive also aims to harmonize some accounting norms and basic banking regulations within the Community. It enumerates a wide range of financial services which can be offered in all member countries, with authorization coming from the country of origin. Thus, this text lays the groundwork which will serve as the basis for the establishment of the truly universal European bank.

It is evident that the principle of mutual recognition will serve to promote the rapid integration of Europe's financial market, but this does not eradicate the full extent of potential difficulties, nor the possibility of discrimination. For example, in accordance with the established criteria of the aforementioned legislation, a French bank, authorized by the French financial authorities, would be capable of offering leasing services in Spain, given that leasing was one of the activities covered by the Second Directive. However, until the beginning of 1988, Spanish banks were not allowed to carry out leasing activities.

In this example it is shown that the Second Directive would considerably quicken the pace of legislative integration among member countries, with the objective that banks of a given country would not be in an unfavourable position relative to banks of other national origins.

The Second Directive gives jurisdiction over the supervision of the banks' solvency and liquidity to the country of origin, while transferring to the receiving country regulation regarding the execution of monetary policy and other practical norms of operation. That is what is meant by the expression 'home country control and host country rules'.

In addition, the Second Directive attempts to establish behavioural guidelines in two specific areas connected with the establishment of banks. The first makes reference to the establishment of a minimum amount of capital, which is set at five million ECU for each bank; branches which are established in foreign countries would not have to comply with this requirement. The EC has left the question of the ratio of total investment to equity in the hands of the Bank for International Settlements (BIS). Compliance with the new regulation approved by the BIS, which is very demanding (an 8 per cent ratio), could cause problems for certain Italian and French banks with weak capital bases.

The second area is the application of the principle of reciprocity for countries outside the Community. In accordance with the basic requirements of the banking licence, a bank of a third country could establish an affiliate in any country in the Community, and from there open branches in any other member country. However, some of the countries might object to this, on the basis that the bank's country of origin does not concede equal treatment to the recipient country's banks. In such cases, the Commission must study whether or not the country of origin offers reciprocal treatment to each and every one of the member countries before granting authorization for investment by third-country enterprises.

Another related aspect of interest regarding the banks of third countries is the discrimination that the legislation establishes between the subsidiaries and branches of a foreign bank. Subsidiaries have to meet the requirements of the country in which they are established, but from there they can expand their activities to other EC countries as if they were really EC member banks. On the other hand, branches of third-country banks, lacking independent jurisdiction, must meet the norms of each country. Among other things, this implies that each of the branches in different countries will have to comply with their minimum capital requirements, which will increase capital costs relative to those banks operating through subsidiaries.

A final relevant point in the Second Directive is the limitation of risks. According to this, banks will be unable to participate in industrial and commercial businesses which exceed 10 per cent of their capital. Furthermore, they will not be able to loan more than 40 per cent of their capital to a single company or group of companies. In countries where there are universal banks, such as in Germany, and where these banks have extensive corporate networks, including substantial shareholdings, such restrictions have not been welcomed.

It is evident then that the legal environment of the banking industry will undergo considerable changes in the next few years, such that the effects of integration will have a greater impact on the banking industry than on many other sectors.

In addition to the legislative changes effected by the Second Directive, complementary modifications affecting the liberalization of capital flows within the European Community are occurring simultaneously, and they can be expected to have important consequences for the European financial services industry.

In effect, the creation of the Single European Market would not be conceivable without the free selling of financial services and the unrestricted movement of capital. This is the real objective of the integration: to provide free access to the financial services they require.

The eradication of barriers to the free circulation of capital within the EC will help to allocate resources from savings more efficiently, and in turn improve investment opportunities for individual savers, creating additional possibilities of financing for companies—at lower costs. Furthermore, the free circulation of capital is one condition for completing the process of financial integration and reaching a greater efficiency within the Single European Market.

With this objective, the Council of Ministers of the European Community met on 14 June 1988 and approved on 24 June the Directive (88/361/EEC) which ensured that the movement of capital within the EC would be completely deregulated as of 1 July 1990. Spain, Ireland, Greece, and Portugal would be given more time to adapt.

The essential elements of this new decree were the following:

(*a*) Object of liberalization. All the restrictions to the capital flows would be removed. In particular, short-term operations and the opening of current accounts in any country would be unrestricted.

(*b*) Period of application. The Directive gave an ordinary period of adaptation until June 1990, and an extraordinary period for the afore-mentioned countries, which will end on 31 December 1992. In addition, Greece and Portugal would be allowed an additional period, until December 1995, in the event that they experience difficulties with regard to their balance of payments.

(*c*) Clauses of guarantee. With the intention of giving the monetary authorities of each country the capacity to react to economic downturns, the Directive contains a clause for which the authorities can modify the regulation of short-term capital movements, up to a period of six months. However, and with the objective of preventing unnecessary protectionist actions, the European Community should approve, even if a posteriori, the adopted measures; also the Council of Ministers could approve or revoke the decision of the Commission.

In addition to this Directive, the European Commission approved in June 1988 a calendar for the fiscal aspects of the liberalization of capital flows, as well as for the implementation of regulatory measures covering financial institutions in the EC.

Succinctly, the Council was especially concerned with three aspects of the new regulations: (1) the co-ordination of the different Community fiscal administrations; (2) the harmonization of certain aspects of corporate taxes; and (3) capital income taxation, in particular, the returns on bank deposits. Among these, the last, without doubt, was the most problematic.

The positive points of the new regulation can be grouped as follows:

(*a*) Financial institutions. As a result of deregulation, competition between financial institutions has intensified. This means that they have had to adapt to the new environment with new strategies, as we shall see in Part III of this study. Logically, these will favour innovation and productivity improvements.

The harmonization of regulations bearing on the operations of financial institutions will help to eliminate the financial privileges of the public sector. Therefore, this should finance the part of the public expenditures not covered by ordinary incomes with funds of equal cost to those utilized by the private sector. In this way, the European financial market offers a restriction on the growth of public spending.

(*b*) Non-financial companies. The advantages of a unified financial market for non-financial companies lie in the availability of a greater number of financial products at lower cost and higher profitability. The reduction of cost of credit and loans comes as much from the reduction of the banks' transfer costs as from greater competition.

(*c*) Savers. Logically, a large European financial market has to offer savers new opportunities for investment, with improved correlations between profitability and risk. An additional advantage of liberalization is the free flow of capital between households, without any type of control.

The complete liberalization of capital movement among EC member countries entails two fundamental threats, however. First, there is the danger of a concentration of power in the hands of a few financial institutions. This threat is real and emerges from one determinant: to attain economies of scale in certain segments requires a certain level of concentration among entities. However, despite this potential drawback, it is clear that the process of liberalization will provide net benefits, and that any negative consequences should be avoided through rigorous measures aimed at protecting competition throughout the Community.

The second threat refers to the impact of liberalization of the capital movements on the European Monetary System (EMS)—as it is actually designed—and the formulation of the monetary policy in each country. These three aspects are broadly interrelated.[11]

Until now, the EMS, generally understood within the context of a monetary regime of fixed exchange rates, has coexisted with autonomous central banks in each of the member countries, which are free to impose monetary discipline at their own discretion, or pursue independent macroeconomic objectives, for example, growth of the GDP, inflation rate, interest rates, etc., which might differ from other countries. However, this coexistence has not been possible due to, among other reasons, controls which existed on the movement of capital in some countries, for example, France, Spain, and Italy.

These controls have served to discourage speculative movements against the weak currencies of the EMS, for example, the French franc and the Italian lira, at times of a lack of coherent economic policies, or at times of instability in the international financial markets (during fluctuations in the value of the dollar, for example).

Therefore, the complete elimination of capital controls could adversely affect the EMS, at least in its present form. As a result, different solutions have been proposed which range from the establishment of temporary exchange controls (which is an alternative included in the Directive of the Commission), to the existence of a dual exchange market (one for purely financial transactions and another for commercial operations), to the establishment of an unequal tax on interest (such as that proposed by J. Tobin), or a greater range of fluctuation for the currencies of the EMS. Of these options, the first three impose high costs

[11] See Wyplosz (1988).

of administration, although the possibility of establishing temporary controls is very interesting. However, to permit a greater interval of currency fluctuations runs contrary to the objectives of the EMS.

It is clear that if a system of fixed exchange rates is to be maintained in Europe, within the context of a system permitting the free circulation of capital, the only valid alternative is a greater co-ordination of the different monetary policies of each country. This would require the gradual creation of a European monetary authority, (for example, the Central Bank of Europe, as suggested in the Delors Plan) comprising some of the attributes of the national central banks, but, more importantly, which would maintain the necessary discipline to establish monetary stability in Europe. What remains clear is that the liberalization of capital movements offers new opportunities for banks, but not without problems requiring quick solutions in order to avoid additional difficulties which could adversely affect the industry.

Regarding the implications for banks' strategies, the liberalization of activities will trigger an increase in price competition, which in turn will lead to a reduction of price discrepancies between countries, as we shall see in Chapter 8.

4. INTERNATIONAL FINANCIAL INSTABILITY

Instability in worldwide financial markets is among the most widely recognized characteristics of the international economy in recent years. This phenomenon stems from different causes, such as the growing integration and globalization of the financial markets, the greater efficiency with which the intermediaries act in these markets, the spectacular increase in the volume of international capital flows, and, finally, the lack of consistency between the economic policies of the principal industrialized countries.

This study will analyse this phenomenon beginning with a brief description of the process of financial globalization which has been responsible for all of these forces and which will continue to shape the industry's future development. We will then look at the primary causes of international financial instability, placing particular emphasis on the system of floating exchange rates. Finally, we will consider what is perhaps the greatest threat to international financial stability: the external debt of developing countries.

The globalization of financial markets and the instability it causes in international markets is particularly relevant since it is undoubtedly one of the most important forces driving the deregulatory process examined in the previous section. In effect, exchange-rate volatility and high interest rates have forced central banks to approve deregulatory measures, and have also served as stimuli for a number of important financial innovations, as we shall see.

The relevance of these issues for banking concerns is evident: present-day international financial instability has increased the risk associated with interest rates and exchange rates, and thus the possibility of bankruptcy. Therefore, an understanding of the forces behind these factors is imperative.

4.1. Internationalization and globalization of financial markets

Globalization represents one of the most significant trends in the international economy in recent years. For the firm, this trend represents a new challenge which must be confronted, although to a lesser or greater extent depending on the sector. There is general consensus, however, that the impact of globalization has been particularly strong in the financial sector and in each of its associated segments.

The fact that the competitive position of a firm in one country will be affected by its position (or lack thereof) in other countries demonstrates that the financial sector has evolved into a truly global industry.[12]

What are the implications for a sector so highly influenced by the trend towards global integration? Undoubtedly, globalization will profoundly affect the industry's long-term structural composition, as well as influence firms' long-range strategic planning. The reason for this is that 'global' infers not just the mere sum of national companies spread throughout different countries, but rather the entire network of competitive (or collusive) relations which these companies maintain in order to reach leadership positions in the international market-place.

International integration of the financial sector began in the 1960s when American commercial banks began offering their services abroad (loans, management of foreign currency, 'clearings', etc.), and was further stimulated by the growth of foreign direct investments and firms' efforts to achieve greater economies of scale. Later, the American investment banks also extended their operations to other countries, starting with the Eurobond market in London. The collection, and later investment (or recycling), of the so-called 'petrodollars', as well as the introduction of a system of floating exchange rates, also contributed to the globalization of financial markets.

In addition, this process was accelerated by the appearance of floating exchange rates and interest rates and, in turn, greater flows of capital between countries—transactions which were of a purely financial or speculative nature, without any real basis to support them. New financial

---

[12] Porter (1986) deals with the concept of global industry in depth. Regarding the banking industry, see Pecchioli (1983).

instruments were created around these transactions, permitting transfer of the power to buy at the time, thereby contributing to the further integration of the sector on a worldwide scale. Such convergence was also aided by the gradual standardization of the demand for financial services throughout the industrialized countries, as well as increases in firms' economies of scale.

A crucial factor in the consolidation of worldwide financial markets was the creation of the 24-hour currency market, in which financial activity continues around the clock in the principal financial centres of the world.[13] Later, this market grew to incorporate transactions in the short-term money markets.

Two additional factors have continued to stimulate international financial integration. The first is the worldwide expansion of the futures markets. The list of firms which they handle grows longer; other assets such as government bonds and shares have been added to the foreign currency. However, the stock-market crash of 1987, caused in part by the extreme volatility generated by the futures markets, has made some governments more cautious over the authorization of deregulation for this type of operation.

The deregulation and modernization of national stock markets comprise another factor of importance. Without a doubt, that which has been of greatest relevance until now is the so-called 'Big Bang' of London. In spite of the technical difficulties with which the Big Bang started, it has had some effects on firms' competitiveness and their ability to create new and innovative products, although there are those who would argue that the benefits have been excessive.

Another factor to add to this list is the progressive dismantling of the controls on the free capital movements between countries, and the financial innovations which permit separation of the currency in which companies operate from that of the country from which they initiate their activities; the obvious result of this is the globalization of international finance.

Despite great progress, the road to a truly integrated global economy remains to be completed as many countries, developed as well as developing, still maintain capital controls and other regulations which act as an impediment to this goal. Some EC countries, such as Spain and Portugal, are good examples of this. But as indicated earlier, many controls within the EC will disappear within the next few years and a unified European financial market will eventually come to be a reality, which will undoubtedly encourage other countries to adopt similar reforms.

What are the implications of a greater level of international integration for financial and non-financial companies? The former can expect

---

[13] Kohlagen (1983) provides a detailed study of this question.

intensified competition and an erosion of the advantages derived from operating internationally, given that this is becoming a common characteristic of the sector. For the latter, globalization suggests three new advantages: a wider offer of financial services, a reduction of general financial costs, and a greater capacity to operate in the international arena.

4.2. Volatility of interest and exchange rates

4.2.1 Floating exchange rates

In March 1973, one month after an approximately 10 per cent devaluation of the US dollar, the principal currencies of the EC were severed from the gold standard, putting an end to the fixed exchange rate system installed through the Bretton Woods accord of 1944. All at once, the Swiss franc, the pound sterling, the lira and the yen began floating independently. The crisis of the dollar was the last in a long series of problems to emerge from the fixed exchange-rate system, beginning in 1959, when the American currency was submitted to strong speculative attacks.

There were primarily three weaknesses in the fixed exchange-rate system.[14] The first relates to the uncertainty the system generated since it was based on a reserve currency—the dollar—convertible to gold. The dollar was exposed to a confidence crisis when the US Government's capacity to convert dollars to gold was brought into question. The second weakness lay in the system's inability to correct disequilibriums in the balance of payments, which delayed the necessary adjustment of exchange rates. Lastly, liquidity was not guaranteed under the existing system; based on the availability of gold, the level of worldwide currency reserves could only increase relative to the availability of the main currency of reserve—the dollar—thereby strengthening the dollar–gold relationship. Thus, the increase in the amount of dollars issued to finance the US deficit caused a crisis of confidence with regard to the value of the dollar and the capacity of the United States to convert dollars to gold if necessary.

In December 1971, the leading industrialized countries subscribed to the Smithsonian Agreement whose objective was to strengthen the system. This agreement was the last effort to avoid the total collapse of the system of fixed exchanges rates. It was preceded by a number of economic measures approved in August 1971 by the US Government and aimed at reducing inflation and correcting the commercial deficit through an increase in tariffs. But the most important measure was undoubtedly the decision to suspend the fixed exchange-rate system. Hence, the dollar

---

[14] See Viñals (1986) and Canals (1988).

was devalued relative to gold by 7.89 per cent (the price of gold exceeded $38 an ounce) and the principal currencies maintained the parity they previously had with gold or were re-evaluated.

The combination of these adjustments provoked a devaluation of the dollar relative to other currencies on the order of 7 per cent. At the same time, the signing governments adopted a couple of additional measures. The first consisted of authorizing an increase in the band of fluctuation from 1 per cent to 2.25 per cent; the second consisted of the promise to begin discussions within the International Monetary Fund to consider the reform of the international monetary system.

Today the declaration of these measures seems surprising since such an action would certainly encourage speculation in the currency markets, as indeed occurred during the weeks following the signing of the Smithsonian accord. Meanwhile, some governments had taken the decision to establish a dual exchange rate: one fixed for commercial transactions and another for financial transactions. In any event, the system devised at Bretton Woods was by now defunct. At the beginning of January 1973, after President Nixon announced Phase III of his plan to control prices and salaries, the devaluation of the dollar was announced, and weeks later the free–flexible exchange-rate system was implemented.

Once the system of fixed exchange was abandoned, the main question was whether or not the variable plan would be 'dirty' (controlled by the central banks) or 'clean' (determined by market forces). This issue was the focal point of the meeting of the Economic Ministers of the Group of Ten in Paris in March 1973. The meeting resulted in a controlled system. According to the official declaration, appropriate government intervention in the exchange markets would help to maintain stability.

The flexible exchange-rate scheme, which put an end to the long period of fixed exchange rates, opened a period of hope that the next system would definitely correct the deficiencies of the international monetary system. At least, this is what can be inferred by the analysis of the characteristics of the floating rates, as we shall see.

In theory, the flexible exchange-rate system offers greater protection from external economic shocks, and, in particular, from inflation resulting therefrom. With a fixed exchange, any inflationary shock from abroad, for example, a large increase in the price of oil, would directly carry over to domestic prices, without the exchange rate acting as a compensating factor. However, within a free-floating system, a currency depreciation would make imports less attractive, raise exports, re-establish external equilibrium, and partially insulate the domestic economy from the effects of the external jolt.

A second advantage of this system lies in the fact that currency fluctuations will theoretically cause automatic adjustments in the balance of payments, eventually bringing it back into equilibrium. This factor presents a dual advantage: greater insulation from external shocks on the one hand, and the possibility of gearing national economic policies towards the attainment of internal equilibrium on the other (economic growth, prices stability, full employment, etc.).

The system can also facilitate the successful implementation of monetary policy. In effect, flexible exchange rates permit the balance of payments always to return to equilibrium; as a result, the variations in financial assets coming from abroad disappear, as they are converted automatically in variations of the money supply, which occurs when the central bank does not act in a compensating manner—one of the important characteristics of fixed exchange rates.

Lastly, it is assumed that within a system of floating rates the speculative crisis experienced by some currencies under the fixed-rate system will disappear. In summary, these crises emerge when some currencies are perceived to be overvalued which in turn causes a mass sale of these currencies to avoid potential losses from an official devaluation; but the same massive sale exacerbates this already critical situation, making a devaluation even more necessary. It is assumed that under a floating-rate system adjustment of overvalued (or undervalued) prices will take place through the market mechanism, an eventuality not provided for by the fixed-rate system.

The hope awakened by the new system appeared justified; however, the results have not always been up to expectation. The floating-rate system actually has not provided the desired insulation from external forces with a negative impact on national economies. It is true that the flexible system offsets a growth of assets abroad, preventing mismanagement of the money supply. However, the flexible exchange has neither compensated completely for international economic distortions, such as the increase in oil prices, nor for financial shocks, such as those related to the US deficit.

The floating-exchange scheme has also failed to achieve another important objective. While it has permitted the financing of the current account deficit by attracting long- and short-term capital (a mechanism which also functioned with fixed rates), it has not facilitated an automatic balance in the exchange of goods and services. Indeed, one of the major sources of economic tension in the world economy stems from trade imbalances between the United States, Japan, Germany, and some of the countries of South-East Asia.

The capacity of monetary policy to improve internal objectives also has not been significantly improved under the new system. In the first place, some central banks, especially the US Federal Reserve, have utilized instruments of monetary policy—principally interest-rate adjustments and market intervention—to sustain currency prices. This has been particularly true in the case of the dollar.

Monetary policy, therefore, cannot be separated from outside market forces. However, there is yet another reason why monetary policy has not been an effective tool for attaining internal objectives, although it does not have to do with exchange rates: the redirection of monetary policy in the fight against inflation in the latter 1970s. The motives behind this policy change dealt with rigidities in the productive system and the high expectations on the part of some agents; both factors thwarted many of the advantages that monetary policy had to offer as an anticyclic instrument. Meanwhile, interest-rate adjustments have been widely utilized to achieve certain objectives in the exchange markets, which has made them less stable.

Also, floating exchange rates have not eliminated distortions related to speculation with the weaker currencies. What happened to the dollar during 1986 and 1987, especially after the stock market crash of October 1987, is a good example. To a certain degree, this need not have occurred. A speculative distortion takes place when it is perceived that the rate of exchange is not sustainable in the medium term, creating the expectation of government intervention to devalue the currency.

In theory, if exchange rates float freely, currency markets will always return to equilibrium, although it is true that price fluctuations do not disappear in the short term. The problem emerges, however, when central banks actively intervene to artificially sustain prices. When this happens, there is widespread speculation with regard to how long the intervention will last, which causes subsequent distortions.

In rather broad terms, this describes what happened in the case of the dollar following the Louvre Accord of February 1987. Through this agreement, the central banks of the USA, Germany, and Japan committed themselves to intervene actively in the foreign-exchange markets in order to devalue the dollar. The aim of the accord was to reduce the value of the dollar and therefore avoid serious trade imbalances. However, for the accord to reach fruition there was the need for a quick change of prior budgetary policies among the countries involved. But the slow correction of the US deficit and the excessive budget discipline of Japan and Germany foiled the objectives of the plan.

For the reasons stated, the floating exchange-rates system has not had entirely positive effects. Yet it is difficult to imagine that a fixed-exchange model would have responded better to the two energy shocks and the increasing US public deficit. However, the floating rates have not been a panacea for resolving the problems of the international economy. Furthermore, and perhaps this is the weakest point, the fluctuations observed between 1973 and 1987 were very large, much more so than expected when the system was implemented in 1973. The cause of these fluctuations is not necessarily a result of the structure of the variable system itself, but rather of the number of variables which can affect

TABLE 1.2. *Monthly changes in nominal effective exchange rates, 1961–1983*

	1961–70	1974–83	1961–83
Canada	0.26	0.82	0.61
France	0.22	1.01	0.64
Germany	0.27	1.00	0.67
Italy	0.13	1.06	0.64
Japan	0.15	1.74	0.95
United Kingdom	0.26	1.32	0.81
United States	0.11	1.35	0.78
Weighted average	0.20	1.18	0.71

Source: IMF (1984).

the rate of exchange. The most severe problem of such fluctuations is not related to their frequency, or magnitude, but to the instability they cause in the business environment, which introduces additional costs for firms engaged in international business.

We will first examine the magnitude and importance of these fluctuations and then analyse their impact on international trade. To measure the volatility of exchange rates there are four different procedures:[15] the bilateral nominal rate of exchange (between two currencies), the nominal effective exchange rate, the real bilateral exchange rate, and the real effective exchange rate. The effective rates are different from the bilateral rates because they represent the average of the bilateral rates between two countries that maintain commercial ties. The real rates are nominal adjusted rates by price levels. Selection among the different types depends on the objective of the analysis: the nominal rate demonstrates more clearly firms' international exposure to exchange risks; the real rate reflects more accurately differences in the competitive position of each country, indicating at the same time changes in the allocation of resources.

We will study the real rate, since it expresses changes in the allocation of resources. In the event that exchange rates are erratic, the system of flexible rates will cause distortions in the allocation of resources and, consequently, instability in the international business environment.

The analysis of exchange-rate values between 1960 and 1983 shows that, in the period of fixed rates (1960–73), variations in real rates were relatively small. However, since 1973, real rates have varied considerably (see Tables 1.2 and 1.3).

[15] See IMF (1984).

TABLE 1.3.　*Monthly changes in the real effective exchange rates, 1961–1983*

	1961–70	1974–83	1961–83
Canada	0.35	0.91	0.63
France	0.38	1.98	0.69
Germany	0.42	1.00	0.76
Italy	0.30	1.00	0.68
Japan	0.68	1.85	1.27
United Kingdom	0.45	1.46	0.97
United States	0.28	1.41	0.87
Weighted average	0.38	1.22	0.83

Source: IMF (1984).

In accordance with the theory of parity purchasing power in a floating exchange-rate regime, the real exchange rate remains constant if conditions in the product and financial markets do not change, while the nominal rate fluctuates to adjust for the differences in the internal and external rates of inflation.

However, in practice, this has not occurred and real rates have fluctuated significantly. The reason is the relative lack of price flexibility in the product markets in response to currency swings, which shows that these swings have a faster transitory effect on real rates than on product prices. This effect is known as 'overshooting' and represents an excessive adjustment of real exchange rates. Thus, in the end, the variability of the real rate has negative consequences for the allocation of resources and raises the level of uncertainty in the business environment.

This variability in the exchange rate can be the cause of a more permanent disconnection between the real rate of exchange on the market and the equilibrium exchange rate (or fundamental rate of exchange). In the literature two principal causes have been indicated for this divergence.[16]

The first is the lack of credibility of a government's monetary and budgetary policies. Such a lack of confidence generates a gap between the real and expected rates of exchange, and the rate of exchange—which we will call theoretical—that would result if the government were to apply monetary and budgetary policies effectively. This divergence between expected and theoretical rates will affect the actual rate in such a way that it will most likely grow farther apart from the ideal value of equilibrium.

---

[16] See Dornbusch (1987).

The second cause lies in speculation. Speculation occurs most when the actual rate of exchange is different from that which it is thought it should be, due to a lack of political credibility or other reasons. Ideally, when this happens, financiers buy or sell in order to cover risks or obtain profits, depending on the case, forcing the exchange rate back into equilibrium. However, financiers can obtain even greater profits when the exchange rate is held at values visibly above or below the perceived level of equilibrium, which tends to stimulate speculation and wider rate fluctuations.

This second cause could well have been behind the erratic behaviour of the dollar between 1983–5 and 1986–7. (It is also one of the factors explaining the 'boom' and later 'crash' of the stock market in October 1987). Yet, if we take into account the lack of credibility of US economic policy since 1985 and the technical effect described previously as 'over-shooting', a more complete picture emerges for understanding the evolution of the dollar and its volatility over the last few years.

The effects of the excessive volatility of the exchange rates on international trade and business are negative. There are many reasons why this is so. Volatility generates costs in terms of greater uncertainty in the business environment, which are derived from the changes imposed on the allocation of resources; it affects the inflation rate and international trade, causes protectionist behaviour and generates high costs for the exchange risk not covered.

The majority of empirical studies presented until now do not take up arguments in sufficient depth.[17] Therefore, caution must be expressed in evaluating the results of these studies; the only thing they affirm is that exchange-rate volatility has not negatively affected exports, since exports have not declined in spite of this phenomenon. What is not explained in these works is the relative importance of other factors affecting the growth of exports, which may have compensated for the adverse effects of fluctuating exchange rates. This remains an open question, still pending a definite answer in empirical work. It should be kept in mind from a theoretical and general point of view, however, that the volatility of exchange rates has a negative influence on international business and, in particular, on the international operations carried out by banks. Banks are especially implicated due to the higher risks associated not only with exchange-rate variability, but also with interest-rate fluctuations, which, as previously noted, have also increased recently.

The volatility of exchange rates has transformed the banking business in two ways. The first involves a higher level of innovation in response to the changing nature of international finance. The second relates to the

[17] See the conclusions of IMF (1984).

globalization of the sector, resulting in part from the increase in capital flows stemming from the variable-rate system.

4.2.2. *The European Monetary System: a zone of monetary stability*

The European Monetary System (EMS) came into being on 13 March 1979. Its immediate antecedent was the 'European serpent', including the currencies of nine countries (which would later form part of the EC-9). However, after a short period (less than two and a half years), France, Britain, Ireland, Sweden, and Italy all abandoned the 'serpent' for different reasons. The currencies of Austria, Switzerland, and Norway had also been included in the system (the first two as official countries with full rights); however, by the end of 1977, the system had fallen almost entirely under the influence of the German mark.

In July 1978, the Council of Ministers of the EC started to study a plan to implement a new monetary regime which would include all the currencies of the Community. This plan was approved in December 1978 and began operating in March 1979.

In the first phase, the principal objective of the new system was to create a zone of monetary stability in Europe; the second phase was characterized by the launching of a kind of European Monetary Fund. As experience shows, this still has not been possible, not only for technical reasons but also due to a lack of political will. The recently approved Delors Plan—in the event that it is fully implemented—will contribute to the creation of a Central European Bank and European currency, reaffirming as such the initial aim of the EMS. In any event, in its 11 years of existence, the EMS has had a generally positive effect, and those who thought a fixed but adjustable system would not work were mistaken.

In addition to the objective of monetary stability—an imperative for reducing financial and commercial instability— the founders of the EMS had three other goals. First, they sought to limit speculative risks within a system of flexible exchange rates: the new scheme does not eliminate speculation—which in periods of crisis is especially significant—but does reduce it. Secondly, they tried to help to establish economic discipline, at least within the realm of EC countries. This was to be done by embarking on a fight against inflation as a first priority of economic policy. Thirdly, they aimed to form a large European economic region comparable in size to the United States in terms of international leadership, though this objective was too ambitious if we take into account the way in which it was to be done.

The European Monetary System establishes a system of fixed parities, with permitted fluctuations within a range of ±2.25 per cent. There is a wider band of ±6 per cent for those countries unable to keep their currencies in a narrower interval. The system is based on the principle of monetary intervention when the rate of exchange reaches the highest or lowest level of permitted fluctuation. In the event that this happens, the financial authorities of the two countries affected by this situation—one with its currency appreciating, the other with its currency depreciating—should buy or sell the weak or strong currency, depending on their respective positions, to re-establish equilibrium. It always involves a type of bilateral intervention.

A second type of arbitrage exists called 'intramarginal' intervention. This type of intervention is guided by a variance indicator which gives the price of each country's currency in relation to the average of all other EC currencies. It expresses valuation from the point at which the price of a currency separates from others in a significant manner. The indicator oscillates between 0 and 1, and intervention is necessary when there is a 75 per cent realignment within these parameters.

Official intervention can be administered in various ways: a readjustment of the prices of the implicated currencies, a strong intervention in the currency markets, a unilateral or co-operative modification of interest rates, or a more radical change in the economic policies of the most affected countries.

Readjustments of prices have been exercised often (12 times in the ten years of the system's existence, the last in January 1990). The principal cause of these changes in prices has been a lack of conviction among certain countries about correcting some basic internal inefficiencies, or a lack of co-ordination of economic policies, especially when some currencies have been under the strain of speculative attacks.

The most useful criterion to evaluate the results obtained by the EMS throughout its 11 years of existence is the level of exchange-rate stability the system has achieved. The importance of the variability of exchange rates emerges from the costs that they generate, especially when the divergences with respect to a given level of equilibrium persist in the medium term, since the associated risks are more difficult to cover.

In analysing the stability of exchange rates within the EMS two separate periods can be distinguished. Until 1983, the EMS was characterized by various degrees of tension and frequent changes in the official rates of the system. In addition, according to some indicators, the variability of rates in this period increased. But in recent years variability has clearly declined, as can be seen in Table 1.4. The main explanation for the manner in which these fluctuations evolved lies in the second oil crisis, which generated a new supply shock in the real sector of the industrialized economies. The lack of a more pronounced macroeconomic adjustment in some countries and the lack of policy co-ordination also help to explain this greater level of instability. In general, the nominal rate of exchange was a factor of adjustment utilized in confronting the balance of payments crisis.

TABLE 1.4. *Monthly exchange-rate variability before and after the EMS*[a]

	1979	1982	1986
EMS currencies (Weighted average)	11.8	15.3	6.1
Dollar	38.6	33.2	34.1

[a] Monthly data variability has been measured as the ratio of variation in effective monthly exchange rates (multiplied by one thousand).

Source: IMF (1984) and estimate.

Since 1983 a greater degree of stability has been registered within the EMS, probably as a result of the strength of the dollar up to 1985—which discouraged heavy speculation in relation to the German mark—and the absence of other notable economic shocks to the EC member countries. However, since 1985, when the Group of Five announced its plan to let the value of the dollar fall, other economic crises have occurred within the EMS: one in April 1986 and the other in January 1987.

These experiences have served to support an explanatory theory of the functioning of the EMS.[18] According to this view, the periods of greatest stability in the EMS coincide with those of a strong dollar (1983–5), while those of greatest instability coincide with those of a weak dollar (1979–82). But a curious phenomenon is evident between 1985 and 1990, a period characterized by a weak dollar, as well as a great deal of uncertainty with regard to its future value. In these circumstances there have not been more than two crises (in more than four years). Perhaps the most sensible explanation is that during this period EC member countries have gone to great efforts to co-ordinate their monetary policies, especially France and Germany, resulting in more stability despite a very unstable financial environment. A complementary explanation is that capital controls, especially in France and Italy, have limited speculative activities in some countries, including those where significant differences in interest rates exist, resulting in greater movement of capital between countries.

4.3. The foreign debt crisis

The foreign debt crisis (see Tables 1.5 and 1.6) began in August 1982 when Mexico declared itself unable to service its debt obligations to

---

[18] See Giavazzi and Giovannini (1986).

TABLE 1.5. *Debt indicators for developing countries, 1980–1987*

Indicator	1980	1981	1982	1983	1984	1985	1986	1987
Debt as a percentage of gross national product	20.6	22.4	26.3	31.4	33.0	35.8	35.4	39.4
Debt as a percentage of exports	90.0	98.0	117.6	134.8	121.2	143.7	144.5	154.7
Interest on debt	16.0	17.5	20.6	19.4	19.5	21.4	22.3	24.4
Interest on debt as a percentage of gross national product	3.7	4.0	4.6	4.5	4.9	5.3	5.5	5.8
Interest on debt as a percentage of exports	6.9	8.3	10.4	10.1	10.3	10.8	10.7	10.8
Private debt as a percentage of total debt	63.1	64.5	65.0	65.8	65.7	63.9	63.5	64.3

Source: Canals (1989).

foreign creditors. Later declarations from other countries followed the Mexican announcement initiating a series of discussions to renew loans, renegotiate interest rates, extend payment periods, and even exchange debt for shares in public companies within the debtor country.

The causes of the debt crisis are numerous: the impact of the oil shocks on the non-oil-producing countries; the fall in exports to the industrialized countries caused by the economic recession during the 1970s; the later decline in oil prices, which seriously affected the oil-producing nations; a lack of competitiveness among the debtor countries; and, finally, the increase in worldwide interest rates beginning in 1979.

Nevertheless, the debt crisis has two other causes: the imprudence of the international banks with regard to their loans to these countries, and the countries' poor economic management, especially in terms of inflation and external equilibrium, and undisciplined debt policies. Together, these factors shed considerable light on the Latin American debt problem.

These two causes, although central in bringing the crisis about, cannot be treated exclusively. As already indicated, each country is subject to different circumstances and, as a result, it is difficult to generalize. It is true that the economic policies of the majority of these countries had been inspired by a model of high investment, the financing of which was

TABLE 1.6.   *External debt indicators, 1987* ($bn.)

Country	Total	Payment 1987–9	Interest in 1987
Argentina	49,400	23,700	7,900
Bolivia	4,600	1,600	600
Brazil	114,500	61,400	20,000
Chile	20,500	9,800	4,200
Colombia	15,100	8,500	3,100
Costa Rica	4,500	2,400	700
Ecuador	9,000	4,300	1,800
Ivory Coast	9,100	4,000	1,700
Jamaica	3,800	1,600	600
Mexico	105,000	44,900	22,200
Morocco	17,300	8,100	2,600
Nigeria	27,000	12,200	3,800
Peru	16,700	7,900	2,200
Philippines	29,000	12,000	4,600
Uruguay	3,800	1,300	700
Venezuela	33,900	15,900	6,600
Yugoslavia	21,800	10,200	3,700

Source: Canals (1989).

insufficiently based on internal savings. The availability of external credit and the proclivity of the international banks to invest in those countries were the necessary complements.

The initial response to the debt crisis was a renegotiation of terms on the part of each country and its creditor banks, accompanied by policies of internal adjustment tied to approval by the International Monetary Fund. This was probably the correct strategy, and it aroused positive expectations with the beginning of the US economic recovery in 1983, and the consequent recovery of prices for some primary goods. Only high interest rates continued to present an obstacle to a solution to the problem.

The results obtained by the debtor countries were less than expected, however. While some countries managed to achieve notable improvements in their current accounts, this was due not to increases in exports, but to the suppression of internal demand and imports. Thus, between 1981 and 1985, aggregate investment in Latin America dropped by as much as 6 per cent.

The difficulties these countries faced in maintaining growth offer one of the keys to understanding the policy changes of the 1984 accords, which were later consolidated by the Baker Plan in 1985. This consisted of an extension of the debtor countries' repayment periods, but with little change in associated interest costs. It was thought at the end of 1985 that this would represent a sufficient reduction of annual service payments. However, this was not the case, and new alternative forms of solution emerged, such as swaps of debt for shares, the indexing of payment schedules in relation to export prices, or an increase in direct investment in the debtor countries—on occasions through the privatization of public companies. In addition to these solutions, creditor banks offered a partial debt relief and increased their provisions for bad loans.[19] This strategy was initiated by Citicorp after Mexico reached a settlement with its creditors in July 1989.

Is debt relief necessary? Are other alternatives available? The answer to these questions is complex. The strategy of gradual payment followed by each country from 1984 until the end of 1987 demonstrated something of considerable importance: that the lending banks could not be allowed to go close to bankruptcy as a result of their loans to developing countries, which provided a stronger base for international economic stability. However, the economic situation of the developing countries has not greatly improved. Their real growth in the last few years has been small and living standards have fallen.

The question of debt condonation is a matter of equity (of help from the favoured to the less-favoured countries), but also of efficiency. In effect, excessive debt relief could further destabilize international financial markets, if all hope of recovering past loans is lost. Furthermore, some countries, such as Turkey and Bolivia, have made serious efforts to comply with their debt payments and effect positive economic change. A third argument against relief from the point of view of efficiency is that developing countries would lose an important external incentive to proceed with needed structural adjustments. Lastly, relief would further erode their financial credibility before the international community.

These problems could be mitigated through appropriate measures, for example, by compromising on the issue of relief. Together with the ongoing reduction of debt, all the countries in this situation need to follow very orthodox economic policies, as much in the macroeconomic realm, especially in the fight against inflation, as in the area of micro-economic management: price setting according to the market, gradual trade liberalization, improvement of management, and privatization of public enterprises. The objective must be to achieve a level of export growth above the average interest on subscribed debts. This is the condition necessary for interest costs not to exceed the principal itself. Logically, these efforts have to be accompanied by other actions on the part of the creditors: no interruptions in the flow of funds, reduction of interest rates, and the dismantling of import barriers.

---

[19] A technical analysis of this solution can be found in Williamson (1988).

In any event, the basis for the success of any accord between debtors and creditors is the desire to co-operate in the solution of global problems, and the willingness to share the economic weight of those measures necessary to bring about a full solution. Without co-operation, any attempt to waive the debt will ignite a war of confrontation which would only further aggravate the problem. And the accord reached by Mexico in 1989 shows that co-operation, along with the provision of appropriate technical solutions, is the only basic philosophy for negotiation.[20]

The implications of widely fluctuating interest and exchange rates, and the debt problems related to the crisis of 1982, are clear. International financial instability has increased significantly in the last decade, as have the risks associated with interest and exchange-rate volatility, threatening the continued well-being of the world's major financial institutions. The management of banks has become much more complicated in recent years, and, in view of the circumstances described above, it will be necessary to contribute to a reduction of international financial instability. This would positively affect both the worldwide economy and the competitive positions of the banks themselves. Also, the creation of more stable financial zones such as the EMS, as well as the negotiation of foreign debt, will certainly contribute to advances in this direction.

5. INFORMATION TECHNOLOGIES IN THE BANKING INDUSTRY

Banks have traditionally made great use of information technology. Without such applications, labour costs would be higher and the speed with which many tasks are executed would be reduced. Information technologies reduce costs and, more importantly, they allow for changes in the traditional manner of operating a bank, since information offers new business opportunities and different ways of managing traditional operations.[21] In effect, technological innovation has generated the development of a very wide range of products, limited only by the level of demand they are capable of sustaining.

The process of large-scale automation started at the beginning of the 1960s, with the incorporation of large computers in the banks' central offices, offering great savings in costs and time. The following decade would see the development of telecommunications and computer tech-

---

[20] The facts have confirmed the validity and strength of the call for co-operation and the international solidarity which Pope John Paul II has expressed with clarity in the encyclical "Sollicitudo rei socialis", as a philosophy to resolve international economic differences.

[21] Mookenjee's work (1988) is of considerable interest with regard to this subject.

TABLE 1.7. *Number of credit cards in circulation, 1987*

Country	Total no.
France	20,380,000
Germany	2,570,000
Italy	3,220,000
Spain	9,450,000
United Kingdom	43,750,000
United States	700,000,000

nology giving users greater flexibility in the execution of withdrawals, deposits, and transferals of funds. These innovations have resulted in significant cost reductions; however, the most serious effect of this second wave of innovation is that it has distanced customers from banks' main offices, prohibiting the possibility of selling new services.

At the present time, we are undergoing a third wave of innovation, characterized by the need for banks to be nearer to their clients. The focal point of this new process is the CIF (customer information file), and its principal advantage is that it permits the central storage of all the information available on a particular client. Naturally, this provides for better knowledge about the client and the ability to separate specific services by age, economic, or professional circumstances. We will return to this subject in Chapter 10.

In any event, an important number of technological innovations have completely transformed classic banking operations, increasing the speed at which certain transactions are executed and thereby reducing costs. Among the new services made possible by technological innovation, the most notable of all is the credit card. The credit card is not just a means of payment, but an instrument of identification which permits the realization of economic transactions without the use of cash and which also facilitates the acquisition of cash when the user needs it.

There are many types of credit cards (see Table 1.7). One is the credit card in the strict sense, which offers the client credit against the issuer of the card. Another type is the debit card, which permits transactions to be charged against the client's account, up to a certain limit. Thirdly, there are transaction cards, which permit certain operations to be effected through the use of automatic bank terminals. Finally, there are commercial cards, which enable their users to acquire goods and pay for services at specific commercial outlets. As a result of the existence of so many different types of card with various applications, at times the generic term 'plastic' is used in place of 'credit'.

TABLE 1.8.   *Volume of credit transactions, 1986*

Country	Credit card transactions[a]	% of the total	Cheque-based transactions[a]	% of the total
France	325	5.4	3,558	59.5
Italy	4	0.5	585	80.1
United Kingdom	452	9.6	2,779	59.0
United States	5,300	11.5	40,000	87.0

[a] In billions of units.

Source: Banco Bilbao Vizcaya, 1988.

TABLE 1.9.   *Installed automatic teller machines, December 1987*

Country	No. of automatic tellers	% increase 1985–7	No. of networks	No. of tellers per 1,000 inhabitants
France	11,895	26	5	214
Germany	5,396	116	1	88
Italy	5,658	166	1	99
Spain	7,873	63	4	203
United Kingdom	12,520	43	11	221
United States	63,865	8	609	262
Rest of the world	320,200	—	—	—

Source: Banco Bilbao Vizcaya, 1988.

As can be observed in Table 1.8, in countries such as Italy, Germany, and Spain, there is room for a substantial increase in the number of cards, especially taking into account the high volume of operations effected through credit cards in such countries as the USA.

Another important technological application is the ATM (automatic teller machine). This terminal permits users to carry out a range of automatic services in conjunction with a variety of operations, such as cash withdrawals and deposits, transfers, and information processing. Also, in recent years the functions of ATMs have been expanded into such interesting areas as the processing of limited credit and mortgage applications. The first automatic tellers were installed in the late 1960s. Their growth in the 1980s has been spectacular and, at the end of 1987, there were more than 400,000 installed around the world, as shown in Table 1.9.

With widespread use of automatic tellers, their applications have been enriched and the service they provide has been much enhanced by an increase in the number of operating hours offered to customers. In this way traditional bank tellers have become a thing of the past and ATMs can now be physically located farther away from central-office locations. Naturally, all of this has had far-reaching consequences for the banking business and has greatly affected firms' strategic planning, as we shall see later on.

Another important innovation is the so-called 'home banking', a system which permits teller-windows to be transferred directly into the home. The terminals utilized tend to consist of a telephone receptor and a television screen, along with a small personal computer. Through the latter piece of equipment it is possible to conduct anything from the simple verification of a bank balance to execution of certain stock-market operations. In spite of all the system's possibilities, its use among the public is less popular than credit cards or automatic tellers. This inno-vation, in spite of its appeal, is being diffused slowly and some banks that were promoting the system have now practically abandoned it (such as Midland and Chemical Bank).

A similar system is the point-of-sales terminal (PST). This permits the execution of on-premise commercial operations without the need for cash. The PST is a card-reader that electronically transmits information from points of sale to a bank database, where authorization is given (or not given) to effect a specific transaction. Perhaps the main virtue of the PST is that it is physically supported by a telephone inside the cash register, with a design that is compatible with all types of credit cards.

Another interesting development is the 'smart card', which is a much more sophisticated type of credit card. It stores information on a micro-chip with memory capacity. Recently, these cards have been designed so that they practically function as pocket computers. This instrument presents notable advances with respect to the traditional card: the capacity to store more data, greater security (the stored information is automatically erased in the event of fraud), proof of transactions (which can be used in the case of a return), and lastly, since individual trans-actions are not linked directly to a central computer, it allows for a much more rational use of information.

The main impediment to the widespread diffusion of these cards is their high cost, estimated to be ten times that of traditional cards. However, it is expected that further advancements in information technology will allow the instrument to be obtained at lower cost within the next five years.

However, as is the case in other sectors of the economy, technological improvements can involve high risk and major operational difficulties. The need for elaborate back-office computers to store documentation—cheques especially—and effect accounting operations requires competent management of sophisticated technology. In effect, banks make large

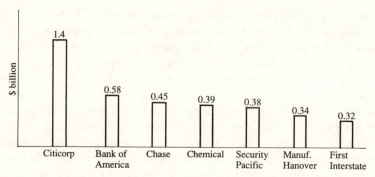

FIG. 1.2   *Investments in information systems by seven large US banks*
Source: The Economist, *1989.*

investments in order to consolidate internal information networks, but
the results are not always satisfactory and mistakes can be costly. Required
investments, as indicated in Figure 1.2, can also be enormous. Such has
been the case for Citicorp, the largest US bank and the one that spends
the most on information systems. Some time ago it bought a system
called TAPS (Trade Analysis and Processing System). It was designed
mainly for investment banks that manage large volumes of financial
assets. The system was good, but too specialized for Citicorp, whose
activities cover such a broad range of areas. As a result, the company
finally divested it—at a loss of $50 million.

Something similar happened to the Bank of America in 1988 when it
bought special software to do certain accounting operations. After having
tried unsuccessfully to apply the new software for five years, it got rid of
it—at a cost of $80 million. As is evident, such errors can be quite
expensive.

Another example is the application of home banking systems. They
deal with expensive transactions that banks offer their clients at little or
no cost, but that are difficult to manage. This has limited their popularity
to the extent that some banks, such as those mentioned previously
(Midland in the UK and Chemical Bank in the US) have discontinued
them.

These anecdotes demonstrate how firms have accepted new technologies without sufficiently evaluating the costs and profit potential of
new applications. In addition, managers must assess whether or not the
business to which the new technology relates is important enough within
the overall strategic goals of the bank to justify its application. Some US
banks require an annual return of 20 per cent on an initial investment in
new computers, which must be reached through related cost reductions.[22]

[22] See *The Economist*, 'Survey on International Banking', 25 Mar. 1989, p. 34.

But this does not mean that banks will stop this flow of investment to improve their internal operations. There are three reasons why this is so. In the first place, the volume of paper the banks handle is extraordinarily high and requires vast resources to manage. This is especially true of cheques. Thus, in the United Kingdom the payment of cheques demands that firms physically present each bank with the original cheques. For this reason, some British banks have transferred the majority of their cheque-handling operations from the bank branches to specialized centres, with a view to liberating space and time at the points of sales. Without a doubt, electronic cheques, which were recently introduced, will provide some relief in this area, since it is estimated that processing costs are 90 per cent less than for traditional cheques.

The second reason is that banks must continue looking for ways to use specific information to complete specific tasks. And the globalization of the banking business adds another requirement: the availability of needed information in any place at any time. Given the importance of infor-mation in the banking sector, the basis for assessing whether or not a bank is global rests on how much information it has available through-out its worldwide network, facilitating the provision of detailed services within a highly integrated international framework.

Thirdly, the use of certain information technologies not only allows banks to reduce processing costs, but also to diversify into new businesses compatible with core activities, offering a broader distribution for the fixed costs characteristic of investments in information equipment. This has important implications for the banking business, as the widespread diffusion of information technologies, which strengthens the relationships between banks and their customers, will help to personalize the sale of financial services. Bank offices will undergo a radical transformation, and will soon consist of more than mere windows at points of sale of financial services.

This will greatly affect banks' operating results since the financial margins—financial revenues minus financial costs—will be reduced, increasing the proportion of revenues from the sale of other financial services.

We will return to these questions later in our analysis of the changing competitiveness of the European banking industry. However, an issue relevant to the present discussion is banks' size. This factor has been identified as a key to the development of competitive advantage as it permits banks to enjoy economies of scale and a wider reach of oper-ations. We will analyse this point later, but what should be emphasized here is that the large banks have the ability to acquire the best infor-mation technology, permitting them to process greater quantities of information more quickly than their smaller competitors.

Empirically, it is also true that large banks spend more on information-processing than the small banks. According to estimates from McKinsey (1988), large banks invested 15 per cent of total expenses in information equipment, while small banks invested only 10 per cent, which represents a very considerable difference. It is evident, therefore, that size is a relevant factor in profitability and costs in relation to the management of information equipment, the rational use of which can offer great competitive advantages. However, some observers have proposed that the smaller banks, with their lower levels of information technology, have an advantage, especially within an intensely competitive market: they are attractive to the larger banks wanting to expand through acquisition and can thus apply new technologies to the acquiring bank.

An important effect of the diffusion of information technology is that bank branches, at least in their present form, are becoming obsolete. For many years, the growth in the number of branches throughout Europe was the result of the desire to obtain a larger market share by being closer to clients. Also, in countries such as Spain, this was the only way banks could increase their size and profitability, especially taking into account that the prime basis of the client–bank relationship was geographic proximity.

Nevertheless, the competitive advantage which physical proximity once provided, can now be achieved through the installation of an ATM, and associated administrative functions can be moved to centralized offices.

The advantages of automated offices are twofold: they allow banks to expand at significantly lower costs; and they help mitigate the problem of excess office space and personnel (more evident in French and Spanish banks). These will become even more acute with the increase in the number of mergers.

6. THE ACCELERATED PROCESS OF FINANCIAL INNOVATION

Together with the forces of change described above, financial innovation has been another factor of great importance, especially in recent years. Financial innovations have profoundly affected the nature of the banking industry, including the distribution of services aimed at different sectors of the economy.

This intensity of innovation within the industry raises several important questions.[23] The first of these relates to the origins and causes of innovation, and the possibility of their identification and control. The second refers to the impact of innovation on the markets, in terms of interest rates, prices of financial assets, and perceptions of risk. A third question

---

[23] See Zecher (1984).

of interest has to do with the effects of innovation on the management of financial entities, and on specific strategies related to innovative activities.

More is known about the second and third questions, as a degree of consensus exists among industry observers and participants on the actual process of financial innovation. In recent years financial institutions have learned that certain changes within the business environment can act as a stimulus to the development of new products, which normally respond to the needs of clients who are not sufficiently satisfied with the new situation.

The first aspect of environmental change which has spurred the innovative process in the banking sector is government regulation. In Section 2 of this chapter we looked at the process of regulation and, later, deregulation, which has taken place throughout the industrialized countries during recent decades. For a financial firm, regulation has two general implications. It creates entry barriers on the one hand, and adds implicit taxes on permitted operations on the other. As a result, banks have sought ways to escape the legal corset forced on them by the authorities of many countries. In any event, and as opposed to other sectors of the economy, the role of the public sector, and, in particular, of the central banks, is crucial to any explanation of the level of innovation, as well as the industry's structure and competitive composition.

A second factor to consider is technology. In the previous section we saw how the application of new technologies in the banking world has provided not only for the reduction of certain operating costs, but also for the creation of new services for clients. Information systems have made available more information, and this allows for better decision-making with regard to investments.

The third aspect to emphasize in this context is the financial environment. The instability of international finance in addition to the internal problems some countries have had to confront in terms of inflation has created an even more difficult operating environment. Many new financial instruments have been introduced under these conditions and they have been a formidable force in promoting financial innovation.

A final aspect worth noting with regard to the changes in the financial industry is the process of disintermediation, in which US and British companies have played leading roles. This has caused significant growth in the issue of commercial paper. As we shall see in Chapter 2, the process of disintermediation has provoked great changes in the structure of banks' assets, and consequently in their activities.

Table 1.10 demonstrates the evolution of bank debt in certain countries. As can be seen, between 1976 and 1984, the bank debt of non-financial firms has fallen considerably. Initially, it was the large companies in search of funds that went to capital markets instead of banks. However, this has also spread to medium-sized firms, whose solvency is less evident

TABLE 1.10.   *Sources of funds in non-financial companies*[a]

Type of debt	United States		France		Germany		Italy	
	1976	1984	1976	1984	1976	1984	1976	1984
Non-bank debt	48.1	48.5	15.9	42.0	16.6	25.8	33.5	56.4
Bank debt	33.4	31.1	58.5	45.7	75.5	64.3	23.9	14.2

[a] As a percentage of total funds.
Source: OECD (1987*c*).

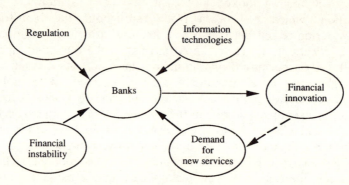

FIG. 1.3   *The process of financial innovation*

for the investor; now they can also issue assets of this type, the means for them being the acquisition of a guarantee of payment by a banking entity.

The innovative process in the financial sector can be explained, therefore, as a process of transformation within the institutions and markets responsible for propagating new types of instruments. This transformation, as we have seen, stems from the interplay of four elements: financial instability, regulation, automation, and the demand for new services, as reflected graphically in Figure 1.3.

6.1 Financial innovation in international banking

Many financial innovations have appeared in the international markets in recent years. Tables 1.11 and 1.12 list some of those which have had an impact in a number of countries.[24] It will be useful to examine some of these in more detail.

---

[24] A treatment of this question can be found in any text on finance. See e.g. Brealey and Meyers (1988).

TABLE 1.11. *Causes and effects of financial innovation*

Causes	Effects	Financial innovations
1. Changes in the financial environment:		
Large public deficits	Growing financial needs of public administrations	Long-term securities with indexed interest rates
		Variable interest rate loans
Financial instability in the international markets (floating exchange rates, oil and debt crises in the developing countries)	Increase in interest-rate volatility	Reduction of medium-term maturity for loans with fixed interest
	Increase in the international flow of funds	Emergence of futures and options markets
	Volatility of exchange rates	Substitutions of loans and credits for financing by negotiable instruments
	Negative impact on banks' solvency	Deposits with money-market-based interest rates
High inflation	Rise of interest rates	Swaps of currencies and interest rates
		Appearance of new competitors
2. Development of information technologies	Reduction of processing costs for financial transactions	
3. Changes in financial regulation (deregulation)	Increase in competition	
	Liberalization of national financial systems	
4. Demand for new services	New investment possibilities	
	Possibility for investors to diversify their portfolios	

First of all, we will look at floating-interest-rate bonds, commonly known as 'floating-rate notes' (FRN). They appeared in the beginning of the 1970s as an answer to high interest rates, and the reluctance of investors to acquire fixed-interest notes, independent of the situation in the market. The advantage of these kinds of instruments is that they transfer the risks associated with interest-rate fluctuations to the lender.

TABLE 1.12. *Financial innovations since 1970*

New fund-collecting techniques (for borrowers)			New risk-management techniques (for investors)	New financial advice and support techniques	Automation of traditional services
Direct debt					
Bonds	Equities	Floating-rate loans			
Floating-rate bonds	Venture-capital finance		Securitized assets	Note-issue facilities	Electronic fund transfer
Zero-coupon bonds	Secondary and tertiary market listings		Mutual funds (securities and money-market instruments)	Credit-enhancing guarantees	Electronic security markets
Eurobonds	Convertibles and other equity-linked bonds		Options (securities and currency)	Leveraged buy-out	Computerized cash management
Junk bonds	Non-voting shares		Futures (securities and currency)		Programme trading
Convertible bonds					Global trading
Commercial paper					
Interest-rate swaps					

Source: OECD (1987*c*).

The growth of this instrument began in 1974, when the large international banks started issuing vast quantities of FRN. The reason was that banks opted for the issue of these titles to attract capital at reasonable rates. Although initially this product became popular in the international markets, little by little it has appeared in the different national financial markets as well, due to its perceived advantages in relation to volatile interest rates.[25]

The FRN market grew to reach a volume of $56 billion dollars in 1985 and began to decline thereafter. The reason lies in the appearance of alternative instruments such as 'swaps' which could attract funds and simultaneously cover interest-rate risks. The rise and fall of the FRN is characteristic of financial innovation. A product emerges in response to certain market needs, but the related circumstances disappear or more sophisticated instruments appear which causes the new product to enter a period of decline, eventually leading to its obsolescence.[26]

Financial futures constitute another innovation of great magnitude.[27] The existence of currency-forward operations between companies and banks goes back to past decades, but the organization of a market with forward contracts occurred in 1972, when these contracts were admitted for the first time on the Chicago Stock Exchange. Three years later, futures on interest rates were introduced, also on the Chicago Stock Exchange. The negotiation of these types of contracts was extended gradually throughout the world. The principal asset which was negotiated was related to contracts on government bonds and, in particular, on US government bonds and three-month Eurobonds.

The great success of futures is owed to the fact that the risk in the international financial operations has grown considerably as a result of the price volatility of certain assets. Also, new technologies provide protection from these risks in an important way: large computers permit the creation of a centralized market in which operations are immediately executed.

What does this contract consist of? It consists of a standard contract to deliver a quantity of predetermined financial assets on an agreed future date. These contracts have some general characteristics which allow them to trade in an organized market at a minimum cost. Only the price remains variable, and it is set through public trading.

The most common future is connected to interest rates. An investor utilizes this instrument by buying bonds, for example, in cash with borrowed money, to sell them in cash and lending money. In this way the

---

[25] See Harrington (1987).

[26] See Silber (1975) and Harrington (1988).

[27] In the Bank for International Settlements (1986) a good presentation of this and other financial instruments specific to international banks is offered. Also of interest is Fernández and Pregel (1989).

investor is able to neutralize risk. The principal characteristics of this type of future are the following: the transactions are carried out in organized markets in a standard way; there is a clearing-house between the buyer and seller; and once the details of the contract are agreed upon, an initial deposit of up to 3 per cent of the value of the contract is required. It is important to note that a futures transaction is not a title or a financial asset, but the contract itself: the promise of delivery (or receipt) of a title associated with a contract is compensated by opposing transactions realized in the same market. The difference between futures and 'forward' interest rates (FIR) consists largely in the uniformity of their contracts, as well as the role played by the clearing-house.

Exchange-rate futures have also played a prominent role in the recent history of financial innovation. They provide a means of covering associated risks in times of market instability and operate in much the same way as interest-rate futures.

Another important financial instrument is the swap of shares and interest rates. Swaps are valued as much for their operational flexibility as for their negotiated volume. Swaps are used in relation to interest rates of specific financial assets in the foreign-exchange markets or on other financial assets. The types which have experienced greatest growth are those related to interest rates and to currencies, since they offer an adequate response to the volatility on the prices of financial assets, and as a result, they help to improve the management of risk associated with interest and exchange rates.

Swaps were devised at the beginning of the 1980s. They grew out of certain practices having to do with international banks effecting parallel or reciprocal loans with the intention of circumventing exchange-rate regulations in certain countries. Their existence in the UK dates back to 1979. A British firm having pounds but wanting to diversify its currency exposure was prohibited from buying foreign exchange directly. But if it found a foreign company with a surplus of treasury in a denomination of interest, both companies could effect a parallel loan. The versatility of this instrument soon made it very useful in certain financial operations which no longer had to do with the circumvention of exchange controls. A modern-day currency 'swap' consists of the exchange of financial assets denominated in different currencies, while an interest-rate swap consists of the exchange of financial assets tied to specified interest rates (fixed or floating).

Specifically, the clearest advantage of swaps is that they make it possible for treasury officers to take advantage of opportunities in a variety of markets, in addition to their added usefulness in covering risk associated with interest and exchange-rate.

The typical case of an interest-rate swap is that in which one party issues a financial asset at a fixed interest rate; the other party issues an

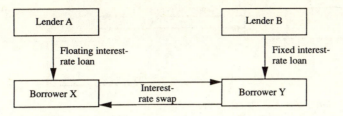

F<small>IG</small>. 1.4 *Diagram of an interest-rate swap*

asset at a variable interest rate, and both exchange payment obligations (see Figure 1.4). What is the advantage obtained by the various parties from this operation? Normally, if one party obtains assets with floating interest rates, desires a fixed compromise on the swap, and cannot accede to the capital-market ideal conditions, he might find it interesting to exchange that part of the debt with another firm which has fixed-interest debt. If the differential in interest rates is sufficiently significant, the firm indebted with floating interest will be interested in effecting a swap. The attraction for the firm indebted with fixed interest is that they can obtain a premium on the interest rate always less than the differential previously indicated.

The amortization of principal is not the object of the interchange and is assumed from the first moment by the corresponding debtor. The periodic payments which each side effects correspond only to interest. In this way, it can occur that the respective borrowers do not know that their lenders have come to an accord to carry out a swap in interest rates.

The role of banks in this operation can be active or passive. It is active when banks intervene as part of the agreement. Banks can be interested in participating in a swap of interest rates to meet the requirements of certain clients or to adapt their financial costs to the structure of their investment. It must be remembered that the banks have a financial cost coming from relatively fixed deposits. The role of the bank can also be passive; this is what occurred especially when these operations began. In those cases, the bank acted as an intermediary with the aim of matching line needs of both parts and bringing about the close of the operation successfully. However, in 1985, the International Swap Dealer Association put forth a number of swap-negotiation clauses for interest rates which have been converted into standard practical rules for these operations. Presently, anyone can know at any moment the price of swaps since they are always available.

We turn now to the swap of foreign currencies. This operation permits a company X which has easy access to currency 1, and is interested in possessing currency 2, to put this at the disposal of company Y with easy access to currency 2. This operation, as we have already said, was in place

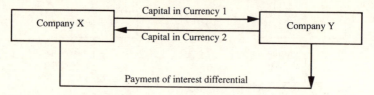

FIG. 1.5   *Diagram of a currency swap*

at the beginning of the 1980s, in the presence of strong controls on the free movement of capital. In areas in which these have been relaxed, swaps will continue to disappear. However, swaps also offer the possibility of covering the long- and medium-term risk of exchange in the presence of strong international economic instability, which was an especially important function of the 1980s (see Figure 1.5).

A currency swap involves three distinct phases: initial exchange of capital, exchange of interests, and final exchange of capital. Once the contracting parties have agreed upon the value of initial principal, they proceed to exchange the payments for interest. The exchange of principal is effected according to the exchange rate at the beginning of the operation which, in principle, is the same as that which terminates the swap at the end of the operation, with an agreement of resale (or repurchase). The exchange of interest is effected on the basis of interest-rate parity: the party which receives the currency with the highest interest rate will pay the differential corresponding to the other party.

The bank's role as intermediary between the participating parties in currency swaps has grown significantly in recent years. Their responsibility consists of satisfying the needs of each side through a simultaneous swap, carried out with each party separately. However, the swap operations have an important advantage for the banks: they deal with operations outside of the balance, as with options and futures. For this reason, the bank ratios bear no relation to these types of operations, and thus they can expand as much as the bank deems opportune.

In this chapter a brief review has been given of some of the consequences of financial innovation on bank management. We will return to this topic in analysing the banking sectors of various countries in the chapters ahead.

6.2. A special example: financial innovation in Spain

The process of financial innovation previously described has had wide-ranging consequences for Spanish banks.[28]

---

[28] Ample treatment of this is provided in Toribio (1984) and Gutiérrez and Chuliá (1988).

TABLE 1.13. *Spanish financial institutions' assets* (% of GDP)

	1975	1981	1983	1985	1987
Public administration	7.4	8.0	9.4	19.6	27.8
Companies and families	81.2	76.5	76.7	68.9	66.3
External sector	10.7	25.9	23.4	23.1	22.7
Inter-bank operations	23.7	28.0	41.9	45.9	41.9

Source: Banco de España, 1988.

Three factors are primarily responsible for innovation in the Spanish banking industry. The first is the public deficit with its associated financial requirements, which has greatly affected financial intermediation and monetary policy during the last few years. As can be seen in Table 1.13, government bond issues quadrupled between 1975 and 1987, while financing for companies and families declined sharply. This could mean that non-financial firms came to rely on alternative means of financing, or that the private sector was driven—at least in part—by the financial needs of the public sector.[29] In any event, the result has been a greater degree of concentration in banking activities for bonds issued by the public sector. In addition, the attraction of these in terms of profitability has led banks to issue a large number of instruments which could compete with these public bonds.

A second factor in the evolution of financial innovation in Spain has been the lack of entry by new competitors. The competitive environment has been conditioned by the starting-point, characterized by heavy regulation of the markets and institutions. Then, the bank crisis of the 1970s and 1980s added a further element of reservation at the point of lifting certain regulations. Lastly, the financial needs of the public deficit have also hindered the deregulatory process and limited competition in a number of ways.

In relation to competitors, the three factors which have most stimulated competition have been the easing of restrictions on savings-bank activities, the entry of foreign banks, and the development of the stock markets (public debt, corporate equity), which had diverted funds that could be put directly into savings, further contributing to disintermediation. Table 1.14 shows the evolution of the level of intermediation in the Spanish financial system, expressed in terms of assets and liabilities in circulation. Between 1978 and 1987, an increase is evident in the proportion of financial assets held by savings banks compared with a reduction of those held by the banks.

---

[29] See Canals (1988) for a discussion of the 'crowding-out' process.

TABLE 1.14. *Evolution of financial intermediation in Spain* (% GDP)

	1975	1981	1983	1985	1987
Financial liabilities of the non-financial residential sectors (total)	123.4	127.6	152.6	193.3	215.1
Of public administration	13.4	23.1	35.0	47.0	48.5
Of companies and families	110.0	104.5	117.6	146.3	166.6
Financial institutions' assets compared with those of non-financial sectors (total)	109.7	105.2	115.6	117.7	117.8
Of the banking system	88.6	84.1	86.1	88.5	94.1
Of the comercial banks	65.9	60.0	60.2	63.9	61.3
Of the savings banks	27.7	24.1	25.9	24.6	32.8

Source: Banco de España, 1988.

The last factor which has influenced the innovative process in Spain has been, as in other countries, regulatory changes. In addition to the restrictions regarding exchange controls, which have adversely affected the internationalization of Spanish banks and the liberalization of interest rates, the most important aspect of financial deregulation is that it has coincided with the reform of direct taxation. The fiscal attributes of certain instruments issued by the public sector have significantly affected the innovative process, in some cases accelerating it by optimizing fiscal advantages.

These three factors clearly affected the process of financial innovation in Spain. Public-sector financial needs prompted in 1982 the issue of short-term bonds and discounted treasury notes which could be acquired directly by any investor. This was the beginning of large-scale financial disintermediation. Financial firms, in the face of intense competition, reacted in order to avoid losses due to lower deposits, not only creating alternative instruments, but also intermediating in the financial flows diverted from savings to public funds: this caused a temporary halt to the issue of treasury notes with the option to repurchase.

Initial public-sector financial disintermediation was followed by that of the large non-financial companies. In recent years, they have gone directly to savers through the issue of bonds and commercial paper which have a lower cost for the issuer. Banks have tried to react to this in the same way that they did to the bonds issued by the public sector: by issuing new competitive instruments, on occasion transforming their credit into negotiable financial assets, a process known as 'securitization'.

The role of financial assets issued by the public sector in the Spanish money markets is very important, especially since the initiation of a new accounting system for public-debt transactions. This system permits transmission and acquisition of shares issued by the public sector, with the physical support of a central accounting facility under the supervision of the Bank of Spain. This suggests an important advance in the modernization of this market because of the flexibility which it offers to participants. The system includes short-term treasury notes as well as long-term and medium-term debt titles.

This innovation process, both facilitated and inhibited at different times by the public sector, has caused tremendous change in the competitive conditions of the Spanish banking industry and is exerting considerable influence on banks' strategies and the structure of the sector.

PART II

2
The Evolution of the European Banking Industry: An Overview

1. INTRODUCTION

The objective of this chapter is to provide an overview of the situation in the banking industry in the five European countries analyzed in this study: Germany, Spain, France, the United Kingdom, and Italy. In the following chapters a more detailed analysis of each country is presented; here we will set up a comparative framework which will help to lay the groundwork for a deeper analysis. We will begin by examining the importance of the banking industry within the economy and later analyse the determinants of the banks' profitability.

While concentrating on the structure of the banking industry we must not lose sight of the forces described in the previous chapter that have had a significant impact on banking operations. Such factors as international financial instability, financial innovation, and the process of deregulation must be taken into account in order to analyse the structure of the industry.

Banks are traditionally classified according to the following categories: commercial banks, universal banks, savings banks, investment banks, co-operatives, and specialized institutions (such as mortgage banks and brokerages). This study will focus primarily on commercial banks.

Bank operations in these countries revolve around two possible models. The first corresponds to that of the 'universal' bank. In theory, a universal bank can legally perform any kind of banking operation. German banks are typical representatives of this model. The second possibility centres on the principle of specialization which requires banks to concentrate on specific financial activities, often dictated by government regulation. British banks provide the best illustration of specialized firms. In reality, these two examples constitute the opposite extremes, though a range of possibilities exists in between throughout the European industry.

In the United Kingdom the large commercial and investment banks have very distinct areas of operation. In Italy, there is also a certain degree of formal specialization, to the extent that a distinction can be made between firms engaged in short- and long-term operations. However, in other countries where historically there has been a degree of specialization, the deregulatory process and increased concentration are

TABLE 2.1.   *General data on the banking industry, 1986 ($bn.)*

	Banks among the 500 largest in the world	%	Total assets	Net profits	Return on assets (%)	Return on equity (%)
Europe	203	40.6	4,047	26.9	0.69	17.6
United States	110	22.0	2,187	17.4	0.47	17.4
Asia	126	25.2	3,345	15.0	0.80	26.0
Others	61	12.2	717	6.9	1.08	10.0
TOTAL	500	10.0	10,296	66.2	—	—

Sources: *The Banker* and Salomon Bros. (1989).

TABLE 2.2.   *Relative share of commercial banks, 1987* (% of assets over GDP)

Country	Banks	Savings banks	Others	Total
France[a]	84.4	15.7	29.8	129.9
Germany	44.0	70.3	32.3	146.4
Italy	88.5	26.7	2.9	118.1
Spain	87.6	43.7	4.5	135.8
United Kingdom	242.7[b]	—	37.6	280.3

[a] 1986 figures
[b] Includes the TSB Group
Source: Lladó *et al*. (1989).

leading to more differentiated activities. Germany, Spain, and France all support, to varying degrees, the development of the universal bank, as we will see in the chapters ahead.

2. THE SCALE OF THE EUROPEAN BANKING INDUSTRY

To facilitate the task of comparison, we will distinguish data corresponding to the savings banks from that of commercial banks. We will measure the relative size of the banking industry with respect to the overall economy using the ratio of financial intermediation, which relates the volume of a country's liquid assets to its GDP. Alternatively, the relationship between total bank assets and the GDP can be used. This indicator constitutes an indirect reference of the banking system's efficiency, explaining the volume of financing necessary to obtain one unit of a product. Table 2.1 presents some joint data, while Table 2.2 gives the value of that ratio for commercial and savings banks for the individual countries.

TABLE 2.3. *The banking industry in the EC economies, 1985*

Country	Employment in the banking industry	% of total employment	Value-added (ECU million)	% GDP
France	448,000	2.6	29,277	4.5
Germany	604,000	2.8	44,417	5.5
Italy	379,000	2.5	26,998	5.6
Spain	292,000	3.9	13,929	6.1
United Kingdom	527,000	2.4	70,240	12.6

Source: Price Waterhouse (1988*a*).

From the figures presented in Table 2.2 those related to the United Kingdom are particularly striking as the proportion of total industry assets is two and a half times GDP, a stark contrast to the figures for the other countries where the next largest is Italy's, with an industry asset value equal to 88.5 per cent of GDP. This discrepancy is explained by the fact that London is a major European financial centre very active in international financing and with a strong foreign banking presence. The relative proportion of the different types of institutions within the various financial systems is also of interest. Particularly noteworthy is the proportion of savings banks in Germany.

Table 2.3 shows some data on the banking industry in terms of total value-added contribution and employment in the different countries. From this perspective, the British banking system appears quite strong, which is probably a result of the circumstances mentioned previously. In the rest of the countries, the proportion of value-added from the industry is around 5 per cent of GDP, not an insignificant figure, especially considering the industry employment figures: the highest (in Spain) at 3.9 per cent of total employment with the others all under 3 per cent.

The study of banks' size is important to understanding the configuration, efficiency, and profitability of these entities. As we shall see in Part III, numerous studies have drawn correlations between profitability and size.[1] In the majority of industrialized countries a high level of concentration is evident, such that a small number of banks represent a large proportion of bank assets and liabilities, as shown in Table 2.4. The most significant concentration is in France where 35.1 per cent of total assets are in the hands of three major banks which, according to Table 2.8, are placed first, second, and fourth respectively in the ranking of the largest European banks. Additionally (from Table 2.5), 10 per cent of the

[1] The correlation is not clear, as shown by the majority of empirical studies. See Revell (1988), Short (1979), Smirlock (1985), and Bourke (1989).

TABLE 2.4. *Market share, 1989* (% over assets)

	Three largest institutions	Five largest institutions	Ten largest institutions
France	36.1	54.5	71.4
Germany	15.3	26.1	37.0
Italy	19.5	36.2	48.7
Spain	30.2	38.5	64.7
United Kingdom	22.5	29.2	42.4

Source: Estimates from *The Banker*, July 1990.

TABLE 2.5. *Nationality of major banks 1987*

10 Largest		500 Largest	
France	2	EC	162
Japan	7	France	20
United States	1	Germany	44
		Italy	33
100 Largest		Japan	107
EC	44	Spain	13
France	10	United Kingdom	15
Germany	11	United States	28
Italy	8	Others	77
Japan	28		
Spain	2		
United Kingdom	5		
United States	11		
Others	17		

Source: *The Banker*, July 1988.

100 largest banks in the world are French, further evidence of how concentrated the country's banking industry is. Conversely, its total assets are less than those of Germany or the UK, as can be seen in Table 2.6, and its relative proportion of GDP is also lower. Spain also shows a fairly high level of concentration. However, in absolute terms its relative presence in the international markets is still very low.

Table 2.6 also shows that the concentration of assets is greatest in the United Kingdom, France, and Germany. In Britain's case this is due to the sophistication of financial systems; in France and Germany, because of the industries' relative proportions of their respective economies.

TABLE 2.6. *Banks' assets 1980–1987* (ECU billion)

	1980	1984	1987
France	493.5	727.9	811
Germany	641	973	1,217
Italy	222.3	352.7	372.7
Spain	184	203.6	228.3
United Kingdom	416.5	983.7	1,031.6

Source: Canals (1990).

Tables 2.7 and 2.8 offer information on the size of institutions. In particular, Table 2.7 presents data regarding stock-market capitalization and return on assets for some of the principal banks in each country. Note the high profitability of the British and Spanish banks and, in contrast, the low profitability of the German and Italian banks. We will return to this issue later.

Table 2.9 summarizes some figures of interest on the size of the banking industry in the EC relative to each country's branch network. Spain stands out with regard to number of branch offices (16,498), easily the highest of the five countries. Next is the United Kingdom with 14,300 offices and yet a much more sophisticated financial system. This figure loses its significance, however, if taken in context with the number of employees per office, in which case Spain falls to the bottom of the list.

To analyse the strategic significance of the number of branch offices, we must look at one more figure. The volume of credit granted per office reveals the productivity of the Spanish bank to be particularly low: 6.4 million ECUs, compared to 17.6 million ECUs for Italy (the country immediately following Spain), and 63.5 million ECUs in loans granted in the United Kingdom in 1986. This provides a clear indicator of the strength of the British financial system. If in place of analysing the volume of credit per office we compare the volume of credit per employee, the picture does not change markedly, the only difference being that Italy reveals an even lower ratio than Spain.

In the following chapters we will analyse in more detail the concentration within each country individually. This introduction is meant only to provide an overview.

3. THE PROFITABILITY OF THE EUROPEAN BANKING INDUSTRY

Comparative analysis of banks' profitability presents two methodological problems. The first is the use of an adequate measure of profitability

TABLE 2.7. *Stock-market indicators of some European banks, December 1988*

	Market capitalization (billion)	Current price[a] (billion)	Book value[a] (billion)	Market value as % of book value	Return on equity 1987 (%)
France					
Banque National de Paris	2.99	318	384	0.83	12.9
Crédit Commerciale de France	1.34	202	96	2.10	10.7
Crédit Lyonnais	2.01	480	695	0.69	13.2
Germany					
Commerzbank	2.90	236	244	0.97	9.2
Deutsche Bank	9.84	510	304	1.68	6.3
Dresdner Bank	4.40	309	245	1.26	6.9
Italy					
Banca Nazionale Lavoro	1.13	11,560	2,653	4.36	−0.8
Banca Nazionale Agricoltura	1.45	13,010	2,259	5.76	9.7
Banca Commerciale Italiana	2.88	3,717	3,930	0.95	7.6
Spain					
Banco Bilbao Vizcaya	7.89	8,099	3,404	2.38	19.0
Banco Central	4.08	4,750	2,069	2.55	15.5
Banco Popular	2.02	8,050	4,429	1.82	20.3
Banco Santander	5.68	6,278	1,972	3.78	19.0
United Kingdom					
Barclays Bank	8.57	4.51	5.16	0.87	17.6
Lloyds Bank	4.95	3.55	3.55	1	23.5
National Westminster	7.65	5.88	6.54	0.9	18.7

[a] Prices and values: France = francs; Germany = DM; Italy = lire; Spain = ptas.; UK = £.

Source: Salomon Bros. (1989).

TABLE 2.8. *The ten largest banks of the European Community in assets, 1987*

	Assets ($m.)	Country	Ranking
1. Crédit Agricole	210,596	France	9
2. BNP	196,955	France	13
3. Barclays	189,368	United Kingdom	14
4. Crédit Lyonnais	178,876	France	15
5. National Westminster Bank	178,505	United Kingdom	16
6. Deutsche Bank	167,133	Germany	20
7. Société Générale	155,483	France	22
8. Dresdner Bank	127,312	Germany	24
9. Paribas	121,617	France	26
10. Midland Bank	100,849	United Kingdom	33

Source: Canals (1990).

incorporating accurate information. The second is the homogenization of data from different countries and, more generally, the need for an accurate reference point from which to compare the results obtained.

The most widely recommended indicator in the financial literature is the return on investment (*ROI*). This treats profit as a ratio instead of an absolute value since profitability is actually a measure of relative and not absolute values. Usually, net profit is used for the numerator, though for the denominator total capital, total assets, or total investments may be used.[2]

There is no contradiction here since the indicator

Return on Equity (*ROE*) = Net profit/Equity

can be broken down in the following manner:

Return on Equity (*ROE*) = Net profit/Assets · Assets/Equity

The expression that we have just obtained indicates that return on equity (*ROE*) is the same as the product Net profit/Assets (*ROI*) times the quotient Assets/Equity (which expresses the debt of the company in question).

Of the two indicators, return on equity and return on assets, the former is preferred for two reasons: it facilitates the comparison of profitability between different industries and takes into account the debt strategy of the bank, very significant in a business such as banking where the level of debt can be quite high. Nevertheless, within a given sector, comparison of returns on assets is also perfectly acceptable; in fact, in the following pages we use both types of indicators.

---

[2] A good methodological study of these issues can be found in Brealey and Myers (1988).

TABLE 2.9. Size of banks in the European Community, 1989

	Inhabitants (millions)	No. of banks	No. of offices	No. of employees	Credits (ECU million)	Inhabitants per office	Employees per office	Credits per office (ECU '000s)	Credits per employee (ECU '000s)
France	56,160	418	9,888	223,805	302,300	5,880	22.6	30,572	1,351
Germany[a]	61,080	308	6,382	6,382	255,390	9,571	29.97	40,017	1,335
Italy	57,525	232	9,864	240,137	210,070	5,832	24.3	21,297	875
Spain	38,888	146	16,623	157,056	141,010	2,339	9.4	8,483	898
United Kingdom	57,236	449	14,600	444,090	1,248,500	3,920	30.4	85,514	2,811
EEC	325,981	2,179	71,743	1,598,020	2,725,300	4,544	22.3	37,987	1,705

[a] Data for Germany refer to 1988.

Source: Canals (1990).

TABLE 2.10.   *Commercial banks: income statement*

| | Financial products |
| Less: | Financial costs |

| | Intermediation margin |
| Plus: | Other products |

| | Gross margin |
| Less: | Operating costs |

| | Operating margin |
| Plus/Less: | Extraordinary expenses/Revenues |

| | Cash flow |
| Less: | Provisions |

| Net profits |

Reserves   Dividends   Taxes

TABLE 2.11.   *Commercial banking: return on equity* (%)

	1981	1982	1983	1984	1985	1986	1987	1988
France	17.57	14.45	13.90	12.78	13.13	14.93	12.7	11.2
Germany	10.63	12.85	14.41	17.34	19.49	16.95	14.8	13.6
Italy	21.85	24.03	17.56	15.72	17.16	13.45	14.9	13.4
Spain	11.54	9.77	10.94	10.96	13.04	14.36	12.9	12.5
United Kingdom	17.42	16.48	16.50	21.9	24.16	21.7	20.4	20.2
United States	17.14	14.97	14.20	13.75	14.66	12.96	10.8	9.7

Source: OECD (1988) and author's estimates.

In order to analyse the profitability of the banking system, we will take into account the classic income statement, as is shown in Table 2.10.[3]

3.1. Bank profitability

We will start by studying the case of commercial banking. In Table 2.11 the evolution of return on equity for each of the reference countries can be seen. In light of these figures, the most profitable banking is the

---

[3] For a deeper explanation of this point see Deutsche Bundesbank (1988).

TABLE 2.12.   *Banks: equity over total assets* (%)

	1981	1986
France	2.80	2.57
Germany	4.06	4.68
Italy	3.22	5.05[a]
Spain	6.48	5.68
United Kingdom	5.5	5.4
United States	5.81	6.19

[a] Corresponding to 1985.
Source: OECD (1988).

TABLE 2.13.   *Commercial banks: return on equity corrected for inflation* (%)

	1981	1982	1983	1984	1985	1986
France	4.3	2.7	4.3	5.4	7.3	12.2
Germany	4.7	7.6	11.1	14.9	17.3	17.2
Italy	0.7	6.2	0.9	4.9	8.0	—
Spain	−3.0	−4.6	−1.2	−0.3	4.2	5.6
United States	6.7	8.9	10.9	9.5	11.2	11.0

Source: OECD (1988).

German, with a return on equity of 16.95 per cent in 1986, although this is lower than the 19.49 per cent figure recorded in 1985. Italian banking also shows high profitability, with a figure of 17.16 per cent in 1985 although, again as with Germany, this represents a decrease in comparison to figures from earlier years, particularly 1981 and 1982. French banking also experienced a decline in profitability between 1981 and 1984, though from 1985 it appears to be recuperating. Somewhat to the contrary, Spanish banking after posting relatively low profitability figures in the first half of the 1980s—undoubtedly due to the gravity of the Spanish economic crisis—experienced a considerable take-off in 1985, rising to 16.69 in 1987, the second highest after Germany. Two factors help to explain the poor results for Spanish banking in the earlier years: greater allocation for insolvency provisions and higher proportion of required reserves (see Table 2.12).

The results change somewhat, however, if the return on equity is corrected for inflation, as shown in Table 2.13. Here we can observe how the high inflation rates in Spain and Italy until 1984 made the banks' results extremely poor, especially in relation to Germany. (We have introduced US data as well here to broaden the comparison.)

Central banks in these countries play an important role in the determination of banks' profitability for two reasons. The first is the execution of monetary policy and its effects on banks' asset and liability interests. The second is the set of ratios banks must observe (for example, on equity) which directly affect the income statement. In studying each of the banking systems in the following chapters, we will look at this in more depth.

The return on equity ratio has a more complex component which has to do with before-tax profits and their generation. Table 2.14 offers data on sources of revenues and expenses from the income statements of private banks, measured against total average assets. Again, information on the United States is also included for comparison. The first point of interest is the intermediation margin—the difference between financial revenues and financial expenses. Spain stands out with a margin of 3.73 per cent in 1986, the highest of the countries we are considering, despite the decline which it experienced about 1980. The historical explanation for this is related to the low remuneration traditionally offered to savings depositors. Increasing financial literacy of clients and competition from other entities, especially foreign banks, has corrected this imbalance so that the margin has decreased slowly over recent years. It is expected that this trend will continue considering the projected competition facing Spanish banks in the coming years from both foreign banks and domestic savings institutions.[4]

There are two additional factors to take into account in analysing the intermediation margin for Spanish banking. The first is that the required reserve ratios to which Spanish banks are subjected are higher than those of other European banks, thus the volume of funds invested at lower than the market rate of profitability is more significant; to the extent that the reserve and investment ratios are reduced, the return on assets will increase. The second factor is the growing interrelationship between interest rates on asset and liability operations and those on the regulatory loans.[5] The anticipated decline of Spanish interest rates in the next few years will affect both sides of the balance sheet, thus its effect on the margin of intermediation is not expected to be significant.

The other European countries present a somewhat lower intermediation margin. Germany is an interesting case because of the recovery of its margin—which goes from 1.89 per cent in 1980 to 2.46 per cent in 1986—as is the United Kingdom because of the deterioration of its

---

[4] Perhaps this is a result of a certain market power of Spanish banking, although the hypothesis is not easy to prove. For more on this issue, see Gual and Ricart (1988).

[5] For a detailed analysis of this point see Termes (1987).

2. Structure of the Banking Industry

TABLE 2.14. *Commercial banks: income statements (% of total assets)*

	1980	1981	1982	1983	1984	1985	1986	1987
Margin of intermediation								
France	2.84	2.79	2.74	2.63	2.58	2.53	2.72	2.79
Germany	1.89	2.07	2.44	2.71	2.54	2.44	2.46	2.48
Italy[a]	3.37	3.47	3.19	3.14	3.08	3.02	2.8	2.72
Spain	4.23	4.15	3.90	3.95	3.77	3.57	3.73	3.79
United Kingdom[b]	4.0	7.56	3.25	3.11	3.06	3.17	3.16	3.04
United States	3.08	3.06	3.16	3.13	3.35	3.47	3.34	3.12
Non-financial products (net)								
France	0.49	0.53	0.53	0.53	0.39	0.41	0.46	0.56
Germany	0.83	0.85	0.89	0.90	0.89	1.05	1.09	1.15
Italy	0.98	1.24	1.32	1.17	1.25	1.31	1.36	1.40
Spain	0.75	0.78	0.88	0.91	0.77	0.83	0.87	0.94
United Kingdom	1.48	1.28	1.54	1.61	1.85	1.67	1.79	1.83
United States	0.87	0.97	1.03	1.13	1.10	1.25	1.42	1.49
Gross margin								
France	3.30	3.32	3.27	3.17	2.97	2.95	3.18	3.35
Germany	2.72	2.92	3.33	3.61	3.43	3.49	3.55	3.63
Italy	4.34	4.70	4.51	4.31	4.34	4.34	4.20	4.12
Spain	4.97	4.93	4.78	4.86	4.54	4.40	4.61	4.73
United Kingdom	5.49	4.84	4.79	4.71	4.92	4.84	4.95	4.87
United States	4.72	4.02	4.20	4.26	4.45	4.72	4.76	4.61
Operating costs								
France	2.24	2.23	2.21	2.14	2.06	2.06	2.12	2.28
Germany	2.04	2.03	2.07	2.19	2.19	2.20	2.24	2.35
Italy	2.82	2.82	2.81	2.97	3.00	2.97	3.07	3.14
Spain	3.49	3.42	3.30	3.09	2.89	2.80	3.00	3.12
United Kingdom	3.77	3.59	3.42	3.29	3.32	3.16	3.23	3.39
United States	2.63	2.76	2.92	2.95	3.04	3.14	3.18	3.34
Provisions								
France	0.60	0.60	0.68	0.68	0.61	0.57	0.68	0.71
Germany	0.24	0.46	0.73	0.82	0.52	0.47	0.52	0.55
Italy	0.90	1.18	0.83	0.61	0.57	0.50	0.65	0.78
Spain	0.75	0.76	0.90	1.12	1.01	0.87	0.79	0.83
United Kingdom	0.37	0.28	0.54	0.61	0.75	0.54	0.54	0.61
United States	0.25	0.26	0.40	0.47	0.57	0.67	0.78	0.91
Profit before taxes								
France	0.49	0.49	0.38	0.34	0.30	0.31	0.38	0.36
Germany	0.44	0.43	0.53	0.60	0.72	0.83	0.79	0.73
Italy	0.90	1.18	0.83	0.61	0.57	0.50	0.48	0.20
Spain	0.73	0.75	0.58	0.65	0.64	0.72	0.81	0.78
United Kingdom	1.35	0.97	0.83	0.81	0.85	1.12	1.18	0.87
United States	1.07	1.00	0.88	0.84	0.84	0.90	0.80	0.36

[a] The data for Italy in 1986 are not compatible with those of the previous years, and thus have been excluded.
[b] The data corresponding to the UK in 1985 and 1986 are derived from only 8 clearing banks.
Source: OECD (1988) and author's estimates.

margin between 1980 and 1984 when it lost almost a point over average total assets. Italy and France maintain relatively stable margins throughout. The small variations between these figures is not very representative, however, since they are aggregates of the overall banking systems for the five countries analysed.

Another important point regarding net results has to do with non-financial products. These correspond to a variety of services that banks offer their clients at fixed commissions. The data reveal that US banks' non-financial products in 1986 were 1.42 per cent, almost a third of their total gross margins. Until 1985, British banks had shown excellent results from banking services. On the other hand, Continental banks are still far behind British banks, although all are rapidly gaining ground, especially in Italy and Germany.

One of the strategic characteristics of American banks in the 1980s was to generate revenue through the development and marketing of high-value services.[6] This trend has been accelerated by increased competition for deposits and the process of disintermediation, as well as the growing demand for more specific financial services. European banks' revenues have been somewhat lower in comparison, although they have been improving over the last few years.

Gross margins are obtained through the sum of the intermediation margin and revenues from non-financial products. The figures presented above show values higher than 4 per cent on average assets for Spain, the United States, Great Britain, and Italy, although they are slightly lower for Germany and France. After the recession of 1982–4, French banks slowly improved their positions. Pre-tax profits are obtained by subtracting operating costs from the gross margin. The data presented earlier show that operating costs are very high in Spain, the United States, and the United Kingdom, although they have fallen considerably in Spain since 1980, while they have actually risen in the United States.

While British and American banks' higher operating costs can be explained partly by a greater specialization of services—which require higher personnel costs, as demonstrated in Table 2.15—the Spanish case is somewhat different due to a relatively larger number of branches compared to other European countries. This is a result of the expansion of offices during the 1970s, which corresponded to the necessity of attracting capital by being closer to savers, since existing regulations prevented banks from doing so through other means. However, the dense network of existing branches has its positive aspects, as it constitutes a considerable entry barrier.

An analysis of Table 2.15 offers a more detailed breakdown of

---

[6] Ballarín (1985) offers a good interpretation of this phenomenon which has transformed the structure of the banking industry.

TABLE 2.15.   *Commercial banks: personnel costs (%)*

	1980	1981	1982	1983	1984	1985	1986
France	1.53	1.49	1.48	1.42	1.34	1.33	1.35
Germany	1.42	1.42	1.43	1.50	1.47	1.45	1.48
Italy	2.13	2.06	1.97	2.17	2.17	2.14	—
Spain	2.41	2.31	2.24	2.08	1.94	1.88	2.10
United Kingdom	2.52	2.35	2.16	2.05	2.01	1.89	1.93
United States	1.39	1.44	1.49	1.49	1.52	1.52	1.51

Source: OECD (1988).

operating costs as it gives specific personnel costs, measured in relation to average total assets. Personnel costs constitute the most important variable in the evolution of operating costs, except for the United States, where they represent approximately half of operating costs. Anyway, these data have to be handled with care, as they are not completely homogeneous, despite efforts by the OECD to standardize them in its studies on 'banking profitability'.

After subtracting the corresponding figure for the provisions, we arrive at the before-tax profits. We can observe two distinct groups of banks. The first, formed by Spain, the United States, Germany, and Great Britain, show a before-tax return on assets of around 0.8 per cent. In the case of the Spanish and German banks there is a trend towards a general increase for this ratio; in Spain, this is explained by a reduction in operating expenses and in Germany by an increase in the gross margin, especially of non-financial products.

On the basis of before-tax returns on assets, we can learn something about the efficiency of the banking systems. Here, Italian and French banks show a definite weakness; in the case of Italy because of high operating expenses and in France because of the lower intermediation margin.

These data on the efficiency of asset and liability management must be kept in perspective, however, as they fail to take into account one crucial aspect of return on equity, as we saw previously—the level of debt. The return on equity is the product of the return on assets for the level of debt, and this last factor, as can be seen in analysing the return on equity, plays an important role. In Spain, this has been especially true since the banks have generally had to maintain a higher ratio of equity than in other countries.

In sum, it can be said that the European banks as a whole present nominal return on equity results that are relatively acceptable and somewhat above that of the American banks. However, they are far from

TABLE 2.16. *Savings banks: return on equity* (%)

	1981	1982	1983	1984	1985	1986
Germany	25.38	32.31	34.88	32.62	27.91	25.06
Italy	20.56	26.51	18.76	18.80	19.63	26.10
Spain	17.10	17.68	18.19	17.87	18.92	16.89
United States[a]	−17.77	−16.00	0.92	3.61	20.34	—

[a] Mutual savings banks.
Source: OECD (1988).

TABLE 2.17. *Savings banks: return on equity corrected for inflation* (%)

	1981	1982	1983	1984	1985	1986
Germany	19.5	27.0	31.6	30.2	25.7	25.3
Italy	2.8	9.9	4.3	8.0	10.4	20.3
Spain	2.5	3.3	6.0	6.6	10.1	8.1
United States	−28.2	−22.1	−2.4	−0.7	26.8	—

Source: OECD (1988).

those of the Japanese banks, which we have not considered here, where levels of return on equity are on the order of 20 per cent.

3.2. Savings banks' profitability

We move on now to consider the case of the savings banks and their profitability. We will deal very little with French or British cases in this analysis, since they do not provide much coherence. The reason for this in the UK is that the savings banks (Trustee Savings Banks) are included within the group of commercial banks for reasons relating to the nature of their operations and also as a result of their privatization. This dates from the approval of the Trustee Savings Bank Bill which completely restructured these organizations. In the case of France, the savings business is very different as the investment of deposits is not decided by the savings banks themselves, but by the Caisse de Dépôts and the SOFEFI, organizations under governmental control.

The results obtained by the savings banks on equity are presented in Table 2.16. In Table 2.17 is the relative profitability data after subtracting each country's corresponding inflation rate. In nominal terms, the high profitability of Germany and Italy stand out, as do the negative figures for

TABLE 2.18.   *Savings banks: income statements* (% over total average assets)

	1980	1981	1982	1983	1984	1985	1986	1987
Margin of intermediation								
Germany	2.93	3.26	3.47	3.62	3.45	3.32	3.18	3.27
Italy	3.62	3.76	3.53	3.83	3.84	3.73	2.78	2.70
Spain	4.31	4.73	4.86	5.28	5.05	4.28	4.68	4.87
United States[a]	0.65	−0.39	−0.41	0.77	1.03	2.13	1.89	1.73
Non-financial products (net)								
Germany	0.35	0.38	0.36	0.38	0.38	0.35	0.35	0.43
Italy	1.35	1.55	1.83	1.32	1.35	1.16	0.92	0.99
Spain	0.29	0.34	0.34	0.36	0.38	0.52	0.66	0.66
United States	0.60	0.81	1.14	0.94	0.71	0.94	0.98	0.99
Gross margin								
Germany	3.28	3.64	3.83	4.00	3.83	3.66	3.54	3.70
Italy	4.96	5.31	5.36	5.15	5.19	4.89	3.71	3.69
Spain	4.60	5.07	5.20	5.64	5.43	5.20	5.35	5.53
United States	1.25	0.42	0.73	1.71	1.74	3.07	2.87	2.72
Operating costs								
Germany	2.22	2.23	2.21	2.21	2.18	2.17	2.17	2.41
Italy	3.12	3.17	3.17	3.46	3.25	3.08	3.16	3.35
Spain	3.47	3.55	3.67	3.61	3.53	3.39	3.83	3.51
United States	1.36	1.43	1.51	1.63	1.50	1.81	1.90	1.94
Provisions								
Germany	0.32	0.55	0.51	0.57	0.47	0.46	0.43	0.48
Italy	1.40	1.58	1.40	1.03	1.12	0.89	0.63	0.75
Spain	0.26	0.49	0.54	0.98	0.85	0.77	0.61	0.81
United States	0.03	0.02	0.03	0.03	0.05	0.12	0.20	0.24
Profits before taxes								
Germany	0.73	0.86	1.11	1.22	1.18	1.03	0.94	0.81
Italy	0.45	0.56	0.79	0.66	0.82	0.93	0.92	−0.41
Spain	0.87	1.03	0.99	1.06	1.04	1.04	0.91	1.22
Untied States	−0.14	−1.03	−0.81	0.00	0.19	1.14	0.77	0.54

[a] Mutual savings banks

Source: OECD (1988) and author's estimates.

the United States during some years (owing to insolvency) and the low profitability of the Spanish savings banks.

Our study of profitability continues with a more detailed examination of the savings banks' income statements (Table 2.18). The intermediation margin is very high in the Spanish savings banks compared with the other three countries and higher also than the margin obtained by the commercial banks studied earlier. This last phenomenon is specific to Spain and, to a lesser extent, Germany. In Spain this is explained by the singular role that the savings banks have had within the Spanish financial system and, in particular, by the historically low remuneration on deposits.

Non-financial products still have relatively little impact on the income statement here in comparison to the effect they have on those of commercial banks. Italy is the exception because savings banks pass costs on to the customer, thus producing higher net incomes. Even there, though, this practice has diminished in recent years as a result of greater competition. Growth in Spain is indicative of two things. The first is that the savings banks have begun to charge their customers for services that previously were free. And the second is that they have started to offer new services as a means of proactively confronting future competition.

Spanish savings banks show a higher gross margin along with a stronger trend toward growth. American savings banks also show signs of growth in terms of gross margin. In contrast to this, Italian savings banks have experienced a significant gross margin decline owing to the decrease in the intermediation margin and a reduction of non-financial products.

Operating expenses in the Spanish and Italian savings banks, with percentages over 3 per cent because of the relatively high provisions for bad debts, make before-tax profits lower than those in the United States but close to those in Germany. In general, the before-tax returns on average assets of the savings banks are somewhat higher than those of the commercial banks, as can be verified by comparing Tables 2.18 and 2.14.

Finally, we can take the debt ratio of the savings banks into account with the return on average assets (see Table 2.19); the decrease of this figure in case of the Spanish and Italian savings banks can be compared to the relative stability in the American banks and the relative growth in the German banks. The greater debt level maintained by the Italian and German savings banks is the reason behind their higher return on equity despite returns on assets that are approximately the same. To the extent that EC legislation tries to standardize equity requirements for financial entities, this problem will diminish. Then the banks which are more strictly regulated today will have the opportunity to create more offensive strategies based on a higher proportion of debt yielding higher returns on equity.

3.3. Some conclusions

The analysis of banks' profitability in some European countries allows us to draw some conclusions about the relative efficiency and competitive capacity of the different banking systems for the years ahead, especially in view of the liberalization and deregulation measures put forth by the Single European Act.

TABLE 2.19.   *Savings banks: equity over total assets* (%)

	1981	1986
Germany	3.39	3.76
Italy	4.73	3.51[a]
Spain	6.04	5.38
United States	5.79	5.62

[a] Data for 1986 are not available. This figure corresponds to 1985.

Source: OECD (1988).

The first interesting conclusion is that, in terms of return on equity, the European banking industry does not differ significantly between countries. Logically, there are banks that behave differently from others but, in general, the results are similar. Thus, in France and Italy, banks have higher debt levels, but their return on assets is lower, and their return on equity is comparable to other countries. In Spain and Germany the opposite occurs: return on assets is higher, and debt is lower due to regulation. At the same time, the higher profitability of Spanish banks is the result of higher intermediation margins, while German and Italian banks earn more from non-financial products.

The second conclusion, which derives from the former, is that banks' strategies have been very conditioned by regulation, which has forced them to adopt different strategies. Greater competition can be expected in the future, as much from the harmonization of regulation as the deeper penetration of American and Japanese banks in Europe.

With regard to the savings banks, the most significant factor is the high profitability of the Italian and German savings banks in relation to their American and Spanish counterparts. It is important to distinguish between the different means by which these results are achieved. Thus, while profits before taxes on total assets are approximately the same in the cases considered here, the lower debt levels of the Italian and German savings banks provide completely different results, with these being more profitable than the Spanish or American savings banks.

These conclusions show two clear facts. The first is that there are no substantial differences regarding profitability among European banks, although this is not the case for the savings banks. The second is that these slight differences do not conceal the diversity of the strategies pursued by the banks of each country, which have been determined partly by government regulation and partly by the banks themselves.

4. THE CREATION OF VALUE: AN ALTERNATIVE FOCUS FOR MEASURING EFFICIENCY IN THE BANKING SECTOR

4.1. A conceptual framework

Throughout the previous sections we have centred our attention on the analysis of the European banking industry's profitability strictly on the basis of accounting results. These types of ratio are useful but are still insufficient and may on occasion cause confusion. They are insufficient because they do not take into account expected future profits; in addition, by themselves they do not show anything about the relationship between earned profits and the cost of capital for the resources employed. They can also cause confusion as a result of deficiencies in accounting systems and procedures or because alone they reveal nothing about contribution to the future development of the bank itself.

To understand better this last dimension—the future development of a bank—and to correct for some of the limitations previously mentioned, modern financial theory has proposed an alternative criterion for evaluating resource allocation: the creation of value. According to this criterion, the profitability of a company will depend on whether it creates value for those that participate in it. If it does, the firm's market value will be greater than its book value (the investment effected by the shareholders). Otherwise, if the market value is less than the book value, the company will be decreasing in value.

The utilization of the market price of a company's shares implies an adequate assessment by investors of the expected return on the firm's assets over a relatively long period of time. In addition, market value is assumed to be a good indicator for measuring the results of a company in relation to other companies in the same industry. On top of this are the interests of the firm's directors, not only in the evolution of share prices but also in the development of policies which lead to an increase in these prices. This behaviour is rational to the extent that it has a long-term perspective and is consistent with the solvency and profitability of the firm. It ceases to be rational when it is focused completely on day-to-day fluctuations since investors take into account not only these decisions but also their consequences in the medium- and long-term.

In order to measure with greater precision the profitability of a company we will use the M/B model (M: Market Value; B: Book Value), which expresses the relationship between the market price of a company and its book value. The numerator of this quotient expresses the perception of investors, while the denominator expresses strictly an accounting figure.[7]

---

[7] This indicator coincides, approximately, with the ratio known as the q of Tobin, although this takes into account the market value of the assets of a company which is different in comparison from the M/B indicator.

The interpretation of this indicator is very simple. When $M/B > 1$, investors value a bank more than the accounting figure indicates which suggests that this bank is creating value, registering a higher profitability than its cost of capital. When $M/B = 1$, a bank neither creates nor destroys value, and the resources generated will compensate for the resources committed. Conversely, when $M/B < 1$ a bank is decreasing the value for the shareholder, as the resources it generates are not sufficient to recover the original investment.

The market value of a firm can be expressed by the following equation:

$$M = \frac{D_1}{(1 + k_c)} + \frac{D_2}{(1 + k_c)^2} + \ldots + \frac{D_n}{(1 + k_c)^n} + \ldots, \tag{1}$$

where: M: market value; D_i: dividends distributed by the company in period i; and k_c: cost of capital

This expression assumes that the company has an indefinite life, and that shareholders do not have any other way of perceiving value than through dividend returns. To make this expression somewhat more operative, we will assume that dividends grow at a constant rate over the long term and that the firm has an indefinite life. In this case, applying the value to a declining geometric series, we obtain that

$$M = \frac{D}{k_c - g}, \tag{2}$$

an expression which coincides with the formula of Gordon–Shapiro.[8]

The ratio g is not only the rate of dividend growth, but also of the possible growth in firms' sales; given a constant level of debt it also equates the necessary growth of capital to finance this.

With after-tax profit (ATP) equivalent to the sum of dividends (D) plus retained profit (R), and dividing equation (2) by the value of a company, we find

$$\frac{M}{B} = \frac{\dfrac{ATP}{B} - \dfrac{R}{B}}{k_c - g} = \frac{ROE - g}{k_c - g}, \tag{3}$$

where ROE is the return on equity and g (rate of dividend growth) is equivalent to the quotient R/B, R being the size of retained profits. By further developing the previous equation (3), we get:

$$\frac{M}{B} = \frac{ROE - g}{k_c - g} = \frac{ROE - k_c}{k_c - g} + \frac{k_c - g}{k_c - g} = \frac{ROE - k_c}{k_c - g} + 1, \tag{4}$$

an expression that indicates that the relationship between the market value of a firm and its book value depends on the difference between return on equity and the cost of capital, according to the following equation:

[8] See Gordon and Shapiro (1956).

$$\frac{1}{k_c - g}.$$

When the rate of growth g is zero (that is, the firm does not grow), the ratio M/B takes on the following expression:

$$M/B = \frac{ROE - k_c}{k_c} + 1. \tag{5}$$

This expression has two important limitations. The first is that it assumes that the cost of capital k_c remains constant over the long term, a rather unreasonable assumption. As a result, the differential between the return on assets and the cost of capital should be taken for brief periods of time. Otherwise, its significance is lost. The second limitation refers to the fact that the rate g is assumed to be constant; in the real world this assumption is hardly realistic.

The concept of cost of capital has been the object of numerous theoretical discussions. Here we define it as the minimum rate of return on an investment project.[9] The cost of capital consists of two components. The first is associated with some type of risk-free interest (that can be likened to the type of interest on long-term public debt) that becomes something like a minimum level of profitability required. The second component is associated with a risk rate or premium. In turn, the risk premium is estimated as the product of two financial indicators: the volatility of the capital generated multiplied by the average risk premium of the market. Analytically:

$$k_c = r + p \tag{6}$$

$$p = \alpha p_m, \tag{7}$$

where k_c: cost of capital; r: risk-free interest rate; p: risk premium; α: volatility of resources generated; and p_m: average risk premium.

The average risk premium (p_m) represents the premium that, on average, is registered in a specific financial market. For its part, the volatility measures the risk of the resources generated; when this ratio equals one, the risk is exactly equal to that of the market, in which case the risk premium of a business is equivalent to the average premium of the market.

In looking at the concept of cost of capital it becomes clear that accounting profits alone are not sufficient to determine whether a particular firm makes efficient use of its resources. A company must create value for all of its stakeholders (directors, shareholders, and employees) which only happens when the average return on equity is greater than the cost of capital. But note that with the accounting criterion it is sufficient

---

[9] This can be found in any finance text, such as that of Brealey and Myers (1988).

for merely the return on assets to be positive. In contrast, with the creation of value criterion that return must be greater than the cost of capital. Consequently, we can see that this second criterion is more complete and also that there is greater use for it.

The expression (4) above has some practical consequences of interest. One of these, commented upon previously, is that for the return on equity to be greater than the cost of capital ($ROE > k_c$), the company must create value ($M/B > 1$). Thus, the creation of value depends on the differential between ROE and k_c. The greater this differential is, the greater the subsequent value created.

The second consequence is that when the differential $ROE - k_c$ is positive growth is a favourable condition since it contributes the generation of more value, as can be seen in equation (4) above. However, we could also turn this around and state that a policy of growth is only valid when the return on assets is greater than the cost of capital. In this case, as we have seen, the company creates value. On the other hand, if the differential is negative growth will have a destructive impact on the value of the company, in which case such a policy would be unreasonable.

We can observe then that the criteria for evaluating profitability with the ratio M/B and with the differential $ROE - k_c$ are perfectly consistent. The business will be generating value when $M/B > 1$ or, alternatively, when $ROE - k_c > 0$. Also, in both cases, growth will contribute to the creation of value. The company does not create value when $M/B < 1$ or, alternatively, when $ROE - k_c < 0$. In these cases, growth will contribute to the destruction of value. When $M/B = 1$ or $ROE - k_c = 0$ the company neither creates nor destroys value and growth will not affect this process.

In Fig. 2.1 the factors that influence the Market Value/Book Value ratio are shown, given a more broken-down account of the possible mechanisms for value-creation within the company. (In Chapter 9 we will return to this question in analysing the value-chain of the bank.) The type of analysis is useful since it presents in a clearly visible way those critical aspects of a company which contribute decisively to creating value.

4.2. An application to the European banking industry

The conceptual model of value-creation presented in the previous section enables us to design a new instrument for evaluating profitability in an industry and of a company within that industry.[10]

A simple way to evaluate the competitive position of businesses in a specific industry is to compare, at a given time, the M/B with the differential $ROE - k_c$. This is what we have done for the main banks of

---

[10] See Wilcox (1984) for a more complete treatment of this topic.

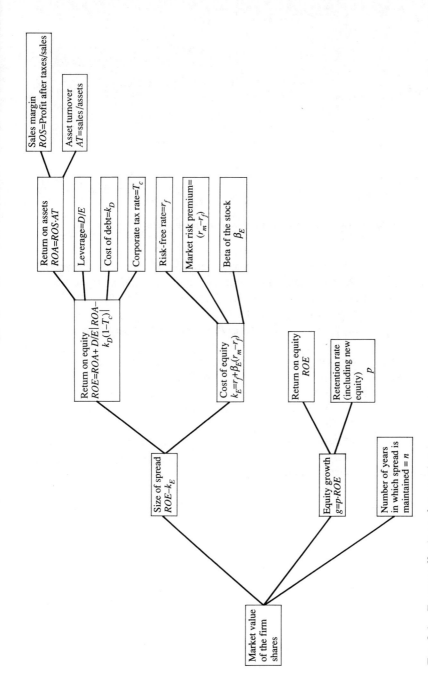

Fig. 2.1 *Factors affecting value-creation*
Source: *Hax and Majluf (1984).*

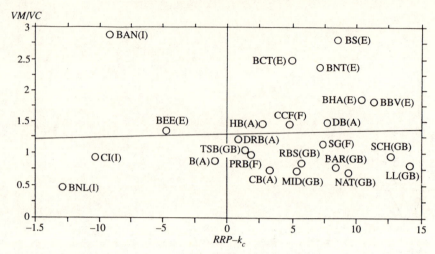

Fig. 2.2   *Market value/book value ratios for European banks*

Germany, Spain, the UK, and Italy respectively in Figures 2.2 through 2.5. (French banks were not included because the major banks were publicly owned and not listed on the stock market as of 31 May 1988, the reference date for the study.) The data for these calculations were taken from the report prepared by Banking Analysis Ltd of London in June 1989.

In Figure 2.2 we have presented the combination of values M/B and $ROE - k_c$ (at the end of 1988) for the principal banks of the five countries. Given the difficulty of obtaining a value for the cost of capital, we have estimated it by using the interest rate of long-term government bonds in each of the five countries.

A necessary clarification is that the data presented here must be kept in perspective, especially when comparing across countries, since capital markets and accounting practices are somewhat different. Also, M is the value taken at 31 March 1988, while B is the value taken at 31 December 1988. Thus, the M/B ratio does not reflect precisely what it was intended to so the results are more for illustrative purposes.

Figure 2.2 shows the adjustment of the observed figures according to the minimum-squares criterion. Actually, the positioning obtained is not that good, with a very small ratio of R^2, undoubtedly from the large number of observations lying significantly far from the line.

The first observation to be made is the low value of the M/B ratio (less than 1) for the British banks. This has two possible explanations. The first is the October 1987 stock-market crash. The second has to do with the increase of lending by British banks to Latin America, many of which were not sufficiently covered by sufficient bad debt provisions. Thus, it is

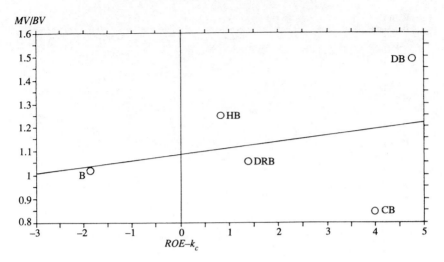

FIG. 2.3   *German banks:* M/B *and* ROE-k$_c$ *relationship*

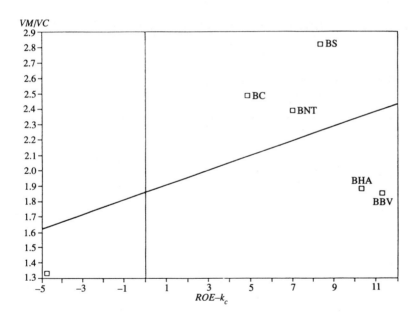

FIG. 2.4   *Spanish banks:* M/B *and* ROE-k$_c$ *relationship*

possible that the market takes a negative view of profits given the pending amortization of debt. In the summer of 1989, one of the largest British banks, NatWest, announced losses for the first half of the year as a result of insufficient provisions for bad debts.

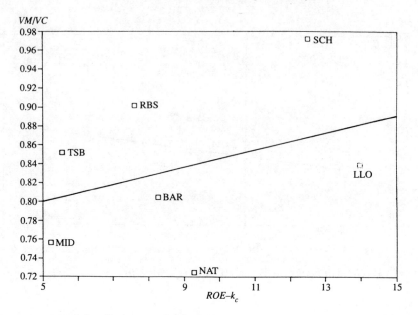

FIG. 2.5   *British banks:* M/B *and* ROE-k_c *relationship*

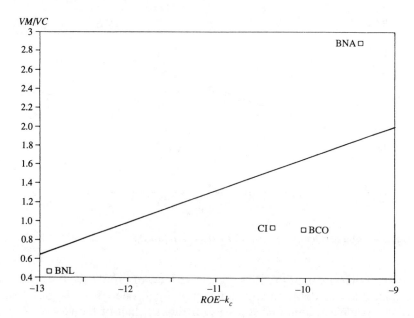

FIG. 2.6   *Italian banks:* M/B *and* ROE-k_c *relationship*

At the other extreme are Spanish banks which show rather high M/B ratios in keeping with significant $ROE - k_c$ differentials. In this case, the market seems to be reacting positively to the quality of management in Spanish banks and their future potential, for example, their capacity to maintain market share and profits. This is consistent with what the theory of options predicts.

Italy shows a poor combination of M/B and $ROE - k_c$ values. In this case, the market places a negative value on the banks' assets because of their poor economic showing. The exception is the Banca Nazionale dell'Agricultura with a negative $ROE - k_c$ differential (-9.35 per cent) giving it an M/B of 2.88, the highest of those considered here, though this seems to be a very optimistic valuation on the part of the market.

In Figures 2.3 to 2.6 we have presented the combination of M/B and $ROE - k_c$ values for the leading banks of each country. The Banco di Roma and the Bank of Scotland have been excluded since both showed a very negative $ROE - k_c$ differential at the end of 1988, significantly lower than the rest.

These results are not necessarily contradictory, but they do bring into question the findings from the previous section and those obtained from purely accounting-based criteria; in particular, they emphasize how the market values not only the banks' present profitability but also their future prospects.

3

The German Banking Industry

1. AN OVERVIEW OF THE GERMAN FINANCIAL INDUSTRY

Since the 1960s the German financial industry has enjoyed much more freedom than those of most other countries. Interest rates were liberalized at the end of the 1960s at the same time that a number of other regulations were eliminated. And this decade has experienced a new era of deregulation, aimed at strengthening a few domestic financial centres.

The core of the German financial system is its commercial banks. Between 1970 and 1985, around 70 per cent of the funds from family and business savings were intermediated by banks.[1] The scale of these firms and their strong ties to industrial companies help explain their dominant position. But the performance of German banks cannot be fully understood without looking at the Deutsche Bundesbank, the German central bank whose prudence in managing monetary policy and financial deregulation has fostered financial innovation in West Germany.

It is possible to distinguish between two large groups of banks. The first is the group of universal banks which make up three-quarters of the sector's total business and includes three types of firms: commercial banks, savings banks, and credit cooperatives. The second group consists of those banks which offer more specialized services: mortgage banks, specialized banks, and the postal banking services.

A study of the data in Tables 3.1 and 3.2 shows the dominance of universal banks over specialized banks, from the point of view of credit investment and shareholder assets. The comparative number of branches (Table 3.4) also confirms this.

2. AN ANALYSIS OF THE FINANCIAL INSTITUTIONS AND MARKETS

2.1. The German banking industry

The German banking industry has expanded considerably in recent years as measured by its growing contribution to the total value-added of the economy, by its volume of activity, and by its contribution to employment.

---

[1] See Francke (1984) or Schneider et al. (1982). Throughout this chapter we will refer to West Germany as simply Germany.

TABLE 3.1. *Structure of the German banking system*

Category/Sub-category	No. of banks	Offices	Total assets (%)	Credit investments (%)	Liabilities (%)
Universal banks	4,392	44,166	74.72	70.23	71.92
Commercial banks	311	6,602	22.43	21.44	17.85
Large banks	6	3,126	8.30	8.07	7.66
Regional and other commercial banks	157	2,971	10.90	11.41	9.02
Branches of foreign banks	59	90	1.79	0.74	0.19
Private banks	89	415	1.44	1.22	0.98
Central institutions of savings banks	12	243	15.07	14.14	15.01
Savings banks	586	17,893	20.79	21.57	23.81
Central institutions of co-operatives	7	42	4.43	1.34	1.09
Credit co-operatives	3,476	19,386	11.99	11.73	14.17
Specialized banks	98	17,774	25.28	29.77	28.08
Mortgage banks	38	70	13.41	17.48	15.87
Private	27	53	8.64	11.92	10.88
Public	11	17	4.76	5.56	4.98
Specialized banks	16	100	6.44	5.28	5.68
Postal services	15	17,554	1.43	1.00	1.96
Credit banks being built	29	50	4.00	6.00	4.58
Total banking system	4,490	61,940	100.00	100.00	100.00

Source: Lladó *et al.* (1989).

TABLE 3.2.   *Major German banks*[a]

Bank	Assets	Deposits	Loans	Equity
Deutsche Bank	167,133	100,648	107,237	6,620
Dresdner Bank	127,312	78,362	78,272	4,332
Commerzbank	99,633	71,631	57,027	2,965
Bayerische Vereinsbank	91,319	28,760	72,244	2,281
Westdeutsche L. Girozentrade	90,727	41,988	44,900	2,325
Bayerische L. Girozentrade	79,291	52,821	38,704	1,717
Hypo-Bank	74,858	29,813	57,190	2,204
Deutsche Genossenschaftsbank	72,319	39,778	32,600	1,345

[a] In $m, as of 31 Dec. 1988.
Source: Business Week, 26 June 1989.

The size of Germany's banking industry doubled between 1978 and 1988. The industry accounted for 4 per cent of the total value-added in 1988 compared with 3.5 per cent in 1978. The number of employees at the end of 1988 reached 615,000, compared with 500,000 in 1978, a figure corresponding to 2.5 per cent of the population. Nevertheless, this information conceals important data.[2] In the last decade competition between banks has increased considerably. The cause lies in that securities companies and other non-banking financial intermediaries have taken away many of the banks' customers. Financial disintermediation has been another influencing factor, especially intense on the asset side of the balance sheet. Many companies have gone away from traditional banking credit.

All of this has spurred banks to offer new financial services such as real estate and mortgages, treasury-bill management, and administration of inheritances in order to compensate for their losses in other business areas. The income from these new services has become an extremely important supplement.

This has meant that banks have had to expand their off-balance sheet operations.[3] For example, between June 1986 and December 1988 the volume of swaps contracted by German banks rose from DM44.5 billion to DM230 billion; currency options and financial futures also showed similar increases.

Another important trend in German banking is its greater presence in other countries over the past five years. At the end of 1988, German

[2] For a study of trends in the German banking industry, see Deutsche Bundesbank (1989).
[3] In this respect, the work published by the Deutsche Bundesbank (1987) is of interest.

TABLE 3.3.   *Evolution of German financial institutions' market share, 1981–1987*

	Total assets		Liabilities		Credit investments	
	1981	1987	1981	1987	1981	1987
Commercial banks	29.8	30.0	25.3	24.8	28.3	30.5
Large banks	11.7	11.1	11.5	10.6	11.3	11.4
Regional and others banks	13.5	14.5	12.0	12.5	13.8	16.2
Branches of foreign banks	2.7	2.3	0.4	0.2	1.4	1.0
Private bankers	1.9	1.9	1.3	1.3	1.6	1.7
Large banks	50.2	48.0	55.9	53.9	53.9	50.8
Savings banks	28.6	27.8	34.4	33.1	31.5	30.7
Central savings banks	21.6	20.1	21.4	20.8	22.4	20.1
Co-operative sector	20.0	21.9	18.7	21.2	17.7	18.6
Credit co-operatives	14.6	16.0	17.2	19.7	15.8	16.7
Central institutions	5.4	5.9	1.5	1.5	1.8	1.9
TOTAL	100.00	100.00	100.00	100.00	100.00	100.00

Source: Deutsche Bundesbank (1989).

banks had 186 foreign offices with a volume of operations totalling DM511 billion, corresponding to 13 per cent of the volume of domestic business.

Table 3.1 illustrates the strength of the universal banks within the German system. Among these the savings banks stand out markedly, as shown in Table 3.3, having relatively high market shares compared to other types of banks. In spite of the fact that they lost market share between 1981 and 1987, they continue to be the most powerful institutions. Commercial banks' share remained roughly equal throughout the period, and co-operatives increased their market share. This increase reflects, among other things, their extensive network of branches and the process of concentration which has taken place in recent years.

Concentration in commercial and savings banks[4] occurred in the 1960s and 1970s, although its impact on the German market is not that significant. The evolution of the industry structure in Germany is shown by the data on changes in the number of firms and branches (see Table 3.4). The number of commercial banks decreased dramatically between 1957 and 1980, rising again at the start of 1990. The number of co-operatives also declined between 1957 and 1987: from 11,814 down to 3,482. In contrast, the number of bank branches has grown for all types of banks—commercial, savings, and co-operative.

[4] See Pöhl (1982).

TABLE 3.4.   *Number of German banks and branches*

	No. of Banks			No. of Offices		
	1957	1980	1987	1957	1980	1987
Commercial banks	364	246	311	2,281	6,160	6,602
Savings banks	885	611	598	9,268	17,814	18,136
Co-operative sector	11,814	4,235	3,482	14,208	19,727	19,428
TOTAL	13,063	5,092	4,391	25,757	43,701	44,166

Source: Lladó *et al*. (1989).

TABLE 3.5.   *Deposit structure by type of bank*

	Current accounts		Savings deposits		Other	
	1978	1988	1978	1988	1978	1988
Commercial banks	33.8	32.5	15.9	14.5	18.5	14.5
Savings banks	41.8	40.3	53.4	51.2	69.7	59.2
Credit co-operatives	24.4	27.2	30.7	34.3	11.8	26.3
TOTAL	100.00	100.00	100.00	100.00	100.00	100.00

Source: Deutsche Bundesbank (1989)

Table 3.5 shows the distribution of liabilities, which indicates the specialization of banks in attracting deposits from savers. The savings banks are strong in a number of areas: current accounts, savings accounts, and other instruments. The commercial banks' strengths lie in current accounts, while the co-operatives maintain a significant presence in various kinds of deposits. However, the figures reveal a relative loss in market share for the commercial and savings banks to the co-operative credit firms. This reflects a strong process of disintermediation, a phenomenon that has been accompanied by an increase in a variety of new financial services.[5]

Also of interest are the productivity and efficiency data for the different institutions. Table 3.6 lists figures for 1987 and shows that the savings banks posted the best results for total activity, as well as credits and liabilities, when the numbers are evaluated relative to the number of employees. However, the number of employees per office is definitely higher in commercial banks.

[5] Deutsche Bundesbank (1989) provides an in-depth analysis of these points.

TABLE 3.6. *Relative productivity data, 1987* (DMm.)

	Commercial banking	Savings banks	Credit co-operatives	Total universal banking
Assets per employee	4,509	5,660	4,335	4,948
Assets per office	132,654	77,227	33,015	66,064
Liabilities per employee	2,500	4,268	2,806	3,319
Liabilities per office	73,554	58,234	21,368	44,307
Credits per employee	2,564	3,352	2,053	2,766
Credits per office	75,435	45,737	15,631	36,933
No. of employees per office	29.4	13.6	7.6	13.4

Source: Lladó *et al.* (1989).

Among the universal banks, savings and large commercial banks stand out the most. They also account for more than a third of the total assets and liabilities in the banking system, the result of a large branch network that grew to a total of 17,893 offices by the end of 1987. Savings banks engage in activities that are very similar to those of commercial banks, with the exception of highly speculative operations such as taking open positions on currencies or shares on their own account. The majority of these are public, belonging to the local or state authorities.

In the commercial banking segment, there are three outstanding institutions: the Deutsche Bank, Dresdner Bank, and Commerzbank. Together they account for almost 40 per cent of all commercial banking and are characterized by being the only banks with subsidiary networks covering practically the entire country. Their importance within the European banking industry is therefore sizeable.

Table 3.7 gives the breakdown of assets and liabilities for the large commercial and savings banks. On the asset side, two things stand out: the strength of treasury bills for commercial banks and the importance of stock portfolios for savings banks. On the liability side, the percentage of equity in commercial banks and the level of deposits in savings banks are significant.

2.2. *The financial markets*

The most important financial market in Germany is the bond market, which accounts for six times the number of transactions executed in the stock market.[6] Curiously, financial innovation in the German markets

[6] See OECD (1986*a*).

TABLE 3.7.   Balance sheet structure, 1987 (%)

Balance	Commercial banks	Large banks	Savings banks
Assets			
Treasury	3.57	5.27	3.17
Inter-bank	24.07	20.00	8.62
Credits	56.87	57.77	61.72
Equity portfolio	12.92	14.75	23.02
Others	2.57	2.22	3.47
TOTAL	100	100	100
Liabilities			
Liabilities	5.50	6.40	3.76
Inter-bank	33.68	21.96	12.55
Deposits	45.78	59.01	78.36
Securities	9.67	5.24	1.43
Others	5.37	7.38	3.90
TOTAL	100	100	100

Source: Deutsche Bundesbank (1989).

during recent years has been much less than in other countries. The conservativeness of the central bank and the close relationships between banks and industrial firms seem to be the cause of this.

Banks play a very important role in the bond market as issuers and holders, particularly savings banks and credit co-operatives. The principal issuers are mortgage banks and the government, the latter in order to reduce its deficit. In contrast, the stock market has played a more modest role in financing the German economy, in part because of the tendency for companies to obtain funds through bank credits. At the end of 1983 only 400 firms were listed on the stock market with a total capitalization of DM224 billion, while in London 2,400 companies were listed with a capitalization of DM640 billion.[7] In addition, of all the shares issued, only those of 30 companies were traded regularly.

A third important market is the money market with its various sub-divisions. The spot market is used by banks to reach their required reserves and lend cash balances at a reasonable rate. The more long-term markets are used to hedge future fixed payments. In spite of the 1987 interest-rate liberalization, non-financial institutions have hardly participated in this market.

A last point of interest in the discussion of financial markets is the asset and liability structure of non-financial institutions (families and

[7] See OECD (1986a).

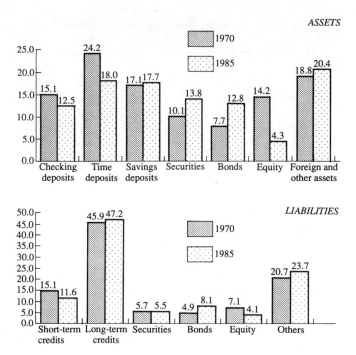

Fig. 3.1   *Financial assets and liabilities for households and companies*
Source: *Deutsche Bundesbank (1989).*

businesses). The most significant phenomenon—similar to that in other countries—is the demand by families and businesses for greater remuneration on different types of financial activities. Figure 3.1 clearly shows the breakdown of assets and liabilities in 1970 and 1985.

Of note is the reduction in bank deposits which went from 56 per cent of the total financial assets owned by families and companies to 49 per cent in 1985. This cannot be explained solely by the difference in interest rates between banks and capital markets; it is also related to the greater sensitivity of savers toward investment decisions.

The above data do not mean that banks' financial intermediation has declined since bonds account for the greatest growth in recent years. As long as banks manage to maintain a role in these transactions—as they have done—their level of intermediation will not be threatened.

With regard to household and company deposits, the role of banks continues to be quite large. Around 60 per cent of the deposits contracted in 1985 were bank credits, about the same as in 1970. And almost half of the total financing for companies and families was in the form of long-term debt. This is indicative of banks' importance in corporate financing.

TABLE 3.8.   *Non-financial companies: debt/equity ratio, 1984*

	Accounting values	Market values
Germany	2.3	1.3
Japan	—	1.3
United Kingdom	1.1	0.6
United States	1.1	0.6

Source: OECD (1986*a*).

Compared to other countries, the proportion of equity used to finance companies is significantly lower and can only be explained by the close relationships which German banks have historically had with other industries.

Table 3.8 compares several debt/equity ratios for different countries. Two sets of values are given: one set reflects accounting figures and the other reflects replacement costs. These numbers clearly show the greater level of debt carried by German banks compared to their American and British counterparts. Only the Japanese have similar debt/equity ratios.

3. THE PROFITABILITY OF THE GERMAN BANKING INDUSTRY

Various indicators suggest that the efficiency of the German banking industry has increased significantly in recent years. Banks' return on equity has gone from 10.63 per cent in 1980 to 16.95 per cent in 1986 (see Table 2.11). Return on total average assets has also increased from 0.49 per cent in 1980 to 0.79 per cent in 1986 due to increases in both the intermediation margin and the number of non-financial products offered. Operating expenses and provisions for bad debts have increased slightly as well (see Table 2.14).[8]

Along with these changes there has been a marked increase in concentration, very significant for co-operatives but somewhat less so for savings banks. Together, the two types of institution accounted for nearly 70 per cent of total assets in the banking system in 1987. At the same time, all the banks, including commercial banks, increased their number of branches by approximately 100 per cent between 1960 and 1987. Along with the greater presence of foreign banks in the German economy—

---

[8] An evaluation of the income statement in 1987 can be seen in a study published by the Deutsche Bundesbank (1988).

TABLE 3.9. *German banks' profitability ratios* (on assets, %)

	1980	1981	1982	1983	1984	1985	1986
Intermediation margin	1.89	2.07	2.44	2.71	2.54	2.44	2.46
Gross margin	2.72	2.92	3.33	3.61	3.43	3.49	3.55
Operating costs	2.04	2.03	2.07	2.19	2.19	2.20	2.24
Profits before taxes	0.44	0.43	0.53	0.60	0.72	0.83	0.79

Source: OECD (1988).

TABLE 3.10. *Banking system: return on equity* (%)

	1981	1982	1983	1984	1985	1986
Banks	10.63	12.85	14.41	17.34	19.49	16.95
Savings banks	25.38	32.31	34.88	32.62	27.91	25.06

Source: OECD (1988).

whose investment in credits equalled about 7.6 per cent of the 1987 total—this has given way to greater competition among the different financial institutions. This competition has often been measured by changes in the margin of intermediation. According to this indicator, however, competition had not increased but diminished since 1980 and in 1986 the margin increased from 1.89 per cent to 2.46 per cent above total average assets (Table 3.9). But the improvement in the intermediation margin is also an indication of the banks' increased exposure to risk over the years. The economic crisis which began at the start of the 1980s would have contributed to this and the banks' increases in provisions for bad debts serves as an indicator of this perceived added risk.

It is also interesting to compare the results of the German banks with those of their affiliates in more liberal markets such as Luxemburg. According to estimations by the OECD, the profitability figures for banks in each area are remarkably similar, for which we can assume that the level of competition in Germany is not at all low.[9]

Table 3.10 offers an overview of the changes in return on equity for German commercial and savings banks. Figures for the latter are substantially greater than those for the former as a result of two things: savings banks' higher return on total assets and their higher leverage. In

[9] See OECD (1986*a*).

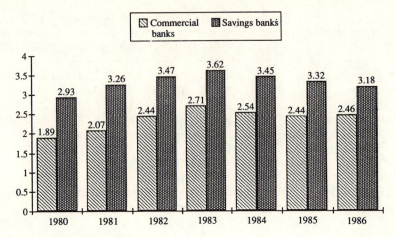

FIG. 3.2   *Intermediation margin*
Source: *OECD (1988).*

1986, the ratio of equity to total assets was 4.06 per cent for commercial banks and 3.39 per cent for savings banks which has had an impact on the return on equity.

The strength of the savings banks is also evident in terms of their market share and credit investments, such as deposits. These factors further explain their higher profitability in relation to commercial banks.

Figure 3.2 shows the change in the intermediation margin for both commercial and savings banks. Although the margin has fallen somewhat in recent years, it continues to be higher for savings banks than for commercial banks. Undoubtedly, this has to do not only with the difference in activities between the two, but also with the greater market shares they each have in different segments of the industry, as was shown in Table 3.3.

It does not look as though the German banking industry will experience serious problems in the years ahead. The existing freedom of capital flows and the proximity of Luxemburg as a competing financial centre mean that some of the challenges of a unified market in 1992 will have less impact in Germany than in other banking industries.

The first thing to be considered with regard to the German banking industry over the next few years is the reduction in the intermediation margin, following its peak in 1983 of 2.17 per cent on total assets. This indicates not only greater competition in asset operations, but also the change in savings from savings accounts to interest-bearing current accounts, bonds, and life insurance. Thus, the reduction of the intermediation margin is significant and the price of financial services is becoming a key variable for evaluating the competitive capacity of a bank. This will ultimately go directly against some of the informal relations which German companies and banks have traditionally maintained.

Closely related to this is the increase in competition, which will come from foreign, regional, and savings banks all offering not only attractive means of placing savings, but also interesting asset operations related to corporate finance. This will amount to an accumulation of very similar products with little differentiation, resulting in a further reduction of the margin of intermediation.[10]

This increase in competition also affects the number of branches. In 1986 the figure for Germany was lower only than that of Spain, which had 8.5 per 10,000 inhabitants, and much higher than that of the next country, France, with 4.6 per 10,000 inhabitants. The number of banks (see Table 3.4) though down from 1960, was still a relatively high 4,391 in 1987.

The explanations for this are varied. One is the classification of the banks into different categories: commercial banks, savings banks, cooperatives, mortgage banks, and mortgage banks with special functions. Secondly, there is a strong regional tradition which has been conducive to the growth of many banks, particularly savings banks and co-operatives. Finally, there are also a large number of banks—again, mainly savings banks and co-operatives—in the hands of the local public sector.

Given the increased placement of savings with insurance companies, a certain degree of movement, at times hostile, at times friendly, between banks and non-financial firms is inevitable. The Deutsche Bank's announcement of its plans to launch its own insurance campaign toward the end of 1988 is clear evidence of this. The reaction of insurance companies such as Allianz (the country's most prominent) might be either to close ranks completely or to ally itself with another large bank to distribute its policies through the bank branches.[11]

There are those, however, who question the wisdom of jointly distributing banking and insurance services, asking whether in place of synergy there will not be more competition. This is a risk, though it appears to be one which the banks are willing to take in the face of the growing proportion of private savings moving toward insurance companies.

Finally, there is the multi-dimensional phenomenon of the single European Market. The first point to consider is that local banks have a

---

[10] See *The Banker*, May 1988, p. 23.
[11] See *The Economist*, 17 Dec. 1988, p. 76.

very powerful presence in the national market in addition to a history of close customer relationships. Foreign banks wishing to conduct business in Germany have only two options: to go to the inter-bank market for capital, which enormously limits operating capacity, or to acquire an existing bank. However, the public ownership of many banks becomes an obstacle with this second option, and the number of banks listed on the stock market is limited, made up primarily of large banks whose size precludes the possibility of acquiring a majority share. Mergers with some of the medium-sized banks may perhaps be the most realistic strategy for foreign firms trying to strengthen their position there.

It is interesting to note here Barclays' acquisition in September 1990 of Bankhaus, Meck, Flinck & Co., a private bank with strong ties with the German financial community. This seems to suggest that careful segmentation of an acquisition target may be another strategy for foreign banks in the future, particularly from the United Kingdom and France.

A different topic is what the German banks plan to do in relation to the rest of Europe. Some of them have already clearly indicated their strategies, such as the Deutsche Bank (see Appendix to this chapter) with its acquisition of the Bank of America in Italy and Morgan Grenfell in the United Kingdom, or the Commerzbank with its 10 per cent participation in the Banco Hispano-Americano. Undoubtedly the Deutsche Bank is positioned above all the rest to play a major role, combining the operations of commercial bank, investment bank, and industrial conglomerate. Evidence of this last motive is its 28 per cent share in Daimler-Benz, Germany's premier industrial group, 25 per cent stake in Karstadd (a retail store chain), and 30 per cent participation in Phillip Holzmann (a large construction company). On top of this, in 1988 the Deutsche Bank accounted for 20 per cent of all stock-market transactions and financed 22 per cent of German companies' exports.[12]

The rest of the large German banks show a much more cautious approach to internationalization, trying first to consolidate their position in the German market through an improvement of their financial services, particularly in life insurance, investment funds, and mergers and acquisitions.

It is likely that other German banks will follow the example of Commerzbank, looking for some type of alliance with foreign banks to offer jointly specific products or financial services. However, as we shall see in Chapter 10, this type of strategy is not without risk, most often due to the lack of a clear leader.

The economic liberalization of Eastern Europe constitutes an enormous opportunity for German banks to expand internationally. The economic union of the two Germanies will further enhance this situation

---

[12] See *New York Times*, 30 July 1989, p. 8F.

although the process will undoubtedly be slow, as much because of their different demands for financial services as East Germany's precarious economic situation. Nevertheless, at least for East Germany, an increase in the presence of West German banks seems imminent, followed by a subsequent wave of acquisitions.

A future strategy geared toward operating as a large bank at the global level will only be viable for a relatively small group of firms. The problem they will face lies in implementation: through alliances, mergers, or acquisitions. However, for the majority of banks the problem lies in how to position themselves in an increasingly competitive market, with the competition coming not so much from foreign rivals as from existing firms in the local market.

APPENDIX

The Plans of the Deutsche Bank AG

The Deutsche Bank AG, the largest German bank in terms of assets, capital, and profits, fits the stereotype of the European bank in search of a truly global strategy. Its strength lies in its volume of transactions in the London Euromarket, although there is a weakness in its international operations: the management of non-German institutional investment funds.

Corporate objectives for the Deutsche Bank since 1986 have been the following:

- To strengthen the bank's position in stock trading through an expansion of the services offered.
- To continue development of classic commercial activities (deposit and loan operations), increasing the bank's presence in other high-growth international markets.
- To strengthen the bank's role in retail banking, entering new business areas and using new distribution channels.
- To develop all business lines with the application of electronic data-processing as an important qualitative and quantitative complement to the group's activity.

To achieve these objectives, the bank's directors have set forth measures for advancement in three specific areas: internal organization, international presence, and product range.

The purpose of the internal reorganization started in 1986 was to achieve more efficient use of all the bank's resources. Toward this end, a policy was enacted to improve the transparency of the group's costs and profits and also the co-ordination between similar strategic units.

In addition, certain areas of middle management went through a process of decentralization designed to free more of the Board of Directors' time for non-operational, strategic management.

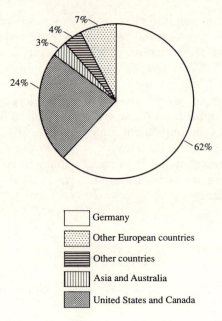

Germany

Other European countries

Other countries

Asia and Australia

United States and Canada

FIG. 3A.1   *Activities of Deutsche Bank by geographic areas, 1988*

The direction of this internal change appears consistent with the bank's second goal: the strengthening of its international presence. In December of 1986 the Deutsche Bank acquired Bank of America's Italian subsidiary, with 98 offices in Italy and a balance of DM8.9 billion. At the end of 1988, it bought a block of shares in Amro Bank from H. Albert de Bary & Co. of Amsterdam. In Portugal, the Deutsche Bank owns 100 per cent of MDM Sociedade de Investimiento, S.A., an important player in the Portuguese capital markets. In March 1989, the bank acquired a majority share of Barcelona's Banco Transatlántico and later that year in November it acquired the British bank Morgan Grenfell. All of this contributes to strengthening the banks' footholds in other EC countries before 1992.

The presence of the Deutsche Bank in countries outside Europe is also growing. During 1988, it reorganized its group in Asia with central offices in Singapore and plans for new offices in India, Indonesia, and Japan. Concurrently, the bank has increased its activity in Australia, the United States (principally in investment banking), and South America, though perhaps not as intensely.

Thus, it is clear that international positioning figures prominently in Deutsche Bank's strategy. At the same time, this is not intended to be at the expense of domestic business as the bank is equally committed to plans for expansion within Germany as well.

The bank's third strategic focus is the broadening of the range of their financial services. To achieve this Deutsche Bank has made a considerable effort to improve services for multinationals, strengthen its merger and acquisition operations

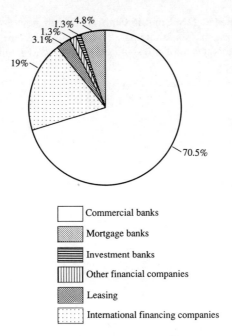

Commercial banks

Mortgage banks

Investment banks

Other financial companies

Leasing

International financing companies

FIG. 3A.2   *Activities of the Deutsche Bank Group, 1988*

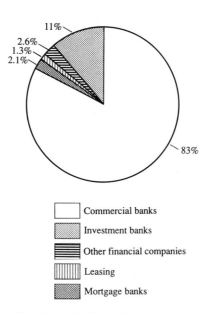

Commercial banks

Investment banks

Other financial companies

Leasing

Mortgage banks

FIG. 3A.3   *Activities of the Deutsche Bank Group: operating results, 1988* (through the Sociedad DB Consult GmbH), launch new financial products for consumers, increase its shareholding in international companies, and become more active in the issuance of titles and other transactions outside of the balance sheet (mainly swaps).

4

The Spanish Banking Industry

1. A HISTORICAL INTRODUCTION

Since 1977, the Spanish financial system has experienced significant legal changes in the way of numerous regulatory and deregulatory rearrangements of the system. From a market point of view, important changes have occurred in banking activities due to the attitudes of families and firms regarding financial intermediation and the deregulation process.[1]

The change has occurred over the course of the last 13 years at varying rates, though a sense of purpose has persisted and ultimately served to create a more modern and open financial system. Monetary policy has been geared toward the struggle against inflation. For this, the Bank of Spain has had to secure the necessary financing required by the authorities, while keeping in mind its objectives regarding inflation. This is important in explaining the changes in the banks' required savings and investment ratios.

The concept of liberalization has been the driving force behind this process of reform and deregulatory activity in the banking industry has been constant. The reason for this is tied not only to confidence in the free-market mechanism for allocating resources, but also to growing competition from abroad. The three most significant aspects of this process have been interest-rate liberalization, bank deregulation, and stock-market reform. The growing openness of the Spanish economy to other European economies, and the interdependence which this generates, has only pushed the process farther along in the same direction.

Financial innovation has also caused changes in the system, creating new products and markets, increasing competition between institutions, and presenting new challenges to the financial authorities.[2]

In addition, the Spanish financial system has experienced a process of disintermediation, through which traditional bank credit has been substituted to some extent by commercial paper and stock-market securities. This has generated greater competition in the banking system, making it more difficult for the banks to satisfy the needs of families and companies.

The modernization of the Spanish financial system began in 1977 with the new economic policy accorded by the government through the

---

[1] With regard to this subject see Cuervo et al. (1987), Trujillo et al. (1988), Ballarín, Ricart, and Gual (1988), and Canals (1990).

[2] Gutiérrez and Chuliá (1988) deal with this topic extensively.

Moncloa Plan.[3] The criteria for redesigning the system were three: reduction of the required ratios, with a view to eradicating them completely; liberalization of interest rates and commissions; and restatement of financial specialization.

With regard to the required bank ratios, there are three currently in existence. The first is the reserve ratio requiring banks to keep a fixed proportion of their deposits (called computable deposits) with the Bank of Spain. The purpose of the ratio is connected to monetary policy and it serves as an intermediate variable in the overall process of monetary control. The Bank of Spain has increased this ratio in recent years, with the objective of avoiding a flood of liquidity in the economy. Thus it went from 8.25 per cent in 1981 to 19.5 per cent in June 1989. Perhaps the booming demand, private as well as public, made it necessary to rely more heavily on the power of monetary policy. In the future, though, the Bank of Spain must trust less in this mechanism if it does not want to diverge sharply from common practice in other European countries.

The second important ratio is on investment. Its purpose is to channel funds toward certain industries through special loans or bonds issued directly by the public sector. The theoretical justification for this is linked to the hypothesis that the market will not channel resources to these industries, despite the fact that they produce social benefits, because their private benefits are considered insufficient. The critical difficulty arises, however, in the actual measurement of these benefits.

In 1985, the investment ratio was substantially reformed according to the criteria of submitting all financial institutions to uniform regulation and bringing the interest rates of such privileged financing progressively closer to those of the market. At the same time, there was the intention of creating a plan for financing the public deficit such that it would not become a deadweight for the financial intermediaries.

The assets in which funds can be invested to comply with this ratio are, according to the law, those issued by the public sector, although they may include assets corresponding to other public necessities, such as export loans or the restructuring of certain industries. In 1989, the maximum limit of the ratio was 35 per cent of computable assets. Along with this, two other limits were set: 15 per cent for short- and medium-term public debt and 25 per cent for the remaining computable assets. Currently, the ratio is 11 per cent and the computable assets correspond to treasury bonds and other titles, such as bonds issued by the autonomous communities and other specified assets to stimulate agricultural development.

Finally, the capital ratio requires banks to maintain a minimum level of capital, the actual amount depending on the risk of their investments.

---

[3] García Alonso (1984) offers an interesting historical vision of this process of modernization.

This ratio ensures the liquidity of the banking system, since banks need to retain adequate levels of capital as this is what allows them to absorb losses during a crisis. At the same time, it is difficult for the market to determine the necessary ratio due to the amount of information this requires. This is the reason for regulation.

Until 1985, the capital ratio in Spain put emphasis on maintaining a set relationship between equity and liabilities. At the same time, legislation included a series of limitations, related to concentration of risk, stock portfolios, the number of branches, etc., on equity. The principal limitation of this ratio was that it did not consider other risks financed from different sources. The new capital ratio also takes these into account and reflects all of a bank's assets, investments, and other commitments.

With regard to interest rates, the Bank of Spain historically had fixed the maximum levels that the banks could receive on financial assets and pay on deposits. The process of liberalization was initiated in 1969, although it did not garner significant support until 1977. That year all bank asset and liability operations not subject to ratios and of term longer than one year were liberalized. This measure was consolidated in 1981 when interest rates on asset operations not subject to investment ratios were liberalized. At the same time interest rates on liability operations were also liberalized. This was still only a partial measure since other restrictive regulations remained. It was in 1987 that all interest rates and commissions were finally deregulated, after numerous set-backs and considerable debate.

Another group of measures was adopted in 1978. Among the most significant were the authorization of foreign banking (July 1978), with its subsequent impact on broadening competition; the granting of a special financial statute for the credit co-operatives and rural savings banks (with more objective norms for market access, official registration, ratios, etc.)

TABLE 4.1. *Major Spanish banks*[a]

Firm	Liabilities	Assets
Banco Bilbao Vizcaya	2,790,400	4,322,000
Banco Central	2,061,400	2,912,200
Banco Español de Crédito	1,984,900	2,517,400
Banco de Santander	1,380,800	2,107,200
Banco Hispano Americano	1,567,600	2,298,900
Banco Popular	760,063	1,160,400
Banco Exterior de España	845,408	2,136,605

[a] In million ptas. as of 31 Dec. 1987.

Source: Canals (1990).

bringing them more in line with other financial firms; the establishment of a new regime for Sociedades de Garantía Recíproca; and a new reorganization of the financial industry. This group of reforms signified tremendous progress in the modernization of the industry and also laid the foundation for greater competition and efficiency.

Most recently, three new reforms have been introduced. The first is the establishment of the Instituto de Credito Oficial in January 1988 as another bank assuming ownership of the shares of Banco de Crédito Agrícola, Banco Hipotecario de España, Banco de Crédito Industrial, and Banco de Crédito Local. The second was the founding of the Sociedades de Crédito Hipotecario, regulated by the law of 25 March 1981. The object of these enterprises is to develop the mortgage market financed through mortgage guarantees which, in turn, can be traded in this market. Finally, although not directly related to credit banks, the new stock market law of 28 July 1988 had a significant impact not only on the overall financial sector, but also on the banking industry in particular.

2. STRUCTURE AND TRENDS IN THE SPANISH BANKING INDUSTRY

The Spanish banking industry consists of two principal types of institution: universal banks and specialized banks as shown in Fig. 4.1.

Within universal banks, of particular note are commercial and savings banks. Table 4.2 shows the structure of these entities and of the credit co-operatives in terms of their respective share of the Spanish banking market, while Table 4.3 compares their relative growth rates between 1983 and 1987. This period is especially important because it spans the time during which the structure of the Spanish banking industry underwent the most dramatic change.

The first observation to be made is the obvious dominance of commercial banks, with 65 per cent of total assets and 57 per cent of liabilities in 1987. However, it is immediately clear that savings banks experienced considerable growth throughout this period, going from 26.9 per cent to 32.2 per cent of total assets, and from 32.7 per cent to 39.4 per cent of liabilities. The data for the credit co-operatives during this period change hardly at all. Thus, as a result of the increased freedom the savings banks enjoyed, together with the aggressive commercial policies some of them employed, they were able to expand their assets, increase their capital, and thereby create more competition for the other banks.

It is also interesting to note that, while the number of branches and employees declined for other types of banks, it actually increased for the savings banks, an unmistakable sign of commercial vitality. An average annual increase of 3.8 per cent in the number of offices, in a country that

TABLE 4.2. *The Spanish banking system: market share by type of bank*

	No. of firms		No. of offices		No. of employees		Total assets (%)		Liabilities (%)		Credits (%)	
	1983	1987	1983	1987	1983	1987	1983	1987	1983	1987	1983	1987
Commercial banks	136	138	16,197	16,642	172,580	156,986	69.8	64.5	63.9	56.9	72.3	66.9
Savings banks	81	78	11,792	13,543	62,178	70,390	26.9	32.2	32.7	39.4	24.8	29.8
Credit co-operatives	147	126	3,197	3,248	10,675	9,805	3.3	3.3	3.4	3.7	2.9	3.3
TOTAL	364	342	31,186	33,433	245,433	237,181	100.0	100.0	100.0	100.0	100.0	100.0

Source: Lladó *et al.* (1989).

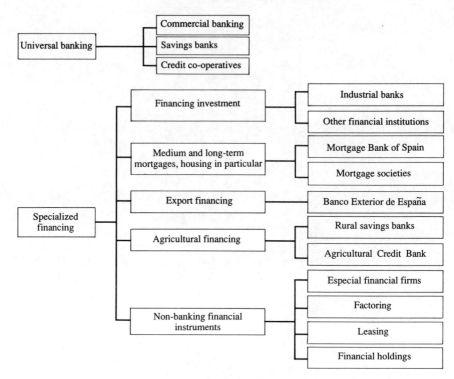

F<small>IG</small>. 4.1   *Structure of the Spanish banking system*

T<small>ABLE</small> 4.3.   *Growth rate according to type of bank, 1983–1987* (%)

	Assets	Liabilities	Credits	Offices	Employees
Commercial banks	11.2	9.9	7.0	1.4	−2.2
Savings banks	18.2	17.7	15.7	3.8	3.3
Credit co-operatives	12.9	15.1	9.7	1.3	−1.4
Total banking system	13.2	12.8	9.3	2.3	−0.7
Government banks	12.5	45.8	12.2	26.3	n.d.

Source: Lladó *et al.* (1989).

already has large numbers of branches, clearly illustrates this point.

A similar trend is demonstrated by annual growth rates for assets and liabilities during the same period (see Table 4.3). Savings banks showed growth of 18.2 and 17.7 per cent in both respectively, while commercial banks grew by only 11.2 and 9.9 per cent respectively. These figures leave little room for question about the thorough penetration of savings banks over recent years.

TABLE 4.4. *Distribution of assets* (% of GDP)

Assets	1975	1981	1983	1985	1987
	88.6	84.1	86.1	88.5	94.1
Government	7.4	8.0	9.4	19.6	27.8
Families and firms	81.2	76.5	76.7	68.9	61.3
Commercial banks	65.9	60.0	60.2	63.9	61.3
Government	6.7	6.2	6.0	15.9	16.9
Families and firms	59.2	53.8	54.2	48.0	44.4
Savings banks	27.7	24.5	25.9	24.6	32.8
Government	5.7	1.8	3.4	3.7	10.9
Families and firms	22.0	22.7	22.5	20.9	21.9

Source: Banco de España, 1988.

Table 4.4 gives information on the distribution of bank assets among commercial and savings banks over a longer time period. The first thing to point out here is the increase in government financing in parallel with the decline, in relative terms (average over GDP) of that for families and companies. Financing for the public sector grew from 9.4 per cent of GDP in 1983 to 27.8 per cent of GDP in 1987, nearly tripling over four years. Meanwhile, financing for families and companies went down from 76.7 per cent of GDP in 1983 to 61.3 per cent of GDP in 1987. There are two major reasons for this decline. The first is the growth in the public sector's demand for financing, primarily between 1983 and 1985, and the subsequent increase in interest rates.[4] The second, more important reason is a direct result of financial intermediation: families and companies, trusting less and less in banks, have gone straight to the market in search of finance. This phenomenon was particularly relevant for the commercial banks, whose financing to families and companies slid from 54.2 per cent of GDP in 1983 to 44.4 per cent of GDP in 1987. In contrast, financing to the public sector increased steadily from 6 per cent of GDP in 1983 to 16.9 per cent in 1987.

In contrast, savings banks adapted much better to the new situation created by the public-sector deficit. In fact, their financing to families and companies did not decline at all during this time, which means that a great deal of effort must have been expended to retain this business. Meanwhile, public-sector financing rose from 3.4 per cent of GDP in 1983 to 10.9 per cent in 1987.

Similar data for deposits broken down in Table 4.5 show the trends in liabilities for savings and commercial banks.

---

[4] A conceptual explanation of this phenomenon can be seen in Canals (1988).

TABLE 4.5. *Structure of liabilities* (%)

	Commercial banks		Savings banks		Total industry[a]	
	1983	1987	1983	1987	1983	1987
Public sector	2.9	4.0	5.4	4.7	3.6	4.2
Private sector	87.3	8.8	92.4	93.5	89.3	90.4
Current accounts	20.2	20.6	10.1	10.4	16.7	16.4
Savings deposits	11.1	10.3	39.8	31.2	21.3	19.3
Time deposits	46.0	28.3	35.9	30.9	42.8	30.1
Loans	3.6	3.6	3.3	5.1	3.4	4.1
Security operations	0.0	0.0	2.1	9.2	0.7	3.6
Government bonds	0.0	19.2	0.0	5.1	0.0	13.0
Other assets	0.0	0.2	0.0	0.0	0.0	0.1
Other credits	6.1	5.3	0.9	1.3	4.2	3.5
Non-residential sector	9.8	8.0	2.1	1.8	7.1	5.4
TOTAL	100.0	100.0	100.0	100.0	100.0	100.0
(billion ptas.)	13,420.9	18,878.6	6,881.1	13,081.2	21,014.8	33,182.3

[a] Includes Credit co-operatives.
Source: Banco de España, 1988.

TABLE 4.6. *Commercial banks: concentration of assets and liabilities*
(billion ptas.)

	1983				1987			
	Capital	%	Assets	%	Capital	%	Assets	%
Banks								
Top 10	9,415.8	70.16	12,027.4	61.96	12,655.9	67.04	19,138.1	61.58
Others	4,005.1	29.84	7,999.3	38.04	6,222.7	32.96	11,940.2	38.42
TOTAL	13,420.9	100.00	21,026.7	100.00	18,878.6	100.00	31,078.3	100.00

Source: Lladó *et al.* (1989).

An important trend in commercial banking has been the decreasing importance of time-deposits, which accounted for 46 per cent of total deposits in 1983. At that time temporarily lent assets appeared as a new and more attractive instrument, and with the promise of repurchase on the part of the banks. With this fall of savings and time deposits, savings banks opted in favour of developing temporarily lent assets and insurance operations, an attractive form of placing savings.

An analysis of commercial and savings banks' asset and liability structures reflects numerous differences among them. Table 4.6 indicates a

TABLE 4.7. *Savings banks: concentration of assets and equity* (billion ptas.)

	1983				1987			
	Capital	%	Assets	%	Capital	%	Assets	%
Savings banks								
Top 10	3,538.5	51.42	4,276.7	52.78	6,922.9	52.92	8,222.9	53.02
Others	3,342.6	48.58	3,825.8	47.22	6,158.3	47.08	7,286.4	46.98
TOTAL	6,881.1	100.00	8,102.5	100.00	13,081.2	100.00	15,509.3	100.00

Source: Lladó *et al.* (1989).

greater trend toward concentration for commercial banks. These figures show a relatively high degree of concentration in Spanish banking, with the ten largest commercial banks (7 per cent of all banks) accounting for 67.04 per cent of liabilities and 61.58 per cent of assets in 1987. Between 1983 and 1987 there was a slight decline in the concentration of liabilities, although not in assets. However, this is not to say that a large Spanish banking presence exists at the international level. Rather this is only an indication of a concentration of activities among a small number of large banks with a strong national presence—a situation similar to that in Germany.

The savings banks also show a fairly high degree of concentration, although less than the commercial banks. The ten largest savings banks, of a total 78, had 52.9 per cent of the liabilities and 53 per cent of assets in 1987, as shown in Table 4.7.

From the perspective of management, there have been three notable changes in Spanish commercial and savings banks' activities in recent years.[5] The first of these is the expansion of off-balance sheet operations—those financial services which are not reflected in the typical accounts of a bank's balance sheet (see Figure 4.2). Some of the causes behind this are similar to those observed in other countries though some are rather different. Among the latter are the high savings and investment ratios for deposits and the rapid growth of the stock market in recent years. Conversely, however, swaps, futures, and options have been very strictly regulated by exchange controls and such activities have been limited by the lack of organized markets for these instruments. The simple nature of off-balance-sheet operations makes it difficult to determine their proportion of banks' overall activities.

An analysis of income statements reveals that intermediation in the government bond market has been the most significant. It is estimated

[5] See Gutiérrez and Chulía (1988) and Gutiérrez (1988).

Fig. 4.2   *Off-balance sheet operations*
Source: *Canals (1990).*

that 90 per cent of the market (8 trillion pesetas in short-term bonds and more than 4 trillion in medium- and long-term bonds) is channelled through the banking system.

Another important aspect with regard to stock intermediation is the introduction of commercial paper, initially developed by foreign banks to attract capital. However, because of its inclusion in the ratios mentioned before, it has not become very significant. Also, operations in pure stock-market intermediation in 1988 grew by approximately a billion pesetas.

An additional area of interest with regard to the changes occurring in the Spanish banking industry is the growing importance of negotiable assets in the banks' balance sheets as much for assets as for liabilities (see Figure 4.3). In both commercial and savings banks, these instruments are more prominent on the asset side of the balance sheet, although negotiable liability instruments have also grown considerably as a result of the process of securitization in the Spanish financial system. As in all countries, financial and fiscal legislation has profoundly influenced the development of this process.

The most utilized financial instruments have been mortgage titles, bonds, letters of credit, credit participations, and temporarily lent assets, particularly those issued by the public sector to finance its deficit. On the other hand, banks have continued to issue other traditional instruments such as treasury and savings bonds, which has contributed to the measurable growth of negotiable liabilities between 1981 and 1985. A period of much greater stability was brought about with the changes in 1985 on the

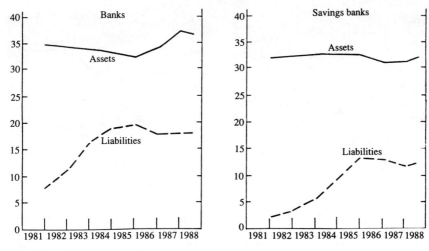

FIG. 4.3   *Negotiable instruments*
Source: *Canals (1990).*

fiscal structure regarding certain financial assets and the standardization of laws for the majority of financial bank innovations dealing with
negotiable stock-market assets.

In analysing asset trends, their increased stability during the 1980s is
somewhat surprising given the growth of liabilities. However, as we saw
in Table 4.4, the composition of negotiable assets changed notably with a
growing proportion issued by the public sector instead of the private
sector. In part, this can be attributed to the process of financial disintermediation. But the combination of these phenomena indicates significant change in the structure of the stock portfolios of commercial and
savings banks.

The regulatory role of the public sector has had a great deal to do with
this through its modification of the investment ratio, prompting banks to
acquire more government bonds. What subsequently occurred was that
commercial and savings banks purchased even more bonds than required
by the change in legislation with the objective of later relocating them, or
temporarily placing them with private investors. Thus, banks have used
this instrument to increase their activities as financial intermediaries,
taking advantage of their extensive branch networks throughout the
country.

The increased proportion of government bonds in bank portfolios has
prompted another important management change. Because government
bonds are negotiable in secondary markets, they are much more liquid
than certain shares and obligations issued by public or private industrial

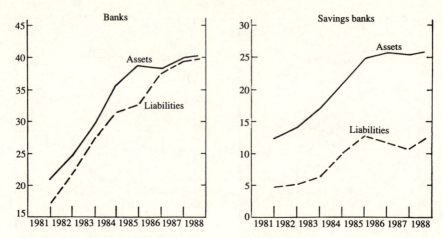

FIG. 4.4 *Instruments' sensitive to interest rates*
Source: *Canals (1990).*

groups. These government bonds have not only increased the liquidity of bank assets, but they have also permitted greater management flexibility.

The third important structural change in the management of banks is related to the greater sensitivity of various financial instruments (assets and liabilities) to interest rates. The primary cause of this is the greater volatility of interest rates throughout the 1980s.[6]

The development of the inter-bank market over the last ten years, and the banks' subsequent use of it for securing capital, is another relevant point which can help to explain interest-rate sensitivity. This is so because the inter-bank market is the first to absorb interest-rate variations before changes are made in monetary policy. In Figure 4.4 data on the changes in these financial instruments over the last nine years show the steady growth banks have experienced in this area. This is true for both assets and liabilities, accounting for 40 per cent of each in the aggregated bank balance sheet. Greater dependence on the inter-bank market has encouraged banks to develop asset instruments which are sensitive to interest rates yet still attractive for the demanders of funds. Among these, credits on variable interest-based titles using the MIBOR as the basic index have been particularly successful.

Savings banks have been less active in the issue of liabilities and variable-rate asset investments. The volume of these increased by roughly 26 per cent of total assets at the end of 1988, in comparison with 40 per cent for commercial banks. The volume of liabilities was even

---

[6] For an international interpretation of this phenomenon, see Canals (1987).

TABLE 4.8. *Commercial banks: income statement* (% total assets)

	Average 1970–4	Average 1975–9	Average 1980–4	1985	1986	1987
Intermediation margin	3.17	3.96	4.00	3.57	3.73	3.89
Non-financial products	0.47	0.48	0.82	0.83	0.87	1.06
Gross margin	3.64	4.34	4.82	4.40	4.61	4.95
Operating costs	2.38	3.17	3.24	2.80	3.00	3.04
Provisions	0.08	0.23	0.91	0.87	0.79	0.91
Profits before taxes	1.17	0.94	0.67	0.72	0.81	1.00

Source: Banco de España, 1988.

TABLE 4.9. *Savings banks: income statement* (% of total assets)

	Average 1970–4	Average 1975–9	Average 1980–4	1985	1986	1987
Intermediation margin	2.52	3.45	4.85	4.68	4.68	4.87
Non-financial products	0.32	0.15	0.34	0.52	0.66	0.66
Gross margin	2.84	3.60	5.19	5.20	5.35	5.53
Operating costs	1.49	2.65	3.57	3.39	3.83	3.51
Provisions	—	0.20	0.62	0.77	0.61	0.81
Profits before taxes	1.35	0.87	1.00	1.04	0.91	1.22

Source: Banco de España, 1988.

lower. The principal reason for this is that savings banks have supported their liability structure over the base of their deposits. The significant accumulation of these has eliminated the need for using the inter-bank market to develop asset operations and, conversely, they have actually acted as lenders in this market.

3. THE PROFITABILITY OF THE SPANISH BANKING SYSTEM

We have already seen that Spanish banks posted returns on total average assets during the 1980s that were higher than those of banks in other European countries. However, higher capital requirements and subsequent lower debt levels have reduced their returns on equity.

Tables 4.8 and 4.9 show income statements for the Spanish commercial and savings banks respectively. To a large extent, their relatively high profitability can be attributed to the industry's protectionist mechanisms directed at controlling competition from both within the industry or abroad. Collusive behaviour between firms is also a possible cause of this higher profitability, at least among the large banks—a predictable result for a sector with substantial entry barriers and asymmetrical information between the demand and supply of financial services.[7] This has been particularly prevalent during recent years, since banks have tended to adopt more or less similar strategies in response to the changes in the financial system (interest-rate liberalization, the increase of government bond issues, stock-market reform, etc.). If we also consider Spain's lethargy in the promotion of financial literacy, it is easier to understand the collusive behaviour of the large banks. Also, the possibility of improved efficiency from certain cost advantages—with deposits, for example—cannot be disregarded in explaining Spanish banks' profitability.

The relative lack of competition in the Spanish banking industry can also be explained, in part, by the financial authorities' regulatory actions, especially at the end of the 1970s and beginning of the 1980s during the banking crisis. The perceived need to protect banks' solvency and stability in the midst of increasing inflation and a growing public deficit relegated the liberalization of interest rates and the reworking of bank ratios to a lower priority. The same was true for commitments to the creation of new banks and to the geographical and operational expansion of savings and foreign banks. Thus, the public sector played a decisive role in sheltering the industry from greater competition.

For many years, in this environment of low competition and undemanding depositors, Spanish banks' strategies were based on the principles of geographic proximity—achieved through expanding their number of branches—and free services, which helped banks to establish closer ties with their customers. Table 4.8 shows how commercial-bank operating costs escalated from 2.38 per cent of assets during 1970–4 to 3.17 per cent in the period 1975–9, when most of this expansion activity was taking place. The corresponding increase in the intermediation margin during the same period did not manage to absorb this increase in operating costs. In addition, the increase of capital allocated for bad debt—which went from 0.08 per cent for 1970–4 to 0.23 per cent in the period 1975–9—did nothing to alleviate this situation. Taking all of these factors into consideration, it becomes clear why commercial banks experienced this second period of decline in their returns on assets (before taxes), which dipped to 0.94 per cent (1975–9) from 1.17 per cent (1970–4).[8]

---

[7] See Caminal, Gual, and Vives (1989).

[8] A detailed analysis of the profitability of the Spanish banking industry can be found in studies by Termes (1987) and Lagares (1987).

During the 1970s and 1980s two events occurred to reshape the Spanish banking industry. First, the banking crisis led to the disappearance of many small and medium-sized banks that could not survive the acute economic crisis and high inflation. Secondly, the more recent liberalization measures approved in 1988 led to more freedom within the sector in terms of price liberalization and foreign-bank authorization.

Foreign banks have had to begin their Spanish operations under the constraint of numerous legal restrictions, which have affected both their geographical freedom and the size of their assets. At the end of 1987 foreign banks' assets accounted for only 7 per cent of Spanish industry. Still, foreign banks have been a powerful stimulus for competition in the industry, especially among the larger companies (national and multi-national, public and private), with innovative financial instruments such as variable-interest-rate credits, treasury accounts, and other attractive vehicles for the placement of savings.

For Spanish banking, the period between 1980 and 1984 brought a significant deterioration in return on assets as gross margins went from 4.34 per cent of total average assets (1975–9) to 4.28 per cent and before-tax net profit dropped from 0.94 to 0.67 per cent. The increase in operating costs and provisions for bad debt were undoubtedly the cause of these declines, which were so substantial that even the significant growth of non-financial products, from 0.48 per cent to 0.82 per cent during the same period did little to offset them.

Since 1985 banking results have improved markedly, owing to intensified efforts to control operating costs and the continued increase in non-financial products. By 1987, returns on assets had once again reached 1 per cent (compared with 1.17 per cent in 1974). The intermediation margin was at 3.89 per cent in comparison to its 1985 value of 3.57 per cent, the lowest in ten years. Financial products represented 1.06 per cent of total assets, and operating costs and provisions for bad debt were at 3.04 per cent and 0.91 per cent of assets, respectively.

In terms of both international comparison and income-statement structure, Spanish banking's strength lies in its high intermediation margin. However, it still has a long way to go with regard to increasing income from non-financial services and reducing operating costs. Spanish banks are very far away from the comparable figures on operating costs for German and French banks, which are around 2.2 per cent of average assets. Much of this is due to their branch office expansion and present lack of automation—both of which have tended to be quite expensive.

We will analyse in greater detail the changes in some relevant income-statement variables. The first is the intermediation margin hitting a minimum of 3.57 per cent in 1985. In general, with the introduction of financial innovation and an increase in competition, the intermediation margin decreases. However, it is feasible for banks to react offensively

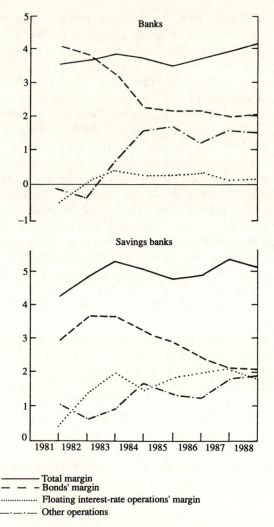

FIG. 4.5   *Structure and evolution of the intermediation margin*
Source: *Canals (1990).*

creating new products that help generate a greater margin. Figure 4.5
reflects the trends for overall financial margins and specific product
margins for both commercial and savings banks.

The fall in the margin for negotiable bond transactions, resulting
from an increase in costs coupled with a dramatic decrease in revenues
(see Table 4.10), has affected commercial and savings banks alike. The
explanation for this is complex. Without doubt, the most important factor
is the drop in tradable asset prices, due to falling interest rates and low

TABLE 4.10. *Commercial banks: bond performance* (% of total assets)

	Revenue		Expenses		Margin	
	Banks	Savings banks	Banks	Savings banks	Banks	Savings banks
1982	4.72	3.99	0.91	0.14	0.91	0.14
1983	4.44	4.19	1.26	0.40	1.26	0.40
1984	4.01	4.01	1.51	0.70	1.51	0.70
1985	3.99	3.99	1.61	0.99	1.61	0.99
1986	3.39	3.39	1.33	1.00	1.33	1.00
1987	3.02	3.02	1.51	0.96	1.51	0.96
1988	3.12	3.12	1.52	1.09	1.52	1.09

Source: Banco de España, 1988.

profits on some government bonds which the banks were required to have. At the same time, the banks' cost of liabilities increased as a result of the greater competition for savings. Figure 4.5 offers an overview of profitability trends for various operations.

To combat this situation, some banks changed their strategy in relation to the bond business such that, despite the decrease in the margin, they increased the number of bond transactions. Through this strategy, they tried to compensate for the lower margins with a greater volume of business to maintain revenues. In addition to this, banks expanded their role as intermediaries of share titles, distributing them to families and companies. This activity is not reflected in the intermediation margin but is evaluated among non-financial products as commissions on the sale of these products and capital gains. Consequently, a new source of revenue for the banks was developed.

Figure 4.6 also shows the drastic drop in instruments sensitive to interest rates, relative to their 1982 levels. The reason for this fall is that although interest rates generally declined between 1982 and 1987, interest rates on deposits did so more slowly than did those on loans, in part because of the greater competition, as can be seen in Table 4.11.

Savings banks experienced an even greater decline in rates compared to the peak reached in 1983. However, in this case the trend is not so significant since savings banks were not as involved in the inter-bank market except as lenders.

Regarding banks' revenues from operations outside of the balance sheet, the substantial growth evident in Figure 4.7 is due mainly to revenues from the sale of shares since 1985; savings banks also experienced a very noticeable increase of activity in this area of business. The

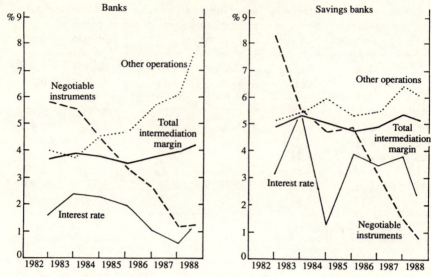

FIG. 4.6   *Average net profit on certain operations*
Source: *Canals (1990).*

TABLE 4.11.   *Operations sensitive to interest rates*

	Average return (%)		Average costs (%)	
	Banks	Savings banks	Banks	Savings banks
1981	—	—	—	—
1982	14.89	14.39	13.32	11.22
1983	13.17	15.46	10.80	10.01
1984	12.47	12.53	10.22	11.30
1985	11.19	11.44	9.25	7.55
1986	10.13	10.83	9.09	7.40
1987	10.84	11.15	10.31	7.33
1988	10.31	10.71	9.23	8.35

Source: Banco de España, 1988.

evolution of commissions is less pronounced. They remained relatively stable, although savings banks do show a gradual increase. This should be taken with caution, however, since behind the apparent stability a certain variability may be concealed as a result of government regulation.

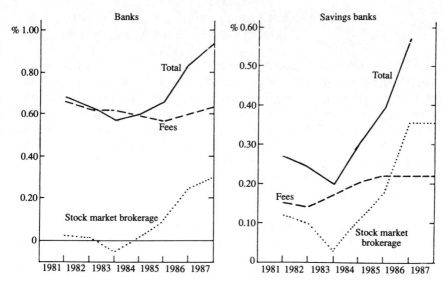

FIG. 4.7   *Off-balance sheet operations income*
Source: *Canals (1990).*

In analysing the savings banks' profitability, Table 4.9 shows the significant increase in the intermediation margin during the last ten years, from 3.45 per cent of total average assets between 1975 and 1979 to 4.85 per cent between 1980 and 1984. In 1987 it was 4.87 per cent, the highest of all the countries analysed. Given this high intermediation margin and the growing revenues for non-financial products, the gross margin on assets for the savings banks remained very high at 5.53 per cent in 1987, the highest since 1970. This margin is also the highest of the European savings banks analysed and is considerably higher than that of the Spanish commercial banks (which was 4.95 per cent in 1987).

The main obstacle for Spanish savings banks (one they share with the Italian banks) is their operating costs, which in 1987 reached 3.51 per cent of average assets. The extraordinarily large personnel base contracted by Spanish savings banks is undoubtedly at the core of the problem.

With this data, a figure of 1.22 per cent before tax on average assets was obtained for 1987, which is somewhat above that of savings banks in other countries and also above the return on assets for commercial banks, which was 1 per cent in 1987. Nevertheless, these figures do nothing to reduce the savings banks' operating costs. This is relevant as these costs are high not only in relation to other foreign savings banks—which will pose little competitive threat—but also relative to the commercial banks, which the savings banks will increasingly resemble in terms of operating structure due to deregulation.

4. FUTURE DEVELOPMENTS IN THE SPANISH BANKING SYSTEM

The liberalization of financial services within the EC means that foreign banks will be free to establish operations in Spain with greater movement than they enjoy now, which will inevitably lead to more competition within the industry. At the same time, savings banks have already received authorization to expand their operations nationally, which will lead to more competition in the retail sector. Finally, the foreseeable movement of banking toward new financial services will also increase the level of competition with other financial intermediaries such as insurance companies and money-market funds.

All of this is going to radically change the future landscape of the Spanish banking industry. In retail banking, the primary sources of intensified competition are foreign and savings banks. With regard to the former, though, they will not have a big impact initially because of entry-barrier deterrents. This is unquestionably true—one has only to consider Spain's unusually high number of branch offices; however, foreign competition is still a factor to be reckoned with. With current operating limitations foreign banking has come to account for 7 per cent of the industry's total assets; as deregulation continues they will be likely to increase their market share, perhaps through acquisitions of small and medium-sized Spanish banks.

Even the possibility of a foreign acquisition of one of the large Spanish banks cannot be disregarded. The efforts of some banks' Boards of Directors to avoid large blocks of public shares from being purchased is proof that this possibility must be considered. Thus, in one form or another, foreign banks must be counted as players in the future configuration of the Spanish banking system.

Savings banks, with their greater freedom to operate on a national basis, could become serious competitors for the commercial banks—due to their extended presence in areas where they did not previously operate, and to their offer of new, more competitive products. In fact, even in recent years, savings banks have managed to gain a larger share of the market at the expense of commercial banks despite the regulations still hindering them.

Within the savings banks segment, greater concentration—mainly through mergers—can be expected because of the relatively small scale of many regional firms; by not consolidating, they risk losing part of their customer base which will switch to other, more competitive banks.

An important phenomenon in Spain in recent years has been a widespread increase in financial literacy which, coupled with an improved standard of living, has generated a demand for more sophisticated financial services. Consequently, the increase in banking competition may well come from the investment banks, primarily foreign banks and new national investment banks. Its result will be considerably greater price competition for various financial services in the Spanish market. A good example of this is the ongoing 'war' for savings—now remunerated at market rates—which was unleashed in 1989 by Banco Santander, and which will continue to have important consequences for banks' profitability as well as the structure of the industry in the years to come.

This war for deposits, as of September 1990, still had not provided any clear results. The Banco Santander, exploiting 'first mover' advantages, was able to increase its deposits by 31 per cent; however, its financial costs rose by 41 per cent and financial products by only 20 per cent, resulting in a drop in its overall intermediation margin by 11 per cent. Other particularly aggressive banks have also lost considerable intermediation margin; in these cases sustained profits came through extraordinary revenues. The results of this struggle were predictable: the new price competition brought with it an increase in market share for the bank that moved first but a drop in margin for almost all banks that followed.

However, the decrease in intermediation margins during 1990 was not related exclusively to the increase in the price of deposits, but also to a credit restriction established by the Bank of Spain in June 1989 which impeded an issue of new loans. In addition, the new reserve ratio—interpreted by some as confiscation of net worth—imposes still another cost on Spanish banks in terms of profits.

The appearance of price competition in Spanish banking will lead to the disappearance of previously implicit price co-ordination. These schemes were made possible by a small number of large banks, price regulation, a high cost of entry into the strategic group of large banks, and the reputation for co-operation which these banks tried to maintain. The deregulatory process, greater competition from savings and foreign banks, and a desire to exploit new opportunities created by the new financial environment (headed by Banco Santander), will drastically change banks' behaviour in the markets. Serious doubt exists as to whether this will positively affect the industry, at least in terms of maintaining high profitability. Until now, the answer has certainly been 'no', as margins have fallen sharply. However, the previous system could not have been maintained indefinitely, and Banco Santander, by breaking it, at least took advantage of the benefits of being the leader. This event seems to have signalled the transformation of Spanish banking from an industry based on client–bank relationships to an industry based on competition, price, and service.

Foreign banks intend to compete heavily in investment banking, given that commercial banking is more complicated, for reasons indicated earlier. This presents additional problems for Spanish commercial banks.

In addition, it must be remembered that investment banks can play a much more active role as stockbrokers, attracting important groups of private investors. However, the Spanish commercial banks are not standing still and are currently putting forth considerable efforts to broaden their product lines and offer better service.

We saw earlier (Figure 4.7) that commercial banks' stock-market intermediation has been a lucrative source of revenue. If a growing part of their business is taken by foreign banks, operating results are bound to suffer. And while this is perhaps not so serious a concern at the moment—with the current growth and expansion of the financial markets— it is an issue that bank strategists must deal with now in anticipation of the time when the market will stabilize.

Regarding commercial banks' activities in the investment banking, things are very different. Many national banks have already created their own investment banks, some with strong ties to Spanish industrial groups. However, the size of these entities, although adequate for certain types of specialized operations, is still small in comparison with national and multinational enterprises. In order to achieve the economies of scale or scope associated with investment banking, considerable scale is necessary, and this may have been the critical issue in the merger between the Banco de Bilbao and the Banco de Vizcaya (see Appendix to this chapter). There are, of course, the obvious motives of avoiding stiffer price competition and attaining greater power for negotiating with the regulatory authorities. However, mergers also present significant problems and it is possible that to achieve alliances of this calibre, agreements or consortiums with foreign banks may prove more successful.[9]

Scale is not the only competitive factor in the banking industry, although it is a necessity for global banks with bases of operations in countries around the world trying to serve companies within those countries. A merger is not the only way to achieve this objective, however. Banco Santander, which has acquired shares in small and medium-sized banks in different EC countries—or Banco Popular and Banco Hispano-Americano, with their distribution agreements with other foreign firms— are clear proof of this. In the last two cases, the co-operation of the foreign firm is critical for the success of these operations and it eliminates many of the operative complications associated with mergers.

[9] See Benston *et al.* (1982) and Revell (1987). For the case of Spain, see Ballarín (1988*b*) and Fanjul and Maravall (1985).

APPENDIX

The Strategy of the Banco Bilbao-Vizcaya: The Desire to be a Universal Bank

After numerous weeks of meetings between the senior directors of the Banco de Bilbao and the Banco de Vizcaya, on 21 January 1988 the merger between the two banks was announced. It was approved by the respective shareholders in June 1988, thus becoming the Banco Bilbao-Vizcaya (BBV).

The merger conditioned the strategy of BBV in 1988 and 1989 and will continue to do so in the years ahead. The management of the merger was generally defined in the Basic Plan of Integration, which established the necessary tasks for achieving an operative merger of both entities.

These basic tasks were grouped into five categories. The first was the design of a new organizational structure, with the definition of functions for each position and the appointment of the corresponding people to fill them. The second was the reassignment of personnel at different levels of the new structure. The third was the redefinition of the branch network aimed at improving attention and service to customers. The fourth was the design of information systems for the integrated management of the businesses of both banks. The last task was the development of the Integration Plans, which included objectives for each of the business units as well the basic outline of their organization.

As a result of the merger, BBV emerged as the leading Spanish bank in terms

TABLE 4A.1. *Subsidiaries as at 13 December 1988*

	Assets (million ptas.)	Cash-flow (million ptas.)	Profits (million ptas.)	No. of employees	No. of offices
Grupo Catalana	1,029,350	24,684	9,107	3,273	383
Induban	299,907	7,124	4,053	588	58
Industrial de Bilbao	119,698	4,909	4,746	19	0
Comercio	270,412	6,024	3,407	961	122
Occidental	39,225	2,376	412	0	0
Mas Sardá	79,267	3,301	1,888	520	55
Bilbao Merchant	15,899	2,111	1,534	166	4
Oeste	36,933	1,293	1,067	4	0
Meridional	54,271	1,813	1,477	358	61
Crédito y Ahorro	65,594	1,959	1,324	223	32
Extremadura	32,869	840	533	176	41
Promobanc	19,856	635	216	5	0
Canaribank	38,700	468	297	142	23
Bankisur	8,576	−133	−144	7	0
TOTAL	2,110,557	57,404	29,917	6,442	779

TABLE 4A.2. *External network to 31 December 1988 ($m)*

	Assets	Credit investment	Share portfolios	Creditors	Permanent resources	No. of offices	No. of employees
Europe	6,020	1,573	930	2,591	459	48	833
Latin America	1,960	724	180	1,295	121	21	329
Others	165	92	39	126	8	22	321
TOTAL	8,145	2,389	1,149	4,012	588	91	1,483

TABLE 4A.3. *Industrial portfolio*

Sector	%
Food	22.9
Energy	22.5
Construction	12.3
Manufacturing	12.3
Oil	9.1
Venture capital	6.7
Other	14.2
TOTAL	100.0

of asset size and capital (see Table 4.1). However, at the worldwide level BBV is still a medium-sized bank, ranking number 68 in terms of assets.

The main points of BBV's strategy are similar to those of the individual banks before the merger. Among them are the following:

- A strong presence in the household segment through continued financial intermediation and a price policy based on market demands.
- A widespread use of electronic banking services, the income from which has grown in the last several years, in credit cards as well as in self-service banking.
- A strengthening of relationships with large commercial companies, directly or through specialized units. (Of note is a network of 12 offices dedicated specifically to these types of client and their financial needs in both pesetas and foreign currencies. Present operations include treasury management, leasing, and factoring, etc.).
- A growing presence in three areas of the capital markets: issue of fixed or variable rate titles, specialized financing, and business mediation services. In these areas, BBV appears to be the clear market-leader in Spain.
- A greater presence in the principal international markets, maximizing the advantages of increased scale. To this end, BBV has created a number of specialized international finance units.
- Capital participation in companies considered to be strategic for the future of the Spanish economy. In the last two years capital investment in non-financial companies has grown considerably.

The objectives of the new bank are tremendously ambitious and involve a high degree of risk. The management of a universal bank will become increasingly complex, requiring exceptional managerial talent to overcome the potential problems inherent in a lack of specialization.

5
The French Banking Industry

1. AN OVERVIEW

The French economy has also gone through a recent process of deregulation and liberalization which has significantly affected its financial system. For many years, this system has been characterized by strong control over the size of total credit and interest rates, strict separation of the different financial markets, the power of the public sector as a buyer of funds, and rigid regulation of capital flows and foreign enterprises.[1]

Table 5.1 presents an overview of savings and investment in the French economy between 1970 and 1985 broken down into various sectors. The low level of corporate savings is evident, as is the high debt level (measured against GDP), although since 1983 it has gradually declined. In addition, the government in 1974 became a demander of funds, a role which increased dramatically in the 1980s with the growth of the public deficit. Finally, the decline of net savings for French families has been another significant factor, dropping from 7.9 per cent of GDP (1979–83) to 6.2 per cent in 1985.

Figure 5.1 shows an international comparison of French corporate indebtedness relative to that of other industrialized countries, measured according to total cash flow over gross investments. This ratio for French companies is around the same as for Japanese and German companies— below 1 over the last 20 years—indicating that firms have needed external financing to cover the needs of their investment projects.

In the case of the French companies, though, the reasons are different from those for German or Japanese firms. First of all, there is the disincentive of higher taxes on profits, making deductible interest expenses more attractive. Additionally, dividends are subject to double taxation, first as profit and second as income, which makes the purchase of shares less attractive. Introduction of more neutral fiscal clauses and recent stock-market reform, however, improved the situation somewhat for private investors.

Another influencing factor in France has been the absence, until only a few years ago, of a large financial market. The problem was not so much the result of a low rate of private savings (though in international terms it is among the lowest), but the fact that a good part of savings were

---

[1] For an overall study of these issues see Fanger *et al.* (1989) and Voisin (1987).

TABLE 5.1.   *Saving and investment by sector* (as % of GDP)

	1970–3	1974–82	1983	1984	1985
Net saving					
Central government	4.0	1.2	−1.0	−1.0	−0.9
Households	9.5	9.3	7.8	7.1	6.2
Non-financial companies	1.9	−1.2	−2.4	−1.0	−0.5
Debts					
Central government	0.8	−1.2	−3.1	−2.9	−2.6
Households	3.6	4.3	4.1	3.8	3.0
Non-financial companies	−4.4	−4.7	−3.5	−2.5	−1.8
Rest of the world	−0.1	+1.1	+1.7	+0.9	+0.8

Source: OECD (1986*b*).

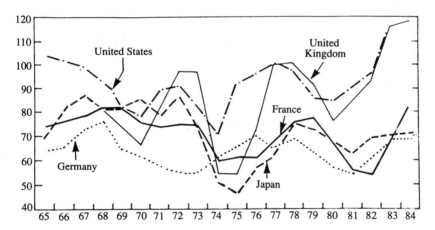

FIG. 5.1   *Debt levels of non-financial companies: cash flow/gross investment*
Source: *OECD (1986).*

directed toward the acquisition of homes either directly or indirectly through preferential credits.

Thirdly, household financial investment for many years was steered only toward traditional bank accounts or government bonds. As shown in Table 5.2, shares represented only 6.2 per cent of families' total assets in 1983. It should also be noted that approximately half of French families' capital was tied up in their homes, a sizeable figure that becomes even more significant when compared to that of other countries.

TABLE 5.2.   *Structure of family wealth* (%)

	1970	1983
Assets	7.4	6.2
Debt	2.5	3.9
Other financial assets	19.7	22.3
Including bank deposits	7.2	5.5
Non-financial assets	70.4	67.5
Including housing	45.0	50.3

Source: OECD (1986*b*).

2. THE STRUCTURE OF THE FRENCH BANKING INDUSTRY

The deregulation of the French banking system began relatively recently and dates back to only 1984. At that time, all previous banking legislation (from just after the Second World War) was eradicated with the purpose of deregulating and modernizing the system.[2] The principal characteristic of the old regulation was that it encouraged specialization within credit activities, establishing different legal requirements for each type of bank (deposit banks, investment banks, and medium- and long-term credit banks). In contrast, the new legislation is based on the principle of universality and establishes standardized directives. These institutions are defined generically as credit firms, and are described as companies that carry out professional banking operations, including receipt of deposits and management of means of payment.

The size of the French banking industry is relatively large. At the end of 1986 there were 2,080 firms, with a network of 25,480 offices, accounting for more than 440,000 employees (see Table 5.3). There are a number of banks that do not directly receive deposits. Among these are the specialized institutions which handle credit for specific public interests, designated by the government and generally related to regional economic development. Their principal operation is the concession of medium- and long-term loans, some of which are ordinary and some of which are subsidized.

This group of banks includes those financial firms that do not receive public deposits for less than two years. Collectively they represent a relatively important network, accounting for 1,049 establishments at the end of 1986. Included in these are credit firms, leasing companies and mortgage banks. These banks, unable to attract resources through deposits, became reliant on the financial markets, in particular the inter-

[2] See Coupaye (1984) and Pastre (1985).

TABLE 5.3. *Structure of the French banking industry*

	No. of banks	Offices	Employees	Total assets (%)	Credit investment (%)	Liabilities (%)
Banks which take deposits	1,000	25,480	409,350	78.82	67.87	83.36
AFB banks	386	9,917	253,877	51.21	42.17	39.67
Co-operative banks	192	11,125	124,183	18.08	21.32	27.96
Savings banks	401	4,391	30,256	9.39	4.13	15.52
Public savings banks	21	47	1,034	0.13	0.24	0.20
Banks which do not take short-term deposits	1,080	n.d.	32,696	21.18	32.13	16.64
Other financial firms	1,049	n.d.	21,917	11.84	15.13	6.83
Specialized financial institutions	31	n.d.	10,779	9.35	17.00	9.82
TOTAL: Banking industry	2,080	25,480	442,046	100.00	100.00	100.00

Source: Lladó *et al.* (1989).

bank market. In 1986 inter-bank financing represented more than 50 per cent of these entities' capital.

Commercial banks account for the bulk of the French banking system in terms of both assets and equity. By the end of 1986, together they held 51 per cent of the industry's total assets, 42 per cent of credit investments, and 39 per cent of deposits. Clearly their position within the French credit system is quite substantial. Under the 1984 law, these are the only firms legally permitted to carry out a full range of banking activities. Thus, they are sometimes referred to as 'AFB banks', which signifies their membership in the Association of French Banks (Association Française des Banques).

The three largest French banks are Banque Nationale de Paris, Crédit Lyonnais, and Société Générale. Table 5.4 provides relative data on the size of French banks in each of the main categories. As is clearly shown, the three leading banks have over 5,000 branches—more than half the total number—along with 41 per cent of assets, 47 per cent of credits, and 53 per cent of the deposits, a scale which is quite impressive even at the international level. According to a report prepared by the IBCA (Banking Analysis Ltd) in 1988, these banks ranked 13th (Banque Nationale de Paris), 15th (Crédit Lyonnais), and 22nd (Société Générale) in the world on the basis of assets.[3]

The second group of banks, in terms of size, consists of the rest of the banks with central offices in Paris that account for more than 1,300 offices, 30 per cent of the banking system's total investment in assets, and 25 per cent of deposits.

Another point of note is the foreign banking presence within the French system.[4] In 1986, out of a total of 386 banks, 145 were from abroad—clearly a very high proportion. The number of foreign banks increased measurably during the 1960s and 1970s when they reached their present levels. Their activity generally tends to be highly segmented. Seventy per cent of their equity comes from money markets, while only 14 per cent comes from public deposits. On the asset side, their investment is mainly in large companies, capital markets, or external trade operations.

A unique characteristic of the French banking system is the strong presence of the public sector. Until 1986—when legislation facilitating privatization of larger public companies was approved—63 per cent of assets, 73 per cent of credits, and 83 per cent of deposits were concentrated in public banks (38 in total).[5] This governmental involvement began in

---

[3] See *Business Week*, 26 June 1989, pp. 63 ff.
[4] This is taken up from the strong international financial integration of France. See the study by the Bank of France (1988).
[5] A treatment of this subject is given by the Ministère de l'Économie (1987).

TABLE 5.4. *Comparative data on French banks, 1986*

	No. of banks	No. of permanent branches	Total assets		Liabilities		Credits	
			FF billion	%	FF billion	%	FF billion	%
BNP, CL, and SG	3	5,609	1,749.5	41.3	788.4	53.3	697.4	47.5
Other banks based in Paris	122	1,308	1,293.1	30.6	377.4	25.5	453.1	30.9
Regional banks	67	2,455	294.2	7.0	176.3	12.0	136.2	9.3
Discount banks	21	24	206.5	5.0	16.0	1.1	7.2	0.5
Banks under foreign control	145	500	674.3	15.9	113.8	7.7	170.7	11.6
Other banks	28	21	14.5	0.3	6.7	0.5	3.3	0.2
TOTAL	386	9,917	4,232.1	100.0	1,478.6	100.0	1,467.9	100.0

Source: Lladó *et al.* (1989).

TABLE 5.5. *Size of major French banks, 1986* (FF million)

	Total assets	Deposits	Credits
Banque Nationale de París	807,092	309,235	314,173
Crédit Lyonnais	773,202	269,436	306,190
Société Générale	673,351	234,179	243,319
Banque Paribas	263,133	59,749	93,404
Banque Indosuez	175,035	49,402	74,823
Crédit Commercial de France	157,492	40,497	58,109
Crédit du Nord	88,912	37,782	40,454
CIC París	60,830	19,604	21,882
Top 3 banks	2,253,645	812,850	863,682
Top 8 banks	2,999,047	1,019,884	1,152,354
Total AFB Banks	5,105,693	1,412,826	1,779,317
Concentration, 3 banks (as % of the total)	44.14	57.53	48.54
Concentration, 8 banks (as % of the total)	58.74	72.19	64.76

1945, the year in which the Bank of France, along with the four most prominent banks at the time—Crédit Lyonnais, Société Général, Banque Nationale pour le Commerce et l'Industrie, and Comptoir Nationale d'Escompte de Paris—passed from private to public hands. These last two merged in 1966 creating the Banque Nationale de Paris. In somewhat of a reversal of this movement, within a programme of nationalization started by the first government of François Mitterrand in 1982, the fourth and fifth largest banks—Banque Paribas and Banque Indosuez—were privatized, along with all banks in possession of resident deposits of greater than one billion French francs.

Table 5.5 provides information on the degree of concentration within the French banking industry. In 1986, the three largest banks accounted for 57 per cent of deposits, 44 per cent of total assets, and 49 per cent of credits. If the list is lengthened to include the eight largest banks, the percentages increase to 72 per cent, 58 per cent, and 64 per cent respectively. These figures suggest an extremely high level of concentration, the most significant among the countries analysed.

This high concentration has not varied noticeably in the last 30 years, which is indicative of a certain degree of oligopoly.[6] This also explains French banks' low bankruptcy rate.

A final point on the nature of French banking is the recent trend in the public sector toward progressive deregulation of the institutions and financial markets. However, some very significant regulations still exist, principally in the area of deposit remuneration. Specifically, current

[6] See Boissieu (1990).

TABLE 5.6.   *Number of branches and employees*

	Offices				Employees		
	1970	1975	1980	1986	1975	1980	1986
AFB banks	6,458	9,528	9,675	9,917	236,700	243,700	253,877
Mutual banks	6,361	8,559	10,341	11,124	86,100	104,316	124,183
Crédit Agricole	2,602	4,098	5,164	5,622	52,600	62,303	72,549
B. Populaires	928	1,363	1,470	1,570	23,900	26,393	27,644
Others	2,831	3,098	3,707	3,932	9,600	15,620	23,990
Savings banks	2,250	2,989	3,630	4,391	13,400	20,034	30,256
TOTAL	15,069	21,076	23,646	25,432	336,200	368,050	408,316

Source: Lladó *et al.* (1989).

accounts (*comptes ordinaires*, *comptes de cheque*, and *comptes courant*) do not earn interest. This means that close to 37 per cent of banks' total capital (1986 figure) was exempt from financial costs. In return, however, the banks are not allowed to charge commissions on deposit-related services, such as cheques or transfers. Other types of savings are also strictly regulated, and, although it is possible to grant some interest, the amount is still either determined or limited by the financial authorities.

These rules on the remuneration of banks' savings affect not only the composition of French banks' operating costs, but also their behaviour. In effect, there have been relatively few alternative products offered to savers; thus, the strategy of the banks has been directed toward other objectives.

The third type of bank, in size, is the mutual bank, or co-operative. What is notable about this group is its number of branches (over 11,000 as opposed to the less than 10,000 for the savings banks), proof of its wide-ranging geographic presence (see Table 5.6). Its participation in intermediation activities is also significant; these transactions accounted for 18 per cent of total activity, 21 per cent of credit investments, and 28 per cent of liabilities in 1986. This last figure reflects the capacity for catering to the household segment. The investment of these banks has been orientated more toward small and medium-sized companies and agricultural concerns, which probably accounts for their large number of branch offices.

Table 5.7 shows statistics on the structure of the French mutual banks. Among these, the largest is Crédit Agricole with 66 per cent of total assets, 71 per cent of credits, and 66 per cent of liabilities in 1986. In 1988 it was the largest in France and in Europe with respect to assets and ninth in the world according to the ranking cited earlier in *Business Week*.

TABLE 5.7. *Mutual banks' size, 1986*

	No. of firms	Offices	Employees	Total assets (%)	Credit investments (%)	Liabilities (%)
Crédit Agricole	95	5,622	72,549	66.71	71.78	66.41
Crédit Populaire	40	1,570	27,644	16.64	13.64	15.62
Crédit Mutuel	23	3,303	18,581	13.23	11.55	15.59
Crédit Mutuel Agricole et Rural	14	448	1,902	1.45	0.75	0.94
Crédit Coopératif	20	182	3,507	1.76	2.28	1.45
Total Mutual Banks and Co-operatives	192	11,125	124,183	100.00	100.00	100.00

The structure of the mutual banks is very unusual. A prime example is provided by Crédit Agricole. This bank is really a combination of savings banks, hierarchically organized on three different levels. On the first level are the local savings banks, which provide the points of sale and customer-contact base. These banks attract deposits and participate in loans.

However, the actual management of deposits and credits are the responsibilities of the regional savings banks, which carry out these activities autonomously. Capital not invested within a certain period of time is transferred to the next level up, the Caisse National du Crédit Agricole. Until 1988, this body was publicly owned but with legislation approved in January 1988 its shares were redistributed to the regional savings banks and employees. The function of the Caisse National is the management of the surplus from the savings and regional savings banks, as well as the allocation of resources for the financing of medium- and long-term investments. It also supervises the regional savings banks, verifying their accounts and naming their senior directors.

The last major group of French banks is the savings banks, which in 1986 had more than 4,300 offices, with 9.4 per cent of total bank assets and 15.5 per cent of total liabilities. These firms are geared toward commercial banking, receiving savings which are directed toward investment projects of national interest. Their present organization was created in July 1983, and also consists of three levels: the savings banks ('Caisses d'Épargne), the 'Sociétés Regionales de Financiament' (SOREFI), and the 'Centre National des Caisses d'Épargne et de Preroyance'.

Savings banks receive capital from the public, although they do not manage or invest it, but rather transfer it to the SOREFI or 'Centre National'. However, they were authorized to grant personal and housing loans and, as of June 1987, some corporate credit.

We will close this structural overview of the French banking system with some figures related to productivity (Table 5.8). The first observation to be made is the higher productivity of the savings banks in terms of assets per employee as well as equity per employee. The banks, on the other hand, show better statistics with regard to assets per office, equity per office, and credits per office.

Another point of interest is the variance of ratios in the credit system. In comparison with the results of the savings banks, the co-operatives present a much poorer showing, although they have the most extensive branch network within the system.

In international terms, the number of offices for each 10,000 inhabitants in France was 4.6 at the end of 1986, which put it right at the middle of the European countries studied—below Spain and Germany and above the UK and Italy. Also for number of employees per office, France ranked in the middle with an average figure of 16.1 employees per office,

TABLE 5.8.   *French banking productivity ratios, 1986* (FF million)

	Banks	Co-operatives	Savings banks	Municipal credit	Total
Assets per employee	16,670	12,032	25,656	10,270	15,911
Assets per office	426,749	134,307	176,781	225,936	255,616
Capital per employee	5,825	8,393	19,127	7,316	7,591
Capital per office	149,110	93,683	131,791	160,957	121,947
Credits per employee	5,782	5,977	4,747	8,248	5,771
Credits per office	148,019	66,723	32,707	181,447	92,714
Employees per office	25.6	11.2	6.9	22.0	16.1

Source: Lladó *et al.* (1989).

above Spain and Germany, and below the UK and Italy. Again, these figures are based on comparisons which are not homogeneous, as the diversity of institutions and operations within each country is significant, and should be treated as such.

3. PROFITABILITY IN THE FRENCH BANKING INDUSTRY

French banks' profitability has suffered since 1981, although in 1985 and 1986 it experienced a slight upturn. In terms of return on equity, the results ranged from 17.57 per cent in 1981 to 14.93 per cent in 1986, which also placed France above Spain and below the other three European countries considered here (see Table 2.9). Two specific time periods are immediately distinguishable from the data: the first, from 1981 to 1984 when return on equity fell to 12.7 per cent, and the second from 1985, when it began to recover.

Following what was discussed earlier in terms of debt, it is not surprising that the French banks emerge as the most highly leveraged, with an equity-to-debt ratio of 2.57 per cent, well below the British banks (the next in line) with 3.9 per cent. Consequently, French banks also show an extremely low return on assets.[7]

Figures 5.2 and 5.3 show the trends between 1980 and 1986 for the gross margin over total assets and profit before tax over total assets ratios. They are shown in comparison with the same ratios for British banks, which showed the best results. French banks' gross margin is

---

[7] The low capitalization of the French banks presents serious problems of limiting resources for the future. For a related discussion of this subject, see *Euromoney*, Sept. 1988.

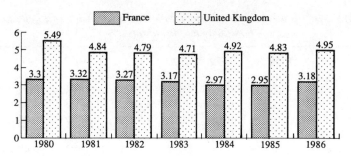

FIG. 5.2   *France and the United Kingdom: gross margin*
Source: *OECD (1988).*

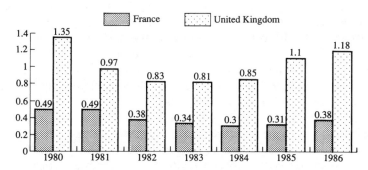

FIG. 5.3   *France and the United Kingdom: return on assets*
Source: *OECD (1988).*

obviously consistently lower than that of the British banks. It dropped between 1980 and 1985, though it showed signs of recovery in 1986.

As we saw before, this figure hides two factors: the intermediation margin and the revenue from non-financial products. The latter has remained relatively stable at around 0.5 per cent of total assets, which shows that intermediation has been kept as an almost exclusive activity of the banks. This is in direct opposition to what we have seen of the banks of other countries which have become increasingly involved in the provision of financial services in an attempt to compensate for the decline of their intermediation business.[8]

The intermediation margin for the same period dropped slightly, from 2.84 per cent on total assets in 1980 to 2.72 per cent in 1986, with

[8] See Banque de France (1989).

TABLE 5.9.   *Bank financing in the total external financing of families and non-financial institutions* (%)

	1978	1979	1980	1981	1982	1983	1984
France	0.77	0.80	0.69	0.77	0.78	0.71	0.62
Germany	0.90	0.95	0.97	0.98	0.79	0.80	0.82
Italy	0.42	0.62	0.59	0.48	0.39	0.30	—
United Kingdom	0.56	0.62	0.68	0.69	0.80	0.62	0.75
United States	0.71	0.76	0.57	0.62	0.42	0.48	0.63

Source: OECD (1986*b*).

significant decreases in 1984 and 1985. During these years, the French margin was only higher than that of the German banks, and was well below those of the Italian, British, and Spanish banks.

Conversely, the French banks show better performance in operating results. The operating costs/average assets ratio improved to 2.12 per cent in 1986 from 2.24 per cent in 1980. In addition to being indicative of falling costs, this figure is also the second lowest among the European banks analysed, after the German banks.

After subtracting the corresponding allocations for bad debt, which are approximately the same as those of other countries, the operating results of the French banks showed a return on assets of 0.38 per cent in 1986, below the 0.49 per cent recorded in 1980. This decline, which persisted over the years, is particularly worrying when compared with the improvement of this ratio in German, British, and Spanish banks. Only the Italian banks experienced such significant deterioration in their return on assets. As indicated previously, the cause of this low profitability is not to be found in operating expenses, which were relatively low, but in the margin of intermediation and the low volume of non-financial products. Both phenomena are related to the strict regulation imposed on French banks by the authorities. These include interest-rate controls, obligatory credit lines to certain industries, and limits on the development of activities not strictly corresponding to banking intermediation.

Nevertheless, the share of bank financing of non-financial institutions has diminished considerably. Table 5.9 shows the change in the percentage of bank financing which went from 77 per cent in 1987 to 62 per cent in 1984. With the exception of Italy, France is the country with the least banking participation in corporate liabilities.

What has happened is that with increasing commercial paper issues companies have relied less on bank credit, a trend which has caused a decline in financial intermediation. This trend is common to other countries as well, but seems to have moved more quickly in France. In

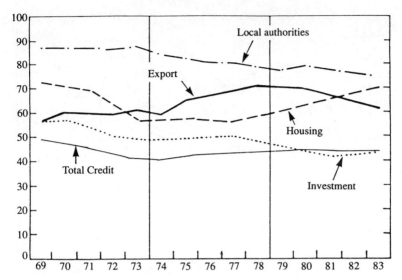

F<small>IG</small>. 5.4 *Special credits with preferential interest on credits*
Source: *OECD (1986b)*.

spite of rigorous regulations, banks tried to deal with this through the creation of subsidiaries that have offered traditional investment banking services in the area of corporate finance.

In conclusion, the French banking industry until recently has been highly regulated, its profitability is low, and few oligopolistic benefits exist. On the other hand, the operating costs and margin of intermediation are both fairly low in comparison with other countries, which indicates that the system is relatively efficient.

Contrary to this argument is the data on concentration. As we saw from Table 5.4, 50 per cent of the French banking business is concentrated among the three largest banks. Generally this would be equated with collusive practices among the participants. However, this is not necessarily so. The level of competition in an industry is dictated, not only by the number of firms in the sector, but also by the competitive behaviour between them. It has been argued that there are situations in which there may be only two large competitors, for example, Boeing and Airbus, but between which the competition is tremendous. The data on margins and costs of French banking appear to support this argument, although there are other interpretations as well: that the competitive stability generated is a consequence of strong public-sector concentration, for instance. And it is true that the public-sector has intervened very actively in the banking system during the past decades, often through

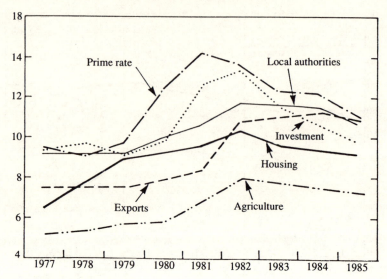

Fig. 5.5   *Selected investments*
Source: *OECD (1986b)*.

such direct means as outright ownership of numerous financial institutions. In this respect, Figures 5.4 and 5.5 are very illustrative of public-
sector intervention, not necessarily in terms of the proportion of assets it
owns but in terms of management. Figure 5.4 expresses the percentage
of total credits conceded at preferential interest rates (fixed by the Bank
of France directly or indirectly). The percentages are surprisingly high:
around 80 per cent in credits to local authorities, 60 per cent in loans for
the acquisition of homes, and around 45 per cent of the total credit.
These figures reveal a fairly high degree of intervention, which would
sustain the hypothesis that the low margins are not a result of competition,
but of the industry's stability due to many years of heavy regulation by
financial authorities.

Figure 5.5 is also a good reference in this respect as it expresses
preferential interest-rate trends in comparison to regular bank interest
rates. All of the preferential rates are far below basic bank rates, although
in recent years the gap has been gradually closing in some cases. The
most extreme cases are agricultural credits (heavily subsidized), home
acquisition, and export support.

This regulatory practice of privileged financing for certain sectors of the
economy is incompatible with the spirit of the Single Act and with EC
banking legislation. Thus, it is to be expected that the process of financial
liberalization initiated in recent years will continue, eventually allowing

for a less regulated, more competitive banking industry based on the principle of price competition.

Already in 1985 there was a reduction in the proportion of credit granted with preferential interest, at the same time that interest rates rose closer to market rates. Between 1984 and 1986, credits granted to industrial sectors with reduced interest rates decreased by FF12,000 billion. Concurrently, the interest rate differential was reduced from four points to only one point.[9]

The process of banking liberalization in France was partially completed in 1990 when capital flows between France and other countries were fully deregulated. This is an important milestone in the strengthening of competition in the French banking industry since it is probable that competition from abroad will increase. This is a logical consequence of French banks' growing expansion in other countries and of the increasing integration of the French and international economies. It is also possible that a number of large French banks will sell some of their regional units to foreign groups, as has happened in other countries.

The two major challenges for French banks in the coming years will be to reach an adequate volume of capital to comply with BIS regulations—which will not be easy, especially for the public banks—and to develop a price framework for their different services, a necessary step for the improvement of operating results. Until very recently, competition in French banking was not based on the price, but on client relationships, due, in part, to the fact that many services were free. The imperative of achieving greater profitability is changing this approach, however, and greater price competition is beginning to appear.

In general, the large French banks such as Banque Nationale de Paris (see Appendix to this chapter) or Crédit Agricole do not think of international expansion in the same way as their English or German counterparts. Their principal focus is the strengthening of domestic operations, to increase profitability by more fully utilizing their extensive branch networks, actively trading in the stock market through intermediary firms, and increasing their product lines to include international investment funds and life insurance. Thus Crédit Agricole has recently created a specialized bank solely for the execution of mergers and acquisitions.

Other French banks are acquiring significant shares of industrial companies in order to expand. In June 1990 the corporate stock held by Crédit Lyonnais, BNP, and Société Générale de la Banque was valued at 9,300 million dollars. In this, they seem to be following the German example of developing close bank–industry relationships. This strategy

---

[9] See OECD (1986*b*).

is difficult to justify in economic terms but understandable from the standpoint of the greater negotiating power it gives in relation to the financial authorities.

A few banks, including Paribas, Indosuez, or Crédit Commercial de France, are following a more international tack on the basis of strengthening certain product lines, such as the issue of ECU bonds and other capital-market operations. This is true of Paribas, which holds minority shares in other European banks and is closely tied to the Groupe Bruxelles Lambert. Indosuez has initiated a similar policy, with analogous participations and with a powerful position in Société Générale de Belgique. All of this constitutes a clear signal that at least some of the French banks are not going to remain tied exclusively to the domestic market, but are planning to internationalize their operations, offering specialized services in specific market segments.

APPENDIX

Changes in Banque Nationale de Paris (BNP)

On 1 July 1986, the BNP celebrated its twentieth birthday. Born of the merger between the Comptoir National d'Escompte de Paris and the Banque Nationale pour le Commerce et l'Industrie, the BNP is a bank with a strong national and international presence.

During recent years, the directors of the BNP have taken a series of decisions to position the bank. In accordance with this, installation of international tele-processing was started in 1985, new offices were opened for share intermediation in Paris and London, and part of the bank's international finance operations were moved to London.

In response to the new initiatives, the BNP decided to increase its volume of capital in 1986, offering shares and voting rights to French and foreign investors. In total, 11 million shares were offered, whose subscription exceeded FF5.3 billion, increasing capital by almost 50 per cent.

The objective of the BNP in the past years has been to concentrate primarily on three areas of activity:

- Commercial banking—in France and abroad, particularly in Europe and Africa.
- Financial markets—participating on its own account as well as for clients.
- Investment banking—directly or through subsidiaries.

During 1988, the BNP made a number of assessments about its competitive ability for the years ahead. The management of BNP in France was reorganized in

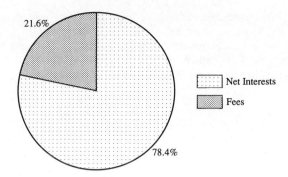

Fig. 5A.1   *BNP Group: structure of results, 1988*

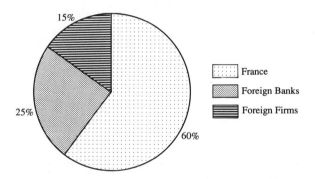

Fig. 5A.2   *BNP Group: geographic breakdown of operating results*

terms of both products and markets. Simultaneously, a direct-marketing unit was created to work in co-ordination with the decentralized services.

Other organizational changes included the establishment of an Office of Internal Development and the restructuring of administrative and auditing services.

The strengthening of the bank's commercial structure has been a guide for future reference. One particularly successful decision was the installation of automatic teller machines which accounted for 60 per cent of all cash withdrawals in 1988—an illustration of the fact that the bank has been able to adopt new direct-marketing techniques with very satisfactory results.

Also central to BNP's strategy is the concentration of effort in EC member countries. In countries in the south of Europe, BNP has managed to introduce a wide range of products and services, while in the northern countries it has followed a product-by-product strategy. In the UK it has strengthened its presence through the initiation of activities in the mortgage market; whereas, in Spain the bank offers a variety of products and services. The European expansion of BNP is not strictly limited to banking operations, but also extends to other activities such as leasing or factoring.

BNP's data-processing equipment for line and share transactions has also been significantly improved, as much for the purpose of improving buy-and-sell operations as providing real time information for clients.

6

The British Banking Industry

1. AN OVERVIEW OF THE BRITISH FINANCIAL SYSTEM

The UK has a long tradition as a premier financial centre. This is a result not only of British political leadership at the international level in the nineteenth century, but also the liberal approach to financial regulation.[1] However, the downturn of the economy and the rise of other international financial centres has caused London to lose some of its prominence.

British authorities have not watched this change passively, but have acted assertively to try to correct this loss of competitiveness. An example of this is the government's 1979 decision to abolish exchange controls which, in spite of initial uncertainty, has contributed greatly to the internationalization of the markets. In response to the rapid development of stock markets in other parts of the world, London proceeded in 1986 with a complete reorganization of the British stock market. The objective of this was the liberalization of operations, but without the financial authority relinquishing its supervisory role, and the strengthening of its international competitive position.[2]

All of this was critical for the United Kingdom, particularly since more than 7 per cent of GDP is generated by banks and insurance companies. The same statistic for other countries is much lower: 4.7 per cent in the United States and 5.7 per cent in France, for example.[3] In addition, the value-added contribution of the financial sector within the overall economy is quite significant, as shown in Table 6.1. The financial sector's value added went from 5.9 per cent in 1975 to 7.4 per cent in 1985; if financial services to companies and leasing are included, the percentages become 10.6 in 1975 and 13.9 in 1985, showing the growth of new financial services. On the other hand, the banking industry has stabilized regarding its participation within the overall financial system which, in turn, demonstrates the growth of other segments.

The recent growth rate of this sector has been dizzying, far above the GDP growth rate as indicated in Figure 6.1. The first section of the graph shows, at constant prices, some trends in the financial services output, including corporate services and leasing. The steady growth in volume of the financial-services business is obvious. The same can be concluded by

---

[1] See Peat Marwick (1985) and the Bank of England (1986c).
[2] See Bank of England (1986a,b) Clarke (1986) and Ritchie (1986).
[3] Figures taken from OECD National Accounts (1986).

TABLE 6.1.   *Value-added generated by the financial sector* (% GDP)

	1975	1980	1985
Banks	2.7	2.6	2.6
Other financial institutions	1.3	1.4	2.2
Securities	1.9	2.3	2.7
Total financial sector	5.9	6.3	7.4
Total including services to companies and leasing	10.6	11.6	13.9

Source: OECD (1987*a*).

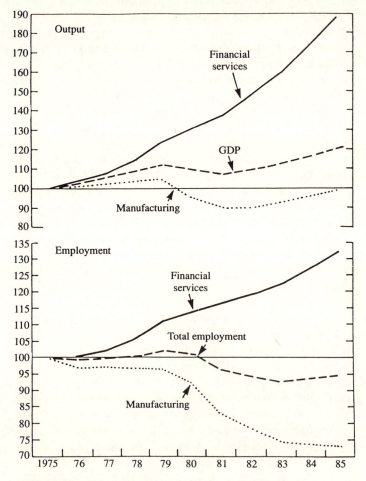

FIG. 6.1   *Production and employment trends for various sectors*
Source: UK National Accounts, *1986*.

TABLE 6.2.   *British financial institutions' foreign-exchange income* (£m.)

	1975	1980	1983	1985
Insurance	450	968	1,962	2,920
Banks	7	160	1,556	2,054
Others	66	202	720	920
Stock Exchange	18	43	71	106
TOTAL	541	1,373	4,309	6,000
Total income from the City	1,029	2,020	5,418	7,130
Balance of services	1,751	1,568	3,969	5,628

Source: UK Balance of Payments, 1986.

analysing the trends for employment. In 1986 financial services provided jobs for nearly 760,000 people, representing 3.6 per cent of total employment and a growth rate of 20.8 per cent between 1978 and 1986.

Another indicator of the financial industry's importance to the British economy is financial institutions' contribution to foreign exchange revenue in the balance of payments. The composition of these revenues is varied and includes interest on loans to foreign companies, export of services to residents abroad, profits from direct investment of British companies with operations outside of the UK, and profits from portfolio investment effected abroad.

Among the financial institutions, insurance companies stand out (see Table 6.2) with an income of £2,920 million in 1985. Commercial banks follow with a total income of £2,054 million for the same year, showing rapid and sustained growth. This growth has been accompanied by a change in the banks' operations from regular and syndicated loans to securitized loans and bond-placement services. These figures are especially relevant in comparison to those of ten years earlier (1975) and to net income from the British balance of services. In 1985, the balance of services total was at £5,628 million, while total income from abroad for all institutions listed was £7,130 million, a clear demonstration of the British financial industry's growing strength.

A focal point of the British financial system is, of course, London, which has retained a significant role in international finance, particularly in Euromarkets. Nearly one-quarter of the international assets for banks with operations abroad are located in London (see Table 6.3).

The international position of the UK has slowly deteriorated since 1973, but it is still dominant over such countries as the United States and Japan.

With respect to its stock markets, though, London is somewhat behind

TABLE 6.3. *Participation in international bank loans* (%)

	1975	1980	1983	1986
United Kingdom	29.3	26.4	24.9	23.3
Japan	6.2	6.2	8.2	15.0
Other centres	15.6	18.4	22.7	20.2
Other European countries	36.4	36.7	27.0	28.2
United States and Canada	12.5	12.3	17.2	13.3

Source: Bank of England (1986*c*).

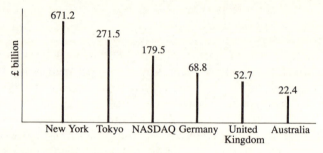

FIG. 6.2   *Stock-market capitalization, 1985*
Source: *OECD (1987).*

the others, specifically, New York, Tokyo, NASDAQ (the American electronic system), and the group of German exchanges, as shown in Figure 6.2.

The strong presence of foreign firms has unquestionably contributed to London's role as an international financial centre. At the end of 1985 there were close to 500 foreign banks operating in London accounting for nearly 60,000 people and representing over 60 countries. Foreign banks in the UK provide a constant source of pressure for improved efficiency in the British banking system as well as a serious threat for the medium-sized domestic banks that cannot compete with the foreign banks' greater access to capital.[4] The financial authorities have not intervened in this increasingly competitive environment but have stimulated it, while at the same time requiring conditions of reciprocity with foreign governments in order to penetrate financial markets abroad.

London's position in financial intermediation at an international level has declined relative to what it was 15 years ago and its stock market

[4] See Bank of England (1986*c*, 1987*b*) and Gilbody (1988).

TABLE 6.4. *Savings and investment by sectors* (% GDP)

	1963–9	1970–9	1980	1981	1982	1983	1984	1985
Households								
Savings	6.3	7.8	10.6	9.2	9.0	8.0	8.3	7.7
Investments	4.1	4.5	4.2	4.0	4.2	4.5	4.5	4.7
Balance	2.2	3.2	6.4	5.2	4.7	3.5	3.8	3.1
Non-financial companies								
Savings	7.9	10.1	7.8	7.5	7.0	8.5	10.2	9.2
Investments	7.5	9.8	7.6	6.6	6.1	6.5	7.5	7.3
Balance	0.4	0.4	0.2	0.9	0.9	2.0	2.7	1.9
Financial sector								
Savings	0.3	1.2	1.5	1.4	1.4	1.6	1.8	2.1
Investments	0.9	1.7	2.3	2.3	2.3	1.8	2.2	2.3
Balance	−0.6	−0.5	−0.8	−0.9	−0.9	−0.2	−0.4	−0.3
Public sector								
Savings	5.9	3.9	1.2	1.6	2.0	1.5	0.7	1.0
Investments	7.9	7.8	5.8	4.8	4.7	5.3	4.8	3.9
Balance	−2.1	−3.8	−4.6	−3.2	−2.7	−3.7	−4.1	−2.9
Domestic sector								
Savings	20.4	23.0	21.0	19.6	19.4	19.6	21.0	20.0
Investments	20.4	23.7	19.8	17.7	17.3	18.1	19.0	18.2
Balance	0	−0.7	1.2	2.0	2.1	1.5	2.0	1.8

Source: OECD (1987*a*).

capitalization is now below that of other counties. While it has grown over the years, this growth has not occurred at the same rate as in New York or Tokyo.

Concerned about London's loss of competitiveness in the stock markets, financial authorities implemented a set of measures for adapting the British stock-market to the new conditions of international competition. This process—which came to be known as the 'Big Bang'—was designed to eliminate existing stock-market entry barriers.[5] At the same time, the London Stock Market reached an agreement with the International Securities Regulatory Organization (ISRO), representative of the larger stock companies, to establish a unified national and international stock market. These steps were taken to secure London's future position with New York and Tokyo in the triad of leading international financial centres. The complete liberalization of capital flows within the EC furthers this objective.

[5] For more information on the Big Bang, see Bank of England (1986*a*).

In relation to the efficiency in a financial system, the channelling of funds from savers to investors is an important factor. The savings rate in the UK has been roughly 20 per cent of GDP since the 1960s. This rate is very high compared to the United States and France, although it is still below savings rates in Japan and Italy. Table 6.4 offers an overview of savings and investment by industry from 1963 to 1985.

An important characteristic of the British economy in recent years is that internal savings have been higher than investment, which has created a surplus in the current account balance and an accumulation of assets higher than that of most other countries. This is the result of various factors such as the high savings rate in non-financial sectors—as high or higher than that of families because of the level of undistributed profits and the high profitability for North Sea oil companies. Contrary to this trend, family savings have suffered a gradual decline since 1980, although they remain at a level roughly equal to that of the 1970s and continue to be a net lender to other sectors of the economy. Also, the public sector has been reducing its budget deficit every year since 1980 (with the exception of 1984) which has helped to compensate for the decline in family savings. All of this has led to an internal savings rate of approximately 20 per cent which has remained relatively stable for a long time.

The information on savings and investment by sectors is complemented by the information in Table 6.5 on financial flows between sectors and the instruments which have been used.[6] The importance of private saving (families plus non-financial companies) is clear, in spite of the family savings decline in 1981 and 1985 relative to 1973–7. The public-deficit decrease is also measurable: from 5.6 per cent of GDP between 1973 and 1977 to 3.3 per cent of GDP between 1981 and 1985. Within the financial sector, however, an important distinction must be made between commercial banks and other financial institutions. Banks are characterized by a positive balance, while the other financial institutions show a deficit somewhat above 1 per cent of GDP.

Within the family sector, significant trends include increased placement of savings in financial institutions and investment in insurance companies and pension funds. Also, as a result of greater economic activity, family home loans increased from 3.2 per cent of GDP (1973–7) to 4.9 per cent (1981–5).

The non-financial sector between 1981 and 1985 recorded an increase in its surplus; this reached the level of 1.8 per cent of GDP, a dramatic rise from the deficit of 0.9 per cent posted between 1973 and 1977. Also of note here is the definite trend away from bank financing for corporations, dropping from 2.8 per cent of GDP (1973–7) to 1.2 per cent in the

---

[6] Regarding the weight of these sectors on the English financial system see Revell (1973) or Rowley (1987).

TABLE 6.5. Acquisition of assets by sectors (% GDP)

	Households		Non-financial companies		Public sector		Banking industry		Other financial institutions		External sector	
	1973–7	1981–5	1973–7	1981–5	1973–7	1981–5	1973–7	1981–5	1973–7	1981–5	1973–7	1981–5
Cash	0.3	0.1	0.4	—	-0.7	-0.2	—	—	—	—	—	—
Bank deposits	1.8	1.3	1.2	1.1	0.2	0.1	-4.9	-13.3	0.5	1.4	1.2	9.3
Deposits in other financial institutions	3.6	3.6	0.1	0.1	—	—	—	—	-3.7	-3.8	0.1	—
Life insurance and pension funds	4.3	5.4	—	—	—	-0.2	—	—	-4.3	-5.2	—	—
Public debt	1.6	1.6	0.2	—	-6.0	-3.4	1.0	-0.3	3.1	1.8	0.1	0.3
Corporate equity	—	-1.0	-0.6	-0.1	—	-0.3	-0.2	-0.3	-0.1	1.1	—	0.6
Foreign equity	-1.4	0.1	0.7	0.6	0.1	—	0.2	1.8	1.0	1.2	0.3	-3.7
Banks loans	-0.4	-1.7	-2.8	-1.2	-0.3	—	3.9	11.8	-0.3	-1.3	—	-7.7
Mortgage loans	-3.2	-4.9	—	—	0.4	—	0.1	1.1	2.7	3.8	—	—
Public sector loans	—	—	-0.1	—	0.6	—	—	—	—	—	-0.3	—
Other loans	-0.1	-0.2	—	—	-0.2	0.2	—	-0.2	0.2	-0.2	0.1	0.1
Foreign investments	—	—	1.3	1.3	—	—	—	—	0.1	—	-1.4	-1.2
Foreign investments in the United Kingdom	—	—	-1.5	-0.6	—	—	—	—	—	—	1.5	0.7
Official financing	—	—	—	—	0.9	-0.2	0.3	—	—	—	-1.2	0.2
Others	-1.0	-0.9	0.4	0.4	-0.5	-0.7	-0.1	0.1	0.1	0.2	1.2	-0.2
Financial surplus or deficit	5.4	3.9	-0.9	1.8	-5.6	-3.3	0.5	0.7	-1.0	-1.2	1.5	-1.2

Source: OECD (1987a).

TABLE 6.6. *Non-financial companies: debt/equity ratio*

	1974–9	1980	1983	1985
United Kingdom	1.38	1.13	1.87	0.70
France	1.33	1.23	1.56	n.d.
Germany[a]	3.36	3.85	3.48	2.39
Japan	3.31	3.14	2.68	1.82
United States	0.96	0.77	0.78	0.83

[a] Excludes real estate.

Source: OECD (1987*a*).

period between 1981 and 1985. Foreign investment is high as well for British firms at 1.3 per cent of GDP, a figure which has remained stable over the last two periods but which is also indicative of the degree of internationalization the system has achieved.

From this comparison of financial and non-financial institutions we can conclude that the volume of disposable funds has been sufficient to support real investment. The problem for the British economy with regard to investment, however, has been a lack of available projects with sufficiently attractive profit potential.[7]

Yet these aggregate figures hide two important facts on the productivity in the British economy. The first is that firms, rather than being demanders of net funds have been offerers, further indicating that there have not been enough projects of interest in which to invest, or that it has been more profitable to invest in financial assets. The second fact is the growing trend of mergers and acquisitions, which was equal to 50 per cent of real total investment in 1986. When a country invests so heavily in simple financial restructurings it may be an indication that corporate directors are more concerned with short-term results than with the long-term objectives.

In general, non-financial firms rely on the markets to raise capital, which has profound implications for the process of financial intermediation. Table 6.6 gives the debt-to-equity ratio for these firms in comparison with their counterparts in other countries. In the UK—as in the USA—firms have had to depend less on debt and more on equity for financing ordinary business activities and investment projects. At the other extreme are companies in countries such as Germany and Japan that are highly dependent on debt financing. Without doubt, this is one of the strengths of corporate Britain. However, this characteristic could also be a symptom of a low level of innovative activity and of risk-aversion,

---

[7] See OECD (1987*a*): *United Kingdom*, where this subject is well documented.

in which case it becomes a definite weakness. Still, lower debt levels demand greater returns on assets in order to reach higher returns on equity. Although these data suggest a greater degree of solvency for British companies, they should be examined critically from the corporate perspective.

British financial authorities have recently approved a number of measures to further liberalize their financial system, intensify competition, and increase incentives for foreign companies. The process started at the beginning of the 1970s with the elimination of direct credit controls and collusive practices that tended to fix interest rates. The quantitative control of credit reappeared at that time, only to be completely eradicated in July 1982 once its ineffectiveness for reaching monetary objectives was proven. One of the more important steps toward liberalization was the removal of controls on capital flows in 1979.

However, the bulk of the reforms were made between 1985 and 1987. The objective at that time was to increase the degree of firms' freedom and the level of inspection and control to better protect private investors. The first set of reforms corresponded to the stock market's Big Bang.[8] The process of change was initiated in 1983 with an examination of the stock-market regulations on competition in the Restrictive Trade Practices Act. This was completed in 1986 when foreigners were permitted to acquire majority positions in companies listed on the London Stock Exchange. The most important measures pertaining to this reform were implemented on 27 October 1986. These included the abolition of the minimum commissions, the elimination of the distinction between stockbrokers and jobbers, a restructuring of the money market, and the introduction of a completely automated trading system.

The second important measure was contained within the Building Societies Act of 1986 and permitted savings banks to grant certain types of credit which were previously prohibited. It also gave them the opportunity to act as commercial banks for the receipt of deposits. Along with these measures, the supervision of these entities was also reinforced.[9]

Thirdly, the Financial Services Act of 1986 brought about a profound revision of investment banking operations. It was intended to improve supervision by the authorities and extend it to international futures, options, and share transactions. These activities became the objective of the Securities and Investment Board (SIB), a private organization that could delegate supervisory responsibility to the self-regulating organizations (SRO), including the Securities Organization—the result of the merger of the Stock Exchange and the International Securities Regulatory

---

[8] See the already cited works of the Bank of England (1986*a,b*).
[9] In Bank of England (1987*a*) there is an excellent analysis of this reform. For an analysis of these entities see Boleat (1982).

Organization. The function of the SRO is to supervise and control the conduct of listed companies. Therefore, investment banks came to depend on the SIB for those operations connected to the stock markets, while still under the control of the Bank of England for traditional operations in financial intermediation.

Finally, and along the lines of reinforcing supervision and control by financial authorities, a new Banking Act was approved in 1987 as a substitute for the 1979 Banking Act. Included in this legislation was a new set of criteria intended to apply universally to all institutions, directed by the Board of Banking Supervision, a council connected to the Bank of England. This body was authorized, according to the new law, to obtain information from banks, companies, or associated individuals to carry out their function of supervision.

It is unusual to have a single set of measures including strong liberalizing objectives together with strict supervisory and control aims. However, this is logical since neither interferes with the other and yet, together, they attempt to guarantee a reasonable level of competition and stability within the system.

2. THE STRUCTURE OF THE BRITISH BANKING INDUSTRY

As a result of London's position in the international finance world, the weight of the British banking sector is considerable, to the point that London has the largest concentration of banks in the world and the largest number of loan operations of all the international financial centres (see Table 6.3).

Financial institutions can be divided into two large groups.[10] The first of these is made up of commercial, merchant, and 'other' British banks, in addition to foreign and consortium banks. The second group can be classified as building societies and national savings societies. Table 6.7 provides information on the number of entities in the British banking industry at the end of 1987 and Table 6.8 breaks these data down by type of bank.

As seen in Table 6.9, the majority of assets and deposits in the British banking system are dominated by foreign banks. Since the greater part of these firms' activities are carried out in foreign currencies rather than pounds, the importance of free-capital flows is clear. Besides foreign banks, retail banks and building societies also play a significant role in the market; between the two, they account for 34 per cent of all assets and 36 per cent of deposits.

---

[10] Revell (1973), Rowley (1987), and Goadres and Curwen (1987) are recommended for further information on the English banking industry.

TABLE 6.7. *Size of the British banking system*

	No. of banks	No. of branches	No. of employees
Retail banks	19		
Merchant banks	35		
Other British banks	206	14,300	403,000
Foreign banks	361		
Consortium banks	22		
Building societies	151	6,952	69,200
TOTAL	784	21,252	472,200

Source: Bank of England (1986*c*).

TABLE 6.8. *Offices, employees, and capital, 1986*

	Banks	Savings banks	Others	Overall
Offices ('000s)[a]	12.7	1.6	7.0	21.3
Offices per 10,000 inhabitants	2.2	0.3	1.2	3.8
Employees ('000s)	383.8	19.2	69.2	472.2
Employees per 1,000 inhabitants	6.8	0.3	1.2	8.3
Employees per office	30.2	12.0	9.9	22.2
Deposits per office ($bn.)[b]	78.3	10.2	26.5	56.2
Deposits per employee ($bn.)[b]	2.6	0.8	2.7	2.5

[a] The figure corresponding to banks is from 1987.
[b] Exchange rate: £1–$1.47.
Source: Lladó *et al.* (1989).

TABLE 6.9. *The British banking industry: assets and deposits by institutions, 1987* (%)

	Assets	Deposits
Retail banks	20.8	20.1
Merchant banks	3.1	2.7
Other British banks	7.1	4.8
Foreign banks	54.1	56.6
Consortium banks	1.7	1.3
Building societies	13.2	15.5

Source: Canals (1990).

TABLE 6.10.   *Principal British banks, 1987*

	Equity (£m.)	Net interest income (£m.)
Barclays Group	4,229	2,671
National Westminster Group	4,959	2,756
Midland Bank	2,685	1,593
Lloyds Bank	2,393	1,778

Source: Salomon Brothers (1989).

The three largest British commercial banks are Barclays Bank, National Westminster Bank, and Midland Bank (see Table 6.10), with dollar assets of 189, 178, and 100 billion respectively at the end of 1988, and ranked 14th, 16th, and 33rd respectively worldwide in terms of assets. Barclays was also the third largest European bank after Crédit Agricole and the Banque Nationale de Paris.

In addition to these three, another five medium-sized banks stand out: Lloyds Bank, Standard Chartered Bank, Royal Bank of Scotland, TSB Group, and Bank of Scotland. They form part of the Committee of London and Scottish Bankers (CLSB) and together this group of eight employs 75 per cent of all banking industry personnel and holds 93 per cent of all bank offices—indicative of a fairly significant degree of concentration.

Commercial banks historically have been known as 'clearing banks' because of the control they have exerted over clearing houses, the institutions handling British credit and cheque compensation, thus facilitating the economy's payment mechanism. Their activities are not limited to domestic operations. The internationalization of London's financial markets has also driven British banks to expand overseas, through the participation in Euromarket operations and the establishment of offices in other countries.

Also among the commercial banks are the 'Trustee Savings Banks' (TSB), which are similar to savings banks with the principal objective of attracting deposits for investment in public debt. In the 1970s, the public sector began to deregulate many of these banks' activities and by 1976 legislation was passed permitting them to grant credit. However, the most important transformation of this sector took place in 1984 with the approval of the Trustee Savings Bank Bill, which had the effect of turning the TSB into a commercial company. At the same time, the TSB holding company (the TSB Group) was privatized and its shares were listed on the stock market. This gave the TSB Group the power to increase its capital, as well as much more operating flexibility.

Another group of institutions is made up of the merchant banks, which carry out their activities mainly in the international arena since their principal operations are in the capital markets and the financing of international trade. Bond issues are a typical merchant bank activity, though recently they have also started their own investment banking operations in such fields as mergers, corporate restructurings, and general corporate finance.

Consortium banks—a group of banks from different countries— process foreign exchange credits and facilitate other transactions sharing a common pool of human resources and materials. This classification is specific to banks within the segment of merchant banking. They comprise medium-sized banks from all over the world that feel the need to be in London in order to meet clients' needs. One way of achieving this more economically is through a 'consortium bank'. Their position with regard to assets and deposits within the system is not all that significant, however: they represented only 1.7 per cent and 1.3 per cent respectively in 1987.

Lastly, there are the building societies, mutual savings banks primarily concerned with mortgages for house-buyers. Their main sources of income are equity and deposits. The shareholders of these societies receive not only dividends, but also interest at a somewhat higher rate than general depositors. These firms have had an important influence on the centralization of savings in the UK. Table 6.8 shows the strong presence which these institutions have in the British financial system.[11] At the end of 1987 they held 13.2 per cent of the system's total assets and 15.5 per cent of deposits. Within their share of assets, they maintain a high proportion of credits for home acquisition: at the end of 1986, 75 per cent of the financing for the purchase of homes had been granted by the building societies. However, building societies do not limit themselves to these operations; a good part of their activity is also in the area of commercial banking. Since 1986, in accordance with the laws cited previously, they have been authorized to carry out similar functions as commercial banks.

Table 6.11 shows some interesting data on banks' deposit levels in comparison with other financial intermediaries such as insurance companies or pension funds. The loss of commercial banks' share of these deposits is substantial. While they accounted for 32 per cent of total deposits in 1970, this figure decreased to 25.8 per cent in 1985, mainly a consequence of strong price competition. To the contrary, market shares of building societies and insurance companies in 1985 were approximately equal to those of 1970. The building societies' maintenance of market share is the result of their segmentation strategy and also of the favourable tax

---

[11] See Boleat (1982).

TABLE 6.11.   *Market share: deposits* (%)

	1974	1980	1983	1985
Banks	32.0	29.8	26.6	25.8
Building societies	17.4	19.7	17.9	17.8
Insurance companies	26.5	25.4	26.2	25.7
Pension funds	15.5	21.0	25.4	26.7
Investment trust	7.3	3.3	3.1	3.1
Finance houses	1.3	0.8	0.8	0.8

Source: Rose (1986).

treatment for housing loans. In contrast, pension funds have grown considerably during this period, from 15.5 per cent of total deposits to 26.7 per cent. These statistics reveal a profound change in the UK's financial intermediation activity.

3. PROFITABILITY AND EFFICIENCY IN THE BRITISH BANKING SYSTEM

The British banking system generates the highest profitability among those recorded in this study. The graph in Figure 6.3 shows the evolution of return on equity for British commercial banks compared with those of German banks, the next on the list in terms of profitability. Clearly, the differences are rather substantial, particularly in recent years. The higher earnings of British banks is primarily due to a high return on assets and

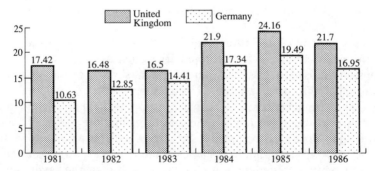

FIG. 6.3   *Return on equity*
Source: *OECD (1988).*

TABLE 6.12. *Before-tax profits/total assets ratio*

	1980	1981	1982	1983	1984	1985	1986
United Kingdom	1.35	0.97	0.83	0.81	0.85	1.12	1.18
Germany	0.44	0.43	0.53	0.60	0.72	0.83	0.79

Source: OECD (1988).

TABLE 6.13. *Intermediation margin* (over assets, %)

	1980	1981	1982	1983	1984	1985	1986
United Kingdom	4.0	3.56	3.25	3.11	3.06	3.17	3.16
Germany	1.89	2.07	2.44	2.71	2.54	2.44	2.46

Source: OECD (1988).

not to a high debt level. In 1986 the equity/assets ratio was 5.4 per cent, a relatively high figure in comparison to other European countries. In effect, the before-tax return on assets is quite high. In 1980 it was 1.35 per cent and in 1986, 1.18 per cent. During these years, this ratio declined substantially, however, falling as low as 0.81 per cent in 1983. Nevertheless, even this figure was quite high in comparison to other European banks (see Table 6.12).

This higher level of profitability was mainly a product of the higher intermediation margin, as shown in Table 6.13. In 1980, this margin stood at 4 per cent, though it fell to 3.16 in 1986 as a result of greater price competition in the British market and a decline in interest rates. Still, this figure remains relatively high for Europe, surpassed only by that of the Spanish banks.

During the 1980s, British banks were also generating a sizeable part of their net income from non-financial products. These represented 1.48 per cent of total average assets in 1980, but by 1986 had grown to account for 1.79 per cent as shown in Table 6.14. In this respect, British banks have moved much more quickly than other European and American banks. The combination of these elements (intermediation margin and non-financial products) yielded a gross margin for English banks which was the highest in Europe for this period. Between 1980 and 1986, however, British banks experienced a downturn, with margins falling from 5.49 per cent to 4.95 per cent, as shown in Table 6.15, where they are compared with those of Spanish banks, ranked second in terms of margin.

TABLE 6.14.　*Non-financial products* (over assets, %)

	1980	1981	1982	1983	1984	1985	1986
United Kingdom	1.48	1.28	1.54	1.61	1.85	1.67	1.79
United States	0.87	0.97	1.03	1.13	1.10	1.25	1.42

Source: OECD (1988).

TABLE 6.15.　*Gross margin* (over assets, %)

	1980	1981	1982	1983	1984	1985	1986
United Kingdom	5.49	4.84	4.79	4.71	4.92	4.84	4.95
Spain	4.97	4.93	4.78	4.86	4.54	4.40	4.61

Source: OECD (1988).

TABLE 6.16.　*Operating and personnel costs* (over assets, %)

	1980	1981	1982	1983	1984	1985	1986
Operating costs							
United Kingdom	3.77	3.59	3.42	3.29	3.32	3.16	3.23
France	2.24	2.23	2.21	2.14	2.06	2.06	2.12
Personnel costs							
United Kingdom	2.52	2.35	2.16	2.05	2.01	1.89	1.93
France	1.53	1.49	1.48	1.49	1.52	1.52	1.51

Source: OECD (1988).

The British banks' main weakness is their high operating costs, which are far above those of other European banks (even the Spanish). It is true that there has been an effort to reduce costs (they have gone from 3.77 per cent in 1980 to 3.23 per cent in 1986 in terms of total average assets), but they are still higher than those of other countries with comparable systems, such as France (see Table 6.16). The major component of total operating costs is personnel costs which, as shown above, are also much higher than that of French banks. However, this has been reduced in recent years, which has somewhat improved overall operating costs.

Another critical factor is the large debt owed by developing countries to British banks. The total volume of debt is sizeable, though still lower than that of American banks. Until now, provisions for bad debt have been low in the UK, but they are expected to rise over the next few years and subsequently affect profitability. As discussed in Chapter 2, it seems that the stock market is anticipating and preparing for this eventuality.

For the British banking industry in the coming years, an increase in competition can be expected given such high levels of return on equity and the current efforts to reduce operating costs, especially those related to personnel.

Figures provided by the European commission indicate that prices for financial services in the UK are above average in comparison with the other countries analysed.[12] Liberalization of financial markets, the entry of the sterling pound into the EMS, and progressive convergence of macroeconomic policy should bring these prices down in the medium term; however, meanwhile the relatively higher earnings may attract new competitors, especially large banking groups from other European countries and the US.

The international strategy of British banks is developing in two directions. On the one hand, the large banks already have established extensive branch networks abroad—under their own names or those of affiliates, as with NatWest and Barclays in Spain (see Appendix to this chapter), or Lloyds in Germany. They will continue to strengthen and expand these, but selectively, taking into account the market situation of individual countries. It would be unreasonable to expect British banks to randomly embark on a series of foreign bank acquisitions; most likely, they will follow a strategy of selective foreign market penetration offering services not yet fully developed by local competitors in target segments, such as pension funds, credit cards, euromarket issues, or corporate restructurings. An exception to this is Barclays, which is determined to become an authentic global bank offering complete service throughout the five continents.

On the other hand, medium-sized banks will have to proceed more cautiously, planning their objectives carefully, perhaps through the distribution of very specific financial products and in alliance with other foreign banks if they are to succeed in international markets. As in other countries, within the medium-sized banking segment the number of mergers is increasing. In the end, however, the outcome of these activities will depend largely on the policies of the Bank of England.

---

[12] See the Cecchini Report (1988).

APPENDIX

The New Strategy of Barclays Bank

At the beginning of 1988, Barclays Banks, the UK's leading bank in terms of assets, redefined its long-term strategy with three general aims:

- To take advantage of their strong presence in the UK and protect their operations from the competition.
- To create a global network of integrated offices that offer a truly global service to large corporations.
- To increase activity in the three major economic centres: Europe, the US, and Japan.

To initiate a new market direction, the directors of the bank made three significant decisions in 1988. The first was to increase capital—by £920 million—in an effort to strengthen international development and in this way provide better customer service.

The eventuality of a single European market in 1992 was behind the bank's second major decision which divided the European division into various market units. Barclays is the British bank with the greatest presence in the European Community, with a total of 250 offices in 10 of the 11 EC countries at the end of 1988. The objective of this restructuring was twofold: to increase the volume of activity in other European countries and co-ordinate the strategies of the distinct operating units in each country.

Finally, the bank approved a development plan for an electronic banking system designed to allow immediate access to all bank services from any office in any country.

Thus, during 1988, Barclays proceeded with the implementation of the organizational changes approved in 1987, which reclassified its UK operations along two broad market lines. This entailed reorganization of the existing branches into 325 'Barclays Business Centres' and 2,400 retail branches. At the same time, the number of regional offices controlling the different branches was reduced, thus decentralizing branch functions.

With this reorganization aimed at achieving greater proximity to the market, Barclays hoped to achieve the following objectives:

- A long-term reduction of operating costs, eliminating those offices not contributing directly to the generation of income.
- Improvement of customer service, by reducing bureaucracy and permitting branch directors to respond more quickly to client needs.
- An increase in the quality of communication between distinct levels of the hierarchy, as well as increased commitment of employees to the institution.

During 1988 the bank also reduced its allocation for bad debt by 23 per cent. This decision was not based on improvement in the country's general economic conditions, but on confidence in better management of the conceded credits resulting from the organizational changes.

Lastly, the bank's internationalization of its activities proves it to be one of the few banks with the vision of becoming a truly global institution. In this context,

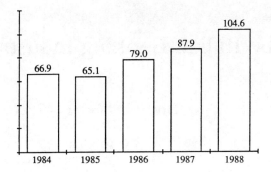

FIG. 6A.1 *Barclays Bank: total assets*

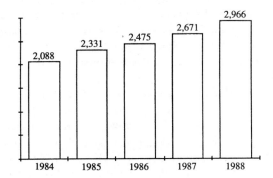

FIG. 6A.2 *Barclays Bank: intermediation margin*

two decisions adopted in 1988 are particularly significant. The first was the restructuring of the treasury bill product line under the name of 'Global Treasury Services' (GTS), with the objective of offering these services worldwide. In connection with this decision, the bank launched a 24-hour intermediation service which allowed clients to use a local telephone number and be transferred immediately to any open GTS centre in any country.

7

The Italian Banking Industry

I. AN OVERVIEW OF THE ITALIAN FINANCIAL SYSTEM

The Italian financial system, like that of other industrialized nations, has experienced very profound changes in recent years.[1] Some of them have been motivated by factors external to the financial system, such as economic adjustment policies in response to the second oil crisis or the high public-sector deficit. Others changes, such as advancements in communications, information technology, and technological innovation, correspond to common developments in the worldwide financial system. However, analysts agree that the pace of change has not been as rapid in Italy as in other European countries. An example of this is the time it is taking for the structure of the banking industry to move from extreme fragmentation toward greater concentration.

The combination of a high private savings rate (for families and firms) and a high public-sector deficit has greatly influenced the Italian financial system. The upper half of Figure 7.1 shows the financial weight of each of the various economic sectors. The lower part of the graph indicates the trends in savings as gross percentages of GDP. Private savings consistently have been about 20 per cent of GDP—one of the highest rates in industrial countries. Corporate savings have remained stable at around 5 per cent of GDP, although non-financial firms experienced a reduction of savings in 1973 and 1979.

The public-sector savings rate has deteriorated markedly since 1970. Together with the increase in public investment, this has generated a great need for financing and, as a result, has caused an increase in interest rates. This, in turn, has brought about changes in the placement of private savings. During the 1970s, a high share of savings was placed in the purchase of homes. Higher interest rates in the 1980s encouraged a movement of savings toward financial assets and government bonds.

Immediately following the two oil crises, financial disequilibrium was at its worst as a result of accelerated inflation and its effect on the price of financial assets. Also significant is the relatively low weight of the financial system within the overall economy. In 1987, assets controlled by commercial and savings banks represented 105 per cent of GDP, as opposed to the much higher percentages found in countries such as the

---

[1] With respect to these see the studies of Onado (1980), Banca d'Italia (1984), and Merussi (1985).

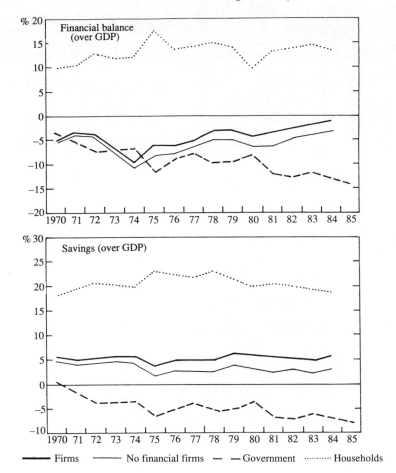

Fig. 7.1   *Financial balance and savings by sectors*
Source: *OECD (1987b).*

United Kingdom and the United States. This figure, however, is somewhat above those of Germany and France (see Table 2.2).

   In any event, this does not seem to have had an impact on any of the
major trends in financial disintermediation, as the same changes that are
occurring in Italy are happening in the rest of the world as well. But they
are coming about with greater force as a result of the inefficiency in the
Italian banking system. Figure 7.2 shows financial institutions' share
of national financial assets. Until 1975 there was a definite increase
in financial intermediation in Italy, but since then the percentage has
dropped from 39 to 30 per cent in 1985. This decline is significant not
only in absolute terms, but also in relative terms compared to other

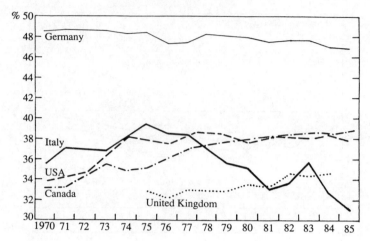

FIG. 7.2   *Financial firms' share in national financial assets*
Source: *OECD (1987b).*

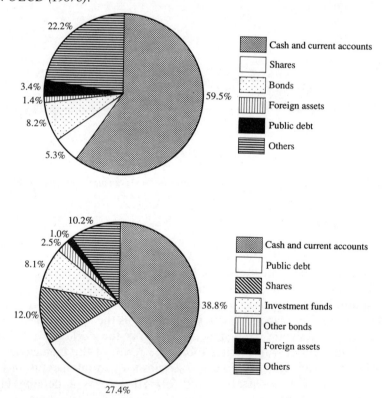

FIG. 7.3   *Households: distribution of financial assets*
Source: *Banca d'Italia (1986a).*

TABLE 7.1.   *Volume of bank deposits* (% GDP)

	1976	1985
Italy	71.7	63.8
France	37.6	37.4
Germany	46.4	50.1
Japan	73.1	84.3
United Kingdom	46.4	50.1
United States	31.5	31.5

Source: Canals (1990).

industrialized countries. In this respect, the steep rise in Italy's public debt has been a major contributory factor to the acceleration of the decline.

Figure 7.3 offers considerable insight with regard to where Italians have been placing their savings in the past decades. In 1975, almost 60 per cent of families' financial assets were placed in cash and bank deposits, 5.3 per cent in shares, and 3.4 per cent in public debt. (Distribution of the remaining percentage is not relevant for this analysis.) Eleven years later, in 1986, the landscape had changed considerably. The percentage of cash and deposits had gone down to 38.8 per cent, while the placement of savings in shares had increased to 12 per cent, and subscription to public debt had also risen to 12 per cent. Investment funds first appeared in 1984, and by 1986 they already accounted for 8 per cent of the financial-assets market.

In spite of this trend, deposits continued to be important in the Italian financial system. In 1985 the volume of deposits was at a level of 64 per cent of GDP, down from 71.7 per cent in 1976. However, this figure was still quite high compared with other countries: in 1985 Germany had only 50.1 per cent of GDP in deposits, while France had a mere 37.4 per cent. Only Japan showed a higher figure of 84 per cent of GDP, as indicated in Table 7.1.

Within the whole of financial disintermediation, it is worth examining the evolution of Italian firms' bank debt (presented in Figure 7.4). Short-term debt grew fairly steadily from 1974 to 1981, undoubtedly due to the widespread lack of liquidity after the two oil crises, then later fell significantly (1981–2), in some cases from close to 40 per cent of total debt to only 30 per cent. This reduction of short-term bank debt was compensated by the issue of other financial instruments. In contrast, the level of long-term debt shows much greater stability, although in recent years it too has tended to be lower than before.

[2] This question is studied in Banca d'Italia (1984).

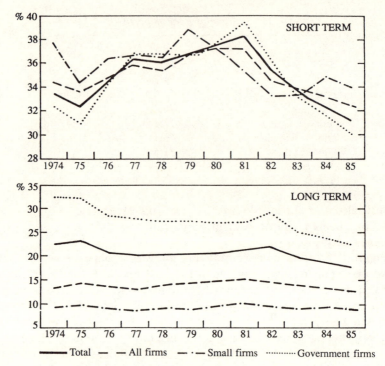

FIG. 7.4   *Corporate bank debt (over total debt)*
Source*: OECD (1987*b*).*

The regulation of the Italian financial system dates back to the Bank Law of 1936, though there have been additional resolutions since, and has had a significant impact on the industry.[2] This law was passed, however, with an important bias: the financial crisis of the 1930s, which helps to explain the law's emphasis on achieving greater stability in the Italian financial system. With this aim the law rejected the possibility of universal banking and instead opted for specialization according to criteria based on credit and deposit maturity terms. Thus, banks had to specialize to accept deposits and grant credits of less than 18 months. Other banks specialized in medium- and long-term loans, in particular those partially subsidized by the public sector. Over time, however, this difference has gradually disappeared and banks are now authorized to carry out all types of operations.

The Bank Law of 1936 also tried to preserve banks' independence with respect to certain industrial groups in an attempt to prevent the banks from being exposed to excessive risk. This was particularly important given that Italy's large industrial groups have more financial resources than many of the banks themselves. The liquidity of some of these larger groups would easily have allowed for majority share acquisitions of specialized banks.

Another objective of this law was to promote a relatively balanced economic development in the different regions of the country. For this reason, the law included strong controls on the opening of new offices. With this restriction the government intended to improve regional development while avoiding the danger of excessive scale in the banking system. However, this regulation also presented costs in terms of inefficiency, ultimately leading to its abolition some years ago.

Another important aspect of Italian banking regulation was exchange control. In accordance with a measure introduced in 1959, transactions between residents and non-residents were highly regulated. During the 1960s, following the directives of the EC, financial transactions with non-residents were progressively liberalized, although the heavy capital outflows at the end of that decade drove the authorities to adopt new controls. Toward this end, they adopted a group of measures which included the supervision of the external conversion of banknotes (between 1972 and 1986); restrictions on foreign-asset acquisition by residents in the form of an obligatory deposit equivalent to 50 per cent of the intended foreign investment (this was finally eliminated in 1987); and restrictions on the charge and payment delays on external trade operations. The authorities did not prohibit foreign loans, though, as a means of obtaining foreign exchange; consequently, given Italy's high interest rates, this became a popular method of financing.

In March 1987, in an effort to move closer to the liberalization of EC capital flows, financial authorities eliminated nearly all restrictions on non-resident financial operations. A number of these restrictions had already fallen into general disuse by way of exception, since companies and individuals could apply for exemption from some of them.

More recently, the Bank of Italy has progressed in its liberalization of the banking system in an attempt to improve efficiency and intensify competition. Some important changes have been introduced such as the progressive deregulation on the opening of new bank branches, liberalization of foreign banking (limited before to operations of foreign firms in Italy or Italian exporters), and the abolition of specialization requirements. Accordingly, some of the regulations controlling the development of investment banks have been modified, allowing them much greater freedom. All of this has subsequently led to a more efficient banking system.

Parallel to these reforms, financial innovation has had an equally important impact on the Italian banking industry. The formation of a market for government bonds and the growth of the public deficit since 1975 are two factors that must be taken into account in relation to this development. Initially, sales on these bonds were outstanding since their interest was tax-exempt. In 1977, the Italian treasury issued a new type of medium-term bond, called a Treasury Certificate, with the added attraction of indexed interest rates. For many years, Italian banks controlled the majority of the transactions on these bonds, and at the same time they held a large number of them because of the attractiveness of the terms they offered. However, in 1984 their fiscal advantage was eliminated and thus the proportion of these assets in banks' portfolios declined notably.

Commercial banks have also been active in the process of financial innovation by creating certificates of deposits, with the aim of circumventing the regulation to which traditional bank deposits were subject. The Bank of Italy also encouraged this development, allowing higher interest rate deposits. The banks themselves started leasing and factoring companies, which meant greater financing possibilities for industrial companies. Nevertheless, financial innovation in Italy has not advanced at the same pace as in the countries already analysed.

There have also been numerous changes in Italy's capital markets over the years. In 1984, the government authorized the creation of investment funds, which grew considerably in terms of both volume and activity, to the extent that in 1986 they already accounted for more than 8 per cent of Italian families' wealth. These funds originally came from Italian banks and the insurance companies, which is the reason behind the high level of concentration in the sector; thus, in 1986 half of the assets of these firms were in the hands of only three institutions. The main vehicles for the investment of these funds have been government bonds and, more recently, purchase of shares.

The appearance of investment funds in Italy, closely linked to an inflow of large sums of capital from American and British institutional investors, propelled an increase in stock-market prices beginning in 1985. As a result, expectations of sustained economic growth and corporate profits

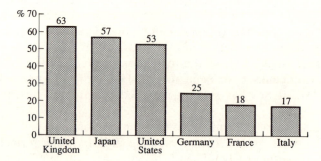

Fig. 7.5   *Stock-market capitalization, 1986 (% GDP)*
Source: *Canals (1990).*

TABLE 7.2. *Structure of the Italian banking industry*

	No. of Firms		Offices		No. of Employees	
	1983	1987	1983	1987	1983	1987
Government banks	6	6	1,766	2,200	67,319	71,443
Banks of national interest	3	3	917	1,272	51,393	49,934
Ordinary credit banks	119	111	2,947	3,416	70,381	71,767
Co-operative banks	148	132	2,408	2,784	41,323	45,016
TOTAL	276	252	8,038	9,672	230,416	238,160
Savings banks	87	84	3,623	4,165	69,463	73,444
Specialized banks	5	5	5	5	1,345	1,442
Rural banks	683	726	1,158	1,392	9,286	11,799
Foreign banks	33	37	54	65	n.d.	n.d.
Others	6	5	36	66	n.d.	n.d.
TOTAL	1,090	1,109	12,914	15,365	310,510	324,845

Source: Lladó *et al*. (1989).

grew, leading companies to issue new shares in the stock market. The volume of total transactions registered between 1984 and 1986 grew almost ten times, although the volume of capitalization at the end of 1986 was still low according to international standards, as can be seen in Figure 7.5.

In 1987, the CONSOB (the national agency in charge of regulating and supervising the stock market) began to study numerous measures aimed at liberalizing the stock market, ultimately permitting banks to conduct stock-market transactions through their own brokerage firms. Thus, a monopoly situation was eradicated and banks were given access to an area of activity to which they previously had been prohibited.

2. THE STRUCTURE OF THE ITALIAN BANKING SYSTEM

As explained in the previous section, the Italian banking industry was highly specialized for many years. However, in recent years there has been considerable deregulation, which, together with the consolidation of activities within individual banks, has led to an erosion of the differences between the various types of institutions. Table 7.2 shows data on the structure of the Italian banking industry and its size with respect to the number of offices and employees.

There are two distinct groups of banks in Italy. The first is made up of four different categories: government-owned banks, national banks,

TABLE 7.3.   *Major Italian banks ($m.)*[a]

Banks	Assets	Deposits	Loans	Capital
Banca Nazionale del Lavoro	87,782	68,027	36,310	2,963
Istituto Bancario San Paolo	79,291	52,821	38,705	3,234
Monte dei Paschi di Siena	66,600	52,488	20,845	3,627
Cassa di Risparmio della Province Lombarde	63,716	42,707	28,876	3,737
Banca Commerciale Italiana	62,739	54,271	23,813	3,370
Crédito Italiano	55,786	50,731	18,262	2,478
Banco di Napoli	55,583	42,119	23,510	946
Banco di Roma	50,432	45,852	18,173	1,308

[a] As at 31 Dec. 1988.

Source: Business Week, 26 June 1989.

commercial banks, and popular co-operative banks.[3] Of these, only the commercial banks are owned by private shareholders. The rest are either publicly owned (totally or majority owned), or are under government control (as in the case of the popular co-operatives).

The government-owned banks are deposit banks and are controlled directly by the government, who designates their senior directors. They comprise six institutions, including Banca Nazionale de Lavoro, the Instituto Bancario San Paolo di Torino, Monte di Paschi di Siena, Banco di Sardegna, Banco di Napoli, and Banco di Sicilia (see Table 7.3). The first three lead in the ranking of Italian banks according to asset size. Their worldwide ranking was 44, 53, and 62 (according to asset size) in 1988. Thus, their international presence is not all that significant.

National banks are companies owned by the state holding IRI, although their shares are listed on the stock exchange, and a minority of their capital is in private hands. There are only three firms within this group, including Banca Commerciale Italiana, Bank of Rome, and Credito Italiano. They are relatively large institutions, occupying fourth, fifth, and sixth place respectively in the ranking of Italian banks; at the end of 1986 they accounted for almost 14 per cent of the credit system's total assets.

Commercial banks are privately held corporations or limited partnerships whose activities are primarily directed at short-term financing. They represent a relatively important group within the industry, although none of them engages in specialized operations.

Lastly, the popular co-operative banks are characterized by the high

---

[3] For a study of these see Peat Marwick (1981) or Merussi (1985).

fragmentation of their shareholders. As can be seen in Table 7.3, they have expanded considerably in recent years, taking away market share from other institutions.

The second group of banks consists of general savings banks, rural savings banks, foreign banks, and special credit institutions. The government also possesses a significant stake in these firms, either in terms of their operating size (in the case of the savings banks), or their property (in the case of specialized credit institutions).

The savings banks are government-owned and the Bank of Italy exercises considerable control over them. Regulations on savings banks have changed quite a bit since 1982, with a view toward liberalizing their operations and expanding their range of business. With the objective of strengthening their solvency, financial authorities have authorized savings banks to diversify their equity bases through the creation of 'risk funds'.

Also of note are some of the characteristics of the special credit institutions. These institutions are authorized to grant financing for more than 18 months, and although in the past banks were authorized to handle these types of operations, they have opted to create their own special credit institutions. Some of the savings banks have done the same. The credit institutions specialize in granting long-term credit to specific sectors of the economy: public works, agriculture, real estate, etc. For many years the government has been subsidizing interest rates on these types of operation, as it deems them to be especially important. In the past it has stopped doing so, however. The credit institutions' main activity is the issue of government bonds (in 1987 they were 76 per cent of their liabilities) and of deposit certificates (which in 1987 represented 20.7 per cent of their liabilities).

Table 7.4 offers insight into each of these segments of the industry with regard to their share of assets, credit investments, and deposits between 1983 and 1987. Of note is the relative strength of the banks with regard to assets and deposits, but not in credit investments, a situation which has remained practically the same throughout the period. The national banks' share in these areas has slowly declined, as has the share of ordinary credit banks in assets and deposits. However, these changes are not very significant. Also, credit banks have a relatively strong position given the considerable role of the public sector and the low degree of concentration in this segment.

The savings banks also account for a substantial part of the market (with 19.9 per cent of assets and 21 per cent of total deposits in 1987), as do the special credit institutions (with 17.6 per cent of assets, 20 per cent of deposits, and 35 per cent of total credits in 1987).

Table 7.5 complements the above information in offering the growth rates by type of institution in each of the considered markets between 1983 and 1987. A very balanced growth can be observed in the case of the

TABLE 7.4.   *Financial institutions: market share*

	Assets (%)		Deposits (%)		Credits (%)	
	1983	1987	1983	1987	1983	1987
Government banks	16.73	16.36	14.14	13.99	11.25	11.87
Banks of national interest	11.18	9.77	8.63	8.20	8.60	8.00
Ordinary credit banks	22.95	20.64	18.10	18.11	17.19	16.60
Co-operative banks	10.52	11.32	12.33	13.14	8.79	9.87
TOTAL	61.38	58.09	53.20	53.44	45.83	46.34
Savings banks	16.86	18.94	21.00	21.02	13.76	15.25
Specialized banks	2.07	2.08	0.05	0.11	1.28	0.80
Rural banks	2.20	3.20	3.39	4.76	2.13	2.54
Special credit firms	17.49	17.68	22.36	20.67	37.00	35.07
TOTAL	100.00	100.00	100.00	100.00	100.00	100.00
(L. trillion)	1,062.15	1,492.17	507.76	723.83	323.29	515.63

Source: Lladó *et al.* (1989).

TABLE 7.5.   *Annual growth rate* (%)

	Assets	Credits	Deposits	Employees	Offices
Banks of public right	9.24	14.04	9.40	1.65	4.78
Banks of national interest	3.90	11.43	8.80	−0.56	7.52
Ordinary credit banks	8.33	12.77	10.02	0.75	3.14
Co-operative banks	12.53	16.33	12.04	3.21	3.12
TOTAL	8.36	13.54	10.12	1.16	3.98
Savings banks	13.50	15.67	10.28	1.50	2.95
Specialized banks	7.82	2.63	31.22	2.36	0.00
Rural banks	21.16	17.22	20.43	6.84	4.31
TOTAL	9.77	13.94	10.66	1.42	3.71

Source: Lladó *et al.* (1989).

banks, while in the case of the other institutions rural savings banks showed the strongest growth.

Throughout this period the increase in the number of offices and especially employees stabilized considerably, with the exception of the rural savings banks. This trend is related to the introduction of information technology and ongoing concentration.

TABLE 7.6.   Concentration of the banking industry (%)

	Assets		Deposits	
	1983	1987	1983	1987
Top eight banks	44.95	44.0	42.95	41.42
Other banks	57.05	58.58	55.05	56.0
TOTAL	100.00	100.00	100.00	100.00

Table 7.6 shows data related to the concentration of the industry, which appears to be highly stable. The first eight banks (of a total of 252) accounted for 44 per cent of total assets and 41 per cent of deposits, with a slow decline having been registered in 1983. However, the 244 remaining banks are considerably smaller, which suggests the need for greater concentration. Presently, many operations are leading in this direction, such as the Instituto San Paolo de Torino's 40 per cent acquisition of the Credop. The result is a much larger bank in which the service to leading companies is complemented with a strong commercial network among second-tier firms. Another case is the recent sale of 51 per cent of the Banco di Santo Spirito (property of IRI) to the Cassa di Risparmio di Roma, which is the second Italian savings bank.[4] In turn, the IRI sold 20 per cent of the Banco di Roma to IMI—a special firm—so that IMI could distribute its products through the close to 400 branches of the Banco di Roma. As can be seen, the trend towards consolidation is notable, and the objective is clear: to achieve a more efficient banking industry.

Finally, we will discuss the productivity of the Italian banking industry. First, it is surprising that the number of branches per thousand inhabitants is lower than in the countries studied: 2.4, while in the following country, Great Britain, the ratio is 3.8 offices. On the other hand, and in part because of this, the number of employees per office is higher than in the other countries considered, with an average of 23.6 employees. (England has an average of 22.2 employees and is the next in ranking.) However, this does not mean that the total number of employees in the industry is very high: in 1986 there were 5.6 employees for each 1,000 inhabitants, while there were 7.4 in France.

Liabilities per bank office were $29 million in 1986, far greater than in Spain, Germany, and France, although lower than in Great Britain. On the other hand, the capital per employee ratio is much lower than for Germany and the UK, although roughly the same as France. Table 7.7

---

[4] See *Financial Times*, 22 Mar. 1989.

TABLE 7.7.   *Some Italian banking industry ratios, 1987* (million Lire)

	Deposits/ office	Deposits/ employee	Credits/ office	Credits/ employee	Employees/ office
Government banks	46,031.1	1,417.5	27,825.8	856.86	32.5
Banks of national interest	46,677.1	1,189.0	32,427.0	826.03	39.3
Ordinary credit banks	37,166.9	1,826.1	24,589.6	1,192.7	20.6
Co-operative banks	34,166.9	2,113.0	18,276.4	1,130.3	16.2
Savings banks	36,537.3	2,072.0	18,885.8	1,071.0	17.6
Specialized banks	160,960.0	558.1	822,740.0	2,852.8	288.4

gives an overview of relevant data for the year 1987, broken down by type of firm.

In general, although the number of offices is not particularly high in relation to other European countries, this is not the case with regard to the number of employees, which probably indicates an excess of personnel and a lack of automation.[5] There is a consensus that in Italy there is a surplus of banks and dearth of offices, which has slowed the process of concentration.

3. THE PROFITABILITY OF THE ITALIAN BANKING INDUSTRY

Italian banks' profitability has deteriorated since 1982 (the year in which it was higher than other banks), reaching a value of 17.16 per cent on equity in 1985, a figure which is higher than that of the French and Spanish banks, but lower than for the German and English banks, as can be seen in Figure 7.6. Their lower degree of leverage helps to explain this fall in profitability.[6] In effect, between 1981 and 1985, the equity/total liabilities ratio went from 3.22 per cent to 5.05 per cent, way above the figures for France and Germany. In addition, the decline in the ratio of profitability over total assets, which was the second highest in Europe in 1980 with a value of 0.9 per cent, became the second worst in 1985, with a ratio of 0.5 per cent, as can be seen in Figure 7.7. This, together with a reduction in leverage, helps to explain the deterioration of profitability over equity.

The deterioration of before-tax profit on assets has various causes. The

[5] See the Bank d'Italia (1988) report.
[6] In Banca d'Italia (1984) a comprehensive analysis can be found in relation to bank concentration and profitability.

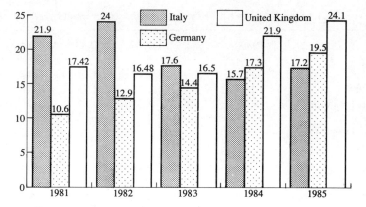

FIG. 7.6   *Banks: return on equity (%)*
Source: *OECD (1988).*

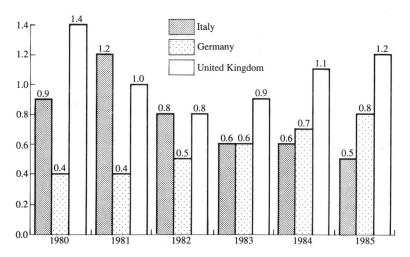

FIG. 7.7   *Profits before taxes on total assets (%)*
Source: *OECD (1988).*

first is the fall of the intermediation margin, which went from 3.37 per cent on assets in 1980 to 3.02 per cent in 1985. Of this figure, 66 per cent represented personnel expenses. Tables 7.8 and 7.9 offer relevant information on these points.

The high intermediation margin in this industry has been explained by the high degree of concentration which tends to increase asset interest

TABLE 7.8.　*Margin of intermediation*　(over assets, %)

	1980	1981	1982	1983	1984	1985
Italy	3.37	3.47	3.19	3.14	3.08	3.02
Germany	1.89	2.07	2.44	2.71	2.54	2.44
United Kingdom	4.0	3.56	3.25	3.11	3.06	3.17

Source: OECD (1988).

TABLE 7.9.　*Operating and personnel costs* (over assets, %)

	1980	1981	1982	1983	1984	1985
Operating costs						
Italy	2.82	2.82	2.81	2.97	3.0	2.97
Germany	2.04	2.03	2.07	2.19	2.19	2.20
United Kingdom	3.77	3.59	3.42	3.29	3.32	3.16
Personnel costs						
Italy	2.13	2.06	1.97	2.17	2.14	2.14
Germany	1.42	1.42	1.43	1.50	1.47	1.45
United Kingdom	2.52	2.35	2.16	2.05	2.01	1.89

Source: OECD (1988).

rates and reduce rates on liabilities.[7] However, as we saw earlier (Table 7.6) the degree of concentration has not changed very much while the intermediation margin has fallen, which can be explained by growing competition for the placement of investment funds.

Italy's high personnel costs can be observed in Table 7.9. In Great Britain and in Germany banks have developed a strategy for reducing personnel costs while in Italy banks have not managed to do so and profits have deteriorated noticeably. In the past, Italian banks have allowed themselves the luxury of having high personnel costs, due to their high intermediation margin. However, in recent years margins have deteriorated considerably, leaving them only two options (which are not mutually exclusive): a reduction of personnel costs, or an increase of revenues from non-financial products. Only through an important and profound change in management objectives, which have suffered from a lack of foreign competition and government control of many institutions, will Italian banks be able to confront the growing competition from abroad.

[7] This is the opinion maintained by various authors and adopted by the OECD (1987*b*).

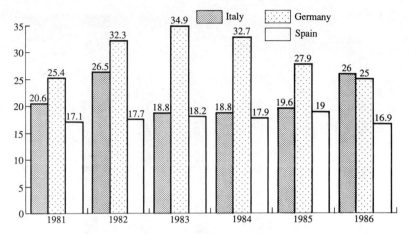

FIG. 7.8   *Savings banks: return on equity (%)*
Source: *OECD (1988).*

With regard to the savings banks, profitability levels are somewhat more flattering; although they share some of the problems common to the banks, their better segmentation has helped them to obtain somewhat superior results. Profitability on equity in 1986 was 26.10 per cent, a figure higher than that of the German and Spanish savings banks, as can be seen in Figure 7.8. However, between 1983 and 1985, they experienced a significant decline in profitability, owing principally to the increase in operating costs which rose from 3.17 of assets in 1982 to 3.46 in 1983.

On the other hand, in 1986, savings banks experienced a substantial decline in the margin of intermediation, which went from 3.37 per cent in 1985 to 2.78 per cent in 1986, causing a drop in the gross margin, from 4.89 per cent to 3.71 per cent. However, this was compensated by a reduction in operating costs and provisions for bad loans, as shown in Figure 7.9.

In the future a considerable increase in the intensity of competition is expected in the Italian banking industry—perhaps to a greater degree than in other countries—for three reasons. The first is the recently initiated deregulatory process, which will lead to a movement away from specialized banking toward greater price competition. The second is privatization, foreseeable in the case of some of the larger banks, which undoubtedly will make them more efficient. Finally, the presence of foreign banks will increase, the natural outcome of deficiencies in the financial services offered by domestic banks, as well as the opening of national borders.

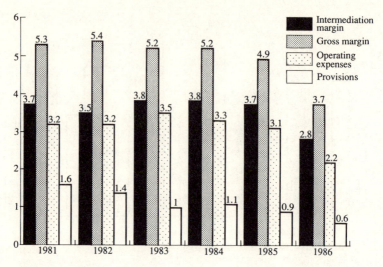

FIG. 7.9   *Savings banks: intermediation margin, operating expenses, and provisions (%)*

Source: *OECD (1988).*

This increase in competition is forcing banks to revise their strategies, especially in order to strengthen their positions in the Italian market. With the exception of such cases as the Banca San Paolo, the majority of Italian banks are not planning internationalization strategies, but rather are trying to consolidate their domestic positions, especially by increasing the quantity and quality of the financial services they offer, as demonstrated by the case of the Banca Nazionale del Lavoro (see Appendix to this chapter).

As in France, some important challenges for Italian banks are: the increase in capital required by European regulation, the charging of commissions on the financial services, and an improvement of the products and services offered. Also, mergers will contribute to a restructuring of the sector and will be likely to improve firms' operating efficiency, as illustrated by the merger between the Banco di Roma and the Cassa di Risparmio de Roma, announced in September of 1990. Given these foreseeable tendencies, banks can be expected to reorientate their strategies toward improving the quality of their management and adequately segmenting targeted markets.[8] On the other hand, mergers—driven in part by the regulatory authorities, with the objective of having Italian banking groups of a similar scale to those in France and

---

[8] These options are shared by specialists and directors of the Italian banking industry. Of related interest are Dini (1988) and Ceccatelli (1989).

England—will have a considerable impact on the industry. Undoubtedly, such changes will positively affect the Italian banking industry and it is to be hoped will improve its efficiency to give it a more prominent role in the international economy.

APPENDIX

Banca Nazionale del Lavoro: An Effort to Adapt

The Banca Nazionale de Lavoro (BNL) is a government-owned bank whose shares are the property of the Treasury and other administrative agencies, such as the Instituto Nazionale delle Assicurazioni (INA) and the Instituto Nazionale della Presidenza Sociale (INPS). It is one of the Italian banks that is most actively trying to adapt (at the EC level) to the more competitive market conditions being brought about by the single European market. In Italy, where the banking industry is experiencing significant changes, partly as a result of its perceived growth potential, such an attitude is truly necessary.

In 1983, the BNL outlined its future strategic objectives consisting of the following points:

(1) Increase the bank's efficiency and growth.
(2) Strengthen and further develop the Division of Special Credits.
(3) Increase the international scale of activities.
(4) Stimulate the growth of quasi-banking activities.
(5) Develop new financial instruments in line with the development of information-processing technology.

To achieve these guidelines, the bank's Board of Directors approved the following policies to achieve them:

(1) Make more effective use of the BNL's strong points, such as its extensive branch network and its presence in non-financial activities to create synergies between bank and non-bank activities; in particular, to offer non-financial products throughout its branch network.
(2) Diversify, with new products and services, developing their own financial innovations.
(3) Increasing management efficiency through improved strategic co-ordination and planning. Improve treasury management, for which it is necessary to expand information systems, which will also provide benefits in other operating areas.
(4) Promote the internationalization of activities, on three different fronts: (*a*) in strong currencies (the dollar, the German mark, and yen); (*b*) in economically important countries; and (*c*) in other markets in which the image of the bank can be used as a competitive advantage.
(5) Improve personnel management, promoting the development of professional skills and taking into account employees' personal preferences.

These objectives were complemented in 1987 with a slogan corresponding to the organization's new strategic directives: 'from a growth oriented strategy to a profit oriented strategy'.

Some of the actions taken include the following: (*a*) a 30 per cent increase in capital; and (*b*) the reorganization of the bank with a view to achieving an improvement in the quality of services and a rationalization of costs.

The second decision entailed a restructuring of the bank in five areas: commercial, financial, resources, administration, and loans. The core of the idea was to achieve an organization (including management) focused on the sale of products and financial services.

PART III

8

The Transformation of the Banking Business: A Corporate Perspective

1. INTRODUCTION

Throughout the previous chapters we analysed some of the changes that have taken place in the banking system of the major European countries. Financial disintermediation, globalization of institutions and markets, deregulation, and international financial instability have been discussed in considerable depth. All of these have influenced banks' strategies and operating results, and will continue to do so in the future.

However, it is important to understand at the bank level which of these factors will most affect banks' short-, medium-, and long-term results, and, in turn, how they will affect their capacity to compete.

It will be useful to provide a synthesis of these factors in order to understand better the implications they will have for individual enterprises. Such an analysis will offer further insights into how they have altered the structure of the banking industry. Structural analysis of industries helps to identify the key elements related to the competitive characteristics of an industry, and to a certain extent, the profitability of the companies that comprise it. Every sector has specific structural and competitive features, which explains why the results of a comparative analysis between two distinct sectors can be vastly different.

The combination of these elements in a bank's strategy will be analysed throughout the following three chapters, on the basis of the conceptual scheme shown in Figure 8.1. According to this figure, there are two determining factors of a competitive strategy. The first has to do with the structural characteristics and, in particular, the evolution of the previously discussed environmental and regulatory changes which have occurred throughout the past years. In order to evaluate their relative weight, in this chapter a simple analytical framework will be used to show some important structural determinants in any industry, such as entry barriers to potential competitors, negotiating power in relation to suppliers and buyers, and the intensity of rivalry.[1] This scheme is shown in Figure 8.2.

A knowledge of the changes that have taken place in the banking sector is not sufficient for formulating a competitive strategy. It is also

[1] Porter (1980) provides this framework. Neven (1990) offers another perspective.

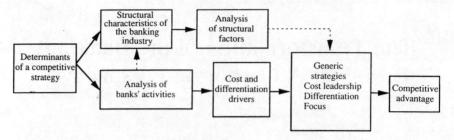

Fig. 8.1   *Determinants of a competitive strategy in the banking industry*

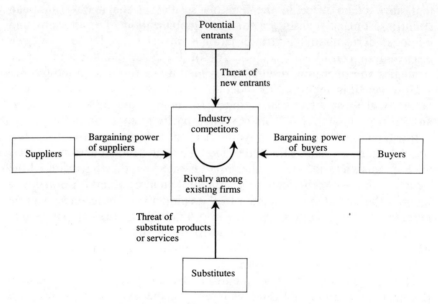

Fig. 8.2   *Structural determinants of an industry*
Source: *Porter (1980).*

necessary to study the activities of value-creation in a bank. Chapter 9 is dedicated to its analysis.

2. AN ANALYSIS OF THE STRUCTURAL DETERMINANTS OF THE BANKING INDUSTRY IN THE EC

Using the conceptual scheme presented in Figure 8.1, this section looks at the forces which condition and determine the structure of the European banking industry, and therefore banks' competitiveness within an integrated market.

2.1. Barriers to entry

Entry barriers refer to those conditions which impede firms outside the sector from gaining access to an industry and, as a result, restrict the number of competitors offering the same product.[2] We will analyse some of these barriers.

Scale. The first entry barrier is related to economies of scale. These appear when the unit cost of production for a specific good falls as output increases. When there are considerable economies of scale, the minimum volume of production to achieve a desired level of profitability (or lower costs) is high, such that potential competitors either have a cost disadvantage, or they have to reach a minimum sales volume equivalent to their output.

Economies of scale are decisive in capital-intensive industries, where the high level of required initial investment discourages the entry of potential competitors. Is this the case in the banking industry?

This issue will be considered in depth in Chapter 10, in examining the relative scale required of banks to compete efficiently. However, prior empirical studies on economies of scale in the banking industry do not seem to show that this is a major structural determinant. Some studies on the US banking industry show the existence of economies of scale for firms with operations of less than $25 million, and diseconomies of scale for firms operating with more than 100 million dollars.[3] One study on the Spanish banking industry shows that economies of scale are somewhat more important than in the American industry.[4] But these studies, despite their relevance, contain certain methodological flaws which makes it impossible to apply them universally.[5] It can be generally affirmed, however, that the existence of economies of scale is doubtful.

Scope. These studies analyse economies of scope as well. Economies of scope occur when production costs for a group of goods or services are less than the sum of costs for each of the individual products.[6] The cause

---

[2] The first and most complete treatment of entry barriers can be found in Bain (1956). For a good description of the different theories that explain the entry of new companies in any industry see Gilbert (1989).

[3] See Benston *et al.* (1982) and Gilligan *et al.* (1984).

[4] See Fanjul and Maravall (1985). For the Italian banking system, see Bank d'Italia (1984) and for the German industry, Pöhl (1982) is of interest.

[5] See the criticism given with this regard in Berger *et al.* (1987).

[6] Panzar and Willing (1981).

of economies of scope is that fixed, or almost fixed costs, are spread through a higher production volume for the different products.

Practical experience in bank management shows that important economies of scope can be reached within the industry. A specific area which has become very important for banks is information systems. For example, in a bank the information that is obtained about a client with regard to his deposits can be extremely useful in determining what other financial services to offer—anything from the selling of shares to the extension of credit.

Large banks have better opportunities for increasing revenues from the sale of financial services which are not strictly bank-related, and this has had a significant effect on operating results. Another factor is new information technologies which allow banks to handle a greater number of activities, products, and clients, offering the potential for important economies of scope. According to recent studies by McKinsey, in the United States large banks spend 15 per cent of their total expenses on information systems, while the smaller banks spend only 10 per cent. Citicorp—the leading American bank—spends 20 per cent of total investments on information technologies.

Nevertheless, economies of scope do not end there. It is evident that they can be extended with relative ease to other aspects of capital costs: for example, branches and associated personnel costs. There is a slow but clear tendency for big banks to create a structure of 'hub and spokes', that is, branches around central offices which carry out specialized operations.

For all these reasons, it seems reasonable to conclude that economies of scope can play an important role in the efficiency and profitability of the banking industry in the future. The arguments in support of this hold considerable weight and the evidence presented allows for a more detailed evaluation of this assumption.

Cost advantages. Another entry barrier is cost advantage independent of scale. This barrier can consist of exceptional managerial talent or access to certain raw materials at lower costs, perhaps as a result of the integration of suppliers within the same company or industrial group. However, a particularly important factor is the experience curve. This shows that with an increase in accumulated total production of a specific good, the unit costs of production will fall considerably. This phenomenon was observed for the first time in the aeronautics sector, and has since been proven in many other industries. Actually, it appears to be an even greater entry barrier than economies of scale. On the other hand, the experience curve requires the accumulation of a production volume obtainable only with the passage of time. Thus, to achieve economies of scale comparable to a company such as IBM might be feasible; however, to reach the same production volume as IBM in computers would be much more difficult.

It is not easy to measure the effects of the experience curve in the banking industry, but they appear to be less influential than in other, more technology-dependent sectors. Still, the experience curve can be a decisive factor in certain specialized operations which will constitute a larger share of banking activity in the future, such as operations in stock or currency markets, and future and option contracts. In these specialized fields, the experience curve could give some banks an important cost advantage.

Capital requirements. The need for capital constitutes another important entry barrier in the banking industry. In general, this barrier is common to those sectors requiring large capital investments, which discourage potential competitors from entering the market if they lack resources and access to adequate financing. Industries with high capital requirements, such as the aeronautics, steel, or automotive sectors, provide good examples of how this kind of barrier works.

In the banking sector this kind of entry barrier has two dimensions. The first and most basic is the minimum capital requirement, which has two different aspects within the EC. The first corresponds to the Committee for Regulation and Banking Supervision of the Bank for International Settlements of Basle. This committee co-ordinates the norms of bank supervision within the different countries of the Group of Ten, focusing especially on banks' international activities. Their accords are not binding, but are only guidelines. At the end of 1987, this committee reached an agreement, known within the banking industry as the Basle Accord, which included three points: the definition of banks' capital ratio, an assessment of the risks included in the capital ratio, and, lastly, the level of the ratio itself.

Leaving legal aspects aside for a moment, the Basle Accord established a capital requirement ratio of 8 per cent on total net investments. However, the Commission of the European Communities has approved an obligatory condition contained in the Second Directive.[7] Accordingly, the minimum capital requirement for creating a credit entity is 5 million ECUs.

Parallel to this requirement, there is another directive, called the Directive of Ratios and Solvency. This requires that EC member banks maintain a minimum level of capital proportional to total assets. Thus, it is somewhat similar to the Basle Accord. The Directive defines the different weights assigned to assets and their risk of insolvency to be included in the ratio; they range from 20 per cent for EC member credit

---

[7] Clarotti (1988) offers a detailed discussion on the terms of this directive.

firms to zero in the case of public credits or bonds of any EC member country. From an economic point of view, however, the required investment cannot be classified as excessive, especially given the entry barriers related to capital requirements in other industries.

The second aspect of such an entry barrier, therefore, is not related to legal requirements, but rather to economic considerations. In effect, investments in modern information technologies necessitate initial capital investments that greatly exceed the legal minimum capital requirement. Thus, legal capital requirements do not by themselves constitute the most significant entry barriers. Rather, barriers result more from the economic necessities for successfully establishing a bank, which can play an important role in discouraging potential competitors from entering the industry.

Regulation. Another entry barrier of importance in the banking industry is government regulation. It was shown previously that the public sector controls minimum capital requirements within the EC. However, the public-sector regulatory role does not end there. Within the realm of the EC, we can identify three broad areas that have been objects of strong control, but which will be increasingly liberalized in the coming years as a result of the financial deregulation that has occurred throughout the 1980s.

The first area has to do with the single banking licence, which encourages each country in the EC to accept the establishment—or the simple sale of services—of any bank from another EC country, as long as the country of origin gives authorization. The operations which can be carried out are described in the Second Directive and guidelines are based on the model of a universal bank. However, the country of origin holds the right to authorize the international expansion of a national bank. For the receiving State, the decision of the country of origin is binding. This is one of the principal points of the Second Directive, which sets a relatively liberal framework for the internationalization of banking activities in the EC.

The second area of deregulation within the EC was the liberalization of capital movements in July 1990, with the exception of Spain, Greece, Ireland, and Portugal, which have a longer transition period. The liberalization of financial services, although already present in the EC's Directive of 28 July 1973, will not be completed until capital flows are entirely deregulated.

The third area of deregulation is not related to the EC, but is manifested at a national level. It has to do with the liberalization of previously restricted banking activities, which the financial authorities of the five countries analysed in this work have adopted in recent years in order to standardize legislation between countries. The objective has been to increase competition and efficiency in domestic banking industries as a mandate for preparing for a more competitive financial system before the challenge of 1992.

As shown in previous chapters, this process of deregulation has been accompanied by an increase in supervision and control in the banking industry. However, the overall effects of this process are positive: despite the increased requirements for information, there is more freedom of movement within the different segments of the industry, which has further eroded barriers to entry.

Product differentiation. Product differentiation, especially with switching costs, or costs a client incurs from changing products between competitors, is an important entry barrier.[8] That cost can be economic—in some sectors the client faces conditions requiring it to buy from a specific company—or simply moral—derived from the loyalty a client has for a specific company it has been doing business with for some time. Yet the overall level of product differentiation within the banking industry is not very strong, with the exception of those products that are exceptionally innovative, such as Merrill Lynch's Cash Management Account. This lack of differentiation means the only discriminating factor is often the geographic proximity of a point of sale or branch office. Thus, an extensive physical distribution network implies not only a significant cost factor for an individual bank, but—and perhaps more importantly—a formidable entry barrier for potential competitors.

For this reason, the excess of branches in such countries as Spain, owing, among other reasons, to previous regulation, can act as a double-edged sword: disadvantageous because of its high cost, but advantageous in that it raises a wall against possible competitors. Certainly, the final utility of this weapon will depend upon how it is used, in other words, its role within the overall strategy of the corresponding bank. For some banks it can represent a considerable disadvantage, but for others, a formidable competitive weapon.

2.2. Relationships with suppliers and buyers

The relationships between a firm and its suppliers and buyers have a significant bearing on the structure and competitive characteristics of any industry. Suppliers can have a great deal of influence in a specific sector if their products are very differentiated, if their prices weigh heavily on buyers' operating results, if the supplier industry is not competitive, and if a change from one supplier to another implies high switching costs. In any of these cases, suppliers can greatly influence a company, both in terms of

---

[8] See this concept and its effects in Klemperer (1987*a*,*b*).

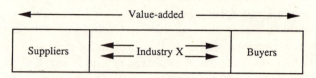

FIG. 8.3   *Value added in a series of transactions*

its structure and results, since prices are determined by suppliers, and buyers have little power to effect changes in the system. In this case, firms are under the dominant influence of their suppliers.

Something similar occurs in the case of buyers. Buyers have more negotiating power if switching costs are low, if the product is standardized, or if its importance with regard to the buyers' operating results is high. In these cases, buyers will desire to derive the greatest possible benefits from supplier relationships, resulting in a reduction of profits and greater competition in the industry as a whole.

These relationships can be visualized in Figure 8.3, which shows the total value-added of a sequence of transactions from suppliers to end clients.

The value-added of a given industry depends on the negotiating power between a firm and its suppliers and buyers. Any given industry will try to gain a greater share of the total value-added: the higher it is, the more attractive the sector will be in terms of profitability, which will naturally attract more competitors.

In the context of the banking industry, some of these ideas can be applied to understand better the industry's structure. In this case the suppliers are the owners of the banks' deposits while the buyers are the buyers of assets.

It is evident that in many countries banks have traditionally enjoyed a notable degree of power over their suppliers of capital, especially depositors, as a result of regulations restricting the remuneration of deposits along with a lack of development of other financial instruments and capital markets. Thus the suppliers of funds were offered few other financial alternatives. However, the increase in price competition, innovation, financial deregulation, and inter-firm rivalry within the sector have made the transfer of funds to other institutions—outside the banking sector—more attractive. This accounts for the process of financial disintermediation which, as shown in previous chapters, is evident in many of the countries analysed in this study.

It is important to note that this increase in competition has to do with the dismantling of certain entry barriers such as legal requirements, for example, the principle of specialization in some countries, and with the difficulty of adequately managing large quantities of information related to consumer-goods purchases. Consequently, new competitors have appeared, such as financial conglomerates (insurance companies, pension funds, and investment banks) and have fragmented the banking market to the detriment of traditional institutions. Additionally, large retail stores (such as Sears Roebuck, Benetton, and Marks & Spencer) are offering consumer credit and other financial products, further eroding the banks' positions. Even clearly defined banking areas such as collection and payment processing are being ceded to information companies which have started to develop these activities. This is the case of Electronic Data System (an affiliate of General Motors) or First Data (a company established by American Express). Lastly, the growing development of the capital markets is another factor which must be considered in this context.

In general, suppliers of funds now have more alternatives outside the banking system, which in turn, increases their power in relation to the banks. Banks can only recover lost ground by being imaginative in new product development and by offering attractive and profitable alternatives for the placement of funds.

From the point of view of assets, considerable changes of equal proportion to those described above have also been taking place. In the financial services industry, commercial banks are facing greater competition from investment banks, and especially from the increasingly sophisticated capital markets, which have attracted a growing flow of funds. Also, the banks' new competitors have an additional advantage: they are not subject to a minimum capital requirement, which allows them to be more competitive.

In sum, the power of negotiation between a bank and its suppliers and buyers of funds can completely change the competitive panorama, strengthening or weakening it depending on the case. The evolution of the relationships between banks and their suppliers and buyers appears to have strengthened the competitive environment. Thus, banks can either adopt a passive attitude and continue competing as best they can in traditional business lines which are increasingly less profitable, or react actively by establishing closer ties with suppliers and demanders of funds in order to raise switching costs. This second path will be more difficult, but is the most interesting for the banks in the near future.

2.3. The threat of substitution by new financial products

Any business is based on the existence of a concrete product, more or less successful, which serves a specific function or fulfils a certain need. Products have a vital cycle consisting of a launching stage, a growth stage, a maturity stage, and a period of decline. The speed of this process depends mainly on the appearance of new competitors and new substitute products. As a result, it is important to understand the function of a product and the need it fulfils for the consumer. Only by answering these questions is it possible to find out whether or not a product is being threatened by substitutes. Naturally, it is not simply a question of pondering only the value which the product provides, but also of relating this value to price.

This is what has occurred in the banking industry over the last decade: it has experienced an avalanche of new substitutes for the more traditional products. Additionally, the majority have come from new competitors, such as those described in the previous section. Others, although fewer, come from within the sector itself. All of this is provoking the fragmentation of something as essential to the sector as the intermediation of payments; the appearance of credit cards is to blame.

New financial products and the gap they have opened in the banking industry have initiated a considerable increase in competition and, therefore, have caused important changes in the structural characteristics of the sector. In addition, they are responsible for significantly altering the essence of traditional banking businesses. These changes will continue to affect the industry and banks will become increasingly engaged in the sale of financial services, offering their clients new and attractive products which will progressively have less to do with standard banking activities. Therefore, the range of competitors and substitute products has grown, and will continue to grow. Only those banks that understand this process, and know how to react with offensive strategies, will be able to confront it successfully.

2.4. *Rivalry in the banking industry*

The economic attraction which an industry has for a firm depends to a certain extent on the degree of rivalry between the companies that operate within that industry. In principle, the more competitive it is, the less attractive the industry will be as its profitability declines. On the other hand, if the degree of competition is low, the industry will be more attractive.

What are the elements that determine the level of rivalry within a sector? We can identify two, at least. The first has to do with a low level of product differentiation. In general, companies have two possible generic competitive strategies in this regard: cost leadership or product differentiation—in the entire market or in a specific segment.[9] When what is offered is a standard product, the only possibility of substituting it successfully is through cost leadership. However, with or without cost leadership, industries with a low level of product differentiation usually are besieged by price wars which corrodes profits tremendously.

[9] See Ghemawat and Nabeluff (1985).

The second factor that affects competition is the weakening of demand. This becomes even more critical depending on the relative weight of fixed costs for the companies in the sector. When these are high, firms need to generate higher incomes in order to break even. The combination of high fixed costs and a weakening of demand will exacerbate the competitive environment as firms fight to stay in the market-place.

A last point of interest relates to the exit barriers corresponding to a particular industry. Exit barriers are a group of obstacles which make it difficult for firms to abandon a sector. Such impediments can be economic—the level of prior investments or the cost of rescinding labour contracts—or simply personal or sentimental reasons on the part of a company's directors. Evidently, the higher the exit barriers, the greater the competition will be in any given field.

There are two principles that define the existing level of rivalry in the banking industry. One is the low level of product differentiation between firms, which has led to increased price competition, especially on the part of the savings and foreign banks. Competition has intensified due to the increase in competitive pricing, but also as a result of the appearance of new banks which have met the demand for innovative products and services. The rate at which these new products and services are being brought to market is increasing; this, in turn, has provoked a radical transformation of the sector, as we have seen in Section 2.1 when analysing entry in the European banking industry.

The transition from competition based on client relationships to price-based competition demonstrates both the dismantling of collusive practices in some countries, and the demand for services based on price and quality, further differentiating banks from one another. Financial deregulation has thus provoked a notable change in the competitive nature of the industry.

3. SOME CONCLUSIONS

In this chapter we have analysed some of the implications of the changes that have occurred during recent years in the respective banking systems of the major European countries. The basis of this analysis was a conceptual scheme which helped to identify the principal competitive forces and determine the structure of the European banking industry at the beginning of the 1990s.

Among the five structural determinants studied here—entry barriers, suppliers' power, buyers' power, differentiation, and degree of rivalry within the sector—the entry of new competitors appears to be an underlying factor resulting from deregulation, globalization, and the sector's attractiveness. At the same time, the increase in price competition has caused significant changes in the structural characteristics of the industry.

Greater competition has threatened the traditional banking industry on all fronts. In the commercial banking industry, new products and financial institutions have emerged. In addition, the role of banks as intermediaries in the processing of payments has deteriorated in favour of such innovations as the credit card.

On the other hand, investment banks—and, in some countries such as Spain, foreign banks—have attracted a greater number of companies to which they offer traditional banking services, in addition to more specific services such as debt underwriting, capital issues, or financial restructuring.

Commercial banks have also suffered from the rise of the capital markets. Deregulation has increased the ability of the capital markets to attract companies in search of direct financing. Thus deregulation and financial disintermediation have considerably diminished the competitive position of the banks in the different countries analysed.

These changes are giving way to a banking industry that will be very different in the future. Financial disintermediation and the appearance of new instruments of payment will leave less room for traditional activities. Simultaneously, and as a result of these economic developments, companies and individuals will increasingly demand more sophisticated financial services.

In effect, and as demonstrated by the previous profitability study of the different banking systems, the margin of intermediation has experienced a general decline in recent years. Conversely, the income from mainstream financial products has increased in proportion to the total assets of the banks. This is a trend confirmed by recent experience and one which will become more accentuated in the future as a result of the greater operating freedom which banks will enjoy in the single market of 1992— once the steps have been taken to liberalize capital movements and liberalize the sale of financial services in any EC country. All these phenomena will not only affect the structure of the sector and the competitive characteristics of the banking industry, but also banks' strategies. The next two chapters will be dedicated to this question.

9

The Value-Creation Process in Banking

1. INTRODUCTION

In Chapter 8 we saw how the transformation of the financial industry within the European Community has affected the evolution of the banking industry on the basis of the determinant forces of any industry. This type of analysis has been especially useful as it has helped to understand better the characteristics of the industry, as well as the relationships between the different agents within the context of the single market of 1992.

The specific purpose of this chapter is to deepen this line of investigation. Until now we have limited ourselves to analysing those factors that have affected the structure, and in turn, the industry as a whole. What we shall try to do is examine the consequences of this at the firm level.

In effect, the greater degree of rivalry in the sector, deregulation, and the increasing power of suppliers and buyers of funds, have significant consequences with regard to banks' results, and influence in one way or another their strategic choices. Thus, a decline in the margin of intermediation should be compensated by either an increase in other income from non-financial products, a reduction of capital costs, or a higher level of lending to maintain or improve profitability. In the following section these questions will be discussed in detail.

2. AN ANALYSIS OF A BANK'S CHAIN OF ACTIVITIES

In the previous pages we have examined recent changes in the banking industry which have greatly affected the results of individual banks. Although intuition alone may provide insight into how these changes have affected banks, we now analyse their effects on banks' operations and creation of value through the value-chain.

The value-chain defines the different steps through which a company's value added rises or falls. Figure 9.1 presents a generic version of the creation of value for an industrial company.[1] In accordance with this concept, the activities of a company can be divided into nine large types of operations which, in turn, are grouped under two generic types of activity: primary and support. The former relate to production and

[1] See Porter (1985).

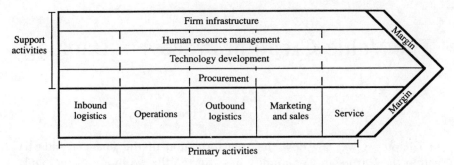

FIG. 9.1   *Value-chain of an industrial firm*
Source: *Porter (1985).*

distribution processes and thus represent the core of value-creation in any company. Support activities are indispensable to the efficient development of primary operations.

The relative weight of each of these activities in the creation of value differs from firm to firm and between industries. Therefore, a firm-level analysis of the process of value-creation requires a study of each stage of the value-chain.

Together with an analysis of the value-chain, it is important to determine cost structures at each of these stages. This is a particularly important step when a firm undertakes a cost-leadership strategy. An analysis of a firm's value-chain is particularly useful in explaining its cost structure—or the comparative differences in its cost structure *vis-à-vis* competitors—which, in turn, allows for better decision-making.

In some stages of the value-chain the determining cost factor will be the experience curve; in others, economies of scale or human resources. It is important to understand which of these is most decisive in the creation of value-added in order to minimize their importance in the company's profit and loss statement. Such a cost analysis is also useful for better understanding of the effects of those costs on the different activities. This analysis allows companies to look for complementary activities or new products which take advantage of existing costs, leading to greater economies of scope.

All of these ideas have considerable application with regard to banks. The value-chain of banks—and, generally, of financial service enterprises—has a much simpler structure. This is basically true of all service-sector companies in which production operations do not exist, and in which most activities are related to documentation and support of primary activities. Figure 9.2 shows the value-chain for a bank.

The preceding illustration, based on the original value-chain, presents

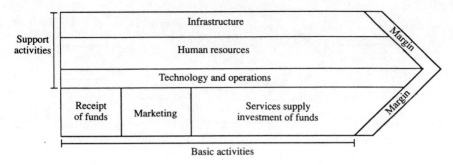

F<small>IG</small>. 9.2 *Value-chain of a bank*

as support activities the 'production' operations of a bank.[2] Without a doubt, effective management of these types of operations can offer important cost savings. However, it seems that in a bank these operations are more support-related than in industrial firms.

Regarding primary activities, three basic functions have been distinguished. The first of these is the receipt of funds, by attracting depositors or operating in the capital markets. The receipt of deposits implies a real sale of services. The marketing function in a bank is not only aimed at the offer of financial services such as loans. It also serves to attract funds, through conditions appealing for savers, and also interesting for banks from the point of view of their financial margin. The third primary function of a bank is to offer financial services in general. Some of them—as with credit—appear in the balance sheet of the company. Others, such as the stock market brokerage, foreign exchange, futures, and swaps, do not appear in the balance sheet, as their results are accounted as simple income from non-financial products. Naturally, the function of marketing also plays an important role in the distribution of these services.

Operations relating to the receipt of capital—especially various forms of deposits—and of the offering of financial services can be—and are—strongly related to a bank's management. The receipt of capital, on occasions, is effected not only on the basis of an attractive combination of profitability and liquidity, but also with the offer of a service (a credit card or life insurance policy) from which the bank can derive additional revenue. In the past, this dimension (the offer of complementary services) was not a basic characteristic in the marketing of financial services. However, the demand for new and better financial services on the part of

---

[2] Ballarín (1988*a*) presents a somewhat different value-chain for a bank. In his version, production would be a primary activity and the commercial activities would have support characteristics. Although it is a different model, it is equally useful for understanding the creation of value in a bank.

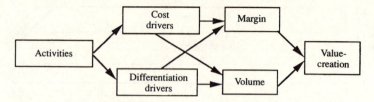

Fɪɢ. 9.3   *Value-chain activities and value-creation*

savers on the one hand, and the financial firms' strategy of obtaining income from non-financial products on the other, has led to a situation in which this type of income is increasingly important for banks. Therefore, it constitutes a key reason for increasing deposits.

3. DRIVING FORCES OF THE VALUE-CHAIN ACTIVITIES

3.1. Introduction

In the previous section it has been asserted that the value-chain is a useful tool for identifying the relevant activities in the creation of value within a bank. But the value-chain allows us to take an additional step. With an understanding of the relevant activities, we can broaden our analysis to the determining forces of these activities, or in other words, the driving forces of value-creation for a given company. When we speak of the drivers of value-creation we refer to certain economic activities within a company. We do not underestimate the role of people which, in the end, act as the determining factor in the long-term success of any organization. However, here we will emphasize the economic aspects specific to a bank.

Within the wide spectrum of forces behind the creation of value two large groups can be distinguished: cost drivers and differentiation drivers. As can be observed in Figure 9.3, these drivers contribute to the creation of value through attaining a certain scale of business, or through the difference between prices and unit costs.

The concept of 'driver' can be defined as a structural factor which connects activities carried out in an organization with their underlying economies.[3] Cost drivers refer to the determinants of cost behaviour within a specific function. Each segment of the value-chain has its particular cost structure. To adequately distinguish between them through a cost analysis, certain criteria must be used. In particular, a value-chain activity can be relevant if its costs are high (or its rate of growth is high),

---

[3] Porter (1985).

if its determining factors are unique, or if the activity plays an important role in the value-chain of competitors.

It is possible to distinguish between several possible cost drivers: economies of scale, learning processes, capacity, connections between a firm's own value-chain and the value-chains of clients or suppliers, interrelations with other operative units of the organization, degree of vertical integration, and localization of activities. This classification is useful in identifying the cost determinants of any activity.

However, the identification of the driving forces of costs is only a preliminary step to achieving competitive cost advantages. In effect, gaining competitive advantage has more to do with controlling these drivers by reconfiguring the value-chain in order to reduce costs. The increasing price competition within the banking industry makes these ideas especially relevant to cost management within specific activities in the sector.

The drivers of differentiation make it possible for a company to offer a product buyers consider to be somehow unique, permitting the firm to charge higher prices. Of course, the price premium must be higher than the cost of differentiation. These factors are found in each one of the activities in the value-chain, and we can distinguish drivers of differentiation such as economies of scale, relations with clients and suppliers, localization of activities, relations with other business units, learning processes, or degree of integration.

In much the same way as with cost drivers, it is necessary to control differentiation drivers because this allows companies to offer additional value to their clients. This value is manifested in two ways: lower purchasing costs or improved results—in a broad sense—for the buyer. To achieve these objectives it is important to know what the effect of the product offer has on the value-chain of its client. Every point of connection between a firm's value-chain and that of its clients creates opportunities for differentiating products and, consequently, for gaining competitive advantage.

3.2. The driving forces of bank activities

As was shown in the previous section, the value-chain offers a useful tool to understand the banks' results. In effect, the two primary activities situated at the respective extremes of the value-chain will determine the gross margin of a bank, defined as the sum of the financial margin (financial income minus financial expenses) plus income from non-financial products.

Let us start with the intermediation margin. In this case, the decisive factors, as previously indicated, are the negotiating power a company has with its suppliers and its buyers of funds. This negotiating power has changed radically in the recent years partly as a result of entry-barrier erosion. As previously indicated, the negotiating power of a traditional bank with its depositors has been diminished by the entry of investment banks, as well as by the development of new financial products, or the capital markets themselves, which have attracted a growing proportion of family savings. Additionally, intensified competition has resulted in a change in the banks' strategies in three different areas: the profit/liquidity ratios (which have been changed increasingly in favour of the investor), new financial innovations (which also provide investors with greater liquidity and profitability), and, finally, non-financial products (which compensate for losses from rising financial costs).

With regard to the buyers of funds, the position of banks has also deteriorated. Capital markets offer increasingly better opportunities for firms in need of funds; and, investment banks also are playing a more active role in this segment. As if this were not enough, competition has increased in the area of consumer credit as a result of the new services offered by large department stores, insurance agencies, and automotive manufacturers. The result of all of this is a further weakening in the banks' position in relation to their biggest clients and the concentration of loans in small and medium-sized companies which have no clear financing alternatives.

In the future, to prevent greater losses, it is imperative that banks offer their clients competitive services in both lending and saving which would permit them to establish more stable relationships with their clients than in the recent past.

A final point with regard to the intermediation margin refers to the maturity of operations and, as a result, the volatility of interest rates. In periods of high stability in interest rates, volatility is not important. However, the high level of financial instability that has characterized the 1980s (as described in Chapter 1) and the growing influence of monetary policy, as much as in the control of inflation as exchange rates, has converted interest rates into a central variable, both for financial and non-financial enterprises.

The management of interest-rate risk appears closely related to the management of financial assets and liability, then maturity and their interest rates. This management technique takes as given certain relationships with clients and providers of funds and aims at obtaining the greatest profitability differential possible from the efficient management of the corresponding assets and liabilities.[4]

The management of assets and liabilities in a bank tends to vary depending on two opposing criteria: minimum risk (trying to group liability

---

[4] Any textbook on bank management offers a description of these concepts, e.g., see Wilson (1986) and Lewis (1987).

operations with profitable asset operations with similar maturities) or maximum profitability. The advantages and disadvantages of both criteria are evident. Investing safely in order to minimize risk offers a high degree of security but, as a compromise, can diminish profitability. As demonstrated by the previous discussion, it is clear that efficient asset and liability management is considerably more important in a bank than in an industrial firm, owing to the potential risk of interest-rate loss.

The second element that has to be emphasized concerning gross margin is the impact of income from non-financial products as a result of their growing importance for banks' operating results. The traditional banking sector has had to pay an additional cost for the appearance of new non-banking intermediaries in the sector.

The introduction of new services implies additional development costs. It is important not only to measure carefully the profitability of new products (once the costs that their services will entail are subtracted), but also to create products that will share pre-existing costs in order to gain greater economies of scope. This is an increasingly complex task, not only because of the growing sophistication of existing financial products and their widespread proliferation (which makes it more difficult to find a successful niche in the market), but also because the capacity to protect innovations from duplication is very low. Nevertheless, the future profitability of a bank will depend increasingly on its ability to create new and different products on the basis of continuous financial innovation. Such a strategy will be more efficient the more it is adapted to existing cost structures.

Other important costs correspond to operating costs, including personnel costs and overheads, which are the highest of all in the banking industry. Operating costs refer to those expenses resulting from support activities and primary activities which are not naturally financial in nature. Personnel costs are closely associated with the size of a bank's network of offices. What activities account for the greatest volume of operating costs? According to a McKinsey study on American banks, the highest operating costs are not related to information processing, but to initiating client relationships.[5] According to McKinsey's estimates, between 70 per cent and 80 per cent of operating costs of the American banks concentrate on this function.

The mechanization of this activity should lower operating costs considerably. Automated tellers and other technological innovations have reduced them, although they have also required large investments, and can be very costly in terms of the low level of efficiency associated with certain technologies. As indicated earlier, high costs can also be incurred

---

[5] See *The Economist*: 'A Survey of International Banking', 25 Mar. 1989.

through the installation of multimillion dollar systems that ultimately are unsuited to a bank's particular needs.

The analysis of entry barriers in Chapter 8 demonstrated the importance of scale in relation to a bank's capabilities in information technology which, if implemented and managed properly, leads to significant economies of scale. The same is true of economies of scope derived, for example, from new services which share storage costs, or the processing costs of a single block of information. However, there is a limit from a management point of view. In effect, the internal processing of information can be automated to the extent that available technology permits. And activities centring around client contact are more problematic: even though they would be possible to automate, it would not always be wise to do so.

The reason for this paradox is surprising: machines are not capable of selling. This activity requires people. And given the growing importance that the sale of financial services has acquired—and will continue to acquire within the sector—it is necessary to count on people capable of understanding their clients well in order to offer them financial services which best meet their family, personal, and economic needs.

The transformation of banking offices into points of sales is not free of problems, however. The first of these is human and consists of professionally retraining branch office employees, who are accustomed to mere administrative tasks, to sell financial products. Another is related to the growth of automation, as it is not easy to attract actual or potential clients to a bank branch if the services it provides can be found at any other automated teller. Banks wishing to reduce operating costs as they strive for complete automation should think twice and carefully consider the costs and benefits of such an approach, otherwise, failures will not be surprising.

Offices that serve as a point of sales for various financial services act as a double-edged sword: on the one hand, they represent an element of considerable costs, and on the other they contribute to the sale of products whose value will be increasingly decisive for the income statement. At the same time, they constitute a significant entry barrier, as we saw earlier.

All of these considerations suggest that with the need to reduce costs— especially in European countries such as Spain and the United Kingdom, where costs are relatively high—efforts to do so cannot be indiscriminate ('we have to reduce all costs'), but carefully selective so as not to damage activities related to the future prosperity of the banking system: the distribution of financial services.

The successful strategic management of a bank refers not only to the reduction of operating costs, but also—and equally important—to those costs derived from the sale of financial services, which also contribute to non-financial and financial revenues. To attract a certain type of client—of high income, for example—can be beneficial for a bank as a means of reaching a segment of the population with a growing demand for financial services. However, a careful cost-benefit analysis must be conducted.

The same can be said of other important sources of revenue: revenue from non-financial products, whose relative importance depends on the firm's innovative capacity. The innovative talent of managers may be capable of generating truly attractive products, but related decisions cannot be taken without simultaneously considering costs.

Throughout this discussion reference has been made several times to firms' costs relative to competitors. This is an important point, since a knowledge of competitors' costs represents an indication for a firm to reduce its own costs. To some extent, costs help to define profitability and, in the end, this is a relative concept, not only in terms of what it measures, but also in relation to how it is viewed within the industry.

Lastly, we will refer to two important aspects of banks' income statement, which do not fit in any of the activities of the value-chain as it has been defined until now. They have to do with provisions to cover the risk of insolvency, and dividend and capital policy.[6]

The risk of debt default represents a very important variable for banks, although it will diminish as financial intermediation decreases. However, this process will occur slowly, and thus allocations to provisions for bad debt will continue to play an important role in determining banks' results. Also, it cannot be forgotten that the regulatory authorities can interfere and set the level of these provisions. Allocations for bad debt have influenced the results of the American and English banks (the latter to a lesser extent) owing to the debt crisis of Latin America. Banks' profits have deteriorated and stock-market capitalization has fallen in some cases. In the summer of 1989, two top-level English banks, Midland and National Westminster Bank, reported losses from exceptional allocations to provisions for bad debt for credits extended to various Latin American countries. As can be seen, the strategic importance of this policy is considerable.

Finally, reference will be made to capital and debt policies, which also are not explicit in the analysis of the value-chain.[7] However, it is difficult to overemphasize their importance.

In effect, dividend and debt policies have an immediate effect on how stock markets perceive a company. Given banks' growing need for capital, as much for carrying out large operations as for acquiring up-to-

---

[6] See Donnelly and Skinner (1989).
[7] This is undoubtedly a limitation of this conceptual tool, which allows for an analysis of the competitive position of a firm in the context of a market, but not in some situations involving financing and dividends. As we have seen in Ch. 2, debt policy can contribute to increase profitability.

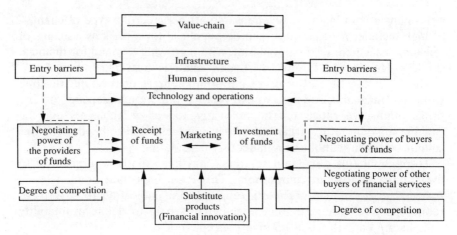

FIG. 9.4   *Value-chain of a bank and structural characteristics of the banking industry*

date information technologies, it is natural for dividend and debt policies to hold more strategic relevance.

4. SOME CONCLUSIONS

This chapter demonstrated some of the intimate connections between the competitive forces of a sector and its structure, as well as their effects on the value-chain in the banking industry. The value-chain facilitates an integrated view of the process by which value is created in a firm. However, this process is closely tied to the structural characteristics of the sector, as shown in Figure 9.4.

What is the virtue of integrating both types of analysis? It offers an organizational link, tying together information relevant to the industry to information about the firm, making it much easier to identify challenging objectives, and the best ways of achieving them.

Lastly, the value-chain is useful for studying cost structures connected to the different services offered by a bank. Intensified price competition requires a greater understanding of how this affects income statement and, indirectly, stock-market prices. All of this makes an analysis of costs on the basis of the value-chain quite useful.

10
Competitive Strategies
of the European Banks

1. BANKS' GENERIC STRATEGIES

The analysis of the value-chain in the banking industry helps to identify key functions that contribute to the creation of value within a bank. Once these have been identified, the ideal strategy should aim at finding concrete ways of strengthening them in order to create competitive advantages.

For a specific function or activity in a firm to represent a real competitive advantage, it must be sustainable over a long period of time. Otherwise, competitors could enjoy similar benefits, thus eroding the firm's competitive position. Figure 10.1 shows the connection between competitive strategies, the characteristics of the sector, the configuration of bank activities, and competitive advantage. Figure 10.1 shows some of the sources of competitive advantage for a bank.

In the literature on strategy it is possible to distinguish between three possible generic strategies for achieving or maintaining competitive advantage: cost leadership, product differentiation, and focus.[1]

Cost leadership historically has been the strategy most used to gain competitive advantage. The reasons are twofold. In the first place, the acceleration of product life-cycles means that innovations become standardized very quickly. As a result, advantages gained from product differentiation disappear and the solution is to innovate again or compete in a mature market through cost reductions in manufacturing and distribution. The second reason has to do with the protection of new technological innovations. This does not refer as much to legal safeguards as to protection derived from the difficulty of imitation. The possibilities of not just copying a product, but of imitating it and improving upon it have grown considerably in recent years due to technological advances. This in turn causes products to mature more rapidly, leaving only two strategic options: renewed innovation or cost leadership.

Cost leadership affects not only one activity within the value-chain of banks, but extends to each and every activity comprising primary as well

---

[1] See Porter (1980), who maintains that a firm will only attain an acceptable profitability through following one of these strategies. For a critical discussion of these ideas see Chrisman *et al.* (1988), Hill (1988), and Thietat (1988).

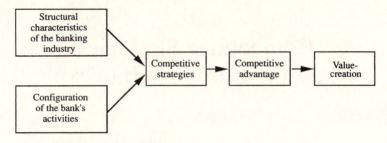

F<small>IG</small>. 10.1   *Value-creation in a bank*

T<small>ABLE</small> 10.1.   *Sources of competitive advantages in banking*

Human resources	Financial resources	Assets	Intangible assets

People
Professional
  training
	Capital
	Deposits
		Branch network
		Information
		  systems
		Telecommunication
		  systems
			Brand image
			Experience
			Managerial
			  talent
			Product quality
			Service quality

as support functions. Yet this does not mean that any bank, considering its personnel structure, resources, and history, cannot concentrate on just one activity or link together value-chain functions to reduce costs. Cost leadership requires, among other things, economies of scale and learning, as well as rigorous cost control throughout the entire value-chain. This strategy has important implications for those banks that follow it. Cost leadership can lead to maximum market share. It also affects financial innovation, whose advantage will not be in design but in lower costs. In turn, economies of scale or learning usually require continuous large-scale investments in plant and equipment.

The advantages of cost leadership are clear. In the first place, cost leadership can give the firm superior benefits compared to competitors, while establishing close relationships with clients. However, the application of this generic strategy is not easy. It is relatively common to commit errors such as concentrating excessively on production costs, forgetting distribution and sales costs, failing to reduce costs in activities which are crucial for the creation of value, or finally, forgoing cost advantages from interrelationships between the distinct value-chain activities. A bank that wants to follow a cost-leadership strategy needs to be aware of these common errors.

The strategy of differentiation consists of offering financial services that potential or actual buyers perceive as being unique. Differentiation can be achieved in every one of the activities.

Differentiation can lead to sustainable competitive advantage to the extent that it permits relative control of the structural forces of the industry: it creates special links with clients, diminishes competition to some extent, and creates barriers to entry while protecting against substitution. Differentiated products allow the firm to fix higher prices compared to standard products, implying higher margins. We have already indicated in Chapter 9 a necessary condition for this: that the cost of differentiation does not exceed its benefits.

The nature of differentiation in many cases makes it incompatible with cost reductions. This is because, in general, differentiation requires higher capital outlays in some activities, such as research and development. A differentiation strategy can result in the inability to gain greater market share, since not all clients will want to pay a premium for the exclusive characteristics that might distinguish a product from mass-produced goods.

There are also potential pitfalls in a differentiation strategy. They include the erroneous pursuit of product characteristics that hold no value for the consumer, differentiating beyond consumer needs, overpricing, and a failure to assess accurately associated costs. In the end, these mistakes have a common denominator: they neglect the role of the product in the value-chain of the client. It is necessary to understand these factors, as well as those criteria relevant to the purchasing decision, in order to pursue such a strategy successfully.

Product differentiation, with all its inherent difficulties, presents important possibilities within the financial service sector. The reason is simple. To the extent that demand for these services increases—and it has in recent years in the industrialized countries due to a rise in the standard of living and the higher level of financial literacy on the part of consumers—so does their sophistication and complexity, characteristics which improve conditions for diversification. Again, the example of Merrill Lynch's Cash Management Account is appropriate for demonstrating that conditions in the banking sector are ideal for specific differentiation strategies as a means of gaining sustainable competitive advantages in the medium term.

Some authors indicate that differentiation contains another additional advantage: the creation of a specific image for the corresponding firm.[2] This image can act as an important entry barrier, offering protection

---

[2] See e.g. Davis (1985), who maintains that corporate image acts as an entry barrier in industries whose products are perceived as being complex.

TABLE 10.2. *Relationship between generic strategies*

		Competitive advantage	
		Cost leadership	Differentiation
Competitive scope	Broad target	1. Cost leadership	2. Differentiation
	Narrow target	3A. Cost focus	3B. Differentiation focus

Source: Porter (1980).

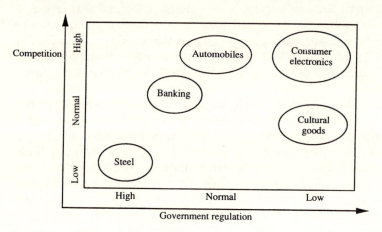

FIG. 10.2 *Dynamics of an industry*

against loss of market share, while also being helpful in the launching of new products.

The third generic strategic option is focus on a specific market segment, a segment which can correspond to specialization in certain product areas (mortgage loans, for example), clients (small companies or families), or geographic areas. In turn, focus can encompass a strategy of cost leadership or differentiation.

The difference between this strategy and the previous two is that a bank can serve its clients more efficiently than its competitors if it concentrates on a particular market segment, following within that segment a strategy of differentiation or cost leadership. The conditions for this strategy to be successful are analogous to those for any company wanting to follow any one of these strategies within the framework of an industry. Table 10.2 presents the relationship between these three strategies.

We have described the three generic strategies available to any bank to build a sustainable competitive advantage *vis-à-vis* competitors. However, the purpose of this book is not so much to discuss the virtues of one strategy or another, but to examine how they should be implemented within the context of a unified market in 1992. In other words, what should banks be considering in order to successfully adapt given the new conditions within the industry?

Before analysing in detail some of the strategies available to European banks within the next few years, in the following section a conceptual scheme will be presented which will help towards a better understanding of the velocity of change within the banking industry from a strategic point of view.

2. A CONCEPTUAL FRAMEWORK FOR UNDERSTANDING THE PACE OF CHANGE IN THE EUROPEAN BANKING INDUSTRY

In order to evaluate the pace of change with more precision, several illustrations will be especially useful, and they are applicable to any industry.[3] Figure 10.2 shows a matrix which relates two relevant variables: the degree of government regulation and the degree of competition within an industry. Together, these variables represent what can be called the 'dynamics' of an industry. According to this matrix, the dynamics of the industry, or its rate of change, will be as high as the intensity of competition, and as low as the level of government regulation.

The variable regulation has been chosen because it is the most common form of government intervention. Other forms, such as public enterprises or subsidies, will tend to disappear in the future, within a political framework in defence of competition in the European Community.

The other variable, the degree of inter-firm competition, has been presented in Chapter 8 in relation to the forces of the industry. It is evident that the greater the level of competition, the quicker the dynamics of the sector. An increase in competition will contribute considerably to the continued liberalization of financial services and the single banking licence in all EC countries. And the progressive dismantling of entry barriers—economic or legal—will contribute to a progressive intensification of competition.

In Figure 10.2 we have situated a group of industries within the given matrix. The consumer electronics industry on the one hand, and the steel industry on the other, show the two opposite poles of dynamics within

---

[3] For a different presentation see Friberg (1988).

the spectrum of European industries. The matrix provides a means for comparing diverse fields such as automobiles and airlines, both of which are subject to intense competition, although in both cases regulation has prevented them from being even more robust. The banking industry has a similar profile to these two sectors, although the level of competition is less fierce.

Nevertheless, the Single European Market of 1992 will present other challenges to the various economic sectors deriving from slightly different variables from those mentioned above. Two are especially relevant in the context of a unified market. The first is related to the new business opportunities which a unified market will offer European companies, opportunities based on three distinct features: a larger market, lower costs resulting from potential economies of scale, and the possibility of creating European-wide brands, leading to a simplification of marketing tasks and reinforcing consumers' perception of a company.

The second challenge arises from the homogeneity of consumption or demand in different European countries. The characteristics of demand for certain products is not the same across countries, since some products contain strong domestic components—as a result of cultural factors, for example—while others are more internationally orientated.

Figure 10.3 shows a matrix which combines both dimensions, grouping into one variable the characteristics which will correspond to the industry in the future. According to this matrix, a sector will have a European (or global) character to the extent that business opportunities in the industry grow due to the economic integration of 1992, and as a result of a broader geographic market or economies of scale. This is the case in both the automobile and banking industries.

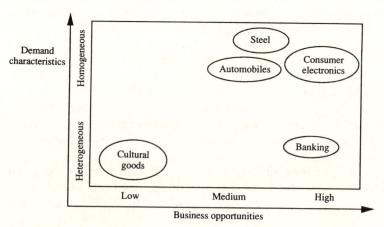

Fig. 10.3 *Geographical scope*

The Euro-global character of an industry will also depend on the characteristics of the demand in that sector. In effect, a more homogeneous type of demand (offering the possibility to market more or less standardized product in all, or at least most EC countries) will experience an immediate geographic increase within the sector. For example, this is the case with the airlines. On the other hand, a more heterogeneous type of demand, which corresponds to the peculiarities of each country, will demonstrate a low propensity for 'Europeanization'. This is the case with the publishing and communications industries, in which language (or other cultural conditions) impedes globalization. Yet this does not prevent firms in these sectors from following global strategies by carefully segmenting the distinct national markets, forming alliances or acquiring companies in foreign markets.

It is interesting to note here that one sector can be in two geographic regions which are completely different and take maximum advantage of the possibilities offered in both depending on the characteristics of demand. In the case of the banking sector, it is possible for firms to market relatively standardized financial services at a global level, while offering more specialized products at a national, or even regional level. In any event, the matrix in Figure 10.4 demonstrates the overall tendencies of the sector, indicating the various possibilities for companies and products.

The adequate combination of the variables just defined—the dynamics of a sector and its geographic range—provides additional insight into the impact that a unified European market will have on any industry. The range that these variables offer can be seen in Figure 10.4. The more

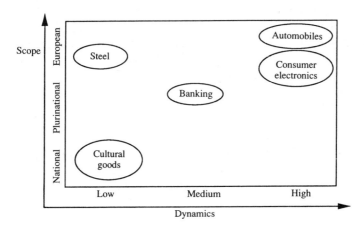

FIG. 10.4 *Transformation of an industry as a result of the Single European Market*

dynamic and 'European' an industry is, the more it will be affected by the integration of the European market. Industries such as consumer electronics and automobiles are already being affected by the changes leading up to 1992. In other cases, such as in banking or airlines, the speed of change is slower, which is the result of geographic considerations and government intervention. In the case of cultural goods, the speed of change is even slower.

These variables allow us to identify three paces of change resulting from European integration. A fast pace of change is characterized by a high degree of dynamism and a geographic spread limited to only one country. In this case, it can be anticipated that the velocity of change within the industry will be very fast due to little government regulation, homogeneous demand, and the possibilities for attaining high economies of scale. Consumer electronics and automobiles are good examples.

A medium rate of change has more heterogeneous characteristics. The clearest cases are those industries which are moderately dynamic and have a global, or at least multinational, scope. In these circumstances, the pace of change resulting from economic integration will not be as fast as in the previous case, owing to such factors as government regulation, competition, demand characteristics, or business opportunities. It seems that the banking sector represents this type of industry mainly because of government regulation, which will eventually be relaxed, although it has undoubtedly contributed to creating a market structure and type of financial institution which will not change easily. The heterogeneity of demand for financial services in Europe is another factor which is preventing the industry sector from becoming decidedly global—one more reason why the rate of change will be slower than in other sectors. In effect, only a few financial products can be considered global: mortgages, mutual funds, and credit cards. Nevertheless, the switching costs for consumers are low and the propensity for change is high. On the other hand, more sophisticated products have more local characteristics.

Finally, we can identify a type of industry with a slower rate of change, either because its natural scope is not European, or because the degree of dynamism is low, a result of the fewer opportunities that come with a larger market.

Once we have identified the variables that determine the rate of change given the new competitive environment, we need to ask which is the best possible strategy for the banks. The answer is complex because each firm needs to devise a strategy which best balances two characteristics: those related to the banks themselves and those related to the new competitive environment.

Without pretending to exhaust all the possibilities, Figure 10.5 presents a topology of strategic possibilities within the banking industry, dis-

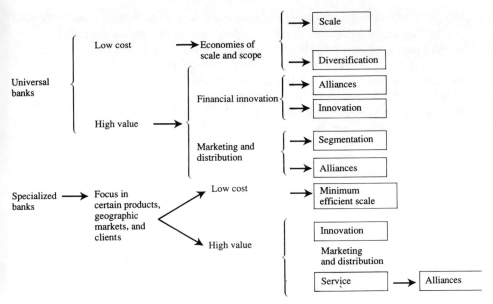

F<small>IG</small>. 10.5   *Banks' strategies for competitive advantage in the Single European Market*

tinguishing between two different types of banks: universal banks and specialized banks. In each of these groups there are different options for taking advantage of the possibilities offered by the unified market in order to confront intensified competition in the years ahead.

By universal banks we mean those firms which develop, or plan to develop, a group of activities including commercial and investment banking in virtually every major market and geographic segment. These include those commercial banks that have clearly opted for the universalization of activities and which, in addition, do not limit themselves geographically. Also included are investment banks which, without pretending to be able to develop a full range of commercial banking activities, do wish to target typical bank clients, such as medium-sized companies and high-income families.

On the other hand, among the specialized commercial banks there are those whose principal activities comprise the distribution of financial banking products typically offered to families and companies, such as the intermediation in the stock market or in the issuance of companies' debt.

Within this context, a set of strategic moves has emerged among banks concentrating their efforts on gaining cost leadership or differentiating products or services within specific market segments; some banks are attempting to do both, and there are many smaller commercial banks that are improving their competitive positions by reducing costs. It is natural for banks to try to pursue both objectives, although one must outweigh the other, as has been recently indicated in the literature.[4]

Within the context of global banking, the first possible strategy in the case of cost leadership consists of looking for a size which offers economies of scale at a business or global level, or greater diversification of activities. The difficulty of achieving such a scale is common to most firms, and only very large firms such as Citicorp or the Deutsche Bank are capable of success in this area. For other companies, growth requires a strategy of mergers, alliances, or capital exchanges which lead to co-operative agreements. The many mergers and alliances between banks in recent years reflect such strategies.

For the development of a differentiation strategy, firms have various options. Within them, and leaving aside the issue of scale which allows for global economies of scope, we can highlight a strategy which places emphasis on financial engineering in new product design, as well as a strategy aimed at generating new clients through the intense use of marketing and distribution by segmenting appropriate markets. According to the latter, a bank can have a global presence, while still segmenting the type of financial markets it wants to operate in, as would be the typical case of investment banks.

In the case of specialized banks, their competitive advantage comes from differentiation in a specific market segment. These banks, however, should attempt to exploit scale economies. The question of distribution is especially important for these banks for two reasons. The first is that it is the most basic sales activity and, consequently, must represent a source of competitive advantage. The second is that in the case of countries such as Spain and Germany, where operations tend to be on too large a scale, the selection of appropriate distribution channels is very important. We cannot forget the apparent contradiction, indicated earlier, between advantages and disadvantages of a dense network of branch offices: the benefits of the entry barrier which such a network creates can be greater than its costs, depending on the strategy the firm follows.

From the brief discussion of these strategic moves, four emerge which deserve special attention: scale, diversification of activities, segmentation, and financial innovation. Given that ample treatment has already been given to financial innovation in Chapters 1 and 2, the next sections present a brief discussion of the other three.

---

[4] See Hill (1988) and Murray (1988).

3. THE IMPORTANCE OF SCALE IN THE BANKING SECTOR

The strategies followed by some of the European banks such as Barclays, NatWest, Dresdner, or Crédit Lyonnais, aim at gaining scale in order to compete more efficiently in the unified European market. This by itself demonstrates that the question of scale is important, or at least that the directors of these companies—who know the sector better than anyone—believe this to be the case.

In general, larger size can help any firm reach a higher level of economies of scale and scope.[5] At the same time, scale permits greater technological innovation and the attainment of greater market power. A bank operates with economies of scale, as we saw in Chapter 8, when operating costs increase proportionally less than production volumes and average unit costs decline as output grows. Conversely, the opposite phenomenon is diseconomies of scale which occurs when costs increase proportionally more than production volumes.

Economies of scope occur when two or more products can be produced or offered simultaneously at a lower cost than if they are produced or offered separately.

Economies of scale or scope are very powerful in shaping the structure of an industry. In the event that these economies really exist, the sector will be made up of large companies which can produce at lower unit costs than smaller companies. In this case, these companies will capture a significant share of the market. On the contrary, if production conditions do not permit economies of scale or scope, the industry will tend to be comprised of small and medium-sized firms, which together will dominate the larger ones.

Two fundamental types of economies of scale can be distinguished. The first consists of those economies which emerge in the production of a specific product, called product-specific economies of scale. The second consists of economies associated not with a single product, but with a range of products, called bank-level economies of scale. Both types of economies of scale are identical in the case of a one-product bank, and can exist in multi-product banks as well. In the latter case, bank-level economies of scale are registered if, in increasing the production of all the bank's products by one unit, total production costs rise at a lower proportion.

Two types of economies of scope can also be distinguished. The first, known as economies of global scope, are obtained when total costs derived from total production are less than the cost of obtaining each

[5] For a thorough discussion of these concepts, see Scherer (1980). Their application to the banking sector has been carried out in a number of works, including Gilligan *et al.* (1984), Clark (1984), Kilbride *et al.* (1986), Berger *et al.* (1987), Clark (1988), and Lawrence (1989). And for an analysis of their effects on pricing policies see Moebs (1986).

product separately. The second type is known as product-specific economies of scope. In this case, the cost of producing a particular product together with other products is less than if it were produced separately.

What are the real sources of economies of scale or scope in a bank? The first lies in investments in computer and telecommunications equipment. Only large banks can have access to large-scale information-processing equipment, capable of handling enormous amounts of information at a very small additional cost per transaction. To the extent that these organizations increase their volume of transactions, the unit cost of processing the information is immediately reduced.

Additionally, the excess of capacity of these computers can be used at low cost to process other kinds of accounts, new products, and even new banks that are the target of an acquisition. The pursuit of economies of scale and scope need not exclude potential benefits gained from differentiation. Differentiation can be achieved at a low cost if, for example, a product shares information related to other financial products within the same network. When the cost of using the same information at various times is less than the cost of obtaining that information separately, the cost savings for the corresponding enterprise can be very significant.

The growing reach and influence of information technology permits banks to use resources more efficiently in the sale of services, which for a distribution-orientated firm is very important.

A second source of economies of scale is the diversification of risk, which implies additional costs in terms of provisions for bad debts. In a large bank these can be proportionally less than in a smaller institution. These advantages also derive from investments in assets, increases in the number of clients, as well as market and geographic diversification, which dilute risk between a larger number of accounts. But they can also appear on the liability side: with a rise in the number of deposits, their variability diminishes as a result of the law of large numbers, reducing in this way the risk of bankruptcy.

Economies of scale or scope can also be gained in distribution activities. A brand image associated with a larger enterprise will facilitate its penetration into other national or international markets. Alternatively, the cost of loan distribution will decline if one branch is capable of offering its clients treasury bonds or pension funds, for instance, thus achieving economies of scope. Finally, the role of advertising cannot be disregarded (as well as the growing importance of advertising costs) in the process of selling banking services. As sales of a particular product grow, the effects of a given advertising campaign on the costs of a specific product line will decline; in addition, the advertising of a large firm supports the sales of all the services it sells.

The previous discussion demonstrates that economies of scale and scope play an important role in the banking industry. In this case, reaching a minimum scale appears to be the lowest requirement for a bank to remain competitive/ Nevertheless, it must be remembered that this is only one road to gaining competitive advantage and that commercial banks do not necessarily have to look for this kind of advantage. Nevertheless, the empirical evidence of economies of scale in the banking industry is not as definitive as the group of theoretical arguments employed here.

Related studies attempt to obtain empirically banks' cost functions, assuming that they depend on the price of production factors and on the level and composition of production. This functional relationship is derived from the dual property between the production function and cost as taught in microeconomics. The most utilized method of conducting these estimates is the translog function; its advantage lies in that it offers accurate information on economies of scale and scope for different levels of production, separately or together. Once the function is selected, it goes on to choose the sample and estimates the parameters of the function, which give measures of the economies of scale or scope, respectively.

Conclusions based on this type of analysis have already been presented in Chapter 8. As previously indicated, firm-level economies of scale occur with relatively small volumes of transactions (less than $25 million) and,

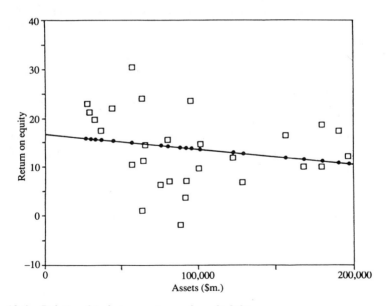

FIG. 10.6   *Relationship between size and profitability*

very large transactions result in diseconomies of scale.[6] In addition, there is no consistent evidence of the existence of complementary production costs.[7]

In Figure 10.6 we have conducted a simple exercise. We have calculated the regression between the size of assets and return on equity in 1988 for the 24 banks analysed in Chapter 2, Section 4.2. The results are quite surprising: the relationship is negative, such that the most profitable banks are the smallest. The fact that the data refer to only one year may account for this irregularity. Thus, in light of these studies, it is not clear that economies of scale result from an increase in scale. The existence of economies of scope is also subject to question, although it is possible to obtain complementary costs in certain product lines.

Nevertheless, these studies must be considered with caution. The first reason is that the majority of the studies refer exclusively to American banks, whose structure and type of business does not adapt perfectly to that of the European countries considered here.[8] The second reason has to do with methodological problems related to the definition of output and costs. Two measures of output tend to be used: the size of assets, and the size of credit and deposit accounts.[9] However, neither of these is completely satisfactory. The third limitation in most of these studies is in the data and their level of aggregation, which is excessive for constructing cost functions.

There is another methodological problem with the cost functions used to measure economies of scale or scope, the translog function, and, in particular, the utilization of these in relation to information obtained from the banks themselves. Most of these studies use data from the 1970s, that is, before the widespread dissemination of large-scale information-processing technology. It has already been suggested that such technology has created enormous possibilities for obtaining economies of scale and scope, and in many cases this has been used to justify the pursuit of larger-scale operations. The case of Citicorp and its annual investments in information technology is very significant. Thus, it is not surprising that data which do not represent the intense process of automation that has taken place over the recent years does not provide empirical evidence of economies of scope or scale.

Nevertheless, scale or scope economies are not the only arguments in favour of pursuing growth strategies in the banking industry. There are a

---

[6] See e.g. Benston *et al.* (1982), Gilligan *et al.* (1984), and Clark (1988).

[7] See Berger *et al.* (1987) and Benston *et al.* (1983).

[8] There are some exceptions to this assertion. Fanjul and Maravall (1985) calculate the existence of economies of scale in the Spanish banking sector and arrive at the conclusion that they can exist in certain circumstances. Recently, Delgado (1989) has observed economies of scale and the branch level.

[9] See Clark (1988).

number of reasons why larger scale is desirable, different from scale or scope economies. First, banks might want to be large to gain market power, defined as the capacity to increase prices above those that would exist in a situation of perfect competition. Market power tends to be more prevalent in very concentrated industries with high entry barriers. This situation allows larger firms to establish monopolistic pricing policies.

According to Cournot's classic model, the difference between price and marginal cost for any given firm is proportional to its market share. Therefore, larger companies can obtain greater market share and, consequently, higher profitability, allowing them to fix prices above marginal costs. This can also be explained using the experience curve. This concept suggests that firms with the largest market shares will have the highest production volumes, resulting in declining costs proportionate to their experience curves. Although the empirical evidence is not conclusive, there are good reasons to believe in the validity of the argument in favour of market power and larger size. For example, the process of globalization on the part of the big European, American, and Japanese banks would appear to support this assertion.

Another justification for a larger scale of operations has to do with diversification. The spectacular development of certain financial markets has diminished the relative proportion of banks' mainstream activities. This fact is quite relevant for banks at a time when financial disintermediation appears to be an irreversible trend.

In light of this trend, European banks have reacted with a growing presence in the stock markets—as brokerage agents or on their own accounts—depending on the regulations of the respective countries. Naturally, this has led to an increase in competition, as indicated earlier.

To enter into these types of operation successfully and diversify successfully into these activities (taking into account that evolution of the financial markets) a larger volume of capital is required than for banking intermediation, which leads once again to the need for a larger scale of activities. Diversification is not exempt from risks, but it is evident that it can provide economies of scope to the extent that the costs associated with new activities can be spread among existing business lines, without incurring additional costs above marginal revenues. In summary, the argument for diversification in some instances justifies the necessity for banks to grow larger, although the high risks associated with initiating stock-market operations cannot be overlooked.

Related to the argument in favour of diversification—which emerges in response to the trend toward financial disintermediation—the question arises as to securitization of credit. In effect, disintermediation has diminished the importance of the traditional credit business developed by European banks, which has in turn made them more vulnerable to some of the stronger foreign concerns, especially those from the United

States and Japan. To compete in this field requires human resources and information technology which only large banks can afford.

Innovation is another argument for large scale. It is acknowledged that large banks tend to innovate more than smaller ones, suggesting that the former have, or can have, greater market power. In general, the process of innovation speeds up when activities related to it derive from economies of scale. But this is not the only benefit. Large banks can better control the results of their research, such that they are not easily copied by competitors; the probability that duplication will not occur is greater if a firm has a larger market share. Concurrently, a bank with substantial market share will tend to use innovation as a means to maintain and increase its position in the market. In the banking industry it can be shown that, in general, larger companies tend to be the most innovative. This assertion does not mean that there are not large institutions that are not innovative, or that there are no small innovative firms. However, the relationship between scale and innovation is one more reason why larger firms enjoy certain competitive advantages.

The liberalization of capital in the EC may convert small and medium-sized firms into acquisition targets for bigger foreign banks. As a consequence, larger scale may offer these firms more effective protection against adversaries in search of take-over targets. Larger scale by itself will not eradicate this threat entirely, although it may make it more difficult for the acquiring companies, as evidenced throughout the last several years in a number of different industries.

In conclusion, it can be said that scale of operations is an important variable for the future development of banks, not just from the point of view of economies of scale and scope, but also from the perspective of a wider range of other potential advantages including greater market power, the improvement of defences against hostile take-overs, and financial benefits derived from friendly take-overs.

However, the pursuit of larger scale presents a serious problem: how actually to obtain it. Mergers, as well as alliances, present significant operational difficulties which only an exceptional management team can overcome. Yet, large-scale growth is not the only possible strategy, nor is it always the best. And although it may be a necessary objective in certain cases, it should be accompanied with an appropriate segmentation strategy.

4. THE DIVERSIFICATION OF ACTIVITIES

Deregulation and disintermediation represent two characteristics of the more advanced financial systems which are driving banks to diversify their activities. Deregulation provides an incentive for entry into segments that were previously prohibited to banks. Disintermediation requires that banks find alternative businesses as a result of their declining position in modern financial markets.[10]

The first advantage of diversification is that by expanding the range of activities, economies of scope can be gained in the distribution of a larger number of financial services. The possibility of sharing certain resources also provides a strong argument in favour of diversification. In addition, diversification permits complementary links between different kinds of services, at the same time as offering clients a wider range of financial options. It also allows for the more efficient use of a network of offices, or for acquiring more sophisticated information-processing equipment whose cost can be spread over a high volume of capital and wider range of businesses.

However, the principal advantage of diversification as a strategy is that it permits banks to reduce the variability of their revenues in relation to a given amount of capital, or expressed differently, a given amount of business which, although more diversified, requires a lower ratio of capital.

In general, banks face three distinct types of risk. The first is related to interest rates, and comes from the simple fact that asset investments have a different period of maturity from savings, and consequently different interest rates. However, this structure can vary as a result of fluctuations in money markets, causing risks known as interest risks. The second type is the risk of solvency, derived from credit granted by banks. The third is liquidity risk, which exists because of the possibility of immediate deposit withdrawals which, in turn, would require the cancellation of asset investment, with all the difficulties this would entail, not to mention the potential losses that would be incurred.

Banks have dealt with these risks by entering into other activities (or targeting other clients) through an appropriate diversification of investment portfolios or a realignment of segments. Some examples will help to illustrate this process. The first of these refers to the reduction of interest-rate risks through loans with variable rates; in the last few years some savings banks and mortgage companies have begun to offer mortgages with variable interest rates. The second refers to the possibility of covering interest-rate risks through operations in the futures markets, allowing companies to protect against an eventual increase in the cost of liabilities. The third has to do with the process of securitization and has been developed especially in the United States. Thanks to this, some

---

[10] For a treatment of the two fundamental theories on the advantages of diversification see Markowitz (1959) and Fama (1976). The diversification of a bank's business portfolio will not be discussed here. Rather, the discussion will focus on product diversification within the bank itself. For a discussion of those aspects not considered here see Daems (1978) and Rumelt (1986).

commercial banks have sold credits in the secondary markets to investment banks which, in turn, have used them to sell shares backed by the credits themselves.

These innovations highlight the importance of the diversification of banks' investments, as it enables them to combine different assets in a single portfolio, thereby spreading risks and providing greater financial stability. This does not mean that the diversification of a bank's activities will necessarily be better for its shareholders, who can also follow the principle of diversification and spread their investments across a variety of instruments. However, the advantages for shareholders are significant if they can invest in a larger number of shares; otherwise, the gains from diversification are not as clear-cut. Thus it is preferable that risks be diversified by the bank itself.

In general, the few empirical studies that take up the question of diversification in financial services show that the benefits banks derived over a specific period could have been more stable had they diversified their activities to an even greater extent.[11] At the same time, these studies show that there are many opportunities for reducing risks by combining banking with non-banking activities.

Within non-banking activities the underwriting of other firms' shares can be distinguished as a particularly attractive business. In effect, in placing shares banks assume the associated risks for only a few days and, in principle, these shares are relatively liquid and easily replaceable. On the other hand, in the case of a loan, the bank assumes risk until the debt reaches maturity and is liquidated.

Until now, we have considered the possibility of the banks reducing their exposure to risks by entering into other non-banking activities. Some authors have studied this approach in a context where banks have combined traditional activities with non-banking activities.[12] Others suggest that only a slight degree of diversification produces noticeable improvements.[13] However, in these cases effective management has been vital to success.

In any event, the experience of the German banks is relevant here. Their high capital participation in industrial companies has been positive, and French and Spanish banks recently have begun to imitate this type of diversification.

Banks' opportunities in the sale of non-banking financial products are also very important. Previously, we highlighted the activity of share underwriting. However, operations such as issuing and underwriting

---

[11] See Heggestad (1975). In general, this type of study tends to compare banks' profits with those obtained by more diversified banks, as well as those obtained by firms in other industries.

[12] See Meinster and Johnson (1979) and Litan (1987).

[13] See Boyd *et al.* (1980).

should not be disregarded, nor should futures, options, currency transactions, or other financial services such as the management of portfolios
and financial assessment.

As a result of financial disintermediation, banks must become capable
of selling new financial services, which is a critical argument for understanding the advantages of, and need for, diversification. Nevertheless,
the simple addition of new activities to already existing ones is not
sufficient for successful diversification. What is really important is to have
related activities which can be sustained simultaneously through the use
of the firm's strategic assets, such as distribution networks or information
systems.

5. A FRAMEWORK FOR THE SEGMENTATION OF THE BANKING INDUSTRY

The segmentation of an industry can be defined as its division into
different markets with the purpose of formulating and implementing a
competitive strategy in each segment. The ultimate reason for segmentation is that different consumers have distinct needs. As a result, products
tend to differ between segments—although they may have common
characteristics—subsequently affecting the competitive conditions in each
one. A careful assessment of the different needs among clients in each
segment is required.[14] These differences are relevant as they always give
way to a different combination of the forces that determine the structural
characteristics of an industry and therefore the intensity of competition in
the sector. Yet, the effects of these differences are not limited to the
structural characteristics of an industry; rather, they extend to the value-
chain of all of its firms. In effect, to the extent that the value-chain
reflects relationships between buyers and sellers, it provides an implicit
message about the forces that affect competition in the industry. As a
result, a change in these forces will affect the activities of the corresponding company. Thus, segmentation entails adapting the value-chain of a
company to the competitive conditions of the targeted segment.

From a strategic point of view, the process of segmentation is crucial,
as it allows for a response to two key questions. The first is the selection
of the segment in which the company chooses to compete. The second is
the way in which the strategy should reflect the characteristics of the
selected segment. In effect, a well-defined segmentation analysis reveals
which forces affect competition in each of the relevant segments and, in
turn, the attractions of each, particularly from the point of view of the
strengths and weaknesses of the company. A study of the competitive

---

[14] See Donnelly *et al.* (1985) for a discussion of this concept.

forces of a specific segment leads to a more precise analysis of a company's value-chain and, in turn, a better strategy.

Segmentation is particularly conducive to one of the three generic strategies presented before: focus activities within a specific industry which, in turn, can be based on the pursuit of cost leadership or product differentiation depending on the characteristics of the given segment.

Existing relationships between segments in which a firm sells its goods or services is another significant aspect of segmentation. A bank can obtain important advantages if it is capable of organizing its activities in such a way that different operating units can share activities or assets and, as a result, costs. This possibility becomes even more important if they are subject to economies of scale or learning.

Spreading activities has its drawbacks, however, such as costs derived from additional organizational requirements and the consequent loss of operational flexibility. However, if the firm is capable of attaining inter-relationships between activities, notable competitive advantages can be gained.

To carry out a segmentation analysis, a number of relevant variables are useful. The first is segmentation in terms of the products offered in each market niche. The second is segmentation according to consumer characteristics and, more specifically, in relation to the role which the product plays in the consumer's value-chain. The third is geographic segmentation, reflecting the areas in which a bank intends to sell its products. A geographic analysis requires the subsequent selection of those consumers which the firm intends to target.

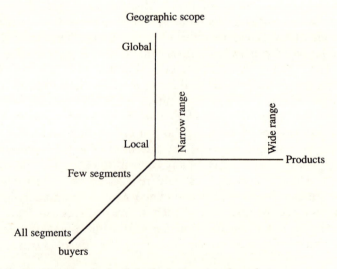

FIG. 10.7   *Positioning of a bank in the international market*

In the following sections we will conduct a segmentation analysis for the banking industry based on three criteria which represent the specific environment in which a firm will compete. These segmentation criteria permit an analysis of the international positioning of a bank as a function of geography, products, and selected customers. Figure 10.7 shows these three dimensions.

Whatever the position of a bank in the financial markets, its strategy must take into account the importance of establishing and maintaining client relationships. In Chapter 1 we emphasized the significance of this for strategic success. However, the importance of competition in price and product differentiation cannot be neglected. As indicated earlier, changes in the macroeconomic environment and in the process of deregulation have made these variables particularly important for the strategic development of a bank.

5.1. Geographic scope and banks' international strategies

The first criterion that can be used to develop a segmentation strategy is geographic. Accordingly, any bank can choose to be global, national, or local. Notice that these types of alternative are somewhat different from those presented in Section 1 of this chapter. There, reference was made to the special characteristics of certain types of industry. But here we are not referring to the spatial area in which a bank desires to develop its activities.

Naturally, both aspects are very related. The scope of any industry will have a powerful influence on companies' strategies. In a global sector, a firm cannot be a leader on the basis of a purely national strategy; unless it changes it will not reach its objectives. In the case of the banking industry, this has been understood by those companies aiming to gain a position of leadership in the international market-place, for which they have adopted (or are changing in order to adopt) truly global strategies. This means that such strategies are designed on the basis of a vision of the entire world as the firm's market, and thus decisions are made using information available not only in specific local markets, but also in the global market-place.

Some banks, such as Citicorp or Barclays, have opted for this type of strategy, forcing them to make large investments in information technology in order to provide clients with information on a worldwide basis, and also exercise minimum control over local branches, which have to operate with a greater degree of autonomy.

Porter presents a two dimensional scheme to characterize an international strategy.[15] The first dimension relates to the configuration of the

---

[15] For conceptual support of these ideas, see Porter (1986).

firm's activities, that is, the placement of each of the primary and support activities in specific geographic locations. The second is co-ordination, which expresses the relationships that exist between diverse activities carried out in different geographic areas.

There are two possibilities with regard to the configuration of firms' activities. The first is concentration on an activity in one country from which it would be offered to the rest of the world. The second option is dispersion, or the placement of activities in many different locations throughout which the company's operations are spread. In the case of total dispersion, different activities would be completely reproduced in each country. In the case of a bank, these two alternatives are clear. Concentration will occur when a company's revenues are generated primarily from a specific market, so that they can be transferred to other geographic areas and invested profitably. Dispersion will be the desired option when a bank actively seeks revenues and investment opportunities in its home market and, simultaneously, is able to invest in other locations where returns are sufficiently attractive. Normally, global banks tend to develop dispersion strategies, which allows them to act actively to attract funds in their corresponding markets.

Regarding co-ordination, the number of possibilities ranges from almost complete autonomy for local subsidiaries to a very broad organization of strategy and operations. It is evident that in the banking industry, co-ordination tends to be extraordinarily high given the global characteristics of the industry and the standardization of money and financial activities.

These considerations can be seen in Figure 10.8. A number of European and American banks are represented as a function of the combination of co-ordination and configuration. Naturally, the common characteristic

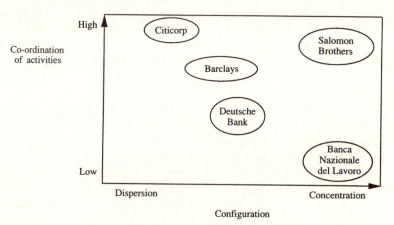

FIG. 10.8   *Dimensions of an international strategy*

among these banks is that they are large banks aiming to develop strong international positions. One of the clearest cases is Citicorp, with dispersed configuration and high co-ordination. Salomon Brothers is the case of an investment bank with concentrated configuration—although it is moving towards dispersion—and high co-ordination. The case of the Deutsche Bank or Barclays is typical of banks which have branches in various countries—with intermediate co-ordination—and are in the midst of carrying out their globalization strategies. Finally, another interesting case is the Banca Nazional de Lavoro, which in 1988 was ranked number 44 in terms of asset size. This bank aims to have an international presence, but at the moment is very concentrated in the Italian market.

Although we are considering the different opportunities available to banks that want to extend their international activities due to a desire to orientate their activities, other opportunities exist, such as concentrating activities exclusively in national or local markets. In both cases the two dimensions of co-ordination and configuration are important, although in the latter case they are much simpler. However, financial markets are becoming much more global and, especially in the European Community, competition will intensify greatly in the next few years as a result of deregulation. This does not invalidate a national or local geographic segmentation strategy, although the increase in competition and decline in financial margins requires that banks opting for such a strategy implement it with an adequate segmentation of their clients and markets.

Globalization as a strategy appears to be a valid option, but it contains a preliminary requirement: scale. Without sufficient size, globalization is not possible, and this is an important consideration in determining the scale of operations that a bank will want to achieve.

5.1.1. A model of internationalization for the banking industry

In Table 10.3 we present some of the international moves of 27 EC banks in recent years. From this comparison a pattern of behaviour emerges that is somewhat different from that presented in the previous section. The 27 cases can be divided into three groups: alliances (eight cases), alliances with strategic interests (four cases), and acquisitions (fifteen cases). In the table they have been categorized according to two criteria: the number of countries in which they are present, and the breadth of their product lines. This classification shows that universal banks tend to be much more international than specialized banks, which may be indicative of the importance of scale for commercial banks.

The first reason for internationalizing operations is to gain economies of scale and scope through the acquisition of a larger size of operation.[16] This objective has been attained primarily through the acquisition of

---

[16] See Tschoegel (1984) and Kogut (1985).

TABLE 10.3.   Alliances and acquisitions in the European banking industry

Category	Target Country	Year	Transaction
Alliances/Minority Interests			
San Paolo Bank (Italy)	UK	1986	Acquired 6% of Hambros Bank
Deutsche Bank (West Germany)	UK		Acquired 5% of Morgan Grenfell
San Paolo Bank (Italy)	France	1987	Acquired 1% of Compagnie Financiere de Suez
Générale de Banque (Belgium)	France	1987	Purchased 1.5% of Compagnie Financiere de Suez
Cariplo (Italy)	Spain	1988	Acquired 1% of Banco Santander
Commerzbank (West Germany)	Spain	1984	Purchased 10% of Banco Hispano-Americano
Proventus (Sweden)	Finland	1988	Intends to acquire 2% of Kansallis Osake Pankki as part of asset swap
Banco Santander (Spain)	UK	1988	Swapped initial 5% shareholding with Royal Bank of Scotland
Strategic Minority Interests			
Hongkong Bank (Hong Kong)	UK	1987	Acquired 15% of Midland Bank
Cartera Central (Kuwait, Spain)	Spain	1988	Purchased 12% of Banco Central
Cariplo (Italy)	Spain	1988	Acquired 30% of Banco Jover from Banco Santander
Banco Santander (Spain)	Italy	1988	Purchased 30% of IBI from Cariplo
Skadinaviska Enskilda Bank (Sweden)	Denmark, Norway Finland	1986	As part of Scandinavian Bank partnership, S-E-Banken acquired 5% of Privatbanken, 6% of Bergen Bank, and 3% of Union Bank of Finland

Kansallis Osake Pankki (Finland)	Sweden	1988	Intends to acquire equivalent of 16% of Gotagruppen from Proventus Holding Company
Cross-Border Acquisitions			
Banca Popolare di Novara (Italy)	France	1988	Acquired 80% of Banque de L'Union Maritime from CCF
National Australia Bank (Australia)	UK, Ireland	1987	Purchased Clydesdale Bank and other interests from Midland Bank
Bank or Ireland (Ireland)	UK	1987	Acquired Bank of America's UK mortgage loan subsidiary
Banque Nationale de Paris (France)	UK	1988	Purchased Chemical Bank's UK mortgage loan subsidiary
Dresdner Bank (West Germany)	UK	1988	Acquired 70% of Thornton Fund Management Group
Deutsche Bank (West Germany)	Italy	1986	Purchased Banca d'America e d'Italia from Bank of America
Banco Santander (Spain)	West Germany	1987	Purchased CC Bank from Bank of America
Citicorp (US)	Spain	1983	Acquired Banco de Levante
Citicorp (US)	Italy	1984	Acquired Banco Centrosud from Banco di Roma
Chase Manhattan (US)	Spain		Purchased Banco de Finanzas from regulatory authorities
Barclays (UK)	Spain	1981	Acquired Banco de Valladolid from regulatory authorities

TABLE 10.3. *Continued*

Category	Target Country	Year	Transaction
Crédit Lyonnais (France)	Netherlands	1987	Purchased Nederlandse Credietbank from Chase Manhattan
Banco Santander (Spain)	Belgium	1988	Acquired Belgian Subsidiary of Crédit du Nord
San Paolo Bank (Italy)	France	1987	Acquired control of Banque Vernes from Suez Group
Lloyds Bank (UK)	West Germany	1984	Acquired failed Schoroeder, Muenchmeyer Hengst merchant bank
National Mergers/Acquisitions			
Westdeutsche Landesbank	West Germany	1988 (proposed)	Projected merger with Hessische Landesbank. Other mergers of Landesbanken expected to follow
Banco de Bilbao/Banco de Vizcaya	Spain	1988	Merger of equals in October 1988
Banco Español de Crédito/Banco Central	Spain	1988 (proposed)	Projected merger, since called off
ABC Bank	Norway	1985	Leading Oslo Savings Bank merged with Central Savings Bank in Norway
Lloyds Bank	UK	1988	Bought 57% of Abbey Life

Source: Salomon Bros. (1989).

TABLE 10.4. *Banks positioning*

	Number of foreign countries with a significant presence in banking		
	0	1–2	>2
Number of products			
Universal banking	9	6	5
Specialized banking	6	1	—

other foreign banks. Table 10.4 shows the more important moves recently registered in the EC.

The strategies followed by some European banks, such as Barclays, Natwest, or the Dresdner Bank, were intended to increase their scale in order to compete more efficiently in the common European market, further suggesting the importance of scale.

The expansion of international activities to increase the scope of services is sometimes intended to create new distribution channels, or exploit deficiencies in markets where products or services may be lacking. These two reasons are evident in a number of specific cases. For instance, German, Belgian, and Italian banks tend to follow their home-country clients abroad to provide them with the same services they would receive in the domestic market. Commerzbank, San Paolo di Torino, and Société de la Banque provide good examples of this. In the following section we will analyse how these strategies have been executed.

The second reason refers to the entry of the American and English banks (Citicorp, Chase Manhattan; Natwest, Barclays) into the Spanish and Italian markets, with the intention of taking advantage of a relative lack of innovation and other deficiencies. If this is particularly true of the Spanish market it is a result of the willingness of the Bank of Spain and the Spanish Government to accept a larger presence of foreign banks. As a result, competition in this market has intensified considerably. The application of information technologies has made this reason especially important in terms of cash management, since centralized management of the treasury brings important benefits in the way of profits and efficiency. This fact adds another differentiating factor for the client.

The third reason for internationalization corresponds to the possibility of transferring resources to other countries. These resources can be tangible (capital) or intangible (financial innovations, human resources, and managerial talent). This provides one of the major explanations

TABLE 10.5.   *Criteria followed by banks in their international activities*

No. of products	Scale	Attention to the client	Transfer of resources
Large	Deutsche	Crédit Lyonnais	Barclays Bank
Small	Banco Bilbao Vizcaya	San Paolo	Générale de la Banque (Belgium)

for the existence of multinational corporations,[17] and it can be clearly appreciated in relation to the banking industry.

Examples of this type of movement abound. The Deutsche Bank, with its policy of acquiring smaller firms such as Morgan Grenfell or the Bank of America in Italy, intends not only to gain larger scale in order to benefit from scale economies, but also to transfer resources (mainly intangibles like managerial talent) to operations which are not well run. Another example is the Bank of Santander, with its acquisition of small banks such as the affiliate of Crédit Nord or the CC Bank.

It is evident that such decisions can affect those factors previously mentioned such as obtaining a specific scale or improving customer service. However, in the cases cited, the reasons are based primarily on the transfer of resources.

An argument that falls midway between the transfer of resources and customer service is the distribution channel of financial services. Some alliances between banks aim to expand distribution networks abroad for products and services traditionally offered in their domestic markets. The problem that emerges is that a bank which facilitates the distribution of an innovation developed by another bank may find it easier to copy that innovation rather than sell it on behalf of the other bank. The lack of information on this subject, however, makes it difficult to draw definite conclusions at this time, although greater insights will be gained as these relationships evolve in the future. Still, the proliferation of these types of agreement in the last few years is striking, which suggests that they do offer considerable advantages. The San Paolo and the Banco Bilbao Vizcaya offer examples of companies that have reached such accords.

In our study we have observed that most banks follow, to some extent, one of the three criteria previously mentioned, as shown in Table 10.5. This does not mean that these enterprises have forgotten the other two, but shows that their international activities are concentrated in one area.

The three dominant causes of internationalization in the banking industry over the last several years (scale, customer service, and transfer

---

[17] See Caves (1982).

TABLE 10.6. *Organizational characteristics of different types of international companies*

Organizational characteristics	Multinational	Global	International	Transnational
Configuration of assets and capabilities	Decentralized and nationally self-sufficient	Centralized and globally scaled	Sources of core competencies centralized, others decentralized	Dispersed, interdependent, and specialized
Role of overseas operations	Sensing and exploiting local opportunities	Implementing parent-company strategies	Adapting and leveraging parent-company competencies	Differentiated contributions by national units to integrated worldwide operations
Development and diffusion of knowledge	Knowledge developed and retained within each unit	Knowledge developed and retained at the centre	Knowledge developed at the centre and transferred to overseas units	Knowledge developed jointly and shared worldwide

Source: Bartlett and Ghoshal (1989).

of resources) are somewhat similar to three factors shown to be present in internationalization strategies in other industries. Bartlett and Ghoshal (1989) highlight the following strategic aspects of international competition: efficiency, responsiveness, and transfer of knowledge. Efficiency can be directly related to scale. Capacity for response is very similar to client service. Transfer of knowledge is a more specific type of transfer of resources.

The empirical evidence also shows a similar conclusion to that of Bartlett and Ghoshal. Banks with better strategic positions at the international level are those which manage better the three elements of scale, client service, and transfer of resources. Thus, the transnational solution proposed by Bartlett and Ghoshal is worth applying (see Table 10.6).

These ideas are especially important for bank management in the global market-place. Until very recently, managers tended to put emphasis on only one of the three elements mentioned above, to the exclusion of the other two. In this way, international bank branches became excessively dependent on—or independent of, depending on the case—the home office. Inefficient co-ordination of these factors is evident in the case of the Banque National de Paris, or Barclays. Thus, it is critical to integrate them within a clear, comprehensive, and well-balanced international plan in order to create sustainable competitive advantage.

Banks have numerous alternatives for internationalizing their operations. Figure 10.9 gives a sequence for international market entry, while Tables 10.7 and 10.8 show different means of entry. The first is expansion into foreign markets with the intention of offering new services to international firms from the banks' home country or offering new financial techniques in the host country. While the first strategy has little impact on banks' strategies, the second can be a significant factor in the internationalization of operations. This was the road which has generally been followed by the North American banks since the 1960s in the European, Asian, and Latin American markets, and which in some cases has enabled them to improve their positions substantially in international markets. In EC countries such as Spain, France, and Italy, there are

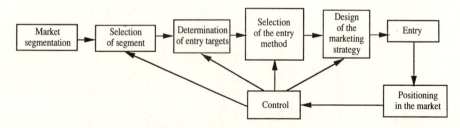

FIG. 10.9 *Decision criteria for entering the internationl market*

TABLE 10.7. *International moves of banks*

---

1. Introduction of new products (credit cards, investment funds) either using their own networks, or from the country of origin (US Banks in Europe)
2. Alliances with or without capital participation (Bayer Hypo, Commerzbank, San Paolo)
3. Alliance with a strategic capital participation (Cariplo, Banco Santander)
4. Majority acquisition (Deutsche Bank, Barclays)
5. Merger of two banks from two different countries
6. Merger between two different banks from the same country

---

significant possibilities for developing new products widely tested in the United States and the United Kingdom, such as risk-management tools, swaps, or investment and pension-fund management. In these countries the commercialization of new products could serve as a good entry strategy.

It is also possible to introduce new products in a given country without a local distribution network through direct sales. The introduction of credit cards was successfully implemented in this way by American Express. The same possibilities exist for international investment funds managed in countries such as Switzerland, Luxemburg, or The Netherlands.

The second option of internationalization is the formation of alliances with other foreign banks. Normally, the first objective of this kind of move is to exploit the strong points of both companies, such as distribution in their respective countries. This makes sense if both firms have competitive advantages in areas such as product quality or distribution. Such is the case of the alliance between Bayer Hypo and the Banco di Trento, in which the former purchased 18 per cent of the latter's shares in 1987, and of the Commerzbank and the Banco Hispano Americano, where the Commerzbank took control of 10 per cent of the Banco Hispano Americano, which, in turn, took a 5 per cent position in the Commerzbank.

This type of alliance has a second objective which focuses on reinforcing the capital of one of the two companies, and/or strengthening the core of those shareholders to deter a hostile take-over. This is the case of Crédit Commercial de France and the Banco Hispano Americano.

A third type of strategy consists of forming alliances through the purchase of a sizeable number of minority shares. The concept is very similar to the one we saw earlier, the difference in this case being the size of capital participation and its strategic goal. The recent moves of the Banco Popular and of Paribas would fit within this category. Figure 10.10

TABLE 10.8. *Presence of the principal European banks in other EC countries, 1988*

	Germany	Spain	France	United Kingdom	Italy	Portugal	Benelux
Germany							
Commerzbank	—	OM Banco Hispano	OM	OM/MV	—	—	OM MV
Deutsche Bank	—	Banco Comercial Transatlántico	OM	OM/MV	OM Bank of America	—	OM MV
Dresdner Bank	—	OM	OM	OM/MV	OM	—	MV
Spain							
Banco Bilbao Vizcaya	OM	—	RO	OM/MV	OM	OM	OM
Banco Central	OM	—	OM	OM	—		OM
Banco Santander	CC Bank	—	RO	OM	Istituto Bancario Italiano		Crédit du Nord

France							
Banque National de Paris	OM	OM	—	OM/MV	OM	OM	OM
Société Générale	OM	—	—	OM/MV	OM/RO	—	OM
Crédit Lyonnais	OM	OM-RO	—	OM/MV	OM	OM/RO	OM/RO
Crédit Commercial de France	OM	OM	—	OM/MV	OM	—	OM
United Kingdom							
Barclays Bank	OM	RO	RO/MV	—	OM	OM	OM
Lloyds Bank	MV	MV	RO	—	—	MV	OM
National Westminster	OM	RO	RO	—	OM/RO	—	OM
Midland Bank	MV	OM	RO	—	OM	—	—
Italy							
Banca Naz. del Lavoro	MV	OM	OM	MV	—	—	—
Banca Naz. dell'Agricoltora	—	—	—	MV	—	—	—
San Paolo	OM	—	OM (Banque Vernes)	MV	—	OM	OM

Notes: MO: Majority office; B: Stock market brokerages; ON: Office networks.
Source: Salomon Bros. (1989).

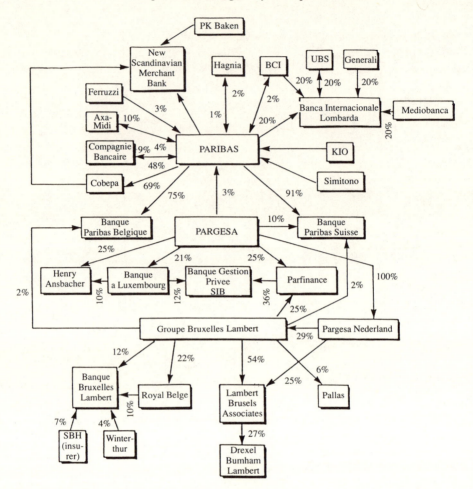

Fig. 10.10   *Shareholders' network of Paribas*
Source: La Actualidad Económica, *21 Nov. 1989.*

shows the network of shareholding relationships which Paribas has with other banks.

An opposing strategy proposes taking a majority position in a foreign bank with the aim of co-ordinating its activities with the global strategy of the acquiring bank. This option is even clearer and has the advantage that it is probably the most effective as a means of gaining a real presence in a foreign market. This assertion is supported by the fact that different firms have followed such a strategy in recent years: Deutsche Bank, which bought Banca d'America e d'Italia in 1986 and Morgan Grenfell in 1989;

Barclays Bank which bought the Banco de Valladolid in 1981; and Crédit Lyonnais which bought Nederlandse Creditbank in 1987.

Another strategic option for internationalization is the merger of two banks from different countries to create a 'supranational' group. However, the difficulties of achieving this type of merger are considerable, as demonstrated by that between Belgium's Societé Genérale de la Banque and The Netherlands' Amro, which failed in September 1989. The reasons for this failure relate not only to those problems typical of any merger, but also to those resulting from differences in financial regulation between countries and, to a certain extent, distinct national and financial cultures.

Finally, a merger between two banks from the same country could also be considered as an initial step toward one of the international strategies discussed above. The reason is that scale could play an important role in the development of international operations. The merger between the Banco de Vizcaya and the Banco de Bilbao appears to correspond to this philosophy, as has been indicated by their directors on numerous occasions.

In general, it can be affirmed that the strategy of mergers and acquisitions functions as a means of increasing the size of a company. The attainment of a larger scale of operations and the reduction of costs are the driving forces behind such a strategy. On the other hand, the other internationalization strategies are geared toward improving the capacity to provide service to clients, while transferring financial innovations to other affiliates. The first strategy has clear objectives, although it entails considerable risk as the possibility of failure is high. The second strategy entails less risk, although its results are also uncertain since alliances with competitors also have a long history of failure. Only time will tell which of these is the better of the two options.

5.2. Segmentation by financial products

An important characteristic of the banking industry is that different financial products create different financial markets. For this reason, in order to study segmentation according to products, it is necessary to look at the specific markets in which these are offered.

It has been reiterated that globalization is an unstoppable trend in modern financial markets. It has important implications for a given country's financial system to the extent that what happens in one market affects all the others. For example, in any country, when prime rates rise, so do other interest rates (credit rates, and later, possibly even the rates on deposits). It is a simple illustration of the integration and growing interdependence of the national financial markets. For this reason, in talking about segmenting markets, the distinct markets can be viewed as separate commercial objectives though they are undoubtedly interconnected.

To segment financial products, various criteria can be used. We will follow a practical approach, which is to analyse the four major markets within the banking business: credit, savings, money, and capital markets.

5.2.1. The credit market

The credit market is where banks and borrowers negotiate the conditions under which credit is given. Together with deposits, this is among the most common of banking businesses.

Within the credit market there are three major activities: company credit, family credit, and mortgages. Each will be analysed in relation to those forces that are causing a radical transformation of the banking system.

In the previous chapters describing the recent evolution of banking systems in the five major European countries, a common tendency has been observed in each: a decline in the importance of credit to firms, which has contributed to financial disintermediation as companies reduce their reliance on bank-related credit.

The causes of this are clear. In the first place, the parallel growth of the capital markets has allowed companies to raise funds directly through the issue of commercial paper. The advantage of this instrument lies in its lower cost, since banks require a higher interest rate on their loans in order to cover operating costs and a lower volume of revenue resulting from their minimum capital ratio requirements.

Together with this phenomenon, there is a second cause of disintermediation which is the dilution of financial margins as a result of intensified competition (entry of foreign competitors, and deregulation affecting the savings and loans institutions).

Finally, we must highlight profitability as an important cause of disintermediation. In effect, the general recovery of corporate profitability amid financial prosperity has allowed companies to rely more on the resources generated by new investments and less on the credit banks.

As a means of actively confronting this process of disintermediation, an almost parallel phenomenon has arisen in the USA and the UK. It is securitization, as mentioned previously. Through this process, loans are divided into parts which are, in turn, divided between investors. Thus, banks manage to reduce risks of insolvency by mobilizing credit and increasing liquidity; in addition, they earn extra revenues. Also, this process is a useful tool in the management of interest-rate risk as it allows the sale of those assets which least adapt to the structure of interest rates.

Generally, the loans which are split and sold later are mortgage loans. This tool is not exclusive to the United States, either; it has been accepted in various European countries through authorization to issue bonds with the support of a mortgage loan.

The typical form of this operation is what is called a 'pass through' loan. This instrument represents a portfolio of loans which is given to the investors in the form of property certificates. The lender takes charge of collecting the principal and interest of the ultimate borrower; with this yield he can offer the investors a reasonable interest rate. This loan is usually carried out through a commercial bank, an investment bank, and some investors. The first sells the loan to the second which, in turn, places it with investors.

For the success of these operations, two conditions are required. The first—not necessary but convenient—is the integration between the commercial bank and the investment bank, exploiting in this way the advantages offered by vertical integration. Obviously, this is not an easy task, since an investment bank is naturally different from a commercial bank, and thus has no reason to know how to adequately manage such an activity. The second condition—absolutely necessary—is the maintenance of a certain level of liquidity in the concentration of these shares. For this, the banks must contribute to maintaining a secondary market in which investors can buy or sell depending on their liquidity situation.

In terms of business, we see that traditional credit will continue losing importance—not rapidly but irreparably. And therefore, banks which, by tradition or vocation, continue offering this type of financial service for companies should think, simultaneously, of the possibilities to 'securitize'. However, as we have indicated with regard to the majority of the banks analysed, this is not easy because the business opportunity will only be present for those that really peruse it.

For other reasons as well, the traditional credit market for companies is not, in the judgement of the managers in the industry, an attractive market. It does not justify the entry of a foreign bank into another national market, unless the bank decides to acquire a national network and contemplates this investment for the long-term, as the large American banks have done in Europe.

A second category of the credit market deals with consumer credit, whose basic end is the financing of families. The most popular form of consumer credit is the credit card. In Chapter 1 we made extensive reference to the development of this credit instrument which has given the banks the possibility of offering a new service to their clients, thereby increasing their incomes, while allowing them at the same time to come into contact with potential clients. The increase in living standards and the rising level of financial literacy on the part of consumers is evidenced by the dramatic increase in the volume of cards in circulation, of their use and sophistication ('intelligent' cards, for example). This will constitute

TABLE 10.9.   *Estimate of the demand for credit cards ($m.)*

Country	Adult population	Potential users		Current Demand (%)	
		Low	High	Potential demand	
				Low	High
France	44,300	44,300	66,450	46.0	30.7
Germany	48,900	48,900	73,350	5.3	3.5
Italy	45,800	45,800	68,700	7.0	4.7
Spain	30,900	30,900	46,350	30.6	20.4
United Kingdom	45,500	45,500	68,250	96.2	64.1

Source: Salomon Bros. (1989).

an important source of income for the banks that understand how continuously to adapt their services to the growing demands of consumers.

On the other hand, growing competition in this area will also come—and is coming—from other non-financial firms such as Sears in the United States, Marks & Spencer in the United Kingdom, and El Corte Inglés in Spain. Again, it is clear that the first to enter the game will emerge as the winners. The business is interesting, but only those institutions that closely follow the evolution of the market and the customers' needs will be in a position to take advantage of it.

The control of the distribution channel is the key for success in the credit-card business. Generally, this distribution can be effected through a branch network or through a sales network. However, there is a third way which has been successful for some credit cards and this is advertising and sale by direct mail.

In the first chapter we have indicated that the potential market in Europe for credit cards is quite large (estimated at around 400 million cards); presently, though, less than one-fourth of this potential market has been realized.

Table 10.9 presents some estimations on potential demand for credit cards in the five countries analysed. With all the reservations an estimation of this sort deserves, it seems clear that the only country showing a degree of saturation is the United Kingdom. In Spain and France the development of these products is growing as a result of the development of information systems which permit acceptance of credit cards for a variety of consumer decisions. Italy, although with relative tardiness, is also advancing in this direction through the installation of similar electronic systems. The case of Germany is different. Its low level of credit-card acceptance is the result of very established cultural and commercial customs, though these too will gradually change in the years to come.

In addition to the credit card, consumer credit has experienced a surge in the last few years also due to the increase in families' living standards along with their higher level of financial sophistication and more positive attitude toward debt. In an atmosphere of expanding consumption, the demand for credit on the part of families has proven to be insensitive to interest rates; thus banks catering to this segment of the market have carried out very profitable operations. The continuation of these conditions in the future will delay innovation of products serving this segment of consumers.

The third type of credit is mortgages. This kind of instrument is very attractive for various reasons—to a greater extent in some countries. The first reason is the recent growth of the home-construction industry, no doubt a result of the favourable economic cycle. Consequently, the demand for property loans has risen. A second reason is that in the past few years mortgage credits have been the object of new legal treatment in the EC, with a view to permitting the securitization process described earlier.

As proposed, the future of this business requires the creation of a secondary market where participation in mortgage loans can be compared and sold easily. Mortgage participation is not the only means of mobilizing mortgage loans; there are also bonds guaranteed by mortgages. However, the former have the advantage for the issuing banks that the risks are not normally computable in calculating the maximum risk as a function of capital. In addition, for the investors there is the benefit that they can recover their investment to the extent that the amortization of the mortgage is produced, without having to wait for the expiration (maturity) of the mortgage.

This characteristic makes these bonds attractive. But for their development it is important to have a sufficiently large secondary market. At present, with the exception of the United Kingdom, in the respective financial systems there is still much ground to be covered in this area. This is the reason that it has become such an attractive business. The key question is whether the phenomenon of 'securitization' will spread in continental Europe in the same way that it has in Britain. Two factors appear to suggest that it will. The first is the interest of the European Commission in the creation of a large market for the financing of the construction industry. The second is the growing demand for discovery of new consumer credit instruments for the different national markets.

The development of a market for bonds of this type in the future will depend on three conditions. The first is the existence of an adequate secondary market. The second has to do with the barriers imposed by different laws on the functioning of these instruments in each country: the legal aspect of the process of securitization, the minimal capital requirements, their fiscal treatment, etc. in addition to the financial practices of the banks themselves, which can favourably influence the development of these operations. Finally, of considerable importance are the economic factors such as interest-rate structure and the size of the financial markets for placing these titles.

5.2.2. Savings instruments

Deposits have constituted, together with the granting of loans, the traditional banking activity. In recent years, three phenomena have powerfully influenced deposits. The first is the fall in the savings rate of households in relation to GDP. During periods of crisis, the fall in savings was due to the fact that profits did not increase (and in some cases fell) while consumption increased, at least in nominal terms. In recent years, the phenomenon has been manifest in the majority of industrialized countries with the increase of economic growth powerfully driving a more than proportional growth of consumption and a decline in savings. This has had subsequent repercussions on the volume of new banking deposits.

The second phenomenon is the growing financial knowledge of the savers, which has induced them to require higher returns on deposits in financial entities. This has had its logical negative effects for banks on the intermediation margin which, as we have verified in Chapter 2, has been reduced in the countries analysed, undoubtedly as a result of this occurrence.

The third factor which has affected deposits is the appearance of new competitors in search of family savings. Within this group, we can distinguish between three categories of competitors. First, there is the public sector. Secondly, there are the non-financial companies which have substituted classic banking credit for commercial paper, attracting to this instrument a certain volume of private savings. Finally, we have the insurance companies, with their money funds and pension funds offering very attractive ways of placing savings, often with the added benefit of a solution to the financial problem of retirement.

In the face of this process, banks have tried to react by offering treasury bonds. However, the possibility of success with this type of instrument depends not only on its intrinsic quality, but also on the actual degree of saturation of the money markets. Definitely public-sector deficits have made possible the expansion of the banking business.

What options remain for banks to confront this phenomenon of decreasing deposits? Principally, there are two. The first is financial innovation and differentiation aimed at offering the client an increasingly sophisticated financial service. If the service is really innovative, banks

TABLE 10.10.   *The insurance market in some European countries* ($m.)

Country	Total policies	Percentage 1985–6	Percentage over GDP	Insurance percentage over GDP	Total insurance
France	14,180	25.7	256.00	1.83	22,260
Germany	25,304	10.0	414.50	2.53	34,381
Italy	2,641	30.4	46.20	0.40	11,811
Spain	1,088	67.9	28.10	0.45	4,231
United Kingdom	27,201	18.6	479.30	4.95	18,603
EC	80,246	16.2			105,480

Source: Salomon Bros. (1989).

will gain the confidence of the savers. This is the case, for example, with current accounts remunerated at market interest rates. The second is diversification of activities. In a previous section we discussed this strategy which now becomes relevant in the context of banking dis-intermediation on the liability side of the balance sheet. Banks must take advantage of their extensive network of offices to offer new products, and with them attempt to compensate for both the reduction in the volume of deposits and the decreasing financial margin. Among the new financial instruments for attracting savings, life insurance and pension funds are noteworthy.

Life insurance has recently attracted banks' attention as much because of the decline of traditional instruments for attracting savings as because of the opportunities provided by this new instrument in Europe. Table 10.10 offers information on the situation of these products at the end of 1986 in the principal European markets.

Banks' interest in these products derives from the possibility of selling them to clients of the bank, relating traditional bank services with life insurance. Also, the tenuous separation existing today between the banks and other non-bank financial institutions drives banks to offer new financial services. Together with this argument, other advantages such as distribution and utilization of branch networks, which can result in the reduction of costs common to both activities, need to be considered. Finally, it cannot be forgotten that the growth potential of these products is considerable, especially in countries such as Spain, Italy, and France.

One of the questions emerging from the offer of life insurance is the legal regulation in each country. In Germany, this product is regulated in such a way that banks cannot sell life insurance directly. What the German banks have done is establish affiliates which distribute these products or form agreements with insurance companies. If there is a lesson to be learned from the recent changes in the system, it is that an entity must offer those products considered useful for its clients and life insurance is one of them.

In Spain, the legal situation is similar to that in Germany. The difference is that the principal Spanish banks have a majority participation in important insurance companies and, those that do not, are creating affiliates that co-ordinate their sales efforts with the respective bank branches.

Conversely, in France banks have affiliated insurance companies and, at the same time, their branches directly commercialize insurance policies in those companies.

The Italian banks are prohibited from distributing insurance policies and the only thing they can do is create affiliates, as the Banco Nazionale del Lavoro recently did.

Finally, in the United Kingdom, according to the Financial Services Act, banks that want to distribute life insurance have to create an affiliated company if they plan to sell these products using their own network of offices. Or they can establish an insurance company, which can utilize the information which the bank can provide on its clients. As can be seen, the diversity of national regulations is considerable and their convergence could take several years.

The second important instrument for savings investment in the years to come is pension funds. The European Commission is contemplating the possibility of allowing an authorized fund in any EC country to operate in any other. This could generate an intensification of competition at the international level, especially in those markets, such as the French and German, with a long tradition in these products but little opening to the outside.

Table 10.11 presents information on the situation of this product in the European countries. As is evident, the circumstances are very different in each country. France is the country with the greatest volume (in value and size) of funds, the result of a number of historical incidents, and of the funds' popularity for placement of individual savings. French banks (owners of a good number of funds) are as concerned about external competition as they are about the possibility of French savings moving to other countries with more favourable fiscal regulations, such as Luxemburg.

The English pension fund market, in the absence of controls on capital movement, finds its strength on the basis of the stock market and favourable fiscal legislation. The proven capacity of its institutions and their experience in the management of international funds predicts an increase in foreign competition in the English market. This in turn could incite English funds to venture into international expansion, either directly or in the form of alliances.

TABLE 10.11. *Pension funds in some European countries ($m.)*

Country	Number of funds	Value
France	3,662	204,000
Germany	162	42,255
Italy	72	50,884
Spain	89	3,245
United Kingdom	1,137	57,937

Source: Salomon Bros. (1989).

Finally, the German, Spanish, and Italian markets also have potential for strong growth, a result of low market saturation, stock-market reform, and the recent liberalization of pension funds.

Again, the deciding factor will be distribution. With direct sales as a possible mechanism, the branch network of banks makes this a potential route for gaining a significant share of the market, especially in the countries with greater predicted growth.

5.2.3. Money market instruments

The money market is made up of a group of agents (government, other financial firms and banks) which handle titles characterized by high liquidity—there is a large secondary market—a shorter maturity, and lower interest rates than those which dominate the capital markets.

The initial money market was the inter-bank market used by banks for temporary placement of inactive funds, which therefore did not generate any profits. At the same time other banks used it to cover the reserve requirements required by financial authorities. However, the important development of variable interest-rate credit which all countries have experienced has led to the growth of the inter-bank market, as this has become a source of additional short-term liquidity.

In recent years, new players have entered the money markets, including the governments and non-financial firms, with their commercial paper issues. All of this has provided a number of liquid financial instruments with relatively high profitability.

What is an advantage for banks from the point of view of their management of assets is ultimately a serious threat from the perspective of attracting deposits, as we have seen in the previous section. On the one hand, government bonds, on occasion accompanied by a number of important fiscal advantages, have captured a considerable volume of private savings. This has had effects not only on attracting deposits, but also on the placement of other types of instruments in the same market. On the other hand, commercial paper has acquired a notable market share over the last few years. This instrument has had a doubly unfortunate effect on the banks. First, it has reduced their relations with companies as a growing number of firms have become familiar with the use of this instrument.

However, what some banks perceive as a threat others have interpreted as an opportunity, offering their services for commercial paper issues. This provides two important advantages. The first consists of the additional revenues which, through commissions, they can obtain through leading an issue. The second is that this offers new business opportunities by providing new services for existing clients or potential ones. What must be kept in mind, however, is that in entering this area of activity, banks start to intervene actively in the specific operations of the investment banks, taking advantage of their own broad capacity for product distribution through numerous points of sale.

From a strategic viewpoint, money markets will play an increasingly important role in the future helping to improve interest-risk management, increasing the revenues from non-financial products, and creating new business possibilities. From this perspective, it is perfectly clear why a number of companies are putting highly qualified directors in charge of these activities.

5.2.4. Other capital markets

The recent growth in the volume of capitalization experienced by the principal international financial centres is, without doubt, one of the important characteristics of the current financial environment. This expansion, in addition, has in some cases been accompanied by legal stock-market reforms aimed at allowing greater operating flexibility and increased competition between brokers and intensifying supervision on the part of the financial authorities. These legal reforms have extended to regulation at the EC level (the Second Banking Directive), which will have additional effects on this type of activity.

In any case, this area of operation encompasses for the banks both a potential advantage and a potential risk. The advantage is that with a bull market and a more liberal framework, the possibilities of acting on it as brokers—or as mere intermediaries—with a minimum of success are quite high.

However, in the case of brokers, potential risks appear as well, which have proven fatal for many companies. Some prominent banks have recently lost money in the British stock market, for example, which has made them much more cautious in playing an active role in the stock markets.[18] As in so many other banking activities, the professionalism of

---

[18] See *Business Week*, 15 May 1989.

TABLE 10.12. *Typology of bank activities according to market segmentation*

	Number of countries	
	One	Many
Product range		
High	Universal national bank	Universal global bank
Low	Specialized national bank	Specialized global bank

the directors and the careful evaluation of the potential risks are essential so that what may be a good business opportunity today does not turn into a fatal disaster later. Thus the segmentation aimed at this market unavoidably carries considerable risks.

The criteria for segmentation already studied (the degree of a bank's internationalization and the number and variety of products it offers) gives way to the type of activities significant for the positioning of a bank which is presented in Table 10.12. According to these criteria any bank has before it the option to be national or global, a direct function of the countries in which it develops its operations. Clearly, immediate alternatives fit somewhere between these extremes. Within each geographic option, a bank can then offer a wide or narrow range of financial products, thereby creating specialized or universal functions. As we have seen in the preceding countries, the positioning of a bank in any of these categories has important strategic implications.

5.3. The specialization of institutions

Traditionally, the specialization of banks has corresponded to the criteria of client selection. The three main types of banking business are commercial banking, investment banking, and merchant banking. We will leave aside merchant banking because of its absence in many European countries, and consider the selection of the customers, distinguishing between families and small and medium-sized companies (served mainly by the commercial banks) and the large corporations (served differently by the larger commercial and investment banks).

5.3.1. Commercial banks

The specific business of commercial banks has always dealt with household savings and credit lending to both families and companies, normally small and medium-sized enterprises. Because large corporations currently have direct access to capital markets, they have stayed at the fringes of this segment. However, with the growing financial knowledge of consumers, this segment has experienced a notable increase in activity over the past few years in Europe. It is attractive for a variety of reasons. The singular activity of banking intermediation alone has traditionally been profitable, as was indicated in Chapter 2 through studying the intermediation margin for the banks. Furthermore, being a commercial bank offers the possibility of providing for their clients a greater number of services, thus increasing income from commissions.

However, it is also true that commercial banking activity is already exploited within the realm of the European Community, making saturation inevitable in certain areas. Given this situation, commercial banks' activities should be characterized by two key variables. The first of these corresponds to an adequate market segmentation, taking into account both clients and products which are being offered to them. Undoubtedly, in this environment, differentiation and focus are critical choices. What seems clear is that few firms will be able to reach the entire population of one country and fulfil all its financial needs; therefore, adequate segmentation by customer-type becomes critical.

In proceeding to this segmentation, there are two clear economic criteria which, as it happens, are also complementary. The first is the potential growth of the population within the segment; and the second refers to the profitability involved in the sales of a product for that particular segment of the population.

Within the arena of investments, a product which will be a star in the coming years is mortgage credit, not only because of its foreseeable growth, but also because it will become the object of a securitization process. Reference has been made to these points in the previous section.

The second aspect to bear in mind, as was highlighted before, is the necessity of broadening the product range offered to customers. The reason is twofold. First, customers will become increasingly demanding with respect to their financial products and if banks do not want to lose their business, they must fulfil the changing needs of customers. Secondly, there is an economic reason: in principle, in increasing the range of products (assuming some proportion of constant fixed costs) revenues will also be increased, enabling banks to generate additional profits. Furthermore, a broader product range can contribute to a complete utilization of the capacity of a bank's branch network, which constitutes a major fixed-cost component for these enterprises, especially in countries where there tends to be an excess capacity of the branch networks.

However, such an addition of new products cannot be chosen randomly, since in this case the possibility of improved operating results actually diminishes. What is really beneficial is to identify products with something in common with an existing product, allowing for synergy. As we saw previously, this is a strictly necessary condition for profiting through diversification. In the end, the message of this section is that, within a particular segment, for the group of commercial banks serving households and small and medium-sized businesses, diversification of activities is a prerequisite for the survival of the bank.

5.3.2. Investment banks

Traditionally, investment banks have specialized in preparing and underwriting new issues of an industrial company's shares, along with other related activities known as corporate finance. More generally, the activity of an investment bank consists of offering services to large companies through which they can gain access to the capital markets in order to obtain the financial resources they need. Their specific segment is large corporations. The process of financial disintermediation has distanced large companies from commercial banks; some commercial banks have reacted by entering the segment of corporate finance more or less successfully. Is this a foreseeable trend for commercial banks?

One important characteristic of investment banks derives from this: by being companies that essentially lend services, the size of their assets relative to the level of business they manage is comparatively small. At first glance the business of investment banks seems more attractive than that of commercial banks, perhaps because of the lower fixed costs required and the greater specialization.[19] However, profitability is highest in those cases where investment banks have acted as insurers of corporate share issues, which necessarily requires considerable capital.

This last factor explains why some of the large American commercial banks (Bank of America is a good example) have opted to abandon investment banking activities when they could not compete with the prices set by giants such as Salomon Brothers or Nomura. Another factor also serves to diminish the attractiveness of this type of business for commercial banks. This has to do with the additional risk a bank incurs in acquiring shares on its own account, as well as the potential conflict that may arise between those responsible for a firm's investment banking on the one hand, and those in charge of its commercial banking on the other, both operating within the same organization.

These considerations (the need for a large capital base and the increase in risk) must be taken seriously, with the advantages and disadvantages carefully weighed if a commercial bank decides to offer its services also as an investment bank.

In countries such as France, Spain, and Italy the development of

---

[19] See Hayes and Hubbard (1989) and the study prepared by Price Waterhouse (1988*b*). Also of interest is Walter and Smith (1989). On the specific problems of management in these enterprises, see Eccles and Crane (1988).

investment banking activities also depends on an additional factor: the existence of broad and flexible capital markets. One of the positive things to come as a result of the public deficit in some European countries has been the development of short-term—and in some cases medium-term— fixed income markets which serve as indicators of price and profitability. Together with this, legal reform of the stock markets initiated in these countries also pushes farther in the same direction, since it liberalizes operations, stimulates competition, and attracts private savings. The same can be said of the growing importance of commercial paper in the money markets in Europe. Concurrently, the process of corporate restructuring recently initiated in Europe will continue in the future, making corporate finance activities particularly appealing in the EC countries throughout the 1990s.

We lack the temporal perspective necessary to measure the possible effects of these changes. However, it is expected that they will be positive for the development of investment banks in those countries in which the industry has not yet reached a period of maturation. Here again, as in other sectors, those who enter first are most likely to emerge as winners.

In the years ahead, established investment banking activities will continue to develop a global shape, though with an emphasis not on the size of growth, but on the degree of profitability. However, considering the forecast increase in domestic competition, it is probable that some banks will react by opting to compete not only on the basis of cost leadership, but also through diversification and specialization of activities. In conclusion, as the competition in this segment grows, so will the inherent business opportunities.

6. SOME CONCLUSIONS

In this chapter we have outlined possible strategies for the commercial banking industry in the face of the single financial market of 1992. Throughout the analysis, we have presented some ideas that help us to understand better the change that is being introduced and how to adapt to it. The first is that the change will be gradual. Naturally banks heading in the same direction (that of deregulated markets and greater national and foreign competition) opt for a winning strategy.

The second concept presented is that the strategy of adaptation will not always be the same; rather, various possibilities exist and they are not mutually exclusive. Those banks that understand how to capitalize on the strong points of one or another, at the same time taking into account the realities within their organizations, will gain an important advantage.

Thirdly, the need to expand the offer of financial products is of increasing importance, though on the basis of a careful segmentation. Demand is growing and will continue to do so in the coming years, both on the part of families (with new forms of consumer credit and savings placement), and companies (with new forms of financing and risk coverage, assessment services for corporate restructuring, etc.).

Fourthly, the special characteristics of the banking industry make it such that the entities already present in the country have a singular advantage for retaining their clients. However, for this to hold true in the future (especially with the growing requirements of consumers), national banks must make efforts to increase their range of products and services to satisfy these increasing financial needs of the customers. Otherwise, traditional relationships will account for very little before a better offer on quality or price from competitors.[20]

Finally we want to emphasize again the importance of adequately utilizing the branch network in rationalizing their resources, and in discovering possibilities for distributing new financial services. The introduction of new information technologies will provide a new challenge to the direction of banks, consistent with the conversion from a personnel orientated toward administrative tasks, to one orientated toward sales tasks. Many other important issues for the banks will depend upon the success of this conversion.

---

[20] This has occurred in the USA, the UK, and Spain. For a detailed study of this subject see Meerschawm (1989).

11
Changes in the Banking Industry and the Challenge of 1992

1. INTRODUCTION

In previous chapters we offered an analysis of the evolution of the banking industry in the EC countries from a three-part perspective: the evolution of the global financial environment, the competitive and profitability positions of banks, and, finally, strategic options for European banks in view of the regulatory changes which the unified financial market will bring. We have tried to set forth a diagnosis of the situation and of the relevant data on the financial environment to examine possible strategies for action.

In this final chapter we intend to pinpoint two aspects of this process of change. The first makes reference to the recommendations to the governments in preparation for the process of banking universalization and the concentration of the industry. This point is crucial, since financial authorities' definitions of legal framework and operating norms condition to a great extent banking activity. The second aspect relates to the transformation of the banking industry and the subsequent urgency of respective corporate strategies.

2. THE DEREGULATORY PROCESS AND THE PROMOTION OF EFFICIENCY IN THE BANKING INDUSTRY

During recent years a double phenomenon has been recognizable within the European banking industry. The first, on the part of governments, has been a broad deregulatory process leading to the elimination of interest-rate controls, the liberalization of certain activities previously restricted to particular banks, and the progressive elimination of required investment ratios. As indicated previously, the deregulatory process has brought about an important increase in price competition, breaking down existing structures for price co-ordination in some countries.

The second, more recent phenomenon stems from the expansion strategies followed by some banks which have led to the acquisition of other banks. All of this leads to a gradual but inevitable concentration of the sector. The next question is whether this will result in increased competition and efficiency within the banking industry.

Financial authorities responsible for the supervision of the banking system have always found themselves with conflicting duties. On the one hand, they have had to fulfil their mission of maintaining solvency in the system and protecting savings. On the other, they have simultaneously attempted to increase the degree of competition, though perhaps not always in a manner consistent with achieving the first objective. The dilemma between regulation and efficiency is classic in public-sector economic intervention.[1]

On top of this, the European Community has within it an additional dichotomy. At one end are the systems designed and approved by national authorities, which have been habitually influenced by the historical trajectories followed by their respective financial systems. At the other end are the norms emanating from the European Commission intent on harmonizing each EC member country's numerous regulations into a minimum of legislative measures or, at least, those relative to the crucial aspects of the banking system.

The principal criterion for intervention by the financial authorities has been that of encouraging competition, to the point of creating the impression that governments have forgotten the other two important objectives in their function as regulators: the stability of the system and the protection of the investor.

The principal instruments used to increase competition have taken various forms: liberalization of interest rates and commissions (as well as an expressed prohibition of price agreements between banks); an elimination of barriers preventing certain financial firms from offering specific services; as a result of the above, the abandonment of specialization criteria to organize the banking industry in favour of the universalization of activities; the opening of the industry to greater competition from abroad; the progressive disappearance of obligatory investment ratios and other regulations which limited free placement of funds in certain assets; the promotion of the innovative process and the modernization of the markets, often as a result of the increase in the public debt; and, finally, an increase in the information requirements on financial entities so that investors can better understand where and how to invest their savings.

This process of deregulation has greatly encouraged another process which has simultaneously been occurring in financial markets: internationalization. In effect, the globalization of this industry has increased the attraction of certain financial centres—those more liberal in their regulations and less demanding from the fiscal point of view. The fear of

---

[1] See Litan (1987) and OECD (1989).

certain governments of losing competitive ground due to an excess of regulation has pushed them to make important reforms in their regulatory codes with the objective of increasing the attractiveness of their financial markets. It is evident that in spite of this process there are still many limitations to complete freedom of operations, some due to historical incidents, others to peremptory necessities such as the articulation of monetary policies or the financing of the public deficit.

Of all these changes, the liberalization of prices and commissions permitting true price competition is the most significant. This has given way to a break with the traditional ties between banks and clients (based on personal relationships), and a move toward relationships based on the binomial price–quality criteria for evaluation of different financial services. In principal, these policies, which permit better resource allocation and increased efficiency of the financial system and the overall economy, have been generally positive. It is true, though, that in some cases price liberalization could give way to price increases in markets with significant entry barriers. This is the risk which exists, particularly in the sector of investment banking, but it can also be partly circumvented through complementary measures encouraging competition.

An example is the abolition of the specialization principle in favour of the universalization of banking activities. This is the second most significant characteristic of the actual process of deregulation. With this has come dedication to the idea that the opening of barriers and markets, previously closed to certain institutions, can be a decisive contribution to the promotion of competition. These measures, in general, have not been coercive since they have permitted those banks wishing to remain specialized to continue as such. However, they have opened the door for new business opportunities, which ultimately signifies a guarantee of greater efficiency owing to the disappearance of legal entry barriers.

Within the context of the return to the principle of universalization, there is one factor that will be critical in the future. Doubt still exists as to whether stock-market operations specific to investment banks should be completely separated. The reason for this doubt is twofold. First, there is the motive of protection of the investors: to the extent that banks become involved in more volatile activities, risk increases. Secondly, there is a potential conflict of interest between the department of shares and the department of loans or deposits (within one bank), which could instil in the investor a lack of confidence in the institution.

Different countries have had to admit, with certain limitations, both functions within the same bank, due to considerations of efficiency.[2]

---

[2] Litan (1987) offers a good discussion of the existing possibilities for dealing with this question. The recent announcement of an important reform in the American banking industry goes along this line.

However, it is also true that many banks have opted for the creation of affiliated entities, with independent social capital and specialized personnel, to carry out investment banking operations. There are two reasons they have done this. First, the naturally radical difference of investment banks. A very consolidated and specialized organization such as a commercial bank cannot easily absorb the requirements of an investment bank. The second reason is that there are requirements for specialists with different qualifications from those of commercial bank employees to carry out such activities.

As a result of the relaxation of regulation, it is expected that these activities will become more competitive. At the same time, however, the regulatory bodies have tried to compensate for this deregulation with increased information requirements for the banks and consequently their supervisory activities.

All of this helps to instil greater confidence in the market mechanisms for regulation of different financial markets. This movement is not unusual, since to a greater or lesser extent it has occurred in all industries throughout the 1980s.[3] Does this mean the disappearance of the government as guarantor of the competition and stability of the financial system? Probably not. The decrease in number of regulatory measures is different from leaving the financial system to the unrestricted free play of the market. The stability of the system and protection of savings still requires some degree of intervention, however little it may be. On the other hand, maintenance of a certain minimum level of competition is a necessary condition for the system to function efficiently, provided this minimum level corresponds to the public-sector directive.

This growing confidence in the markets does have two important exceptions. The first refers to the co-operative agreements between financial authorities and monetary authorities on questions such as the organization and management of the money or capital markets; the restructuring of the payments system; the deposit system and liquidation of operations in the stock markets. Along with these are the co-operative agreements between banks to share automatic tellers, the services of a specific credit card, or the distribution of certain financial products. To the extent that these agreements are not collusive, they are clearly beneficial for improving the efficiency of the financial system. A reasonable combination of free market and co-operative agreement can be positive for the overall system.

The second exception to unlimited confidence in the markets comes from the system adopted by the Second Directive to recognize banks in countries different from their origins. This deals with the principle of reciprocity, to which we referred previously. In an area as important to

[3] See OECD (1987).

the promotion of competition as the relative access of foreign entities to a national financial market, the decision is left to the authorities of the country of the bank's origin. This point, which in practice would require 12 distinct laws, constitutes an important limitation to the principle of stimulating competition and a certain restriction on the intrinsic market mechanisms.

Another important factor in this process of deregulation has been the internationalization of the industry—also studied in Chapter 1—and its opening of barriers to foreign entities, thus increasing competition. In countries such as Spain or France, the entry of foreign banking has been an important determinant to stir competition in the sector.

In relation to this phenomenon, a new goal has emerged: it aims at maintaining or increasing the competitiveness of a country's financial system relative to other countries' systems. This is important not only for countries with large financial centres, but also for financially less important countries, such as France (with the Paris Stock Market), Italy (with the Milan Stock Market), and Spain (with the reform of its stock market), who have proceeded to develop measures aimed at increasing the attractiveness of their financial systems. These measures are extremely varied, but they start from the common deregulatory measures previously indicated, continue until the development of an efficient technological infrastructure (payment systems, deposit systems, general telecommunications networks, etc.), and pass through an elimination of legal entry barriers or a fiscal reform.

Within the EC, the liberalization of capital movements which has been established in the majority of countries since July 1990, will unquestionably contribute to a strengthening of the trend toward internationalization, although, as we have seen in Chapter 1, it will not be exempt from certain risks, in particular, relative price stability for currencies within the free market. However, it is evident that without such freedom of capital movements, true internationalization and opening of the different national financial markets to foreign competition would be impossible.

In summary, and from a historical perspective, this process of deregulation is a substantive contribution to improving the efficiency of the financial system. In addition, to guarantee a certain degree of stability and investor protection, monetary authorities have opted not to increase regulation that distorts banking activities, but to increase information requirements and intensify the specific activities of supervision in an effort to make some of the norms clearer and simpler. In some cases, proper functioning of the markets through the strengthening of financial entities' competitiveness has been considered the best possible solution, not only for the efficiency of the banking system, but also for its security and stability.

3. CONCENTRATION AND COMPETITION IN THE BANKING INDUSTRY

In the previous pages we have seen some criteria for banking reform by the financial authorities. Accordingly, entry barriers impeding access to certain activities have been eliminated making way for the appearance of universal banks which, when of considerable size, are becoming comparable to financial supermarkets.

In parallel with this phenomenon, a wave of mergers and acquisitions has prevailed throughout Europe: many motivated by the intention of creating an advantageous position in the unified financial market of 1992, and all with a view to benefiting from greater scale, an aspect which we analysed in the previous chapter. However, this greater concentration generates a new problem: to what point does free competition not constitute a threat? For years, the principal objective of policies promoting competition in many countries has been orientated toward avoidance of greater concentration, assuming that this would result in a lessening of competition.

Recently, these policies have changed. The reasoning lies in the perception that a greater degree of concentration does not necessarily lead to greater inefficiency. In any event, there are two considerations to be taken into account here: promoting the construction of competitive corporate groups in an increasingly global economy, and the growing lack of confidence in regulatory policies. Historically, it was believed that a government, using its power for the common good, should invest in certain economic industries to guarantee their efficiency or to direct distribution. The experience of so many years, however, shows that governments do not generally possess any better information than companies for intervention in the market-place and, therefore, their effects on efficiency and means of distribution are questionable.[4]

All of this has led to general acknowledgement that the central objective of the financial authorities should be the creation of a policy intended to promote competition rather than limit concentration.[5]

This phenomenon of competition is a decisive factor to improve the banking industry's competitive capacity in any country. According to Porter (1990), the factors that play a critical role in the creation of competitive advantage in a global industry are the following: the structure and intensity of rivalry in that sector, the conditions of demand,

---

[4] For a critical discussion of the efficiency of the industrial policy see OECD (1987) or Canals (1989).

[5] The concept of 'contestable markets', which has appeared in the literature of industrial organization in the 1980s, is an attempt to explain the conditions in which a high level of competition is given, although the number of competing firms is not very big. For a related study see Baumol, Panzar, and Willig (1982).

production factors (basic or created), and related industries. For the service sector, the two primary factors (competition and demand) are particularly decisive. Public policy should be orientated towards the promotion of savings and competition.

Still, potential market power is not the only consideration that governments must take into account in analysing the process of banking concentration. There are other issues of significance such as greater efficiency derived from economies of scale; more efficient management; increased financial innovation; and the possibility that national banks may be absorbed by foreign banks. Clearly these points are all relevant, and any of them can justify a process of concentration as long as the final result does not lead to a market position of absolute dominance. In this respect, the role of the public sector should be understood as one of an agent committed to the maintenance of competition through the abolition of entry barriers, ultimately leading to a more efficient allocation of resources.

4. STRATEGIC ALTERNATIVES FOR COMMERCIAL BANKS

We have stressed the importance of three elements which are altering the structure of the European banking industry: the general process of deregulation, which is responsible for the elimination of entry barriers and thus the appearance of new competitors, the phenomenon of financial disintermediation, and the globalization of the industry. These trends have promoted a major modification of the traditional banking activity through decreased emphasis on the basic banking functions (deposits and loans) and increased emphasis on the offer of a wide range of financial services.

It is interesting to note here the anticipated increase in competition in the EC from the United States and Japan. For a number of American banks (Citicorp, Morgan), the European market offers opportunities, and while no one predicts an invasion of American banks similar to that of the 1960s, it would be naïve to think that these banks will remain inactive. The Japanese as well, with their large financial resources, will want to establish a presence before the threat of a closed European market. Therefore, it is to be expected that they too will substantially increase their operations in Europe.

This has tremendous implications for banking activities and structure. Banks have already lost their dominant position in the credit market and risk losing what they still have in the payment system. For this reason, the intermediation margin will account for less and less on banks' income statements, whereas revenues from sales of financial services will account for more and more. Parallel to this, standard credit and deposit transactions will also decrease in importance as a result of financial disinter-mediation and of the break with traditional client relationships.

Nevertheless, the offer of new financial services is not easy for banks. First, there is the question of competition: numerous non-banking firms (insurance companies, brokerages, etc.) have been providing similar services for some years now and have attained a notable degree of specialization and experience. The demand for financial services is growing and will continue to do so as living standards increase, which means the market will grow. But to take advantage of this, banks will have to change their structure and strategy in order effectively to compete in selling financial products. The first condition for success in this task is the professional retraining of employees. Of course this is not a simple process, but it will prove to be a very necessary one for banks wishing to secure their future competitiveness.

A second difficulty arises from the fact that to sell financial services with high value, it is necessary to attract customers to the point of sale. And in this sense, machines are not good salesmen. To sell, people must instil trust. In recent years banks have gone through an automation revolution which has had the effect of distancing customers from branch offices. Now banks are faced with reversing the process, though not indiscriminately; it will not be necessary to attract all current-account holders, but rather only those potential purchasers of specialized products.

As we have argued previously, it is vital that banks' strategies be dictated according to customer-service criteria. This will ensure a stable position in the domestic market and also act as a means for penetration in foreign markets.

The option of selling financial products seems to provide an effective response to the phenomena of deregulation and financial disinter-mediation. In this sense, there is not much difference between the future for American and European banks. Both must observe, however, a certain degree of prudence in entering these potentially attractive, but inherently riskier, business areas (as was clearly demonstrated by the October 1987 stock-market crash). This is very much the case with such activities as the underwriting of shares or bonds, much more familiar to the investment banks specialized in these operations.

What then distinguishes the European banking industry? Without doubt, it is the creation of a unified financial market, characterized by three measures: the single banking licence, the free offer of financial services, and the free circulation of capital. These factors make the impact of the single European market of 1992 much stronger in the banking industry than in other sectors.

In Chapter 10 we evaluated the process of change currently under way in Europe, and pointed out that while the desire itself for change is strong, the actual implementation of that change is slower. This is certainly true with respect to the EC bureaucracy, which has been slow to take decisions on a number of issues. Even with the approval of the Second Directive, change will not be instantaneous. The reasons for this are varied. The first and most important is that the supposed single financial market will not be a uniform market. Differences between the most advanced countries and the least advanced countries within the Community are substantial as is evidenced by their many different profitability levels, consumer profiles, savings characteristics, payment customs, intensities of demand, and, certainly, socio-cultural norms. This diversity is significant and although with time it will tend to lessen, it will not disappear completely. This will make changes in the European banking industry much slower than is aimed at by the simple edicts of the Second Directive.

At the same time, in each of the EC countries there is a group of banks with a strong national presence and considerable market power. While their situation is not one of monopoly (and while there are many opportunities for segmentation and differentiation, particularly in certain countries), competition will still increase less dramatically than would be expected with the creation of a truly unified market. As a result, the potential benefits of the unified financial-services market as reviewed in the Cecchini Report will be realized more slowly and possibly to a lesser extent than is indicated there.

These factors do not invalidate the current processes of globalization in the banking industry. They simply mean that change will not occur as quickly as it might and that distinct national markets will maintain their identities. Nevertheless, this is not to say that banks should not be prepared. The industry offers recent, very clear examples of the direction of change such as the acquisition of a Dutch Chase Manhattan affiliate by Crédit Lyonnais; the purchase of the Italian Bank of America affiliate by Deutsche Bank; the alliance between Commerzbank, Crédit Lyonnais, and Banco Hispano Americano; and, finally, the merger between the Banco de Bilbao and the Banco de Vizcaya in Spain.

What does all this mean? The industry will become increasingly more global in nature, with a higher presence of American and Japanese competitors. As a result, for banks wishing to compete in this environment, scale is a significant element: without considerable resources it will be impossible to enter certain areas or acquire the technology necessary to conduct specific banking activities efficiently. Therefore, the pursuit of scale becomes a logical strategy.

A greater international presence can be achieved not only through mergers and acquisitions, but also through alliances between banks. In fact, some banks have already started down this road and so the problems associated with alliances are known. On the one hand, it is necessary for one party to excercise leadership, though it is not always easy for the other party to accept it, causing a conflict of interests for each of the participating companies. This conflict leads to unclear decision-making processes and extremely complex management systems. This helps to explain why certain alliances in the past, such as those between Dunlop and Pirelli or between Citroën and Fiat, ended badly.

Some executives of the industry say that alliances should not be discarded as a means to achieve a greater presence in the international financial markets. This is especially true in cases where links can be established between banks and non-bank firms (for example, department stores or travel agencies) or manufacturing companies with wide distribution networks (for example, automotive companies with their dealership networks).[6] With more synergy between partners it may be possible to resolve some of the conflicts with which alliances historically have been plagued. Alliances geared toward distribution agreements could also prove successful to the extent that they help both companies to strengthen their positions in the national market. Alliances of these kinds serve to fill in the gaps within individual organizations thus corresponding to a more product-orientated and thereby customer-orientated strategy.

On the downside, mergers, alliances, and internationalization all present new management problems, as indicated in Chapter 10. Attention to a combination of factors such as operational efficiency, customer service, adaptation to local markets, and international transfer of resources or products will become necessary for the improvement of bank management within an international context. Only in this way will banks be able to avoid local market fragmentation and reach a point where management is a source of sustainable competitive advantage.

Mergers, acquisitions, and alliances thus become viable tools for expansion, especially in the European financial markets. However, as we saw before, they are not the only possibilities for successful strategy in this new environment. Scale is important in many cases, but it is not a necessary prerequisite for all, nor is it sufficient in itself to guarantee success (not even for those companies determined on a global strategy). Of equal importance is the efficient maintenance and utilization of a national branch network which can be a considerable source of competitive advantage.

There are other alternatives as well, such as diversification of activities and differentiation, within a clear and well-designed segmentation strategy. The financial services industry has a wide variety of clients, operates in many different markets, and projects a diverse geographic

---

[6] See the lecture of H. Norrington (1988), 'A UK Bank Planning for the Internal Market', London, in which he defends a strategy of globalization based on acquisition, although without discarding the possibility of establishing strategic alliances with other non-bank enterprises or manufacturing or distribution companies.

focus. Therefore, what is particularly relevant in segmentation is a precise, well-designed product–client combination which also takes into account the process of financial disintermediation and, consequently, the need to offer services which satisfy the new needs of consumers. Strategies based on new product introduction, in both domestic and international markets, will generate the successes of the future.

In terms of size, it should be clear that greater scale is not incompatible with a proper segmentation strategy. All banks should have some type of segmentation plan for those areas of activity in which they decide to establish a presence. Future strategies will be based on identifying profitable market segments and designing appropriate products to offer customers the services they demand and shareholders the return they expect.

Segmentation, therefore, is a necessary condition for the success of large banks as well, since the idea of the bank that can do it all and do it all well seems more relevant to theoretical speculation than reality. One reason for this is that companies do not want to deal with banks that do everything, but, principally, with banks that know how to offer efficiently the services they require. In fact, large companies tend to have relationships with four or five banks and, before making specific decisions, generally test the market to see who can provide the best overall conditions. Thus, segmentation and specialization both are necessary parts of a successful strategy.

The new competitive requirements present a novel and important challenge to banks: the improvement of the qualifications of their personnel and the efficient use of new information technologies. These two aspects of bank management, together with an improvement of the management process, are decisive for gaining competitive advantage.

Also of considerable importance is the control that the stock markets have come to excercise over banks. Confronted with increasing capital requirements, banks have had to follow the verdicts of the market in order to effect the necessary increases in capital. These, in turn, have required them to demonstrate acceptable levels of profitability and maintain sufficiently high prices. All this constitutes a powerful incentive for the banks to cut costs and increase revenues, abandoning unprofitable activities and intensifying their presence in those segments where success is more certain.

Given these pressures, if the European banks share a common characteristic in the next decade it is that they will all have clear segmentation strategies, some in a global context with larger clients, others in a local context and with smaller clients. Achievement of a larger scale will lack significance unless it is undertaken in conjunction with an analysis of potential new sevices to be offered or the improvement of existing ones.

What can be learned from all of this is that banks must try to be the best not in all areas of service (an unrealistic objective), but rather in certain areas or with certain types of customers. In this sense, the banking industry resembles an athletic competition, in which one does not try to win in every event, but to win in one of the events, and at the very least, to compete with dignity.

References

Abad, F. J. (1987), 'La banca extranjera en España', *Papeles de Economía Española*, 32: 333–4.

Aliber, R. Z. (1984), 'International Banking: A Survey', *Journal of Money, Credit and Banking*, 16(4): 661–95.

Aspinwall, R. C., and Eisenbeis, R. A. (1985), *Handbook for Banking Strategy* (New York: John Wiley).

Bain, J. S. (1956), *Barriers to New Competition* (Cambridge, Mass.: Harvard University Press).

Ballarín, E. (1985), *Estrategias competitivas para la banca* (Barcelona: Ed. Ariel).

—— (1988a), 'Distribución de servicios financieros en España: implicaciones competitivas', *Papeles de Economía Española. Suplementos sobre el sistema financiero*, 23: 145–56.

—— (1988b), 'El proceso de concentración de la banca española: Teoría y práctica', *Boletín*, Círculo de Empresarios, 41–54.

—— Gual, J., and Ricart, J.E. (1988), 'Rentabilidad y competitividad en el sector bancario español: Un estudio sobre la distribución de servicios financieros en España' (IESE, Barcelona).

Baltensperger, E., and Dermine, J. (1987), 'Banking Deregulation in Europe', *Economic Policy*, 44: 64–109.

Banca d'Italia (1984), *Italian Credit Structures: Efficiency, Competition and Controls* (London: Euromoney Publications).

—— (1986a), *Annual Report* (Rome).

—— (1986b), 'L'attuazione della Directiva comunitaria 77/780 in materia di constituzione de enti creditizi', *Bolletino economico*: 1–7.

—— (1988), *White Paper on the Payment System in Italy* (Rome).

Banco de España (1988), 'Análisis comparativo de la rentabilidad del sistema bancario español', *Boletín económico* (March), 11–20.

Bank for International Settlements (1986), *Recent Innovations in International Banking* (Basle).

Bank of England (1986a), 'City Regulation after Big Bang', *Quarterly Bulletin* (Mar.), 71–3.

—— (1986b), 'The UK Approach to Financial Regulation', *Quarterly Bulletin* (Mar.), 48–50.

—— (1986c), 'International Banking in London', *Quarterly Bulletin* (Sept.), 367–8.

—— (1986d), 'New Issue Costs and Methods in the UK Equity Market', *Quarterly Bulletin* (Dec.), 532–42.

—— (1987a), 'Building Societies: A Changing Role', *Quarterly Bulletin* (Aug.), 400–3.

—— (1987b), 'Japanese Banks in London', *Quarterly Bulletin* (Nov.), 518–25.

Banque de France, 'Statistiques monétaires définitives', various years (Paris).

—— (1988), 'L'integration financiére interne et externe', *Cahiers économiques et monétaires*, 31: 119–227.

—— (1989), 'La Titrisation des crédits bancaires', *Bulletin trimestriel*, 45–60.

Bartlett, C. A., and Ghoshal, S. (1989), *Managing Across Borders* (Boston: Harvard Business School Press).

Baumol, W., Panzar, J., and Willig R. (1982), *Contestable Markets and the Theory of Industry Structure* (New York: Brace Jovanovich).

Bengoechea, J., and Lerena, L. A. (1988), 'La dimensión como condicionante de la estrategia bancaria', *Papeles de Economía Española*, 36: 77–89.

Benston, G. J., Hanweck, G. A., and Humprey, D. B. (1982), 'Scale Economies in Banking: A Restructure and Reassessment', *Journal of Money, Credit and Banking*, 14: 435–56.

—— Berger, A. N., Hanweck, G. A., and Humprey, D. (1983) 'Economies of Scale and Scope', Conference on Bank Structure and Competition, Proceedings, Federal Reserve Bank of Chicago, Chicago.

Berger, A. N., Hanweck, G. A., and Humprey, D. B. (1987), 'Competitive Viability in Banking: Scale, Scope and Product Mix Economies', Research Paper, Federal Reserve Bank, Washington, DC.

Boissieu, C. de (1990), 'The French Banking Sector in the Light of European Financial Integration', in J. Dermine, *European Banking in the 1990s* (Oxford: Blackwell).

Boleat, M. (1982), *The Building Society Industry* (London: Allen & Unwin).

Bourke, P. (1989), 'Concentration and Other Determinants of Bank Profitability in Europe, North America and Australia', *Journal of Banking and Finance*, 13: 65–79.

Boyd, J. H., Hanweek, G. A., and Pithyachariyakul, P. (1980), 'Bank Holding Company Diversification', Conference on Bank Structure and Competition, Proceedings, Federal Reserve Bank of Chicago, Chicago.

Brealey, R., and Myers, S. (1988), *Principles of Corporate Finance* (New York: McGraw-Hill).

Caminal, R., Gual, J., and Vives, X. (1989), *Competition in Spanish Banking*, Paper no. 89–03, FEDEA, Madrid.

Canals, J. (1987), 'La evolución de los tipos de interés en la década de los 80', Working Paper 13, Caixa de Pensions, Barcelona.

—— (1988*a*), 'La coordinación de políticas económicas: Una solución a los problemas de la economía internacional?', Working Paper 16, Caixa de Pensions, Barcelona.

—— (1988*b*), 'Déficit público, deuda pública y tipos de interés', *Información comercial española*, (Mar.), 19–36.

—— (1989), *El entorno económico de los negocios internacionales* (Bilbao: Ed. Deusto).

—— (1990), Estratigias du sector bancario en Europa (Barcelona: Editorial Ariel).

Campbell, T. S. (1982), *Financial Institutions, Markets and Economic Activity* (New York: McGraw-Hill).

Caves, R. E. (1982), *Multinational Enterprise and Economic Analysis* (Cambridge, Mass., Harvard University Press).

Ceccatelli, E. (1989), 'Risks and Opportunities for the Major Italian Banks in a Climate of Closely Integrated Financial Markets', mimeo, Rome.

Cecchini, P. (1988), *Europa 1992: Una apuesta de futuro* (Madrid: Alianza Editorial).

Chrisman, J. J., Hofer, C. W., and Boulton, W. R. (1988), 'Toward a System for Classifying Business Strategies', *Academy of Management Journal*, 13(3): 413–28.

Clark, J. A. (1984), 'Estimation of Economies of Scale in Banking using a Generalized Functional Form', *Journal of Money, Credit and Banking* (Feb.), 53–68.

—— (1988), 'Economies of Scale and Scope at Depository Financial Institutions: A Review of the Literature', *Economic Review*, Federal Reserve Bank of Kansas City (Sept.–Oct.), 16–33.

Clarke, M. (1986), *Regulating the City* (Philadelphia: Open University Press).

Clarke, W. M. (1986), *How the City of London Works* (London: Waterlow).

Clarotti, P. (1988), 'La armonización de las exigencias requeridas a las entidades de crédito en la Comunidad Europea', *Papeles de Economía Española. Suplementos sobre el sistema financiero*, 24: 7–18.

Cooper, K., and Fraser, D. R. (1986), *Banking Deregulation and the New Competition in Financial Services* (Cambridge: Ballinger).

Coupaye, P. (1984), 'Les Banques en France', *Notes et études documentaires*, no. 4759 (Paris).

Crane, D. B., and Hayes, S. L., III (1982), 'The New Competition in World Banking', *Harvard Business Review*, 60(4): 88–94.

—— —— (1983), 'The Evolution of International Banking, Competition and Its Implications for Regulation', *Journal of Banking Research*, 14(1): 39–53.

Cuervo, A., Rodríguez, L., and Parejo, J. A. (1987), *Manual del sistema financiero* (Barcelona: Ed. Ariel).

Daems, H. (1978), *Holding Company and Corporate Control* (Boston, Mass.: Martinus Nijhoff).

Danker, D. J., and McLaughlin, M. M. (1985), 'Profitability of Insured Commercial Banks in 1984', *Federal Reserve Bulletin* (Nov.), 845–9.

Davis, S. I. (1985), *Excellence in Banking* (New York: St Martin's Press).

Delgado, F. L. (1989), 'Economías de escala en el sector bancario español', mimeo, Madrid.

Dermine, J. (1986), 'Measuring the Market Value of a Bank: A Primer', Working Paper 86/35, INSEAD, Fontainebleau.

Deutsche Bundesbank (1986), 'The Profitability of German Banks in 1985', *Monthly Report* (Aug.), 15–32.

—— (1988), 'New Off-balance Sheet Financial Instruments and Their Implications for Banks in the Federal Republic of Germany', *Monthly Report* (Apr.), 23–7.

—— (1989), 'Longer-term Trends in the Banking Sector and Market Position of the Individual Categories of Banks', *Monthly Report* (Apr.), 13–22.

Dini, L. (1988), 'The Italian Financial System in the Perspective of 1992', *Banca Nazionale del Lavoro Quarterly Review*, 4: 441–9.

Dixit, A. (1979), 'Quality and Quantity Competition', *Review of Economic Studies*, 46: 587–99.

Donnelly, J. H., Berry, L. L., and Thompson, T. (1985), *Marketing Financial Services* (Homewood, Ill.: Dow-Jones Irwin).

—— and Skinner, S. (1989), *The New Banker* (Homewood, Ill.: Dow-Jones Irwin).

Dornbusch, R. (1987), 'Exchange Rate Economics: 1986', *Economic Journal*, 97(1): 1–18.

Eccles, R. B., and Crane, D. B. (1988), *Doing Deals: Investment Banks at Work* (Boston, Mass.: Harvard Business School Press).

Fama, E. F. (1976), *Foundations of Finance: Portfolio Decisions and Securities Prices* (New York: Basic Books).

Fanjul, O., and Maravall, F. (1985), *La eficiencia del sistema bancario español* (Madrid: Alianza Universidad).

Faugere, J. P., and Voisin, C. (1989), *Le Systéme financier français* (Paris: Éditions Nathan).

Fernández, P., and Pregel, G. (1989), *Bonos convertibles en España* (Barcelona: Ed. IESE).

Francke, H. H. (1984), *Banking and Finance in West Germany* (Bromley, Kent: Croom Helm).

Friberg, E. G. (1988), 'The Challenge of 1992', *McKinsey Quarterly* (Autumn), 27–40.

García Alonso, J. M. (1984), 'La evolución reciente del sistema financiero español', *Papeles de Economía Española*, 18: 60–85.

Ghemawat, P., and Nabeluff, B. (1985), 'Exit', *Rand Journal of Economics* 16(2): 184–94.

Giavazzi, F., and Giovannini, A. (1986), 'The EMS and the Dollar', *Economic Policy*, 2: 456–85.

Gilbert, R. A. (1984), 'Bank Market Structure and Competition', *Journal of Money, Credit and Banking*, 16(4): 617–45.

—— (1986), 'Requiem for Regulation Q: What It Did and Why It Passed Away', *Federal Reserve Bank of St Louis Review* (Feb.), 22–37.

—— (1989), 'The Role of Potential Competition in Industrial Organization', *Journal of Economic Perspectives*, 3(3): 107–27.

Gilbody, J. (1988), *The UK Monetary and Financial System* (London: Routledge).

Gilligan, T., Smirlock, M., and Marshall, W. (1984), 'Scale and Scope Economies in the Multi-Product Banking Firm', *Journal of Monetary Economics*, 13: 393–405.

Goacher, D. J., and Curwen, P. J. (1987), *British Non-Bank Financial Intermediaries* (London: Allen & Unwin).

Gordon, M. J., and Shapiro, E. (1956), 'Capital Investment Analysis: The Required Rate of Profit', *Management Science*, 3: 102–10.

Gual, J., and Ricart, J. E. (1988), 'Poder de mercado en la captación de depósitos a plazo del sector bancario español', Working Paper 151, IESE, Barcelona.

Gutiérrez, F. (1988), 'La innovación financiera en España', *Papeles de Economía Española. Suplementos sobre el sistema financiero*, 21: 103–18.

—— and Campoy, J. A. (1988), 'Eficiencia y competencia en el sistema bancario español', *Boletín económico*, (Dec.), 51–63.

—— and Chuliá, C. (1988), 'La innovación financiera y las operaciones del sistema bancario', *Boletín económico*, (Nov.), 21–33.

Harrington, R. L. (1987), *Asset and Liability Management by Banks* (Paris: OECD).

—— (1988), 'Innovación financiera y banca internacional', *Papeles de Economía Española. Suplementos sobre el sistema financiero*, 23: 10–20.

Hax, A., and Majluf, N. (1984), *Strategic Management: An Integrative Perspective* (Eglewood Cliffs, NJ: Prenticé-Hall).

Hayes, S. L., Spence, A. M., and Marks, D. V. P. (1983), *Competition in the Investment Banking Industry* (Cambridge, Mass.: Harvard University Press).

—— and Hubbard, P. M. (1989), *Investment Banking. A Tale of Three Cities* (Boston: Harvard Business School Press).

Heggestad, A. (1975), 'Riskiness of Investments in Nonbank Activities by Bank Holding Companies', *Journal of Economics and Business*, 27(1): 219–23.

Hernsi, F. (1985), *Transformazioni della banca pubblica* (Bologna: Il Mulino).

Hill, C. W. L. (1988), 'Differentiation versus Low Cost or Differentiation and Low Cost: A Contingency Framework', *Academy of Management Review*, 13(3): 401–12.

IMF (1984), 'Exchange Rate Volatility and World Trade', Occasional Paper 28 (Washington, DC: IMF).

Institute of Bankers (1986), *Bank Strategies for the 1990s* (London).

Kennedy, S. (1973), *The Banking Crisis of 1933* (Lexington: University of Kentucky Press).

Kilbride, B. J., McDonald, B., and Miller, R. E. (1986), 'A Reexamination of Economies of Scale in Banking Using a Generalized Functional Form', *Journal of Money, Credit and Banking* (Nov.), 19–26.

Klemperer, P. (1987a), 'Markets with Consumer Switching Costs', *Quarterly Journal of Economics*, 102: 375–94.

—— (1987b), 'The Competitiveness of Markets with Switching Costs', *Rand Journal of Economics*, 18(1): 138–50.

Kogut, B. (1985), 'Designing Global Strategies: Comparative and Competitive Value-Added Chains', *Sloan Management Review* (Summer), 15–28.

Kohlagen, S. (1983), 'Overlapping National Investment Portfolios', in R. Hawkings, R. Levich, and C. Wihlborg (eds.), *Research in International Business and Finance* (Greenwich, New York: JAI Press).

Korsvick, W. J., and Juris, H. A. (1989), *The New Frontiers in Bank Strategies* (Homewood, Ill.: Dow-Jones Irwin).

Kreps, D., and Wilson, R. (1982) 'Reputation and imperfect information', *Journal of Economic Theory*, 27: 253–79.

Krugman, P., and Obtsteld, M. (1988), *International Economics* (New York: Scott, Foreman).

Lagares, M. J. (1987), 'Los resultados de las cajas de ahorros', *Papeles de Economía Española*, 32: 292–313.

—— (1988), 'Cajas de ahorros: Los retos del futuro', *Papeles de Economía Española*, 36: 157–173.

Lawrence, C. (1989), 'Banking Costs, Generalized Functional Forms and Estimation of Economies of Scale and Scope', *Journal of Money, Credit and Banking* (Aug.), 368–80.

Lewis, M. K. (1987), *Domestic and International Banking* (Oxford: Philip Allan Publishers).

Lindenberg, E. B., and Ross, S. A. (1981), 'Tobin's Q Ratio and Industrial Organization', *Journal of Business*, 54(1): 1–32.

Litan, R. E. (1987), *What Should Banks Do?* (Washington, DC: Brookings Institution).

Lladó, A., Gracia, A., Soler, M., and Isern, J. (1989), 'Los sistemas bancarios de los principales países de la CEE', Working Paper 18, Caixa de Pensions, Barcelona.

Madroñero, A. (1988), 'La regulación bancaria en la CEE', *Papeles de Economía Española. Suplementos sobre el sistema financiero*, 24: 19–22.

Markowitz, H. (1959), *Portfolio Selection: Efficient Diversification of Investments* (New York: John Wiley).

Meerschawm, D. (1991), *Breaking Local Boundaries in Finance* (Boston: Harvard Business School).

Meinster, D. R., and Johnson, R. D. (1979), 'Bank Holding Company Diversification and the Risk of Capital Impairment', *Bell Journal of Economics*, 10(3): 683–94.

Menéndez, M. A. (1988), 'Los problemas estadísticos derivados de la innovación financiera: Especial referencia a los mercados monetarios y de valores', *Boletín económico* (Nov.), 35–46.

Merussi, F. (1986), 'La tutela dei depositanti: dalla riserva obbligatoria al Fondo Interbancario di Garanzia', *Note Economiche*, 3–4, 233–8.

Milgrom, P., and Roberts, J. (1982), 'Predation, Reputation and Entry Deterrance', *Journal of Economic Theory*, 27: 280–312.

Ministère de l'Économie (1987), 'Privatisation', Paris.

Mookerjee, A. S. (1988), 'Global Electronic Wholesale Banking', unpublished doctoral thesis, Boston: Harvard University, Graduate School of Business Administration.

Murray, A. I. (1988), 'A Contingency View of Porter's Generic Strategies', *Academy of Management Review*, 13(3): 390–400.

Neven, D. J. (1990), 'Structural Adjustment in European Retail Banking: Some Views from Industrial Organization', in J. Dermine (ed.), *European Banking in the 1990s* (Oxford: Blackwell).

OECD (1986a), *Germany* (Paris: OECD).

—— (1986b), *France* (Paris: OECD).

—— (1987a), *United Kingdom* (Paris: OECD).

—— (1987b), *Italy* (Paris: OECD).

—— (1987c), *Structural Adjustment and Economic Performance* (Paris: OECD).

—— (1988), *Bank Profitability: Statistical Supplement* (Paris: OECD).

—— (1989), *Competition in Banking* (Paris: OECD).

Onado, M. (1980), *Il sistema finanziario italiano* (Bologna: Il Mulino).

Ontiveros, E. (1987), 'El proceso de innovación en los mercados financieros internacionales', *Papeles de Economía Española*, 32: 194–226.

Panzar, J. C., and Willig, R. D. (1981), 'Economies of Scope', *American Economic Review*, 71(2): 268–72.

Pastre, O. (1985), 'La Modernisation des banques françaises' (Paris: La Documentation française).

Pavel, C., and Rosenblum, J. (1985), 'Banks and Non-banks: The Horse Race Continues', *Federal Reserve Bank of Chicago Economic Perspectives* (May–June), 3–17.

Peat Marwick (1981), *Banking in Italy* (Milan).

—— (1982), *Banking in Germany* (Frankfurt).

—— (1985), *Banking in the United Kingdom* (London).

Pecchioli, R. M. (1983), *The Internationalization of Banking* (Paris: OECD).

Phillips, A. (1985), 'Changing Technology and Future Financial Activity', in Aspinwall and Eisenbeis (1985).

Phillips, L. W., Chang, D. R., and Buzzell, R. D. (1983), 'Product Quality, Cost Position and Business Performance: A Test of Some Key Hypotheses', *Journal of Marketing*, 47(2): 26–43.

Pöhl, M. (1982), *Konzentration in Deutschen Bankwesen* (Frankfurt: Fritz Knapp Verlag).

Porter, M. E. (1980), *Competitive Strategy* (New York: Free Press).

—— (1985), *Competitive Advantage* (New York: Free Press).

—— (ed.) (1986), *Competition in Global Industries* (Boston: Harvard Business School Press).

—— (1990), *The Competitive Advantage of Nations* (New York: Free Press).

Pregel, G., Suñol, R., and Nueno, P. (1989), *Nuevos instrumentos financieros al servicio de la empresa* (Bilbao: Ed. Deusto).

Price Waterhouse (1988*a*), *El coste de la no integración europea en mercados financieros* (Madrid).

—— (1988*b*), *Investment Banking: A Survey* (London).

Ramsler, M. (1982), 'Strategic Groups and Foreign Market Entry in Global Banking Competition', unpublished doctoral thesis, Boston: Harvard University, Graduate School of Business Administration.

Rappaport, A. (1986), *Creating Shareholder Value: The New Standard for Business Performance* (New York: Free Press).

Revell, J. R. S. (1973), *The British Financial System* (London: Macmillan Press).

—— (1980), *Costs and Margins in Banking: An Internatinal Survey* (Paris: OECD).

—— (1987), 'Mergers and the Role of Large Banks', Research Monograph in Banking and Finance, 2 (Bangor: Institute of European Finance).

Ritchie, N. (1986), *What Goes on in the City?* (Cambridge: Woodhead Faulkner).

Rose, H. (1986), 'Change in Financial Intermediation in the United Kingdom', *Oxford Review of Economic Policy*, 2(4).

Rowley, E. E. (1987), *The Financial System Today* (Manchester: Manchester University Press).

Rumelt, R. P. (1986), *Strategy, Structure and Economic Performance* (Boston: Harvard Business School Press).

Salomon Brothers (1989), 'European Banking Integration in 1992. The Competitive Challenges Facing US Multinational Banks' (New York).

Sánchez Asiaín, J. A. (1984), 'Algunas reflexiones sobre la banca del futuro', *Papeles de Economía Española*, 18: 175–94.

Scherer, F. M. (1980), *Industrial Market Structure and Economic Performance*, 2nd edn. (Boston: Houghton Mifflin).

Schneider, H., Hellwing, H. J., and Kingsman, D. J. (1982), *The German*

Banking System (Frankfurt: Fritz Knapp Verlag).

Shaw, E. R. (1981), *The London Money Market* (London: Heinemann).

Short, B. K. (1979), 'The Relation between Commercial Bank Profit Rates and Banking Concentration in Canada, Western Europe and Japan', *Journal of Banking and Finance*, 3.

Silber, W. (ed.) (1975), *Financial Innovation* (Lexington: Lexington Books).

Smirlock, M. (1985), 'Evidence on the (Non) Relationship Between Concentration and Profitability in Banking', *Journal of Money, Credit and Banking*, 17(1).

Termes, R. (1987), 'Los resultados de la banca privada', *Papeles de Economía Española*, 32: 274–91.

Thietart, R. A. (1988), 'Success Strategies for Businesses that Perform Poorly', *Interfaces*, 18(3): 32–45.

Toribio, J. J. (1984), 'La innovación financiera en España', *Papeles de economía española*, 18: 100–35.

Trujillo, J. A., Cuervo-Arango, C., and Vargas, F. (1988), *El sistema financiero español* (Barcelona: Ed. Ariel).

Tschoegel, A. E. (1987), 'International Retail Banking as a Strategy: An Assessment'. *Journal of International Business Studies*, 18(2): 67–88.

US Congress (1984), *Effects of Information Technology on Financial Services System* (Washington, DC: OTA).

Viñals, J. (1986), 'Hacia una mayor flexibilidad de los tipos de cambio en el sistema monetario internacional?', *Papeles de Economía Española*, 28: 2–22.

Vives, X. (1988), 'Concentración bancaria y competitividad', *Papeles de Economía Española*, 36: 62–76.

Voisin, C. (1987), 'Le Système bancaire français: Vingt ans de mutations', (Paris: La Documentation française).

Walter, I. (1988), *Global Competition in Financial Services* (Cambridge: Ballinger).

—— and Smith, R. (1989), *Investment Banking in Europe after 1992* (Oxford: Blackwell).

Wilcox, J. W. (1984), 'The P/B–ROE Valuation Model', *Financial Analysts Journal*, 1: 58–66.

Williamson, J. (1988), 'Voluntary Approaches to Debt Relief' (Washington, DC: Institute for International Economics).

Wilson, J. S. G. (1986), *Banking Policy and Structure: A Comparative Analysis* (Bromley, Kent: Croom Helm).

Wyplosz, C. (1988), 'Capital Flows Liberalisation and the EMS', *European Economy*, 36: 85–103.

Zecher, J. R. (1984), 'Financial Innovation in the 1980s', in Federal Reserve Bank of St Louis, *Financial Innovations* (Boston: Klywer-Nijhoff), 151–75.

Index

Madrid
August '00

Notas de prensa
Obra periodística 5
(1961-1984)

Notas de prensa

Obra periodística 5
(1961-1984)

GABRIEL GARCÍA MÁRQUEZ

MONDADORI

© 1991, 1999 Gabriel García Márquez
© 1991, 1999 de la presente edición para España:
MONDADORI (Grijalbo Mondadori, S.A.)
Aragó, 385. 08013 Barcelona
www.grijalbo.com
Derechos exclusivos únicamente para España. Prohibida su venta
en los demás países del área idiomática de lengua castellana.
Primera edición
ISBN: 84-397-0431-3
Depósito legal: M. 33.645-1999
Impreso y encuadernado en Artes Gráficas Huertas, S.A.,
Fuenlabrada (Madrid), sobre papel AMABULK, offset volumen ahuesado,
de Papelera de Amaroz, S.A.

ÍNDICE

UN HOMBRE HA MUERTO DE MUERTE NATURAL

Esta vez parece ser verdad: Ernest Hemingway ha muerto. La noticia ha conmovido, en lugares opuestos y apartados del mundo, a sus mozos de café, a sus guías de cazadores, a sus aprendices de torero, a sus choferes de taxi, a unos cuantos boxeadores venidos a menos y a algún pistolero retirado.

Mientras tanto, en el pueblo de Ketchum, Idaho, la muerte del buen vecino ha sido apenas un doloroso incidente local. El cadáver permaneció seis días en cámara ardiente, no para que se le rindieran honores militares, sino en espera de alguien que estaba cazando leones en África. El cuerpo no permanecerá expuesto a las aves de rapiña, junto a los restos de un leopardo congelado en la cumbre de una montaña, sino que reposará tranquilamente en uno de esos cementerios demasiado higiénicos de los Estados Unidos, rodeado de cadáveres amigos. Estas circunstancias, que tanto se parecen a la vida real, obligan a creer esta vez que Hemingway ha muerto de veras, en la tercera tentativa.

Hace cinco años, cuando su avión sufrió un accidente en el África, la muerte no podía ser verdad. Las comisiones de rescate lo encontraron alegre y medio borracho, en un claro de la selva, a poca distancia del lugar donde merodeaba una familia de elefantes. La propia obra de Hemingway, cuyos héroes no tenían derecho a morir antes de padecer durante cierto tiempo la amargura de la victoria, había descalificado de antemano aquella clase de muerte, más bien del cine que de la vida.

En cambio, ahora, el escritor de sesenta y dos años, que en la pasada primavera estuvo dos veces en el hospital tratándose una enfermedad de viejo, fue hallado muerto en su habitación con la cabeza destrozada por una bala de escopeta de matar tigres. En favor de la hipótesis de suicidio hay un argumento técnico: su experiencia en el manejo de las armas descarta la posibilidad de un accidente. En con-

Publicado originalmente el 9 de julio de 1961.

tra, hay un solo argumento literario: Hemingway no parecía pertenecer a la raza de los hombres que se suicidan. En sus cuentos y novelas, el suicidio era una cobardía, y sus personajes eran heroicos solamente en función de su temeridad y su valor físico. Pero de todos modos, el enigma de la muerte de Hemingway es puramente circunstancial, porque esta vez las cosas ocurrieron al derecho: el escritor murió como el más corriente de sus personajes, y principalmente para sus propios personajes.

En contraste con el dolor sincero de los boxeadores, se ha destacado en estos días la incertidumbre de los críticos literarios. La pregunta central es hasta qué punto Hemingway fue un grande escritor, y en qué grado merece un laurel que a él mismo le pareció una simple anécdota, una circunstancia episódica en la vida de un hombre.

En realidad, Hemingway sólo fue un testigo ávido, más que de la naturaleza humana, de la acción individual. Su héroe surgía en cualquier lugar del mundo, en cualquier situación y en cualquier nivel de la escala social en que fuera necesario luchar encarnizadamente no tanto para sobrevivir cuanto para alcanzar la victoria. Y luego, la victoria era apenas un estado superior del cansancio físico y de la incertidumbre moral.

Sin embargo, en el universo de Hemingway la victoria no estaba destinada al más fuerte, sino al más sabio, con una sabiduría aprendida de la experiencia. En ese sentido era un idealista. Pocas veces, en su extensa obra, surgió una circunstancia en que la fuerza bruta prevaleciera contra el conocimiento. El pez chico, si era más sabio, podía comerse al grande. El cazador no vencía al león porque estuviera armado de una escopeta, sino porque conocía minuciosamente los secretos de su oficio, y por lo menos en dos ocasiones el león conoció mejor los secretos del suyo. En *El viejo y el mar* −el relato que parece ser una síntesis de los defectos y virtudes del autor− un pescador solitario, agotado y perseguido por la mala suerte, logró vencer al pez más grande del mundo en una contienda que era más de inteligencia que de fortaleza.

El tiempo demostrará también que Hemingway, como escritor menor, se comerá a muchos escritores grandes, por su conocimiento de los motivos de los hombres y los secretos de su oficio. Alguna vez, en una entrevista de prensa, hizo la mejor definición de su obra al compararla con el iceberg de la gigantesca mole de hielo que flota en la superficie: es apenas un octavo del volumen total, y es inexpugnable, gracias a los siete octavos que la sustentan bajo el agua.

La trascendencia de Hemingway está sustentada precisamente en la oculta sabiduría que sostiene a flote una obra objetiva, de estructura directa y simple, y a veces escueta inclusive en su dramatismo. Hemingway sólo contó lo visto por sus propios ojos, lo gozado y padecido por su experiencia, que era al fin y al cabo lo único en que podía creer. Su vida fue un continuo y arriesgado aprendizaje de su oficio, en el que fue honesto hasta el límite de la exageración: habría que preguntarse cuántas veces estuvo en peligro la propia vida del escritor, para que fuera válido un simple gesto de su personaje.

En ese sentido, Hemingway no fue nada más, pero tampoco nada menos, de lo que quiso ser: un hombre que estuvo completamente vivo en cada acto de su vida. Su destino, en cierto modo, ha sido el de sus héroes, que sólo tuvieron una validez momentánea en cualquier lugar de la tierra, y que fueron eternos por la fidelidad de quienes los quisieron.

Ésa es, tal vez, la dimensión más exacta de Hemingway. Probablemente, éste no sea el final de alguien sino el principio de nadie en la historia de la literatura universal. Pero es el legado natural de un espléndido ejemplar humano, de un trabajador bueno y extrañamente honrado, que quizá se merezca algo más que un puesto en la gloria internacional.

DESVENTURAS DE UN ESCRITOR DE LIBROS

Escribir libros es un oficio suicida. Ninguno exige tanto tiempo, tanto trabajo, tanta consagración en relación con sus beneficios inmediatos. No creo que sean muchos los lectores que al terminar la lectura de un libro, se pregunten cuántas horas de angustias y de calamidades domésticas le han costado al autor esas doscientas páginas y cuánto ha recibido por su trabajo. Para terminar pronto, conviene decir a quien no lo sepa, que el escritor se gana solamente el diez por ciento de lo que el comprador paga por el libro en la librería. De modo que el lector que compró un libro de veinte pesos sólo contribuyó con dos pesos a la subsistencia del escritor. El resto se lo llevaron los editores, que corrieron el riesgo de imprimirlo, y luego los distribuidores y los libreros. Esto parecerá todavía más injusto cuando se piense que los mejores escritores son los que suelen escribir menos y fumar más, y es por tanto normal que necesiten por lo menos dos años y veintinueve mil doscientos cigarrillos para escribir un libro de doscientas páginas. Lo que quiere decir en buena aritmética que nada más en lo que se fuman se gastan una suma superior a la que van a recibir por el libro. Por algo me decía un amigo escritor: «Todos los editores, distribuidores y libreros son ricos y todos los escritores somos pobres».

El problema es más crítico en los países subdesarrollados, donde el comercio de libros es menos intenso, pero no es exclusivo de ellos. En los Estados Unidos, que es el paraíso de los escritores de éxito, por cada autor que se vuelve rico de la noche a la mañana con la lotería de las ediciones de bolsillo, hay centenares de escritores aceptables condenados a cadena perpetua bajo la gota helada del diez por ciento. El último caso espectacular de enriquecimiento con causa en los Estados Unidos es el del novelista Truman Capote con su libro *In Cold Blood*, que en las primeras semanas le produjo medio millón de dólares en regalías y una cantidad similar por los derechos para el

Publicado originalmente en julio de 1966.

cine. En cambio, Albert Camus, que seguirá en las librerías cuando ya nadie se acuerde del estupendo Truman Capote, vivía de escribir argumentos cinematográficos con seudónimo, para poder seguir escribiendo sus libros.

El Premio Nobel –que recibió pocos años antes de morir– apenas fue un desahogo momentáneo para sus calamidades domésticas, acarrea consigo unos cuarenta mil dólares más o menos, lo que en estos tiempos cuesta una casa con un jardín para los niños. Mejor aunque involuntario, fue el negocio que hizo Jean-Paul Sartre al rechazarlo, pues con su actitud ganó un justo y merecido prestigio de independencia, que aumentó la demanda de sus libros.

Muchos escritores añoran al antiguo mecenas, rico y generoso señor que mantenía a los artistas para que trabajaran a gusto. Aunque con otra cara, los mecenas existen. Hay grandes consorcios financieros que a veces por pagar menos impuestos, otras veces por disipar la imagen de tiburones que se ha formado de ellos la opinión pública, y no muchas veces por tranquilizar sus conciencias, destinan sumas considerables a patrocinar el trabajo de los artistas. Pero a los escritores nos gusta hacer lo que nos da la gana, y sospechamos, acaso sin fundamento, que el patrocinador compromete la independencia de pensamiento y expresión, y origina compromisos indeseables. En mi caso, prefiero escribir sin subsidios de ninguna índole, no sólo porque padezco de un estupendo delirio de persecución, sino porque cuando empiezo a escribir ignoro por completo con quién estaré de acuerdo al terminar. Sería injusto que a la postre estuviera en desacuerdo con la ideología del patrocinador, cosa muy probable en virtud del conflictivo espíritu de contradicción de los escritores, así como sería completamente inmoral que por casualidad estuviera de acuerdo.

El sistema de patrocinio, típico de la vocación paternalista del capitalismo, parece ser una réplica a la oferta socialista de considerar al escritor como un trabajador a sueldo del estado. En principio la solución socialista es correcta, porque libera al escritor de la explotación de los intermediarios. Pero en la práctica hasta ahora y quién sabe por cuánto tiempo, el sistema ha dado origen a riesgos más graves que las injusticias que ha pretendido corregir. El reciente caso de dos pésimos escritores soviéticos que han sido condenados a trabajos forzados en Siberia, no por escribir mal sino por estar en desacuerdo con el patrocinador, demuestra hasta qué punto puede ser peligroso el oficio de escribir bajo un régimen sin la suficiente madurez para admitir la verdad eterna de que los escritores somos unos facinerosos a quienes los corsés doctrinarios, y hasta las disposiciones legales nos aprietan más que los zapatos. Personalmente, creo que el escritor, como tal, no tiene otra obligación revolucionaria que la de escribir bien. Su inconformismo, bajo cualquier régimen, es una condición esencial que no tiene remedio, porque un escritor conformista muy probablemente es un bandido, y con seguridad es un mal escritor.

Después de esta triste revisión, resulta elemental preguntarse por qué escribimos los escritores. La respuesta, por fuerza, es tanto más melodramática cuanto más sincera. Se es escritor simplemente como se es judío o se es negro. El éxito es alentador, el favor de los lectores es estimulante, pero éstas son ganancias suplementarias, porque un buen escritor seguirá escribiendo de todas maneras aun con los zapatos rotos, y aunque sus libros no se vendan. Es una especie de deformación que explica muy bien la barbaridad social de que tantos hombres y mujeres se hayan suicidado de hambre, por hacer algo que al fin y al cabo, y hablando completamente en serio, no sirve para nada.

MIS DOS RAZONES CONTRA ESTA REVISTA

De modo que aquí está otra vez *Alternativa*. Vuelve después de un receso de casi cuatro meses que por supuesto nos sirvió para trabajar menos, para perder menos plata y tal vez para equivocarnos menos, pero también para reflexionar, como los curas de otros tiempos, sobre el destino de nuestras almas. Sin embargo, volvemos a salir otra vez como semanario y esta vez a veinte pesos. Lo que quiere decir que los retiros espirituales nos ayudaron a resolver muchos problemas, menos los dos que a mi modo de ver son la desgracia de esta revista: la frecuencia y el precio.

Quienes propugnábamos porque *Alternativa* se convirtiera en diario seguimos creyendo tener la razón. También siguen creyendo tenerla los compañeros que sustentaban la opinión contraria. Son ellos los que ganan, sin embargo, por la razón de peso completo de que ni los unos ni los otros, ni todos juntos, tenemos la plata que haría falta para hacer un diario. Es decir: no hay campanas. Era por ahí, desde luego, por donde hubiéramos debido empezar.

La revista ha sido un género desdichado en Colombia. Todas, de cualquier clase, han tenido el destino de los amores de verano y de los ministros de Educación: intenso y fugaz. La única que ha logrado perdurar por más de sesenta años a través de los azares de las peluquerías y los infartos mortales de los cambios de dueños, parece más bien una advertencia de Dios para escarmiento de ingenuos y temerarios. Tal vez sea que los colombianos no sabemos hacer revistas. Tal vez sea que no las sabemos leer. Pero tal vez sea solamente que el lapso de ocho días es un reto descomunal para la mala memoria histórica de los colombianos: cuando llega el sábado ya los lectores se han olvidado de la que fue su revista favorita el sábado pasado, de modo que ésta tiene que cautivar cada semana una clientela nueva que ni siquiera recordaba haber sido la misma clientela fugitiva de la semana anterior. Es triste pero cierto: cada semana compramos una

Publicado originalmente en mayo de 1977.

revista diferente con la misma ilusión efímera e irrepetible con que cada cuatro años elegimos un presidente de la república. Así las cosas, es muy difícil implantar un semanario, y retener el interés de un público numeroso y comprensivo, y sensible además a una propuesta política distinta, mientras no se pueda competir todos los días, y en condiciones similares, con los órganos de opinión que tienen en sus manos todos los poderes del poder. Las peleas de los sábados –los borrachos lo sabemos muy bien– no son más que pleitos de cantina. El otro problema esencial es el precio. Sin grandes anuncios –que no queremos y que además nadie nos daría–, sin un partido político que nos sustente ni un centro mundial de poder que nos mantenga, ni una agencia central de inteligencia que nos subsidie para después poderlo decir, esta revista huérfana de padre y madre no se puede vender a menos precio y la amarga verdad, duélale a quien le duela, es que los lectores con posibilidades de gastarse veinte pesos en una revista no son los que más nos interesan. De manera que nos queremos dirigir a un público y en realidad llegamos a otro. Hacemos una revista para pobres que muchos pobres no pueden comprar. Tratamos de crear una conciencia popular, pero a nuestra clientela más accesible no le interesa tanto la justicia social como las vacaciones en Miami.

A pesar de eso, con la temeridad profesional y política que nos distingue de otros mortales más felices, aquí está otra vez *Alternativa*. Yo sigo estando en ella como siempre desde aquel septiembre casual y ya remoto de su fundación, porque creo que a pesar de sus dos problemas mayores es un órgano indispensable en las condiciones actuales del país y de la prensa de izquierda. Lo único nuevo es que no estaré siempre en toda la revista, sino que cada quince días estaré dentro de las cuatro paredes de esta columna personal, para decir lo que me dé la gana por mi propia cuenta. Hoy, por desgracia, no he tenido mucho tiempo para decirlo.

TONTO ÚTIL, PARA SERVIRLE A USTED

Los amigos políticos y personales del exterior se preguntan cómo es posible que *Alternativa* diga las cosas que dice, y que el gobierno las tolere. Se preguntan cómo explicamos el contrasentido de que la revista diga que en Colombia no hay libertad de prensa, si la sola afirmación impresa es una prueba de que sí la hay. Se preguntan, en fin, qué clase de país es éste donde todavía pueden suceder semejantes cosas mientras el resto del continente es una selva de gorilas.

Aunque parezca increíble, la respuesta a esas preguntas la dio el propio presidente López Michelsen, hace poco más de un año, cuando el ministro de Defensa, general Abraham Varón Valencia, le pidió en un Consejo de Ministros que clausurara *Alternativa* por las cosas que decía sobre las fuerzas armadas. La negativa del presidente fue inmediata:

–Yo no le hago a la revista ese favor político.

Doña María Elena de Crovo, que entonces era ministra del Trabajo, temió que la píldora presidencial le resultara demasiado amarga al general Varón Valencia, y se la endulzó con el consuelo de que no se preocupe, mi general, que al fin y al cabo, según ella, *Alternativa* estaba en las últimas por falta de plata. Su cálculo era muy simple: en un país donde la prensa grande está subsidiada por los avisos del gran capital y los favores del poder, no le parecía posible que una revista pudiera subsistir sin más apoyo que el de sus lectores. Le faltó pensar que en el mundo hay gente que ve la vida de otro modo, y que es capaz de arriesgar no sólo su plata sino inclusive el pellejo por la defensa de una causa buena, casi con tanta temeridad como ella lo estaba arriesgando en el Ministerio del Trabajo por la defensa de una causa mala. Un año y medio después, mal que bien, la revista sigue aquí mientras que doña María Elena de Crovo no sigue ya ni mal ni bien en ninguna parte.

Publicado originalmente en julio de 1977.

A las pocas semanas de aquella reunión estalló una bomba en la redacción de *Alternativa* y más tarde otra en la casa de su director actual. Era evidente que alguien con más prisa y menos visión histórica estaba tratando de hacernos el favor político que no quiso hacernos el presidente. Y de hecho nos lo hizo, porque los dos bombazos alertaron a los lectores, aumentaron la circulación de la revista y promovieron expresiones de solidaridad en el exterior. Un telegrama del futuro presidente de Francia, François Mitterrand, provocó una de las rabias mitológicas del presidente López Michelsen. Primero, porque no podía decir que Mitterrand fuera comunista, si en realidad es un socialista de avanzada, como habría podido serlo López Michelsen si no se le hubieran interpuesto los negocios. Y segundo, porque era precisamente esa clase de ruidos internacionales los que había de impedir cuando se negó a cerrar la revista.

Por supuesto, estas repercusiones no le hubieran importado mucho al presidente, o habría tenido que afrontarlas, si *Alternativa* constituyera un peligro real a corto plazo para el gobierno, o a un plazo más largo para el sistema. Pero tanto él como nosotros sabemos que todavía no lo es. Sabe, como lo sabemos nosotros, que en el archipiélago de incomunicación e irrealidad de las izquierdas colombianas, no hay un criterio unánime en relación con nada, pero mucho menos en relación con esta revista. En las universidades hay semanas en que la leen en voz alta, entre aplausos y gritos de entusiasmo, y hay otras semanas en que los mismos que la aplaudieron la queman en ceremonias públicas aprendidas de los fascistas. En cambio, *El Tiempo*, cuya inercia de poder es tan grande que ya nadie se toma el trabajo de quemarlo, no puede permitirse el lujo de tener redactores que piensen con su propia cabeza porque el gobierno no se lo permite ni a la propia dirección del periódico, que tantas veces tuvo la vanidad de coronarse a sí misma como el campeón mundial del derecho a disentir. Tampoco a *El Tiempo*, desde luego, el presidente le haría el favor político de cerrarlo. Pero le asusta con la amenaza peor de renunciar, o sea, de cerrarse a sí mismo si no lo dejan gobernar como él quiere.

En síntesis: al presidente no le faltaba razón cuando no quiso hacernos aquel favor político. Pero le faltó agregar, para ser perfecto, que al negarnos el favor a nosotros se hacía a sí mismo un favor mucho más grande, que a él le importaba mucho mantener fuera del país una imagen de liberal que ya muy pocos le creen en el interior. Tal vez sea triste, pero de eso vivimos. Hasta el punto de que muchos amigos del exterior se preguntan a menudo si esta revista de oposición no será más bien el tonto útil del presidente López Michelsen.

Mi respuesta es sencilla: me tiene sin cuidado ser el tonto útil del gobierno, si las cosas que hacemos son de todos modos más útiles para la izquierda. Por supuesto, tendrán que pasar todavía muchos años antes de que se sepa a ciencia cierta para cuál de los dos señores resultó más útil el tonto en la cuenta final. Pero no hay prisa. Al fin y al cabo, lo último que un tonto pierde no es la cabeza sino el optimismo.

LOS IDUS DE MARZO DE LA OLIGARQUÍA

El país se está llenando de unos enormes carteles rojos que dicen: «Turbay próximo presidente liberal». Esto quiere decir, para quienes sepan leer bien, que antes del doctor Julio César Turbay puede haber otros presidentes que no sean liberales. Semejante pronóstico involuntario es una prueba grande de las muchas trampas malditas que se esconden dentro del arte de escribir. A menos que sea algo peor: una traición del subconsciente de quien escribió el anuncio. Porque las cosas en Colombia no están para afirmar nada, y menos algo tan oscuro como el nombre del presidente que sucederá al actual.

El apoyo de *El Tiempo* a la candidatura del doctor Lleras Restrepo es una bomba de ruido con un cierto poder estremecedor, pero no es probable que sea decisiva en la contabilidad electoral. El periódico más grande del país, justo es reconocerlo, no está ya para ganar elecciones con la misma facilidad de antes en estos tiempos del transistor. En cambio, su decisión es muy reveladora del estado de ánimo de la vieja oligarquía, rutinaria y un poco perezosa, pero muy segura de lo que hace a la hora de escoger entre un oligarca de nación como el doctor Lleras Restrepo, y un liberal abnegado como lo ha sido siempre el doctor Turbay, pero que llega al concurso del poder con una pandilla de nuevos ricos y una manada de lobos.

El doctor Turbay tenía todo el derecho a suponer que sería el candidato de la oligarquía. Tenía a su favor veinte años de sacrificios personales en nombre de todos los lugares comunes de la clase en el poder: la unidad del partido, la concordia entre los colombianos, la supervivencia de las instituciones democráticas, y tantas otras vainas que se dicen en los discursos y que ya nadie cree. Habían sido veinte años que le permitieron llegar hasta las goteras del poder por la escalera del buen servicio, tragándose las rabias sin contestar, sin pelearse con nadie, aguantándose las burlas de sus anfitriones de sangre azul por no saber a qué temperatura debe estar el vino para acom-

Publicado originalmente en agosto de 1977.

pañar los espárragos ni cuántos siglos se necesitan en realidad para leer siete mil libros, y todo eso con una paciencia que de todos modos parecería oriental aunque no lo fuera.

Sartre dijo alguna vez que la conciencia de clase empieza cuando uno descubre que es imposible cambiar de clase. El doctor Turbay, por supuesto, no leyó a tiempo esta sentencia, como no leyó a tiempo tantas otras, pero va a empezar a vivirla. Por primera vez en su larga carrera de oligarca asimilado tendrá que salir a pelear en la calle, a hacer populismo aunque sea del más barato, tratando de recuperar en seis meses tantos años perdidos, sólo por no haber leído a Sartre a su debido tiempo. En todo caso, es casi una burla de su destino personal, y un nuevo infortunio para la historia patria, que un Turbay se vea ahora obligado por pura urgencia demagógica a recoger del suelo después de treinta años las banderas polvorientas de Jorge Eliecer Gaitán. Lo malo para todos es que no es muy probable que alguien crea en el cambio de clase del doctor Turbay. Lo más seguro, por el contrario, es que a la hora del crujir de dientes, muchos liberales preferirán votar por el godo con piel de cordero que ya empieza a balar de júbilo entre los geranios. Ésta debía ser la inquietud que yacía en el subconsciente del autor de los carteles rojos, cuando pronosticó que el próximo presidente liberal será el doctor Turbay, pero no dice cuándo.

«POR QUÉ NO LE CREO NADA, SEÑOR TURBAY»

En febrero de este año le escribí una carta al señor Julio César Turbay Ayala, y era una carta inspirada por un ánimo tan respetuoso y redactada en términos tan comedidos, que la misma prensa oficial la interpretó como un elogio. Yo proponía en esa carta –ampliando una idea original del ministro de Defensa– que fueran invitados tres periodistas de renombre internacional para que esclarecieran de una vez y divulgaran por el mundo entero la situación real de los derechos humanos en Colombia. Tuve el cuidado de no afirmar nada, y mucho menos de formular algún cargo pues no conocía en aquel momento ninguna evidencia terminante. Además, fui muy explícito en el sentido de que no compartía la creencia generalizada de que existen dentro del gobierno de Colombia otras fuerzas capaces de entorpecer los mejores designios del presidente de la república.

Transcurridos cinco meses sin ninguna respuesta, me considero con derecho a creer que el señor Turbay Ayala no rendirá a los dioses de la buena educación el tributo elemental de contestarme. Otra persona menos comprensiva que yo podría considerarlo como un desaire, y peor aún: un desaire inmerecido. Sin embargo, una de mis pocas virtudes es saber colocarme en la situación de los otros para entender sus actos, y esa virtud me basta para entender la incorrección del señor Turbay Ayala. Yo tampoco, si fuera él, hubiera sabido cómo contestar esa amable carta.

La conciencia limpia

Por el contrario, tal vez deba agradecerle al señor Turbay Ayala que me haya dado este tiempo para reflexionar. Ahora sé, en primer término, que una persona que no contesta las cartas no sólo no merece el tono que yo adopté en la mía, sino que no merece siquiera que

Publicado originalmente en junio-julio de 1979.

se le escriba. Ahora sé, en término segundo, que no era necesario invitar a los tres periodistas, pues las denuncias de atropellos y torturas que se hacían contra el gobierno no eran calumnias de la oposición. Por último, ahora hay más razones para creer, tal como yo lo creía desde entonces, que no hay en Colombia ningún poder por encima del señor Turbay Ayala, sino que él mismo es el responsable más alto de un reducido grupo de militares frenéticos que están arrastrando por el suelo el buen nombre de las fuerzas armadas.

El señor Turbay Ayala ha adoptado un sistema simple frente a las denuncias: negarlas sin apelación no sólo dentro de Colombia, sino también fuera de ella, en un viaje costoso y triste que no tenía otra finalidad de fondo que la de seguir negando a los cuatro vientos, y cuyo resultado real fue poner al país en ridículo a través de medio mundo. En París, ya en el delirio de la negación, el señor Turbay Ayala llegó a decir que él era el único preso político de Colombia, y lo dijo sin saber que el derecho de mamar gallo con asuntos tan graves es un privilegio reservado a quienes tenemos la conciencia limpia.

El señor Turbay Ayala niega sin parpadear contra toda evidencia. No contra el testimonio desgarrador de miles de hombres y mujeres –culpables o inocentes– que han sido sacados de sus casas y maltratados como perros en las cárceles militares. Entre ellos, el muy ilustre don Luis Vidales, cuyo cautiverio infame son las quince horas de mayor oprobio no sólo de este gobierno sino del propio destino personal del señor Turbay Ayala. Niega contra la evidencia de investigaciones que él mismo había promovido atropellos comprobados por comisiones del Congreso Nacional y el Concejo de Bogotá, de las averiguaciones serenas y pruebas abrumadoras del Foro de los Derechos Humanos, y aun contra el estupor de su ministro de Gobierno, que vio en sus propias oficinas los estragos causados por la tortura en el cuerpo de un detenido. Niega, en fin, a pesar de que sabe tan bien como todo el mundo que un universitario desarmado fue muerto a tiros y su cuerpo escamoteado por una patrulla militar, y que casi dos años después de su arresto por la fuerza pública no existe ningún rastro de la estudiante Omaira Montoya.

La palabra de las víctimas

Estamos pues en presencia de un poder personal y absoluto, convencido de que no existe en el mundo ninguna otra novedad distinta de su palabra suprema. Ante esta realidad tenebrosa, a los colombianos sin amparo no nos queda otro recurso que decidir con la conciencia de qué lado está la razón. De un lado están los relatos dramáticos de los torturados y sus familias, y aun de los niños arrestados como rehenes. Del otro lado está la negativa impertérrita del señor Turbay Ayala. Yo no vacilo un instante: les creo a las víctimas. Por esta convicción se regirán todos mis actos a partir de ahora, en relación con el estado de los derechos humanos en Colombia.

Supongo, por supuesto, que todo esto le importa muy poco al impávido señor Turbay Ayala. Lo creo así porque hace poco él declaró a un periodista español: «A García Márquez le tengo más admiración como intelectual que como defensor de los derechos humanos en Colombia». La admiración como intelectual, por venir de quien viene, no puede menos que conmoverme. Pero la otra admiración me interesa más, porque es mucho más útil para el país. De modo que el señor Turbay Ayala puede estar seguro que consagraré todas mis fuerzas a conseguirla y merecerla.

EL FANTASMA DEL PREMIO NOBEL (1)

Todos los años, por estos días, un fantasma inquieta a los escritores grandes: el Premio Nobel de Literatura. Jorge Luis Borges, que es uno de los más grandes y también uno de los candidatos más asiduos, protestó alguna vez en una entrevista de prensa por los dos meses de ansiedad a que lo someten los augures. Es inevitable: Borges es el escritor de más altos méritos artísticos en lengua castellana, y no pueden pretender que le excluyan, sólo por piedad, de los pronósticos anuales. Lo malo es que el resultado final no depende del derecho propio del candidato, y ni siquiera de la justicia de los dioses, sino de la voluntad inescrutable de los miembros de la Academia Sueca.

No recuerdo un pronóstico certero. Los premiados, en general, parecen ser los primeros sorprendidos. Cuando el dramaturgo irlandés Samuel Beckett recibió por teléfono la noticia de su premio, en 1969, exclamó consternado: «¡Dios mío, qué desastre!». Pablo Neruda, en 1971, se enteró tres días antes de que se publicara la noticia, por un mensaje confidencial de la Academia Sueca. Pero la noche siguiente invitó a un grupo de amigos a cenar en París, donde entonces era embajador de Chile, y ninguno de nosotros se enteró del motivo de la fiesta hasta que los periódicos de la tarde publicaron la noticia. «Es que nunca creo en nada mientras no lo vea escrito», nos explicó después Neruda con su risa invencible. Pocos días más tarde, mientras comíamos en un fragoroso restaurante del Boulevard Montparnasse, recordó que aún no había escrito el discurso para la ceremonia de entrega, que tendría lugar 48 horas después en Estocolmo. Entonces volteó al revés la hoja de papel del menú, y sin una sola pausa, sin preocuparse por el estruendo humano, con la misma naturalidad con que respiraba y la misma tinta verde, implacable, con que dibujaba sus versos, escribió allí mismo el hermoso discurso de su coronación.

Publicado originalmente el 8 de octubre de 1980.

La versión más corriente entre escritores y críticos es que los académicos suecos se ponen de acuerdo en mayo, cuando se empieza a fundir la nieve, y estudian la obra de los pocos finalistas durante el calor del verano. En octubre, todavía tostados por los soles del sur, emiten su veredicto. Otra versión pretende que Jorge Luis Borges era ya el elegido en mayo de 1976, pero no lo fue en la votación final de noviembre. En realidad, el premiado de aquel año fue el magnífico y deprimente Saul Bellow, elegido deprisa a última hora, a pesar de que los otros premiados en las distintas materias eran también norteamericanos.

Lo cierto es que el 22 de septiembre de aquel año –un mes antes de la votación–, Borges había hecho algo que no tenía nada que ver con su literatura magistral: visitó en audiencia solemne al general Augusto Pinochet. «Es un honor inmerecido ser recibido por usted, señor presidente», dijo en su desdichado discurso. «En Argentina, Chile y Uruguay se están salvando la libertad y el orden», prosiguió, sin que nadie se lo preguntara. Y concluyó impasible: «Ello ocurre en un continente anarquizado y socavado por el comunismo». Era fácil pensar que tantas barbaridades sucesivas sólo eran posibles para tomarle el pelo a Pinochet. Pero los suecos no entienden el sentido del humor porteño. Desde entonces, el nombre de Borges había desaparecido de los pronósticos. Ahora, al cabo de una penitencia injusta, ha vuelto a aparecer, y nada nos gustaría tanto a quienes somos al mismo tiempo sus lectores insaciables y sus adversarios políticos que saberlo por fin liberado de su ansiedad anual.

Sus dos rivales más peligrosos son dos novelistas de lengua inglesa. El primero, que había figurado sin mucho ruido en años anteriores, ha sido ahora objeto de una promoción espectacular de la revista *Newsweek*, que lo destacó en su portada del 18 de agosto como el gran maestro de la novela; con mucha razón. Su nombre completo es nada menos que Vidiadhar Surajprasad Naipaul, tiene 47 años, nació aquí al lado, en la isla de Trinidad, de padre hindú y madre caribe, y está considerado por algunos críticos muy severos como el más grande escritor actual de la lengua inglesa. El otro candidato es Graham Greene, cinco años menor que Borges, con tantos méritos y también con tantos años de retraso como él para recibir ese laurel senil.

En el otoño de 1972, en Londres, Naipaul no parecía muy consciente de ser un escritor del Caribe. Se lo recordé en una reunión de amigos y él se desconcertó un poco; reflexionó un instante, y una sonrisa nueva iluminó su rostro taciturno. «Good claim», me dijo. Graham Greene, en cambio, que nació en Berkhamsted, ni siquiera vaciló cuando un periodista le preguntó si era consciente de ser un novelista latinoamericano. «Por supuesto», contestó. «Y me alegro mucho, porque en América Latina están los mejores novelistas actuales, como Jorge Luis Borges.» Hace algunos años, hablando de todo, le expresé a Graham Greene mi perplejidad y mi disgusto de que a un autor como él, con una obra tan vasta y original, no le hubieran dado el Premio Nobel.

«No me lo darán nunca», me dijo con absoluta seriedad, «porque no me consideran un escritor serio.»

La Academia Sueca, que es la encargada de conceder el Premio Nobel de Literatura, sólo ése, se fundó en 1786, sin pretensiones mayores que la de parecerse a la Academia Francesa.* Nadie se imaginó entonces, por supuesto, que con el tiempo llegaría a adquirir el poder consagratorio más grande del mundo. Está compuesta por dieciocho miembros vitalicios de edad venerable, seleccionados por la propia academia entre las figuras más destacadas de las letras suecas. Hay dos filósofos, dos historiadores, tres especialistas en lenguas nórdicas, y sólo una mujer. Pero no es ése el único síntoma machista; en los ochenta años del premio, sólo se lo han concedido a seis mujeres, contra 69 hombres. Este año será concedido por una decisión impar, pues uno de los académicos más eminentes, el profesor Lindroth Sten, murió el pasado 3 de septiembre: hace quince días.

Cómo proceden, cómo se ponen de acuerdo, cuáles son los compromisos reales que determinan sus designios, es uno de los secretos mejor guardados de nuestro tiempo. Su criterio es imprevisible, contradictorio, inmune incluso a los presagios, y sus decisiones son secretas, solidarias e inapelables. Si no fueran tan graves, podría pensarse que están animadas por la travesura de burlar todos los vaticinios. Nadie como ellos se parece tanto a la muerte.

Otro secreto bien guardado es dónde está invertido un capital que produce tan abundantes dividendos. Alfred Nobel (con acento en la *e* y no en la *o*), creó el premio en 1895 con un capital de 9.200.000 dólares, cuyos intereses anuales debían repartirse cada año, a más tardar el 15 de noviembre, entre los cinco premiados. La suma, por consiguiente, es variable, según haya sido la cosecha del año. En 1901, cuando se concedieron los premios por primera vez, cada premiado recibió 30.160 coronas suecas. En 1979, que fue el año de intereses

---

* Los otros cuatro premios son: Física y Química, concedidos por la Real Academia de Ciencias; Medicina o Fisiología, concedido por el Comité Nobel del Instituto Carolino, y el de la Paz, concedido por el Comité Nobel del Parlamento de Noruega.

más suculentos, recibió cada uno 160.000 coronas (2.480.000 pesetas). Dicen las malas lenguas que el capital está invertido en las minas de oro de África del Sur y que, por consiguiente, el Premio Nobel vive de la sangre de los esclavos negros. La Academia Sueca, que nunca ha hecho una aclaración pública ni respondido a ningún agravio, podría defenderse con el argumento de que no es ella, sino el Banco de Suecia, quien administra la plata. Y los bancos, como su nombre lo indica, no tienen corazón.

El tercer enigma es el criterio político que prevalece en el seno de la Academia Sueca. En varias ocasiones, los premios han permitido pensar que sus miembros son liberales idealistas. Su tropiezo más grande, y más honroso, lo tuvieron en 1938, cuando Hitler prohibió a los alemanes recibir el Premio Nobel, con el argumento risible de que su promotor era judío. Richard Khun, el alemán que aquel año había merecido el Nobel de Química, tuvo que rechazarlo. Por convicción o por prudencia, ninguno de los premios fue concedido durante la segunda guerra mundial. Pero tan pronto como Europa se repuso de sus quebrantos, la Academia Sueca cometió la que parece ser su única penosa componenda: le concedió el premio de Literatura a sir Winston Churchill sólo porque era el hombre con más prestigio de su tiempo, y no era posible darle ninguno de los otros premios, y mucho menos el de la paz.

Tal vez las relaciones más difíciles de la Academia Sueca han sido con la Unión Soviética. En 1958, cuando el premio le fue concedido al muy eminente Boris Pasternak, éste lo rechazó por temor de que no se le permitiera regresar a su país. Las autoridades soviéticas consideraron el premio como una provocación. Sin embargo, en 1965, cuando el premiado fue Mikhail Sholokhov, el más oficial de los escritores oficiales soviéticos, las propias autoridades de su país lo celebraron con júbilo. En cambio, cinco años más tarde, cuando se lo concedieron al disidente mayor, Alexander Solzhenitsyn, el gobierno soviético perdió los estribos y llegó a decirse que el Premio Nobel era un instrumento del imperialismo. A mí me consta, sin embargo, que los mensajes más cálidos que recibió Pablo Neruda con motivo de su premio provenían de la Unión Soviética, y algunos de muy alto nivel oficial. «Para nosotros», me dijo, sonriendo, un amigo soviético, «el Premio Nobel es bueno cuando se lo conceden a un escritor que nos gusta, y malo cuando sucede lo contrario.» La explicación no es tan simplista como parece. En el fondo de nuestro corazón todos tenemos el mismo criterio.

El único miembro de la Academia Sueca que lee en castellano, y muy bien, es el poeta Artur Lundkvist. Es él quien conoce la obra de nuestros escritores, quien propone sus candidaturas y quien libra por ellos la batalla secreta. Esto lo ha convertido, muy a su pesar, en una deidad remota y enigmática, de la cual depende en cierto modo el destino universal de nuestras letras. Sin embargo, en la vida real es un anciano juvenil, con un sentido del humor un poco latino, y con una casa tan modesta que es imposible pensar que de él dependa el destino de nadie.

Hace unos años, después de una típica cena sueca en esa casa —con carnes frías y cerveza caliente—, Lundkvist nos invitó a tomar el café en su biblioteca. Me quedé asombrado. Era increíble encontrar semejante cantidad de libros en castellano, los mejores y los peores revueltos, y casi todos dedicados por sus autores vivos, agonizantes o muertos en la espera. Le pedí permiso al poeta para leer algunas dedicatorias, y él me lo concedió con una buena sonrisa de complicidad. La mayoría eran tan afectuosas, y algunas tan directas al corazón, que a la hora de escribir las mías me pareció que hasta la sola firma resultaba indiscreta. Complejos que uno tiene, ¡qué carajo!

EL FANTASMA DEL PREMIO NOBEL (Y 2)

Se ha dicho muchas veces que los más grandes escritores de los últimos ochenta años se murieron sin el Premio Nobel. Es una exageración, pero no demasiado grande. Leon Tolstoi, cuya novela *Guerra y paz* es, sin duda, la más importante en la historia del género, murió en 1910, a la edad muy nobiliaria de 82 años, cuando ya el Premio Nobel se había adjudicado diez veces. Su libro magistral llevaba ya 45 años de gloria, con numerosas traducciones y reimpresiones en el mundo entero, y ningún crítico dudaba de que estaba destinado a existir para siempre.

En cambio, de los diez escritores que obtuvieron el Premio Nobel mientras Tolstoi vivía, el único que permanece vivo en la memoria es el inglés Rudyard Kipling. El primero que lo obtuvo fue el francés Sully-Prudhomme, que era muy famoso en su tiempo, pero cuyos libros no se encuentran ahora sino en librerías muy especializadas. Más aún, si uno busca su nombre en un diccionario francés, se encuentra con una definición previa que parece una mala jugada del destino: «Prototipo moderno de la nulidad satisfecha y la trivialidad magistral». Otro de los diez primeros laureados fue el polaco Henryck Sienkiewicz, que se había colado de contrabando en la gloria con su ladrillo inmortal, *Quo Vadis*. Otro había sido Federico Mistral, un poeta provenzal que escribió en su lengua vernácula y que tuvo el triste honor de compartir el premio con uno de los dramaturgos más deplorables que parió la madre España: don José Echegaray, ilustre matemático a quien Dios tenga en su santo reino.

En los dieciséis años siguientes murieron sin obtener el premio otros cinco de los grandes escritores de todos los tiempos: Henry James, en 1916; Marcel Proust, en 1922; Franz Kafka, en 1924; Joseph Conrad, en el mismo año, y Rainer Maria Rilke, en 1926. También durante esos años estaban sentados en el escaño de los genios nadie menos que G. K. Chesterton, que murió sin su premio en 1936, y

Publicado originalmente el 9 de octubre de 1980.

James Joyce, que murió en 1941, cuando su *Ulysses* había cambiado el curso de la novela en el mundo, diecinueve años después de su publicación.*

En cambio, de los catorce autores que lo obtuvieron en esa mala época, sólo cuatro perduran: el belga Maurice Maeterlinck, los franceses Romain Rolland y Anatole France, y el irlandés George Bernard Shaw. El indio Rabindranath Tagore, a quien debemos tantas lágrimas de caramelo, fue arrastrado por los vientos de la justicia del carajo. Knut Hamsun, el noruego que obtuvo el premio en 1920 en el apogeo de la gloria, ha corrido la misma suerte, aunque menos merecida. Dos años después, la Academia Sueca sufrió su segundo accidente mortal en lengua castellana: el inefable don Jacinto Benavente, a quien Dios tenga lo más cerca posible de don José Echegaray hasta el fin de los siglos. Con mayores o menos méritos, ninguno de los premiados de este lapso lo merecieron tanto como los que se murieron mereciéndolo.

La omisión de Kafka y Proust es comprensible. En 1917, cuando el Premio Nobel fue compartido por dos ilustres conocidos en su casa –Karl Gjellerup y Henryk Pontoppidan–, Franz Kafka tuvo que retirarse de la compañía de seguros donde trabajaba, y murió siete años después aniquilado por la tuberculosis en un hospital de Viena. *La metamorfosis*, su obra maestra, había sido publicada poco antes en una revista alemana. Sólo en 1926 –como se sabe tal vez demasiado–, su amigo Max Brod contrarió la voluntad del muerto y publicó sus dos novelas geniales: *El castillo* y *El proceso*. Ese año le concedieron el Premio Nobel a la italiana Grazia Deledda, quien vivió todavía diez años más para creerlo.

También Marcel Proust murió sin conocer su gloria. En 1916, el primer tomo de su obra máxima había sido rechazado por varios editores, y entre ellos Gallimard, por decisión de su consejero literario, André Gide, quien por cierto había de ser el muy justo Premio Nobel de 1947. Fue publicado más tarde por cuenta del propio autor. Luego, en 1919, publicó el segundo volumen –*A la sombra de las muchachas en flor*–, que le valió un prestigio inmediato, y la distinción mayor de las letras francesas: el Premio Goncourt. Pero hay que ser justos: sólo un poder adivinatorio real hubiera podido prever lo que sería el espléndido monumento literario de este siglo: *A la búsqueda del tiempo perdido*, sólo publicada en su totalidad después de la muerte del autor.

---

* Otros que merecían el Premio Nobel: Thomas Hardy, Aldous Huxley, Virginia Woolf, Henri de Montherlant y, por supuesto, André Malraux.

En la misma conversación que cité aquí ayer, Graham Greene me dijo que sus dos influencias decisivas habían sido las de Henry James y Joseph Conrad, ambos considerados en vida como dos clásicos de la lengua inglesa. El año en que murió Henry James, el Premio Nobel fue el sueco Verner von Heidenstam. El año en que murió Conrad lo fue otro escritor nacido en Polonia —como él—: Wladyslaw Reymont. Ninguno de los dos era un genio oculto, como sin duda lo son el griego Giorgios Seferis, premiado en 1963, y el norteamericano Isaac B. Singer, premiado en 1978.

Al contrario de Kafka y de Proust, Conrad había vivido su gloria. Había publicado dieciséis novelas y numerosos cuentos, la mayoría de ellos magistrales; estaba reconocido como uno de los más grandes escritores de su tiempo y se había dado el lujo de rechazar el título de caballero del imperio británico. Acababa de cumplir 67 años, que entonces era una buena edad para morirse tranquilo.

Marie Curie obtuvo el Premio Nobel de Física en 1903, compartido con su esposo Pierre, y obtuvo, luego, el de Química, ella sola, en 1911. También el norteamericano John Bardeem compartió el premio de Física en 1956, por descubrir los efectos del transistor, y volvió a compartirlo en 1972, por su aporte al desarrollo de la teoría de la superconductividad. Por último, el profesor Linus Carl Pauling, que obtuvo el premio de Química en 1954, repitió con el de la Paz en 1962. Einstein, en cambio, mereció dos veces el premio de Física, y sólo se lo dieron una vez. Los encargados de adjudicarlo fueron previsivos: temiendo que la teoría de la relatividad resultara falsa, le concedieron el premio por el descubrimiento de la ley de los fenómenos fotoeléctricos.

La Academia Sueca no incurre en esas frivolidades. Al contrario: si una virtud hay que reconocerle es su carácter drástico. No tiene miedo de equivocarse —y se equivoca mucho, por supuesto—, concede el premio una sola vez por una obra de toda la vida, y parece considerar que quien es bueno en una ciencia no puede serlo también en el arte de las letras. La única inconsecuencia en que ha incurrido —y tal vez no lo vuelva a hacer— fue adjudicar un premio póstumo, en 1931, al poeta más popular de Suecia, Erik Axel Karlfeldt, que había muerto seis meses antes. Más raro aún: Karlfeldt había declinado el premio en 1918, y en consecuencia fue declarado desierto ese año. Uno no se explica entonces por qué no se hizo lo mismo cuando lo rechazaron Boris Pasternak, en 1958, y Jean-Paul Sartre, en 1964, sino que se les siguió considerando premiados contra su voluntad.

En todo caso, una superstición muy difundida entre escritores pretende que el Premio Nobel de Literatura es siempre un homenaje póstumo: de setenta y cinco premiados, sólo doce están vivos. Conozco varios escritores grandes que por estos días no sienten la ansiedad de Borges, sino todo lo contrario, un terror metafísico, porque cada vez prospera más la creencia de que nadie sobrevive siete años al Nobel de las letras. Las estadísticas no lo prueban, pero tampoco lo desmienten: veintidós han muerto dentro de ese plazo.

El mal ejemplo lo dieron los primeros. Sully-Prudhomme murió seis años después de recibirlo. El alemán Theodor Mommsen murió al cabo de un año. El noruego Björnstjerne Björnson murió a los siete años. El récord del primer decenio lo batió el poeta italiano Giosué Carducci, que recibió el Premio Nobel en noviembre de 1906 y murió en febrero del año siguiente. Sin embargo, el récord actual lo conserva el gran poeta inglés John Galsworthy, quien recibió el premio en 1932 y murió sesenta días después del hecho.

Quienes no creen en supersticiones, por supuesto, tienen la explicación lógica: la edad promedio a que se adjudica el premio es de 64 años, de modo que es una probabilidad estadística que los premiados mueran dentro de los siete años siguientes. Lo demuestran por la negativa con los premiados más jóvenes: Rudyard Kipling, el más joven de todos, que lo recibió a los 42 años, murió a los 76; Sinclair Lewis, que lo obtuvo a los 45, murió a los 66; Pearl S. Buck, la bien olvidada, que lo obtuvo a los 46, murió a los 81, y Eugene O'Neill, que lo recibió a los 48, murió a los 73. La excepción bien triste fue Albert Camus, que obtuvo el premio a los 44 años, en el esplendor de su gloria y su talento, y murió dos años después, en el accidente de un automóvil conducido por un destino que tal vez no era el suyo.

Sin embargo, la vida siempre encuentra la manera de estar contra la lógica. Para demostrarlo está la lista de los tres premiados más viejos: el alemán Paul Heyse, con 80 años; Bertrand Russell, con 78, y Winston Churchill, con 79. Heyse, que en este caso es la excepción al revés, murió cuatro años después del premio. Pero Churchill sobrevivió once años, fumándose una caja de puros y bebiéndose dos botellas de coñá al día, y Bertrand Russell batió todas las marcas mundiales: murió veinte años después de recibir el premio, a los 98 años de su edad.

El caso más extraño, y fuera de todo cálculo, fue el de Shmuel Y. Agnon y Nelly Sachs, que compartieron el premio en 1966. Agnon había nacido en Polonia en 1888, pero emigró a Israel con su familia, y adquirió la nacionalidad israelí. Fue, sin duda, el más grande novelista hebreo. Nelly Sachs, que fue una gran poeta y muy buena autora de teatro, había nacido en Berlín en 1891, también en el seno de una familia hebrea, pero conservó siempre la nacionalidad alemana. Al principio de la segunda guerra mundial escapó de la persecución nazi y se radicó en Suecia. El 17 de febrero de 1970, a la edad de 82 años, Agnon murió en Jerusalén, cuatro años después de recibir el Premio Nobel. Ochenta y cuatro días después, el 12 de mayo, y a los setenta años de su edad, Nelly Sachs murió en Estocolmo.

Jean-Paul Sartre no dio nunca ninguna muestra de creer en estos misterios de los números. Salvo una; cuando un periodista le preguntó si estaba arrepentido de haber rechazado el Premio Nobel, contestó: «Al contrario, eso me salvó la vida». Lo inquietante es que murió seis meses después de decirlo.

SEAMOS MACHOS: HABLEMOS DEL MIEDO AL AVIÓN

El único miedo que los latinos confesamos sin vergüenza, y hasta
con un cierto orgullo machista, es el miedo al avión. Tal vez porque
es un miedo distinto, que no existe desde nuestros orígenes, como
el miedo a la oscuridad o el miedo mismo de que se nos note el
miedo. Al contrario: el miedo al avión es el más reciente de todos,
pues sólo existe desde que se inventó la ciencia de volar, hace apenas 77 años. Yo lo padezco como nadie, a mucha honra, y además
con una gratitud inmensa, porque gracias a él he podido darle la
vuelta al mundo en 82 horas, a bordo de toda clase de aviones,
y por lo menos diez veces.

No; al contrario de otros miedos que son atávicos o congénitos,
el del avión se aprende. Yo recuerdo con nostalgia los vuelos líricos
del bachillerato, en aquellos aviones de dos motores que viajaban
por entre los pájaros, espantando vacas, asustando con el viento de
sus hélices a las florecitas amarillas de los potreros, y que a veces se
perdían para siempre entre las nubes, se hacían tortillas, y había que
salir a medianoche a buscar sus cenizas del modo más natural: a lomo
de mula.

Una vez, siendo reportero de un diario de Bogotá, en una época irreal en que todo el mundo tenía veinte años, me mandaron con
el fotógrafo Guillermo Sánchez a perseguir una mala noticia en uno
de aquellos Catalinas anfibios que habían sobrado de la guerra. Volábamos sobre la plena selva de Urabá sentados en bultos de escobas,
porque asientos no había en aquel sepulcro volante, ni una azafata
de consolación a quien pedirle el número de su teléfono en el paraíso, y de pronto el avión se metió a tientas por donde no era y se
extravió en un aguacero bíblico. No sólo llovía afuera, sino también
adentro. Agarrándose a duras penas, el copiloto nos llevó un periódico para que nos tapáramos la cabeza, y vimos, con asombro, que
apenas si podía hablar y le temblaban las manos.

Publicado originalmente el 26 de octubre de 1980.

Ese día aprendí algo muy alentador: también los pilotos tienen miedo, sólo que a ellos, como a los toreros, no se les nota tanto en el temblor de las manos como en las supersticiones. Un amigo español –tan temeroso del avión que nunca viajaba sentado– lo descubrió una mala noche de invierno en que lo invitaron a presenciar el decolaje en la cabina de mando. Era en Nueva York, durante una tormenta de nieve, y la tripulación permaneció muy serena en la cabeza de la pista, hasta que le dieron la orden de decolar. Entonces, como si fuera un requisito técnico insalvable, todos se persignaron al unísono. Mi amigo, comprendiendo que en el fondo de su alma también los pilotos tenían miedo, le perdió para siempre el miedo al avión.

Yo tuve una prueba todavía más sutil volando por entre las estrellas sobre el océano Atlántico. Hablando de todo, le pregunté al comandante por otro piloto amigo que había sido mi compañero de escuela. Yo ignoraba, por supuesto, que se había estrellado en el aeropuerto de Tenerife cuando trataba de aterrizar en medio de la borrasca. El comandante me lo dijo de otro modo, pero más revelador:

–Se retiró de la compañía hace tres años, en las islas Canarias.

Sin embargo, el buen miedo al avión no tiene nada que ver con las catástrofes aéreas. Picasso lo dijo muy bien: «No le tengo miedo a la muerte, sino al avión». Más aún: hubo muchos temerosos que perdieron el miedo al avión después de sobrevivir a un desastre. Yo lo contraje como una infección incurable volando a medianoche de Miami a Nueva York, en uno de los primeros aviones a reacción. El tiempo era perfecto y el avión parecía inmóvil en el cielo, llevando a su lado esa estrella solitaria que acompaña siempre a los aviones buenos, y yo la contemplaba por la ventanilla con la misma ternura con que Saint-Exupéry veía las fogatas del desierto desde su avión de aluminio. De pronto, en la lucidez de la vigilia, tuve conciencia de la imposibilidad física de que un avión se sostuviera en el aire, y me juré que nunca volvería a volar.

Lo cumplí durante diez años, hasta que la vida me enseñó que el verdadero temeroso del avión no es el que se niega a volar, sino el que aprende a volar con miedo. Es una especie de fascinación. De todos los temerosos insignes que conozco, el único que de verdad no vuela es el arquitecto brasileño Oscar Niemeyer. En cambio, su compatriota Jorge Amado, que es un timorato aéreo de los más grandes, ha tenido la audacia poética de volar en Concord desde París hasta Nueva York, para allí tomar un barco que lo llevara a Río de Janeiro. El escritor venezolano Miguel Otero Silva y el director de cine brasileño Ruy Guerra, por distintos caminos, han llegado a la conclusión de que la única manera de combatir el miedo al avión es volando con miedo, y lo combaten casi todos los meses. Carlos Fuentes, que no voló durante quince años y hacía unos viajes épicos de ocho días, cambiando de trenes, desde México hasta Nueva York, no sólo ha vuelto a volar, sino que la semana pasada fue a dictar una conferencia en la Universidad de Indiana, en una avioneta de un solo motor. Sin embargo, entre los grandes especialistas del miedo al avión no hay ninguno mejor que don Luis Buñuel, que a los ochenta años sigue volando impávido, pero muerto de miedo. Para él, el verdadero terror empieza cuando todo anda perfecto en el vuelo y, de pronto, aparece el comandante en mangas de camisa y recorre el avión a pasos lentos, saludando a cada uno de los pasajeros con una sonrisa radiante.

Mi madre no ha volado más de dos veces en su larga vida. Nunca ha sentido miedo, pero conoce muy bien el de sus hijos –que son doce–, de modo que mantiene siempre una vela encendida en el altar doméstico para proteger a cualquiera de nosotros que se encuentre en el aire. Su fe es tan cierta, que a uno de sus hijos –que es ingeniero de caminos– se le cayó hace poco un bulldozer en una cuneta. Mi madre oyó decir que el rescate podía costar más de 100.000 pesos, y le dijo a mi hermano que no gastara ni un céntimo, pues ella iba a encender una vela para sacar el bulldozer. Mi hermano le reprendió: «Sólo a ti se te ocurre que una vela pueda sacar un bulldozer de una cuneta». Mi madre, impasible, le replicó:

–¡Cómo no va a sacarlo, si sostiene un avión en el aire!

EL ALQUIMISTA EN SU CUBIL

Con las primeras cerezas de 1972, en la vitrina de la galería Pyramid, de Washington, se exhibió un cuadro que causó un escándalo fácil entre las señoras de sombreros floridos que llevaban a cagar a sus perros en el parque cercano. Parecía ser la fotografía demasiado realista de una mujer en cueros, derrumbada en un mecedor vienés y abierta de piernas frente a los transeúntes sin el menor recato, si bien la expresión de su sexo era más desolada que libertina. La policía ordenó retirar el cuadro, pero su ímpetu se quedó sin razones cuando le demostraron que no era una fotografía, sino un dibujo. El arte tiene sus privilegios, y el más raro de ellos es que se le toleren ciertos excesos que no están permitidos a la vida.

El autor de aquel dibujo tan perfecto que hasta la policía de Washington lo confundía con una foto era un colombiano de veintiocho años que sobrevivía a duras penas en un cuarto de servicio del barrio de Saint Michel, en París. Su nombre no le decía nada a nadie: Darío Morales. Su esposa, Ana María, estaba peor que él, porque además estaba encinta. Pagaban el alquiler del cuarto limpiando a gatas las escaleras del decrépito edificio de seis pisos. De noche, Ana María dividía el espacio con una manta para poder dormir, con su niña dormida en el vientre, mientras su esposo pintaba hasta el amanecer. Como no tenía bastante luz, Darío Morales oprimía con cinta pegante el interruptor regulado de la escalera, de modo que no se apagara cada minuto, como estaba previsto, sino que permaneciera encendido toda la noche mientras él pintaba. En Francia hay delitos más graves que ése, por supuesto, pero ningún otro les duele tanto a los franceses.

Alguien había tratado de convencer a Darío Morales de la inutilidad de aquellas miserias, y le había aconsejado volver a Cartagena de Indias, la fragorosa ciudad del Caribe donde nació y donde le sería más fácil subsistir. Darío Morales rechazó el consejo con un

Publicado originalmente el 4 de noviembre de 1980.

argumento hermético: «Dondequiera que yo vaya seguiré siendo el mismo». En París tenía, al menos, eso que los escritores lánguidos suelen llamar alimento espiritual: la posibilidad perpetua de ver en carne y hueso la mejor pintura del mundo. Además, según había leído por esos días en un periódico de la tarde, sólo en el Barrio Latino había más de 11.000 pintores anónimos del mundo entero, viviendo en las mismas condiciones que él. Ninguno, hasta donde recordaban las estadísticas, se había muerto de hambre. La noticia le había hecho sentirse menos solo, que es algo muy alentador cuando se es joven y no se tiene nada que comer en París.

Sin embargo, una de esas tardes de lluvias oscuras en que a uno se le vuelve de cenizas el corazón, escribió una carta a Colombia pidiendo que le mandaran el boleto de avión de la derrota. Pero la carta no llegó nunca, por una razón que se ha venido repitiendo a través de los siglos desde el principio de la humanidad: Ana María no la puso al correo. Fue una decisión sabia. Antes de un año, la vida del pintor, de la mujer clarividente que no puso la carta, y de la bella Estefanía, que nació en abril, se había resuelto de pronto. La primera exposición individual de óleos de Darío Morales en la galería Pyramid, de Washington, en junio de 1973, fue un acontecimiento artístico y comercial. Si hubiera aceptado todos los encargos que le hicieron esa vez, habría tenido que pintar, a su ritmo de orfebre, durante más de 116 años. Pintando lo mismo: esa mujer sin identidad, con el sexo afligido, en una habitación escueta donde no vive nadie y con muy pocos objetos dispersos que ya no sirven para nada.

¿Quién es esa mujer?

Tal vez Darío Morales daría algo de su propia vida por saberlo, aunque no volviera a pintar más cuando lo supiera. Después de todo, eso parece ser lo único que busca con el delirio de su arte, desde que empezó a pintar, a los doce años, en su casa natal del barrio de la Manga. Era una casa grande y vacía, con una terraza de baldosas ajedrezadas y un patio de sombras frescas con palos de mango y matas de guineo, donde cantaban hasta reventar de gozo las chicharras del calor. La vida andaba suelta por las calles ardientes, en la peste de pescados muertos de la bahía, en el almendro solitario de la esquina del Trébol, donde en otro tiempo amanecían los borrachos ahorcados por amor. Pero Darío Morales no parecía ver la vida de dentro ni la vida de fuera, sino sólo el universo ilusorio del baño de servicio a través de un agujero que había taladrado en el muro. Era lo único que pintaba. Tanto, que uno se preguntaba desde entonces si no se daría cuenta de que en el mundo había también mujeres vestidas. Su abuela, que fue su primer crítico, se lo dijo:

—¡No sabes pintar nada más que tetas y pan!

Ahora, a los 36 años, Darío Morales sigue tratando de rescatar aquellas ilusiones de su paraíso perdido. Sus cuadros son cada vez más grandes y más ansiosa la búsqueda de sus verdades milimétricas, tal vez con la esperanza de que un milagro de su alquimia termine por implantar sus nostalgias en la realidad.

No es cierto, como se dice con tanta facilidad, que Darío Morales sea un realista. No: sus cuadros no se parecen a la vida, sino a los sueños recurrentes. No tienen el color, ni el clima, ni la luz de la vida, sino el color y el clima y la luz de la ilusión. Darío Morales se ha hecho retratar frente a alguno de ellos, y no se sabe muy bien dónde termina él y dónde empieza la pintura. Pero es demasiado evidente que se sentiría mejor si estuviera de veras dentro del cuadro. Hay una foto suya tomada frente a su autorretrato, y el Darío Morales pintado se parece más a él que el Darío Morales de la realidad. Hay también un cuadro insólito en su obra, donde se ve a Ana María —vestida— cosiendo en la máquina de otros cuadros. De la habitación contigua sólo se ve un ángulo iluminado, con otra máquina de coser y otro mecedor vacío, y uno sabe, por la naturaleza de la luz, que esa otra máquina y ese mecedor ineludible no existen ni siquiera en la realidad de la pintura, sino que Darío Morales los está soñando en algún lugar de la casa. Son los muebles de su obsesión, y por eso se sabe que volveremos a encontrarlos en otros cuadros. Pero su misterio volverá a cambiar por completo en cada ocasión, según su tiempo y su lugar, como sucede con los sueños que se repiten a sí mismos durante toda la vida.

Yo entendí esa alquimia secreta de Darío Morales hace muy pocos años, cuando fui por primera vez a su estudio de París. Abrió la puerta él mismo, con su barba de bebé enorme y una chaqueta y una gorra de lobo de mar, más parecido que nunca a un personaje de Melville. Al final de una escalera empinada había una habitación amplia, de techos muy altos, con cristales lluviosos por donde sólo se veía el cielo de ceniza. Más que un estudio de pintor, aquello era un taller de fabricar recuerdos. Allí estaban la máquina de coser de la hermana que se quedó esperando en la ventana al que nunca volvió, la estufa de carbón de los tiempos del ruido, la lámpara colgada del techo cuyo cordón se estiraba y se encogía a voluntad sobre la mesa de comer. Dispersos por el suelo, en gran desorden, estaban los miembros descuartizados de la mujer del sueño: el torso sin corazón, la pierna helada, la mano muerta para siempre, parecían los estragos de un accidente pavoroso, pero no tenían ni un rastro de sangre, como sólo puede ocurrir en las catástrofes de las pesadillas. Yo sabía desde entonces que Darío Morales había hecho una pausa de pintor para aventurarse en la escultura. Sin embargo, no pude reprimir un leve escalofrío al descubrir otra vez a la mujer recurrente en el fondo del estudio, tumbada en el mecedor, intacta, pero no pintada en un lienzo, sino esculpida en materia tangible: ya casi viva.

—El mejor desnudo es el de la escultura —me dijo Darío Morales—, por una razón muy simple: se puede tocar.

Me volví a mirarlo con un cierto estupor: estaba radiante. Yo, en cambio, me sentí de pronto extraviado dentro de un destino ajeno, como si hubiera dejado atrás mi propia vida y hubiera empezado a formar parte de las nostalgias de Darío Morales. Tal es la magia de su mundo.

DEL MALO CONOCIDO AL PEOR POR CONOCER

Siempre se ha dicho que entre dos males hay que escoger el menos malo, y que en el arte de preferir a los hombres es más seguro el malo conocido que el bueno por conocer. Estados Unidos –al término de una campaña electoral que durante casi un año mantuvo al mundo con el último aliento– ha hecho dos veces lo contrario en una sola vez: eligió al peor desconocido.

Fue un cataclismo arrasador con muy pocos precedentes en la vida de ese país asombroso, cuyo inmenso poder creativo le ha servido para hacer muchas de las cosas más grandes de este siglo, y también algunas de las más abyectas, pero no le ha servido para escoger un presidente digno de su tamaño. Después de su fracaso en playa Girón, hace diecinueve años, el efímero John F. Kennedy dijo una frase hermosa: «La victoria tiene muchos padres, pero la derrota es huérfana». Se la atribuyó a un clásico griego que nunca quiso identificar, y hasta hoy nadie ha podido averiguar quién era. De modo que el presidente James Carter podría repetirla con igual derecho. Pues nada le hace tanta falta como una frase histórica, ahora que su imagen parece destinada a ser la más patética de estos tiempos difíciles: una estrella fugaz lanzada al mercado de la gloria con el poder de seducción de una nueva pasta dentífrica, y de la cual no quedará nada más que la mala memoria de estos cuatro años empedrados de buenas intenciones.

La elección de Ronald Reagan, sin embargo, no es lo más significativo de este desastre. Es apenas un símbolo. Lo esencial es la absoluta falta de misericordia con que los electores de todos los niveles y todos los colores han repudiado a los políticos mansos que durante más de veinte años trataban de gobernar con palabras. Ninguno se ha salvado. Cuatro gobernadores y siete senadores eternos de los más liberales fueron aniquilados, y entre éstos, dos de los más idealistas, los que siempre tuvimos como los abuelitos buenos de la América

Publicado originalmente el 11 de noviembre de 1980.

Latina: George McGovern, candidato demócrata en 1972 y senador sucesivo durante dieciocho años, y Frank Church, que durante veinticuatro años había luchado por establecer la buena imagen de su patria en el mundo. Pero hay algo más expresivo: el senador Jacob Javit, un republicano de Nueva York que se distinguió durante más de veinte años por su corazón liberal, ha sido desplazado por un reaccionario de su propio partido.

El mismo senador Edward Kennedy debe considerarse como un sobreviviente de esta espantosa carnicería política. Fue una votación pasional, cuya explicación parece ser que los electores no sólo votaron en favor de Reagan y en contra de Carter, sino contra varias generaciones de hombres de buena fe que predicaban una cosa, pero no la hacían, o no querían, o no podían hacerla en la vida real.

Esto no quiere decir, por supuesto, que los electores hayan acertado. También los pueblos se equivocan, y la historia de la humanidad está llena de ejemplos atroces. Como todos los países, y más los ricos que los pobres, Estados Unidos tiene instintos primitivos muy fáciles de despertar; es natural que éstos hayan favorecido al candidato más viejo que hubo jamás en la historia de su país, un antiguo pistolero de cine sin ningún defecto ni ninguna virtud que no estén pasados de moda, desde su ideología de las cavernas hasta el copete de vaselina. Otros debieron votar contra su propio deseo, por la rabia de los sueños contrariados y la nostalgia del tiempo perdido, y hasta por la esperanza siempre verde de que un simple cambio de nombres sea también un cambio a favor de los precios del mercado. Muchos latinos, negros y judíos sin corona abandonaron a Carter por desilusión. De modo que las razones del voto pueden ser muy diversas, y todas distintas del amor. Lo grave sería que, tratando de castigar al malo conocido, Estados Unidos se haya aventurado sin quererlo por el callejón sin salida de su desgracia.

Si es así, la América Latina no puede apagar las luces para dormir. Durante la campaña electoral, ninguno de los candidatos se ocupó de ella como algo esencial en la vida de su país, y no mereció ni siquiera una triste mención honorífica en su duelo de bobos de la televisión. Roger Fontain, que es el consejero de Reagan para la América Latina y seguirá siéndolo sin duda durante su presidencia, no tuvo mucho que hacer en la campaña: en la plataforma electoral de ambos partidos apenas si nos tomaban en cuenta. No es raro. Henry Kissinger —que es uno de los dioses tutelares de Reagan— sólo le consagró a América Latina unas sesenta páginas, de las casi 1.600 que tienen los dos mamotretos de sus memorias.

A pesar de eso, para nadie es tan peligrosa esta mala elección como para América Latina. Durante su campaña, Reagan demostró ser tan flexible como le convenga, y es seguro que seguirá siéndolo en la presidencia. Tiene fama de ser más duro de palabra que de obra. Con absoluta seguridad no hará el papel de vaquero de la justicia contra los pieles rojas de la Unión Soviética, ni va a meter a su país en otro pantano de guerra como el de Vietnam. A fin de cuentas, es un republicano de los grandes, y ya sabemos que a los presidentes republicanos se les ha ido la vida tratando de terminar las guerras que empezaron los presidentes demócratas.

Sin embargo, en alguna parte del mundo tiene que acreditar la imagen de gendarme sin corazón que le consiguió tantos votos, y en ninguna le resulta más fácil que en América Latina, este traspatio inmenso y solitario, por el cual nadie distinto de nosotros mismos está dispuesto a sacrificar la felicidad. Peor aún: Reagan no tendrá siquiera que hacer nada. Bastará su sola presencia en la Casa Blanca para que los gorilas militares y civiles se sientan tranquilos en su trono de sangre. Nuestro destino, aunque el propio Reagan no lo quiera, está escrito en la palma de su mano. Por fortuna, estas Américas desdichadas, incluido Estados Unidos, son mucho más grandes y más nobles que sus propios instintos primitivos. Si Ronald Reagan no lo sabe, hay que esperar que se lo enseñen a tiempo sus consejeros, antes de que la realidad —como lo hizo con el presidente Carter— termine por enseñárselo a golpes.

LA COMISIÓN DE BABEL

Hace cuatro años, la Conferencia General de la Unesco, reunida en Nairobi, le pidió a su director general, Amadou-Mahtar M'Bow, que emprendiera un estudio a fondo sobre la comunicación y la información en el mundo contemporáneo. El señor M'Bow delegó el espinoso mandato en una comisión de dieciséis personas, escogidas por él mismo según su propio criterio, en distintos países del mundo, bajo la presidencia del honorable Sean MacBride, antiguo ministro de Relaciones Exteriores de Irlanda, Premio Nobel de la Paz y Premio Lenin de la Paz, y quien —al margen de todo esto— ha venido haciendo una gestión silenciosa y vana para liberar a los rehenes norteamericanos en Teherán.

Los miembros de esta comisión imprevisible no representaban a ningún gobierno ni a ninguna persona pública, ni privada, y no tenían que obedecer a nadie más que a su propia razón. Además del presidente, había un norteamericano, un francés, un zairota, un soviético, un indonesio, un tunecino, un japonés, un nigeriano, un yugoslavo, un egipcio, un holandés, un hindú, dos latinoamericanos y sólo una mujer sola, como siempre: Betty Zimmerman, de Canadá. Uno de los dos latinoamericanos —para mal de mis días— era yo.

La comisión trabajó durante los últimos dos años en ocho sesiones nómadas: cuatro en París, una en Estocolmo, una en Dubrovnik, una en Nueva Delhi y una en Acapulco. Para mí, que soy un cazador solitario de las palabras, escribir un libro junto con otras quince personas, y además tan distintas, era una aventura inquietante. Resultó ser la más extraña: nunca me había aburrido tanto ni me había sentido tan inútil; pero creo que nunca había aprendido tanto en tan poco tiempo. Al final, sólo me quedó la amargura de no haber logrado demostrar que la telepatía, los presagios y los sueños cifrados son medios de comunicación naturales que es necesario rescatar del oscurantismo científico.

Publicado originalmente el 21 de noviembre de 1980.

La comprobación terminante del drama de la comunicación en este mundo la tuvimos desde el primer día en torno de nuestra mesa. Nos entendíamos en tres idiomas oficiales –inglés, francés y ruso–, pero la mayoría pensábamos en nueve lenguas maternas, algunas tan extrañas como el bahasa, que es una de las incontables de Oceanía; el swahili, que se sigue hablando en muchos países de África, y el hindi, que es uno de los doce idiomas oficiales y los ochocientos dialectos de la India. Michio Nagai, el miembro japonés de la comisión, sintió una alarma legítima cuando se encontró hablando en inglés con Mochtar Lubis, el miembro indonesio, que nació a pocas horas de vuelo del Japón, y en la misma orilla. «El imperio romano continúa», protestó Michio Nagai. Con igual derecho hubiéramos podido protestar los dos latinoamericanos. En efecto, el rozagante y tenaz Juan Somavía, que es chileno, hablaba siempre en inglés durante las sesiones, y yo le escuchaba en francés por el sistema de traducción simultánea. Aunque las intérpretes eran eficaces y bellas en sus jaulas de vidrio, yo tenía la impresión de entender todo lo que los otros decían, pero no lo que pensaban.

Lo único en que todos estuvimos de acuerdo desde el principio fue en la certidumbre de que el flujo de la información de este mundo circula en un solo sentido: de los más fuertes hacia los más débiles. La mayoría pensábamos –y yo lo sigo pensando– que la información y la comunicación se han convertido en instrumentos de dominio de los países ricos sobre los países pobres, y esto causa otra desigualdad universal que es necesario corregir. En los dos extremos del dilema, por supuesto, se encontraban el colega de Estados Unidos y el colega de la Unión Soviética.

El norteamericano era Elie Abel, un gringo inteligente y cordial, que durante muchos años fue decano de la escuela de periodismo de Columbia University, considerada la mejor del mundo. El soviético era el director de la agencia Tass, Sergei Losev, cuyo estado de tensión permanente le impedía parecer tan simpático y con tanto sentido del humor como lo era en realidad. Para Abel era imposible concebir cualquier intento de intervención estatal en la información. Para Losev era imposible concebir la más mínima intervención privada. El jamón de este sándwich sin solución éramos los nativos del Tercer Mundo. Unos y otros parecíamos convencidos de la urgencia de democratizar la información; pero era evidente que, alrededor de la mesa, había dieciséis maneras distintas de entender la democracia.

El resultado no podía ser otro: un informe de compromiso, que el señor M'Bow entregó la semana pasada a la Conferencia General de la Unesco, reunida esta vez en Belgrado. «No es un trabajo siste-

mático en la exposición de los diferentes temas que aborda, y a
veces le falta un estilo plenamente coherente y metódico», como lo
dejamos establecido en una nota personal los dos miembros latinoamericanos. Pero, con todos sus defectos y sus enormes posibilidades de controversia, es el mejor informe posible sobre el drama de
la comunicación sin regreso y la información pervertida en esta olla
de grillos caníbales del mundo contemporáneo. No es, no podía ser,
ni pretendía serlo, una fórmula mágica para salvar el alma, sino una
guía espiritual para santos y forajidos. Su validez tendrá que ser distinta de acuerdo con quien la lea, y según su tiempo y su lugar. A fin
de cuentas, la democracia es urgente en todas partes; pero no será
igual en ninguna, tal como lo sentíamos de dieciséis maneras distintas en aquellas jornadas quiméricas de nuestra mesa de Babel.

TELEPATÍA SIN HILOS

Un notable neurólogo francés, investigador de tiempo completo, me contó la otra noche que había descubierto una función del cerebro humano que parece ser de una gran importancia. Sólo tiene un problema: no ha podido establecer para qué sirve. Yo le pregunté, con una esperanza cierta, si no había alguna posibilidad de que ésa fuera la función que regula los presagios, los sueños premonitorios y la transmisión del pensamiento. Su única respuesta fue una mirada de lástima.

Yo había visto esa misma mirada dieciocho años antes, cuando le hice una pregunta similar a un muy querido amigo, que es también investigador del cerebro humano en la Universidad de México. Mi opinión, ya desde entonces, era que la telepatía y sus medios diversos no son cosas de brujos, como parecen creerlo los incrédulos, sino simples facultades orgánicas que la ciencia repudia, porque no las conoce, como repudiaba la teoría de la redondez de la Tierra cuando se creía que era plana. Mi amigo admitía, si no recuerdo mal, que es muy reducida el área del cerebro cuyas funciones están comprobadas a plenitud, pero se negaba a admitir que en el resto de aquellas tinieblas hubiera un lugar para anticiparse al porvenir.

Yo le hacía bromas telepáticas que él descalificaba como casualidades puras, a pesar de que algunas parecían demasiado evidentes. Una noche lo llamé por teléfono para que fuera a comer a nuestra casa, y sólo después me di cuenta de que no había cosas bastantes en la cocina. Volví a llamarle para pedirle que me llevara una botella de vino de una marca que no era usual, y un pedazo de salchichón. Mercedes me gritó desde la cocina que le pidiera también un jabón para lavar platos. Pero ya había salido de su casa. Sin embargo, en el momento de colgar el teléfono, tuve la impresión nítida de que, por un prodigio imposible de explicar, mi amigo había recibido el mensaje. Entonces lo escribí en un papel, para que él no fuera a dudar

Publicado originalmente el 25 de noviembre de 1980.

de mi versión, y por puro virtuosismo poético agregué que llevara también una rosa. Poco después, su esposa y él llegaron con las cosas que les habíamos pedido, inclusive el jabón de la misma marca que usábamos en casa. «El supermercado estaba abierto por casualidad, y decidimos traerles estas cosas», nos dijeron, casi excusándose. Sólo faltaba la rosa. Aquel día mi amigo y yo iniciamos un diálogo distinto que todavía no ha terminado. La última vez que le vi, hace seis meses, estaba dedicado por completo a establecer en qué lugar del cerebro se encuentra la conciencia.

La vida, más de lo que uno cree, está embellecida por este misterio. La víspera del asesinato de Julio César, su esposa Calpurnia vio con terror que todas las ventanas de la casa se abrían de golpe al mismo tiempo, sin viento y sin ruidos. Siglos después, el novelista Thornton Wilder le atribuyó a Julio César una frase que no está en sus memorias de guerra ni en las crónicas fascinantes de Plutarco y Suetonio, pero define mejor que nada la condición humana del emperador: «Yo, que gobierno tantos hombres, soy gobernado por pájaros y truenos». La historia de la Humanidad —desde que el joven José descifraba los sueños en Egipto— está llena de estas ráfagas fabulosas. Conozco dos gemelos idénticos a quienes les dolió la misma muela al mismo tiempo en ciudades distintas, y que cuando están juntos tienen la sensación de que los pensamientos del uno interfieren a los del otro. Hace muchos años, en una vereda de la costa del Caribe, conocí un curandero que se preciaba de sanar un animal a distancia si le daban la descripción precisa y el lugar en que estaba. Yo lo comprobé con estos ojos: vi una vaca infectada, cuyos gusanos se caían vivos de las úlceras, mientras el curandero rezaba una oración secreta a varias leguas de distancia. Sin embargo, sólo recuerdo una experiencia que haya tomado en serio estas facultades en la historia de hoy. La hizo la Marina de Estados Unidos, que no tenía medios para comunicarse con los submarinos nucleares que navegaban bajo la corteza polar, y decidió intentar la telepatía. Dos personas afines, una en Washington y otra a bordo del submarino, intentaron establecer un sistema para intercambiar mensajes pensados. Fue un fracaso, por supuesto, pues la telepatía es imprevisible y espontánea, y no admite ninguna clase de sistematización. Es su defensa. Todo pronóstico, desde los presagios matinales hasta las centurias de Nostradamus, viene cifrado desde su concepción y sólo se comprende cuando se cumple. De no ser así, se derrotaría de antemano a sí mismo.

Hablo de esto con tanta propiedad porque mi abuela materna fue el sabio más lúcido que conocí jamás en la ciencia de los presagios.

Era una católica de las de antes, de modo que repudiaba como artificios de malas artes todo lo que pretendiera ser adivinación metódica del porvenir. Así fueran las barajas, las líneas de la mano o la evocación de los espíritus. Pero era maestra de sus presagios. La recuerdo en la cocina de nuestra casa grande de Aracataca, vigilando los signos secretos de los panes perfumados que sacaba del horno.

Una vez vio el 09 escrito en los restos de la harina, y removió cielo y tierra hasta encontrar un billete de la lotería con ese número. Perdió. Sin embargo, la semana siguiente se ganó una cafetera de vapor en una rifa, con un boleto que mi abuelo había comprado y olvidado en el bolsillo del saco de la semana anterior. Era el número 09. Mi abuelo tenía diecisiete hijos de los que entonces se llamaban naturales —como si los del matrimonio fueran artificiales—, y mi abuela los tenía como suyos. Estaban dispersos por toda la costa, pero ella hablaba de todos a la hora del desayuno, y daba cuenta de la salud de cada uno y del estado de sus negocios como si mantuviera una correspondencia inmediata y secreta. Era la época tremenda de los telegramas que llegaban a la hora menos pensada y se metían como un viento de pánico en la casa. Pasaba de mano en mano sin que nadie se atreviera a abrirlo, hasta que a alguien se le ocurría la idea providencial de hacerlo abrir por un niño menor, como si la inocencia tuviera la virtud de cambiar la maldad de las malas noticias.

Esto ocurrió una vez en nuestra casa, y los ofuscados adultos decidieron poner el telegrama al rescoldo, sin abrirlo, hasta que llegara mi abuelo. Mi abuela no se inmutó. «Es de Prudencia Iguarán para avisar que viene», dijo. «Anoche soñé que ya estaba en camino.» Cuando mi abuelo volvió a casa no tuvo ni siquiera que abrir el telegrama. Volvió con Prudencia Iguarán, a quien había encontrado por casualidad en la estación del tren, con un traje de pájaros pintados y un enorme ramo de flores, y convencida de que mi abuelo estaba allí por la magia infalible de su telegrama.

La abuela murió de casi cien años sin ganarse la lotería. Se había quedado ciega y en los últimos tiempos desvariaba de tal modo que era imposible seguir el hilo de su razón. Se negaba a desvestirse para dormir mientras la radio estuviera encendida, a pesar de que le explicábamos todas las noches que el locutor no estaba dentro de la casa. Pensó que la engañábamos, porque nunca pudo creer en una máquina diabólica que permitía oír a alguien que estaba hablando en otra ciudad distante.

EL NUEVO OFICIO MÁS VIEJO DEL MUNDO

El otoño de París empezó de pronto y tarde este año, con un viento glacial que desplumó a los árboles de sus últimas hojas doradas. Las terrazas de los cafés se cerraron al mediodía, la viada se volvió turbia y el verano radiante que se había prolongado más de la cuenta pasó a ser una veleidad de la memoria. Parecía que en pocas horas hubieran pasado varios meses. El atardecer fue prematuro y lúgubre, pero nadie lo lamentó de veras, pues este tiempo de brumas es el natural de París, el que más le acompaña y el que mejor le sienta.

La más bella de las mujeres de alquiler que hacen su carrera de rutina en las callejuelas de Pigalle era una rubia espléndida que en un lugar menos evidente se hubiera confundido con una estrella de cine. Llevaba el conjunto de chaqueta y pantalón negros, que eran la fiebre de la moda, y a la hora en que empezó el viento helado se puso un abrigo legítimo de nucas de visón. Así estaba, ofreciéndose por doscientos francos frente a un hotel de paso de la calle Dupere, cuando un automóvil se detuvo frente a ella. Desde el puesto del volante, otra mujer hermosa y bien vestida le disparó de frente siete tiros de fusil. Esa noche, cuando la policía encontró al asesino, ya aquel drama de arrabal había retumbado en los periódicos, porque tenía dos elementos nuevos que lo hacía diferente. En efecto, ni la víctima ni el victimario eran rubias y bellas, sino dos hombres hechos y derechos, y ambos eran de Brasil.

La noticia no hizo sino poner en evidencia lo que ya se sabe de sobra en Europa: la prostitución callejera de las grandes ciudades es ahora un oficio de hombres, y los más codiciados de entre ellos, los más caros y los mejor vestidos son jóvenes latinoamericanos disfrazados de mujer. Según datos de prensa, de doscientos travestidos callejeros que hay en Francia, por lo menos la mitad ha llegado de Brasil. En España, Inglaterra, Suiza o Alemania Federal, donde el negocio

Publicado originalmente el 2 de diciembre de 1980.

parece ser todavía más fructífero, el número es mucho mayor y la nacionalidad más variada. El fenómeno tiene matices diversos en cada país, pero en todos se presenta como un cambio de fondo en el oficio más antiguo y conservador del mundo.

Cuando estuve en Europa por primera vez, hace unos veinticinco años, la prostitución era una industria próspera y ordenada, con categorías exactas y territorios muy bien repartidos. Yo llevaba todavía la imagen idílica de los burdeles del Caribe, aquellos patios de baile con guirnaldas de colores en los almendros, con gallinas impávidas que andaban picoteando por entre la música y bellas mulatas sin desbravar que se prostituían más por la fiesta que por la plata y que a veces incurrían en la descomunal inocencia de suicidarse por amor. A veces, uno se quedaba con ellas, no tanto por la vagabundina —como decía mi madre— como por la dicha de sentirlas respirar dormidas. Los desayunos eran más caseros y tiernos que los de la casa, y la verdadera fiesta empezaba a las once de la mañana, bajo los almendros apagados.

Educado en una escuela tan humana, no podía sino deprimirme el rigor comercial de las europeas. En Ginebra merodeaban por las orillas del lago, y lo único que las distinguía de las perfectas casadas eran las sombrillas de colores que llevaban abiertas con lluvia o con sol, de día o de noche, como un estigma de clase. En Roma se las oía silbar como pájaros entre los árboles de la Villa Borghese, y en Londres se volvían invisibles entre la niebla y tenían que encender luces que parecían de navegación para que uno encontrara su rumbo. Las de París, idealizadas por los poetas malditos y el mal cine francés de los años treinta, eran las más inclementes. Sin embargo, en los bares de desvelados de los Campos Elíseos se les descubría de pronto el revés humano: lloraban como novias ante el despotismo de los chulos inconformes con las cuentas de la noche. Costaba trabajo entender semejante mansedumbre de corazón en mujeres curtidas por un oficio tan bárbaro. Fue tal mi curiosidad que, años después, conocí a un chulo y le pregunté cómo era posible dominar con puño de hierro a mujeres tan bravas, y él me contestó, impasible: «Con amor». No volví a preguntar nada, por temor de entender menos.

La irrupción de los travestidos en aquel mundo de explotación y de muerte no ha conseguido sino hacerlo más sórdido. Su revolución consiste en hacer los dos oficios al mismo tiempo: el de prostitutas y el de chulos de sí mismos. Son autónomos y fieros. Muchos territorios nocturnos que las mujeres habían abandonado por su peligrosidad han sido ocupados por ellos a mano armada. Pero en la mayoría de las ciudades se han enfrentado a las mujeres y a sus chulos a golpes de mazo, y están ejerciendo su derecho de conquista en las mejores esquinas de Europa. El hecho de que muchos latinoamericanos estén participando en esta apoteosis del machismo no nos quita ni nos agrega ninguna gloria. Es una prueba más de nuestras perturbaciones sociales y no tiene por qué alarmarnos más que otras más graves.

La mayoría, por supuesto, son homosexuales. Tienen bustos espléndidos de silicón, y algunos terminan por realizar el sueño dorado de una operación drástica que los deja instalados para siempre en el sexo contrario. Pero muchos no lo son, y se han echado a la vida con sus armas prestadas —o usurpadas a golpes— porque es una mala manera de ganársela bien. Algunos son tranquilos padres de familia que hacen de día algún empleo de caridad y por la noche, cuando los niños se duermen, se van para la calle con las ropas dominicales de su mujer. Otros son estudiantes pobres que han resuelto de este modo la culminación de su carrera. Los más diestros se ganan en una buena noche hasta quinientos dólares. Lo cual —según dice mi esposa, aquí a mi lado— es mejor que escribir.

EL CUENTO DE LOS GENERALES QUE SE CREYERON
SU PROPIO CUENTO

Cuando el general Charles de Gaulle perdió su último plebiscito, en 1969, un caricaturista español lo dibujó frente a un general Francisco Franco minúsculo y ladino que le decía, con un tono de abuelo: «Eso te pasa por preguntón». Al día siguiente, el que fuera el hombre providencial de Francia estaba asando castañas en su retiro de Colombey-les-deux-Églises, donde poco después había de morirse de repente y solo mientras esperaba las noticias frente a la televisión. El periodista Claude Mauriac, que estuvo muy cerca de él, describió las últimas horas de su vida y su poder en un libro magistral, cuya revelación más sorprendente es que el viejo general estaba seguro de perder la consulta popular. En efecto, desde la semana anterior había hecho sacar sus papeles personales de la residencia presidencial y los había mandado en varias cajas a unas oficinas que tenía alquiladas de antemano. Más aún: algunos de sus allegados piensan ahora que De Gaulle había convocado aquel plebiscito innecesario sólo para darles a los franceses la oportunidad que querían de decirle que ya no más, general, que el tiempo de los gobernados es más lento e insidioso que el del poder, y que era venido el tiempo de irse, general, muchas gracias. Su vecino, el general Francisco Franco, no tuvo la dignidad de preguntarles lo mismo a los españoles, y poco antes de su mala muerte convocó a los periodistas que su propio régimen mantuvo amordazados durante cuarenta años y también a los que su propio régimen pagaba para que lo adularan, y los sorprendió con una declaración fantástica: «No puedo quejarme de la forma en que siempre me ha tratado la prensa».

Por preguntones acaba de ocurrirles lo mismo que a De Gaulle a los militares turbios y sin gloria que gobiernan con mano de hierro a Uruguay. Pero lo que más intriga de este descalabro imprevisto es por qué tenían que preguntar nada en un momento en que parecían

Publicado originalmente el 9 de diciembre de 1980.

dueños de todo su poder, con la prensa comprada, los partidos políticos prohibidos, la actividad universitaria y sindical suprimida y con media oposición en la cárcel o asesinada por ellos mismos, y nada menos que la quinta parte de la población nacional dispersa por medio mundo. Los analistas, acostumbrados a echarle la culpa de todo al imperialismo, no sólo de lo malo, sino también de lo bueno, piensan que los gorilas uruguayos tuvieron que ceder a la presión de los organismos internacionales de crédito para mejorar la imagen de su régimen. Otros, aún más retóricos, dicen que es la resistencia popular silenciosa, que, tarde o temprano, terminará por socavar la tiranía. No hay menos de veinte especulaciones distintas, y es natural que algunas de ellas sean factores reales. Pero hay una que corre el riesgo de parecer simplista, y que a lo mejor es la más próxima de la verdad: los gorilas uruguayos —al igual que el general Franco y al contrario del general De Gaulle— terminaron por creerse su propio cuento.

Es la trampa del poder absoluto. Absortos en su propio perfume, los gorilas uruguayos debieron pensar que la parálisis del terror era la paz, que los editoriales de la prensa vendida eran la voz del pueblo y, por consiguiente, la voz de Dios, que las declaraciones públicas que ellos mismos hacían eran la verdad revelada, y que todo eso, reunido y amarrado con un lazo de seda, era de veras la democracia. Lo único que les faltaba entonces, por supuesto, era la consagración popular, y para conseguirla se metieron como mansos conejos en la trampa diabólica del sistema electoral uruguayo. Es una máquina infernal tan complicada que los propios uruguayos no acaban de entenderla muy bien, y es tan rigurosa y fatal que, una vez puesta en marcha —como ocurrió el domingo pasado—, no hay manera de detenerla ni de cambiar su rumbo.

Sin embargo, lo más importante de esta pifia militar no es que el pueblo haya dicho que no, sino la claridad con que ha revelado la peculiaridad incomparable de la situación uruguaya. En realidad, la represión de la dictadura ha sido feroz, y no ha habido una ley humana ni divina que los militares no violaran ni un abuso que no cometieran. Pero en cambio se encuentran dando vueltas en el círculo vicioso de su propia preocupación legalista. Es decir: ni ellos mismos han podido escapar de una manera de ser del país y de un modo de ser de los uruguayos, que tal vez no se parezcan a los de ningún otro país de América Latina. Aunque sea por un detalle sobrenatural: Uruguay es el único donde los presos tienen que pagar la comida que se comen y el uniforme que se ponen, y hasta el alquiler de la celda.

En realidad, cuando irrumpieron contra el poder civil, en 1973, los *gorilas* uruguayos no dieron un golpe simple, como Pinochet o Videla, sino que se enredaron en el formalismo bobo de dejar un presidente de fachada. En 1976, cuando a éste se le acabó el período formal, buscaron otra fórmula retorcida para que el poder armado pareciera legal durante otros cinco años. Ahora trataban de buscar una nueva legalidad ficticia con este plebiscito providencial que les salió por la culata. Es como si la costumbre de la democracia representativa –que es casi un modo de ser natural de la nación uruguaya– se les hubiera convertido en un fantasma que no les permite hacer con las bayonetas otra cosa que sentarse en ellas.

SÍ: LA NOSTALGIA SIGUE SIENDO IGUAL QUE ANTES

Ha sido una victoria mundial de la poesía. En un siglo en que los vencedores son siempre los que pegan más fuerte, los que sacan más votos, los que meten más goles, los hombres más ricos y las mujeres más bellas, es alentadora la conmoción que ha causado en el mundo entero la muerte de un hombre que no había hecho nada más que cantarle al amor. Es la apoteosis de los que nunca ganan.

Durante cuarenta y ocho horas no se habló de otra cosa. Tres generaciones –la nuestra, la de nuestros hijos y la de nuestros nietos mayores– teníamos por primera vez la impresión de estar viviendo una catástrofe común, y por las mismas razones. Los reporteros de la televisión le preguntaron en la calle a una señora de ochenta años cuál era la canción de John Lennon que le gustaba más, y ella contestó, como si tuviera quince: «La felicidad es una pistola caliente». Un chico que estaba viendo el programa dijo: «A mí me gustan todas». Mi hijo menor le preguntó a una muchacha de su misma edad por qué habían matado a John Lennon, y ella le contestó, como si tuviera ochenta años: «Porque el mundo se está acabando».

Así es: la única nostalgia común que uno tiene con sus hijos son las canciones de los Beatles. Cada quien por motivos distintos, desde luego, y con un dolor distinto, como ocurre siempre con la poesía. Yo no olvidaré nunca aquel día memorable de 1963, en México, cuando oí por primera vez de un modo consciente una canción de los Beatles. A partir de entonces descubrí que el universo estaba contaminado por ellos. En nuestra casa de San Ángel, donde apenas si teníamos dónde sentarnos, había sólo dos discos: una selección de preludios de Debussy y el primer disco de los Beatles. Por toda la ciudad, a toda hora, se escuchaba un grito de muchedumbres: «Help, I need somebody». Alguien volvió a plantear por esa época el viejo tema de que los músicos mejores son los de la segunda letra del catálogo: Bach, Beethoven, Brahms y Bartók. Alguien volvió a de-

Publicado originalmente el 16 de diciembre de 1980.

cir la misma tontería de siempre: que se incluyera a Bozart. Álvaro Mutis, que como todo gran erudito de la música tiene una debilidad irremediable por los ladrillos sinfónicos, insistía en incluir a Bruckner. Otro trataba de repetir otra vez la batalla en favor de Berlioz, que yo libraba en contra porque no podía superar la superstición de que es un *oiseau de malheur*, es decir, un pájaro de mal agüero. En cambio, me empeñé, desde entonces, en incluir a los Beatles. Emilio García Riera, que estaba de acuerdo conmigo y que es un crítico e historiador de cine con una lucidez un poco sobrenatural, sobre todo después del segundo trago, me dijo por esos días: «Oigo a los Beatles con un cierto miedo, porque siento que me voy a acordar de ellos por todo el resto de mi vida». Es el único caso que conozco de alguien con bastante clarividencia para darse cuenta de que estaba viviendo el nacimiento de sus nostalgias. Uno entraba entonces en el estudio de Carlos Fuentes, y lo encontraba escribiendo a máquina con un solo dedo de una sola mano, como lo ha hecho siempre, en medio de una densa nube de humo y aislado de los horrores del universo con la música de los Beatles a todo volumen.

Como sucede siempre, pensábamos entonces que estábamos muy lejos de ser felices, y ahora pensamos lo contrario. Es la trampa de la nostalgia, que quita de su lugar a los momentos amargos y los pinta de otro color, y los vuelve a poner donde ya no duelen. Como en los retratos antiguos, que parecen iluminados por el resplandor ilusorio de la felicidad, y en donde sólo vemos con asombro cómo éramos de jóvenes cuando éramos jóvenes, y no sólo los que estábamos allí, sino también la casa y los árboles del fondo, y hasta las sillas en que estábamos sentados. El Che Guevara, conversando con sus hombres alrededor del fuego en las noches vacías de la guerra, dijo alguna vez que la nostalgia empieza por la comida. Es cierto, pero sólo cuando se tiene hambre. En cambio, siempre empieza por la música. En realidad, nuestro pasado personal se aleja de nosotros desde el momento en que nacemos, pero sólo lo sentimos pasar cuando se acaba un disco.

Esta tarde, pensando todo esto frente a una ventana lúgubre donde cae la nieve, con más de cincuenta años encima y todavía sin saber muy bien quién soy, ni qué carajos hago aquí, tengo la impresión de que el mundo fue igual desde mi nacimiento hasta que los Beatles empezaron a cantar. Todo cambió entonces. Los hombres se dejaron crecer el cabello y la barba, las mujeres aprendieron a desnudarse con naturalidad, cambió el modo de vestir y de amar, y se inició la liberación del sexo y de otras drogas para soñar. Fueron los años fragorosos de la guerra de Vietnam y la rebelión universitaria. Pero, sobre todo, fue el duro aprendizaje de una relación distinta entre los padres y los hijos, el principio de un nuevo diálogo entre ellos que había parecido imposible durante siglos.

El símbolo de todo esto —al frente de los Beatles— era John Lennon. Su muerte absurda nos deja un mundo distinto poblado de imágenes hermosas. En *Lucy in the Sky*, una de sus canciones más bellas, queda un caballo de papel periódico con una corbata de espejos. En *Eleanor Rigby* —con un bajo obstinado de chelos barrocos— queda una muchacha desolada que recoge el arroz, en el atrio de una iglesia donde acaba de celebrarse una boda. «¿De dónde vienen los solitarios?», se pregunta sin respuesta. Queda también el padre MacKensey escribiendo un sermón que nadie ha de oír, lavándose las manos sobre las tumbas, y una muchacha que se quita el rostro antes de entrar en su casa y lo deja en un frasco junto a la puerta para ponérselo otra vez cuando vuelva a salir. Estas criaturas han hecho decir que John Lennon era un surrealista, que es algo que se dice con demasiada facilidad de todo lo que parece raro, como suelen decirlo de Kafka quienes no lo han sabido leer. Para otros, es el visionario de un mundo mejor. Alguien que nos hizo comprender que los viejos no somos los que tenemos muchos años, sino los que no se subieron a tiempo en el tren de sus hijos.

ESTAS NAVIDADES SINIESTRAS

Ya nadie se acuerda de Dios en Navidad. Hay tantos estruendos de cornetas y fuegos de artificio, tantas guirnaldas de focos de colores, tantos pavos inocentes degollados y tantas angustias de dinero para quedar bien por encima de nuestros recursos reales que uno se pregunta si a alguien le queda un instante para darse cuenta de que semejante despelote es para celebrar el cumpleaños de un niño que nació hace 2.000 años en una caballeriza de miseria, a poca distancia de donde había nacido, unos mil años antes, el rey David. 954 millones de cristianos creen que ese niño era Dios encarnado, pero muchos lo celebran como si en realidad no lo creyeran. Lo celebran además muchos millones que no lo han creído nunca, pero les gusta la parranda, y muchos otros que estarían dispuestos a voltear el mundo al revés para que nadie lo siguiera creyendo. Sería interesante averiguar cuántos de ellos creen también en el fondo de su alma que la Navidad de ahora es una fiesta abominable, y no se atreven a decirlo por un prejuicio que ya no es religioso sino social.

Lo más grave de todo es el desastre cultural que estas Navidades pervertidas están causando en América Latina. Antes, cuando sólo teníamos costumbres heredadas de España, los pesebres domésticos eran prodigios de imaginación familiar. El niño Dios era más grande que el buey, las casitas encaramadas en las colinas eran más grandes que la virgen, y nadie se fijaba en anacronismos: el paisaje de Belén era completado con un tren de cuerda, con un pato de peluche más grande que un león que nadaba en el espejo de la sala, o con un agente de tránsito que dirigía un rebaño de corderos en una esquina de Jerusalén. Encima de todo se ponía una estrella de papel dorado con una bombilla en el centro, y un rayo de seda amarilla que había de indicar a los Reyes Magos el camino de la salvación. El resultado era más bien feo, pero se parecía a nosotros, y desde luego era mejor que tantos cuadros primitivos mal copiados del aduanero Rousseau.

Publicado originalmente el 24 de diciembre de 1980.

La mistificación empezó con la costumbre de que los juguetes no los trajeran los Reyes Magos –como sucede en España con toda razón–, sino el niño Dios. Los niños nos acostábamos más temprano para que los regalos llegaran pronto, y éramos felices oyendo las mentiras poéticas de los adultos. Sin embargo, yo no tenía más de cinco años cuando alguien en mi casa decidió que ya era tiempo de revelarme la verdad. Fue una desilusión no sólo porque yo creía de veras que era el niño Dios quien traía los juguetes, sino también porque hubiera querido seguir creyéndolo. Además, por pura lógica de adulto, pensé entonces que también los otros misterios católicos eran inventados por los padres para entretener a los niños, y me quedé en el limbo. Aquel día –como decían los maestros jesuitas en la escuela primaria– perdí la inocencia, pues descubrí que tampoco a los niños los traían las cigüeñas de París, que es algo que todavía me gustaría seguir creyendo para pensar más en el amor y menos en la píldora.

Todo aquello cambió en los últimos treinta años, mediante una operación comercial de proporciones mundiales que es al mismo tiempo una devastadora agresión cultural. El niño Dios fue destronado por el Santa Claus de los gringos y los ingleses, que es el mismo Papá Noel de los franceses, y a quienes todos conocemos demasiado. Nos llegó con todo: el trineo tirado por un alce y el abeto cargado de juguetes bajo una fantástica tempestad de nieve. En realidad, este usurpador con nariz de cervecero no es otro que el buen san Nicolás, un santo al que yo quiero mucho porque es el de mi abuelo el coronel, pero que no tiene nada que ver con la Navidad, y mucho menos con la Nochebuena tropical de la América Latina. Según la leyenda nórdica, san Nicolás reconstruyó y revivió a varios escolares que un oso había descuartizado en la nieve, y por eso le proclamaron el patrón de los niños. Pero su fiesta se celebra el 6 de diciembre y no el 25. La leyenda se volvió institucional en las provincias germánicas del norte a fines del siglo XVIII, junto con el árbol de los juguetes, y hace poco más de cien años pasó a Gran Bretaña y Francia. Luego pasó a Estados Unidos, y éstos nos lo mandaron para América Latina, con toda una cultura de contrabando: la nieve artificial, las candilejas de colores, el pavo relleno, y estos quince días de consumismo frenético al que muy pocos nos atrevemos a escapar. Con todo, tal vez lo más siniestro de estas Navidades de consumo sea la estética miserable que trajeron consigo: esas tarjetas postales indigentes, esas ristras de foquitos de colores, esas campanitas de vidrio, esas coronas de muérdago colgadas en el umbral, esas canciones de retrasados mentales que son los villancicos traducidos del inglés, y

tantas otras estupideces gloriosas para las cuales ni siquiera valía la
pena haber inventado la electricidad.

Todo eso, en torno a la fiesta más espantosa del año. Una noche
infernal en que los niños no pueden dormir con la casa llena de borrachos que se equivocan de puerta buscando dónde desaguar, o
persiguiendo a la esposa de otro que acaso tuvo la buena suerte de
quedarse dormido en la sala. Mentira: no es una noche de paz y
de amor, sino todo lo contrario. Es la ocasión solemne de la gente
que no se quiere. La oportunidad providencial de salir por fin de los
compromisos aplazados por indeseables: la invitación al pobre ciego
que nadie invita, a la prima Isabel que se quedó viuda hace quince
años, a la abuela paralítica que nadie se atreve a mostrar. Es la alegría por decreto, el cariño por lástima, el momento de regalar porque nos regalan, o para que nos regalen, y de llorar en público sin
dar explicaciones. Es la hora feliz de que los invitados se beban todo
lo que sobró de la Navidad anterior: la crema de menta, el licor de
chocolate, el vino de plátano. No es raro, como sucede a menudo,
que la fiesta termine a tiros. Ni es raro tampoco que los niños
—viendo tantas cosas atroces— terminen por creer de veras que el
niño Jesús no nació en Belén, sino en Estados Unidos.

CUENTO DE HORROR PARA LA NOCHEVIEJA

Llegamos a Arezzo un poco antes del mediodía, y perdimos más de dos horas buscando el castillo medieval que el escritor Miguel Otero Silva había comprado en aquel recodo idílico de la campiña toscana. Era un domingo de principios de agosto, ardiente y bullicioso, y no era fácil encontrar a alguien que supiera algo en las calles invadidas por los turistas. Al cabo de muchas tentativas inútiles volvimos al automóvil, abandonamos la ciudad por un sendero sin indicaciones viales, y una vieja pastora de gansos nos indicó con precisión dónde estaba el castillo. Antes de despedirse nos preguntó si pensábamos dormir allí, y le contestamos —como lo teníamos previsto— que sólo íbamos a almorzar. «Menos mal», dijo ella, «porque esa casa está llena de espantos.» Mi esposa y yo, que no creemos en aparecidos a pleno sol, nos burlamos de su credulidad. Pero los niños se pusieron dichosos con la idea de conocer un fantasma de cuerpo presente.

Miguel Otero Silva, que además de buen escritor es un anfitrión espléndido y un comedor riguroso, nos esperaba con un almuerzo de nunca olvidar. Como se nos había hecho tarde, no tuvimos tiempo de conocer el interior del castillo antes de sentarnos a la mesa, pero su aspecto desde fuera no tenía nada de pavoroso, y cualquier inquietud se mitigaba con la visión completa de la ciudad desde la terraza de verano donde estábamos almorzando. Era difícil creer que en aquella colina de casas encaramadas, donde apenas cabían 90.000 personas, hubieran nacido tantas de genio perdurable, como Guido de Arezzo, que inventó una escritura para cantar, o el espléndido Vasari y el deslenguado Aretino, o Julio II y el propio Cayo Clinio Mecenas, los dos grandes padrinos de las artes y las letras de su tiempo. Sin embargo, Miguel Otero Silva nos dijo con su sentido del humor habitual que tan altas cifras históricas no eran las más insignes de Arezzo. «El más importante», nos dijo, «fue Ludovico.» Así, sin apellidos: Ludovico, el

Publicado originalmente el 30 de diciembre de 1980.

gran señor de las artes y de la guerra que había construido aquel castillo de su desgracia.

Miguel Otero Silva nos habló de Ludovico durante todo el almuerzo. Nos habló de su poder sin medida, de su amor desgraciado y de su muerte espantosa. Nos contó cómo fue que, en un instante de locura del corazón, había apuñalado a su dama en el lecho donde acababan de amarse, y luego azuzó contra sí mismo a sus feroces perros de guerra, que lo despedazaron a dentelladas. Nos aseguró, muy serio, que a partir de la medianoche el espectro de Ludovico deambulaba por su castillo de tinieblas, tratando de conseguir un instante de sosiego para su purgatorio de amor. Sin embargo, a pleno día, con el estómago lleno y el corazón contento, aquello no podía parecer sino una broma como tantas otras de Miguel Otero Silva para entretener a sus invitados.

El castillo, en realidad, era inmenso y sombrío, como pudimos comprobarlo después de la siesta. Sus dos pisos superiores y sus 82 cuartos habían padecido toda clase de mudanzas de sus dueños sucesivos. Miguel Otero Silva había restaurado por completo la planta baja y se había hecho construir un dormitorio moderno con suelos de mármol e instalaciones para sauna y cultura física, y la terraza de flores intensas donde habíamos almorzado. «Son cosas de Caracas para despistar a Ludovico», nos dijo. Yo había oído decir, en efecto, que lo único que confunde a los fantasmas son los laberintos del tiempo.

La segunda planta estaba sin tocar. Había sido la más usada en el curso de los siglos, pero ahora era una sucesión de cuartos sin ningún carácter, con muebles abandonados de diferentes épocas. La planta superior era la más abandonada de todas, pero se conservaba en ella una habitación intacta, por donde el tiempo se había olvidado de pasar. Era el dormitorio de Ludovico. Fue un instante mágico. Allí estaba la cama de marquesina, con cortinas bordadas en hilos de oro y el sobrecamas de prodigios de pasamanería todavía salpicado con la sangre de la amante sacrificada. Estaba la chimenea con las cenizas heladas y el último leño convertido en piedra, el armario con sus armas bien cebadas y el retrato al óleo del caballero pensativo, pintado por algunos de los maestros florentinos que no tuvieron la fortuna de sobrevivir a su tiempo. Sin embargo, lo que más me impresionó fue el olor a fresas recientes que permanecía, sin explicación posible, en el ámbito de la habitación.

Los días del verano son largos y parsimoniosos en la Toscana, y el horizonte se mantiene en su sitio hasta las nueve de la noche. Después de mostrarnos el interior del castillo, Miguel Otero Silva nos llevó a ver los frescos de Piero della Francesca, en la iglesia de San Francisco; luego nos tomamos un café bien conversado bajo las pérgolas de la plaza embellecidas por los primeros aires de la noche, y cuando volvimos al castillo para recoger las maletas encontramos la cena servida. De modo que nos quedamos a comer. Mientras lo hacíamos, los niños prendieron más antorchas en la cocina y se fueron a explorar las tinieblas en los pisos de arriba. Desde la mesa oíamos sus pasos de caballos cerreros por las escaleras, el crujido lúgubre de las puertas, los gritos felices llamando a Ludovico en los cuartos abandonados. Fue a ellos a quienes se les ocurrió la mala idea de que nos quedáramos a dormir. Miguel Otero Silva los apoyó encantado, y nosotros no tuvimos el valor civil de decirles que no.

Al contrario de lo que yo temía, dormimos muy bien; mi esposa y yo, en un dormitorio de la planta baja, y mis hijos, en el cuarto contiguo. Mientras trataba de conseguir el sueño conté los doce toques insomnes del reloj de péndulo de la sala, y por un instante me acordé de la pastora de gansos. Pero estábamos tan cansados que nos dormimos muy pronto, en un sueño denso y continuo, y desperté, después de las siete, con un sol espléndido. A mi lado, Mercedes navegaba en el mar apacible de los inocentes. «Qué tontería», me dije, «que alguien siga creyendo en fantasmas por estos tiempos.» Sólo entonces caí en la cuenta —con un zarpazo de horror— que no estábamos en el cuarto donde nos habíamos acostado la noche anterior, sino en el dormitorio de Ludovico, acostados en su cama de sangre. Alguien nos había cambiado de cuarto durante el sueño.

Surinam —como no todo el mundo lo sabe— es un país independiente sobre el mar Caribe, que fue hasta hace pocos años una colonia holandesa. Tiene 163.820 kilómetros cuadrados y un poco más de 384.000 habitantes de origen múltiple: indios de la India, indios locales, indonesios, africanos, chinos y europeos. Su capital, Paramaribo —que en castellano pronunciamos como palabra grave y que los nativos pronuncian como esdrújula—, es una ciudad fragoroso y triste, con un espíritu más asiático que americano, en la cual se hablan cuatro idiomas y numerosos dialectos aborígenes, además de la lengua oficial —el holandés—, y se profesan seis religiones: hinduista, católica, musulmana, morava, holandesa reformada y luterana. En la actualidad, el país está gobernado por un régimen de militares jóvenes, de los cuales se sabe muy poco, inclusive en los países vecinos, y nadie se acordaría de él si no fuera porque una vez a la semana es la escala de rutina de un avión holandés que vuela de Amsterdam a Caracas.

Había oído hablar de Surinam desde muy niño, no por Surinam mismo —que entonces se llamaba Guayana Holandesa—, sino porque estaba en los límites de la Guayana Francesa, en cuya capital, Cayena, estuvo hasta hace poco la tremenda colonia penal conocida, en la vida y en la muerte, como la Isla del Diablo. Los pocos que lograron fugarse de aquel infierno, que lo mismo podían ser criminales bárbaros que idealistas políticos, se dispersaban por las islas numerosas de las Antillas hasta que conseguían volver a Europa o se establecían con la identidad cambiada en Venezuela y la costa caribe de Colombia. El más célebre de todos fue Henri Charrier, autor de *Papillon*, que prosperó en Caracas como promotor de restaurantes y otros oficios menos diáfanos, y que murió hace pocos años en la cresta de una gloria literaria efímera, pero tan meritoria como inmerecida. Esa gloria, en realidad, le correspondía, con mejores

Publicado originalmente el 6 de enero de 1981.

títulos, a otro fugitivo francés que describió mucho antes que Papillon los horrores de la Isla del Diablo, y sin embargo no figura hoy
en la literatura de ninguna parte, ni su nombre se encuentra en las
enciclopedias. Se llamaba René Belbenoit, había sido periodista en
Francia antes de ser condenado a cadena perpetua por una causa que
ningún periodista de hoy ha podido recordar, y siguió siéndolo en
Estados Unidos, donde consiguió asilo y donde murió de una vejez
honrada.

Algunos de estos prófugos se refugiaron en el pueblo del Caribe
colombiano donde yo nací, en los tiempos de la *fiebre del banano*,
cuando los cigarros no se encendían con fósforos, sino con billetes
de cinco pesos. Varios se asimilaron a la población y llegaron a ser
ciudadanos muy respetables, que se distinguieron siempre por su habla difícil y el hermetismo de su pasado. Uno de ellos, Roger Chantal, que había llegado sin más oficio que el de arrancador de muelas
sin anestesia, se volvió millonario de la noche a la mañana sin explicación alguna. Hacía unas fiestas babilónicas —en un pueblo inverosímil que tenía muy poco que envidiarle a Babilonia—, se emborra-
chaba a muerte y gritaba en su feliz agonía: «Je suis l'homme le plus
riche du monde». En medio del delirio le aparecieron unas ínfulas
de benefactor que nadie le conocía hasta entonces, y le regaló a la
iglesia un santo de yeso de tamaño natural, que fue entronizado con
una parranda de tres días. Un martes cualquiera llegaron en el tren
de las once tres agentes secretos que fueron de inmediato a su casa.
Chantal no estaba ahí, pero los agentes hicieron una requisa minuciosa en presencia de su esposa nativa, que no opuso ninguna resistencia, salvo cuando quisieron abrir el enorme escaparate del dormitorio. Entonces los agentes rompieron los espejos y encontraron
más de un millón de dólares en billetes falsos escondidos entre el
cristal y la madera. Nunca más se supo de Roger Chantal. Más tarde
circuló la leyenda de que el millón de dólares falsos había entrado al
país dentro del santo de yeso, que ningún agente de aduana había
tenido la curiosidad de registrar.

Todo esto me volvió de golpe a la memoria poco antes de la
Navidad de 1957, cuando tuve que hacer una escala de una hora en
Paramaribo. El aeropuerto era una pista de tierra aplanada con una
caseta de palma, en cuyo horcón central había un teléfono de aquellos de las películas de vaqueros, con una manivela que se hacía girar con fuerza y muchas veces hasta obtener la respuesta. El calor era
abrasante, y el aire, polvoriento e inmóvil, tenía el olor de caimán
dormido con que se identifica el Caribe cuando uno llega de otro
mundo. En un taburete apoyado en el horcón del teléfono estaba

una negra muy bella, joven y maciza, con un turbante de muchos colores como los que usan las mujeres en algunos países del África. Estaba encinta, a punto de dar a luz, y fumaba un tabaco en silencio y como sólo he visto hacerlo en el Caribe: con el fuego dentro de la boca y echando humo por el cabo, como una chimenea de buque. Era el único ser humano en el aeropuerto.

Al cabo de un cuarto de hora llegó un jeep decrépito envuelto en una nube de polvo ardiente, del cual descendió un negro de pantalones cortos y casco de corcho con los papeles para despachar el avión. Mientras atendía los trámites, hablaba por teléfono, dando gritos en holandés. Doce horas antes yo estaba en una terraza marítima de Lisboa, frente al inmenso océano portugués, viendo las bandadas de gaviotas que se metían en las cantinas del puerto huyendo del viento glacial. Europa era entonces una tierra decrépita cubierta de nieve, los días de luz no tenían más de cinco horas, y era imposible imaginar que de veras existiera un mundo de sol canicular y guayabas podridas, como aquél donde acabábamos de descender. Sin embargo, la única imagen que persistió de aquella experiencia, y que aún conservo intacta, fue la de la hermosa negra impasible, que tenía en las piernas una canasta con rizomas de jengibre para vendérselas a los pasajeros.

Ahora, viajando otra vez de Lisboa a Caracas, volví a aterrizar en Paramaribo, y mi primera impresión fue que nos habíamos equivocado de ciudad. La terminal del aeropuerto es ahora un edificio luminoso, con grandes ventanales de vidrio, con un aire acondicionado muy tenue, oloroso a medicinas para niños, y esa música enlatada que se repite sin misericordia en todos los lugares públicos del mundo. Hay tiendas de artículos de lujo sin impuestos, tan abundantes y bien surtidas como en el Japón, y una cafetería multitudinaria donde se encuentran revueltas y en ebullición las siete razas del país, sus seis religiones y sus lenguas incontables. Aquel cambio no parecía de veinte años, sino de varios siglos.

Mi profesor Juan Bosch, autor, entre otras muchas cosas, de una historia monumental del Caribe, dijo alguna vez en privado que nuestro mundo mágico es como esas plantas invencibles que renacen debajo del cemento, hasta que lo cuartean y lo desbaratan, y vuelven a florecer en su mismo sitio. Esto lo comprendí mejor que nunca cuando salí por una puerta imprevista del aeropuerto de Paramaribo y encontré una fila de viejas mujeres sentadas impávidas, todas negras, todas con turbantes de colores y todas fumando con la brasa dentro de la boca. Vendían frutas y artesanía del lugar, pero ninguna hacía el menor esfuerzo por convencer a nadie. Sólo

una de ellas, que no era la mayor, vendía raíces de jengibre. La reconocí al instante. Sin saber por dónde empezar ni qué hacer en
realidad con aquel hallazgo, le compré un puñado de raíces. Mientras lo hacía, recordando su estado de la primera vez, le pregunté
sin preámbulos cómo estaba su hijo. Ni siquiera me miró. «No es
hijo, sino hija», dijo, «y acaba de darme mi primer nieto a los veintidós años.»

HAY QUE SALVAR A EL SALVADOR

Un amigo, que volvió tarde a su casa después del trabajo, encontró a la esposa viendo en la televisión un espectáculo bárbaro. Era una muchedumbre de hombres, mujeres y niños masacrados por la fuerza pública en el atrio de una iglesia. Muchos estaban muertos, otros agonizaban como gusanos en una ciénaga de sangre y los últimos vivos se dispersaban espantados bajo el fuego implacable de la metralla. Parecía una imitación barata de la carnicería de Odessa en *El acorazado Potemkin*, la película memorable de Sergei Eisenstein. Sólo que más feroz y sin ninguna consideración artística. Mi amigo, que se horroriza con el cine de horror, le reprochó a su mujer que estuviera viendo semejante película. Pero ella le contestó impasible: «No es una película, sino las noticias de El Salvador». Esto ocurrió a principios del año pasado. Desde entonces, hasta el diciembre que acaba de pasar, 10.000 personas murieron en aquella masacre continua. Como si el año bisiesto hubiera sido más bisiesto en El Salvador que en el resto del mundo.

Ronald Reagan dijo hace poco que ésa es una guerra civil en tres direcciones. Sin duda quería decir que de un lado está la dictadura militar, del otro están las pandillas de criminales de la extrema derecha y del otro lado las fuerzas de la revolución. Pero la aritmética social de El Salvador es más simple. En verdad, esta guerra civil, que ya es la más sangrienta de América Latina en toda su historia, tiene sólo dos bandos: la aristocracia feudal, de un lado, y el resto de la nación, del lado contrario. El 90 % de la población del país —cuya densidad demográfica es una de las más altas del mundo— son indios y mestizos. Sólo el 10 % son blancos, pero también son ellos quienes controlan desde siempre y con puño de hierro la totalidad del poder económico y político. La proporción de las víctimas es igual: el 90 % de los muertos del terrible año bisiesto que acaba de pasar eran del bando de los pobres. Tanto de los pobres en armas como de los inermes, inclusive el arzobispo primado. Es decir, que, a diferen-

Publicado originalmente el 13 de enero de 1981.

cia de Nicaragua, donde el Frente Sandinista logró concertar a los antisomocistas de todos los tamaños y todos los niveles, las tensiones sociales de El Salvador se han resuelto en una irremediable confrontación de clases. Eso explica en gran parte la polarización radical de esta guerra, su ferocidad insaciable y la resolución de exterminio de ambos lados, con episodios tan bestiales que ya resultan insoportables hasta en la televisión.

Es una guerra antigua. Entre 1931 y 1944, el país padeció la dictadura del general Maximiliano Hernández Martínez, un déspota con ínfulas de teósofo cuyo defecto más notable era que estaba loco. Había inventado un péndulo mágico que suspendía sobre los alimentos para averiguar, según su inclinación, si estaban envenenados. En una ocasión trató de conjurar una epidemia de escarlatina cubriendo con papel rojo el alumbrado público del país. Estas fantasías folclóricas que, después de todo, no molestaban a nadie, tuvieron una expresión brutal en 1932, cuando las fuerzas armadas se enfrentaron a tiros a una vasta insurrección agraria y mataron a 31.000 campesinos. Lo repito con todas sus letras: treinta y un mil campesinos. Desde entonces, El Salvador ha pasado por todos los matices del poder militar y, de una u otra forma, la guerra desigual entre los ricos y los pobres no ha tenido un instante de tregua. El inconformismo secular se expresa hoy a través de los movimientos armados, las organizaciones de masas y los partidos políticos de oposición que han logrado por fin una fórmula de unidad, cuya cara pública es el Frente Democrático Revolucionario. Se supone que tienen unos 5.000 hombres sobre las armas, inclusive con piezas de artillería media, y unos 30.000 reservistas dispuestos para la ofensiva final anunciada para estos días. Pero quienes han ido a El Salvador en los últimos tiempos saben que es un ejército popular infiltrado por todas partes. Sus soldados aparecen vestidos de meseros en los restaurantes, de camareras en los hoteles, de choferes en los taxis y hasta de curas con sotanas en los confesionarios. En estas condiciones, el asalto decisivo de estos días podría no ser tan ilusorio como algunos anteriores.

El poder feudal, por su parte, cuenta con el apoyo de Estados Unidos y con unas fuerzas armadas muy bien armadas. Cuenta con bandas de asesinos a sueldo que hacen el trabajo sucio que el gobierno no se atreve a hacer para que no se le vea la cara verdadera. Cuenta, en fin, con una fracción de la Democracia Cristiana que se olvidó de Cristo y parece dispuesta a no dejar ningún cristiano vivo. A esta fracción pertenece el actual presidente de la República, Napoleón Duarte, que no fue elegido por nadie, sino nombrado por los militares en un momento de apuro para tener una pantalla civil.

No es casual esta actitud de la Democracia Cristiana. Al contrario, forma parte de una estrategia global, cuyo paladín en América Latina es el nuevo presidente de Venezuela, Luis Herrera Campins, y cuya finalidad inmediata es torcer los avances democráticos en el Caribe y América Central, con el pretexto de contrarrestar la influencia cubana. Herrera Campins, de quien se dice que come caramelos todo el día y hace la siesta después del desayuno, no se ha dormido en este empeño. Primero, porque corresponde a su ideología, y segundo, porque corresponde a su obsesión de deshacer todo lo que hizo su antecesor, Carlos Andrés Pérez, contra el cual sigue haciendo oposición desde la presidencia. Hasta ahora ha logrado poner a los gobiernos del Pacto Andino contra la liberación de El Salvador, con excepción del presidente de Ecuador, Jaime Roldos, cuya vocación progresista es indiscutible. Pero su obra maestra fue conseguir que Napoleón Duarte fuera invitado como presidente legítimo al sesquicentenario de Simón Bolívar, al tiempo que no se invitó al dictador de Bolivia, como si éste fuera más dictatorial, más sanguinario y de origen menos ilegítimo que el régimen de El Salvador. En todo caso, y con igual derecho que Napoleón Duarte, hubieran podido invitar al presidente de Guatemala, general Romero Lucas, que por lo menos subió al poder mediante una farsa electoral.

Éste es el panorama que encontrará Ronald Reagan la semana entrante, cuando se siente en la silla presidencial de Estados Unidos. El presidente de México, José López Portillo, le mandó a decir en público, y pensando, sin duda, en El Salvador, que no intervenga en América Latina, que respete la voluntad de los países que buscan definiciones nuevas, que son mayores de edad y capaces de ocuparse solos de sus propios asuntos. Es probable, además –conociendo el carácter de López Portillo–, que se lo haya repetido en privado, y ya con nombre propio, en su reciente entrevista de la frontera.

Sin embargo –de acuerdo con un memorando del Departamento de Estado que divulgó *The New York Times* hace un mes–, la intervención de Estados Unidos en El Salvador está ya preparada hasta en sus ínfimos detalles políticos y militares. La ha preparado el presidente Carter, y el presidente Reagan sólo tendría que apretar un botón. Tal como lo hizo John F. Kennedy hace veinte años, cuando llegó al poder y se encontró con el plan de invasión a Cuba preparado por Eisenhower. Dice el refrán que a ningún perro lo capan dos veces, pero lo peligroso en este caso es que se trata de dos perros distintos.

25.000 MILLONES DE KILÓMETROS CUADRADOS
SIN UNA SOLA FLOR

Cuando Neil Armstrong desembarcó en la superficie lunar, hace ahora once años, el animador de la televisión exclamó emocionado: «Por primera vez en la historia, el hombre ha puesto un pie en la Luna». Un niño que estaba con nosotros, y que había seguido con ansiedad los pormenores del desembarco, gritó sorprendido:

—¿Pero es la primera vez? ¡Qué tontería!

Su desencanto era comprensible. Para un niño de su tiempo, acostumbrado a vagar todas las noches por el espacio sideral de la televisión, la noticia del primer hombre en la Luna era como un regreso a la Edad de Piedra. A mí me dejó también una sensación de desaliento, pero por motivos más simples. Estábamos pasando el verano en la isla de Pantelaria, en el extremo sur de Sicilia, y no creo que exista en el mundo un lugar más apropiado para pensar en la Luna.

Recuerdo como en un sueño las llanuras interminables de roca volcánica, el mar inmóvil, la casa pintada de cal viva hasta los sardineles, desde cuyas ventanas se veían en las noches sin viento las aspas luminosas de los faros de África. Explorando los fondos dormidos alrededor de la isla, habíamos descubierto una ristra de torpedos amarillos encallados desde la última guerra; habíamos rescatado un ánfora con guirnaldas petrificadas que todavía tenía dentro los rescoldos de un vino inmemorial carcomido por los años, y nos habíamos bañado en un remanso humeante cuyas aguas eran tan densas que casi se podía caminar sobre ellas.

Yo pensaba con una cierta nostalgia premonitoria que así debía ser la Luna. Pero el desembarco de Armstrong aumentó mi orgullo patriótico: Pantelaria era mejor.

Para quienes perdemos el tiempo pensando en estas cosas, hay desde entonces dos lunas. La Luna astronómica, con mayúscula, cuyo valor científico debe ser muy grande, pero que carece por completo

Publicado originalmente el 20 de enero de 1981.

de validez poética. La otra es la Luna de siempre que vemos colgada en el cielo; la Luna única de los licántropos y los boleros, y a la cual —por fortuna— nadie llegará jamás.

Hasta ahora, la conquista del espacio parece condenada a esta clase de desilusiones. La más triste es que, después del viaje asombroso del *Voyager I*, se puede ya afirmar sin ninguna duda que al menos en esta minúscula provincia del sistema solar no existe la vida como nosotros la entendemos. Venus y Mercurio, los dos planetas más cercanos al Sol, estaban descalificados desde hace mucho tiempo como dos pelotas incandescentes sin ningún valor comercial. Los canales de Marte, que suponíamos excavados por nuestros primos del espacio, no parecen ser mucho más que una pura ilusión. Júpiter, 317 veces más grande que la Tierra, es un bobo gigantesco con doscientos grados bajo cero. Después de la fructífera exploración de Saturno, sólo nos falta conocer a Urano, Neptuno y Plutón, los tres ancianos solitarios de los suburbios solares, cuyas órbitas son tan desmesuradas que el último de ellos se demora más de 248 años de los nuestros para terminar una vuelta alrededor del Sol.

La utilidad científica de estos descubrimientos es incalculable, pero una cosa queda en claro: allá no hay nadie. Es una inmensa noche glacial de 25.000 millones de kilómetros cuadrados donde hay océanos de nitrógeno líquido, vientos diez veces más devastadores que los tifones de Sumatra, y tempestades apocalípticas que pueden durar hasta 30.000 años. Pero no hay una sola flor. Ni siquiera una rosa miserable como esta de mi escritorio, que se aburre quizá por no ser más de lo que es, sin saber que ella sola es un prodigio irrepetible en el universo.

Luciano de Samosata —según dice Jorge Luis Borges en su prólogo a *Crónicas marcianas*, de Bradbury— escribió que los selenitas hilaban y tejían los metales y el vidrio, se quitaban y se ponían los ojos, y bebían extractos del aire. Es una cita como casi todas las de Borges, a la vez deslumbrante y sospechosa, pero ilustra muy bien sobre la imagen que se tenía en el siglo segundo de los seres extraterrestres. Con los progresos de la ciencia y el refinamiento de la imaginación, la visión no ha mejorado, sino todo lo contrario. Los escritores de ficción científica describen a nuestros parientes siderales como criaturas pavorosas con orejas de murciélago, antenas en vez de cuernos, membranas interdigitales y ventosas en los sentidos. Todo lo que tiene que ver con ellos es de naturaleza viscosa e infame, y su única ventaja sobre nosotros son sus armas luciferinas y su prodigiosa inteligencia para la maldad. El cine no había logrado nunca un terror más intenso que el de las películas del espacio.

Tal vez la desilusión del vecindario celeste nos sirva para corregir este grave e injusto malentendido universal. Tal vez, al cabo de tantos milenios de fantasías mezquinas, empecemos a comprender que los aborígenes de los otros planetas no pueden estar donde tanto los buscamos, porque están aquí desde mucho antes que nosotros: son los microbios. Llevan milenios viviendo en nuestra vida, navegando nuestra sangre, durmiendo en nuestras heridas, naciendo y muriendo con nosotros, y todavía, ni ellos ni nosotros sabemos quiénes somos. Su naturaleza diversa les impide hacer lo que quisieran, y nos impide hacer lo que quisiéramos, que es sentarnos a comer juntos en la misma mesa, jugar a las barajas y contarles a los niños las verdades del universo para que no vayan al cine a ver tantas calumnias del espacio.

En cambio de eso, andamos a la greña desde el principio de la creación, ellos tratando de exterminarnos y nosotros tratando de exterminarlos a ellos, empeñados en una guerra a muerte de la cual no sabemos ni siquiera contra quién la libramos. Pues es muy probable que nuestros microbios, al igual que nosotros, tampoco sepan dónde están, ni por qué han venido. «Hay otros mundos, pero están en éste», dijo Paul Éluard. Otro grande escritor de nuestro tiempo que tal vez no crea en los marcianos, lo dijo de un modo más brutal: «La Tierra es el infierno de otros planetas».

LA POESÍA, AL ALCANCE DE LOS NIÑOS

Un maestro de literatura le advirtió el año pasado a la hija menor de un gran amigo mío que su examen final versaría sobre *Cien años de soledad*. La chica se asustó, con toda la razón, no sólo porque no había leído el libro, sino porque estaba pendiente de otras materias más graves. Por fortuna, su padre tiene una formación literaria muy seria y un instinto poético como pocos, y la sometió a una preparación tan intensa que, sin duda, llegó al examen mejor armada que su maestro. Sin embargo, éste le hizo una pregunta imprevista: ¿qué significa la letra al revés en el título de *Cien años de soledad*? Se refería a la edición de Buenos Aires, cuya portada fue hecha por el pintor Vicente Rojo con una letra invertida, porque así se lo indicó su absoluta y soberana inspiración. La chica, por supuesto, no supo qué contestar. Vicente Rojo me dijo cuando se lo conté que tampoco él lo hubiera sabido.

Ese mismo año, mi hijo Gonzalo tuvo que contestar un cuestionario de literatura elaborado en Londres para un examen de admisión. Una de las preguntas pretendía establecer cuál era el símbolo del gallo en *El coronel no tiene quien le escriba*. Gonzalo, que conoce muy bien el estilo de su casa, no pudo resistir la tentación de tomarle el pelo a aquel sabio remoto, y contestó: «Es el gallo de los huevos de oro». Más tarde supimos que quien obtuvo la mejor nota fue el alumno que contestó, como se lo había enseñado el maestro, que el gallo del coronel era el símbolo de la fuerza popular reprimida. Cuando lo supe me alegré una vez más de mi buena estrella política, pues el final que yo había pensado para ese libro, y que cambié a última hora, era que el coronel le torciera el pescuezo al gallo e hiciera con él una sopa de protesta.

Desde hace años colecciono estas perlas con que los malos maestros de literatura pervierten a los niños. Conozco uno de muy buena fe para quien la abuela desalmada, gorda y voraz, que explota a la

Publicado originalmente el 27 de enero de 1981.

cándida Eréndira para cobrarse una deuda es el símbolo del capitalismo insaciable. Un maestro católico enseñaba que la subida al cielo de Remedios la Bella era una transposición poética de la ascensión en cuerpo y alma de la virgen María. Otro dictó una clase completa sobre Herbert, un personaje de algún cuento mío que le resuelve problemas a todo el mundo y reparte dinero a manos llenas. «Es una hermosa metáfora de Dios», dijo el maestro. Dos críticos de Barcelona me sorprendieron con el descubrimiento de que *El otoño del patriarca* tenía la misma estructura del tercer concierto de piano de Béla Bartók. Esto me causó una gran alegría por la admiración que le tengo a Béla Bartók, y en especial a ese concierto, pero todavía no he podido entender las analogías de aquellos dos críticos. Un profesor de literatura de la Escuela de Letras de La Habana destinaba muchas horas al análisis de *Cien años de soledad* y llegaba a la conclusión –halagadora y deprimente al mismo tiempo– de que no ofrecía ninguna solución. Lo cual terminó de convencerme de que la manía interpretativa termina por ser a la larga una nueva forma de ficción que a veces encalla en el disparate.

Debo ser un lector muy ingenuo, porque nunca he pensado que los novelistas quieran decir más de lo que dicen. Cuando Franz Kafka dice que Gregorio Samsa despertó una mañana convertido en un gigantesco insecto, no me parece que eso sea el símbolo de nada, y lo único que me ha intrigado siempre es qué clase de animal pudo haber sido. Creo que hubo en realidad un tiempo en que las alfombras volaban y había genios prisioneros dentro de las botellas. Creo que la burra de Balaam habló –como lo dice la Biblia– y lo único lamentable es que no se hubiera grabado su voz, y creo que Josué derribó las murallas de Jericó con el poder de sus trompetas, y lo único lamentable es que nadie hubiera transcrito su música de demolición. Creo, en fin, que el licenciado Vidriera –de Cervantes– era en realidad de vidrio, como él lo creía en su locura, y creo de veras en la jubilosa verdad de que Gargantúa se orinaba a torrentes sobre las catedrales de París. Más aún: creo que otros prodigios similares siguen ocurriendo, y que si no los vemos es en gran parte porque nos lo impide el racionalismo oscurantista que nos inculcaron los malos profesores de literatura.

Tengo un gran respeto, y sobre todo un gran cariño, por el oficio de maestro, y por eso me duele que ellos también sean víctimas de un sistema de enseñanza que los induce a decir tonterías. Uno de mis seres inolvidables es la maestra que me enseñó a leer a los cinco años. Era una muchacha bella y sabía que no pretendía saber más de lo que podía, y era además tan joven que con el tiempo ha terminado por ser menor que yo. Fue ella quien nos leía en clase los primeros poemas que me pudrieron el seso para siempre. Recuerdo con la misma gratitud al profesor de literatura del bachillerato, un hombre modesto y prudente que nos llevaba por el laberinto de los buenos libros sin interpretaciones rebuscadas. Este método nos permitía a sus alumnos una participación más personal y libre en el prodigio de la poesía. En síntesis, un curso de literatura no debería ser mucho más que una buena guía de lecturas. Cualquier otra pretensión no sirve para nada más que para asustar a los niños. Creo yo, aquí en la trastienda.

LA ENFERMEDAD POLÍTICA DE REZA PAHLAVI

Pierre Salinger —el sagaz periodista norteamericano que fue asesor de prensa de John F. Kennedy— ha dicho, al parecer, en un programa de televisión que David Rockefeller y Henry Kissinger son en último análisis los responsables del triste drama de los rehenes de Teherán, porque fueron ellos quienes lograron que el sha Mohamed Reza Pahlavi fuera admitido en Estados Unidos para una operación quirúrgica que bien podía hacerse en México. Esta declaración pone en primer plano el cuento tenebroso de la enfermedad del sha, cuyos misterios de doble fondo —más políticos que médicos— no son nada fáciles de descifrar.

En realidad, en septiembre de 1979, los médicos mexicanos que se ocupaban del sha estaban preparados para operarlo, y él estaba de acuerdo. En su refugio primaveral de Cuernavaca —una inmensa mansión de seis millones de dólares, muy bien disimulados entre los ruiseñores y las buganvillas—, el sha había sido sorprendido a finales de agosto por unas raras fiebres crepusculares. Después de un examen a fondo, los médicos mexicanos encontraron indicios de una anemia perniciosa, que es una enfermedad endémica en Irán, y que el propio sha creía haber contraído desde que era cadete militar. El doctor Georges Flandrin, del hospital Saint Louis, de París, que en 1973 había detectado por primera vez la grave enfermedad del sha, fue llamado de urgencia. Por su parte, el sha solicitó a su amigo y banquero David Rockefeller que le mandara un especialista norteamericano. Rockefeller —según el periodista Mark Bloom lo contó en la revista *Science*— envió a Cuernavaca al doctor Benjamin Kean, jefe de medicina tropical del New York Hospital, y profesor de parasitología de la Universidad de Cornell. Desde su primera visita, el doctor Kean llegó a la conclusión simple de que los médicos mexicanos habían confundido los parásitos de la malaria con la precipitación de unos cristales de tinte en el análisis de sangre. Estuvo de acuerdo, sin em-

Publicado originalmente el 4 de febrero de 1981.

bargo, en que debía extirpar la vesícula cuanto antes y hacer exploraciones del colédoco, pero consideró que esto sólo era posible en el New York Hospital. Fue en base a ese informe que Rockefeller y Kissinger solicitaron el ingreso del sha en Estados Unidos.

El propio sha parecía tener otros planes. Por instrucciones suyas, el doctor Flandrin había hecho ya gestiones para que le operaran en México, en cuyos médicos confiaban ambos, y cuyos recursos técnicos les parecían suficientes. Por otra parte, el doctor Even Dustin –secretario de Estado adjunto para asuntos médicos– pidió otra opinión calificada, además de la del doctor Kean, antes de dar el visto bueno al ingreso del sha en Estados Unidos. Enterado de esto, David Rockefeller visitó en Washington a su amigo Cyrus Vance, secretario de Estado y antiguo director de la Fundación Rockefeller, y consiguió la autorización de la visa sin más trámites. El sha llegó a Estados Unidos el 22 de octubre de 1979, ingresó de inmediato en el New York Hospital y fue operado sin contratiempos. Dos semanas después, un grupo de universitarios iraníes asaltó la embajada de Estados Unidos en Teherán, como protesta por la presencia del sha en Nueva York, y tomó como rehenes a los 52 empleados que habían de permanecer en cautiverio mucho más que Reza Pahlavi en este mundo.

Éstas debieron ser las razones que Pierre Salinger tomó en cuenta para decir lo que dicen que dijo. Pero el enigma político de la enfermedad del sha tuvo otros episodios secretos, cuyo protagonista más visible fue el doctor Kean, en nombre de David Rockefeller.

El capítulo de Panamá fue tal vez el más intrigante. El sha llegó a ese país el 15 de diciembre de 1979, porque el presidente James Carter le pidió al general Omar Torrijos que le recibiese por pocas semanas, en una tentativa de desempatar las diligencias de liberación de los rehenes. Llegó acompañado por su esposa, Farah Diba; por el coronel Jahnbini, su jefe de seguridad; por la doctora Pernio, su médico personal; por su *valet* iraní, y por Roberto Armao, hombre de confianza de los Rockefeller, que era quien tomaba las decisiones mayores y manejaba el dinero. Llevaba dos perros afganos y siete baúles enormes con documentos secretos. Desde el primer examen a que le sometieron los médicos panameños, se estableció que era urgente extirpar el bazo, cosa que no se había hecho en Nueva York por las malas condiciones del paciente; pero, al cabo de dos meses de cuidados intensos, había aumentado veinticuatro libras, su estado general era el mejor posible y los médicos panameños consideraron que era el momento de operarlo. El doctor Flandrin, que volvió de Francia, estuvo de acuerdo. «Una operación como ésa no es técnicamente difícil», me dijo más tarde el doctor Carlos García Aguilera, uno de los médicos del sha, «y se realizaba sin problemas en todos los hospitales de Panamá.» Sin embargo, por las condiciones y la significación del paciente, se tomaron precauciones excepcionales. Se hizo llevar incluso un separador de sangre fabricado por la IBM sobre un diseño de los médicos del hospital M. D. Anderson, de Houston, cuya función es separar los distintos elementos de la sangre y permitir una transfusión selectiva. Cuando el doctor Kean se enteró de estos aprestos, se dirigió al Departamento de Estado y al comando médico de la zona del canal, para que la máquina no fuera despachada a Panamá. Más tarde pidió la participación del doctor Michael de Bakey, jefe de cirugía cardiovascular del hospital metodista de Houston y una de las estrellas más brillantes de la cirugía mundial. Esto ocurrió el día 7 de marzo de 1980. Sin embargo, el propio doctor De Bakey le contó a un periodista de su país que el doctor Kean le había contratado para operar al sha desde el 4 de marzo, o sea, tres días antes de su reunión con los médicos panameños.

El viernes 14 de marzo, el sha ingresó en el hospital de Paitilla, en Panamá, para ser operado. El equipo médico local estaba dispuesto. Esa misma tarde llegaron en un avión particular los miembros del equipo norteamericano, incluso el doctor De Bakey con sus asistentes personales. Sin embargo, en lugar de proceder a la operación, los dos equipos se encarnizaron en debate tormentoso, que terminó por despojarse de todo su carácter científico. Los compañeros tenían la impresión de que los norteamericanos no tenían otro propósito que el de impedir la operación mediante toda clase de maniobras dilatorias para forzar su regreso a Estados Unidos. En efecto, consiguieron aplazarla.

De regreso a su casa de Contadora —una casa tropical sobre una colina, desde la cual se divisa el océano Pacífico—, el sha no parecía ser ya el dueño de su destino. Su esposa ocupaba el dormitorio de la planta baja. El sha ocupaba el de la planta alta, junto con los siete baúles. En sus últimos días sólo leyó un libro: *La caída del trono del pavo real*, de William Forbis, que era la historia de su propia desgracia. Las únicas visitas que recibía eran las de sus médicos, y a todos les impuso el protocolo real, de modo que tenían que tratarlo como a un monarca reinante. «Lo hacíamos por compasión», me dijo uno de ellos. El 23 de marzo —contra el criterio del propio presidente Carter, pero al parecer con la anuencia de Rockefeller— se fugó a Egipto en un avión privado. Sus administradores no pagaron el alquiler de la casa, ni la cuenta de los médicos, ni los gastos de su seguridad personal. Ocho días después, en Egipto, el doctor Michael de Bakey le hizo por fin la operación del bazo.

LA MUJER QUE ESCRIBIÓ UN DICCIONARIO

Hace tres semanas, de paso por Madrid, quise visitar a María Moliner. Encontrarla no fue tan fácil como yo suponía: algunas personas que debían saberlo ignoraban quién era, y no faltó quien la confundiera con una célebre estrella de cine. Por fin logré un contacto con su hijo menor, que es ingeniero industrial en Barcelona, y él me hizo saber que no era posible visitar a su madre por sus quebrantos de salud. Pensé que era una crisis momentánea y que tal vez pudiera verla en un viaje futuro a Madrid. Pero la semana pasada, cuando ya me encontraba en Bogotá, me llamaron por teléfono para darme la mala noticia de que María Moliner había muerto. Yo me sentí como si hubiera perdido a alguien que sin saberlo había trabajado para mí durante muchos años.

María Moliner –para decirlo del modo más corto– hizo una proeza con muy pocos precedentes: escribió sola, en su casa, con su propia mano, el diccionario más completo, más útil, más acucioso y más divertido de la lengua castellana. Se llama *Diccionario de uso del español*, tiene dos tomos de casi 3.000 páginas en total, que pesan tres kilos, y viene a ser, en consecuencia, más de dos veces más largo que el de la Real Academia de la Lengua, y –a mi juicio– más de dos veces mejor. María Moliner lo escribió en las horas que le dejaba libre su empleo de bibliotecaria, y el que ella consideraba su verdadero oficio: remendar calcetines. Uno de sus hijos, a quien le preguntaron hace poco cuántos hermanos tenía, contestó: «Dos varones, una hembra y el diccionario». Hay que saber cómo fue escrita la obra para entender cuánta verdad implica esa respuesta.

María Moliner nació en Paniza, un pueblo de Aragón, en 1900. O, como ella decía con mucha propiedad: «En el año cero». De modo que al morir había cumplido los ochenta años. Estudió Filosofía y Letras en Zaragoza y obtuvo, mediante concurso, su ingreso al Cuerpo de Archiveros y Bibliotecarios de España. Se casó con don Fer-

Publicado originalmente el 10 de febrero de 1981.

nando Ramón y Ferrando, un prestigioso profesor universitario que enseñaba en Salamanca una ciencia rara: base física de la mente humana. María Moliner crió a sus hijos como toda una madre española, con mano firme y dándoles de comer demasiado, aun en los duros años de la guerra civil, en que no había mucho que comer. El mayor se hizo médico investigador, el segundo se hizo arquitecto y la hija se hizo maestra. Sólo cuando el menor empezó la carrera de ingeniero industrial, María Moliner sintió que le sobraba demasiado tiempo después de sus cinco horas de bibliotecaria, y decidió ocuparlo escribiendo un diccionario.

La idea le vino del *Learner's Dictionary*, con el cual aprendió el inglés. Es un diccionario de uso; es decir, que no sólo dice lo que significan las palabras, sino que indica también cómo se usan, y se incluyen otras con las que pueden reemplazarse. «Es un diccionario para escritores», dijo María Moliner una vez, hablando del suyo, y lo dijo con mucha razón. En el diccionario de la Real Academia de la Lengua, en cambio, las palabras son admitidas cuando ya están a punto de morir, gastadas por el uso, y sus definiciones rígidas parecen colgadas de un clavo. Fue contra ese criterio de embalsamadores que María Moliner se sentó a escribir su diccionario en 1951. Calculó que lo terminaría en dos años, y cuando llevaba diez todavía andaba por la mitad. «Siempre le faltaban dos años para terminar», me dijo su hijo menor. Al principio le dedicaba dos o tres horas diarias, pero a medida que los hijos se casaban y se iban de la casa le quedaba más tiempo disponible, hasta que llegó a trabajar diez horas al día, además de las cinco de la biblioteca. En 1967 –presionada sobre todo por la Editorial Gredos, que la esperaba desde hacía cinco años– dio el diccionario por terminado. Pero siguió haciendo fichas, y en el momento de morir tenía varios metros de palabras nuevas que esperaba ver incluidas en las futuras ediciones. En realidad, lo que esa mujer de fábula había emprendido era una carrera de velocidad y resistencia contra la vida.

Su hijo Pedro me ha contado cómo trabajaba. Dice que un día se levantó a las cinco de la mañana, dividió una cuartilla en cuatro partes iguales y se puso a escribir fichas de palabras sin más preparativos. Sus únicas herramientas de trabajo eran dos atriles y una máquina de escribir portátil, que sobrevivió a la escritura del diccionario. Primero trabajó en la mesita de centro de la sala. Después, cuando se sintió naufragar entre libros y notas, se sirvió de un tablero apoyado sobre el respaldar de dos sillas. Su marido fingía una impavidez de sabio, pero a veces medía a escondidas las gavillas de fichas con una cinta métrica, y les mandaba noticias a sus hijos. En una ocasión les contó que el diccionario iba ya por la última letra, pero tres meses después les contó, con las ilusiones perdidas, que había vuelto a la primera. Era natural, porque María Moliner tenía un método infinito: pretendía agarrar al vuelo todas las palabras de la vida. «Sobre todo las que encuentro en los periódicos», dijo en una entrevista. «Porque allí viene el idioma vivo, el que se está usando, las palabras que tienen que inventarse al momento por necesidad». Sólo hizo una excepción: las mal llamadas malas palabras, que son muchas y tal vez las más usadas en la España de todos los tiempos. Es el defecto mayor de su diccionario, y María Moliner vivió bastante para comprenderlo, pero no lo suficiente para corregirlo.

Pasó sus últimos años en un apartamento del norte de Madrid, con una terraza grande, donde tenía muchos tiestos de flores, que regaba con tanto amor como si fueran palabras cautivas. Le complacían las noticias de que su diccionario había vendido más de 10.000 copias, en dos ediciones, que cumplía el propósito que ella se había impuesto y que algunos académicos de la lengua lo consultaban en público sin ruborizarse. A veces le llegaba un periodista desperdigado. A uno que le preguntó por qué no contestaba las numerosas cartas que recibía le contestó con más frescura que las de sus flores: «Porque soy muy perezosa». En 1972 fue la primera mujer cuya candidatura se presentó en la Academia de la Lengua, pero los muy señores académicos no se atrevieron a romper su venerable tradición machista. Sólo se atrevieron hace dos años, y aceptaron entonces la primera mujer, pero no fue María Moliner. Ella se alegró cuando lo supo, porque le aterrorizaba la idea de pronunciar el discurso de admisión. «¿Qué podía decir yo», dijo entonces, «si en toda mi vida no he hecho más que coser calcetines?»

EL KISSINGER DE REAGAN

Al presidente de Estados Unidos, Ronald Reagan, hay que reconocerle el mérito de haber acabado en menos de treinta días con la vieja contradicción entre el Departamento de Estado y el Pentágono. Lo consiguió de un solo plumazo, al nombrar como maestro de su diplomacia al general Alexander Haig, un militar inteligente y feroz que ha leído a los clásicos griegos y latinos, que ama a los perros bravos y la buena cocina europea, y que sería capaz de manejar los dos ministerios más importantes de su país como si fuera una sola cosa. Es el Kissinger de Reagan, pero con la ventaja adicional de sus cuatro estrellas.

La primera noticia pública del general Alexander Haig la dio Henry Kissinger en sus memorias. No eludió ningún recurso literario para llamar la atención sobre aquel militar con ínfulas de intelectual europeo que fue tal vez la persona más cercana a Nixon durante el drama de Watergate. Kissinger dejó sentado en su libro que fue él quien sobrellevó el delicado encargo de vigilar a Nixon en los últimos días de su infortunio para impedir que hiciera una locura final, inclusive la que podría parecer menos verosímil en Estados Unidos: un golpe de Estado contra la soberanía del Congreso.

A quienes habíamos leído con la atención merecida aquel libro, a la vez fascinante y abominable, no podía sorprendernos que el presidente Reagan hubiera encomendado al general Haig el alto honor de devolverle a Estados Unidos su maltrecho prestigio mundial.

Tampoco nos equivocamos en la profecía fácil de que el esfuerzo había de comenzar por América Latina. Lo han demostrado esos primeros treinta días del presidente Reagan, con las tentativas inaugurales de una diplomacia de mano dura, y un renovado aliento de guerra —destapada o encubierta— en América Central y el Caribe. El Salvador está en llamas. El embajador de Estados Unidos en ese país, un realista llamado White, había sido condenado a muerte por la

Publicado originalmente el 21 de febrero de 1981.

extrema derecha sólo por haber dicho en público muchas veces que una intervención de su país en El Salvador sería favorable a la extrema derecha. Al general Haig, que es un reaccionario químicamente puro, no podían asustarle estos pronósticos. Al contrario: la decisión de asistir en su agonía a la Junta de Gobierno de El Salvador con toda clase de recursos mortíferos, sin intentar ninguna solución intermedia, es un acto machista que define muy bien el estilo de la nueva diplomacia militar.

Nicaragua no ha vuelto a dormir tranquila. A pesar de su decisión reiterada y evidente de implantar un sistema de gobierno independiente de todo centro mundial de poder, la hostilidad de Estados Unidos se ha visto recrudecida en estos treinta días. Las incursiones de bandas somocistas por sus fronteras son cada vez más frecuentes. En este fin de semana se tenían noticias confidenciales de una invasión más grande desde Guatemala.

Cuba, por su parte, está otra vez en pie de guerra. No sólo desde el día de la posesión del presidente Reagan, sino desde la misma noche de su elección. Allí nadie ha tomado a la ligera las amenazas de la campaña electoral, que los primeros actos del general Haig no han hecho sino confirmarlas. Es injusto, sobre todo porque este estado de tensión nacional impone a Cuba una distracción extraordinaria de sus recursos civiles, en un año de gracia en el que el racionamiento de la comida y la ropa se han resuelto en la práctica, y el país se prepara para una de las zafras más fructíferas de este siglo, gracias a una conducta económica más realista, a un invierno frío y sin aguas y a un precio favorable del azúcar en el mercado mundial.

Sin embargo, el gobierno de Panamá es el primero que ha conocido de un modo directo la vocación imperial y el estilo rupestre del general Haig, en una nota verbal inconcebible que le hizo llegar la semana pasada. El embajador de Estados Unidos en Panamá, un liberal simpático que habla el catalán sin acento y un castellano riguroso con todas sus zetas inútiles, vivió dos veces el mal rato de transmitirla de viva voz, en audiencias separadas al presidente Arístides Royo y al general Omar Torrijos. Es una nota tan inadmisible que no puedo resistir a la tentación de una infidencia.

El general Haig —según la nota verbal— se congratula con las buenas relaciones entre Panamá y Estados Unidos. Se congratula con el hecho de que el gobierno panameño hubiera celebrado elecciones en 1980 y con el proyecto de hacer otras en 1984. Entiende que existan relaciones entre Cuba y Panamá, pero le preocupa que sean tan buenas, y le preocupa sobre todo que el intercambio comercial de los dos países esté contribuyendo a romper el bloqueo impuesto por Estados Unidos desde hace más de veinte años. Le preocupa la presencia creciente de personal en la embajada cubana en Panamá y la influencia que estos funcionarios ejercen sobre el gobierno panameño. Le preocupa la presencia de una flota pesquera cubana en las aguas territoriales panameñas y que Cuba utilice a Panamá para mandar armas y gentes entrenadas a El Salvador.

El presidente Royo, que es un hombre inteligente y culto, le dictó al embajador una respuesta serena a cada una de las preocupaciones del general Haig. «Me preocupa mucho la ejecución de los tratados Torrijos-Carter sobre el canal interoceánico, porque Estados Unidos no le está dando estricto cumplimiento», dijo para comenzar. Precisó una vez más que Panamá es un país no alineado, y con una política exterior independiente. Confirmó que, en efecto, Panamá está en contra del bloqueo a Cuba, «porque estamos en contra de bloquear una nación hermana sólo por tener un sistema distinto del nuestro»; dijo que es cierto que Panamá mantiene muy buenas relaciones con Cuba, pero precisó que son mucho menos importantes que las que mantiene Estados Unidos con China y la Unión Soviética.

«En Panamá no nos preocupan los comunistas, y creemos en el pluralismo ideológico», prosigue el presidente Royo, «y el día en que empecemos a perseguir tendencias e ideologías volveremos a tener violencia.» Reiteró que la flota pesquera de Cuba –que en realidad está compuesta por dos barcos– sólo se dedica a pescar, y rechazó por falsa la afirmación de que fuera cada vez mayor la presencia de personal cubano en la embajada de Panamá. Rechazó también la afirmación de que los cubanos influyeran sobre el gobierno de Panamá, «pues lo impide el respeto mutuo entre los dos gobiernos».

«Es falso», dijo por último el presidente Royo, «que se utilice a Panamá para enviar tropas y armas a El Salvador. Ningún país latinoamericano lo ha hecho. El único país que contra nuestra voluntad ha utilizado nuestro territorio para incursiones en El Salvador es Estados Unidos.» El presidente se refería, por supuesto, al envío de recursos al gobierno de El Salvador desde las bases que todavía tiene Estados Unidos en la zona del canal, y al entrenamiento de tropas salvadoreñas en los territorios que todavía no han sido recuperados por Panamá.

El general Omar Torrijos oyó la lectura de la nota masticando su puro y tratando de someter el mechón rebelde que siempre le cae sobre la frente. Al final, pidió a su secretaria una hoja de papel y escribió la respuesta de su puño y letra: «Doy este mensaje como no recibido por haberse equivocado de destinatario. Debió ser enviado a Puerto Rico».

REMEDIOS PARA VOLAR

Una vez más he hecho el disparate que me había propuesto no repetir jamás, que es el de dar el salto del Atlántico de noche y sin escalas. Son doce horas entre paréntesis dentro de las cuales se pierde no sólo la identidad, sino también el destino. Esta vez además fue un vuelo tan perfecto que por un instante tuve la certidumbre de que el avión se había quedado inmóvil en la mitad del océano e iban a tener que llevar otro para transbordarnos. Es decir, siempre me había atormentado el temor de que el avión se cayera, pero esta vez concebí un miedo nuevo. El miedo espantoso de que el avión se quedara en el aire para siempre.

En esas condiciones indeseables comprendí por qué la comida que sirven en pleno vuelo es de una naturaleza diferente de la que se come en tierra firme. Es que también el pollo —muerto y asado— va volando con miedo, y las burbujas de la champaña se mueren antes de tiempo, y la ensalada se marchita de una tristeza distinta. Algo semejante ocurre con las películas. He visto algunas que cambian de sentido cuando se vuelven a ver en el aire, porque el alma de los actores se resiste a ser la misma y la vida termina por no creer en su propia lógica. Por eso no hay ninguna posibilidad de que sea buena ninguna película de avión. Más aún: cuanto más largas sean y más aburridas, más se agradece que lo sean, porque uno se ve forzado a imaginarse más de lo que ve y aun a inventar mucho más de lo que se alcanza a ver, y todo eso ayuda a sobrellevar el miedo.

Semejantes remedios son incontables. Tengo una amiga que no logra dormir desde varios días antes de embarcarse, pero su miedo desaparece por completo cuando logra encerrarse en el excusado del avión. Permanece allí tantas horas como le sean posibles, leyendo en un sosiego sólo comparable al del ojo del huracán, hasta que las autoridades de a bordo la obligan a volver al horror del asiento. Es raro, porque siempre he creído que la mitad del miedo al avión se debe

Publicado originalmente el 24 de febrero de 1981.

a la opresión del encierro, y en ninguna parte se siente tanto como en los servicios sanitarios. En los excusados de los trenes, en cambio, hay una sensación de libertad irrepetible. Cuando era niño, lo que más me gustaba de los viajes en los ferrocarriles bananeros era mirar el mundo a través del hueco del inodoro de los vagones, contar los durmientes entre dos pueblos, sorprender los lagartos asustados entre la hierba, las muchachas instantáneas que se bañaban desnudas debajo de los puentes. La primera vez que subí a un avión –un bimotor primitivo de aquellos que hacían mil kilómetros en tres horas y media– pensé, con muy buen sentido, que por el hueco de la cisterna iba a ver una vida más rica que la de los trenes, que iba a ver lo que ocurría en los patios de las casas, las vacas caminando entre las amapolas, el leopardo de Hemingway petrificado entre las nieves del Kilimanjaro. Pero lo que encontré fue la triste comprobación de que aquel mirador de la vida había sido cegado y que un acto tan simple como soltar el agua implicaba un riesgo de muerte.

Hace muchos años superé la ilusión generalizada de que el alcohol es un buen remedio para el miedo al avión. Siguiendo una fórmula de Luis Buñuel, me tomaba un martillazo de Martini seco antes de salir de la casa, otro en el aeropuerto y un tercero en el instante de decolar. Los primeros minutos del vuelo, por supuesto, transcurrían en un estado de gracia cuyo efecto era contrario al que se buscaba. En realidad, el sosiego era tan real e intenso que uno deseaba que el avión se cayera de una vez para no volver a pensar en el miedo. La experiencia termina por enseñar que el alcohol, más que un remedio, es un cómplice del terror. No hay nada peor para los viajes largos: uno se calma con los dos primeros tragos, se emborracha con los otros dos, se duerme con los dos siguientes, engañado con la ilusión de que en realidad está durmiendo, y tres horas después se despierta con la conciencia cierta de que no ha dormido más de tres minutos y que no hay nada más en el futuro que un dolor de cabeza de diez horas.

La lectura –remedio de tantos males en la tierra– no lo es de ninguno en el aire. Se puede iniciar la novela policíaca mejor tramada, y uno termina por no saber quién mató a quién ni por qué. Siempre he creído que no hay nadie más aterrorizado en los aviones que esos caballeros impasibles que leen sin parpadear, sin respirar siquiera, mientras la nave naufraga en las turbulencias. Conocí uno que fue mi vecino de asiento en la larga noche de Nueva York a Roma, a través de los aires pedregosos del Ártico, y no interrumpió la lectura de *Crimen y castigo* ni siquiera para cenar, línea por línea, página por página; pero a la hora del desayuno me dijo con un suspiro: «Parece un libro interesante». Sin embargo, el escritor uruguayo Carlos Martínez Moreno puede dar fe de que no hay nada mejor que un libro para volar. Desde hace veinte años vuela siempre con el mismo ejemplar casi desbaratado de *Madame Bovary*, fingiendo leerlo a pesar de que ya lo conoce casi de memoria, porque está convencido de que es un método infalible contra la muerte.

Siempre pensé que no hay un recurso más eficaz que la música, pero no la que se oye por el sistema de sonido del avión, sino la que llevo en un magnetofón con auriculares. En realidad, la del avión produce un efecto contrario. Siempre me he preguntado con asombro quiénes hacen los programas musicales del vuelo, pues no puedo imaginarme a nadie que conozca menos las propiedades medicinales de la música. Con un criterio bastante simplista, prefieren siempre las grandes piezas orquestales relacionadas con el cielo, con los espacios infinitos, con los fenómenos telúricos. «Sinfonías paquidérmicas», como llamaba Brahms a las de Bruckner. Yo tengo mi música personal para volar, y su enumeración sería interminable. Tengo mis programas propios, según las rutas y su duración, según sea de día o de noche, y aún según la clase de avión en que se vuele. De Madrid a Puerto Rico, que es un vuelo familiar a los latinoamericanos, el programa es exacto y certero: las nueve sinfonías de Beethoven. Siempre pensé —como he dicho antes— que no había un método más eficaz para volar hasta esta semana de mi infortunio, en que un lector de Alicante me ha escrito para decirme que ha descubierto otro mejor: hacer el amor tantas veces como sea posible en pleno vuelo. De esto —como en las telenovelas— vamos a hablar la semana entrante.

EL AMOR EN EL AIRE

Los viajes —como el poder— son afrodisíacos. Si las crónicas de navegantes y los cuadernos de bitácora dijeran toda la verdad, y no sólo
la verdad, serían textos ejemplares de literatura prohibida. Es por eso
que en las cubiertas de los barcos de pasajeros es imposible encontrar
de noche un rincón sin luz, y los expertos en cruceros de turismo,
sobre todo en el Caribe, aconsejan a los principiantes llevar consigo
una llave inglesa para romper focos. Los legendarios trenes europeos
fueron durante muchos años hoteles de placer sobre ruedas.

El *Orient Express*, además de haber sido escenario de crímenes sin
solución y laboratorio de espías, fue un paraíso nocturno donde se
concibieron en alcobas sin fronteras más de tres testas coronadas. En
el metro de Ciudad de México, por el mismo motivo, y a pleno día,
ha sido preciso establecer vagones separados para hombres y mujeres,
y no a la hora de menor afluencia, sino todo lo contrario, en las más
concurridas.

Los aviones, en cambio, estuvieron considerados durante muchos
años como espacios vedados al amor. Hasta el punto de que el cinturón del asiento nos parece todavía un sustituto compasivo del cinturón de castidad. Tal vez como reacción contra ese castigo surgió
la leyenda mundial de las azafatas fáciles, a quienes nuestras fantasías
juveniles atribuyeron toda clase de virtudes concupiscentes. Hace muchos años, en Barranquilla, corrió la voz de que en el barrio más
elegante de la ciudad se había abierto una casa de citas donde vendían sus gracias las más exquisitas servidoras del aire de las compañías internacionales. Esa misma noche fuimos todos, desde el señor
gobernador con su gabinete en pleno hasta los periodistas peor pagados. Y, en efecto, encontramos una escudería de bellas muchachas
de uniforme acreditadas con las insignias de todos los cielos del mundo: las suecas de la SAS, las alemanas de Lufthansa, las amazonas
universales de la Pan American. Era tal nuestra ilusión de que fuera

Publicado originalmente el 4 de marzo de 1981.

verdad tanta mentira que muchos fingimos no darnos cuenta de
que todas eran tan mulatas como las nuestras, y hablaban el castellano sin acento, con la cadencia inefable de la fábrica de sueños de
Pilar Ternera.

La primera vez en que oí hablar con buen derecho de la posibilidad de hacer el amor en un avión fue también en Barranquilla, bebiendo ron blanco con cáscaras de limón con un veterano piloto
alemán que se había retirado cuando inventaron las turbinas, pues
no podía entender que los aviones volaran sin hélices. Fue él quien
me contó que en los *Constellations* de línea había camas plegadizas
como en los camarotes de los trenes, y que nadie preguntaba qué
hacían en ellas los pasajeros que las alquilaban para dormir. En realidad, habían sido diseñadas por Howard Hughes, el creador del *Constellation* para su uso personal con las estrellas de cine que también di-
señaba. Habían de pasar muchos años antes de que una película se
atreviera a mostrar un acto de amor a bordo de un avión. Se vio por
primera vez en *Emmanuelle*, y fue un acto de amor tan difícil y des-
corazonador que parecía más bien una prueba de que era imposible
hacerlo en pleno vuelo.

En la actualidad, sin embargo, la gente del jet-set lo tiene como
cosa corriente, y lo hacen con tanta frecuencia y tanta naturalidad
como en la vida real. En Estados Unidos existe una sociedad civil
llamada el Mile High Club, en la cual son admitidos quienes puedan demostrar que han hecho el amor a más de una milla de altura.
Sus socios son muchos; todos coinciden en que en esta materia,
como en tantas otras, lo único difícil es empezar. También hay un
vuelo nocturno de Los Ángeles a Miami, o de Los Ángeles a Nueva York, cuyo nombre demasiado obvio es el *Red Eyes Express*,
o sea, el *Expreso de los ojos rojos*. El vuelo dura siete horas, pero
lo único que nadie se permite es dormir, de modo que los pasajeros llegan a su destino con los ojos enardecidos por los fragores de
la noche.

La diferencia entre el *Red Eyes Express* y los vuelos comerciales
de siempre —además de los precios del billete, que son muy bajos—
es que en aquél no hay vigilancia de ninguna clase. No hay más
autoridad que la de los pilotos, que viajan encerrados con aldabas en
la cabina, para que no los salpique la tentación de su propio invento. Los pasajeros llevan su comida y su bebida, sus drogas y su música personales, y cada quien es dueño absoluto de su cuerpo. Es decir: cada uno va en otro viaje dentro del viaje. Nadie les pregunta
quién es quién, ni por dónde, pues en aquellos vuelos babilónicos
de luces apagadas el sexo es lo de menos.

Un error muy común cuando se habla de estas cosas es pensar en los servicios sanitarios del avión. Existe inclusive un manual ilustrado, en el cual se indican las diferentes maneras acrobáticas de hacer el amor en los retretes de las grandes líneas. Los dibujos indican los puntos de apoyo según la edad y los gustos, y se han establecido unas 162 posibilidades al modo occidental. La sola manija de seguridad donde uno se agarra para no caerse durante el uso tradicional del retrete sirve para otras 74 cosas distintas, según el manual. Esto quiere decir que el excusado de los aviones tiene más utilidad demográfica que los automóviles, aunque las estadísticas demuestran que es cada día mayor el número de niños inteligentes y sin fracturas que se conciben en los automóviles, muchos de ellos en marcha.

Sin embargo, los expertos consideran que los servicios sanitarios de los aviones son tan convencionales para hacer el amor como lo son las camas para los senadores de la República. El sitio ideal son los asientos, después de levantar el brazo que los separa. La demostración excesiva la hizo Arnold Schwarzenegger, el desolado míster Universo –de quien, por cierto, se dijo alguna vez que era del otro equipo– que hace unos tres años viajó con su novia en un vuelo nocturno de Los Ángeles a Nueva York y al parecer no la dejó dormir ni un instante. La azafata que debía atenderlos declaró después a la prensa: «Durante todo el vuelo, lo único que vi de ellos fueron los pies».

De modo que a lo mejor tiene razón el lector de Alicante que me escribió para decirme que el amor es el remedio más drástico para el miedo al avión. En efecto, los científicos dicen que no hay mejor tranquilizante que el orgasmo. Además, si uno lo piensa bien, nada demuestra que esté prohibido intentarlo en los aviones. Está prohibido fumar durante el decolaje y el aterrizaje, en algunas áreas del avión y, sobre todo, en los servicios sanitarios, y por eso hay un letrero que se enciende y se apaga para recordarlo. Esto permite pensar que si estuviera prohibido hacer el amor habría también un letrero similar. Más aún: en mis miedos indómitos sobre todos los océanos nocturnos he tenido la paciencia de leer muchas veces el texto microscópico del contrato de vuelo impreso en los billetes y no he encontrado cláusula alguna que se oponga a ninguna función natural. De modo que si usted no lo hace debe ser simplemente por un malentendido. ¡Adelante, pues, y feliz viaje!

UN DOMINGO DE DELIRIO

Un editor de Barcelona hizo la semana pasada una escala en Cartagena de Indias para almorzar conmigo. Después de una comida criolla bien conversada, lo llevé a conocer la ciudad antigua, que, con toda razón, le pareció una de las más bellas del mundo. Lo invité más tarde a tomar un café en casa de mis padres, que tienen 54 nietos, y muchos de ellos habían ido a saludarlos. Por último, sin saber cómo, terminamos en una recepción oficial, y lo trataron con tanta amabilidad que tuvo que escuchar seis discursos y se tomó once vasos de whisky en tres cuartos de hora. Al atardecer, todavía medio aturdido por tantas novedades juntas, se fue con la impresión de haber vivido una de las experiencias más raras de su vida. «No has inventado nada en tus libros», me dijo al despedirse. «Eres un simple notario sin imaginación.» En realidad, estaba preparado para pasar un domingo tranquilo, a salvo de las nieves que había dejado el día anterior en el otro lado del mundo, y se encontró de pronto y sin previo aviso enredado en los hechizos del Caribe.

El delirio empezó en el mismo aeropuerto. Yo nunca había observado, hasta que él me lo hizo notar, que las puertas de abordaje y desembarco son imposibles de distinguir. En efecto, hay una con un letrero que dice: «Salida de pasajeros», y por ella salen los que van a abordar los aviones. Hay otra puerta con otro letrero que dice lo mismo: «Salida de pasajeros», y es por allí por donde salen los pasajeros que llegan. Lo peor es que ambos letreros son correctos, porque por ambas puertas se sale. Por otra parte, hay también una sala de espera que no es para esperar a los que llegan, sino para que esperen la salida del avión los que se van. Allí hay, por supuesto, varias hileras de sillas muy ordenadas y limpias, frente a una serie de puertas numeradas bajo un letrero general: «Salida de vuelos nacionales». Pero esas puertas no se usan. En cambio, los pasajeros que llegan en los vuelos nacionales no salen por ninguna de tantas puertas, sino por la

Publicado originalmente el 10 de marzo de 1981.

salida internacional, que está en otro edificio apartado; sin embargo, cuando una cálida voz de mujer solicita por los altavoces que salgan por la puerta de salida los pasajeros que se van, nadie sufre un tropiezo: «Es que no hay que hacerle caso a los letreros», nos explicó un agente de policía de turismo. «Aquí todo el mundo sabe por dónde se entra y por dónde se sale.»

Para mí, el rincón más nostálgico de Cartagena de Indias es el muelle de la Bahía de las Ánimas, donde estuvo hasta hace poco el fragoroso mercado central. Durante el día, aquélla era una fiesta de gritos y colores, una parranda multitudinaria como recuerdo pocas en el ámbito del Caribe. De noche, era el mejor comedero de borrachos y periodistas. Allí estaban, frente a las mesas de comida al aire libre, las goletas que zarpaban al amanecer cargadas de marimondas y guineo verde, cargadas de remesas de putas *biches* para los hoteles de vidrio de Curazao, para Guantánamo, para Santiago de los Caballeros, que ni siquiera tenía mar para llegar, para las islas más bellas y más tristes del mundo. Uno se sentaba a conversar bajo las estrellas de la madrugada, mientras los cocineros maricas, que eran deslenguados y simpáticos y tenían siempre un clavel en la oreja, preparaban con mano maestra el plato de resistencia de la cocina local: filete de carne con grandes anillos de cebolla y tajadas fritas de plátano verde. Con lo que allí escuchábamos mientras comíamos, hacíamos el periódico del día siguiente.

Mi amigo editor recordaba muy bien el lugar, porque lo conoció descrito en *El otoño del patriarca*, como el remanso nocturno donde monseñor Demetrio Aldus, auditor de la Sagrada Congregación del Rito y promotor y postulador de la fe, se peleaba a trompadas con los marineros. Lo recordaba, digo, pero no lo reconoció cuando lo llevé a conocerlo en la realidad, porque el mercado público fue demolido, y el muelle fue desmantelado, y en su lugar se construye un esperpento descomunal, que será todo lo contrario de la ciudad: el edificio más feo del mundo.

El Centro Internacional de Convenciones –inspirado, como hasta su nombre lo indica, en el Convention Hall de Miami– costará 1.500 millones de pesos, que equivalen a siete veces el presupuesto municipal. Mi amigo, que sabe de números como buen editor catalán, comprendió entonces lo que quiere decir el realismo mágico. En efecto, 3.000 convencionistas necesitan por lo menos diez jumbos de los más grandes para llegar a la ciudad, y por lo menos un mes para salir con la capacidad actual de las siete puertas del aeropuerto. Será necesario paralizar un día completo el tráfico de la ciudad para llevarlos desde sus hoteles hasta el centro de convenciones; y otro día completo, para el viaje contrario, y aun así se formará un embotellamiento apocalíptico con sus propios vehículos.

Por otra parte, la mayoría de los convencionistas, si en realidad valen la pena, serán hombres de empresa que deberán estar en contacto permanente con sus centros financieros. Pero el servicio telefónico de Cartagena es tan rudimentario que, para hablar por teléfono, hay que dejar la ventana abierta, porque lo que uno dice se oye más por la ventana que por el teléfono. Sólo para conseguir que las operadoras de larga distancia les contesten a 3.000 convencionistas agónicos se necesitarán 32 años. Antes que mi amigo, estos cálculos los había hecho una comisión de expertos, internacionales, que consideraron el proyecto como un disparate homérico. Pero los promotores locales se empeñaron en hacerlo con un argumento magistral: «La ciudad lo necesita para coronar todos los años a la reina de la belleza».

Agobiado por tanto realismo fantástico, mi amigo me agradeció, como una pausa de alivio, que lo invitara a tomarse un café en casa de mis padres. Más le hubiera valido no aliviarse. En efecto, como creo haberlo dicho otras veces, mi padre acaba de cumplir ochenta años, y mi madre setenta y seis. Pero no hay manera de sentarlos a descansar. Mi padre se va a pie todos los días, bajo el sol de fuego, hasta el centro de la ciudad, y no hemos logrado disuadirlo de una excursión que quiere hacer por la selva amazónica. Mi madre se ha empeñado toda la vida en hacer los oficios de la casa, y quiere inclusive acabar de lavar los platos que la lavadora eléctrica deja mal lavados. Mi amigo le preguntó si alguien la ayudaba, y ella le contestó con su lenguaje propio: «Tengo dos secretarias». Mi amigo le preguntó desde cuándo, y ella le volvió a contestar: «Desde hace quince días». El secreto de ambos es que nunca se han puesto a pensar en la edad. Hace poco, mi padre compró unos bonos que serán liquidados en el año 2000. Es decir, cuando él tenga cien años. Uno de mis hermanos le reprochó su falta de sentido, y él replicó impasible: «No los compré para mi beneficio, sino para asegurarle a tu madre una vejez tranquila».

Mientras conversábamos, llegó una nieta a contarnos que la noche anterior se había desdoblado. «Cuando regresé del baño», me dijo, «me encontré conmigo misma que todavía estaba en la cama.» Poco después llegaron tres hermanas y dos hermanos, de los dieciséis que somos en total. Una de ellas, que fue monja hasta hace poco, se enredó en un diálogo sobre religiones comparadas con un hermano que es mormón. Otro hermano había mandado hacer una tabla sobre medida, pero cuando la volvió a medir en la casa resultó ser más corta que en la carpintería. «Es que en el Caribe no hay dos metros iguales», dijo. En efecto, midió un metro con el otro, y a uno de los dos le faltaba un centímetro. Otra hermana tocaba al piano la serenata del cuarteto número cinco de Haydn. Le hice ver que la tocaba tan rápido que parecía una mazurca. «Es que sólo toco el piano cuando estoy acelerada», me dijo, «lo hago para tratar de calmarme, pero lo único que consigo es acelerar también al piano.» En ésas estábamos cuando tocó a la puerta una hermana de mi madre, la tía Elvira, de 84 años, a quien no veíamos desde hacía quince años. Venía de Riohacha, en un taxi expreso, y se había envuelto la cabeza con un trapo negro para protegerse del sol. Entró feliz, con los brazos abiertos, y dijo para que todos la oyéramos: «Vengo a despedirme, porque ya casi me voy a morir». Mi amigo no soportó más. Al atardecer, camino del aeropuerto, me costó trabajo convencerlo de que ésa era nuestra vida real de todos los días, y de que yo no había preparado —sólo por impresionarlo— cada uno de los episodios de aquel domingo de delirio.

LA LARGA NOCHE DE AJEDREZ
DE PAUL BADURA-SKODA

A su paso por Bogotá, hace dos semanas, el genial pianista austríaco Paul Badura-Skoda sorprendió a un grupo de sus amigos con una pasión más compulsiva que la música: el ajedrez. Había tenido un concierto muy difícil la noche anterior. A las once de la mañana, después de tres horas de ensayo, se sometió con un rigor asombroso a un programa de televisión de casi cuatro horas, y terminó estragado por la tensión, por las luces y el calor, por las interrupciones y las repeticiones constantes. Pasadas las cuatro de la tarde, sin tiempo para cambiarse de ropa, asistió a un almuerzo con los platos más exquisitos y bárbaros de la cocina criolla, y no sólo comió con un buen apetito de músico, sino que se dejó seducir por los vinos abundantes. Al final, cuando sus anfitriones suponían que estaba al borde del desmayo, preguntó si era posible encontrar a alguien que le hiciera el favor de jugar con él una partida de ajedrez.

La vitalidad de los grandes pianistas es ejemplar. «Tocar una sonata de Mozart es como meter un camión de carga por el ojo de una aguja», me dijo uno de ellos. Hace unos cinco años vi a Arturo Rubinstein en un restaurante de Barcelona, cenando con un grupo de amigos, a las dos de la madrugada, y todavía con el traje de etiqueta del concierto que acababa de ejecutar. Ya había cumplido los 84 años, pero se comió una tortilla de chorizos con fríjoles blancos, como si tuviera dieciocho, y ayudó a sus compañeros de mesa a despachar seis botellas de champaña. Paul Badura-Skoda, aunque sólo tiene 53 años, parece hecho de la misma materia. De modo que cuando dijo que quería jugar al ajedrez, sus amigos llamaron por teléfono al maestro Boris de Greiff —que es una de las estrellas mayores del mundo— y éste no se hizo repetir dos veces la solicitud. Eran las cinco de la tarde, Boris de Greiff prometió que a las siete de la noche recogería a Badura-Skoda en su hotel.

Publicado originalmente el 17 de marzo de 1981.

Cuando el pianista supo cuál era el tamaño de su adversario, pidió que lo dejaran solo en su habitación. Sus amigos pensaron con muy buen sentido que iba a descansar dos horas. Sin embargo, poco después lo llamaron de la empresa de televisión para arreglar las cuentas del programa, y él se negó a ocuparse de un asunto tan trivial. «Ahora no puedo», dijo, «porque estoy preparando una partida de ajedrez.» Era cierto. Cuando Boris de Greiff llegó a recogerlo, lo encontró estudiando en un tablero magnético que lleva siempre en su maleta. Las fichas estaban colocadas en la posición final de la última partida inconclusa de la semifinal que jugaron en enero de este año, en Merano (Italia), el alemán Robert Huber y el disidente soviético Víctor Korchnoi. Esto le dio a Boris de Greiff una idea inquietante de la categoría de su adversario. «Aquella habitación parecía más de un ajedrecista que de un músico», dijo Boris de Greiff. «No había una sola partitura.» En cambio, en la mesa había un libro de ajedrez en inglés y otro en alemán. Ambos muy especializados. Y había además muchos recortes de la sección de ajedrez del *Times* de Londres y del *New York Times*.

Mientras Badura-Skoda practicaba en su tablero magnético, los dueños de la casa donde se iba a jugar la partida, previendo que ésta sería rápida y alegre, se fueron a comprar lo necesario para improvisar una cena. Desde la tienda llamaron a casa para avisar al servicio que tal vez sus invitados iban a llegar antes que ellos, y ordenaron que los atendieran con todos los honores. Cuando volvieron a la casa, un poco antes de las ocho, encontraron en la puerta un Mercedes radiante con dos antenas de televisión, reflectores rojos y verdes y sirenas de alarma. En realidad eran los invitados de una fiesta vecina, que se habían equivocado de puerta, pero las criadas los habían recibido de acuerdo con las órdenes de los dueños de casa. Badura-Skoda y Boris de Greiff llegaron justo en el momento en que se aclaraba el equívoco, pero estaban tan excitados con la inminencia de la partida, que no se dieron cuenta de nada.

La larga noche empezó a las ocho. Por una cortesía pura con sus anfitriones, Badura-Skoda tocó en el piano, sin un punto de inspiración, la tercera *partita* de Juan Sebastián Bach. Estaba en un estado de tensión que no había padecido la noche anterior, en el concierto, ni esa mañana, ante las cámaras. Sólo cuando se sentaron frente al tablero pareció sumergirse en una ciénaga de serenidad. Boris de Greiff contó que en la Olimpiada mundial de Leipzig, en 1960, no se había usado un timbre como señal de partida, sino el aria para la cuerda de sol de la *suite* para orquesta número 3, de Bach. A Badura-Skoda le pareció bien que se usara en aquel momento, y el dueño de casa, que es el más compulsivo fanático de Bach y del sonido electrónico en

Colombia, puso el disco a un volumen prudente. Boris de Greiff, jugando con las blancas, abrió con el peón del rey. Badura-Skoda le replicó con la defensa siciliana. En ese instante terminó el aria para la cuerda de sol, y siguió una gavota. Los testigos tuvieron la impresión real de que a Badura-Skoda se le pusieron los pelos de punta. «Me gusta mucho Bach y me gusta mucho el ajedrez, pero no los soporto juntos», dijo, con su buena educación exquisita. Entonces quitaron el disco, desconectaron el teléfono y el timbre de la puerta, y encerraron los perros amordazados en el dormitorio. Los dueños de casa y la esposa de Boris de Greiff se encerraron con una botella de whisky en el comedor vecino, y la casa y el barrio, y la ciudad entera quedaron sumergidos en un silencio sobrenatural.

La guerra duró seis horas. Badura-Skoda se concentró hasta el punto de que sólo dijo tres veces la misma palabra en alemán después de tres de sus propias jugadas. Boris de Greiff entendió que decía: «Muy mal». Y, en efecto, lo dijo siempre después de las tres jugadas que determinaron su derrota. No levantó la vista del tablero un solo instante, y sólo movió la mano para jugar. «Desde el principio me di cuenta que debía empeñarme a fondo», dice Boris de Greiff, «pues no podía hacer un papelón frente a un jugador tan serio.» Aunque De Greiff es un fumador intenso, y fuma siempre mientras juega, esta vez se abstuvo por consideración a la austeridad de su adversario.

Jugaron cuatro partidas. Badura-Skoda perdió tres, y la cuarta quedó en tablas. No quedó satisfecho, por supuesto. A las tres de la madrugada se empeñó en analizar las partidas, hasta que Boris de Greiff le ayudó a establecer cuáles fueron sus errores decisivos. Luego, cuando le acompañó al hotel, le pidió que subiera al cuarto para explicarle el sistema especial de notación del redactor de ajedrez del *Times*, y siguió hablando de ajedrez hasta que la ciudad amaneció en las ventanas. A todos los testigos de esa noche irrecuperable les quedó la impresión de que Badura-Skoda –que es uno de los pianistas más notables de nuestro tiempo– es en realidad un ajedrecista que sólo toca el piano para vivir.

Pregunta sin respuesta

El señor Hans Knospe, un lector alemán, me dice lo siguiente en una carta: «Usted dice en la página 239 de *Cien años de soledad*: "Y cuando llevaba toda su ropa a casa de Petra Cotes, Aureliano Segundo se quitaba cada tres días la ropa que llevaba puesta y esperaba en calzoncillos a que estuviera limpia". Pregunto: ¿Cuándo se cambiaba y lavaba Aureliano Segundo los calzoncillos?».

EL RÍO DE LA VIDA

Por lo único que quisiera volver a ser niño es para viajar otra vez en un buque por el río Magdalena. Quienes no lo hicieron en aquellos tiempos no pueden ni siquiera imaginarse cómo era. Yo tuve que hacerlo dos veces al año —una vez de ida y otra de vuelta— durante los seis años del bachillerato y dos de la universidad, y cada vez aprendí más de la vida que en la escuela, y mejor que en la escuela. En la época en que era bueno el caudal de las aguas, el viaje de subida duraba cinco días de Barranquilla a Puerto Salgar, donde se tomaba el tren hasta Bogotá. En tiempos de sequía, que eran los más y los más divertidos para viajar, podía durar hasta tres semanas.

El tren de Puerto Salgar subía como gateando por las cornisas de rocas durante un día completo. En los tramos más empinados se descolgaba para tomar impulso y volvía a intentar el ascenso resollando como un dragón, y en ocasiones era necesario que los pasajeros se bajaran y subieran a pie hasta la cornisa siguiente, para aligerarlo de su peso. Los pueblos del camino eran helados y tristes, y las vendedoras de toda la vida ofrecían por la ventanilla del vagón unas gallinas grandes y amarillas, cocinadas enteras, y unas papas nevadas que sabían a comida de hospital. A Bogotá se llegaba a las seis de la tarde, que desde entonces era la hora peor para vivir. La ciudad era lúgubre y glacial, con tranvías ruidosos que echaban chispas en las esquinas, y una lluvia de agua revuelta con hollín que no escampaba jamás. Los hombres vestidos de negro, con sombreros negros, caminaban deprisa y tropezando como si anduvieran en diligencias urgentes, y no había una sola mujer en la calle. Pero allí teníamos que quedarnos todo el año, haciendo como si estudiáramos, aunque en realidad sólo esperábamos a que volviera a ser diciembre para viajar otra vez por el río Magdalena.

Eran los tiempos de los barcos de tres pisos con dos chimeneas, que pasaban de noche como un pueblo iluminado, y dejaban un re-

Publicado originalmente el 25 de marzo de 1981.

guero de músicas y sueños quiméricos en los pueblos sedentarios de
la ribera. A diferencia de los buques del Misisipí, la rueda de impulso de los nuestros no estaba en la borda, sino en la popa, y en
ninguna parte del mundo he vuelto a ver otros iguales. Tenían nombres fáciles e inmediatos: *Atlántico, Medellín, Capitán de Caró, David
Arango.* Sus capitanes, como los de Conrad, eran autoritarios y de
buen corazón, comían como bárbaros, y nunca durmieron solos en
sus camarotes remotos. Los tripulantes se llamaban *marineros* por su
extensión, como si fueran del mar. Pero en las cantinas y burdeles
de Barranquilla, adonde llegaban revueltos con los marineros de
mar, los distinguieron con un nombre inconfundible: vaporinos.

Los viajes eran lentos y sorprendentes durante el día, los pasajeros nos sentábamos por la terraza a ver pasar la vida. Veíamos los cai-
manes que parecían troncos de árboles en la orilla, con las fauces
abiertas, esperando que algo les cayera adentro para comer. Se veían
las muchedumbres de garzas que alzaban el vuelo asustadas por la
estela del buque, las bandadas de patos silvestres de las ciénagas inte-
riores, los cartumenes interminables, los manatíes que amamantaban
a sus crías y gritaban como si cantaran en los playones. A veces, una
tufarada nauseabunda interrumpía la siesta, y era el cadáver de una
vaca ahogada, inmensa, que descendía casi inmóvil en el hilo de la
corriente con un gallinazo solitario parado en el vientre. A lo largo
de todo el viaje, uno despertaba al amanecer, aturdido por el alboroto de los micos y el escándalo de las cotorras.

Ahora es raro que uno conozca a alguien en los aviones. En los
buques del río Magdalena, los pasajeros terminábamos por parecer
una sola familia, pues nos poníamos de acuerdo todos los años para
coincidir en el viaje. Los Eljach se embarcaban en Calamar, los Pena
y los Del Toro –paisanos del hombre caimán– se embarcaban en Plato; los Estorninos y los Vinas, en Magangue; los Villafañes, en el Banco. A medida que el viaje avanzaba, la fiesta se hacía más grande.
Nuestra vida se vinculaba de un modo efímero, pero inolvidable, a la
de los pueblos de las escalas, y muchos se enredaron para siempre con
su destino. Vicente Escudero, que era estudiante de Medicina, se metió sin ser invitado en un baile de bodas en Gamarra, bailó sin permiso con la mujer más bonita del pueblo, y el marido lo mató de un
tiro. En cambio, Pedro Pablo Guillén se casó en una borrachera homérica con la primera muchacha que le gustó en Barrancabermeja, y
todavía es feliz con ella y con sus nueve hijos. El irrecuperable José
Palencia, que era un músico congénito, se metió en un concurso de
tamboreros en Tenerife, y se ganó una vaca que allí mismo vendió por
cincuenta pesos: una fortuna de la época. A veces el buque encallaba

hasta quince días en un banco de arena. Nadie se preocupaba, pues la fiesta seguía, y una carta del capitán sellada con el escudo de su amigo servía como justificación para llegar tarde al colegio.

Una noche, en mi último viaje de 1948, nos despertó un lamento desgarrador que llegaba a la ribera. El capitán Climaco Conde Abello, que era uno de los grandes, dio orden de buscar con reflectores el origen de semejante desgarramiento. Era una hembra de manatí que se había enredado en las ramas de un árbol caído. Los vaporinos se echaron al agua, le amarraron con un cabestrante, y lograron desencallarla. Era un animal fantástico y enternecedor, de casi cuatro metros de largo, y su piel era pálida y tersa, y su torso era de mujer, con grandes tetas de madre amantísima, y de sus ojos enormes y tristes brotaban lágrimas humanas. Fue al mismo capitán Conde Abello a quien le oí decir por primera vez que el mundo se iba a acabar si seguían matando a los animales del río, y prohibió disparar desde su barco. «El que quiera matar a alguien, que vaya a matarlo en su casa», gritó. «No en mi barco.» Pero nadie le hizo caso. Trece años después —el 19 de enero de 1961—, un amigo me llamó por teléfono en México para contarme que el vapor *David Arango* se había incendiado y convertido en cenizas en el puerto de Magangue. Yo colgué el teléfono con la impresión horrible de que aquel día se había acabado mi juventud, y que todo lo último que quedaba de nuestro río de nostalgias se había ido al carajo.

Se había ido, en efecto. El río Magdalena está muerto, con sus aguas envenenadas y sus animales exterminados. Los trabajos de recuperación de que ha empezado a hablar el gobierno desde que un grupo de periodistas concentrados pusieron de moda el problema, es una farsa de distracción. La rehabilitación del Magdalena sólo será posible con el esfuerzo continuado e intenso de por lo menos cuatro generaciones conscientes: un siglo entero.

Se habla con demasiada facilidad de la reforestación. Esto significa, en realidad, la siembra técnica de 59.110 millones de árboles en las riberas del Magdalena. Lo repito con todas sus letras: cincuenta y nueve mil ciento diez millones de árboles. Pero el problema mayor no es sembrarlos, sino dónde sembrarlos. Pues la casi totalidad de la tierra útil de las riberas es propiedad privada, y la reforestación completa tendría que ocupar el 90 % de ellas. Valdría la pena preguntar cuáles serían los propietarios que tendrían la amabilidad de ceder el 90 % de sus tierras sólo para sembrar árboles y renunciar en consecuencia al 90 % de sus ingresos actuales.

La contaminación, por otra parte, no sólo afecta al río Magdalena, sino a todos sus afluentes. Son alcantarillados de las ciudades y los pueblos ribereños que arrastran y acumulan, además, desechos industriales y agrícolas, animales y humanos, y desembocan en el inmenso mundo de porquerías nacionales de bocas de ceniza. En noviembre del año pasado, en Tocaima, dos guerrilleros se arrojaron en el río Bogotá huyendo de las fuerzas armadas. Lograron escapar, pero estuvieron a punto de morir infectados por las aguas. De modo que los habitantes del Magdalena, sobre todo la parte baja, hace mucho tiempo que no toman ni usan agua pura ni comen pescados sanos. Sólo reciben —como dicen las señoras— mierda pura.

La tarea es descomunal, pero esto es tal vez lo mejor que tiene. El proyecto completo de lo que hay que hacer está en un estudio realizado hace algunos años por una comisión mixta de Colombia y Holanda, cuyos treinta volúmenes duermen el sueño de los injustos en los archivos del Instituto de Hidrología y Meteorología (IMAT). El subdirector de ese estudio monumental fue un joven ingeniero antioqueño, Jairo Murillo, que consagró a él media vida, y antes de terminar le entregó la que le quedaba: murió ahogado en el río de sus sueños. En cambio, ningún candidato presidencial de los últimos años ha corrido el riesgo de ahogarse en esas aguas. Los habitantes de los pueblos ribereños —que en los próximos días van a estar en las primeras líneas de la intención nacional con el viaje de la *Caracola*— deberían ser conscientes de eso. Y recordar que desde Honda hasta las Bocas de Ceniza, hay suficientes votos para elegir un presidente de la República.

BREVE NOTA DE ADIÓS AL OLOR DE LA GUAYABA

El señor presidente de la República de Colombia inició su discurso del lunes pasado con las siguientes palabras: «No abrigo, como seguramente ustedes tampoco, ninguna duda acerca de que a lo largo del accidentado recorrido republicano de nuestro país, jamás la subversión se había comprometido con tanta saña y persistencia en el criminal propósito de hacerse con el control del aparato del Estado, como lo ha venido intentando durante la Administración que presido».

Al parecer, los plumíferos del señor presidente están necesitando repasar de buena fe la historia de Colombia. Si algo bueno tiene este país es que siempre ha tenido fuerzas capaces de alzarse contra la injusticia y la desigualdad, y ninguna de esas fuerzas lo ha hecho con tanta persistencia como el propio Partido Liberal del señor presidente de la República. A lo largo del siglo XIX, el país padeció ocho guerras civiles generales, catorce locales, tres golpes de cuartel y, por último, la guerra de los mil días, encabezadas por liberales tan esclarecidos como Rafael Uribe y Benjamín Herrera. Sólo en esta última perecieron no menos de 80.000 colombianos. En tiempos más recientes, el mismo Partido Liberal se lanzó a una subversión justa contra un régimen conservador sanguinario y despótico. El propio ministro de Gobierno actual, doctor Germán Zea, tuvo el honor de ser perseguido por orientar una emisora clandestina.

De modo que sí ha habido más saña y persistencia en otras rebeliones anteriores a la de ahora. Esta exageración de la retórica presidencial no tendría mayor importancia, por supuesto, si no fuera porque muchas de las determinaciones oficiales de los últimos días parecen pecar de la misma ligereza, entre ellas –también por supuesto– la precipitada ruptura de relaciones con Cuba, sin más fundamento público que la declaración de un prisionero supuesto, cuya identidad no se conoce a ciencia cierta, y cuyas acusaciones sincopadas y elusivas no convencieron a nadie. No se sabe ni siquiera

Publicado originalmente el 3 de abril de 1981.

que el gobierno colombiano hubiera intentado obtener una explicación de un gobierno amigo, como ha demostrado serlo el de Cuba en los últimos años. La forma en que Ecuador acaba de manejar y resolver un incidente incluso más grave en sus relaciones con Cuba hace resaltar por contraste la precipitud lamentable de nuestro gobierno. En otra parte de su discurso dijo el presidente que se ha sabido por confesión de los guerrilleros que fueron entrenados en Cuba y que las armas capturadas son del mismo país.

Como esto último no lo había dicho el prisionero, aun los más crédulos pensarían que el gobierno daría otras pruebas. Sin embargo, al día siguiente del discurso presidencial, las autoridades militares hicieron ante los periodistas una exhibición cinematográfica de las armas capturadas a los guerrilleros. Pero no mostraron ninguna prueba de su procedencia. Peor aún: lo único que trató de presentarse como una evidencia del origen de las armas fue la captura de varias pistolas iguales a una que, según dicen los propios militares, le regaló Fidel Castro al comandante del antiguo Frente de Liberación Nacional, Fabio Vásquez Castaño, y que fue ocupada por las fuerzas armadas en 1967.

Hace años, después del robo de las armas del cantón Norte, centenares de presuntos miembros del M-19 denunciaron torturas y malos tratos. Una comisión de Amnistía Internacional comprobó que muchas de las denuncias eran ciertas. El gobierno del doctor Turbay Ayala, sin embargo, negó de plano todos los cargos, fundándose sobre todo en la falta de crédito de los guerrilleros. No obstante, bastó con que un solo prisionero declarara que fue entrenado en Cuba para que el gobierno rompiera sus relaciones con ese país, como si se tratara de la verdad revelada.

Con igual inconsecuencia se procedió a absolver de toda responsabilidad al gobierno de Panamá, cuando el mismo prisionero dijo que era allí donde se habían armado y embarcado. Me une una amistad personal muy antigua, muy seria y muy entrañable con el general Omar Torrijos, lo mismo que con el presidente Arístides Royo, y no abrigo ni la menor sombra de duda de que son ajenos a este incidente. Pero una amistad semejante me une con el presidente Fidel Castro, y con muchos otros dirigentes de la revolución cubana, y no puedo entender que el crédito que sirve para unos no sirva para los otros.

Se ha señalado como una prueba de la serenidad de nuestro gobierno el hecho de que las relaciones con Cuba no hayan sido rotas, sino suspendidas. La verdad es que en la práctica no hay ninguna diferencia. El canciller Lemos Simonds, que es un hombre culto y responsable, tuvo la mala suerte de enredarse en este sofisma en su primera actuación pública. Según dijo al periodista Yamit Amat, en una entrevista radial, la suspensión deja abierta la posibilidad de que otro país se encargue de nuestros asuntos en Cuba, y viceversa. En cambio, según dijo el canciller colombiano, la ruptura excluiría esa posibilidad. Es un error.

En realidad, la distinción no existe en Derecho internacional. Fue el doctor Alberto Lleras, un hombre que conoce muy bien las sutilezas del idioma, quien le mencionó por primera vez cuando su gobierno rompió relaciones con Cuba: el 9 de diciembre de 1961. «No hemos roto relaciones con Cuba», dijo en su discurso de esa ocasión. «Sólo hemos suspendido las existentes con el régimen de Castro.» De modo que los asesores del presidente Turbay le hicieron repetir una simple y muy precisa fórmula literaria como si fuera una figura del Derecho internacional.

Después de veinticinco años, tenía el propósito firme y grato de vivir en mi país. Pero en este ambiente de improvisación y equivocaciones, recibí una información muy seria de que había una orden de detención contra mí, emanada de la justicia militar. No tengo nada que ocultar ni me he servido jamás de un arma distinta de la máquina de escribir, pero conozco la manera como han procedido en otros casos semejantes las autoridades militares, inclusive con alguien tan eminente como el poeta Luis Vidales, y me pareció que era una falta de respeto conmigo mismo facilitar esa diligencia. Las autoridades civiles, entre quienes tengo muy buenos y viejos amigos, me dieron toda clase de seguridades de que no se intentaba nada contra mí. Pero en un gobierno donde algunos dicen una cosa y otros hacen otra muy distinta, y donde los militares guardan secretos que los civiles no conocen, no es posible saber dónde está la tierra firme. Una prueba de eso es que el canciller Lemos Simonds —con quien yo tenía prevista una cita amistosa para el próximo lunes— se refirió a mi persona en términos muy cordiales a través de la radio, y en cambio el comunicado de su propia cancillería dijo que mi decisión de abandonar el país bajo la protección de la embajada de México es una maniobra más en la campaña internacional de desprestigio contra el actual gobierno de Colombia, es decir: al cargo concreto y más gratuito, que no se encontraba el día anterior. Así las cosas, con el dolor de mi alma, me he visto precisado a seguir apacentando, quién sabe por cuánto tiempo más, mi persistente y dolorosa nostalgia del olor de la guayaba.

PUNTO FINAL A UN INCIDENTE INGRATO

Nunca, desde que tengo memoria, he dado las gracias por un elogio escrito ni me he contrariado por una injuria de prensa. Es justo: cuando uno se expone a la contemplación pública a través de sus libros y sus actos, como yo lo he hecho, los lectores deben disfrutar del privilegio de decir lo que piensan, aunque sean pensamientos infames. Por eso renuncié hace mucho tiempo al derecho de réplica y rectificación –que debía considerarse como uno de los derechos humanos– y, desde entonces, en ningún caso y ni una sola vez en ninguna parte del mundo he respondido a ninguno de los tantos agravios que se me han hecho, y de un modo especial en Colombia.

Me veo obligado a permitirme ahora una sola excepción, para comentar los dos argumentos únicos con que el gobierno ha querido explicar mi intempestiva salida de Colombia la semana pasada. Distintos funcionarios, en todos los tonos y en todas las formas, han coincidido en dos cargos concretos. El primero es que me fui de Colombia para darle una mayor resonancia publicitaria a mi próximo libro. El segundo es que lo hice en apoyo de una campaña internacional para desprestigiar al país. Ambas acusaciones son tan frívolas, además de contradictorias, que uno se pregunta escandalizado si de veras habrá alguien con dos dedos de frente en el timón de nuestros destinos.

La única desdicha grande que he conocido en mi vida es el asedio de la publicidad. Esto, al contrario de lo que creo merecer, me ha condenado a vivir como un fugitivo. No asisto nunca a actos públicos ni a reuniones multitudinarias, no he dictado nunca una conferencia, no he participado ni pienso participar jamás en el lanzamiento de un libro, les tengo tanto miedo a los micrófonos y a las cámaras de televisión como a los aviones, y a los periodistas les consta que cuando concedo una entrevista es porque respeto tanto su oficio que no tengo corazón para decirles que no.

Publicado originalmente el 8 de abril de 1981.

Esta determinación de no convertirme en un espectáculo público me ha permitido conquistar la única gloria que no tiene precio: la preservación de mi vida privada. A toda hora, en cualquier parte del mundo, mientras la fantasía pública me atribuye compromisos fabulosos, estoy siempre en el único ambiente en que me siento ser yo mismo: con un grupo de amigos. Mi mérito mayor no es haber escrito mis libros, sino haber defendido mi tiempo para ayudar a Mercedes a criar bien a nuestros hijos. Mi mayor satisfacción no es haber ganado tantos y tan maravillosos amigos nuevos, sino haber conservado, contra los vientos más bravos, el afecto de los más antiguos. Nunca he faltado a un compromiso, ni he revelado un secreto que me fuera confiado para guardar, ni me he ganado un centavo que no sea con la máquina de escribir. Tengo convicciones políticas claras y firmes, sustentadas, por encima de todo, en mi propio sentido de la realidad, y siempre las he dicho en público para que pueda oírlas el que las quiera oír. He pasado por casi todo el mundo. Desde ser arrestado y escupido por la policía francesa, que me confundió con un rebelde argelino, hasta quedarme encerrado con el papa Juan Pablo II en su biblioteca privada, porque él mismo no lograba girar la llave en la cerradura. Desde haber comido las sobras de un cajón de basuras en París, hasta dormir en la cama romana donde murió el rey don Alfonso XIII. Pero nunca, ni en las verdes ni en las maduras, me he permitido la soberbia de olvidar que no soy nadie más que uno de los dieciséis hijos del telegrafista de Aracataca. De esa lealtad a mi origen se deriva todo lo demás: mi condición humana, mi suerte literaria y mi honradez política.

He dicho alguna vez que todo honor se paga, que toda subvención compromete y que toda invitación se queda debiendo. Por eso he sido siempre tan cuidadoso en mi vida social. Nunca he aceptado más almuerzos que los de mis amigos probados. Hace muchos años, cuando era crítico de cine y estaba sometido a la presión de los exhibidores, conservaba siempre el pase de favor para demostrar que no había sido usado, y pagaba la entrada. No acepto invitaciones de viajes con gastos pagados.

El boleto de nuestro vuelo a México de la semana pasada —a pesar de la gentil resistencia de la embajadora de aquel país en Colombia— lo compramos con nuestro dinero. Pocos días antes, sin consultarlo conmigo, un amigo servicial le había pedido al alcalde de Bogotá que hiciera cambiar el horario del racionamiento eléctrico en mi casa, pues coincidía con mi tiempo de trabajo, y tengo un estudio sin luz natural y una máquina de escribir eléctrica. El alcalde le contestó, con toda la razón, que Balzac era mejor escritor que yo y, sin embargo, escribía con velas. Al amigo que me lo contó indignado le repliqué que el señor alcalde cumplió con su deber, y que contestó lo que debía contestar.

La gente que me conoce sabe que ésta es mi personalidad real, más allá de la leyenda y la perfidia, y que si quedé mal hecho de *fábrica* ya es demasiado tarde para volverme a hacer nuevo. De modo que no, ilustres oligarcas de pacotilla: nadie se construye una vida así, con las puras uñas, y con tanto rigor minuto a minuto, para salir de pronto con el chorro de babas de asilarse y exiliarse sólo para vender un millón de libros, que además ya estaban vendidos.

El segundo cargo, de que me fui de Colombia con el único propósito de desprestigiar al país, es todavía menos consistente. Pero tiene el mérito de ser una creación personal del presidente de la República, aturdido por la imagen cada vez más deplorable de su gobierno en el exterior. Lo malo es que me lo haya atribuido a mí, pues tengo la buena suerte de disponer de dos argumentos para sacarlo de su error.

El primero es muy simple, pero quiero suplicar que lo lean con la mayor atención, porque puede resultar sorprendente. Es éste: en ninguna de mis ya incontables entrevistas a través del mundo entero —hasta ahora— no había hecho nunca ninguna declaración sobre la situación interna de Colombia ni había escrito una palabra que pudiera ser utilizada contra ella. Era una norma moral que me había impuesto desde que tuve conciencia del poder indeseable que tenía entre manos, y logré mantenerla, contra viento y marea, durante casi treinta años de vida errante. Cada vez que quise hacer un comentario sobre la situación interna de Colombia lo vine a hacer dentro de ella o a través de nuestra prensa. El que tenga una evidencia contra esta afirmación le suplico que la haga conocer de inmediato, de un modo serio e inequívoco y con pruebas terminantes. Pues también suplico a mis lectores que si esas pruebas no aparecen, o no son convincentes, lo consideren y proclamen desde ahora y para siempre como un reconocimiento público de mi razón.

El segundo argumento es todavía más simple, y no ha dependido tanto de mí como de la fatalidad. Es éste: tengo el inmenso honor de haberle dado más prestigio a mi país en el mundo entero que ningún otro colombiano en toda su historia, aun los más ilustres, y sin excluir, uno por uno, a todos los presidentes sucesivos de la República. De modo que cualquier daño que le pueda hacer mi forzosa decisión lo habría derrotado yo mismo de antemano, y también a mucha honra.

En realidad, el gobierno se ha atrincherado en esas dos acusaciones pueriles, porque en el fondo sabe que mi sentido de la responsabilidad me impedirá revelar los nombres de quienes me previnieron a tiempo. Sé que la trampa estaba puesta y que mi condición de escritor no me iba a servir de nada, porque se trataba precisamente de demostrar que para las fuerzas de represión de Colombia no hay valores intocables. O como dijo el general Camacho cuando apresaron a Luis Vidales: «Aquí no hay poeta que valga». Mauro Huertas Rengifo, presidente de la Asamblea del Tolima, declaró a los periodistas y se publicó en el mundo entero que el Ejército me buscaba desde hacía diez días para interrogarme sobre supuestos vínculos con el M-19. El único comentario que conozco sobre esa declaración lo hizo un alto funcionario en privado: «Es un loquito». En cambio, el primer guerrillero que se declaró entrenado en Cuba provocó, de inmediato, la ruptura de relaciones con ese país. Pero hay algo no menos inquietante: a la medianoche del miércoles pasado, cuando mi esposa y yo teníamos más de seis horas de estar en la embajada de México en Bogotá, el gobierno colombiano fue informado de nuestra decisión, y de un modo oficial, a través del secretario general de la cancillería colombiana, el coronel Julio Londoño. A la mañana siguiente, cuando la noticia se divulgó contra nuestra voluntad, los periodistas de radio entrevistaron por teléfono al canciller Lemos Simonds y éste no sabía nada. Es decir: casi ocho horas después aún no había sido informado por su subalterno. El ministro de Gobierno, aún más despalomado, llegó hasta el extremo de desmentir la noticia.

La verdad es que las voces de que me iban a arrestar eran de dominio público en Bogotá desde hacía varios días y –al contrario de los esposos cornudos– no fui el último en conocerlas. Alguien me dijo: «No hay mejor servicio de inteligencia que la amistad». Pero lo que me convenció por fin de que no era un simple rumor de altiplano fue que el martes 24 de marzo, en la noche, después de una cena en el palacio presidencial, un alto oficial del ejército la comentó con más detalles. Entre otras cosas dijo: «El general Forero Delgadillo tendrá el gusto de ver a García Márquez en su oficina, pues tiene algunas preguntas que hacerle en relación con el M-19». En otra reunión diferente, esa misma noche, se comentó como una evidencia comprometedora un viaje que Mercedes y yo hicimos de Bogotá a La Habana, con escala en Panamá, del 28 de enero al 11 de febrero. El viaje fue cierto y público, como los tres o cuatro que hacemos todos los años a Cuba, y el motivo fue una reunión de escritores en la Casa de las Américas, a la cual asistieron también otros colombianos. Aunque sólo hubiera sido por la suposición escandalosa de que ese viaje tuvo alguna relación con el posterior desembarco de guerrilleros, habría tomado precauciones para no dejarme manosear por los militares. Pero hay más, y estoy seguro de que el tiempo lo irá sacando a flote.

La forma en que la prensa oficial ha tratado el incidente está ya sacando algunas, y más de lo que parece.

Ha habido de todo para escoger. Jaime Soto —a quien siempre tuve como un buen periodista y un viejo amigo a quien no veo hace muchos años— explicó mi viaje en la forma más boba: «El que la debe la teme». Sin embargo, el comentario más revelador se publicó en la página editorial de *El Tiempo*, el domingo pasado, firmado con el seudónimo de *Ayatolá*. No sé a ciencia cierta quién es, pero el estilo y la concepción de su nota lo delatan como un retrasado mental que carece por completo del sentido de las palabras, que deshonra el oficio más noble del mundo con su lógica de oligofrénico, que revela una absoluta falta de compasión por el pellejo ajeno y razona como alguien que no tiene ni la menor idea de cuán arduo y comprometedor es el trabajo de hacerse hombre.

A pesar de su propósito criminal, es una nota importante, pues en ella aparece por primera vez, en una tribuna respetable de la prensa oficial, la pretensión de establecer una relación precisa, incluso cronológica, entre mi reciente viaje a La Habana y el desembarco guerrillero en el sur de Colombia. Es el mismo cargo que los militares pretendían hacerme, el mismo que me dio la mayoría de mis informantes, y del cual yo no había hablado hasta entonces en mis numerosas declaraciones de estos días. Es una acusación formal. La que el propio gobierno trató de ocultar, y que echa por tierra, de una vez por todas, la patraña de la publicidad de mis libros y la campaña de desprestigio internacional. Ahora se sabe por qué me buscaban, por qué tuve que irme y por qué tendré que seguir viviendo fuera de Colombia, quién sabe hasta cuándo, contra mi voluntad.

No puedo terminar sin hacer una precisión de honestidad. Desde hace muchos años, el tiempo ha hecho constantes esfuerzos por dividir mi personalidad: de un lado, el escritor que ellos no vacilan en calificar de genial, y del otro lado, el comunista feroz que está dispuesto a destruir a su patria. Cometen un error de principio: soy un hombre indivisible, y mi posición política obedece a la misma ideología con que escribo mis libros. Sin embargo, el tiempo me ha consagrado con todos los elogios como escritor, inclusive exagerados, y al mismo tiempo me ha hecho víctima de todas las diatribas, aun las más infames, como animal político.

En ambos extremos, el tiempo ha hecho su oficio sin que yo haya intentado nunca ninguna réplica de ninguna clase, ni para dar las gracias ni para protestar. Desde hace más de treinta años, cuando todos éramos jóvenes y creíamos –como yo lo sigo creyendo– que nada hay más hermoso que vivir, he mantenido una amistad fiel y afectuosa con Hernando y Enrique Santos Castillo –a quienes quiero bien a pesar de nuestra distancia, porque he aprendido a entenderlos bien– y con Roberto García Peña, a quien tengo por uno de los hombres más decentes de nuestro tiempo. Quiero suplicarles que digan a sus lectores si alguna vez les he hecho un reclamo por las injurias de su periódico, si alguna vez he rectificado en público o en privado cualquiera de sus excesos, o si éstos han alterado de algún modo mi sentido de la amistad. No; he tenido la buena salud mental de tratarlos como si ellos no tuvieran nada que ver con un periódico que siempre he visto como un engendro sin control que se envenena con sus propios hígados. Sin embargo, esta vez el engendro ha ido más allá de todo límite permisible y ha entrado en el ámbito sombrío de la delincuencia. Me pregunto, al cabo de tantos años, si yo también no me equivoqué al tratar de dividir la personalidad de sus domadores.

De modo que todo este ingrato incidente queda planteado, en definitiva, como una confrontación de credibilidades. De un lado está un gobierno arrogante, resquebrajado y sin rumbo, respaldado por un periódico demente cuyo raro destino, desde hace muchos años, es jugárselas todas por presidentes que detesta. Del otro lado estoy yo, con mis amigos incontables, preparándome para iniciar una vejez inmerecida, pero meritoria. La opinión pública no tiene más que una alternativa: ¿A quién creer? Yo, con mi paciencia sin término, no tengo ninguna prisa por su decisión. Espero.

MITTERRAND, EL OTRO: EL ESCRITOR

Hace algunos años, al término de una cena oficial en la embajada de Francia en México, fuimos invitados por nuestros anfitriones a tomar el café frente a la chimenea. Era una reunión muy reducida de franceses solos, a la cual yo asistí por una amable sugerencia del visitante de honor, François Mitterrand, candidato a la presidencia de la República en aquel momento. La conversación en la mesa había tenido el sabor complaciente, pero efímero, de las cenas mundanas, y era evidente que los anfitriones habían propuesto el café frente a la chimenea en busca de un ambiente más propicio para que Mitterrand se decidiera a hablarnos de los asuntos actuales de Francia y el mundo. Parsimonioso y sonriente, como siempre, él ocupó el sillón favorito del dueño de la casa, y todos nos sentamos alrededor para no perder ni una gota de sus palabras. Entonces, Mitterrand, dirigiéndose a mí, dijo:

—Muy bien; hablemos de literatura.

El ángel de la desilusión se aposentó en la sala. La mayoría pensó que Mitterrand, que es un político con las espuelas muy bien puestas, había recurrido a aquel artificio para eludir el asunto central. Pero al cabo de breves minutos todos estábamos fascinados por la sabiduría y el encanto de aquel maestro que se paseaba con un aire propio a través de los grandes nombres y las desdichas eternas de las letras universales.

Aquel día lo descubrí. Lo había conocido unos años antes, después de que Pablo Neruda le habló de mí y le llevó algunos de mis libros traducidos al francés y le dijo tantas cosas enormes sobre nuestra amistad. Cuando nos encontramos por primera vez, ya parecíamos amigos muy antiguos. Pero yo no había podido superar el prejuicio de que Mitterrand era antes que nada un político, y tenía la tendencia a hablarle sólo de política, como lo hace sin remedio la inmensa mayoría de los políticos. Aquella noche, en México, caí en la cuen-

Publicado originalmente el 14 de abril de 1981.

ta de que el equivocado era yo, y que Mitterrand era en efecto un hombre de letras, en el sentido reverencial y un poco fatalista en que sólo los franceses lo entienden.

En realidad no sólo es Mitterrand un escritor excelente, sino de los que escriben todos los días de su vida, como lo hacen los más grandes. En todos sus libros, pero en especial en *La paille et la graine*, como tantas veces en la vida real, él ha dicho que nunca ha tenido intención de escribir sus memorias. Es comprensible: las memorias son un género al cual recurren los escritores cansados cuando ya están a punto de olvidarlo todo. El propósito de Mitterrand es el contrario: escribe para no olvidar, y su buena costumbre nos ayuda a que tampoco nosotros olvidemos. «Yo tomo notas como demonio sobre algún papel que pierdo más a menudo de lo que me llegan a servir.» Son, como él mismo lo dice, anotaciones fugaces escritas a golpes de emoción, y a las cuales acuerda una importancia por razones variables y casi siempre subjetivas. No hay escritor que no lo comprenda. Todos llevamos esas notas escritas en el revés de los sobres, en esquinas de periódicos, en tiques de autobuses usados y aun sin usar, donde hemos escrito una frase que en un momento nos pareció una nueva revelación del mundo, o del alma humana, y que luego volveremos a encontrar convertidos en pelotitas de cartón piedra, molidos por las aspas de la lavadora eléctrica, macerados por el jabón y petrificados por la plancha. Mitterrand lo sabe y lo dice: «Es una ilusión lírica». Y lo dice con toda razón, porque esas notas fugaces son como los versos que a veces conocemos en sueños, que nos trastornan mientras dormimos, como si fueran la esencia misma de la poesía, y al despertar comprobamos que no era más que una frase de publicidad en la radio de la casa vecina. Era, en efecto, una ilusión lírica. Pero Mitterrand sabe, como todos los escritores, que de esos minúsculos y continuos fracasos está hecha la buena literatura.

A mí me parece que su visión del mundo, más que la de un político, es la de un hombre abrasado por la fiebre de la literatura. Por eso he pensado siempre que sería —¿será?— un gobernante sabio. Es un hombre que se interesa por todas las cosas de la vida, aun las más simples, y lo hace con una pasión, con un gusto y una lucidez que constituyen su mejor virtud. Un hombre al cual le llama la atención, leyendo el diario de los hermanos Goncourt, lo que tal vez a otro lector menos inteligente podía parecerle una frivolidad: que la sociedad protectora de animales, creada en 18.., se anticipó en tres años a la liberación de los esclavos. Cuando visitó a Violete Trefusis, en la casa del Ombrellino de Florencia, lo que más le impresionó, y que había de marcar aquel instante para siempre, fue el eco de sus propios pasos en la inmensa galería de la entrada. De su entrevista con Golda Meir, de quien sabemos que no era bella, nos dejó el testimonio de que era una madre severa y tierna.

De todos los recuerdos que se han escrito sobre Salvador Allende, el de Mitterrand me parece el más revelador. Era en 1971, y el presidente le conducía a través de las galerías del palacio de la Moneda, en Santiago de Chile, cuando se detuvo frente a un busto de José Manuel Balmaceda. «Este hombre era un conservador elegido por la derecha de su época», le dijo. «Pero este conservador era también un legalista que no pudo soportar las agresiones al derecho: se suicidó.» El presidente Allende concluyó: «Ahora todos los chilenos respetan su memoria. Su acción heroica pertenece a la conciencia de nuestro pueblo». Yo estoy seguro de que Mitterrand no podía quitarse de la mente aquel episodio, una mañana en que desayunábamos en México con las hijas del presidente Allende, apenas un año después de su muerte. «Fue preciso movilizar la aviación», anotó en sus papelitos de bolsillo, «y destruir La Moneda, sólo para asesinarlo».

De esas notas quedará una visión de nuestro tiempo sin duda mucho más fiel de lo que suponen sus lectores distraídos. De Georges Pompidou ha escrito: «Tiene la ambición más alta que su poltrona». Como buen escritor, Mitterrand debe saber que nuestras palabras nos persiguen no sólo hasta la muerte, sino hasta mucho más allá de la muerte. Pero, también como a buen escritor, no le teme a ese destino. Un día, mientras almorzaba solo en la *brasserie* Lipps, el propietario se le acercó y le dijo al oído: «Dicen que el presidente ha muerto». El presidente era Georges Pompidou. Recordando aquel día, Mitterrand escribió más tarde que, de todos modos, no pudo evitar una cierta piedad por ese muerto olvidado desde antes de que lo sepultaran. De Valéry Giscard d'Estaing, de quien ha dicho tantas cosas, ha dicho una terrible: «Nadie duda que él posea, en el grado más alto, el arte de explicar los fracasos de los cuales derivan sus triunfos». Sin embargo, ninguna indignación me pareció nunca más lúcida que la suya cuando le dieron el Premio Nobel de la Paz a Henry Kissinger. «No tengo nada en su contra», escribió entonces. Pero consideró que darle a Kissinger el Premio de la Paz por haber puesto término a una guerra que él mismo había enardecido era como dárselo a Sukarno porque no mató más comunistas indonesios después de haber matado 300.000, o como dárselo a Papadópulus, el coronel griego, porque cerró las cámaras de tortura que él mismo había instaurado y abrió al turismo las playas de sus islas de presidiarios; o a Idi Amin Dada, porque no volvió a masacrarle el cráneo a ninguno de sus ministros en los últimos años. «No pongo más ejemplos», escribió, «porque no pienso enemistarme con la mitad del mundo.»

A Julio César, que también era un escritor grande, Thornton Wilder le atribuyó esta frase feliz: «Yo, que gobierno tantos hombres, soy gobernado por pájaros y truenos». El escritor Mitterrand no podía estar a salvo de estas pequeñas supersticiones que hacen más misteriosa y bella la vida de los hombres. La suya, de acuerdo con numerosas anotaciones en sus libros, es la superstición del mes de mayo. El mes de las flores y de las vírgenes que suben al cielo en cuerpo y alma, y en el que le han ocurrido a él las peores y las mejores cosas. Hace unos tres meses cuando apenas se vislumbraba la posibilidad de su candidatura, alguien habló de esto en un almuerzo que nos ofreció Mitterrand en París. «La reelección del actual presidente es probable», dijo, «pero la mía es posible.» No sé por qué tuve entonces la impresión de que Mitterrand contaba en aquel momento en los innumerables factores que determinan la victoria de una elección, pero que entre ellos no descartaba uno que era tal vez el menos extraño a su corazón de buen escritor: el mes de mayo.

LA ÚLTIMA Y MALA NOTICIA SOBRE HAROLDO CONTI

A Haroldo Conti, que era un escritor argentino de los grandes, le advirtieron en octubre de 1975 que las fuerzas armadas lo tenían en una lista de agentes subversivos. La advertencia se repitió por distintos conductos en las semanas siguientes y, a principios de 1976, era ya de dominio público en Buenos Aires. Por esos días, me escribió una carta a Bogotá, en la cual era evidente su estado de tensión. «Martha y yo vivimos prácticamente como bandoleros», decía, «ocultando nuestros movimientos, nuestros domicilios, hablando en clave.» Y terminaba: «Abajo va mi dirección, por si sigo vivo». Esa dirección era la de su casa alquilada en el número 1205 de la calle Fitz Roy, en Villa Crespo, donde siguió viviendo sin precauciones de ninguna clase hasta que un comando de seis hombres armados la asaltó a medianoche, nueve meses después de la primera advertencia, y se lo llevaron vendado y amarrado de pies y manos, y lo hicieron desaparecer para siempre.

Haroldo Conti tenía entonces 51 años, había publicado siete libros excelentes y no se avergonzaba de su gran amor a la vida. Su casa urbana tenía un ambiente rural: criaba gatos, criaba palomas, criaba perros, criaba niños y cultivaba en canteros legumbres y flores. Como tantos escritores de nuestra generación, era un lector constante de Hemingway, de quien aprendió además la disciplina de cajero de banco. Su pensamiento político era claro y público, lo expresaba de viva voz y lo exponía en la prensa, y su identificación con la revolución cubana no era un misterio para nadie.

Desde que recibió las primeras advertencias tenía una invitación para viajar a Ecuador, pero prefirió quedarse en su casa. «Uno elige», me decía en su carta. El pretexto principal para no irse era que Martha estaba encinta de siete meses y no sería aceptada en avión. Pero la verdad es que no quiso irse. «Me quedaré hasta que pueda, y después Dios verá», me decía en su carta, «porque, aparte de escribir,

Publicado originalmente el 21 de abril de 1981.

y no muy bien que digamos, no sé hacer otra cosa.» En febrero de 1976, Martha dio a luz un varón, a quien pusieron el nombre de Ernesto. Ya para entonces, Haroldo Conti había colgado un letrero frente a su escritorio: «Éste es mi lugar de combate, y de aquí no me voy». Pero sus secuestradores no supieron lo que decía ese letrero, porque estaba escrito en latín.

El 4 de mayo de 1976, Haroldo Conti escribió toda la mañana en el estudio y terminó un cuento que había empezado el día anterior: *A la diestra*. Luego se puso saco y corbata para dictar una clase de rutina en una escuela secundaria del sector, y antes de las seis de la tarde volvió a casa y se cambió de ropa. Al anochecer ayudó a Martha a poner cortinas nuevas en el estudio, jugó con su hijo de tres meses y le echó una mano en las tareas escolares a una hija del matrimonio anterior de Martha, que vivía con ellos: Myriam, de siete años. A las nueve de la noche, después de comerse un pedazo de carne asada, se fueron a ver *El padrino II*. Era la primera vez que iban al cine en seis meses. Los dos niños se quedaron al cuidado de un amigo que había llegado esa tarde de Córdoba y lo invitaron a dormir en el sofá del estudio.

Cuando volvieron, a las 12.05 horas de la noche, quien les abrió la puerta de su propia casa fue un civil armado con una ametralladora de guerra. Dentro había otros cinco hombres, con armas semejantes, que los derribaron a culatazos y los aturdieron a patadas.

El amigo estaba inconsciente en el suelo, vendado y amarrado, y con la cara desfigurada a golpes. En su dormitorio, los niños no se dieron cuenta de nada porque habían sido adormecidos con cloroformo.

Haroldo y Martha fueron conducidos a dos habitaciones distintas, mientras el comando saqueaba la casa hasta no dejar ningún objeto de valor. Luego los sometieron a un interrogatorio bárbaro. Martha, que tiene un recuerdo minucioso de aquella noche espantosa, escuchó las preguntas que le hacían a su marido en la habitación contigua. Todas se referían a dos viajes que Haroldo Conti había hecho a La Habana. En realidad, había ido dos veces –en 1971 y en 1974–, y en ambas ocasiones como jurado del concurso de la Casa de las Américas. Los interrogadores trataban de establecer por esos dos viajes que Haroldo Conti era un agente cubano.

A las cuatro de la madrugada, uno de los asaltantes tuvo un gesto humano, y llevó a Martha a la habitación donde estaba Haroldo para que se despidiera de él. Estaba deshecha a golpes, con varios dientes partidos, y el hombre tuvo que llevarla del brazo porque tenía los ojos vendados. Otro que los vio pasar por la sala, se burló: «¿Vas a bailar con la señora?». Haroldo se despidió de Martha con un beso. Ella se dio cuenta entonces de que él no estaba vendado, y esa comprobación la aterrorizó, pues sabía que sólo a los que iban a morir les permitían ver la cara de sus torturadores. Fue la última vez que estuvieron juntos. Seis meses después del secuestro, habiendo pasado de un escondite a otro con su hijo menor, Martha se asiló en la embajada de Cuba. Allí estuvo año y medio esperando el salvoconducto, hasta que el general Omar Torrijos intercedió ante el almirante Emilio Massera, que entonces era miembro de la Junta de Gobierno Argentina, y éste le facilitó la salida del país.

Quince días después del secuestro, cuatro escritores argentinos —y entre ellos los dos más grandes— aceptaron una invitación para almorzar en la casa presidencial con el general Jorge Videla. Eran Jorge Luis Borges, Ernesto Sábato, Alberto Ratti, presidente de la Sociedad Argentina de Escritores, y el sacerdote Leonardo Castellani. Todos habían recibido por distintos conductos la solicitud de plantearle a Videla el drama de Haroldo Conti. Alberto Ratti lo hizo, y entregó además una lista de otros once escritores presos. El padre Castellani, entonces tenía casi ochenta años y había sido maestro de Haroldo Conti, pidió a Videla que le permitiera verlo en la cárcel. Aunque la noticia no se publicó nunca, se supo que, en efecto, el padre Castellani lo vio el 8 de julio de 1976 en la cárcel de Villa Devoto, y que lo encontró en tal estado de postración que no le fue posible conversar con él.

Otros presos, liberados más tarde, estuvieron con Haroldo Conti. Uno de ellos rindió un testimonio escrito, según el cual fue su compañero de presidio en el campo de concentración de la Brigada Goemez, situada en la autopista Richieri, a doce kilómetros de Buenos Aires por el camino de Ezeiza. «En mayo de 1976», dice el testimonio, «Haroldo Conti se encontraba en una celda de dos metros por uno, con piso de cemento y puerta metálica. Llegó el día 20. Dijo haber estado en un lugar del ejército, donde lo pasó muy mal. Dijo que se había quedado encerrado en un baño, donde se desmayó. Apenas si podía hablar y no podía comer. El día 21 pudo comer algo. Se ve que andaba muy mal porque le dieron una manta y lo iban a ver con frecuencia. En la madrugada del día 22 lo sacaron de la celda. Parece que lo iban a revisar o algo así. Estaba muy mal y no retenía orines.» El testigo no lo volvió a ver en la prisión.

No ha habido gestión, ni derecha ni torcida, que la esposa y los amigos de Haroldo Conti no hayamos hecho en el mundo entero para esclarecer su suerte. Hace unos dos años sostuve una entrevista en México con el almirante Emilio Massera, que ya entonces estaba retirado de las armas y del gobierno, pero que mantenía buenos contactos con el poder. Me prometió averiguar todo lo que pudiera sobre Haroldo Conti, pero nunca me dio una respuesta definitiva.

En junio de 1980, la reina Sofía de España viajó a Argentina al frente de una delegación cultural que asistió al aniversario de Buenos Aires. Un grupo de exiliados le pidió a algunos miembros de la comitiva que intercedieran ante el gobierno argentino para la liberación de varios presos políticos prominentes. Yo, en nombre de la Fundación Habeas, y como amigo personal de Haroldo Conti, les pedí una gestión muy modesta: establecer de una vez y para siempre cuál era su situación real. La gestión se hizo, pero el gobierno argentino no dio ninguna respuesta. Sin embargo, en octubre pasado, cuando ya estaba decidido su retiro de la presidencia, el general Jorge Videla concedió una entrevista a una delegación de alto nivel de la agencia Efe, y respondió algunas preguntas sobre los presos políticos. Por primera vez habló entonces de Haroldo Conti. No hizo ninguna precisión de fecha, ni de lugar ni de ninguna otra circunstancia, pero reveló sin ninguna duda que estaba muerto. Fue la primera noticia oficial, y hasta ahora la única. No obstante, el general Videla les pidió a los periodistas españoles que no la publicaran de inmediato, y ellos cumplieron. Yo considero, ahora que el general Videla no está en el poder, y sin haberlo consultado con nadie, que el mundo tiene derecho a conocer esa noticia.

¿QUIÉN CREE A JANET COOKE?

Todo empezó el día en que Janet Cooke, reportera del *Washington Post*, le dijo a su jefe de redacción que había oído hablar de un niño de ocho años que se inyectaba heroína con la complacencia de su madre. «Encuentre a ese niño», le dijo el jefe de redacción. «Será un reportaje de primera página.» En octubre del año pasado, en efecto, el relato revelador y tremendo —bajo el título de «El mundo de Jimmy»— estremeció a Estados Unidos. Hace dos semanas, con sólo tres años en el oficio y veintiséis de edad, Janet Cooke mereció el honor más codiciado del periodismo de su país: el Premio Pulitzer. Aunque sólo por pocas horas, pues el escrutinio inclemente de sus jefes y la presión de su propia alma la obligaron a confesar que el reportaje era inventado y que el pequeño Jimmy sólo había existido en su imaginación.

Este incidente plantea, una vez más, el drama del periodismo de Estados Unidos, cuyo rigor casi puritano lo ha convertido en el mejor del mundo, pero cuyas contradicciones traumáticas lo han convertido también en el más peligroso. De allí que toda nota falsa, como la que Janet Cooke acaba de contar, termine por provocar sin remedio una crisis de conciencia nacional.

Yo tuve una prueba personal de ese rigor, hace unos cuatro años, cuando la revista *Harper*, de Nueva York, me pidió un artículo exclusivo sobre el golpe militar en Chile y el asesinato de Salvador Allende. Uno de los editores principales de la revista llamó por teléfono de Nueva York a París cuando leyó los originales, y me sometió a un interrogatorio casi policial de más de una hora sobre el origen de mis datos. No aspiraba, por supuesto, a que yo le revelara mis fuentes confidenciales, pero quería estar seguro de que yo estaba seguro de ellas, y de que me encontraba en condiciones de defenderlas. Más tarde vi personificada esa moral en mi amigo Elie Abel —el antiguo director de la escuela de periodismo de la

Publicado originalmente el 29 de abril de 1981.

Universidad de Columbia–, con quien trabajé en la comisión especial de la Unesco que hizo un estudio sobre la comunicación y la información en el mundo actual. Elie Abel y yo estábamos a una distancia política de siglos, pero la claridad y la entereza con que se batía por sus principios en aquellas reuniones soporíferas me recordaban a los predicadores iluminados de su compatriota Nathaniel Hawthorne.

Por eso es más sorprendente que un periodismo con fundamentos morales tan drásticos sea también capaz de llegar a extremos inconcebibles de manipulación y falsedad. Hace dos años –por ejemplo–, la revista *Time* publicó a media página la fotografía de algo que parecía ser dos pantallas de radar implantadas en una colina. El texto decía que había sido tomada en secreto en el interior de Cuba, y que eran unos dispositivos soviéticos muy refinados para captar toda clase de mensajes originados en Estados Unidos. Yo lo creí, y me pareció un recurso ordinario en la guerra sin cuartel de la información. Pero mis hijos, que se interesan más que yo en la ficción científica, me hicieron caer en la cuenta de que habíamos visto esas pantallas muchas veces en nuestros tantos viajes a Cuba. No debían ser tan secretas si millares de turistas extranjeros podían verlas y fotografiarlas viajando por carretera desde La Habana hacia el oriente del país. La semana siguiente, en efecto, el encargado de la oficina de intereses de Cuba en Washington aclaró en una carta que aquellas pantallas habían sido instaladas allí desde antes de la revolución por una empresa de comunicaciones de Estados Unidos. Veinte años después, a pesar del bloqueo, de los sabotajes y de los desembarcos armados, las pantallas continuaban en su puesto, todavía al servicio de la misma empresa transnacional norteamericana, y bajo su responsabilidad absoluta. La revista *Time* publicó esta aclaración de una pulgada en la sección de cartas, y quedó en paz con su conciencia. Nunca rectificó.

Más infame y persistente fue la guerra de información contra Vietnam, hace dos años. La prensa occidental, instigada por la de Estados Unidos, hizo creer al mundo que el gobierno vietnamita estaba mandando a morir en alta mar a los residentes chinos. Muy pocos nos tomamos el trabajo de ir a Vietnam a conversar con todo el mundo, inclusive con los chinos que se querían ir, como tanta gente se quiere ir de todas partes. Lo que entonces averiguamos parece hoy muy simple: la solidaridad mundial que Vietnam había conseguido durante la guerra militar, seguía siendo un dolor de cabeza para Estados Unidos, y se propusieron aniquilarla con la otra guerra feroz de la información. Lo lograron, por supuesto.

En todo caso, más allá de la ética y la política, la audacia de Janet Cooke, una vez más, plantea también las preguntas de siempre sobre las diferencias entre el periodismo y la literatura, que tanto los periodistas como los literatos llevamos siempre dormidas, pero siempre a punto de despertar en el corazón. Debemos empezar por preguntarnos cuál es la verdad esencial en su relato. Para un novelista lo primordial no es saber si el pequeño Jimmy existe o no, sino establecer si su naturaleza de fábula corresponde a una realidad humana y social, dentro de la cual podía haber existido. Este niño, como tantos niños de la literatura, podría no ser más que una metáfora legítima para hacer más cierta la verdad de su mundo. Hay por lo menos un punto a favor de esta coartada literaria: antes de que se descubriera la farsa de Janet Cooke, varios lectores habían escrito a su periódico para decir que conocían al pequeño Jimmy, y muchos decían conocer otros casos similares. Lo cual hace pensar –gracias a los dioses tutelares de las bellas letras– que el pequeño Jimmy no sólo existe una vez, sino muchas veces, aunque no sea el mismo que inventó Janet Cooke.

Lo malo es que en periodismo un solo dato falso desvirtúa sin remedio a los otros datos verídicos. En la ficción, en cambio, un solo dato real bien usado puede volver verídicas a las criaturas más fantásticas. La norma tiene injusticias de ambos lados: en periodismo hay que apegarse a la verdad, aunque nadie la crea, y en cambio en literatura se puede inventar todo, siempre que el autor sea capaz de hacerlo creer como si fuera cierto. Hay recursos intercambiables. Si un escritor dice que vio volar un rebaño de elefantes, no habrá nadie que se lo crea, porque el buen periodismo le ha hecho creer al mundo que los elefantes no vuelan. Pero no faltará quien se lo crea si apela al recurso periodístico de la precisión y dice que los elefantes que volaban eran 326. Yo oí contar muchas veces, siendo muy niño, la historia de un cura rural que levitaba en el momento de apurar el cáliz. Intenté contarlo en una novela, pero no conseguía creerlo yo mismo, hasta que cambié el vino por una taza de chocolate, y el cura se elevó como un ángel a dos centímetros sobre el nivel del suelo. Algo de esto debe ser el alcalde de Washington, Marion Barry, pues fue el primero que denunció la falsedad del relato de Janet Cooke. Y no porque creyera que el niño no existía, sino porque le pareció imposible que la madre permitiera inyectarle heroína delante de un reportero.

John Hersey, que era un buen novelista, escribió un reportaje sobre la ciudad de Hiroshima devastada por la bomba atómica, y es un relato tan apasionante que parece una novela. Daniel Defoe, que era también un gran periodista, escribió una novela sobre la ciudad de Londres devastada por la peste, y es un relato tan sobrecogedor que parece un reportaje. En esa línea de demarcación invisible pueden estar los ángeles que Janet Cooke necesita para la salvación de su alma. Pues no habría sido justo que le dieran el Premio Pulitzer de periodismo, pero en cambio sería una injusticia mayor que no le dieran el de literatura.

«MARÍA DE MI CORAZÓN»

Hace unos dos años, le conté un episodio de la vida real al director mexicano de cine Jaime Humberto Hermosillo, con la esperanza de que lo convirtiera en una película, pero no me pareció que le hubiera llamado la atención. Dos meses después, sin embargo, vino a decirme sin ningún anuncio previo que ya tenía el primer borrador del guión, de modo que seguimos trabajándolo juntos hasta su forma definitiva. Antes de estructurar los caracteres de los protagonistas centrales, nos pusimos de acuerdo sobre cuáles eran los dos actores que podían encarnarlos mejor: María Rojo y Héctor Bonilla. Esto nos permitió además contar con la colaboración de ambos para escribir ciertos diálogos, e inclusive dejamos algunos apenas esbozados para que ellos los improvisaran con su propio lenguaje durante la filmación.

Lo único que yo tenía escrito de esa historia —desde que me la contaron muchos años antes en Barcelona— eran unas notas sueltas en un cuaderno de escolar, y un proyecto de título: «No: yo sólo vine a hablar por teléfono». Pero a la hora de registrar el proyecto de guión nos pareció que no era el título más adecuado, y le pusimos otro provisional: *María de mis amores*. Más tarde, Jaime Humberto Hermosillo le puso el título definitivo: *María de mi corazón*. Era el que mejor le sentaba a la historia, no sólo por su naturaleza, sino también por su estilo.

La película se hizo con la aportación de todos. Creadores, actores y técnicos aportamos nuestro trabajo a la producción, y el único dinero líquido de que dispusimos fueron dos millones de pesos de la universidad veracruzana; es decir, unos 80.000 dólares, que, en términos de cine, no alcanzan ni para los dulces. Se filmó en dieciséis milímetros y en color, y en 93 días de trabajos forzados en el ambiente febril de la colonia Portales, que me parece ser una de las más definitivas de la ciudad de México. Yo la conocía muy bien, porque hace más de veinte años trabajé en la sección de armada de una imprenta

Publicado originalmente el 5 de mayo de 1981.

de esa colonia, y por lo menos un día a la semana, cuando terminábamos de trabajar, me iba con aquellos buenos artesanos y mejores amigos a bebernos hasta el alcohol de las lámparas en las cantinas del barrio. Nos pareció que ése era el ámbito natural de *María de mi corazón*. Acabo de ver la película ya terminada, y me alegré de comprobar que no nos habíamos equivocado. Es excelente, tierna y brutal a la vez, y al salir de la sala me sentí estremecido por una ráfaga de nostalgia.

María –la protagonista– era en la vida real una muchacha de unos veinticinco años, recién casada con un empleado de los servicios públicos. Una tarde de lluvias torrenciales, cuando viajaba sola por una carretera solitaria, su automóvil se descompuso. Al cabo de una hora de señas inútiles a los vehículos que pasaban, el conductor de un autobús se compadeció de ella. No iba muy lejos, pero a María le bastaba con encontrar un sitio donde hubiera un teléfono para pedirle a su marido que viniera a buscarla. Nunca se le habría ocurrido que en aquel autobús de alquiler, ocupado por completo por un grupo de mujeres atónitas, había empezado para ella un drama absurdo e inmerecido que le cambió la vida para siempre.

Al anochecer, todavía bajo la lluvia persistente, el autobús entró en el patio empedrado de un edificio enorme y sombrío, situado en el centro de un parque natural. La mujer responsable de las otras las hizo descender con órdenes un poco infantiles, como si fueran niñas de escuela. Pero todas eran mayores, demacradas y ausentes, y se movían con una andadura que no parecía de este mundo. María fue la última que descendió sin preocuparse de la lluvia, pues de todos modos, estaba empapada hasta el alma. La responsable del grupo se lo encomendó entonces a otras, que salieron a recibirlo, y se fue en el autobús. Hasta ese momento, María no se había dado cuenta de que aquellas mujeres eran 32 enfermas pacíficas trasladadas de alguna otra ciudad, y que en realidad se encontraba en un asilo de locas.

En el interior del edificio, María se separó del grupo y preguntó a una empleada dónde había un teléfono. Una de las enfermeras que conducía a las enfermas, la hizo volver a la fila mientras le decía de un modo muy dulce: «Por aquí, linda, por aquí hay un teléfono». María siguió, junto con las otras mujeres, por un corredor tenebroso, y al final entró en un dormitorio colectivo donde las enfermeras empezaron a repartir las camas. También a María le asignaron la suya. Más bien divertida con el equívoco, María le explicó entonces a una enfermera que su automóvil se había descompuesto en la carretera y sólo necesitaba un teléfono para prevenir a su marido. La enfermera fingió escucharla con atención, pero la llevó de nuevo a su cama, tratando de calmarla con palabras dulces.

«De acuerdo, linda», le decía, «si te portas bien, podrás hablar por teléfono con quien quieras. Pero ahora no, mañana.»

Comprendiendo de pronto que estaba a punto de caer en una trampa mortal, María escapó corriendo del dormitorio. Pero antes de llegar al portón, un guardia corpulento le dio alcance, le aplicó una llave maestra, y otros dos le ayudaron a ponerle una camisa de fuerza. Poco después, como no dejaba de gritar, le inyectaron un somnífero. Al día siguiente, en vista de que persistía en su actitud insurrecta, la trasladaron al pabellón de las locas furiosas, y la sometieron hasta el agotamiento con una manguera de agua helada a alta presión.

El marido de María denunció su desaparición poco después de la medianoche, cuando estuvo seguro de que no se encontraba en casa de ningún conocido. El automóvil –abandonado y desmantelado por los ladrones– fue recuperado al día siguiente. Al cabo de dos semanas, la policía declaró cerrado el caso, y se tuvo por buena la explicación de que María, desilusionada de su breve experiencia matrimonial, se había fugado con otro.

Para esa época, María no se había adaptado aún a la vida del sanatorio, pero su carácter había sido doblegado. Todavía se negaba a participar en los juegos al aire libre de las enfermas, pero nadie la forzaba. Al fin y al cabo, decían los médicos, así empezaban todas, y tarde o temprano terminaban por incorporarse a la vida de la comunidad. Hacia el tercer mes de reclusión, María logró por fin ganarse la confianza de una visitadora social, y ésta se prestó para llevarle un mensaje a su marido.

El marido de María la visitó el sábado siguiente. En la sala de recibo, el director del sanatorio le explicó en términos muy convincentes cuál era el estado de María y la forma en que él mismo podía ayudarla a recuperarse. Le previno sobre su obsesión dominante –el teléfono– y le instruyó sobre el modo de tratarla durante la visita, para evitar que recayera en sus frecuentes crisis de furia. Todo era cuestión, como se dice, de seguirle la corriente.

A pesar de que él siguió al pie de la letra las instrucciones del médico, la primera visita fue tremenda. María trató de irse con él a toda costa, y tuvieron que recurrir otra vez a la camisa de fuerza para someterla. Pero poco a poco se fue haciendo más dócil en las visitas siguientes. De modo que su marido siguió visitándola todos los sábados, llevándole cada vez una libra de bombones de chocolate, hasta que los médicos le dijeron que no era el regalo más conveniente para María, porque estaba aumentando de peso. A partir de entonces, sólo le llevó rosas.

COMO ÁNIMAS EN PENA

Hace ya muchos años que oí contar por primera vez la historia del viejo jardinero que se suicidó en Finca Vigía, la hermosa casa entre grandes árboles, en un suburbio de La Habana, donde pasaba la mayor parte de su tiempo el escritor Ernest Hemingway. Desde entonces la seguí oyendo muchas veces en numerosas versiones. Según la más corriente, el jardinero tomó la determinación extrema después de que el escritor decidió licenciarlo, porque se empeñaba en podar los árboles contra su voluntad. Se esperaba que en sus memorias, si las escribía, o en uno cualquiera de sus escritos póstumos, Hemingway contara la versión real. Pero, al parecer, no lo hizo.

Todas las variaciones coinciden en que el jardinero, que lo había sido desde antes de que el escritor comprara la casa, desapareció de pronto sin explicación alguna. Al cabo de cuatro días, por las señales inequívocas de las aves de rapiña, descubrieron el cadáver en el fondo de un pozo artificial que abastecía de agua potable a Hemingway y a su esposa de entonces, la bella Martha Gelhorm. Sin embargo, el escritor cubano Norberto Fuentes, que ha hecho un escrutinio minucioso de la vida de Hemingway en La Habana, publicó hace poco otra versión diferente y tal vez mejor fundada de aquella muerte tan controvertida. Se la contó el antiguo mayordomo de la casa, y de acuerdo con ella, el pozo del muerto no suministraba agua para beber, sino para nadar en la piscina. Y a ésta, según contó el mayordomo, le echaban con frecuencia pastillas desinfectantes, aunque tal vez no tantas para desinfectarla de un muerto entero. En todo caso, la última versión desmiente la más antigua, que era también la más literaria, y según la cual los esposos Hemingway habían tomado el agua del ahogado durante tres días. Dicen que el escritor había dicho: «La única diferencia que notamos era que el agua se había vuelto más dulce».

Ésta es una de las tantas y tantas historias fascinantes —escritas o habladas— que se le quedan a uno para siempre, más en el corazón

Publicado originalmente el 12 de mayo de 1981.

que en la memoria, y de las cuales está llena la vida de todo el mundo. Tal vez sean las ánimas en pena de la literatura. Algunas son perlas legítimas de poesía que uno ha conocido al vuelo sin registrar muy bien quién era el autor, porque nos parecía inolvidable, o que habíamos oído contar sin preguntarnos a quién, y al cabo de cierto tiempo ya no sabíamos a ciencia cierta si eran historias que soñamos. De todas ellas, sin duda la más bella, y la más conocida, es la del ratoncito recién nacido que se encontró con un murciélago al salir por primera vez de su cueva, y regresó asombrado, gritando: «Madre, he visto un ángel». Otra, también de la vida real, pero que supera por muchos cuerpos a la ficción, es la del radioaficionado de Managua que, en el amanecer del 22 de diciembre de 1972, trató de comunicarse con cualquier parte del mundo para informar que un terremoto había borrado a la ciudad del mapa de la Tierra. Al cabo de una hora de explotar un cuadrante en el que sólo se escuchaban los silbidos siderales, un compañero más realista que él le convenció de desistir. «Es inútil», le dijo, «esto sucedió en todo el mundo.» Otra historia, tan verídica como las anteriores, la padeció la orquesta sinfónica de París, que hace unos diez años estuvo a punto de liquidarse por un inconveniente que no se le ocurrió a Franz Kafka: el edificio que se le había asignado para ensayar sólo tenía un ascensor hidráulico para cuatro personas, de modo que los ochenta músicos empezaban a subir a las ocho de la mañana, y cuatro horas después, cuando todos habían acabado de subir, tenían que bajar de nuevo para almorzar.

Entre los cuentos escritos que lo deslumbran a uno desde la primera lectura, y que uno vuelve a leer cada vez que puede, el primero para mi gusto es *La pata de mono*, de W. W. Jacobs. Sólo recuerdo dos cuentos que me parecen perfectos: ése, y *El caso del doctor Valdemar*, de Edgar Allan Poe. Sin embargo, mientras de este último escritor se puede identificar hasta la calidad de sus ropas privadas, del primero es muy poco lo que se sabe. No conozco muchos eruditos que puedan decir lo que significan sus iniciales repetidas sin consultarlo una vez más en la enciclopedia, como yo lo acabo de hacer: William Wymark. Había nacido en Londres, donde murió en 1943, a la modesta edad de ochenta años, y sus obras completas en dieciocho volúmenes –aunque la enciclopedia no lo diga– ocupan 64 centímetros de una biblioteca. Pero su gloria se sustenta completa en una obra maestra de cinco páginas.

Por último, me gustaría recordar –y sé que algún lector caritativo me lo va a decir en los próximos días– quiénes son los autores de dos cuentos que alborotaron a fondo la fiebre literaria de mi juventud. El primero es el drama del desencantado que se arrojó a la calle desde un décimo piso, y a medida que caía iba viendo a través de las ventanas la intimidad de sus vecinos, las pequeñas tragedias domésticas, los amores furtivos, los breves instantes de felicidad, cuyas noticias no habían llegado nunca hasta la escalera común, de modo que en el instante de reventarse contra el pavimento de la calle había cambiado por completo su concepción del mundo, y había llegado a la conclusión de que aquella vida que abandonaba para siempre por la puerta falsa valía la pena de ser vivida. El otro cuento es el de dos exploradores que lograron refugiarse en una cabaña abandonada, después de haber vivido tres angustiosos días extraviados en la nieve. Al cabo de otros tres días, uno de ellos murió. El sobreviviente excavó una fosa en la nieve, a unos cien metros de la cabaña, y sepultó el cadáver. Al día siguiente, sin embargo, al despertar de su primer sueño apacible, lo encontró otra vez dentro de la casa, muerto y petrificado por el hielo, pero sentado como un visitante formal frente a su cama. Lo sepultó de nuevo, tal vez en una tumba más distante, pero al despertar al día siguiente volvió a encontrarlo sentado frente a su cama. Entonces perdió la razón. Por el diario que había llevado hasta entonces se pudo conocer la verdad de su historia. Entre las muchas explicaciones que trataron de darse al enigma, una parecía ser la más verosímil: el sobreviviente se había sentido tan afectado por su soledad que él mismo desenterraba dormido el cadáver que enterraba despierto.

La historia que más me ha impresionado en mi vida, la más brutal y al mismo tiempo la más humana, se la contaron a Ricardo Muñoz Suay en 1947, cuando estaba preso en la cárcel de Ocaña, provincia de Toledo, España. Es la historia real de un prisionero republicano que fue fusilado en los primeros días de la guerra civil en la prisión de Ávila. El pelotón de fusilamiento lo sacó de su celda en un amanecer glacial, y todos tuvieron que atravesar a pie un campo nevado para llegar al sitio de la ejecución. Los guardias civiles estaban bien protegidos del frío con capas, guantes y tricornios, pero aun así tiritaban a través del yermo helado. El pobre prisionero, que sólo llevaba una chaqueta de lana deshilachada, no hacía más que frotarse el cuerpo casi petrificado, mientras se lamentaba en voz alta del frío mortal. A un cierto momento, el comandante del pelotón, exasperado con los lamentos, le gritó:

—Coño, acaba ya de hacerte el mártir con el cabrón frío. Piensa en nosotros, que tenemos que regresar.

LA CONDUERMA DE LAS PALABRAS

Mi amigo Argos ha observado que en *Crónica de una muerte anuncia-da* hay tres expresiones que no son de comprensión inmediata en Colombia. La observación es digna de un interés muy especial, no sólo por venir de quien viene, sino porque hay indicios muy serios de que la novela transcurre en este país. Uno de ellos es la nacionalidad del autor. Otro, más significativo aún, es que cerca del pueblo sin nombre donde sucede el drama hay una ciudad de Colombia muy conocida en el mundo entero —Cartagena de Indias—, que fue fundada 374 años antes de que Madrid se convirtiera en la capital de España, y un poco más lejos hay otra ciudad también colombiana —Riohacha— que fue fundada 64 años antes de que el navegante in-glés Henry Hudson explorara el lugar donde había de fundarse la ciudad de Nueva York. De modo que era razonable esperar que todas las expresiones del lenguaje de la novela fueran también co-lombianas.

Sin embargo, Argos sabe tan bien como todo buen escritor que la guerra cotidiana con las palabras no respeta fronteras. Un pobre hombre solitario sentado seis horas diarias frente a una máquina de escribir con el compromiso de contar una historia que sea a la vez convincente y bella agarra sus palabras de donde puede. La guerra es más desigual aún si el idioma en que se escribe es el castellano, cuyas palabras cambian de sentido cada cien leguas, y tienen que pa-sar cien años en el purgatorio del uso común antes de que la Real Academia les dé permiso para ser enterradas en el mausoleo de su diccionario.

Las tres expresiones observadas son *conduerma, cruda* —entendida como el malestar que se padece al día siguiente de la noche ante-rior— y *hacerse bolas*. Las dos últimas, en efecto, son originarias de México. La primera, según el diccionario de americanismos de Al-fredo Neves, y también según el Vox y el de la Real Academia, es

Publicado originalmente el 19 de mayo de 1981.

un venezolanismo. Las tres son de uso corriente en sus patrias originales.

Sin embargo, yo no aprendí la palabra *conduerma* en ningún diccionario foráneo con pretensiones transnacionales, sino en la casa de mis abuelos, a los cinco años de edad, y con un significado mucho más intenso. Cuando me empeñaba en conseguir algo con una cantaleta invencible de días y días enteros −como lo sigo haciendo de viejo−, mi abuela terminaba por reventar: «Carajo, esta criatura es una conduerma». Así que más que modorra o sueño pesado −que tienen algo de metafórico− la conduerma de mi infancia era un tormento continuado e ineludible, como la amenaza de la muerte, que es el sentido que tiene en mi novela. Con todo, tuve buen cuidado de no decirla yo como narrador, sino que la puse en boca de un personaje, y todo el mundo sabe que los protagonistas de las novelas son los dueños de sus palabras.

Es probable, por supuesto, que aquella conduerma errante viniera de Venezuela. De niño aprendí otras muchas palabras que más tarde volví a encontrar en aquel país, pues pasaban de contrabando de un lado al otro, como las sedas de China y los perfumes de Francia, por una frontera que por aquellos tiempos era de dominio público. Lo que debemos preguntarnos es si al cabo de cincuenta años −y quién sabe cuántos más anteriores− las palabras emigrantes no pueden cambiar de nacionalidad con tanto derecho como cambian de sentido.

La palabra *cruda*, por supuesto, la conocí en México. En Colombia se dice *guayabo*, pero yo preferí la mexicana, porque la nuestra tiene además una connotación de añoranza que me estorbaba en el texto. Con ese sentido escribí hace ya muchos años, en otra novela, que un personaje se sentía atormentado por «el fragante y agusanado *guayabal* de amor que iba arrastrando hacia la muerte». En la *Crónica de una muerte anunciada*, la palabra *guayabo* también aparece en otra parte con el sentido de cruda, pero no está dicha por el narrador, sino por un protagonista, al cual le preguntan por qué está tan pálido, y él contesta: «Imagínese, con este *guayabo*». Por cierto que revisando la versión inglesa encontré que cruda había sido traducida en forma correcta −*hangover*−, que es como si uno siguiese todavía colgado de la noche anterior. En cambio, *guayabo* había sido traducido por error como *hullabaloo*, que no tiene nada que ver con nada, tal vez porque el traductor pisó sin darse cuenta una de las trampas frecuentes y peligrosas del sentido común. En todo caso, si escogí cruda fue por puras razones de gusto personal, pues ningún otro estado del ánimo tiene tantos nombres para escoger en castellano: resaca en España (como en Brasil), ratón en Venezuela, perseguidora en Cuba, chuchaque en Ecuador. Es un verdadero dolor de cabeza, no tanto para los sobrevivientes de la pachanga, sino también para los sabios lingüistas de agua mineral.

El traductor al inglés no entendió tampoco la expresión «hacerse bolas», y se lo preguntó en una carta al escritor Pedro Gómez Valderrama, quien le resolvió de un modo certero no sólo ése, sino otros varios enigmas de la misma novela. El término, en efecto, lo aprendí en México, y no me costó trabajo entenderlo, porque es casi igual a otro colombiano que quiere decir lo mismo y que no yace todavía en ningún diccionario oficial: *embolatarse*. En la novela preferí el mexicano, porque me pareció más expresivo, y también más fácil de descifrar por sentido común.

Pensando en todo esto, caí en la cuenta de que en la misma novela hay otros mexicanismos, además de los que señala Argos. Se dice: «habladas de borrachos», «mulatas destrampadas», «un poco al desgarriate». No sé de dónde venga *habladas*, con el sentido de bravuconadas, pero lo aprendí en México, y no encontré otra palabra más feliz en Colombia. *Destrampadas* viene de *destrampe*, que es la pachanga de delirio en la que todo está permitido. Hacer las cosas *al desgarriate*, es hacerlas de la peor manera posible, y me cuesta trabajo imaginarme una palabra que se parezca tanto a lo que quiere decir.

Los colombianos, que en los últimos tiempos hemos ganado tan mala fama en el mundo por tantas razones distintas, tenemos desde hace años la de hablar el castellano más puro. Dormimos en falsos laureles, pues en realidad hablamos por la calle una lengua muy bella, rica y útil, pero la que nos ha dado la fama no es ésa, sino la que recitan como loros nuestros académicos polvorientos y nuestros presidentes embalsamados.

Para mí, el mejor idioma no es el más puro, sino el más vivo. Es decir: el más impuro. El de México me parece el más imaginativo, el más expresivo, el más flexible. Tal vez porque es la lengua de emergencia de una nación que olvidó los idiomas nacionales antiguos, y al mismo tiempo aprendió mal el que trajo Hernán Cortés. La síntesis logra a veces dimensiones mágicas. Sólo un botón de muestra: en México existe, con su significado completo, la palabra *mendigo*. Pero hay otra, que es la misma, pero pronunciada como esdrújula: *méndigo*. Suele usarse más como adjetivo, y significa, más o menos, miserable. Los mexicanos tienen para las dos una explicación deslumbrante: «Mendigo es el que pide limosna, y méndigo es el que no la da».

MITTERRAND, EL OTRO: EL PRESIDENTE

No fue nada fácil reunir en París a los invitados latinoamericanos del presidente François Mitterrand. En parte porque nadie sabe muy bien en qué lugar del mundo se encuentran en cada momento nuestros escritores y artistas, y en parte porque la decisión de invitarlos se tomó 72 horas antes de la posesión del nuevo presidente. Aún la víspera no estaba todavía muy claro quiénes alcanzarían a llegar y quiénes se estaban excusando con telegramas que serían recibidos cuando ya hubieran sido arriadas las banderas de júbilo y las calles de París estuvieran barridas de la parranda multitudinaria más alegre y ruidosa de que se tuviera memoria desde otro mayo histórico: el de 1968.

El escritor Carlos Fuentes, de quien nadie daba noticias, estaba dictando una conferencia en una remota universidad de Estados Unidos, y tuvo el valor civil de tomar en Washington el Concorde que venía de México para llegar a tiempo a la fiesta. Matilde Neruda, la esposa del inmortal poeta chileno, se preparaba para volar a Buenos Aires cuando le avisaron que el nuevo presidente de Francia, que había sido amigo personal y lector perpetuo de Pablo Neruda, quería tenerla a su lado el día de la posesión. Miguel Otero Silva, en Caracas, tuvo que vencer su dudoso miedo al avión por tercera vez en lo que va de año, y llegó justo en el instante en que empezaban a tocar *La Marsellesa* bajo el Arco de Triunfo.

Otros dos invitados notables, ambos brasileños, no lograron vencer el miedo al avión: el escritor Jorge Amado y el arquitecto Oscar Niemeyer. El cardenal Pablo Evaristo Arns, que era el tercer brasileño invitado, no llegó por motivos distintos. El profesor Juan Bosch, antiguo presidente de Santo Domingo, quien sabía que el nuevo presidente de Francia no había incluido a ningún político en su lista de invitados personales, se sorprendió al recibir el telegrama. Sólo entonces se enteró de que no había sido invitado como político,

Publicado originalmente el 26 de mayo de 1981.

sino como escritor. Doña Hortensia Allende, la viuda del presidente asesinado en Chile, estaba en París una semana antes de la segunda vuelta electoral, y no se quedó desde entonces porque no estaba muy convencida del triunfo de Mitterrand. Ocho días después tuvo que tomar de nuevo el avión de regreso para atravesar el Atlántico por sexta vez en lo que va de año. Julio Cortázar fue el que llegó más fácil: tomó el metro en la esquina de su casa y salió en la estación de la Concorde, a veinte pasos del palacio del Elíseo.

Yo estaba en México, soñando que iba en un tren cargado de guacamayas, cuando sonó el teléfono infame de la mesa de noche. Era Monique Lang, la esposa del nuevo ministro de la Cultura, que había calculado mal la diferencia de horas, y me transmitió la muy amable invitación a las cuatro de la madrugada. Menos mal, porque a las seis debía viajar a las selvas de Chiapas en busca de un lugar con guacamayas silvestres para filmar una película, y no habría estado al alcance de nadie durante una semana.

De modo que el jueves 21 de mayo, a la una de la tarde, la mayoría de los invitados latinoamericanos estábamos en el comedor del palacio del Elíseo, respirando el aire enrarecido de los gobelinos grandilocuentes, pero con un menú inspirado en la inventiva sobria y original de la nueva cocina francesa, como si fuera una señal de un estilo distinto de gobierno. Había unos doscientos comensales, pero los treinta invitados extranjeros del presidente de la República ocupábamos las dos mesas centrales. Una presidida por el propio presidente, y la otra presidida por su esposa, Danielle. A la derecha del presidente, no por disposición del protocolo, sino por voluntad del nuevo dueño de la casa, se sentó doña Hortensia Allende. No se necesitaba demasiada perspicacia para darse cuenta de que aquella deferencia tan especial tenía una significación política muy importante para los invitados latinoamericanos. Poco después, cuando tomábamos el café en los jardines nublados, el presidente se acercó a los distintos grupos para despedirse. Le dije: «Los latinoamericanos tenemos por primera vez la impresión de tener en Francia un presidente nuestro». Mitterrand sonrió. «Sí», dijo, «pero ¿cuáles latinoamericanos?»

Esa mañana, a las 9.30 horas, había tomado posesión de la presidencia en un acto sin invitados. Luego recorrió los Campos Elíseos, de pie en un automóvil descubierto bajo el eterno cielo encapotado de París y aclamado por una muchedumbre interminable, y depositó una ofrenda de rosas vivas en la tumba del soldado desconocido. La última vez que lo había visto fue el 18 de enero anterior, cuando era candidato reciente por tercera vez, y muy pocos creíamos en su victoria. Le había hecho saber que me iba a Colombia, y él me hizo la distinción de citarme una vez más a su despacho de la calle Solferino para despedirnos. No fue por premonición, ni por ilusión, sino por una realidad demasiado evidente, que en esa ocasión me pareció que actuaba como si ya fuera presidente de Francia.

La impresión fue distinta el jueves pasado, cuando ya lo era en realidad, y estaba escuchando *La Marsellesa* bajo el Arco de Triunfo, frente a la llama eterna del soldado muerto. Estaba más pálido que de costumbre, con los ojos fijos en el horizonte de su destino y tratando de reprimir, para que nadie se los notara, los latidos del corazón. Había consagrado toda su vida a merecer aquel instante, había fracasado en dos tentativas anteriores sin dejarse vencer por el óxido de la derrota, y era, por fin, el presidente de su patria desde hacía tres horas, pero estaba tan bien instalado dentro de su piel que daba la impresión de haberlo sido durante toda la vida. A las seis de la tarde, bajo una llovizna tierna, atravesó solo y a paso lento la plaza del Panteón con dos rosas rojas en la mano. Los coros de la Orquesta de París, dirigida por Daniel Barenboim y con altavoces desmesurados en los extremos de la plaza, cantaban el «Himno de la alegría», de la *Novena Sinfonía* de Beethoven. El presidente entró solo en el ámbito helado del Panteón, caminando erguido y sin prisa por entre las losas funerarias de los muertos más ilustres de Francia, y depositó una rosa en cada una de las tumbas de dos mártires grandes: Jean Jaurès, un dirigente socialista asesinado a cuchillo en 1914 por su decidida oposición a la guerra, y Jean Moulin, dirigente de la resistencia durante la segunda guerra, mutilado y muerto por sus torturadores alemanes. La muchedumbre guardaba un silencio inmenso que sólo podía entenderse como el pasmo inexorable ante el misterio sin fondo de la poesía. Luego estalló en un cataclismo de júbilo que se inició en el Barrio Latino y terminó por contagiar a la ciudad entera. Por primera vez desde el mayo de gloria de 1968, el torrente incontenible de la juventud estaba en la calle, pero esta vez no se había desbordado para repudiar el poder, sino embriagado por el delirio de que una época feliz había comenzado. En medio de las músicas confundidas, de los bailes frenéticos, de los teatros de esquina, de los amores públicos de aquella noche enloquecida en que todo París era una sola rumba, yo pensaba que semejante paroxismo de la esperanza era tan emocionante como peligroso. No: yo hubiera querido estar entonces en cualquier parte menos durmiendo dentro del pellejo de François Mitterrand.

VIDES DE PERROS

Subíamos en silencio por la vieja escalera mecánica, erguidos y en orden, como siempre he pensado que se debe subir al cielo, cuando se oyó un chillido espantoso, una explosión como la de una piñata cuando se revienta en una fiesta infantil, y todos corrimos sin saber qué pasaba, pero con el instinto certero de que pasaba algo grave. En la ráfaga de pánico alcancé a ver una señora con un pobre abrigo de primavera salpicado de sangre todavía caliente, y otra que trataba de limpiar las piernas de su hijo embadurnadas de una materia espesa. Sólo entonces nos dimos cuenta de lo que ocurría: la escalera mecánica había oprimido entre dos peldaños un perrito pequinés, lo había reventado, y sus vísceras dispersas habían salpicado a los que estaban más cerca. En la escalera vacía sólo quedó el dueño del perrito, paralizado de espanto, mirando con la boca abierta la traílla rota que le quedó colgando en la mano. Esto sucedió el jueves de la semana pasada en un almacén de París, y es uno de los episodios más raros y estremecedores que he visto en mi vida. Lo más raro, sin embargo, fue la reacción del público. Tan pronto como pasó el pánico, todos soltaron una carcajada un poco histérica, y se pusieron a hablar de perros con una torcedura de la conciencia que no pude entender. El dueño del perrito muerto, por su parte, tuvo que ser atendido de urgencia por el servicio médico del almacén. Más tarde supimos que había hecho desmontar un sector de la escalera mecánica para rescatar hasta el último fragmento del perrito despedazado, y se los había llevado dentro de una caja de zapatos. Tal vez esa misma tarde fue al cementerio para perros, bajo la lluvia desolada de aquel mal jueves de primavera, con su gabardina escuálida y sus zapatos de perdulario, y enterró la cajita con lo poco que quedó de su perro entre las tumbas opulentas de los perros más amados y mejor tenidos de París. Estoy seguro, sin embargo, de que ese pobre hombre no volverá a ser nunca más el mismo de antes.

Publicado originalmente el 2 de junio de 1981.

París es una ciudad de perros privilegiados. En las calles, inclusive en los Campos Elíseos, que tienen la reputación de ser la avenida más bella del mundo, hay que caminar a saltos para no pisar la inconcebible cantidad de caca de perro que se encuentra por todas partes. También en Nueva York es familiar la imagen de los vecinos que sacan a sus perros al atardecer para que hagan sus necesidades en la calle, pero llevan un bastón especial, con una mano mecánica, como la que usaron los astronautas para recoger piedras en el suelo de la Luna. Con esa mano de ficción científica recogen lo que el perro deja, lo echan en una bolsita de plástico, que las tiendas especializadas venden para eso, y lo depositan en el tanque de la basura de la próxima esquina.

En París, donde el arte de amar a los perros no ha alcanzado semejante refinamiento, los animales dejan sus residuos en cualquier parte y de cualquier modo. Se calcula que en toda la ciudad, incluidos los suburbios, se recoge todos los días casi una tonelada de caca de perros, cuyo aprovechamiento industrial no está todavía resuelto. Las autoridades del municipio tienen años de estar buscando una solución desesperada, pero ninguna ha resultado eficaz. En las aceras han pintado la silueta de un perro, y una flecha que indica dónde deben cumplir con su deber los perros de la realidad. La señal está en un sitio por donde pasa al atardecer un arroyo artificial inventado por los ingenieros municipales con el propósito único de arrastrar hasta las alcantarillas la caca de los perros. Pero son muy pocos los que obedecen las señales, y se orinan siempre, como se dice, fuera del tiesto. De modo que no hay remedio: en París siempre hay alguien en una visita con un pegote de perro en la suela de un zapato. En mi tierra dicen que eso trae buena suerte. Si esto es así, en ninguna parte del mundo hay gente tan afortunada como en Francia.

Al parecer, los franceses quieren tener sus perros domésticos por encima de todos los problemas que puedan causar. Sólo en París hay más de un millón. Es decir: un perro por cada diez habitantes. Pero las estadísticas reales no son tan cuadradas, porque muchos franceses tienen ya dos perros en su casa y están tratando de tener tres, mientras el gobierno adelanta una campaña inútil para que tengan un tercer hijo. El control voluntario de la natalidad es tan severo —al contrario de lo que ocurre en el Tercer Mundo— que las autoridades empiezan a preocuparse en serio. «Entre nosotros, un nacimiento es ahora todo un acontecimiento», ha dicho un especialista. «Las familias con tres hijos se consideran como numerosas: las que tienen más de cinco se hacen acreedoras a medallas especiales, y las que tienen siete o más se vuelven célebres en la prensa regional.» El promedio nacional son dos hijos en cada matrimonio. Las madres que se encuentran en este punto, y que trabajan para contribuir al presupuesto familiar, tienen la oferta oficial de ganar el mismo sueldo en la casa si tienen un tercer hijo. No son muchas las que se dejan tentar. En cambio, es cada vez mayor la tendencia a tener un tercer perro.

Todo esto me parece asombroso y digno de ser escrito, porque pertenezco al abundante y respetable grupo de mortales que hubiéramos querido tener doce hijos, como los tuvo mi madre, y en cambio no nos entendemos con los perros. Más aún: les tengo terror. Al parecer, los perros lo saben, porque cuando llego a una casa donde los hay, éstos desprecian de un modo olímpico a quienes los aman y quieren acariciarlos, y en cambio tratan de subirse encima de mí y agobiarme con besos de seducción. Siempre he pensado que los dueños de perros no son conscientes de cuánto sufrimos con los perros quienes no los queremos. Cuando está a punto de sucumbir al horror, los dueños se toman la molestia de decir: «*Dandy*, quédate quieto». *Dandy*, por supuesto, no hace el menor caso, y cuando ya parece haber desaparecido, uno siente en la mesa un rumor, después un calor sobrenatural que se desliza por debajo del mantel y, enseguida, un hocico que surge por entre nuestras piernas. Grito: «¡Aquí está otra vez!». Es siempre un grito exagerado, con la esperanza de que los comensales se rían, de que los dueños se avergüencen, de que haya alguien que sienta un poco de lástima por quien tuvo la mala suerte de no querer a los perros, pero nunca obtengo los resultados previstos. Los dueños de casa apenas si interrumpen la conversación para decir, sin la menor pretensión de autoridad: «Sal de ahí, *Nerón*». Pero *Nerón* sigue ahí, gozando con el espectáculo de Roma a merced de las llamas, y allí sigue hasta el final de la cena. Entonces es para mí la hora providencial en que vuelvo a mi casa llena de las tortugas de la buena suerte, del loro que canta las arias entrañables de Puccini y de las rosas del alma que perfuman la casa sin ladrar, que no muerden, que no se le trepan encima a nadie, a las cuales no hay que sacarlas a pasear todas las tardes para que ensucien la ciudad con sus gracias fragantes.

Esto no quiere decir, por supuesto, que esté contra los perros. Estoy contra muchos dueños de perros que se derriten de ternura con ellos, y en cambio son capaces de cualquier crueldad con los seres humanos. O de los que son víctimas de una confusión de sentimientos cuyos estragos son imprevisibles. Recuerdo, hace muchos años, una señora cubierta casi por completo de zorros azules, a quien las autoridades sanitarias de Nueva York no le dejaban desembarcar su perro. Lo llevaba en una canasta de mimbre, con un abrigo tejido a mano y un lazo de organza en la cabeza, y no podía soportar la idea de que lo sometieran a cuarentena. Al final, ante la intransigencia de las autoridades, escogió la que le parecía la fórmula más humanitaria: echó el perrito al agua, y lo vio ahogarse ante el asombro de todos, con los ojos llenos de lágrimas, pero feliz de que el animalito de su corazón estuviera para siempre a salvo del mal de rabia de los perros humanos.

FANTASÍA Y CREACIÓN ARTÍSTICA

Según el diccionario de la Real Academia de la Lengua, la *fantasía* es «una facultad que tiene el ánimo de reproducir por medio de imágenes». Es difícil concebir una definición más pobre y confusa que esa primera acepción. En su segunda acepción dice que es «una ficción, cuento o novela, o pensamiento elevado o ingenioso», lo cual no hace sino infundir mayor desconcierto en el ya creado por la definición inicial.

De la palabra *imaginación*, el mismo diccionario dice que es «aprensión falsa de una cosa que no hay en la realidad o no tiene fundamento». Por su parte, don Joan Corominas, ese gran detective de las palabras castellanas —cuya lengua materna no era, por cierto, el castellano, sino el catalán—, estableció que fantasía e imaginación tienen el mismo origen y que en última instancia puede decirse sin mucho esfuerzo que son la misma cosa.

Uno de mis mayores defectos intelectuales es que nunca he logrado entender lo que quieren decir los diccionarios, y menos que cualquier otro el terrible esperpento represivo de la Academia de la Lengua. Por una vez que he tenido la curiosidad de volver a él para establecer las diferencias entre fantasía e imaginación, me encuentro con la desgracia de que sus definiciones no sólo son muy poco comprensibles, sino que además están al revés. Quiero decir que, según yo lo entiendo, la fantasía es la que no tiene nada que ver con la realidad del mundo en que vivimos: es una pura invención fantástica, un infundio, y por cierto de un gusto poco recomendable en las bellas artes, como muy bien lo entendió el que le puso el nombre al chaleco de fantasía. Por muy fantástica que sea la concepción de que un hombre amanezca convertido en un gigantesco insecto, a nadie se le ocurriría decir que la fantasía sea la virtud creativa de Franz Kafka, y en cambio no cabe duda de que fue el recurso primordial de Walt Disney. Por el contrario, y al revés de lo que dice

Publicado originalmente el 9 de junio de 1981.

el diccionario, pienso que la imaginación es una facultad especial que tienen los artistas para crear una realidad nueva a partir de la realidad en que viven. Que, por lo demás, es la única creación artística que me parece válida. Hablemos, pues, de «la imaginación en la creación artística en América Latina» y dejemos la fantasía para uso exclusivo de los malos gobiernos.

En América Latina y el Caribe, los artistas han tenido que inventar muy poco, y tal vez su problema ha sido el contrario: hacer creíble su realidad. Siempre fue así desde nuestros orígenes históricos, hasta el punto de que no hay en nuestra literatura escritores menos creíbles y al mismo tiempo más apegados a la realidad que nuestros cronistas de Indias. También ellos –para decirlo con un lugar común irreemplazable– se encontraron con que la realidad iba más lejos que la imaginación.

El diario de Cristóbal Colón es la pieza más antigua de esa literatura. Empezando porque no se sabe a ciencia cierta si el texto existió en la realidad, puesto que la versión que conocemos fue transcrita por el padre Las Casas de unos originales que dijo haber conocido. En todo caso, esa versión es apenas un reflejo infiel de los asombrosos recursos de imaginación a que tuvo que apelar Cristóbal Colón para que los Reyes Católicos le creyeran la grandeza de sus descubrimientos.

Colón dice que las gentes que salieron a recibirlo el 12 de octubre de 1492 «estaban como sus madres los parieron». Otros cronistas coinciden con él en que los caribes, como era natural en un trópico todavía a salvo de la moral cristiana, andaban desnudos. Sin embargo, los ejemplares escogidos que llevó Colón al palacio real de Barcelona estaban ataviados con hojas de palmeras pintadas y plumas y collares de dientes y garras de animales raros. La explicación parece simple: el primer viaje de Colón, al revés de sus sueños, fue un desastre económico. Apenas si encontró el oro prometido, perdió la mayor de sus naves y no pudo llevar de regreso ninguna prueba tangible del valor enorme de sus descubrimientos, ni nada que justificara los gastos de su aventura y la conveniencia de continuarla. Vestir a sus cautivos como lo hizo fue un truco convincente de publicidad. El simple testimonio oral no hubiera bastado, un siglo después de que Marco Polo había regresado de China con realidades tan novedosas e inequívocas como los espaguetis y los gusanos de seda, y como lo habían sido la pólvora y la brújula.

Toda nuestra historia, desde el descubrimiento, se ha distinguido por la dificultad de hacerla creer. Uno de mis libros favoritos de siempre ha sido el *Primer viaje en torno del globo*, del italiano Antonio

Pigafetta, que acompañó a Magallanes en su expedición alrededor del mundo. Pigafetta dice que vio en Brasil unos pájaros que no tenían colas, otros que no hacían nidos porque no tenían patas, pero cuyas hembras ponían y empollaban sus huevos en la espalda del macho y en medio del mar y otros que sólo se alimentaban de los excrementos de sus semejantes. Dice que vio cerdos con el ombligo en la espalda y unos pájaros grandes cuyos picos parecían una cuchara, pero carecían de lengua. También habló de un animal que tenía cabeza y orejas de mula, cuerpo de camello, patas de ciervo y cola y relincho de caballo. Fue Pigafetta quien contó la historia de cómo encontraron al primer gigante de la Patagonia, y de cómo éste se desmayó cuando vio su propia cara reflejada en un espejo que le pusieron enfrente.

La leyenda del Dorado es, sin duda, la más bella, la más extraña y decisiva de nuestra historia. Buscando ese territorio fantástico, Gonzalo Jiménez de Quesada conquistó casi la mitad del territorio de lo que hoy es Colombia, y Francisco de Orellana descubrió el río Amazonas. Pero lo más fantástico es que lo descubrió al derecho —es decir, navegando de las cabeceras hasta la desembocadura—, que es el sentido contrario en que se descubren los ríos. El Dorado, como el tesoro de Cuauhtémoc, siguió siendo un enigma para siempre. Como lo siguieron siendo las 11.000 llamas, cargadas cada una con cien libras de oro, que fueron despachadas desde el Cuzco para pagar el rescate de Atahualpa, y que nunca llegaron a su destino. La realidad fue otra vez más lejos hace menos de un siglo, cuando una misión alemana encargada de elaborar el proyecto de construcción de un ferrocarril transoceánico en el istmo de Panamá concluyó que el proyecto era viable, pero con una condición: que los rieles no se hicieran de hierro, que era un metal muy difícil de conseguir en la región, sino que se hicieran de oro.

Tanta credulidad de los conquistadores sólo era comprensible después de la fiebre metafísica de la Edad Media y del delirio literario de las novelas de caballería. Sólo así se explica la desmesurada aventura de Alvar Núñez Cabeza de Vaca, que necesitó ocho años para llegar desde España a México a través de todo lo que hoy es el sur de Estados Unidos, en una expedición cuyos miembros se comieron unos a otros, hasta que sólo quedaron cinco de los seiscientos originales. El incentivo de Cabeza de Vaca, al parecer, no era la búsqueda del Dorado, sino algo más noble y poético: la fuente de la eterna juventud.

Acostumbrado a unas novelas donde había ungüentos para pegarles las cabezas cortadas a los caballeros, Gonzalo Pizarro no podía dudar cuando le contaron en Quito, en el siglo XVI, que muy cerca de allí había un reino con 3.000 artesanos dedicados a fabricar muebles de oro, y en cuyo palacio real había escalera de oro macizo y estaba custodiado por leones con cadenas de oro. ¡Leones en los Andes! A Balboa le contaron un cuento semejante en Santa María del Darien y descubrió el océano Pacífico. Gonzalo Pizarro no descubrió nada especial, pero el tamaño de su credulidad puede medirse por la expedición que armó para buscar el reino inverosímil: 800 españoles, 4.000 indios, 150 caballos y más de 1.000 perros amaestrados en la caza de seres humanos.

EN CHILE COMO EN CHICAGO

Calama es un pueblo remoto de la provincia de Arica, en medio del desierto desolado de Acatama y a unos 300 kilómetros de Santiago de Chile. El único acontecimiento que había trastornado la rutina rural en los años recientes fue el asalto armado a la sucursal del Banco del Estado en el último diciembre. Por eso el gerente de este establecimiento, Guillermo Martínez, y su cajero, Sergio Yáñez, entendieron como algo muy natural que dos funcionarios de la seguridad local quisieran tomar medidas especiales para evitar futuros asaltos. Eso fue, en efecto, lo que les dijeron los dos hombres que entraron en sus oficinas el pasado 9 de marzo, a las 8.30 de la mañana, y les pidieron en consecuencia que colaboraran con ellos para hacer un asalto fingido. Los dos hombres eran el jefe local de la Central Nacional de Información (CNI), Gabriel Hernández Anderson, y su segundo, Eduardo Villanueva Márquez. Los funcionarios del banco los conocían bien desde hacía tiempo, porque todo el mundo se conocía en el pueblo, pero los conocían mejor desde el diciembre anterior, porque fueron ellos los encargados de investigar el asalto, sin ningún resultado.

Lo primero que hicieron fue sacar todo el dinero del cofre de seguridad para ponerlo en varias cajas de manzanas que llevaron en su camioneta sin insignias: 45 millones de pesos chilenos, equivalentes a un millón de dólares. Luego cargaron las cajas en la camioneta y los cuatro hombres salieron por una puerta posterior del banco, a la hora en que entraban los primeros clientes por la puerta principal. Sólo uno dijo haberlos visto, y aseguró que los empleados bancarios iban con las manos encadenadas. Pero fue una declaración ilusoria, pues la verdad era que ambos salieron sin resistencia, convencidos de que estaban jugando al cine en la vida real.

Los cuatro hombres llegaron a un lugar desierto en las afueras de Calama, y los funcionarios de seguridad les pidieron a los emplea-

Publicado originalmente el 23 de junio de 1981.

dos bancarios que se colocaran de espaldas a un muro de arena, para simular el episodio en que serían muertos por los asaltantes. Los bancarios obedecieron, y en efecto fueron fusilados, pero no con balas de fogueo, como les hicieron creer, sino con varias ráfagas de metralleta. Sus cuerpos fueron sepultados en la arena. Esa misma noche, cuando sus familiares notaron la ausencia, los autores del crimen dejaron escapar la hipótesis de que los dos funcionarios se habían alzado con los fondos del banco y habían huido del país.

La Central Nacional de Información es un cuerpo secreto de represión política. Hay otra policía civil, que es un cuerpo abierto, al cual le correspondió investigar el caso. Entre los dos cuerpos existe una rivalidad a muerte. A los pocos días de cometido el crimen, los autores comprendieron que la policía civil andaba sobre pistas que la llevarían sin remedio al esclarecimiento de la verdad. De modo que volvieron al lugar del fusilamiento, desenterraron los cadáveres en descomposición y los despedazaron con varias cargas de dinamita. La explosión fue tan poderosa que removió diecisiete toneladas de tierra, bajo las cuales quedaron sepultados los miembros dispersos.

Fue inútil, pues la policía civil esclareció el crimen pocos días más tarde. Los cadáveres fueron rescatados a pedazos y recuperado el dinero, que había sido escondido en diferentes lugares del pueblo. Una parte apareció en una quebrada, donde la había escondido un hermano del autor principal del crimen. Otra parte apareció en casa de su suegro, que era nada menos que el flamante alcalde municipal. Más de diez personas fueron arrestadas en conexión con el asalto y el asesinato y, de un modo u otro, todas tenían algún vínculo con el gobierno.

La policía creía haber terminado su labor de limpieza la semana pasada cuando ocurrió un episodio imprevisto. Un mayor del ejército chileno y jefe de la Central Nacional de Información en la ciudad de Arica, Juan Dalmas, apareció muerto en su coche en medio del desierto, con un tiro en la cabeza. A su lado estaba el revólver calibre 38 del cual había salido la bala. El mayor Juan Dalmas, que ejercía su cargo con el nombre falso de Carlos Vargas, fue señalado por la prensa como el autor intelectual del asalto y el crimen de Calama. La policía, también por supuesto, informó que se había suicidado.

La Central Nacional de Información es en esencia la misma DINA, el cuerpo de represión política que se encargó del exterminio de la oposición después del asalto al poder por el general Augusto Pinochet. Le cambiaron el nombre y se le hicieron algunos cambios de forma en 1979, cuando la policía de Estados Unidos estableció que fue ese organismo de terror el que organizó desde Chile el asesinato de Orlando Letelier, antiguo ministro de Defensa de Salvador Allende. Una bomba mortal había sido colocada debajo del asiento de su automóvil y detonada a control remoto cuando circulaba por el centro de Washington. La secretaria de Letelier, Romy Mofit, también murió en el atentado. Un norteamericano, Michel Townley, y cinco cubanos fueron identificados por la policía norteamericana como los autores materiales del crimen. Townley y tres de los cubanos fueron condenados a penas menores, que fueron reducidas hace pocas semanas. Los otros dos no fueron encontrados nunca dentro de Estados Unidos. El director de la DINA, general Manuel Contreras, fue reclamado por la justicia norteamericana para que respondiera por la concepción del atentado. El gobierno chileno, por supuesto, negó la extradición, pero le cambió el nombre a la DINA para que pareciera que había dejado de existir. El general Contreras se fue para su casa, en uso de buen retiro.

Fue, por supuesto, un retiro aparente. El general Contreras es hoy uno de los hombres más poderosos de Chile y un rival implacable del general Humberto Gordon Rubio, director general del CNI, cuyos subalternos cometieron el crimen de Calama. Es, además, el director de una agencia privada de protección bancaria que no depende de ningún organismo oficial y a cuyo cargo está la custodia forzosa o voluntaria de todos los bancos de Chile. El general Contreras tiene también una gran influencia en el cuerpo de policía civil, que esclareció el crimen de Calama, y es, por último, uno de los propietarios de *La Tercera*, un periódico amarillo que destapó y divulgó a grandes voces la culpabilidad de la CNI. Todo perfecto, como en el Chicago de Al Capone.

Este sórdido acertijo parece tener una pieza suelta. Hace algún tiempo fueron encontrados por casualidad numerosos restos humanos en una mina de Lonquen, a pocos kilómetros de Santiago, y en una región solitaria muy cerca de Concepción, que es la segunda ciudad de Chile. A pesar de las versiones torcidas de las autoridades, nadie puso en duda que eran cuerpos de prisioneros políticos ejecutados por las fuerzas de seguridad. Ambos hallazgos tenían una cosa en común: los cadáveres parecían destrozados por una explosión.

ALGO MÁS SOBRE LITERATURA Y REALIDAD

Un problema muy serio que nuestra realidad desmesurada plantea a la literatura es el de la insuficiencia de las palabras. Cuando nosotros hablamos de un río, lo más lejos que puede llegar un lector europeo es a imaginarse algo tan grande como el Danubio, que tiene 2.790 kilómetros. Es difícil que se imagine, si no se le describe, la realidad del Amazonas, que tiene 5.500 kilómetros de longitud. Frente a Belén del Pará no se alcanza a ver la otra orilla, y es más ancho que el mar Báltico. Cuando nosotros escribimos la palabra «tempestad», los europeos piensan en relámpagos y truenos, pero no es fácil que estén concibiendo el mismo fenómeno que nosotros queremos representar. Lo mismo ocurre, por ejemplo, con la palabra «lluvia». En la cordillera de los Andes, según la descripción que hizo para los franceses otro francés llamado Javier Marimier, hay tempestades que pueden durar hasta cinco meses. «Quienes no hayan visto esas tormentas», dice, «no podrán formarse una idea de la violencia con que se desarrollan. Durante horas enteras los relámpagos se suceden rápidamente a manera de cascadas de sangre y la atmósfera tiembla bajo la sacudida continua de los truenos, cuyos estampidos repercuten en la inmensidad de la montaña.» La descripción está muy lejos de ser una obra maestra, pero bastaría para estremecer de horror al europeo menos crédulo.

De modo que sería necesario crear todo un sistema de palabras nuevas para el tamaño de nuestra realidad. Los ejemplos de esa necesidad son interminables. F. W. Up de Graff, un explorador holandés que recorrió el alto Amazonas a principios de siglo, dice que encontró un arroyo de agua hirviendo donde se hacían huevos duros en cinco minutos, y que había pasado por una región donde no se podía hablar en voz alta porque se desataban aguaceros torrenciales. En algún lugar de la costa caribe de Colombia, yo vi a un hombre rezar una oración secreta frente a una vaca que tenía gusanos en la

Publicado originalmente el 1 de julio de 1981.

oreja, y vi caer los gusanos muertos mientras transcurría la oración. Aquel hombre aseguraba que podía hacer la misma cura a distancia, siempre que le hicieran la descripción del animal y le indicaran el lugar en que se encontraba. El 8 de mayo de 1902, el volcán Mont Pele, en la isla Martinica, destruyó en pocos minutos el puerto de Saint-Pierre y mató y sepultó en lava a la totalidad de sus 30.000 habitantes. Salvo uno: Ludger Sylvaris, el único preso de la población, que fue protegido por la estructura invulnerable de la celda individual que le habían construido para que no pudiera escapar.

Sólo en México habría que escribir muchos volúmenes para expresar su realidad increíble. Después de casi veinte años de estar aquí, ya podría pasar todavía horas enteras, como lo he hecho tantas veces, contemplando una vasija de fríjoles saltarines. Nacionalistas benévolos me han explicado que su movilidad se debe a una lana viva que tienen dentro, pero la explicación parece pobre: lo maravilloso no es que los fríjoles se muevan porque tengan una lana dentro, sino que tengan una lana dentro para que puedan moverse. Otra de las extrañas experiencias de mi vida fue mi primer encuentro con el ajolote (*axolotl*). Julio Cortázar cuenta en uno de sus relatos que conoció el ajolote en el Jardin des Plantes de París, un día en que quiso ver los leones. Al pasar frente a los acuarios, cuenta Cortázar, «soslayé los peces vulgares hasta dar pronto con el *axolotl*». Y concluye: «Me quedé mirándolo por una hora, y salí, incapaz de otra cosa». A mí me sucedió lo mismo, en Pátzcuaro, sólo que no lo contemplé por una hora, sino por una tarde entera, y volví varias veces. Pero había allí algo que me impresionó más que el animal mismo, y era el letrero clavado en la puerta de la casa: «Se vende jarabe de ajolote».

Esa realidad increíble alcanza su densidad máxima en el Caribe, que, en rigor, se extiende (por el norte) hasta el sur de Estados Unidos, y por el sur, hasta Brasil. No se piense que es un delirio expansionista. No: es que el Caribe no es sólo un área geográfica, como por supuesto lo creen los geógrafos, sino un área cultural muy homogénea.

En el Caribe, a los elementos originales de las creencias primarias y concepciones mágicas anteriores al descubrimiento, se sumó la profusa variedad de culturas que confluyeron en los años siguientes en un sincretismo mágico cuyo interés artístico y cuya propia fecundidad artística son inagotables. La contribución africana fue forzosa e indignante, pero afortunada. En esa encrucijada del mundo, se forjó un sentido de libertad sin término, una realidad sin Dios ni ley, donde cada quien sintió que le era posible hacer lo que quería sin límites de ninguna clase: y los bandoleros amanecían convertidos en reyes, los prófugos en almirantes, las prostitutas en gobernadoras. Y también lo contrario.

Yo nací y crecí en el Caribe. Lo conozco país por país, isla por isla, y tal vez de allí provenga mi frustración de que nunca se me ha ocurrido nada ni he podido hacer nada que sea más asombroso que la realidad. Lo más lejos que he podido llegar es a trasponerla con recursos poéticos, pero no hay una sola línea en ninguno de mis libros que no tenga su origen en un hecho real. Una de esas trasposiciones es el estigma de la cola de cerdo que tanto inquietaba a la estirpe de los Buendía en *Cien años de soledad*. Yo hubiera podido recurrir a otra imagen cualquiera, pero pensé que el temor al nacimiento de un hijo con cola de cerdo era la que menos probabilidades tenía de coincidir con la realidad. Sin embargo, tan pronto como la novela empezó a ser conocida, surgieron en distintos lugares de las Américas las confesiones de hombres y mujeres que tenían algo semejante a una cola de cerdo. En Barranquilla, un joven se mostró en los periódicos: había nacido y crecido con aquella cola, pero nunca lo había revelado, hasta que leyó *Cien años de soledad*. Su explicación era más asombrosa que su cola. «Nunca quise decir que la tenía porque me daba vergüenza», dijo, «pero ahora, leyendo la novela y oyendo a la gente que la ha leído, me he dado cuenta de que es una cosa natural.» Poco después, un lector me mandó el recorte de la foto de una niña de Seúl, capital de Corea del Sur, que nació con una cola de cerdo. Al contrario de lo que yo pensaba cuando escribí la novela, a la niña de Seúl le cortaron la cola y sobrevivió.

Sin embargo, mi experiencia de escritor más difícil fue la preparación de *El otoño del patriarca*. Durante casi diez años leí todo lo que me fue posible sobre los dictadores de América Latina, y en especial del Caribe, con el propósito de que el libro que pensaba escribir se pareciera lo menos posible a la realidad. Cada paso era una desilusión. La intuición de Juan Vicente Gómez era mucho más penetrante que una verdadera facultad adivinatoria. El doctor Duvalier, en Haití, había hecho exterminar los perros negros en el país, porque uno de sus enemigos, tratando de escapar de la persecución del tirano, se había escabullido de su condición humana y se había convertido en perro negro. El doctor Francia, cuyo prestigio de filósofo era tan extenso que mereció un estudio de Carlyle, cerró a la República del Paraguay como si fuera una casa, y sólo dejó abierta una ventana para que entrara el correo. Antonio López de Santa Anna enterró su propia pierna en funerales espléndidos. La mano cortada de Lope de Aguirre navegó río abajo durante varios días, y quienes la veían pasar se estremecían de horror, pensando que aun en aquel estado aquella mano asesina podía blandir un puñal. Anastasio Somoza García, en Nicaragua, tenía en el patio de su casa un jardín zoológico con jaulas de dos compartimientos: en uno, estaban las fieras, y en el otro, separado apenas por una reja de hierro, estaban encerrados sus enemigos políticos.

Martínez, el dictador teósofo de El Salvador, hizo forrar con papel rojo todo el alumbrado público del país, para combatir una epidemia de sarampión, y había inventado un péndulo que ponía sobre los alimentos antes de comer, para averiguar si no estaban envenenados. La estatua de Morazán que aún existe en Tegucigalpa es en realidad del mariscal Ney: la comisión oficial que viajó a Londres a buscarla, resolvió que era más barato comprar esa estatua olvidada en un depósito, que mandar hacer una auténtica de Morazán.

En síntesis, los escritores de América Latina y el Caribe tenemos que reconocer, con la mano en el corazón, que la realidad es mejor escritor que nosotros. Nuestro destino, y tal vez nuestra gloria, es tratar de imitarla con humildad, y lo mejor que nos sea posible.

MR. ENDERS ATRAVIESA EL ESPEJO

El gobierno colombiano se encuentra emparedado entre dos dolores de cabeza. De un lado, el temor de un conflicto internacional por sus múltiples desacuerdos de fronteras, y del otro lado, la realidad de una insurrección interna que no es para enfrentar con pistolas de agua. En efecto, Colombia tiene conflictos de límites con Venezuela, por un pleito que empezó hace muchos años y que se encuentra hoy más empantanado que al principio. Tiene conflictos con Nicaragua, que está reclamando en mal momento, y con muy poca lucidez política, unos islotes desérticos del archipiélago de San Andrés y Providencia que son colombianos desde siempre. Tendrá en cualquier momento conflictos con Brasil, cuya penetración por el sur, silenciosa, pero implacable, inquieta a los militares colombianos mucho más que las guerrillas domésticas. Un oficial de alto rango me hacía notar hace pocos años que en los grandes ríos meridionales de Colombia ya no se hace el comercio en castellano, sino en portugués.

Las guerrillas, por otra parte, no están tan derrotadas como lo quisieran los mismos militares, que en los últimos treinta años han proclamado once veces la muerte en combate del mismo comandante guerrillero. Todo lo contrario: este año —según se desprende de las propias informaciones militares— las guerrillas están más fuertes, más extendidas y mejor armadas que el año anterior, y han conseguido una imagen popular que no tuvieron en otro tiempo. De modo que las fuerzas armadas de Colombia necesitan más armas para conjurar ambas amenazas. Con una condición difícil, sin embargo: que sean armas suficientes para ganar la guerra interior, pero no tantas ni tan destructoras como para alarmar a los vecinos, que tienen más dinero con que comprarlas en mayor cantidad.

Esta necesidad ineludible de caminar por la cuerda floja se hizo evidente el 10 de junio pasado, después de que los miembros del estado mayor de las fuerzas armadas de Colombia visitaron a su pre-

Publicado originalmente el 8 de julio de 1981.

sidente, el doctor Turbay Ayala, un *géminis* perfecto por su sensualidad y su desidia, que aquel día cumplió sus primeros 65 años de felicidad. Al salir de la fiesta, el ministro de la Defensa, general Camacho Leyva, hizo una declaración pública que era al mismo tiempo un prodigio de acrobacia militar. «Cada país», dijo el ministro, «debe tener el armamento que requiere para combatir la subversión interna y la que se trata de traer de otra nación, pero nada más.» Luego agregó: «Se debe tener lo normal». Pero lo agregó sin explicar, y tal vez sin explicarse a sí mismo, qué se entiende por cantidades normales cuando se habla de hierros para matar. Para cualquier buen entendedor de los misterios colombianos, por supuesto, el ministro de la Defensa estaba pensando que el enemigo inmediato son las guerrillas, pero que el enemigo principal es Venezuela.

Estas filtraciones de la mala vecindad entre Colombia y su vecino rico se habían notado demasiado durante la visita que hizo a Bogotá, también en junio, el ministro de Relaciones Exteriores de Guyana, señor Rashleig E. Jackson. No creo que haya muchos colombianos —además del canciller— que sepan a ciencia cierta dónde queda Guyana, ni que nadie en Colombia tenga algo que ver con ella. Creo, además, que la indiferencia es recíproca. En el comunicado conjunto que firmaron los dos cancilleres, en efecto, se nota también demasiado que aquélla no fue una reunión entre los personeros de dos países amigos, sino de dos gobiernos que se sienten enemigos del mismo enemigo. Sólo así se entiende que hayan insistido tanto en los principios fundamentales de la integridad territorial y la solución pacífica de las disputas internacionales, y que se hayan comprometido «a continuar los esfuerzos para mantener la soberanía y la exclusiva jurisdicción de cada país en las áreas marinas y submarinas adyacentes a sus costas, de acuerdo con las normas establecidas en el Derecho internacional». Todo esto carecería de sentido entre dos países que no tienen fronteras comunes, ni idiomas ni religiones comunes, ni una historia común, pero en cambio, ese asunto de *blanco es gallina* lo pone entre dos países que mantienen desde hace tiempo sus respectivas disputas territoriales con Venezuela.

Ambos, por cierto, están bien correspondidos. Hace algunas semanas, a raíz de una carta que le escribió el secretario de Estado de Estados Unidos, general Alexander Haig, el presidente de Venezuela reunió al estado mayor de sus fuerzas armadas para explicarles la necesidad de que Estados Unidos le vendiera más armas a Colombia para defenderse de una supuesta agresión de Cuba. Sin embargo, los servicios de inteligencia militar de Venezuela no saben menos que los de Colombia. En el mes de marzo, cuando el ejército colombiano capturó a un grupo de guerrilleros del M-19 infiltrados por el sur del país, los militares venezolanos les pidieron compartir sus informaciones para ver si Venezuela estaba incluida en los planes pretendidos de Cuba. Ahora saben que no lo estaban, entre otras cosas, porque esos planes no existían sino en la imaginación fantástica de los sorprendidos y asustados militares colombianos.

Los venezolanos saben además —como lo saben ahora los propios militares de Colombia, aunque no lo digan por conveniencia política— que si Cuba tuvo algo que ver con el entrenamiento de aquellos guerrilleros no tuvo, en cambio, nada que ver con el barco en que llegaron ni con las armas con que desembarcaron. Más aún: saben muy bien quién los embarcó y quién los armó. De modo que la socorrida amenaza de Cuba no les pareció más que un pretexto de Colombia para armarse contra Venezuela, y así se lo dijeron a su presidente, de la manera más delicada de que fueron capaces. Para ellos, el enemigo principal no es Cuba, sino Colombia.

Dentro de este espejo de suspicacias no hay que ser clarividentes para suponer qué vino a buscar en Colombia y Venezuela el secretario de Estado adjunto de Estados Unidos, Thomas Enders, hace dos semanas. Lo único que se proponía era conciliar los intereses de ambos países, para que cada uno de ellos permitiera que se armara el otro, y de ese modo equilibrar contra Cuba el potencial militar en el Caribe. En Colombia, Enders trató de convencer al presidente Turbay Ayala —en vísperas de su cumpleaños feliz— de que el enemigo principal no es Venezuela como piensan sus militares, sino la pérfida Cuba. Tal vez el presidente no lo creyó, porque hace ya mucho tiempo que no cree en nada, pero debió hacer cara de que lo creía, como hace siempre, para no tener que pensarlo. Estados Unidos necesita que se calme la disputa entre Colombia y Venezuela, pero le conviene que se agudice la que mantiene Venezuela con Guyana —a la cual consideran como un aliado potencial de Cuba— y que se agudice también la que mantienen Colombia y Nicaragua, que Estados Unidos considera como una simple proyección de Cuba en tierra firme. Por esto, Enders le dijo al presidente Turbay Ayala que su país veía con inquietud el comunicado conjunto de Guyana y Colombia, que afecta las relaciones de ésta con Venezuela. Y, en cambio, le aseguró que el Departamento de Estado tratará de que el Congreso de Estados Unidos ratifique el acuerdo que asegure a Colombia la soberanía sobre los cayos del archipiélago de San Andrés, porque ese acuerdo ha de deteriorar todavía más las relaciones de Colombia con Nicaragua. Al presidente de Venezuela, por otra parte, le dijo lo mismo que al de Colombia, pero al revés. En realidad, al pasar de un país a otro, Enders debió sentir lo mismo que sintió la pequeña Alicia al atravesar el espejo.

Lo más triste de este ir y venir del señor Enders, y de todo el proyecto del presidente Reagan en América Central y el Caribe, es que parece elaborado con materiales de archivo. Hace veinte años, los Enders del presidente Eisenhower, y después los del presidente Kennedy, estaban haciendo lo mismo que hacen los Enders de hoy a través de los espejos ilusorios del presidente Reagan. No es que el tiempo no pase para ellos –como ocurre en los retratos–, sino que pasa al revés. Como en el interior de los espejos, por supuesto.

¿UNA ENTREVISTA? NO, GRACIAS

En el curso de una entrevista, un reportero me hizo la pregunta eterna: «¿Cuál es su método de trabajo?». Permanecí pensativo, buscando una respuesta nueva, hasta que el periodista me dijo que si la pregunta me parecía demasiado difícil podía cambiarla por otra. «Al contrario», le dije, «es una pregunta tan fácil y tantas veces contestada por mí que estoy buscando una respuesta distinta.» El periodista se disgustó, pues no podía entender que yo explicara mi método de trabajo de un modo diferente para cada ocasión. Sin embargo, así era. Cuando se tiene que conceder un promedio de una entrevista mensual durante doce años, uno termina por desarrollar otra clase de imaginación especial para que todas no sean la misma entrevista repetida.

En realidad, el género de la entrevista abandonó hace mucho tiempo los predios rigurosos del periodismo para internarse con patente de corso en los manglares de la ficción. Lo malo es que la mayoría de los entrevistadores lo ignoran, y muchos entrevistados cándidos todavía no lo saben. Unos y otros, por otra parte, no han aprendido aún que las entrevistas son como el amor: se necesitan por lo menos dos personas para hacerlas, y sólo salen bien si esas dos personas se quieren. De lo contrario, el resultado será un sartal de preguntas y respuestas de las cuales puede salir un hijo en el peor de los casos, pero jamás saldrá un buen recuerdo.

La introducción es siempre la misma, y casi siempre por teléfono. «He leído todas las entrevistas que le han hecho a usted, y todas son iguales», dice una voz amable y muy segura de sí misma. «Lo que yo quiero hacerle es algo distinto.» Es inútil replicar que todos dicen lo mismo. Además, no lo hago de ningún modo, porque siempre me he considerado un periodista, por encima de todo, y cuando otro periodista me solicita una entrevista me siento en un callejón sin salida: a la vez víctima y cómplice. De modo que termino siempre

Publicado originalmente el 15 de julio de 1981.

por aceptar, con ese hilo de suicida irremediable que todos llevamos dentro.

En dos de cada tres casos, el resultado es el mismo: no resulta una entrevista distinta, porque las preguntas son las de siempre. Incluso la última: «¿Quisiera decirme una pregunta que nunca le hayan hecho y quisiera contestar?». La respuesta es siempre la más desoladora: «Ninguna». Tal vez los entrevistadores no se den cuenta de hasta qué punto nos duele su fracaso a los entrevistados, pues en la realidad no es un fracaso de ellos solos, sino, sobre todo, un fracaso nuestro. Siempre me quedo con la impresión sobrecogedora de que el domingo próximo, cuando los lectores abran el periódico, se dirán con un gran desencanto, y quizá con una rabia justa, que allí está otra vez la misma entrevista de siempre, del escritor de siempre, que ya se encuentran hasta en la sopa, y pasarán con toda razón y todo derecho a la página providencial de las historietas cómicas. Tengo la esperanza de que en un día no muy lejano nadie volverá a comprar los periódicos donde se publiquen entrevistas conmigo.

Hay entrevistadores de diversas clases, pero todos tienen dos cosas en común: piensan que aquélla será la entrevista de su vida, y están asustados. Lo que no saben —y es muy útil que lo sepan— es que todos los entrevistados con sentido de la responsabilidad están más asustados que ellos. Como en el amor, por supuesto. Los que creen que el susto sólo lo tienen ellos, incurren en uno de los dos extremos: o se vuelven demasiado complacientes, o se vuelven demasiado agresivos. Los primeros no harán nunca nada que en realidad valga la pena. Los segundos no consiguen nada más que irritar al entrevistado. «Eso es bueno», me dijo un excelente entrevistador de radio. «Si uno logra irritar al entrevistado, éste terminará por gritar la verdad de pura rabia.» Otros emplean el método de los malos maestros de escuela, tratando de que el entrevistado caiga en contradicciones, tratando de que diga lo que no quiere decir, y tratando, en el peor de los casos, de que diga lo que no piensa. He tenido que enfrentarme algunas veces a esta clase de entrevistadores, y los resultados han sido siempre los más deplorables. Debo reconocer, sin embargo, que, en otro género de entrevistas, el método puede conducir a una explosión deslumbrante. Éste fue el caso, hace algunos años, en una conferencia de prensa sobre temas económicos que concedió el presidente de Francia Valéry Giscard d'Estaing. Fue un espectáculo radiante, en el cual los periodistas disparaban con cargas de profundidad, y el entrevistado respondía con una precisión, una inteligencia y un conocimiento asombrosos. De pronto, una periodista preguntó con el mayor respeto: «¿Sabe usted, señor presidente, cuánto cuesta un billete de metro?». El señor presidente, por supuesto, no lo sabía.

En esta clase de entrevistas, que tal vez debían llamarse entrevistas de guerra, el nombre culminante es el de mi admirada Oriana Fallaci. Otros periodistas que creen conocerla –pero que sin duda no la quieren– tienen reservas en relación con su método. Dicen que en efecto no altera ni una sola palabra de lo que dijo el entrevistado frente al micrófono, pero en cambio acomoda a su antojo el orden en que fue dicho, y, sobre todo, cambia y retoca sus propias preguntas como mejor le conviene No me consta, y es muy probable que quienes lo dicen no lo sepan tampoco de primera mano. A fin de cuentas, no creo que ese método sea menos sospechoso que el empleado en la actualidad por las revistas norteamericanas *Time* y *Newsweek*, que graban una conversación de varias horas y luego no utilizan sino el material de una página, sin preguntarse si las omisiones no alteran de algún modo el sentido del texto original. En todo caso, el resultado del método de Oriana Fallaci es casi siempre revelador y fascinante, y muy pocas personalidades de este mundo han resistido a la vanidad de concederle una entrevista. A ella, por su parte, sólo se le ha ablandado el corazón frente a monseñor Helder Cámara. El propio Henry Kissinger admitió en sus memorias que la entrevista de Oriana Fallaci fue la más catastrófica que le habían hecho jamás. Es fácil comprender, porque en ninguna otra había quedado tan descubierto por dentro y por fuera, y de cuerpo entero. Como sólo puede lograrse, desde luego, con los recursos mágicos de la ficción.

Un buen entrevistador, a mi modo de ver, debe ser capaz de sostener con su entrevistado una conversación fluida, y de reproducir luego la esencia de ella a partir de unas notas muy breves. El resultado no será literal, por supuesto, pero creo que será más fiel, y sobre todo más humano, como lo fue durante tantos años de buen periodismo antes de ese invento luciferino que lleva el nombre abominable de magnetófono. Ahora, en cambio, uno tiene la impresión de que el entrevistador no está oyendo lo que se dice, ni le importa, porque cree que el magnetófono lo oye todo. Y se equivoca: no oye los latidos del corazón, que es lo que más vale en una entrevista. No se crea, sin embargo, que estas desdichas me alegran. Al contrario: al cabo de tantos años de frustraciones, uno sigue esperando en el fondo de su alma que llegue por fin el entrevistador de su vida. Siempre como en el amor.

A propósito

Después de terminar la nota anterior me encontré con una entrevista a Mario Vargas Llosa publicada por la revista *Cromos*, de Bogotá, con el siguiente título: «Gabo publica las sobras de *Cien años de soledad*». La frase, entre comillas, quiere decir, además, que es una cita literal. Sin embargo, lo que Vargas Llosa dice en su respuesta es lo siguiente: «A mí me impresiona todavía un libro como *Cien años de soledad*, que es una suma literaria y vital. García Márquez no ha repetido semejante hazaña porque no es fácil repetirla. Todo lo que ha escrito después es una reminiscencia, son las sobras de ese inmenso mundo que él ideó. Pero creo que es injusto criticárselo. Es injusto decir que la *Crónica* no está bien porque no es como *Cien años de soledad*. Es imposible escribir un libro como ése todos los días». En realidad –ante una pregunta provocadora del entrevistador–, Vargas Llosa le dio una buena lección de cómo se debe entender la literatura. El titulador, por su parte, ha dado también una buena lección de cómo se puede hacer el mal periodismo. A propósito de esto, creo que alguna vez tendremos que hablar sobre otro de los aspectos más sucios de una entrevista: la manipulación.

EDÉN PASTORA

Tengo una foto de Edén Pastora con uniforme y botas de campaña, dormido cuan largo es sobre un mesón de cuartel. «Es una foto histórica», me dijo un compañero suyo, «porque Edén Pastora no duerme nunca.» Edén Pastora, por su parte, me dijo: «Cómo será la confianza que te tengo, que me atreví a quedarme dormido delante de ti». Ambos comentarios definen muy bien la personalidad de este comandante sandinista de 45 años, duro y receloso, que ha resuelto renunciar a las vanidades del poder terrenal para irse a tirar tiros en otras tierras, como algunos reyes medievales lo abandonaban todo y se iban para Jerusalén a rescatar el santo sepulcro.

La foto fue tomada al amanecer del 25 de agosto de 1978, en una guarnición de Panamá, pocas horas después de que nos conocimos. Edén Pastora había asaltado el martes anterior el Palacio Nacional de Managua al frente de un comando de veinticinco sandinistas resueltos, y tomó en rehenes a la Cámara de Diputados en pleno. Al cabo de dos días de negociaciones intensas, el dictador Anastasio Somoza liberó sesenta prisioneros políticos, que volaron a Panamá junto con el comando. Edén Pastora tenía entonces cincuenta horas sin dormir, y los otros dos responsables de la acción habían dormido muy poco, pero todos aceptaron contarme su historia de inmediato para un reportaje mundial.

El comandante Hugo Torres, que era el segundo responsable de la acción y es en la actualidad secretario general del Consejo de Estado de Nicaragua, dobló la cabeza sobre el mesón, al cabo de seis horas, y se durmió a fondo. Un momento después, Edén Pastora sufrió una especie de colapso demoledor, como si le hubieran dado en la cabeza con un mazo de picapedrero, y se hundió en un sueño sobrenatural, boca arriba, entre los platos sin lavar y los restos de comida de soldados que quedaban sobre el mesón. La única que consiguió mantenerse en pie fue Dora María Téllez, la bella, que era la

Publicado originalmente el 22 de julio de 1981.

única mujer del comando y su responsable tercera, y que ocupa en la actualidad un cargo superior en las fuerzas armadas de Nicaragua con su grado de comandante conquistado en la guerra. Ella siguió contándome el final de la historia con un sentido sorprendente de la narración, humano y minucioso, hasta que el bravo sol de agosto se encendió en las ventanas. Entonces despertamos al fotógrafo, que dormía en una banca apartada, y fue él quien tomó la foto histórica del guerrero derrumbado.

La impresión que me causó Edén Pastora desde aquella noche en que le vi por primera vez es que era un sandinista distinto. Yo conocía desde antes a casi todos los dirigentes del movimiento. Había tenido algo que ver con las negociaciones largas y difíciles que culminaron con la unificación de los tres grupos en que estaban divididos, y de ahí surgió una relación que fue más allá de la política y la guerra, y terminó por ser una amistad buena que todavía se mantiene. Algunos de ellos aparecían en mi casa de México a la hora menos pensada, y se preparaban en la cocina unas comidas rápidas que «siempre parecían de campaña», y a veces pasábamos las noches en vela, hablando de todo, mientras esperábamos las noticias que no nos dejaban dormir. Allí les pusimos aquel nombre genérico, en clave, que muy pronto se hizo público: los muchachos. Algunos se iban con su mochila al hombro, se despedían con la mano hasta el mes entrante, cuando lograran la victoria, y pocos días después nos llegaba la noticia de su muerte. Era una realidad de la guerra a la cual no logramos acostumbrarnos nunca. En todo caso, lo que más me admiraba de ellos, y que sigo admirando, eran las dos virtudes mayores que todavía no han perdido en el poder: el realismo y la paciencia. Eran buenos guerreros, pero a casi todos se les notaba que su mundo no era el de la guerra, y que estaban en él por necesidad. En cambio, tenían una inteligencia política que fue casi tan útil para ganar la guerra como su sentido militar, y que les ha servido en el poder para resistir a toda clase de injurias y provocaciones. Edén Pastora, en cambio, me pareció desde el primer momento un guerrero puro.

Así lo describí: «Un hombre de 42 años, con veinte de militancia muy intensa, y con una decisión de mando que no logra disimular con su excelente buen humor». Cuando entró en el salón azul de la Cámara de Diputados soltó al aire una ráfaga de metralleta y gritó: «Todo el mundo a tierra». Un testigo presencial me contó después que más de sesenta diputados habían obedecido de inmediato, no tanto por el pavor del plomo como por la devastadora autoridad de su voz. Es de baja estatura, ancho y macizo, con unos ojos intensos y una barba tan dura y montaraz que uno tiene la impresión de verla crecer tan pronto como se acaba de afeitar. Desde la acción del Palacio Nacional hasta los meses inciertos en que fue responsable de la guerra en el frente sur, aquella cara de labrador arisco que heredó de sus abuelos sicilianos se volvió popular en el mundo entero, bajo su estrella solitaria de comandante. El nombre, que es auténtico aunque parezca mentira, y que es apenas uno más de los tantos nombres líricos de su familia, parecía completar su predestinación para la leyenda. Tiene ángel. Un ángel raro, terrestre, que es sin duda el ángel insaciable de la guerra.

Un hombre así no podía acostumbrarse al tiempo parsimonioso del poder. Hace un año le volví a ver en una recepción diplomática en Managua, y no cabía dentro del pellejo: quería irse de sí mismo. Pocas horas antes había desfilado ante la muchedumbre que presenció el primer aniversario de la victoria, al frente de las milicias populares que él había formado y entrenado, y parecía el comandante en jefe de las fuerzas armadas del universo. Era, pues, evidente que no soportaba la paz precaria de su país, amenazada por los cuatro costados, y con la mala conciencia de no ser ya un guerrillero de la justicia, sino un hombre de gobierno que debía tomar en cuenta razones de estado, contemplaciones diplomáticas, criterios de política fugaz. Edén Pastora carece de esa vocación, y ya sabemos que el carácter de los dirigentes es un elemento inevitable de los procesos históricos, de modo que no me cabe ninguna duda de que se va por lo que dice: porque no puede soportar, detrás de un escritorio, que otros estén sufriendo sin él.

Sin embargo, se va en un mal momento. Creo que en política no hay un error más grave que ser inoportuno, y la determinación de Edén Pastora lo es para el destino de su país. Es un plato demasiado suculento y gratuito para sus enemigos. No habrá en el mundo una fuerza de disuasión capaz de convencer a nadie de que no es ésta la primera grieta mortal en el interior del sandinismo en el poder. La analogía con el Che Guevara es inevitable, y es inevitable la repetición de las mismas tergiversaciones. En su número de esta semana, *Time* va más lejos: desliza la suposición perversa de que los 50.000 hombres que tiene Nicaragua sobre las armas —y que necesita para la defensa de sus fronteras— podrían ser muy útiles si uno de sus propios tácticos, ansioso de pelear, decide de pronto ponerse al frente de la contrarrevolución.

Si de algo han dado muestras los sandinistas en el poder, es de su inteligencia para sortear toda clase de provocaciones en la búsqueda de la felicidad para su pueblo, mediante una fórmula original fundada en sus realidades propias. Sin embargo, será también imposible impedir que sus enemigos interpreten la decisión de Edén Pastora como una forma sandinista de intervenir en los asuntos de sus vecinos y de exportar su revolución. Por esto se han hecho tantos esfuerzos para convencer al comandante díscolo de que vuelva a casa. Uno de los argumentos más convincentes, sin duda, es que si de veras quiere pelear por una causa justa no tiene por qué ir tan lejos. Este 19 de julio, al celebrar el segundo aniversario de la victoria, Nicaragua es ya un país bloqueado por Estados Unidos, casi tanto como lo es Cuba desde hace veinte años. Es decir: está siendo sometida a un castigo de país socialista, pero sin serlo. En la frontera de Honduras hay más de tres mil antiguos guardias somocistas dispuestos a invadir Nicaragua en cualquier momento, con los recursos de la potencia militar más agresiva del mundo. Cuando eso ocurra —y el día esté lejano— Edén Pastora va a saber cuánta falta le hace a su país su estrella solitaria.

MI HEMINGWAY PERSONAL

Lo reconocí de pronto, paseando con su esposa, Mary Welsh, por el bulevar de Saint Michel, en París, un día de la lluviosa primavera de 1957. Caminaba por la acera opuesta en dirección del jardín de Luxemburgo, y llevaba unos pantalones de vaquero muy usados, una camisa de cuadros escoceses y una gorra de pelotero. Lo único que no parecía suyo eran los lentes de armadura metálica, redondos y minúsculos, que le daban un aire de abuelo prematuro. Había cumplido 59 años, y era enorme y demasiado visible, pero no daba la impresión de fortaleza brutal que sin duda él hubiera deseado, porque tenía las caderas estrechas y las piernas un poco escuálidas sobre sus bastos. Parecía tan vivo entre los puestos de libros usados y el torrente juvenil de la Sorbona que era imposible imaginarse que le faltaban apenas cuatro años para morir.

Por una fracción de segundo —como me ha ocurrido siempre— me encontré dividido entre mis dos oficios rivales. No sabía si hacerle una entrevista de prensa o sólo atravesar la avenida para expresarle mi admiración sin reservas. Para ambos propósitos, sin embargo, había el mismo inconveniente grande: yo hablaba desde entonces el mismo inglés rudimentario que seguí hablando siempre, y no estaba muy seguro de su español de torero. De modo que no hice ninguna de las dos cosas que hubieran podido estropear aquel instante, sino que me puse las manos en bocina, como Tarzán en la selva, y grité de una acera a la otra: «Maeeeestro». Ernest Hemingway comprendió que no podía haber otro maestro entre la muchedumbre de estudiantes, y se volvió con la mano en alto, y me gritó en castellano con una voz un tanto pueril: «Adioooos, amigo». Fue la única vez que lo vi.

Yo era entonces un periodista de veintiocho años, con una novela publicada y un premio literario en Colombia, pero estaba varado y sin rumbo en París. Mis dos maestros mayores eran los dos nove-

Publicado originalmente el 29 de julio de 1981.

listas norteamericanos que parecían tener menos cosas en común.
Había leído todo lo que ellos habían publicado hasta entonces, pero
no como lecturas complementarias, sino todo lo contrario: como
dos formas distintas y casi excluyentes de concebir la literatura. Uno
de ellos era William Faulkner, a quien nunca vi con estos ojos y a
quien sólo puedo imaginarme como el granjero en mangas de camisa que se rascaba el brazo junto a dos perritos blancos, en el retrato célebre que le hizo Cartier Bresson. El otro era aquel hombre
efímero que acababa de decirme adiós desde la otra acera, y me había dejado la impresión de que algo había ocurrido en mi vida, y
que había ocurrido para siempre.

No sé quién dijo que los novelistas leemos las novelas de los
otros sólo para averiguar cómo están escritas. Creo que es cierto.
No nos conformamos con los secretos expuestos en el frente de la
página, sino que la volteamos al revés, para descifrar las costuras. De
algún modo imposible de explicar desarmamos el libro en sus piezas
esenciales y lo volvemos a armar cuando ya conocemos los misterios
de su relojería personal. Esa tentativa es descorazonadora en los
libros de Faulkner, porque éste no parecía tener un sistema orgánico para escribir, sino que andaba a ciegas por su universo bíblico
como un tropel de cabras sueltas en una cristalería. Cuando se logra
desmontar una página suya, uno tiene la impresión de que le sobran
resortes y tornillos y que será imposible devolverla otra vez a su estado original. Hemingway, en cambio, con menos inspiración, con
menos pasión y menos locura, pero con un rigor lúcido, dejaba sus
tornillos a la vista por el lado de fuera, como en los vagones de ferrocarril. Tal vez por eso Faulkner es un escritor que tuvo mucho que
ver con mi alma, pero Hemingway es el que más ha tenido que ver
con mi oficio.

No sólo por sus libros, sino por su asombroso conocimiento del
aspecto artesanal de la ciencia de escribir. En la entrevista histórica
que le hizo el periodista George Plimpton para *Paris Review* enseñó
para siempre −contra el concepto romántico de la creación− que la
comodidad económica y la buena salud son convenientes para escribir, que una de las dificultades mayores es la de organizar bien las
palabras, que es bueno releer los propios libros cuando cuesta trabajo escribir para recordar que siempre fue difícil, que se puede escribir en cualquier parte siempre que no haya visitas ni teléfono, y que
no es cierto que el periodismo acabe con el escritor, como tanto se
ha dicho, sino todo lo contrario, a condición de que se abandone a
tiempo. «Una vez que escribir se ha convertido en el vicio principal
y el mayor placer −dijo−, sólo la muerte puede ponerle fin.» Con

todo, su lección fue el descubrimiento de que el trabajo de cada día sólo debe interrumpirse cuando ya se sabe cómo se va a empezar al día siguiente. No creo que se haya dado jamás un consejo más útil para escribir. Es, ni más ni menos, el remedio absoluto contra el fantasma más temido de los escritores: la agonía matinal frente a la página en blanco.

Toda la obra de Hemingway demuestra que su aliento era genial, pero de corta duración. Y es comprensible. Una tensión interna como la suya, sometida a un dominio técnico tan severo, es insostenible dentro del ámbito vasto y azaroso de una novela. Era una condición personal, y el error suyo fue haber intentado rebasar sus límites espléndidos. Es por eso que todo lo superfluo se nota más en él que en otros escritores. Sus novelas parecen cuentos desmedidos a los que les sobran demasiadas cosas. En cambio, lo mejor que tienen sus cuentos es la impresión que causan de que algo les quedó faltando, y es eso precisamente lo que les confiere su misterio y su belleza. Jorge Luis Borges, que es uno de los grandes escritores de nuestro tiempo, tiene los mismos límites, pero ha tenido la inteligencia de no rebasarlos.

Un solo disparo de Francis Macomber contra el león enseña tanto como una lección de cacería, pero también como un resumen de la ciencia de escribir. En algún cuento suyo escribió que un toro de lidia, después de pasar rozando el pecho del torero, se revolvió «como un gato volteando una esquina». Creo, con toda humildad, que esa observación es una de las tonterías geniales que sólo son posibles en los escritores más lúcidos. La obra de Hemingway está llena de esos hallazgos simples y deslumbrantes, que demuestran hasta qué punto se ciñó a su propia definición de que la escritura literaria –como el iceberg– sólo tiene validez si está sustentada debajo del agua por los siete octavos de su volumen.

Esa conciencia técnica será sin duda la causa de que Hemingway no pase a la gloria por ninguna de sus novelas, sino por sus cuentos más estrictos. Hablando de *Por quién doblan las campanas*, él mismo dijo que no tenía un plan preconcebido para componer el libro, sino que lo inventaba cada día a medida que lo iba escribiendo. No tenía que decirlo: se nota. En cambio, sus cuentos de inspiración instantánea son invulnerables. Como aquellos tres que escribió en la tarde de un 16 de mayo en una pensión de Madrid, cuando una nevada obligó a cancelar la corrida de toros de la feria de San Isidro. Esos cuentos –según él mismo le contó a George Plimpton– fueron *Los asesinos*, *Diez indios* y *Hoy es viernes*, y los tres son magistrales.

Dentro de esa línea, para mi gusto, el cuento donde mejor se condensan sus virtudes es uno de los más cortos: *Un gato bajo la lluvia*. Sin embargo, aunque parezca una burla de su destino, me parece que su obra más hermosa y humana es la menos lograda: *Al otro lado del río y entre los árboles*. Es, como él mismo reveló, algo que comenzó por ser un cuento y se extravió por los manglares de la novela. Es difícil entender tantas grietas estructurales y tantos errores de mecánica literaria en un técnico tan sabio, y unos diálogos tan artificiales y aun tan artificiosos en uno de los más brillantes orfebres de diálogos de la historia de las letras. Cuando el libro se publicó, en 1950, la crítica fue feroz. Porque no fue certera. Hemingway se sintió herido donde más le dolía, y se defendió desde La Habana con un telegrama pasional que no pareció digno de un autor de su tamaño. No sólo era su mejor novela, sino también la más suya, pues había sido escrita en los albores de un otoño incierto, con las nostalgias irreparables de los años vividos y la premonición nostálgica de los pocos años que le quedaban por vivir. En ninguno de sus libros dejó tanto de sí mismo ni consiguió plasmar con tanta belleza y tanta ternura el sentimiento esencial de su obra y de su vida: la inutilidad de la victoria. La muerte de su protagonista, de apariencia tan apacible y natural, era la prefiguración cifrada de su propio suicidio.

Cuando se convive por tanto tiempo con la obra de un escritor entrañable, uno termina sin remedio por revolver su ficción con su realidad. He pasado muchas horas de muchos días leyendo en aquel café de la Place de Saint Michel que él consideraba bueno para escribir, porque le parecía simpático, caliente, limpio y amable, y siempre he esperado encontrar otra vez a la muchacha que él vio entrar una tarde de vientos helados, que era muy bella y diáfana, con el pelo cortado en diagonal, como un ala de cuervo. «Eres mía y París es mío», escribió para ella, con ese inexorable poder de apropiación que tuvo su literatura. Todo lo que describió, todo instante que fue suyo, le sigue perteneciendo para siempre. No puedo pasar por el número 112 de la calle del Odeón, en París, sin verlo a él conversando con Sylvia Beach en una librería que ya no es la misma, ganando tiempo hasta que fueran las seis de la tarde por si acaso llegaba James Joyce. En las praderas de Kenya, con sólo mirarlas una vez, se hizo dueño de sus búfalos y sus leones, y de los secretos más intrincados del arte de cazar. Se hizo dueño de toreros y boxeadores, de artistas y pistoleros que sólo existieron por un instante, mientras fueron suyos. Italia, España, Cuba, medio mundo está lleno de los sitios de los cuales se apropió con sólo mencionarlos. En Cojímar, un pueblecito cerca de La Habana donde vivía el pescador solitario de *El viejo y el mar*, hay un templete conmemorativo de su hazaña con un busto de Hemingway pintado con barniz de oro. En Finca Vigía, su refugio cubano donde vivió hasta muy poco antes de morir, la casa está intacta entre los árboles sombríos, con sus libros disímiles, sus trofeos de caza, su atril de escribir, sus enormes zapatos de muerto, las incontables chucherías de la vida y del mundo entero que fueron suyas hasta su muerte, y que siguen viviendo sin él con el alma que les infundió por la sola magia de su dominio. Hace unos años entré en el automóvil de Fidel Castro −que es un empecinado lector de literatura− y vi en el asiento un pequeño libro empastado en cuero rojo. «Es el maestro Hemingway», me dijo. En realidad, Hemingway sigue estando donde uno menos se lo imagina −veinte años después de muerto−, tan persistente y a la vez tan efímero como aquella mañana, que quizá fue de mayo, en que me dijo adiós, amigo, desde la acera opuesta del bulevar de Saint Michel.

BREVE NOTA DE ADIÓS AL OLOR DE LA GUAYABA
DE FELIZA BURSZTYN

Si alguien le hubiera avisado a tiempo que iba a ser detenida, la escultora colombiana Feliza Bursztyn habría podido asilarse en una embajada antes de que la manosearan los militares. El gobierno habría dicho entonces que no había nada contra ella, y que sólo se asilaba para hacerle propaganda a sus juguetes de chatarra o para contribuir a la campaña de descrédito de Colombia en el exterior. Pero nadie le avisó, a pesar de las buenas relaciones de Feliza Bursztyn en el alto mundo político de su país, y antes de asilarse tuvo que padecer la humillación previa de un asalto a su casa, a las cinco de la madrugada, por dieciocho militares disfrazados de civil, y vivir todo un viernes de tinieblas con los ojos vendados y contestando preguntas imbéciles en una caballeriza militar. Por muy cultas que sean las fuerzas armadas de Colombia, no se les puede exigir a sus torturadores que sean especialistas en las bellas artes, pero deben saber al menos quiénes son sus torturados, para no perder el tiempo haciéndoles preguntas que se pueden aprender en la escuela primaria.

Esta vez, además de haber cometido un atropello, los militares colombianos han hecho algo peor: el ridículo. No se necesita ser adivino para saber lo que andaban buscando estos Sherlock Holmes de pacotilla, que han esculcado por toda la ciudad en busca de un mortero con el cual fueron disparadas tres granadas contra el palacio presidencial. Alguien debió ser tan bruto como para pensar que Feliza Bursztyn tiene un enorme taller de fundición en el cual han sido forjadas las armas más mortíferas de la escultura nacional, y que con los mismos recursos se hubiera podido fundir un mortero. Lo más ridículo de todo es que, mientras los servicios de seguridad buscaban el cañón secreto en el galpón de hierros viejos de Feliza Bursztyn, el hombre que concibió, y tal vez dirigió, el bombardeo al palacio presidencial estaba muerto de risa en un barrio cercano,

Publicado originalmente el 5 de agosto de 1981.

hablando para una entrevista forzosa con una de las periodistas más bonitas de Colombia, Alexandra Pineda, y con mi muy querido amigo Fernando González Pacheco, que es tal vez el animador más feo de la TV.

La única vez en que Feliza Bursztyn ha conspirado fue en 1958, y lo hizo junto con las damas más perfumadas de la oligarquía nacional, que se sentaron en medio de la calle para derribar la dictadura del general Gustavo Rojas Pinilla, instigadas por los dirigentes de los partidos tradicionales. Desde entonces, Feliza no ha hecho nada más subversivo que convertir en obras de arte los accidentes de tránsito, con una temeridad que le ha costado una limitación pulmonar muy seria por los vapores tóxicos de la fundición, una limitación, dicho sea de paso, que le ha causado trastornos respiratorios, pero que no le ha quitado alientos para disparar las palabras del más grueso calibre en las visitas de sociedad. No son ésos, por cierto, sus únicos quebrantos de salud. Se diría que tiene huesos de vidrio. Hace unos meses se fracturó la columna vertebral y tuvo que ponerse un chaleco ortopédico que parecía un cinturón de castidad fabricado por ella misma, y cuya llave se le perdía en cada pachanga. Salvo por su lengua sin control, no conozco a nadie más inofensivo que ella, ni a nadie más indefenso, ni a nadie a quien sus amigos lo quieran más. Tiene sus ideas políticas, desde luego, y además muy bien puestas en su lugar. Tenemos, además de muchas otras, una convicción común: repudiamos el terrorismo como instrumento de lucha política, así sean las bombas sin corazón o el atentado personal de los guerrilleros desesperados, o los allanamientos militares de madrugada y el cautiverio con los ojos vendados en nombre de cualquier ley. Sin embargo, lo más disparatado del atropello incalificable de que se ha hecho víctima a Feliza Bursztyn es que su exilio va a quitarle el voto más seguro y fervoroso que tenía el candidato conservador a la presidencia de la República, don Belisario Betancur.

No se trata, por supuesto, de una equivocación. La misma noche en que Feliza Bursztyn era detenida, volvieron a tumbarle la puerta al anciano poeta don Luis Vidales, y su casa fue sometida a una requisa tan encarnizada como infructuosa. La única diferencia entre esta vez y la anterior, fue que entonces se lo llevaron vendado a las caballerizas militares, y allí lo mantuvieron varios días, en el que ha de quedar para la historia como el episodio más sombrío no sólo de la presidencia del doctor Turbay Ayala, sino de su propio destino personal. El poeta Luis Vidales no recibió nunca una explicación satisfactoria del atropello. La repetición del allanamiento de su casa tiene, sin duda, una finalidad muy precisa: dejar establecido de una vez por todas que ninguno de los dos atropellos fueron por error −como se hizo creer después del primero−, sino que corresponden al pensamiento expresado por el señor ministro de la Defensa: aquí no hay poeta que valga. Ni escritores, ni músicos, ni damas escultoras, por muy inocentes que sean y muy enfermas que estén. Es una guerra abierta contra los intelectuales y los artistas que tengan la temeridad de pensar, y cuya solidaridad con las causas más justas no dejan dormir tranquilo a un presidente que alguna vez declaró haber leído una biblioteca completa de 5.000 volúmenes. Hace pocos meses, un muy querido amigo mío, que lo es también del presidente Turbay Ayala, y además su partidario incondicional, le dijo en mi presencia al general Omar Torrijos que a todos los gobernadores que rompen relaciones con la inteligencia se los lleva el carajo. No sé si sea ya demasiado tarde para que se lo diga también a Turbay Ayala.

Todo esto ocurre en un momento crítico de Colombia, en que la lucha armada contra el poder establecido está más fuerte y extendida que nunca. Ya no son las bandas dispersas con escopetas de fisto que se paseaban a todo lo largo y todo lo ancho de nuestra historia, sino un verdadero ejército marginal, con *bazookas* y morteros capaces de tronar frente al propio dormitorio del presidente de la República. Varias veces, en los últimos tiempos, las fuerzas armadas han proclamado la victoria final sobre la subversión, pero la realidad se ha encargado de demostrar al día siguiente que la guerra continúa cada vez más intensa y que amenaza con ser sangrienta y sin término.

Todos sabemos por qué se llegó a este punto. Hace menos de dos años, el movimiento M-19 expresó su propósito de deponer las armas para tomar parte en la contienda política, a cambio de una amnistía real y completa. El gobierno tuvo entonces la oportunidad de instaurar una paz civil que tal vez hubiera sido la única verdadera y estable en los últimos treinta años. Pero el presidente Turbay Ayala desoyó las voces de sus consejeros más altos, prestó oídos sordos al clamor nacional, rechazó inclusive un proyecto aceptable de sus propios parlamentarios y se embarcó en una ley de amnistía que no olvidaba nada, sino todo lo contrario. Nadie se acogió a ella. Con razón: era la amnistía del embudo, con la cual el gobierno pretendía resolver su problema sin preocuparse por el de sus adversarios. No era una ley concebida por los asesores jurídicos del presidente, sino por sus asesores militares, que no han podido ganar la guerra contra las guerrillas en veinticinco años y no se resignan a perderla otra vez en el papel. El M-19 declaró en esa ocasión: «O es la amnistía o es la guerra». Puesto que no fue la amnistía, fue la guerra.

Los militares piensan que cuando se publican los partes de batalla, los únicos que lamentan los muertos militares son los miembros de su familia y los propios militares. Se equivocan. No creo ser una excepción entre los adversarios de este gobierno y de este sistema que lamentamos la muerte de los militares, sobre todo la de los soldados que tienen la misma edad que nuestros hijos. Lo lamentamos más porque somos conscientes de que los militares que mueren en combate con las guerrillas no están allí por su voluntad ni por su ideología, sino por cuenta de los muy pocos que toman decisiones políticas dentro de las fuerzas armadas y escogen la prolongación de una guerra aún por simples consideraciones de machismo profesional. Son éstos, en última instancia, los que menos piensan con la mano en el corazón que también los militares se mueren en la guerra.

Las cartas están otra vez sobre la mesa. La oposición armada ha vuelto a hacer una propuesta de paz, en los mismos términos que la anterior, y en un momento en que no es posible decir que lo haga por debilidad, el gobierno tiene otra vez la palabra, pero es de esperar que esta vez no la tenga él solo. Que la tenga todo el país, sin exceptuar, por supuesto, a los militares que piensan distinto y a los curas que todavía crean en Dios. Los candidatos presidenciales de todos los partidos no podrán continuar sus campañas electorales sin definir con toda certeza cuál es su posición frente a un asunto de tanta gravedad para el destino de la nación. Todo esto, a fin de cuentas, le conviene al presidente Turbay Ayala: hace ya bastante tiempo que se aburrió del oficio estéril de gobernar sin soluciones, y lo único que desea es que se le acabe el empleo lo más pronto posible, para retirarse a gozar de sus glorias pasadas, a salvo para siempre del óxido implacable del poder y de las estatuas de pesadilla de Feliza Bursztyn. Termine, pues, su funesto mandato con un capítulo de paz, y no con esa mala imagen de sarraceno enardecido que no perdona ni a las bellas artes.

TORRIJOS

El general Omar Torrijos y un grupo reducido de amigos estábamos invitados a una cena el pasado 20 de julio, en Panamá. Poco antes de la hora en que debíamos irnos, su secretaria de turno interrumpió la conversación informal, que se había prolongado desde el almuerzo, y le recordó al general que además de la cena, a las ocho y media, estaba invitado al cumpleaños de un ministro, a las once de la noche, y que al día siguiente, muy temprano, debía asistir a un acto oficial. El general se volvió hacia mí, masticando el cigarro, y dijo de buen humor: «Ya están tratando de organizarnos a ti y a mí, que somos unos anárquicos». Y luego, dirigiéndose a los otros amigos, precisó: «Dije anárquicos, no anarquistas». Esa tarde había expresado varias veces su entusiasmo por la cena, que era en honor del antiguo presidente de Colombia Alfonso López Michelsen, pero en aquel instante comprendí que no asistiría a ella ni a ninguno de los otros actos programados. Así fue. Poco después ordenó que le tuvieran listos un avión y un helicóptero en el cercano aeropuerto de Paitilla para despegar en cualquier momento. Esto quería decir que aún no había tomado una decisión sobre su rumbo inmediato, pues el avión sólo lo usaba de noche para volar a la isla de Contadora o a la base militar de Farallón, que tiene servicios de aterrizaje nocturno, y el helicóptero podía servirle para cualquiera de los dos sitios o para el centro agrícola de Coclecito, un lugar remoto en las montañas del norte, donde solía apartarse del mundo entre los campesinos. Lo dijo aquella noche como tantas otras: «Lo que más me gusta es que nunca sé dónde voy a dormir». Ni lo sabía nadie. Sólo en el momento en que el avión o el helicóptero estaban listos para despegar le indicaba al piloto el lugar de destino. Esta vez no fue una excepción. Cuando regresé de la cena, encontré la casa iluminada, pero desierta y silenciosa, y comprendí que él se había ido hacía muy poco tiempo, pues en el aire refrigerado estaba todavía el olor de su

Publicado originalmente el 12 de agosto de 1981.

cigarro. Nunca supe para dónde se fue, pero ahora sé que desde aquella noche no volvería a verlo jamás.

Yo había llegado de México dos días antes. Viajaba a Panamá dos o tres veces al año, sólo para estar con él y con los amigos comunes, y siempre iba a un hotel. Esta vez me quedé en un cuarto de su casa de la capital, donde él aparecía muy pocas veces. «Las cosas no están como para andar por ahí», le dije. Esta frase le llamó tanto la atención que la repitió varias veces aquel fin de semana. En realidad, él era consciente de que la situación en América Central y el Caribe no era como para vivir sin precauciones, y procedía en consecuencia. Sus servicios de seguridad habían empezado a tomar medidas excepcionales y él mismo, que era el hombre más imprevisible que he conocido, había adoptado un comportamiento más imprevisible que nunca. Mi impresión es que muy pocas veces, en los tiempos de su poder, tuvo un instante de sosiego, y esto había creado en torno suyo una disponibilidad permanente para cambiar de lugar. Hace unos años, después de una reunión de seis presidentes sobre los tratados del canal de Panamá, varios amigos suyos lo convencimos de quedarse una noche más en Bogotá. Su avión, como siempre, estaba listo para partir en cualquier momento. La fiesta empezaba apenas a calentarse cuando su escolta le informó que el aeropuerto local estaría cerrado por reparaciones desde las doce de la noche hasta las seis de la mañana. El general se sentía tan a gusto que no le dio importancia, pero a las diez de la noche saltó de la silla y ordenó: «Nos vamos». En el camino del aeropuerto me confesó que no hubiera podido estar allí durante las seis horas en que no le sería posible irse de inmediato para donde le diera la gana.

Ya sabemos que cada palabra de alguien, cada gesto anterior, y aun sus actos más naturales cobran una significación espectral después de su muerte. Tal vez por eso tengo la impresión de que nunca como en esta última vez había hablado tanto de la muerte con el general Torrijos, y sobre todo de la que siempre nos amenaza durante el vuelo. Conocía muy bien mi miedo a volar, y siempre lo tomaba en cuenta con un gran respeto. Cuando yo estaba a bordo impartía a los pilotos instrucciones suplementarias para que eludieran los cielos tormentosos, y ordenaba que me subieran una cantimplora de whisky. «No hay nada mejor para volar», decía. «Si a los aviones les echaran whisky en los tanques en vez de gasolina, nunca más se volverían a caer.» La misma noche en que llegué, hace dos semanas, a Panamá, fuimos en helicóptero a la isla de Contadora. El cielo estaba sembrado de estrellas marinas, y el aire era fragante y diáfano sobre el Pacífico. Torrijos me miró de pronto, con sus ojos clarividentes, y me encontró impasible con el vaso de whisky en la mano. Entonces se volvió hacia su esposa, Raquel, con quien yo nunca había volado, y le dijo: «La única persona con quien Gabriel vuela tranquilo es conmigo». Dos días después se lo repitió al antiguo presidente de Venezuela, nuestro amigo Carlos Andrés Pérez, cuando regresábamos en avión a la ciudad de Panamá, sólo que entonces añadió una frase más: «Gabriel sabe que conmigo no puede pasarle nada». El avión en que volábamos entonces, para un trayecto de veinte minutos, era el bimotor *Twin Otter*, de la fuerza aérea panameña, en que Torrijos había de morir el viernes siguiente, en circunstancias que no me parecen del todo accidentales.

Fue un fin de semana alegre y raro en el paraíso de Contadora. El domingo 19 de julio, Gabriel Lewis Galindo, que fue embajador de Panamá en Washington durante el tiempo más difícil de las negociaciones del canal, invitó a un grupo de amigos a navegar en torno de la isla. No invitó a Torrijos, pues todos sabíamos que carecía por completo de vocación náutica. Sin embargo, a última hora conseguimos embarcarlo, y así vivió de muy buen humor el segundo día de mar de su vida. Mientras navegábamos, lo miré varias veces y lo encontré impasible, con su vaso de whisky en la mano, y no pude eludir la suposición de que él debía sentirse en el mar como yo me sentía en el aire. A la hora de las fotos caí en la cuenta de que nunca nos habíamos tomado una juntos, y se lo dije. Entonces él la hizo tomar, y es quizá la foto en que nos parecemos menos a nosotros mismos: en traje de baño. Pero me parece que fue la última de su vida.

Siempre tuve la impresión de que Torrijos corría muchos más riesgos de los que podía permitirse un hombre acechado por tantas amenazas. Aceptaba a duras penas las normas de seguridad, tal vez porque era el ser humano más desconfiado que se podía concebir, y en última instancia no confiaba en nadie ni en nada más que en sus intuiciones misteriosas y certeras. Era su única orientación en las tinieblas del azar. No creo que exista nadie capaz de decir a ciencia cierta qué era lo que pensaba en realidad, ni cuál era el secreto de sus sueños ni el sentido último de sus presagios. Su única debilidad era el corazón, y había conseguido amaestrarlo. «El que se aflige se afloja», decía. Los aviones en que volaba casi todos los días desde hacía muchos años eran buenos y muy bien mantenidos, y sus pilotos rigurosos eran los únicos que tomaban las decisiones del vuelo. Sin embargo, tal vez Torrijos no se daba cuenta de que aquella servidumbre a su intuición sobrenatural, que tal vez le salvó la vida muchas veces, terminó a la larga por ser su flanco más vulnerable, pues al final le daba tantas oportunidades a la fatalidad como a sus enemigos. Cualquiera de los dos pudo causarle la muerte. Pero es imposible no relacionar esta catástrofe con otras similares ocurridas en poco más de un año. En junio de 1980, el avión en que volaba el vicepresidente electo de Bolivia, Jaime Paz Zamora, se precipitó a tierra envuelto en llamas. Se pensó entonces, aunque nunca pudiera comprobarse, que le habían echado azúcar en el tanque de la gasolina. Después fue la tragedia del presidente de Ecuador, Jaime Roldós; más tarde, la del jefe del Estado Mayor de Perú, general Luis Hoyos Rubio, y ahora la del general Omar Torrijos, el hombre providencial e irreemplazable de Panamá. Cuatro personalidades progresistas, cuya desaparición sólo podía favorecer a las tendencias más tenebrosas de las Américas. No es fácil creer que tantos desastres sucesivos sean casuales, porque no es tan selectivo el índice de la muerte y hasta las mismas casualidades tienen sus leyes inexorables.

En todo caso, no era ésta la clase de final que Torrijos esperaba, ni la que deseaba y merecía. Siempre tuve la impresión de que se había reservado el privilegio de escoger el modo y la ocasión de su muerte, y que la tenía reservada como la carta última y decisiva de su destino histórico. Era una vocación de mártir que tal vez fuera el aspecto más negativo de su personalidad, pero también el más espléndido y conmovedor. El desastre, accidental o provocado, le frustró ese designio, pero la muchedumbre dolorida que asistió a sus funerales iba sin duda movida por la sabiduría secreta de que aquella muerte impertinente y sin grandeza es una de las formas más dignas del martirio. Yo no estaba allí, por supuesto. Nunca he tenido corazón para enterrar a los amigos.

Dos muchachos y dos muchachas que viajaban en un Renault 5 recogieron a una mujer vestida de blanco que les hizo señas en un cruce de caminos poco después de la medianoche. El tiempo era claro, y los cuatro muchachos —como se comprobó después hasta la saciedad— estaban en su sano juicio. La dama viajó en silencio varios kilómetros, sentada en el centro del asiento posterior, hasta un poco antes del puente de Quatre Canaux. Entonces señaló hacia delante con un índice aterrorizado, y gritó: «Cuidado, esa curva es peligrosa», y desapareció en el acto.

Esto ocurrió el pasado 20 de mayo en la carretera de París a Montpellier. El comisario de esa ciudad, a quienes los cuatro muchachos despertaron para contarle el acontecimiento espantoso, llegó hasta admitir que no se trataba de una broma ni una alucinación, pero archivó el caso porque no supo qué hacer con él. Casi toda la prensa de Francia lo comentó en los días siguientes, y numerosos parapsicólogos, ocultistas y reporteros metafísicos concurrieron al lugar de la aparición para estudiar sus circunstancias, y fatigaron con interrogatorios racionalistas a los cuatro elegidos por la dama de blanco. Pero al cabo de pocos días, todo se echó al olvido, y tanto la prensa como los científicos se refugiaron en el análisis de una realidad más fácil. Los más comprensivos admitieron que la aparición pudo ser cierta, pero aun ellos prefirieron olvidarla ante la imposibilidad de entenderla.

A mí —que soy un materialista convencido— no me cabe ninguna duda de que aquél fue un episodio más, y de los más hermosos, en la 'muy rica historia de la materialización de la poesía. La única falla que le encuentro es que ocurrió de noche, y peor aún, al filo de la medianoche, como en las peores películas de terror. Salvo por eso, no hay un solo elemento que no corresponda a esa metafísica de las carreteras que todos hemos sentido pasar tan cerca en el curso de un viaje, pero ante cuya verdad estremecedora nos negamos a

Publicado originalmente el 19 de agosto de 1981.

rendirnos. Hemos terminado por aceptar la maravilla de los barcos fantasmas que deambulan por todos los mares buscando su identidad perdida, pero les negamos ese derecho a las tantas y pobres ánimas en pena que se quedaron regadas y sin rumbo a la orilla de las carreteras. Sólo en Francia se registraban hasta hace pocos años unos doscientos muertos semanales en los meses más frenéticos del verano, de modo que no hay por qué sorprenderse de un episodio tan comprensible como el de la dama de blanco, que sin duda seguirá repitiéndose hasta el fin de los siglos, en circunstancias que sólo los racionalistas sin corazón son incapaces de entender.

Siempre he pensado, en mis largos viajes por tantas carreteras del mundo, que la mayoría de los seres humanos de estos tiempos somos sobrevivientes de una curva. Cada una es un desafío al azar. Bastaría con que el vehículo que nos precede sufriera un percance después de la curva para que se nos frustrara para siempre la oportunidad de contarlo. En los primeros años del automóvil, los ingleses promulgaron una ley –*The Locomotive Act*– que obligaba a todo conductor a hacerse preceder de otra persona de a pie, llevando una bandera roja y haciendo sonar una campana, para que los transeúntes tuvieran tiempo de apartarse. Muchas veces, en el momento de acelerar para sumergirme en el misterio insondable de una curva, he lamentado en el fondo de mi alma que aquella disposición sabia de los ingleses haya sido abolida, sobre todo una vez, hace quince años, en que viajaba de Barcelona a Perpiñán con Mercedes y los niños a cien kilómetros por hora, y tuve de pronto la inspiración incomprensible de disminuir la velocidad antes de tomar la curva. Los coches que me seguían, como ocurre siempre en esos casos, nos rebasaron. No lo olvidaremos nunca: eran una camioneta blanca, un Volkswagen rojo y un Fiat azul. Recuerdo hasta el cabello rizado y luminoso de la holandesa rozagante que conducía la camioneta. Después de rebasarnos en un orden perfecto, los tres coches se perdieron en la curva, pero volvimos a encontrarlos un instante después los unos encima de los otros, en un montón de chatarra humeante, e incrustados en un camión sin control que encontraron en sentido contrario. El único sobreviviente fue el niño de seis meses del matrimonio holandés.

He vuelto a pasar muchas veces por ese lugar, y siempre he vuelto a pensar en aquella mujer hermosa que quedó reducida a un montículo de carne rosada en mitad de la carretera, desnuda por completo a causa del impacto, y con su bella cabeza de emperador romano dignificada por la muerte. No sería sorprendente que alguien la encontrara un día de estos en el lugar de su desgracia, viva y entera, haciendo las señales convencionales de la dama de blanco de

Montpellier, para que la sacaran por un instante de su estupor y le dieran la oportunidad de advertir con el grito que nadie lanzó por ella: «Cuidado, esa curva es peligrosa».

Los misterios de las carreteras no son más populares que los del mar, porque no hay nadie más distraído que los conductores aficionados. En cambio, los profesionales –como los antiguos arrieros de mulas– son fuentes infinitas de relatos fantásticos. En las fondas de carreteras, como en las ventas antiguas de los caminos de herradura, los camioneros curtidos, que no parecen creer en nada, relatan sin descanso los episodios sobrenaturales de su oficio, sobre todo los que ocurren a pleno sol, y aun en los tramos más concurridos. En el verano de 1974, viajando con el poeta Álvaro Mutis y su esposa por la misma carretera donde ahora apareció la dama de blanco, vimos un pequeño automóvil que se desprendió de la larga fila embotellada en sentido contrario, y se vino de frente a nosotros a una velocidad desatinada. Apenas si tuve tiempo de esquivarlo, pero nuestro automóvil saltó en el vacío y quedó incrustado en el fondo de una cuneta. Varios testigos alcanzaron a fijar la imagen del automóvil fugitivo: era un Skoda blanco, cuyo número de placas fue anotado por tres personas distintas. Hicimos la denuncia correspondiente en la inspección de policía de Aix-en-Provence, y al cabo de unos meses la policía francesa había comprobado sin ninguna duda que el Skoda blanco con las placas indicadas existía en realidad. Sin embargo, había comprobado también que a la hora de nuestro accidente estaba en el otro extremo de Francia, guardado en su garaje, mientras su dueño y conductor único agonizaba en el hospital cercano.

De éstas, y de otras muchas experiencias, he aprendido a tener un respeto casi reverencial por las carreteras. Con todo, el episodio más inquietante que recuerdo me ocurrió en pleno centro de la ciudad de México, hace muchos años. Había esperado un taxi durante casi media hora, a las dos de la tarde, y ya estaba a punto de renunciar cuando vi acercarse uno que a primera vista me pareció vacío y que además llevaba la bandera levantada. Pero ya un poco más cerca vi sin ninguna duda que había una persona junto al conductor. Sólo cuando se detuvo, sin que yo se lo indicara, caí en la cuenta de mi error: no había ningún pasajero junto al chofer. En el trayecto le conté a éste mi ilusión óptica, y él me escuchó con toda naturalidad. «Siempre sucede», me dijo. «A veces me paso el día entero dando vueltas, sin que nadie me pare, porque casi todos ven a ese pasajero fantasma en el asiento de al lado.» Cuando le conté esta historia a don Luis Buñuel, le pareció tan natural como al chofer. «Es un buen principio para una película», me dijo.

EL CUENTO DEL CUENTO

Poco antes de morir, Álvaro Cepeda Samudio me dio la solución final de la *crónica de una muerte anunciada*. Yo había vuelto de Europa después de un viaje muy largo, y estábamos en su casa de domingos, frente al mar miserable de Sabanilla, cocinando su legendario sancocho de mojarras de a 2.000 pesos.

«Tengo una vaina que le interesa», me dijo de pronto: «Bayardo San Román volvió a buscar a Ángela Vicario.»

Tal como él lo esperaba, me quedé petrificado. «Están viviendo juntos en Manaure», prosiguió, «viejos y jodidos, pero felices.» No tuvo que decirme más para que yo comprendiera que había llegado al final de una larga búsqueda.

Lo que esas dos frases querían decir era que un hombre que había repudiado a su esposa la noche misma de la boda había vuelto a vivir con ella al cabo de veintitrés años. Como consecuencia del repudio, un grande y muy querido amigo de mi juventud, señalado como autor de un agravio que nunca se probó, había sido muerto a cuchilladas en presencia de todo el pueblo por los hermanos de la joven repudiada. Se llamaba Santiago Nasar y era alegre y gallardo, y un miembro prominente de la comunidad árabe del lugar. Esto ocurrió poco antes de que yo supiera qué iba a ser en la vida, y sentí tanta urgencia de contarlo, que tal vez fue el acontecimiento que definió para siempre mi vocación de escritor.

A quienes primero se lo conté fue a Germán Vargas y Alfonso Fuenmayor, unos cinco años después, en el burdel de alcaravanes de la negra Eufemia. Para entonces ya había resuelto ser escritor, y mi padre me había dicho: «Comerás papel». Durante años soñé que rompía resmas enteras y me las comía en pelotitas, y nunca era el papel sobrante de los periódicos donde trabajaba entonces, sino un muy buen papel de 36 gramos, áspero y con marcas de agua, tamaño carta, del que seguí usando siempre desde que tuve dinero para com-

Publicado originalmente el 26 de agosto de 1981.

prarlo. Sin embargo, Alfonso Fuenmayor y Germán Vargas coincidieron en que la historia del crimen era digna de ser escrita, aunque fuera comiendo papel. «No importa que sea inventada», me dijo Alfonso Fuenmayor; «así las inventaba Sófocles, y fíjese lo bien que le quedaban.» Más tarde, cuando regresó graduado de Columbia University, Álvaro Cepeda Samudio estuvo de acuerdo, pero me previno sin reticencias. «Lo único peligroso», me dijo, «es que a esa historia le falta una pata.»

En efecto, le faltaba el final imprevisible que él mismo me contó veintitrés años después del crimen, pero entonces era imposible imaginarlo. Germán Vargas, con su prudencia congénita, me aconsejó que esperara uno o dos años hasta que tuviera la historia mejor pensada. Yo no esperé ni uno ni dos, sino treinta años más.

No fue una demora excepcional, pues nunca he escrito una historia antes de que pasaran, por lo menos, veinte años desde su origen. Pero en este caso la razón era más consciente: seguía buscando, en la imaginación, la pata indispensable que le faltaba al trípode, tratando de inventarla a la fuerza, sin pensar siquiera que también la vida lo estaba haciendo por su cuenta y con mejor ingenio. Fue don Ramón Vinyes quien me dio la fórmula de oro: «Cuéntela mucho», me dijo. «Es la única manera de descubrir lo que una historia tiene por dentro.»

Por supuesto, seguí el consejo. Durante muchos años conté la historia al derecho y al revés, por todas partes, con la esperanza de que alguien le encontrara la falla. Mercedes, que la recordaba a pedazos desde muy niña, la volvió a armar por completo de tanto oírla, y terminó por contarla mejor. Luis Alcoriza se la hizo grabar en su casa de México, en una época en que todo el mundo era joven. A Ruy Guerra se la conté durante seis horas en un pueblo remoto de Mozambique, una noche en que los amigos cubanos nos dieron de comer un perro de la calle haciéndonos creer que era carne de gacela, y ni aun así pudimos descubrir el elemento que le faltaba. A Carmen Balcells, mi agente literario, se la conté muchas veces durante muchos años, en trenes y aviones, en Barcelona y en el mundo entero, y siempre lloró como la primera vez, pero nunca pude saber si lloraba porque la emocionaba o porque yo no la escribía. Al único amigo cercano a quien no se la conté nunca fue a Álvaro Mutis, por una razón práctica: él ha sido siempre el primer lector de mis originales, y me cuido mucho de que los lea sin ninguna idea preconcebida.

La revelación de Álvaro Cepeda Samudio en aquel domingo de Sabanilla me puso el mundo en orden. La vuelta de Bayardo San

Román con Ángela Vicario era, sin duda, el final que faltaba. Todo estaba entonces muy claro: por mi afecto hacia la víctima, yo había pensado siempre que ésta era la historia de un crimen atroz, cuando en realidad debía ser la historia secreta de un amor terrible. Sólo que estuve a punto de no conocer nunca sus pormenores ocultos, porque Álvaro y yo nos desbarrancamos dos horas después en el camión del Catatumbo de Alejandro Obregón, y no nos matamos de milagro. «¡Puta vida», pensaba, mientras caíamos hacia el fondo de aquel mar perdulario; «tanto buscar este final, para morirme sin contarlo!» Tan pronto como me restablecí, sobre todo del susto, me fui a buscar a Bayardo San Román y Ángela Vicario en su casa feliz de Manaure, para que me contaran los secretos de su reconciliación increíble. Fue un viaje más revelador de lo que yo pensaba, y por mejores motivos, porque a medida que trataba de escudriñar la memoria de los otros, me iba encontrando con los misterios de mi propia vida.

Hay dos pueblos cercanos, pero muy distintos, que se llaman Manaure. El uno es una sola calle muy ancha, con casas iguales, en una meseta verde de un silencio sobrenatural. Allí llevaban a mi madre a temperar cuando era niña. Tanto me habían hablado de ese pueblo medicinal en casa de mis abuelos, que cuando lo vi por primera vez me di cuenta de que lo recordaba como si lo hubiera conocido en una vida anterior. No era allí donde vivía el matrimonio feliz, pero Rafael Escalona, el sobrino del obispo, se equivocó de camino cuando íbamos para el otro Manaure. Estábamos tomando una cerveza helada en la única cantina del pueblo cuando se acercó a nuestra mesa un hombre que parecía un árbol, con polainas de montar y un revólver de guerra en el cinto. Rafael Escalona nos presentó, y él se quedó con mi mano en la suya, mirándome a los ojos.

—¿Tiene algo que ver con el coronel Nicolás Márquez? —me preguntó.

—Soy su nieto.

—Entonces —dijo él—, su abuelo mató a mi abuelo.

No me dio tiempo de asustarme, porque lo dijo de un modo muy cálido, como si también ésa fuera una forma de ser parientes. Era un contrabandista de la estirpe legendaria de los Amadises y, lo mismo que ellos, era un hombre derecho y de buen corazón. Estuvimos de parranda tres días y tres noches en sus camiones de doble fondo, bebiendo brandy caliente y comiendo sanchocho de chivo en memoria de los abuelos muertos. Me llevó a distintos pueblos, hasta el interior de la península Guajira, para que conociera a diecinueve de los hijos incontables que el coronel Nicolás Márquez había dejado dispersos durante la última guerra civil. Al cabo de una semana me dejó en el otro Manaure: un pueblo de salitre frente a un mar en llamas. Se detuvo ante una casa que yo hubiera reconocido de todos modos por lo mucho que había oído hablar de ella. «Ahí es», me dijo.

En la ventana de la sala, bordando a máquina en la hora de más calor, había una mujer de medio luto con antiparras de alambre y canas amarillas, y sobre su cabeza estaba colgada una jaula con un canario que no paraba de cantar. Al verla así, dentro del marco idílico de la ventana, no quise pensar que fuera ella, porque me resistía a creer que la vida terminara por parecerse tanto a la mala literatura. Pero era ella: Ángela Vicario, veintitrés años después del drama.

EL CUENTO DEL CUENTO
(Conclusión)

Me doy cuenta de que el lugar en que se cometió el crimen ha sido idealizado por la nostalgia. Era inevitable: allí pasé los años de mi adolescencia, que fueron los más libres de mi vida, hasta que la familia tuvo que cambiar de aires. Después volví dos veces, siempre en relación con el proyecto del libro. La primera fue unos quince años más tarde, tratando de rescatar de la memoria de la gente las numerosas piezas desperdigadas del rompecabezas del crimen, y tratando sobre todo de encontrar el final que todavía la vida no había resuelto. No me pareció que el tiempo hubiera sido demasiado severo con nadie, ni con nada, salvo con la casa de placer de María Alejandrina Cervantes, que había sido transformada en escuela de monjas. Fue una experiencia perturbadora ver un tropel de niñas con uniformes celestiales entrando por el mismo portón de trinitarias por donde toda mi generación había entrado a perder la virginidad.

La segunda vez que volví fue a escribir esta crónica. Fui inducido por el embeleco, tan común entre los realistas teóricos, de capturar en caliente para escribirla, la misma vida que se está viviendo. Escribí en calzoncillos de nueve de la mañana a tres de la tarde durante catorce semanas sin treguas, sudando a mares, en la pensión de hombres solos donde vivió Bayardo San Román los seis meses que estuvo en el pueblo. Era un cuarto escueto con una cama de hierro, una mesa coja que debía nivelar con cuñas de papelitos en las patas, y una ventana por donde se metían los moscardones aturdidos por el calor y la pestilencia de las aguas muertas del puerto antiguo. Ésa fue la única contribución de la vida circundante a mis esfuerzos de escritor comprometido. A medida que escribía me daba cuenta de que la realidad inmediata no tenía nada que ver con la que yo trataba de escribir, ni tal vez tampoco con la que recordaba, y estaba tan con-

Publicado originalmente el 2 de septiembre de 1981.

fundido que llegué a preguntarme si la vida misma no era también una invención de la memoria.

El doctor Dionisio Iguarán, primo hermano de mi madre y nuestro único médico en la época del drama, murió entre esas dos visitas. Su prestigio bien ganado queda repartido entre varios médicos nuevos, y en especial el doctor Cristóbal Bedoya, a quien llamábamos Cristo, que había hecho el tercer año de Medicina en el momento del crimen, y que es un protagonista ejemplar de esta crónica. Fue el amigo íntimo que acompañó a Santiago Nasar hasta unos minutos antes de su muerte, y el único de los 20.000 habitantes del pueblo que se propuso y estuvo a punto de impedir que lo mataran. Sus testimonios fueron los más inteligentes y entrañables. Fue él quien me recordó, al término de nuestras evocaciones incansables, uno de los datos más raros de esta desgracia: la autopsia de Santiago Nasar no la hizo un médico, sino el cura de la parroquia.

Se llamaba Carmen Amador, se preciaba de haber nacido en un risco de Galicia donde nunca se habla la lengua castellana, y bastaba con oírselo decir para saber que era cierto. Yo lo recordaba con cierta amargura porque siendo muy niño me hacía repetir de memoria los falsos poemas de Gabriel y Galán y fue quien me dijo más tarde que Dios había prohibido leer a Gil Vicente. Fue nuestro único párroco hasta donde me alcanza la memoria, pero cuando volví de adulto por primera vez se había ido sin dejar rastros.

Nunca traté de encontrarlo. Sin embargo, durante un verano que pasé hace doce años en la playa de Calafell, muy cerca de Barcelona, alguien me habló de un cura retirado en la tenebrosa casa de salud del lugar, que decía haber perdido media vida en mi tierra. Lo reconocí de inmediato, aunque sólo hubiera sido por sus ojos de ternero de vientre y su castellano rupestre con cadencias del Caribe. Hablamos mucho y muchas veces hasta el final del verano, y era evidente que no había logrado asimilar el mal recuerdo de aquella autopsia.

Un año después de que Álvaro Cepeda Samudio me dio la clave final, el libro estaba listo para ser escrito. Sin embargo, por algunos de esos motivos demasiado simples que los escritores no logramos entender, pasó todavía mucho tiempo sin que lo escribiera. Más aún: hubo una época en que lo olvidé por completo. De pronto, en el otoño de 1979, Mercedes y yo estábamos en la sala oficial del aeropuerto de Argel, esperando que nos llamaran para embarcar, cuando entró un príncipe árabe con la túnica inmaculada de su alcurnia y con un halcón amaestrado en el puño. Era una hembra espléndida de halcón peregrino, y en vez del capirote de cuero de la cetrería clásica llevaba uno de oro con incrustaciones de diamantes. Por supuesto, me acordé de Santiago Nasar, que había aprendido de su padre las bellas artes de la altanería, al principio con gavilanes criollos y luego con ejemplares magníficos trasplantados de la Arabia feliz. En el momento de su muerte tenía en su hacienda una halconera profesional, con dos primas y un torzuelo amaestrados para la caza de perdices, y un neblí escocés adiestrado para la defensa personal.

Sin embargo, la evocación de Santiago Nasar no fue tan comprensible como me pareció cuando vi entrar al monarca del desierto con su animal de volatería coronado de oro. Fue más bien un zarpazo del destino. En el avión de regreso comprendí que la historia tantas veces diferida había vuelto esta vez a quedarse para siempre, y que no podría seguir viviendo un solo instante sin escribirla. La sentía entonces con tanta intensidad como no la había sentido nunca en 32 años, desde el lunes infame en que María Alejandrina Cervantes irrumpió desnuda en el cuarto donde yo continuaba dormido a pesar de las campanas de incendio, y me despertó con su grito de loca: «Me mataron a mi amor».

A propósito: George Plimpton, en su entrevista histórica para *The Paris Review*, le preguntó a Ernest Hemingway si podría decir algo acerca del proceso de convertir un personaje de la vida real en un personaje de novela. Hemingway contestó: «Si yo explicara cómo se hace eso algunas veces sería un manual para los abogados especialistas en casos de difamación».

LA DESGRACIA DE SER ESCRITOR JOVEN

En mi doble destino de periodista y escritor, sólo recuerdo hasta ahora dos cosas de que arrepentirme, y es haber ganado dos concursos literarios. El primero fue en 1954, patrocinado por la Asociación de Escritores de Colombia, cuyo secretario de entonces me suplicó que participara con un cuento inédito, porque no se había presentado ninguna obra que valiera la pena y temían que el certamen fuera un fracaso. Le entregué un cuento sin terminar –*Un día después del sábado*–, y pocos días más tarde apareció jadeante en mi oficina para decirme, como si fuera un milagro ajeno a su diligencia, que me habían concedido el primer premio. No recuerdo cuánto representaba en dinero, pero estoy seguro de que apenas me alcanzó para celebrar la victoria. La ganancia mayor, por supuesto, fue la resonancia en la prensa. Pero yo no era ya demasiado sensible a esa gloria instantánea, a pesar de que sólo tenía veinticinco años, porque llevaba cinco de ganarme la vida escribiendo columnas firmadas en periódicos de provincia, y en aquel momento era reportero de planta de *El Espectador*. La impresión que me quedó después de la premiación solemne, en la cual pusieron flores en el estrado y se pronunciaron discursos trémulos, fue la muy desapacible de haberme prestado a una farsa pública.

El segundo concurso fue todavía más triste. Lo había convocado en 1962 la filial colombiana de una empresa de petróleo de Estados Unidos, y el premio era la publicación de la obra, y nada menos que 3.000 dólares de la época. Yo vivía en México, y ni siquiera había tenido noticias de aquella propuesta tentadora, pero sus patrocinadores mandaron con todos los gastos pagados a mi querido amigo el maestro Guillermo Angulo para que me convenciera de participar en el concurso. El motivo de la diligencia era el mismo: nadie había mandado ninguna obra que valiera la pena, y los patrocinadores temían que el concurso fuera un fracaso.

Publicado originalmente el 9 de septiembre de 1981.

Yo había terminado desde hacía más de un año una novela que no me había preocupado por publicar, pues el placer en aquellos tiempos no era ése, sino el más puro y simple de escribir. Tenía los originales enrollados y amarrados con una corbata en el fondo de un baúl, y se los entregué a Guillermo Angulo tal como estaban, con corbata y todo, sin tomarme el trabajo de volverlos a leer ni de pensar un título. Sólo cuando la novela iba a ser impresa le encontré uno adecuado: *La mala hora*. Con los 3.000 dólares compré un automóvil de segunda mano y pagué los gastos del nacimiento de mi hijo menor, que de aquel modo había traído su propio pan bajo el brazo. Pero no viajé a Bogotá a recibir el premio con todos los gastos pagados, porque tenía la sensación ingrata de haberme prestado una vez más a la promoción de una empresa que no tenía nada que ver con la literatura.

Veinte años después, evocando aquellos tiempos difíciles y viendo cómo proliferan ahora los concursos literarios, sigo pensando que mis escrúpulos de entonces estaban bien fundados. Sin embargo, el entusiasmo casi pueril con que concursan los escritores que hoy tienen la misma edad y las mismas ilusiones que yo tenía entonces me hace pensar que ellos no comparten mis recelos, sino todo lo contrario, y que muchos no escriben por una necesidad ineludible, como debe ser, sino sólo para ganarse un concurso, y eso es algo que debe alarmarlos, como a mí me alarma, si en realidad están dispuestos a entrar con pie derecho en el infierno de los escritores grandes.

En realidad, los concursos literarios patrocinados por las casas editoriales no favorecen a nadie más que a ellas mismas. Los editores piensan, como lo han pensado desde el día de infortunio en que fueron inventados, que son ellos quienes les hacen a los escritores el favor de publicarles sus libros, sobre todo a los escritores nuevos, y que, por consiguiente, son éstos quienes deben pagarles por la publicación. Un editor me dijo hace poco que la industria editorial no la hacen los escritores, ni los escritores y los editores juntos, sino sólo los editores. Yo le dije que eso era tanto como pretender que la industria petrolera la hacen solas las compañías petroleras, sin la modesta colaboración del petróleo. Esta concepción mesiánica de su propio destino es sin duda lo que ha inducido a las casas editoriales a la patraña de los concursos literarios. Los organizan y convocan con ínfulas de benefactores de la humanidad y arcángeles de la cultura, cuando lo único que hacen en realidad es promover el nombre de sus propias empresas a costa de los escritores que no tienen quién los publique, como no lo han tenido al principio de sus vidas ninguno de los escritores grandes que en el mundo han sido.

Se me ocurren estas reflexiones a propósito de la batalla justa y solitaria que está librando un joven novelista colombiano contra la agencia de la Editorial Plaza & Janés en Colombia, que ha instaurado en ese país un concurso anual de novela. El premio, según dicen las bases del concurso, es de 300.000 pesos colombianos: 6.000 dólares. Pero la verdad es que no hay premio, porque al ganador lo hacen firmar un contrato en el cual se establece que esa suma es un anticipo sobre los derechos de autor, y por consiguiente será descontada de la liquidación periódica. Pero hay más. Los derechos de autor que recibe el favorecido no son del 10 % sobre el precio del libro, que es lo normal, sino sólo del 7 %. Esto quiere decir que con la venta de una edición de 3.000 ejemplares —que se vende sin dificultad, por la resonancia del concurso— el editor se paga todos los gastos, y se gana, además, la propaganda gratis que la prensa le hace al concurso, con el apoyo de los organismos culturales y la participación jubilosa de los otros escritores y artistas. Un cálculo conservador de lo que Plaza & Janés se gana en propaganda gratis con su concurso anual es de unos tres millones de pesos. Es decir, que no sólo comete un atraco contra el escritor novato, sino que es éste el que le sirve al editor para enriquecerse más con el menor esfuerzo. Pero como si eso no fuera bastante, el contrato que le hacen firmar le asegura a la editorial los derechos del libro para toda la vida, y no por cinco años, como debe ser, y no sólo para una parte del territorio, sino para todos los países de lengua castellana. Sin embargo, tan pronto como se agota la edición del lanzamiento, el libro se queda prisionero del contrato y sin la esperanza de una nueva impresión, y no circula en todo el territorio contratado, sino a duras penas en el país donde ganó el concurso.

Ése es el caso: no hay desgracia más grande en este mundo que la de ser escritor joven. Sobre todo en estos tiempos infaustos en que está de moda ser famoso. Antes, cuando los escritores jóvenes escribíamos porque no nos quedaba más remedio, teníamos además la ventaja de que los editores no nos hacían caso. Yo necesité cinco años para encontrar quien me publicara la primera novela, y el que encontré fue un pobre editor aficionado y sin recursos que se fugó del país huyendo de los acreedores. Eduardo Zalamea Borda, que era mi verdadero papá literario, llamó entonces a sus libreros amigos para que compraran el libro en los depósitos de la imprenta, y mis otros amigos de siempre escribieron las notas de prensa para que se supiera que estaba ya a la venta. Hace unos pocos años descubrí que los ejemplares sobrantes de aquella edición indigente de mil ejemplares se estaban vendiendo en las calles de Bogotá a un peso cada uno, y compré todos los que pude, con la impresión de comprar las piltrafas sueltas de mi propio pasado. Me gusta contar estas cosas no por la obsesión de hablar de nosotros mismos que tenemos los escritores, sino con la esperanza de que les sirva de algo a los que vienen detrás y creen todavía que es imposible vivir sin los editores. Un día —que ojalá no esté lejano— se convencerán no sólo de que es posible, sino de que es todo lo contrario: son los editores los que no pueden vivir sin nosotros. «Los pobrecitos editores.»

Más de trescientos intelectuales de la América Latina y el Caribe y un grupo de observadores de España se reunieron cuatro días en La Habana la semana pasada para conversar en familia. Había de todo: escritores, pintores, músicos, profesores universitarios, y lo mismo se encontraba un comunista indignado por las agresiones al clero de su país que un sacerdote dispuesto a explicar las conveniencias del socialismo. Pero el tema era uno sólo: los peligros que amenazan la soberanía y la identidad cultural de nuestras naciones, en estos momentos en que un vaquero de película se ha metido a caballo en la Casa Blanca.

Siempre he tenido un prejuicio contra los intelectuales, entendiendo por intelectual a alguien que tiene un esquema mental preconcebido y trata de meter dentro de él, aunque sea a la fuerza, la realidad en que vive. Graham Greene, que al parecer tiene el mismo prejuicio, explicó alguna vez que los novelistas no somos intelectuales, sino emocionales, y ese esclarecimiento me puso la conciencia en orden. Con todo, nunca había asistido a una reunión de intelectuales, y menos a una de trescientos, porque me parecía que era como asistir a un aquelarre de trescientos partidos políticos contrapuestos.

Al menos esta vez me equivoqué. El encuentro de intelectuales por la soberanía de los pueblos de nuestra América fue un certamen compacto y serio, en el cual se pasó por encima de tantas diferencias secundarias y se consiguió un acuerdo unánime en torno de una preocupación que era la mayor de todas. La prensa extranjera, y sobre todo las agencias de Estados Unidos, pusieron sus recursos enormes al servicio del silencio. Que nada se supiera de nosotros, que nunca nos habíamos visto, y que quienes vinimos no éramos los mejores. La verdad es que fueron muy pocos los que faltaron —muchos de ellos porque no pudieron eludir otros compromisos. El do-

Publicado originalmente el 16 de septiembre de 1981.

cumento final, corto, sobrio, sereno, no sólo es reflejo fiel del espíritu que prevaleció en estos cuatro días, sino que muestra muy bien el grado de madurez del encuentro.

Una novedad notable fue la reincorporación de los amigos del Brasil a un diálogo interrumpido desde hacía mucho tiempo. Para mí, y de un modo muy particular, éste es un motivo grande de alegría: de algún modo difícil de explicar, los brasileros, que llevan a todas partes un grano de locura que les da una dimensión nueva a las cosas, son portadores de la buena suerte. Fue la delegación extranjera más numerosa: 36 iluminados, precedidos por un terremoto, la actriz Ruth Escobar, que vino a reivindicar, con su hermosa voz de navegante, los derechos de la mujer. Tenía razón: una falla del encuentro era la escasa participación femenina. Ruth Escobar lo hizo notar desde la primera sesión. Yo, que a pesar de mis esfuerzos constantes no he logrado superar el machismo intravenoso que me inyectaron desde la cuna, me quedé pensando que en la realidad sólo hay algo en que el hombre es superior a la mujer, y es en la ternura. No lo hice por desconcertar a Ruth Escobar: lo creo. Como creo que la mujer dispone de una fuerza de poder de que carecemos los hombres: su falta de indulgencia.

Un periodista europeo, sorprendido por la unanimidad de este encuentro, andaba preguntando entre los asistentes si seríamos capaces de sostener un diálogo tan fluido con los intelectuales europeos. A título personal le contesté que no, por una razón que ni los europeos ni nosotros nos hemos decidido a aceptar: nuestra concepción de América Latina parte de dos análisis distintos. Durante la década de los sesenta, los intelectuales europeos se colocaron en la primera línea de la solidaridad con nosotros, nos desbordaron con un alborozo idealista que, sin embargo, no resistió el primer embate serio de la realidad. Su análisis tenía, y sigue teniendo, un rezago colonial: sólo ellos se creen depositarios de la verdad. Para ellos sólo es bueno lo que ha probado serlo en su propia experiencia. Todo lo demás es extraño, y, por consiguiente, inaceptable y corruptor. En la actualidad les resulta casi imposible hacer cualquier análisis del mundo sin tomar como punto de referencia la intervención soviética en Afganistán o la marmita a alta presión de Polonia. Para ellos nada ocurre en nuestro ámbito que no sea un designio tenebroso de la Unión Soviética. Una tentativa de imponer su modelo. Tal vez sin darse cuenta y, por supuesto, sin desearlo, los intelectuales europeos coinciden en esas concepciones con las del gobierno del presidente Reagan.

Trescientos intelectuales de este lado del Atlántico hemos discutido nuestros asuntos durante cuatro días sin apelar a ningún punto de referencia que no sea de nuestra realidad propia. Para nosotros, por encima de cualquier otro, el riesgo mayor e inminente es la forma en que nos concibe el gobierno de Reagan, cuyo altavoz hacia el mundo es la muy tenebrosa Kirkpatrick —embajadora ante las Naciones Unidas—, quien declaró hace poco que el régimen del general Pinochet es un ejemplo de la democracia autoritaria que nos hace falta para ser felices, y que los nicaragüenses estaban menos oprimidos con el general Anastasio Somoza que con el gobierno actual.

La paradoja mayor es que me parece imposible convencer de su error a los intelectuales europeos, y en cambio creo posible y urgente convencer a los norteamericanos. Siempre he creído que éstos carecen de la arrogancia que distingue a los europeos, y están animados por un deseo de entender que los aproxima más a nosotros.

En realidad, la opinión pública de Estados Unidos es mucho más sensible a nuestro drama, está más dispuesta a escuchar y admitir nuestras razones. Además, está tan penetrada por las corrientes profundas de nuestra cultura que ya casi nada de lo nuestro le parece extraño. En este encuentro, la tendencia fue casi unánime: vamos a conversar, lo más propio posible, con los intelectuales de Estados Unidos.

LA LARGA VIDA FELIZ
DE MARGARITO DUARTE

He vuelto a ver a Margarito Duarte. Apareció de pronto en una de las callecitas apacibles de la Roma antigua, y no sólo me sorprendió por su aspecto irreversible de romano viejo, sino por su tenacidad irracional. La última vez en que lo había visto, hace más de veinte años, conservaba todavía la ropa funeraria y la conducta sigilosa de los funcionarios públicos de los Andes. Ahora, sus maneras me parecieron las de alguien que ya no pertenece a nadie más que a sí mismo. Al cabo casi de dos horas de recordaciones nostálgicas en uno de los cafecitos del Trastévere, me atreví a hacerle la pregunta que más me ardía por dentro.

—¿Qué pasó con la santa?

—Ahí está —me contestó—, esperando.

Sólo el tenor Rafael Rivero Silva y yo podíamos entender la tremenda carga humana de la respuesta. Conocíamos tanto su drama que durante muchos años pensé que Margarito Duarte era el personaje en busca de autor que todos los novelistas esperamos durante toda la vida, y si nunca tomé la decisión de dejarme encontrar fue porque el final de su historia era imprevisible y casi imposible de inventar. Todavía lo sigue siendo.

Margarito Duarte llegó a Roma en el verano de 1954. Era la primera vez que salía de su remota aldea de los Andes, y no necesitaba decirlo para que uno lo supiera a primera vista. Se había presentado una mañana en el consulado de su país en Roma, con aquella maleta de piano lustrado que por su tamaño y su forma parecía el estuche de un violonchelo, y le había planteado al cónsul el motivo asombroso de su viaje. El cónsul llamó entonces a su amigo, el tenor colombiano Rafael Rivero Silva, para que éste le consiguiera una habitación a Margarito Duarte en la pensión donde ambos vivíamos. Así nos conocimos.

Publicado originalmente el 23 de septiembre de 1981.

Ese mismo día nos contó su historia. No había pasado de la escuela primaria, pero su vocación por las bellas letras le había inducido a darse una formación cultural más alta, mediante la lectura concienzuda y un poco apasionada de cuanto material impreso pasaba a su alcance.

A los dieciocho años se había casado con la muchacha más bella de su provincia, que murió dos años después, y de ella le quedó una hija más bella aún, que había muerto poco después a la edad de siete años. Margarito Duarte era el registrador de instrumentos públicos de su municipio desde que terminó la escuela primaria. Y siguió siéndolo hasta que una trastada de su destino lo embarcó en aquel viaje demente que había de torcer para siempre el rumbo de su vida. Todo había empezado seis meses antes de su llegada a Roma, cuando hubo que cambiar de lugar el cementerio del pueblo para construir una empresa. Margarito Duarte, como todos los habitantes de la región, desenterró sus muertos para la mudanza y aprovechó la ocasión para ponerles urnas nuevas. La esposa estaba convertida en polvo al cabo de doce años, pero la niña estaba intacta. Tanto, que cuando destaparon la caja se sintió el olor de las rosas frescas con que la habían enterrado. Las voces que proclamaron el milagro se oyeron de inmediato hasta mucho más allá de su provincia, y durante toda la semana acudieron al pueblo los curiosos menos pensados. No había duda: la incorruptibilidad del cuerpo ha sido siempre uno de los síntomas más visibles de la santidad, y hasta el obispo de la diócesis estuvo de acuerdo en que la noticia de aquel acontecimiento debía llegar hasta el Vaticano para que la sagrada congregación del rito rindiera su veredicto. Fue así como se hizo una colecta pública para que Margarito Duarte viajara a Roma a batallar por una causa que ya no era sólo suya, ni del ámbito estrecho de su pueblo, sino un asunto de la patria.

En el curso de su relato, Margarito Duarte abrió el candado y luego la tapa del baúl primoroso que parecía un estuche de violonchelo, y entonces el tenor Rivero Silva y yo participamos del milagro. No era una momia marchita como las que se ven en tantos museos del mundo, pues aquella hubiera podido confundirse con una criatura que seguía dormida al cabo de doce años bajo la tierra. No había color de miel en reposo, y los ojos abiertos eran diáfanos y vivos, y causaban la impresión irresistible de que nos estaban viendo desde la muerte. La niña había sido vestida de novia virgen para ser sepultada, de acuerdo con una costumbre muy antigua de su región, y le habían puesto en las manos un ramo de rosas.

Pero el raso y los azahares falsos de la corona no habían soportado el rigor del tiempo con tan buena salud como la piel. Con todo, lo más sorprendente era que el cabello no había cesado de crecer y le llegaba hasta los pies. Lo mismo había ocurrido con las uñas, pero se las habían cortado por decisión unánime del pueblo. Pues hasta los intérpretes más puros estuvieron de acuerdo en que era un espectáculo contrario a la santidad. De todos modos, Margarito Duarte llevaba en un frasco las uñas cortadas. Por si hacían falta como prueba adicional del prodigio.

Margarito Duarte empezó sus gestiones en los últimos meses de aquel verano ardiente y ruidoso. Al principio dispuso de una cierta ayuda de las autoridades diplomáticas de su país, pero muy pronto quedó a merced de su propia inspiración. Pío XII, que era el Papa de entonces, no dio nunca señal alguna de haber tenido noticias del milagro. Más aún: la Secretaría de Estado no contestó nunca la carta manuscrita de casi sesenta folios que escribió y entregó Margarito Duarte en persona. En el verano siguiente desistió del concurso inservible de sus diplomáticos, y fue solo a Castelgandolfo con el estuche de la santa para mostrarla al Papa, pero no le fue posible porque el sumo pontífice no circulaba por entre los turistas que llegaban de todo el mundo para verle, sino que apareció en un balcón del patio interior, y desde allí pronunció seis veces el mismo discurso en seis idiomas. Pero ni aquella frustración inicial ni las incontables y muy descorazonadoras que ha padecido desde entonces han logrado quebrantar su determinación.

Invencible Margarito Duarte. La semana pasada, mientras conversaba en el cafecito del Trastévere, me hizo caer en la cuenta de que han pasado ya de cuatro los papas desde que él esperaba, de modo que hay razones para creer que sus posibilidades, hablando en términos estadísticos, son cada vez mayores. Después de eso no tengo ya ninguna duda: el santo es él. Sin darse cuenta, a través del cadáver incorrupto de su hija, Margarito Duarte lleva más de veinte años de estar luchando en la vida por la causa legítima de su propia canonización.

«LOS IDUS DE MARZO»

He vuelto a leer esta semana *Los idus de marzo*, la hermosa novela de
Thornton Wilder que leí por primera vez hace unos veinticinco
años en una traducción apresurada, y que he releído muchas veces
desde entonces con el primer placer. Cuando estaba escribiendo
El otoño del patriarca, como era natural, la tuve siempre a la mano
como una fuente deslumbrante de la grandeza y las miserias del poder. La he comprado muchas veces en distintos idiomas para compartir mi estremecimiento con amigos del mundo entero, y no recuerdo a ninguno que no hubiera sucumbido ante aquel manantial
de belleza. Ahora la he vuelto a leer cuando menos lo pensaba, en
un vuelo apacible de cuatro horas y en un ejemplar ajeno, y sólo
ahora he descubierto cuánto ha tenido que ver con mi vida esa novela magistral.

Mi preocupación por los misterios del poder tuvo origen en un
episodio que presencié en Caracas por la época en que leí por primera vez *Los idus de marzo*, y ahora no sé a ciencia cierta cuál de las
dos cosas ocurrió primero. Fue a principios de 1958. El general
Marcos Pérez Jiménez, que había sido dictador de Venezuela durante diez años, se había fugado para Santo Domingo al amanecer. Sus
ayudantes habían tenido que izarlo hasta el avión con una cuerda,
pues nadie tuvo tiempo de colocar una escalera, y en las prisas de la
huida olvidó su maletín de mano, en el cual llevaba su dinero de
bolsillo: trece millones de dólares en efectivo. Pocas horas después,
todos los periodistas extranjeros acreditados en Caracas esperábamos
la constitución del nuevo gobierno en uno de los salones suntuosos
del palacio de Miraflores. De pronto, un oficial del ejército en uniforme de campaña, cubriéndose la retirada con una ametralladora
lista para disparar, abandonó la oficina de los conciliábulos y atravesó el salón suntuoso caminando hacia atrás. En la puerta del palacio
encañonó un taxi, que le llevó al aeropuerto, y se fugó del país. Lo

Publicado originalmente el 30 de septiembre de 1981.

único que quedó de él fueron las huellas de barro fresco de sus botas en las alfombras perfectas del salón principal. Yo padecí una especie de deslumbramiento: de un modo confuso, como si una cápsula prohibida se hubiera reventado dentro de mi alma, comprendí que en aquel episodio estaba toda la esencia del poder. Unos quince años después, a partir de ese episodio y sin dejar de evocarlo, o sin dejar de evocarlo de un modo constante, escribí *El otoño del patriarca*. Mi primer texto para aprender a descifrar el misterio fue *Los idus de marzo*. Como lo saben quienes la han leído, la novela es la reconstrucción literaria de los últimos años de la República Romana y de la propia vida de su dictador, Julio César. El pretexto del relato, en torno del cual se construye, es una fiesta ruidosa que Clodia Pulcher y su hermano ofrecían en honor de dos varones ilustres: Julio César y el poeta Cayo Valerio Catulo. Es una licencia literaria, porque el año de la fiesta, que era el 45 antes de Cristo, Catulo debía tener unos ocho años de muerto. Pero un escritor grande como Thornton Wilder no podía detenerse en esas menudencias racionalistas. Fue mucho más lejos. En la novela, el dictador, ataviado con sus mejores galas, abandonó la recepción descomunal que la reina Cleopatra le ofrecía aquella noche, y fue a velar a Catulo en su lecho de moribundo. «Toda la noche estuvimos oyendo las orquestas y viendo el cielo iluminado por los fuegos artificiales», dijo un testigo supuesto. El autor atribuyó el relato de aquella velación a una carta que la mujer de Cornelio Nepote le escribió a su hermana Postumia, y concluyó que César, para consolar al moribundo, no hizo más que hablarle de Sófocles. «Cayo murió con un coro de *Edipo en Colono*», decía el relato.

Antes de *Los idus de marzo*, lo único que yo había leído sobre Julio César eran los libros de texto del bachillerato, escritos por los hermanos cristianos, y el drama de Shakespeare, que, al parecer, le debe más a la imaginación que a la realidad histórica. Pero a partir de entonces me sumergí en las fuentes fundamentales: el inevitable Plutarco, el chismoso incorregible de Suetonio, el árido Carcopino y los comentarios y memorias de guerra del propio Julio César. Todos ellos se refieren, por supuesto, a la diligencia frenética con que los augures oficiales descuartizaban animales y escudriñaban la naturaleza para averiguar el porvenir. El primero de septiembre del 45 antes de Cristo —según cuenta Thornton Wilder—, el dictador recibió de sus adivinos más de quince informes, entre ellos el de un ganso que tenía manchas en el corazón y en el hígado, y un pichón siniestro que tenía un riñón fuera de lugar, el hígado hinchado y de color amarillo y una piedrecita de cuarzo en el buche. «Yo, que gobierno tantos hombres, soy gobernado por pájaros y truenos», dijo César, aturdido por tantos y tan confusos presagios. No sé dónde leí que había terminado por clausurar el colegio de augures, y escribió contra ellos un libro de protesta cuyo solo título era un poema: *Auguralia*. Lo busqué durante muchos años, hasta que el crítico Ernesto Volkenin, que es la persona que más sabe de eso en este mundo, me dijo de un modo severo y para siempre: «Ese libro no existió nunca».

A fin de cuentas, *Los idus de marzo* es sólo una hipótesis sobre la personalidad de César. Pero es una hipótesis que tal vez supere la realidad. «Todos comprendemos muy bien al cocinero de César que se quitó la vida cuando se le incendió el fogón», cuenta un Cornelio Nepote inventado por Thornton Wilder. Dice que había invitados importantes cuando ocurrió el percance, y el mayordomo, asustado, obligó al cocinero a que se lo contara a César. Pero éste no se inmutó cuando lo supo, sino que le pidió de muy buen modo al cocinero que le llevara dátiles y ensalada para sustituir la cena perdida. Entonces el cocinero salió al jardín y se degolló con el cuchillo de las verduras.

Veinte siglos después de ese suicidio, circuló en España una historia que ilustraba tan bien como aquélla sobre la fatalidad del poder. Según esa historia, una nieta del generalísimo Franco, de unos siete años, dio muestras de disgusto en casa de un ministro cuando vio una atractiva anunciadora en la televisión. «Es una pesada», dijo la niña. Entonces le preguntaron por qué lo decía, y ella dijo: «Porque mi abuelito dice que es una pesada». Aquélla fue la última vez en que se vio a la atractiva anunciadora en la televisión.

El 15 de marzo del año 44 antes de Cristo, todo el mundo en Roma sabía que a César le iban a matar. Todo el mundo menos él mismo. Plutarco cuenta que el griego Artemidoro, profesor de elocuencia helénica, se abrió paso a través de la muchedumbre que aclamaba al dictador cuando iba para el Senado, y le entregó un papel escrito de su puño y letra, con la advertencia de que lo leyera de inmediato. César solía entregar a sus secretarios los muchos papeles que le daban en la calle, pero aquél lo retuvo en la mano izquierda para leerlo en la primera oportunidad.

Allí estaban contados los pormenores de la conspiración y la forma en que César sería asesinado. Pero él no lo leyó nunca, pues un instante después entró en el Senado y fue muerto de veintitrés puñaladas. Suetonio termina su relato de este modo: «Antisio, el médico, dijo que de todas aquellas heridas sólo la segunda en el pecho debió haber sido mortal». Cualquier parecido con cualquier otra historia, viva o muerta, será pura coincidencia.

¿QUIÉN LE TEME A LÓPEZ MICHELSEN?

Un amigo abusivo −que también los hay, y muy buenos− hizo públicas, la semana pasada, algunas reflexiones que yo hice en privado sobre la candidatura de Alfonso López Michelsen para la presidencia de Colombia, y además me atribuyó conclusiones prematuras y simples que nunca se me hubieran ocurrido en mi sano juicio. Antes de que yo hiciera la aclaración debida, ya muchos amigos me habían llamado por teléfono para expresarme su alborozo o su contrariedad. La cosa no terminó ahí, sin embargo. El domingo siguiente, por primera vez desde su iniciación, hace un año exacto, esta columna semanal no apareció en *El Espectador*, de Bogotá. A pesar de que el periódico explicó a sus lectores que eso se debió a un inconveniente telegráfico, muchos amigos me volvieron a llamar indignados, pues creían que la censura interna de *El Espectador* me había suprimido el artículo por ser favorable a López Michelsen. A todos les expliqué que en realidad el artículo era sólo una evocación a Thornton Wilder, un novelista norteamericano criado en Hong Kong que no tuvo nada que ver con la política colombiana. También les recordé que desde hace más de treinta años estoy vinculado a *El Espectador* por motivos más duraderos que el oficio y la política; que, por consiguiente, no era probable que sus dueños le torcieran el cuello a mi cisne descarriado, y que en un caso extremo no lo harían sin decirles la verdad a nuestros lectores.

El incidente no valdría ni siquiera un recuerdo dominical, si no fuera porque revela de un modo inquietante el grado de sensibilización de los colombianos frente a la candidatura de López Michelsen. No parece haber términos medios, ni la menor incertidumbre de la razón, sino que todos creemos saber ya dónde estamos, y parecemos decididos a estar allí hasta la muerte. Desde que tengo conciencia no recuerdo una candidatura que provocara de inmediato semejante polarización pasional. Ni siquiera la de Alfonso López Pumarejo,

Publicado originalmente el 7 de octubre de 1981.

padre del Alfonso actual, en su segundo asalto al poder en 1942. Tal vez el único antecedente hubiera sido la candidatura de Laureano Gómez, en 1949, pero ésta ocurrió en un momento en que la inmensa mayoría de los colombianos estábamos sometidos por el terror desde el poder, y nadie pudo expresar su repudio.

Tampoco, desde que tengo memoria, había visto al país en un estado de postración como éste, que tiene todos los visos de una encrucijada final. Lo que nos hace falta ahora no es un presidente como tantos otros, sino un salvador providencial. Nunca hemos estado peor. Para empezar, por primera vez, no hay gobierno. Vivimos una especie de sálvese quien pueda nacional, donde cada quien tiene que valerse de sí mismo, aun para los actos más simples de la vida. A partir de 1948, cuando el régimen conservador asesinó a 450.000 colombianos con su política de tierra arrasada, los sobrevivientes creíamos, al menos, en el poder medicinal de las elecciones. Entonces había una oligarquía intacta, inteligente y recursiva, que se daba inclusive el lujo de una cierta sensibilidad social. Hoy esa oligarquía se está desbaratando a pedazos en su silla de ruedas, y amenaza con caernos encima con todo el peso de sus culpas. Nadie cree, no sin cierta razón, que las elecciones sirvan para algo, y en los núcleos más atrasados de las fuerzas armadas, como en el corazón de muchos civiles, germina la ilusión de curarnos con un purgante militar. De modo que la hora no es para retozos democráticos, y la forma emocional en que los partidarios y los enemigos de López Michelsen están considerando su candidatura no concuerda con la seriedad del momento.

Lo menos serio de todo, desde luego, fue la manera como se decidió. Claro que, dentro de las normas que se impuso a sí mismo el partido liberal, la decisión fue legítima. El argumento de que la convención que lo decidió estaba manipulada no tiene ningún valor, porque no hay en este mundo ninguna convención que no sea manipulada: para eso se inventaron. Lo que me parece un escándalo en un país tan menesteroso de seriedad es que la designación se haya hecho por pura politiquería de vereda, y en último caso, por intereses de amistad y simpatía personal, y que haya favorecido a un hombre imprevisible que hasta 48 horas antes y durante muchos meses no se había cansado de reiterar que no quería ser candidato. Menos serio aún es que todo esto haya sucedido sin que López Michelsen dijera qué es lo que nos propone y se propone, y cuáles son las armas secretas de que dispone para conjurar los ventarrones siniestros del apocalipsis. Sin embargo, lo que tenemos enfrente ahora es la realidad irreversible de un hecho cumplido, con el cual tuvieron que ver muy pocos, pero que nos concierne a todos, y creo que es por ahí por donde tenemos que empezar.

Desde hace muchos años mantengo una muy buena amistad personal con Alfonso López Michelsen. Le conocí en la Universidad Nacional de Bogotá, en 1947, cuando él era el mejor profesor de Derecho Constitucional, a los treinta y cuatro años, y yo era su peor alumno, a los diecinueve. Llegaba siempre con una puntualidad irritante, con unas magníficas chaquetas de casimir, cortadas en Londres por el sastre de su padre, que según se decía era el mejor del mundo, y dictaba la clase sin mirar a nadie, con ese aire celestial de los miopes inteligentes, que siempre parecen andar por entre los sueños ajenos. Era una clase muy aburrida, como lo eran para mí todas las clases que no fueran sobre poesía, pero su modo de hablar con una sola cuerda tenía la fascinación de un encantador de serpientes. Ésa es todavía, a los sesenta y ocho años, su virtud más notable. Tiene una cultura literaria que ya quisiéramos muchos escritores, aunque sólo fuera para los domingos; tiene un encanto personal peligroso, tiene una lucidez casi mágica para descubrir de inmediato las segundas intenciones de las gentes, sobre todo, de las que quiere menos, y un talento especial para que sus juicios parezcan certeros, aunque no lo sean. Por todo esto le tengo una admiración personal inmensa, y la he proclamado siempre sin estrecheces de ninguna clase, pero también he conseguido, a lo largo de muchos años, que no interfiera mi juicio político.

No sólo no le vi ni una sola vez mientras fuera presidente de la república, sino que siempre compartí y sigo compartiendo la responsabilidad de la oposición encarnizada —no siempre suficiente y no siempre justa— que le hizo la revista *Alternativa* en sus cuatro años de mal gobierno. Fui, pues, su adversario merecido, y estoy preparado para serlo otra vez con el mismo ardor y ojalá con mejores recursos, si su segunda presidencia resulta como la primera. Pero también estoy preparado para ayudarle hasta donde pueda, siempre desde una oposición democrática y sin complacencias, si resulta ser, al contrario de su padre, el hombre providencial de la segunda oportunidad.

Yo creo, contra toda evidencia, que López Michelsen tiene condiciones de sobra para pagarnos con intereses los tres millones de votos que nos quedó debiendo de su primer mandato, y que no es imposible encontrar una fórmula que nos proteja de una reincidencia. Creo, por pura intuición poética, que tiene la inteligencia, la formación y la autoridad para encontrarle una solución política a la guerra civil larvada que padecemos todos, de ambos lados, y creo que puede promover un proceso de justicia social y recuperación democrática, y recoger del suelo los vidrios rotos de la política internacional. Sólo con eso bastaría para cumplir con un destino histórico. Pero lo que nadie sabe es si López Michelsen lo quiere, porque al parecer no existe nadie en este mundo que sepa a ciencia cierta lo que quiere hacer este hombre insondable, aparte de ser otra vez presidente de la república.

ALLÁ POR AQUELLOS TIEMPOS
DE LA COCA-COLA

Los cubanos han demostrado, entre otras muchas cosas, que se puede vivir sin Coca-Cola a noventa millas de Estados Unidos. Fue el primer producto que se acabó con el bloqueo, y hoy no queda ningún vestigio de su pasado en la memoria de las nuevas generaciones. Como en todos los países capitalistas, pero de un modo especial en la vieja Cuba pervertida por un turismo sin corazón, el refresco más famoso del mundo había terminado por convertirse en un ingrediente esencial de la vida. Su implantación se inició bajo la dictadura feroz del general Gerardo Machado, en aquella segunda década del siglo nacida bajo el signo de la frivolidad, cuando todavía no estaban inventadas las tapas de corona metálica y las botellas de gaseosa se cerraban con una bolita de cristal amarradas a presión con un alambre, como los corchos de champaña. Fue un injerto difícil, tal vez por un inconveniente cultural que nadie había tomado en cuenta: la Coca-Cola no tiene un sabor latino. Sin embargo, una presión publicitaria insidiosa logró abrir poco a poco una grieta de complacencia en los núcleos sociales más influidos por el gusto de Estados Unidos, hasta que el nuevo sabor sajón desplazó en el mercado a la limonada doméstica de limón de verdad y a todos los venerables refrescos de bolita heredados de la España provinciana, y derrotó a los aguerridos chicles Wrigley's como el símbolo de un modo ajeno de vivir.

Se supone que quien toma una botella de Coca-Cola todos los días a una misma hora sucumbe al hechizo de una adicción semejante a la del cigarrillo o el café. Se supone que eso se debe a un ingrediente secreto. Según ciertos entendidos, la Coca-Cola contenía cocaína hasta 1903, y sus orígenes permiten suponer que es cierto. Fue inventada como medicina y no como refresco a finales del siglo pasado por un cierto doctor W. Pamberton, un boticario de Alaba-

Publicado originalmente el 14 de octubre de 1981.

ma, Georgia, que la envasaba en frascos con etiquetas y las vendía ya con su nombre premonitorio para curar dolores menstruales, espasmos de vientre y cólicos de madrugada. El nombre y la época permiten pensar que en realidad se elaboraba con hojas de coca, que es de donde se extrae la cocaína, y que por aquellos tiempos de la belladona y el elixir paregórico era de uso corriente para aliviar los dolores domésticos. El doctor Pamberton vendió la fórmula en 1910 a la empresa de refrescos que había de lanzarla a la conquista mundial, y sólo porque tenía un ingrediente misterioso cobró por ella una cantidad fabulosa para la época: quinientos dólares. No obstante, las autoridades de Perú comprobaron en 1970 que no contenía cocaína, y hubieran podido prohibirla si lo hubieran querido, porque su nombre hacía creer al público que contenía algo que en realidad no tenía. En Francia, donde todo producto debe advertir si contiene un ingrediente de uso delicado, las botellas de Coca-Cola tienen impresa la advertencia de que contienen cafeína. La leyenda dice que sólo dos personas en el mundo conocen la fórmula secreta, y que nunca viajan juntas en un mismo avión.

Durante el Festival de la Juventud de 1957, en Moscú, lo primero que sorprendió a los visitantes occidentales fue que en cuatro días inmensos a través de Ucrania vimos establos solitarios con vacas asomadas por las ventanas, y pueblos ásperos con carretas cargadas de flores y hombres indescifrables que salían en pijama a recibir el tren en las estaciones, pero no vimos en ninguna parte bajo el cielo ardiente del verano ni un solo anuncio de Coca-Cola. Era demasiado notable para nuestras mentes saturadas por la publicidad occidental. Al cabo de varios días de intimidad, una intérprete ansiosa de conocer los encantos del capitalismo se atrevió a preguntarme a qué sabía la Coca-Cola, y yo le contesté con mi verdad: «Sabe a zapatos nuevos». Ya entonces había médicos que la recomendaban como hidratante para los niños con disentería, y otros que la aconsejaban para restaurar los ánimos del corazón, y quienes afirmaban por experiencia propia que tomada con aspirina tenía poderes alucinógenos. Mi dentista, por su parte, asegura sin parpadear que un diente sumergido en un vaso de Coca-Cola se disuelve en 48 horas.

Al triunfo de la revolución en Cuba, el mercado de la Coca-Cola tenía pocas posibilidades de expansión. Sus promotores habían logrado llevarla más allá de sus posibilidades como refresco, al inventar el cubalibre —que es una mezcla de Coca-Cola con ron cubano. Pero aun así, de seis millones de cubanos sólo 900.000 estaban en condiciones de comprarla de un modo regular. Cuando los obreros cubanos se tomaron la embotelladora de La Habana, no pudieron

seguir fabricando la Coca-Cola, porque el ingrediente básico llegaba de Estados Unidos y había muy poco almacenado en la fábrica.
Lo único que quedaba, disperso por todo el país, era un millón de
botellas vacías.

Los más extremistas fueron contrarios a intentar la sustitución de
un producto que era el símbolo de todo cuanto los cubanos querían
olvidar. Pero el Che Guevara, con su asombrosa claridad política, les
replicó que el símbolo del imperialismo no era la bebida en sí misma sino la forma de la botella. Tal vez él no lo supo nunca, pero en
realidad la botella sólo había sido diseñada en 1915, casi veinte años
después de la invención del doctor Pamberton y cuando la Coca-
Cola sólo tenía vida propia dentro de Estados Unidos. Fue a partir
del cambio de la botella cuando se atrevieron a mandarla a caminar
sola por el mundo.

Fue el mismo Che Guevara, como ministro de Industria, quien
decidió que se tratara de fabricar un sustituto como complemento
del cubalibre. Las mentes más cuadradas pensaron en destruir las botellas existentes para exterminar el germen. Sin embargo, un cálculo
más sereno demostró que las fábricas de botellas de Cuba tardarían
varios años en sustituirlas por otras de forma menos perversa, y los
revolucionarios más crudos tuvieron que resignarse a utilizar la botella maldita hasta su extinción natural. Sólo que la usaron en toda clase de refrescos, menos con el que improvisaron para el cubalibre. Los
visitantes del mundo capitalista, hasta hace muy pocos años, padé-
cíamos una cierta confusión mental al bebernos una limonada transparente en una botella de Coca-Cola.

Los propios cubanos fueron los primeros en admitir que la imitación de la Coca-Cola no fue uno de sus éxitos mayores. Una broma callejera muy popular que los propios químicos celebraban era
que cada botella tenía un sabor distinto, lo cual convertía al nuevo
producto en el más original del mundo. Cuando le presentaron la
primera muestra al Che Guevara, éste la probó, la saboreó con una
seriedad de buen catador, y dijo sin ninguna duda: «Sabe a mierda».
Más tarde dijo en la televisión que sabía a cucaracha. Pero aun así se
abrió paso.

El nuevo producto, que se llama refresco de cola, sin más arande-
las, acabó por encontrar un color que se parece mucho al original,
con un sabor que no es ni de mierda ni de cucaracha, y que desde
luego carece de su regusto sajón. Es un poco más dulce, menos ga-
seoso y con un raro fondo de chocolate, y es bueno para la sed y el
calor, y mezclado con el ron cubano legítimo disimula mucho más
su catadura de advenedizo. Por otra parte, el mal uso deliberado aca-

bó con las botellas antiguas mucho antes del tiempo previsto, y el símbolo se disolvió en la memoria social y no alcanzó a las nuevas generaciones. Quince años después de iniciado el bloqueo, un escritor cubano de paso en París encontró por casualidad una botella de Coca-Cola extraviada de Marruecos, con el célebre logotipo en caracteres árabes. El escritor compró la botella por curiosidad para llevársela a La Habana, y al llegar se la mostró alborozado a su hija de quince años. La niña miró perpleja la botella sin comprender los aspavientos de su padre. «Mírala bien», le dijo él, «es una botella de Coca-Cola con letras árabes.» La niña, todavía más perpleja, preguntó: ¿Y qué es Coca-Cola?

BOGOTÁ 1947

En aquella época todo el mundo era joven. Pero había algo peor: a pesar de nuestra juventud inverosímil, siempre encontrábamos a otros que eran más jóvenes que nosotros, y eso nos causaba una sensación de peligro y una urgencia de terminar las cosas que no nos dejaban disfrutar con calma de nuestra bien ganada juventud. Las generaciones se empujaban unas a otras, sobre todo entre los poetas y los criminales, y apenas si uno había acabado de hacer algo cuando ya se perfilaba alguien que amenazaba con hacerlo mejor. A veces me encuentro por casualidad con alguna fotografía de aquellos tiempos y no puedo reprimir un estremecimiento de lástima, porque no me parece que en realidad los retratados fuéramos nosotros, sino que fuéramos los hijos de nosotros mismos.

Bogotá era entonces una ciudad remota y lúgubre, donde estaba cayendo una llovizna inclemente desde principios del siglo XVI. Yo padecí esa amargura por primera vez en una funesta tarde de enero, la más triste de mi vida, en que llegué de la costa con trece años mal cumplidos, con un traje de manta negra que me habían recortado de mi padre, y con chaleco y sombrero, y un baúl de metal que tenía algo del esplendor del santo sepulcro. Mi buena estrella, que pocas veces me ha fallado, me hizo el inmenso favor de que no exista ninguna foto de aquella tarde.

Lo primero que me llamó la atención de esa capital sombría fue que había demasiados hombres deprisa en la calle, que todos estaban vestidos como yo, con trajes negros y sombreros, y que, en cambio, no se veía ninguna mujer. Me llamaron la atención los enormes percherones que tiraban de los carros de cerveza bajo la lluvia, las chispas de pirotecnia de los tranvías al doblar las esquinas bajo la lluvia, y los estorbos del tránsito para dar paso a los entierros interminables bajo la lluvia. Eran los entierros más lúgubres del mundo, con carrozas de altar mayor y caballos engringolados de terciopelo y mo-

Publicado originalmente el 21 de octubre de 1981.

rriones de plumones negros, y cadáveres de buenas familias que se sentían los inventores de la muerte. Bajo la llovizna tenue de la plaza de las Nieves, a la salida de un funeral, vi por primera vez una mujer en las calles de Bogotá, y era esbelta y sigilosa, y con tanta prestancia como una reina de luto, pero quedé para siempre con la mitad de la ilusión, porque llevaba la cara cubierta con un velo infranqueable.

La imagen de esa mujer, que todavía me inquieta, es una de mis escasas nostalgias de aquella ciudad de pecado, en la que casi todo era posible, menos hacer el amor. Por eso he dicho alguna vez que el único heroísmo de mi vida, y el de mis compañeros de generación, es haber sido jóvenes en la Bogotá de aquel tiempo. Mi diversión más salaz era meterme los domingos en los tranvías de vidrios azules, que por cinco centavos giraban sin cesar desde la plaza de Bolívar hasta la avenida de Chile, y pasar en ellos esas tardes de desolación que parecían arrastrar una cola interminable de otros muchos domingos vacíos. Lo único que hacía durante el viaje de círculos viciosos era leer libros de versos y versos y versos, a razón quizá de una cuadra de versos por cada cuadra de la ciudad, hasta que se encendían las primeras luces en la lluvia eterna, y entonces recorría los cafés taciturnos de la ciudad vieja en busca de alguien que me hiciera la caridad de conversar conmigo sobre los versos y versos y versos que acababa de leer. A veces encontraba a alguien, que era siempre un hombre, y nos quedábamos hasta pasada la medianoche tomando café y fumando las colillas de los cigarrillos que nosotros mismos habíamos consumido, y hablando de versos y versos y versos, mientras en el resto del mundo la humanidad entera hacía el amor.

Una noche en que regresaba de mis solitarios festivales poéticos en los tranvías, me ocurrió por primera vez algo que merecía contarse. Ocurrió que en una de las estaciones del norte había subido un fauno en el tranvía. He dicho bien: un fauno. Según el diccionario de la Real Academia Española, un fauno es «un semidiós de los campos y las selvas». Cada vez que releo esa definición desdichada lamento que su autor no hubiera estado allí aquella noche en que un fauno de carne y hueso subió en el tranvía. Iba vestido a la moda de la época, como un señor canciller que regresara de un funeral, pero lo delataban sus cuernos de becerro y sus barbas de chivo, y las pezuñas muy bien cuidadas por debajo del pantalón de fantasía. El aire se impregnó de su fragancia personal, pero nadie pareció advertir que era agua de lavanda, tal vez porque el mismo diccionario había repudiado la palabra lavanda como un galicismo para querer decir agua de espliego.

Los únicos amigos a quienes yo les contaba estas cosas eran Álvaro Mutis, porque le parecían fascinantes aunque no las creyera, y Gonzalo Mallarino, porque sabía que eran verdad aunque no fueran ciertas. En una ocasión, los tres habíamos visto en el atrio de la iglesia de San Francisco a una mujer que vendía unas tortugas de juguete cuyas cabezas se movían con una naturalidad asombrosa. Gonzalo Mallarino le preguntó a la vendedora si esas tortugas eran de plástico o si estaban vivas, y ella le contestó:

—Son de plástico, pero están vivas.

Sin embargo, la noche en que vi al fauno en el tranvía ninguno de los dos estaba en su teléfono, y yo me sofocaba con las ansias de contárselo a alguien. De modo que escribí un cuento —el cuento del fauno en el tranvía—, y lo mandé por correo al suplemento dominical de *El Tiempo*, cuyo director, don Jaime Posada, no lo publicó nunca. La única copia que conservaba se incendió en la pensión donde yo vivía el 9 de abril de 1948, día del bogotazo, y de ese modo la historia patria hizo un favor por partida doble: a mí y a la literatura.

No he podido eludir estos recuerdos personales leyendo el libro encantador que Gonzalo Mallarino acaba de publicar en Bogotá: *Historias de caleños y bogoteños*. Gonzalo y yo estábamos al mismo tiempo en la facultad de Derecho de la Universidad Nacional, pero no éramos tan asiduos en las clases como en el cafetín universitario, donde sorteábamos el sopor de los códigos intercambiando versos y versos y versos de la vasta poesía universal que ambos podíamos decir de memoria. Al final de las clases, él se iba a su casa familiar, que era grande y apacible entre los eucaliptus. Yo me iba a mi pensión lúgubre de la calle de Florián, con mis amigos costeños, con mis libros prestados y mis tumultuosos bailes de los sábados. En realidad, nunca se me ocurrió preguntarme qué hacía Gonzalo Mallarino en las muchas horas en que no estábamos en la universidad, dónde carajo estaba mientras yo daba la vuelta completa a la ciudad leyendo versos y versos y versos en los tranvías. He necesitado más de treinta años para saberlo, leyendo este libro ejemplar, donde él revela con tanta sencillez y tanta humanidad esa otra mitad de su vida de aquellos tiempos.

Al principio fue el pánico: alguien soltó el rumor de que la delegación de Estados Unidos iba a pedir un minuto de silencio en memoria del presidente Sadat. La delegación de Argelia hizo saber de inmediato, mediante otro rumor en sentido contrario, que abandonaría la conferencia con un portazo que se iba a sentir en el mundo entero. La noticia saturó poco antes de la cena inaugural el vestíbulo hermético del hotel Sheraton, de Cancún, donde esa noche había más jefes de Estado por metro cúbico que en cualquier otro lugar del planeta, y se interpretó como un mal presagio. Menos mal que al canciller de México, Jorge Castañeda, que es un hombre sereno y con sentido común, se le ocurrió la solución más simple, que era la que a nadie se le había ocurrido: preguntarle al secretario de Estado de Estados Unidos, Alexander Haig, si la versión era cierta. Haig, para alivio de todos, le contestó que no. Fue así como al día siguiente empezó sin ningún tropiezo esta reunión internacional sobre cooperación y desarrollo, en la que once países ricos y once países pobres se sentaron a conversar en torno a una mesa redonda en cuyo centro, a falta de una idea menos lúgubre, se sembró un jardín de crisantemos.

Fue de veras una reunión secreta. No sólo por las medidas de seguridad, que convirtieron el confuso hotel Sheraton en un recinto sagrado, sino porque se cumplió sin desmayos el acuerdo de que nadie distinto de los escogidos pudiera entrar en la sala de la disputa, para que todos dijeran lo que les diera la gana. Más aún: en el ámbito del inmenso hotel de 323 habitaciones sólo podían penetrar algunos miembros selectos de las distintas comitivas con una credencial inequívoca. No hubo una sola excepción, en un país de América Latina donde casi siempre las reglas sólo sirven para confirmar las excepciones. Los propios automóviles oficiales que regresaban al hotel eran sometidos a una requisa meticulosa. Desde el

Publicado originalmente el 28 de octubre de 1981.

otro lado del jardín, en uno de los mares más bellos del mundo, tres barcos de guerra vigilaban el horizonte, y tres helicópteros militares sobrevolaban la ciudad y venían a posarse en estas playas de harina de trigo que parecen anteriores al descubrimiento de América.

El secreto de las sesiones fue propuesto por México con el propósito de que se estableciera un diálogo real y no una competencia de discursos para lucirse ante la prensa. No se permitió una sola grabadora en el interior del recinto, ni nadie tomó notas oficiales para la historia. Desde el punto de vista notarial, en síntesis, esta conversación a veintidós voces no existió nunca.

Los periodistas, por supuesto, aunaron la inventiva profesional para burlar el cerco. Las conferencias de prensa de las distintas delegaciones, después de cada sesión, no sólo resultaban sospechosas de parcialidad, sino que tenían el sabor de las comidas enlatadas. De modo que todo el que pudo hizo lo suyo por atrapar un pájaro vivo. El primer día, los servicios de seguridad encontraron un micrófono oculto entre los crisantemos. Un micrófono inalámbrico, de una sofisticación exquisita, cuyo punto de destino no fue posible establecer. Una tarde, dos asesores presidenciales acreditados para asistir a la reunión fueron sorprendidos con grabadoras ocultas. Un muy alto funcionario mexicano se introdujo en una cabina de traducción simultánea, y fue invitado a salir de inmediato por los servicios de seguridad. De todos modos, entre los jefes de Estado con sus asesores, los intérpretes simultáneos y los fotógrafos de prensa que alcanzaban a oír más que unas frases sueltas durante los breves minutos en que les permitían entrar, quedaron numerosas piezas sueltas de un inmenso rompecabezas que los 1.200 periodistas acreditados trataban de armar por completo al término de la reunión.

El presidente norteamericano Ronald Reagan, desde luego, fue el ejemplar más curioso de este raro cardumen presidencial. Muchos que no lo entendieron antes se dieron cuenta ahora de por qué puso tanto empeño en que Cuba no asistiera a esta reunión. En realidad, el presidente Reagan era consciente de que asistía a su primera representación pública internacional, y de que una confrontación con Fidel Castro no era lo que más le convenía a su imagen de vaquero de cine de casi ochenta años. Su deterioro físico, visto de cerca y bajo las luces despiadadas de los reflectores, es mucho más notable de lo que se puede imaginar. Los camarógrafos de cine y televisión no se equivocan en las cosas de su oficio, y muchos de ellos coinciden en que el maquillaje del presidente Reagan, inclusive con toques de colorete, es completo y laborioso desde el desayuno hasta la cena, y no es de una técnica de estos tiempos, sino del cine mudo. Sus intervenciones fueron pocas, repetitivas y más bien elementales. Pero varios testigos presenciales estuvieron de acuerdo en que su posición fue moderada, e hizo todo lo posible por evitar cualquier tropiezo con alguno de sus interlocutores desfavorecidos. Inclusive, no se sabe si por cálculo o descuido, no aceptó el whisky que el presidente López Portillo le ofreció en el aeropuerto, a su llegada, sino que pidió, en presencia de medio mundo, su trago favorito: vodka ruso con agua tónica.

La reunión corría el riesgo de fracasar en el pantano florido de los convencionalismos y las buenas maneras. La primera mañana, en una prueba sin precedentes, se despacharon los veintidós discursos de cada uno de los delegados, gracias a la severidad con que el maestro de ceremonias, Pierre Trudeau, primer ministro de Canadá, hizo respetar los minutos asignados. Pero aun en la reunión de la tarde las cosas parecían seguir sin cambios notables, hasta las 16.20 horas, cuando el presidente Reagan se arriesgó en la defensa febril de la propiedad privada en la producción agrícola. Dijo que si todos los países del mundo cultivaran la tierra en la forma en que lo hace Estados Unidos, y con iguales recursos, no sólo quedaría resuelto el drama mundial del hambre, sino que se reducirían en gran medida las tierras necesarias para la agricultura. Lo que quiere decir, en términos más simples, que si todos los países fueran tan desarrollados como Estados Unidos no habría problemas de alimentación en el planeta. Esto sirvió de base para que el presidente de Tanzania, Julius Nyerere, convirtiera por fin en un diálogo verdadero lo que hasta entonces había sido un sartal de monólogos. Con una voz cortante y un inglés pedregoso, el carismático Nyerere puso las cosas en su puesto al recordarle a Reagan que el rendimiento agrícola no es sólo un problema técnico, sino también cultural. De paso, planteó un interesante problema de lenguaje. «Tanzania», dijo, «no es un país en vías de desarrollo, como se dice ahora por una traducción fácil del inglés (*developping country*), sino un país subdesarrollado que ni siquiera ha empezado a encontrar las vías para dejar de serlo.» En realidad, aunque Nyerere no lo dijo, la pobreza de su país es una de las más penosas y originales: su único producto de exportación son las semillas de marañón, y apenas si le alcanzan para comprar la mitad del petróleo que consume.

Con todo, la opinión casi unánime, al término de esta conversación múltiple, es que los resultados fueron mejores de lo que todos esperaban. El documento final, que no estaba previsto, es una carta de identificación de los problemas pendientes entre los ricos del

Norte y los pobres del Sur, y una orientación que puede ser útil para intentar una solución global. En síntesis, se piensa que valió la pena una iniciativa en la que el presidente López Portillo empeñó a fondo su inteligencia y su bien ganado prestigio internacional, que mantuvo en vilo durante 48 horas la atención de medio mundo, y que le costó a México la bicoca de quince millones de dólares.

GEORGES BRASSENS

Hace algunos años, en el curso de una discusión literaria, alguien preguntó cuál era el mejor poeta actual de Francia, y yo contesté sin vacilación: Georges Brassens. No todos los que estaban allí habían oído antes ese nombre –unos por demasiado viejos y otros por demasiado jóvenes–, y algunos que lo menospreciaban porque era autor de discos y no de libros dieron por hecho que yo lo decía por desconcertar. Sólo mis compañeros de generación, los que gozaron y padecieron a París en los años ingratos de la guerra de Argelia, sabían no sólo que yo hablaba en serio, sino que además tenía razón.

Para ellos, más que para el resto del mundo, Georges Brassens ha muerto la semana pasada a los sesenta años, frente al voluble mar de Sète que tanto amaba, y donde tenía su casa llena de flores y de gatos que se paseaban sin romperse entre la vida real y sus canciones. Sólo que no murió en ella. Su discreción legendaria era tan cierta, que se fue a morir en la casa de un amigo para que nadie lo supiera. Y la mala noticia no se conoció hasta 72 horas después por una llamada anónima, cuando ya un reducido grupo de parientes y amigos íntimos lo habían enterrado en el cementerio local. No podía ser de otro modo: para un hombre como él, la muerte era el acto personal más secreto de la vida privada.

Así fue siempre. Había nacido en 1921, en la casa de pobres de un albañil que deseaba para su hijo el mismo oficio. Como todos los niños con vocación vital, el pequeño Georges detestaba la escuela por lo que ésta tenía de cuartel. Una maestra desesperada acabó de rematarlo: lo encerró con llave en un ropero durante varias horas, y cuando por fin lo liberaron habían germinado en su corazón, para siempre, las semillas de la anarquía. Su odio a la autoridad y a toda norma establecida fueron el sustento de sus canciones más hermosas. Para él no había más luz en aquellas tinieblas que la independencia personal y el amor. Una vez cantó: «Morir por las ideas, de acuerdo;

Publicado originalmente el 11 de noviembre de 1981.

pero de muerte lenta». El Partido Comunista francés puso el grito en el cielo en nombre de tantos compatriotas muertos de muerte rápida durante la resistencia.

En realidad, Georges Brassens carecía por completo de instinto gregario. Llevaba una vida tan reservada, que todo lo que tenía que ver con él andaba confundido con la leyenda, y uno se preguntaba a veces si de veras existía. Aun en su época de mayor esplendor, hacia la mitad de los años cincuenta, era un hombre invisible. Nadie sabe cómo lo convenció René Clair de que actuara en una película, y él lo hizo muy mal, abrumado por la vergüenza de ser el centro de la atención; pero en cambio cantó una ristra de canciones originales que se quedaban resonando en el corazón. El tiempo –decía en una de ellas– era un bárbaro de la misma calaña de Atila, y por donde su caballo pasaba no volvía a crecer jamás el amor.

Lo vi en persona una sola vez cuando su primera presentación en el Olympia, y ése es uno de mis recuerdos irremediables. Apareció por entre las bambalinas como si no fuera la estrella de la noche, sino un tramoyista extraviado, con sus enormes bigotes de turco, su pelo alborotado y unos zapatos deplorables, como los que usaba su padre para pegar ladrillos. Era un oso tierno, con los ojos más tristes que he visto nunca y un instinto poético que no se detenía ante nada. «Lo único que no me gustan son sus malas palabras», decía su madre. En realidad, era capaz de decir todo y mucho más de lo que era permisible, pero lo decía con una fuerza lírica que arrastraba cualquier cosa hasta la otra orilla del bien y del mal. Aquella noche inolvidable en el Olympia cantó como nunca, agonizando por su miedo congénito al espectáculo público, y era imposible saber si llorábamos por la belleza de sus canciones o por la compasión que nos suscitaba la soledad de aquel hombre hecho para otros mundos y otro tiempo. Era como estar oyendo a François Villon en persona, o a un Rabelais desamparado y feroz. Nunca más tuve oportunidad de verlo, y aun sus amigos más cercanos lo perdían de vista. Poco antes de morir, alguien le preguntó qué estaba haciendo durante las jornadas de mayo de 1968, y él contestó: «Tenía un cólico nefrítico». La respuesta se interpretó como una irreverencia más de las tantas que soltó en la vida. Pero ahora se sabe que era cierto. Sin que casi nadie lo supiera, había empezado a morirse en silencio desde hace más de veinte años.

En 1955, cuando era imposible vivir sin las canciones de Brassens, París era distinto. Los parques públicos se llenaban por las tardes de ancianos solitarios, los más viejos del mundo; pero las parejas de enamorados eran dueñas de la ciudad. Se besaban en todas partes con besos interminables, en los cafés y en los trenes subterráneos, en el cine y en plena calle, y hasta paraban el tránsito para seguirse besando, como si tuvieran conciencia de que la vida no les iba a alcanzar para tanto amor. El existencialismo había quedado atrás; sepultado en las cuevas para turistas de Saint Germain-des-Près, y lo único que quedaba de él era lo mejor que tenía: las ansias irreprimibles de vivir. Una noche, a la salida de un cine, una patrulla de policías me atropelló en la calle, me escupieron la cara y me metieron a golpes dentro de una camioneta blindada. Estaba llena de argelinos taciturnos, recogidos a golpes y también escupidos en los cafetines del barrio. También ellos, como los agentes que nos habían arrestado, creyeron que yo era argelino. De modo que pasamos la noche juntos, embutidos como sardinas en una jaula de la comisaría más cercana, mientras los policías, en mangas de camisa, hablaban de sus hijos y comían barras de pan ensopadas en vino. Los argelinos y yo, para amargarles la fiesta, estuvimos toda la noche en vela, cantando las canciones de Brassens contra los desmanes y la imbecilidad de la fuerza pública.

Ya para entonces, Georges Brassens había hecho su testamento cantado, que es uno de sus poemas más hermosos. Lo aprendí de memoria sin saber lo que significaban las palabras, y a medida que pasaba el tiempo y aprendía el francés iba descifrando poco a poco su sentido y su belleza, con el mismo asombro con que hubiera ido descubriendo, una tras otra, las estrellas del universo. Ahora, transcurridos veinticinco años, ya nadie se besa en las calles de París, y uno se pregunta asustado qué fue de tantos que se amaban tanto y que ahora no se ven en el mundo. Georges Brassens ha muerto, y alguien tendrá que poner en la puerta de su casa, como él lo pedía en su testamento, un letrero simple: «Cerrado por causa de entierro».

UN DICCIONARIO DE LA VIDA REAL

Hace cuatro años fue llevado a París el cuerpo momificado del faraón egipcio Ramsés II para ser sometido a un examen médico que determinara la naturaleza y el remedio de una floración parasitaria que amenazaba con destruirlo. Puesto que era el cadáver del monarca de un país con el que Francia tiene buenas relaciones, el presidente de entonces, Valéry Giscard d'Estaing, lo recibió en el aeropuerto con honores militares. Pero no fue ése el problema más difícil que planteó el examen del cuerpo, sino otro menos convencional y tal vez sin solución: las vísceras estaban rellenas con una especie de aserrín de diversas materias vegetales, y entre ellas, picadura de hojas de tabaco.

Aquel descubrimiento parecía un disparate histórico. En efecto, Ramsés II murió en 1235 antes de Cristo. Es decir, hace 3.000 años, y es una verdad aceptada por todo el mundo que el tabaco fue descubierto por Cristóbal Colón y llevado por él a Europa después del descubrimiento de América. El hecho de que un faraón milenario lo tuviera en las vísceras, sin embargo, ha puesto a pensar en la posibilidad de que los egipcios conocieran el tabaco, pero no para fumarlo, sino para usos medicinales, y muy en concreto para embalsamar a esos faraones que creían seguir vivos mientras se conservara su cuerpo.

Esta información sorprendente, que no recuerdo haber leído en la prensa, la he encontrado en un diccionario a la vez curioso y divertido que compré hace poco por casualidad. Se llama *¿Desde cuándo?*, y es el catálogo del origen de ochocientos objetos y costumbres de la vida cotidiana, escrito por el francés Pierre Germa. Alguna vez oí decir que Aldous Huxley había leído hoja por hoja los casi treinta volúmenes de la enciclopedia británica, y durante años soñé con repetir esa proeza agotadora y fructífera. Ahora he tenido un premio de consolación: en una noche he leído este diccionario de la vida

Publicado originalmente el 18 de noviembre de 1981.

diaria con la misma tensión y el mismo placer con que se lee una novela de misterio.

En la escuela primaria me llamaba la atención que los maestros atribuían a los chinos la invención de las cosas más fantásticas, además de la pólvora y la brújula. He vuelto a recordarlo porque los sabios que estudiaron la momia de Ramsés II supieron que tal vez el tabaco había llegado a Egipto desde China, y que fue de allí de donde pasó a nuestras Américas. En cambio, el diccionario de orígenes dice que los cristales para corregir los defectos de la visión fueron enunciados en el año 990 por el físico árabe Ibn al Haytam, pero que no fueron tallados para anteojos hasta 1285 por los vidrieros italianos. Sin embargo —y tal vez por una deformación inculcada por mis maestros de la escuela primaria— yo estaba convencido de que también los anteojos habían sido inventados en China. No tengo a la mano *El libro de las maravillas del mundo*, de Marco Polo, pero me parece que era él quien lo decía, y su viaje de veinte años por el Oriente remoto terminó en 1292.

Los datos más interesantes se refieren al progreso de la ciencia, y sobre todo de la medicina. Es bueno saber que Juno, la esposa de Júpiter, en su Olimpo fue la primera protagonista de un parto sin dolor, gracias a las virtudes narcóticas de la lechuga. También es bueno recordar una vez más que la operación de cesárea no se llama así por Cayo Julio César, como tantas veces se ha dicho sin fundamento. En realidad, se practicaba desde tiempos inmemoriales en mujeres que morían cuando estaban a punto de dar a luz, y de ese modo se salvaba la vida del hijo. La primera cesárea en una mujer viva la hizo en el año 1500 un castrador de cerdos de Shiegerhasen, en Thurgovia, suizo, después de que los médicos y parteras del lugar declararon que el parto de su esposa era imposible. El hombre, que se llamaba Jacques Nufer, le abrió el vientre con su cuchillo de castrador, la remendó con hilos de coser, sin ninguna clase de anestesia y tanto ella como el hijo vivieron muchos años.

En 1667 —cuenta este diccionario alegre— el colegio de medicina de Londres le pagó veinte chelines a un loco para que se dejara hacer una transfusión de sangre de cordero. No era la primera vez que se intentaba, pero las transfusiones habían sido prohibidas pocos años antes en Inglaterra, porque eran muy pocos quienes sobrevivían. Sin embargo, el loco no sólo asimiló muy bien la sangre del cordero, sino que un testigo de la época declaró que la transfusión le había transformado en un hombre diferente.

Uno de los artículos más notables es el de los métodos anticonceptivos. Se habla allí de una receta encontrada en un papiro egipcio, que es un emplasto a base de caca de cocodrilo y goma arábiga, y cuya eficacia era absoluta si se le colocaba bien en el fondo de la vagina. Este método me recordó al más primitivo que encontré cuando tuve que ponerlo al servicio de un personaje de novela. Eran unas cataplasmas de mostaza cuyos vapores debían ser recibidos en la vagina poco antes de hacer el amor, y que al parecer se usaban más de lo que uno se cree en América Latina por los tiempos de las guerras civiles del coronel Aureliano Buendía, cuatro siglos después de que el anatomista italiano Fallopio perfeccionó el preservativo con tripas de cordero. También leyendo esto recordé un cuento que circuló en Cuba por la década de los sesenta, y cuya veracidad no he logrado comprobar en mis frecuentes viajes a ese país. Se dice que Cuba le compró a China varios millones de preservativos, pero que éstos eran tan pequeños que los cubanos se los ponían muertos de risa en el dedo meñique. Al parecer, muy pronto fueron retirados del comercio, y por último los pintaron de colores y los usaron inflados como globos para las fiestas de carnaval.

En fin, el diccionario de orígenes nos cuenta con precisión y gracia quién inventó la máquina de lavar, dónde se construyó el primer faro, en qué mar navegó el primer petrolero, desde cuándo se usa el aceite de ricino, quién fue el primer hombre que se lanzó en paracaídas, y tantas cosas más que apenas caben en su orden alfabético. A los escritores les gustará saber, por ejemplo, que una de las máquinas de escribir construidas en el siglo pasado se llamaba «el piano de escribir» y que su cliente más entusiasta fue el escritor Mark Twain. Se preguntarán sin duda –porque el diccionario no lo dice– qué se hizo de la máquina de escribir en chino, que según se dijo hace muchos años había sido inventada por el escritor americanizado Lin Yutang. Les gustará saber que el corsé de varillas de acero fue muy popular en el siglo XIX, a pesar de que era tan incómodo y peligroso que en algunos casos podía causar la muerte. Pero hay que decir –señala el diccionario– que las mujeres de Estados Unidos no dejaron de usarlo por ese riesgo, sino como respuesta a un llamado que les hizo el gobierno en 1917 para que contribuyeran con sus varillas metálicas al esfuerzo patriótico de la primera guerra mundial. De ese modo se recuperaron 28.000 toneladas de acero, que alcanzaron para construir dos acorazados de la época.

NICARAGUA ENTRE DOS SOPAS

Estados Unidos pensaba que los comunistas iban a tomarse el poder en Italia mediante las elecciones generales de abril de 1948. La Agencia Central de Inteligencia (CIA), que acababa de crearse, contribuyó a impedirlo con todo un sistema de maquinaciones truculentas que fueron reveladas hace poco por el escritor norteamericano Thomas Powers en un libro muy bien documentado: *The man who kept secrets*. Leído ahora, es asombroso cómo ese episodio de la Italia de aquellos tiempos se parece a otros que ocurrieron en Chile cuando Salvador Allende era el presidente constitucional, y a los que ocurren casi a diario por estos días en América Central y el Caribe.

Powers cuenta, en efecto, que el presidente Harry S. Truman aspiraba a una intervención directa de Estados Unidos en Italia en caso de una victoria comunista. George Kennan, que era el director de previsiones políticas del Departamento de Estado, dirigió a los diplomáticos norteamericanos en Europa la circular siguiente: «Italia es el país clave del continente. Si los comunistas ganaran las elecciones, nuestra posición en el Mediterráneo, y tal vez en toda Europa, se debilitaría de un modo considerable». No se le puede reprochar a este mensaje la moderación de sus términos. Pero basta con sustituir el Mediterráneo por el mar Caribe para sorprenderse con las analogías. Es como si Estados Unidos, en 35 años y después de haber llegado a la Luna, no hubiera cambiado ni un ápice su sistema de análisis ni sus métodos de intervención. Dice Powers, en efecto, que la CIA hizo circular cartas falsas y documentos supuestos del partido comunista para deteriorar su imagen pública. «Hizo publicaciones anónimas», agrega, «que evocaban de un modo impresionante los excesos cometidos por el Ejército Rojo en Alemania, donde los soldados rusos saqueaban y violaban sin pudor, y pronosticaba para Italia una suerte igual a la de Polonia y Checoslovaquia, que apenas unas semanas antes habían pasado al dominio comunista.»

Publicado originalmente el 25 de noviembre de 1981.

Maniobras como éstas han sido puestas en práctica por los servicios secretos de Estados Unidos, e inclusive de un modo abierto por sus autoridades más altas, para justificar una intervención directa contra Cuba y Nicaragua y para facilitar un golpe de mano en El Salvador. Con el mismo desparpajo con que lo hubieran denunciado si fuera verdad, Estados Unidos dijo hace unos dos meses que había asesores cubanos entre las guerrillas salvadoreñas. Poco después, el propio secretario de Estado de Estados Unidos, el general Alexander Haig, hizo saber a varios gobiernos amigos que Cuba había mandado entre quinientos y seiscientos soldados de su ejército para reforzar a las fuerzas armadas de Nicaragua. El gobierno de Cuba desmintió en ambos casos las afirmaciones de Estados Unidos y pidió que se publicaran las pruebas. Nunca, por supuesto, serán publicadas, pero el efecto ya fue conseguido: en el ánimo de muchos quedará la duda para siempre.

Cuba y Nicaragua, que interpretaron estos y muchos otros infundios similares como indicios ciertos de que algo grande preparaba Estados Unidos contra ellas, tomaron precauciones elementales. Más aún: Cuba empezó a tomarlas, con una previsión de gato escaldado, desde la llegada de Ronald Reagan al poder. Entrenó a medio millón de civiles para la defensa del país ante una invasión eventual, con medio millón de armas compradas de urgencia en el único país que se las vende: la Unión Soviética. Esto significó, por supuesto, un enorme sacrificio para la economía cubana y un nuevo golpe para su producción. Es decir, que con sólo amenazar a Cuba Estados Unidos consigue ocasionarle perjuicios irreparables.

Nicaragua, por su parte, está viviendo una situación de novela fantástica: todos sus esfuerzos, desde la victoria contra Somoza, se han orientado en el sentido de establecer una democracia pluralista, y Estados Unidos no ha hecho nada más que impedírselo. Amenazada desde el primer día por 3.000 guardias somocistas concentrados en la frontera con Honduras, el gobierno nicaragüense salió a buscar armas para defenderse, y donde primero las solicitó fue en Estados Unidos. Se las negaron. De modo que fue a buscarlas donde se las quisieron dar, y encontraron un poco en distintos lados, inclusive en la Unión Soviética y los países socialistas. Esto fue tomado como pretexto por Estados Unidos para castigar a Nicaragua por su flanco más débil, que es la economía: le cancelaron un crédito de setenta millones de dólares, que en el fondo no resolvía sus penas, y le suspendieron sin previo aviso un despacho de trigo, apenas 48 horas antes de que el país se quedara sin pan. México, cuya solidaridad ha sido constante, auxilió a Nicaragua con un cargamento de trigo de emergencia. Otro país se apresuró a ayudarle, en esta ocasión sin condiciones políticas de ninguna clase: Bulgaria.

Hace poco, un grupo de comunistas nicaragüenses fue encarcelado por incitar a la población a que exigiera la imposición inmediata de un régimen socialista. Casi al mismo tiempo, un grupo de empresarios privados, que cometieron el mismo delito desde el extremo contrario, fueron también arrestados. El hecho tuvo una vasta repercusión internacional, como prueba supuesta de una persecución a la iniciativa privada. En cambio, el encarcelamiento de los comunistas, del cual apenas si se habló en la prensa internacional, fue considerado como una farsa del gobierno de Nicaragua, a la cual se habrían prestado los propios extremistas presos.

No es, pues, extraño que un gobierno acosado de un modo tan injusto desde los dos extremos pierda a veces la paciencia. Es esto lo que parece haberle ocurrido con el periódico *La Prensa*, que se complace en hostilizarlo, y no siempre con argumentos justos y oportunos. En dos ocasiones recientes, el gobierno ha suspendido su publicación por un máximo de dos días, al cabo de los cuales ha reaparecido el diario con ínfulas de mártir y con la circulación aumentada. Es comprensible, pero no justificable, que el gobierno de Nicaragua no tenga bastante serenidad y madurez para tolerar los excesos de un periódico que puede hacerle más daño con su desaparición que con sus provocaciones.

Conozco a los más destacados dirigentes de la Nicaragua de hoy desde mucho antes de que estuvieran en el poder, y sé que sus objetivos no están escritos en ningún esquema anterior, sino en uno propio y original, acorde con las condiciones de un país cuyo carácter no tiene muchas cosas en común con sus vecinos. «No queremos hacer una nueva Cuba», han dicho ellos muchas veces, «sino una nueva Nicaragua.» Sin embargo, soy el primero en reconocer que en dos años han tenido que hacer muchas cosas en sentido contrario del que ellos hubieran querido. Lo han hecho obligados por la tozudez de Estados Unidos, que se empeña en empujarlos en brazos de la Unión Soviética sólo para demostrar que no hay sino dos sopas en este mundo, y que los países desamparados no tienen más opción que escoger una de las dos o morirse de hambre.

LOS DOLORES DEL PODER

El estado de salud de una persona es un asunto de su vida privada, salvo si se da la casualidad de que esa persona sea presidente de la república. François Mitterrand lo sabe, y por eso decidió, sin que ninguna ley se lo exigiera, publicar cada seis meses un informe minucioso sobre su estado de salud. El primero de ellos, que se hizo público el 20 de mayo pasado, apenas unos días después de su posesión, era más que satisfactorio para un hombre de 64 años que no se priva de ninguno de los placeres sanos de la vida, y que fue visto hace poco, muy cerca de la medianoche, comiendo arenques ahumados en un restaurante del Barrio Latino.

En Estados Unidos es donde menos secretos se guardan sobre las enfermedades de los presidentes, y de sus parientes y colaboradores. En 1961, el joven y deportivo John F. Kennedy sufrió una torcedura lumbar en el curso de una entrevista con Nikita Jruschov, en Viena, y regresó a su país caminando con un par de muletas, que le valieron una cierta aureola de veterano de guerra. Hace pocos años, uno de los hijos de su hermano Edward sufrió la amputación de una pierna como consecuencia de un cáncer de los huesos, y el hecho fue celebrado por la prensa como una prueba más del valor familiar. Por la misma época, la esposa del presidente Gerald Ford y la esposa de Nelson Rockefeller, gobernador de Nueva York, fueron mutiladas por el rigor de sus cirujanos, y su desdicha mereció el homenaje de las primeras páginas. No parecen cosas de descendientes de ingleses, para quienes es de muy mal gusto hablar en público de los hijos, de dinero y de enfermedades. Pero es, en cambio, una muy sana costumbre, no sólo para anticiparse a toda clase de especulaciones, sino, también, en algunos casos, una manera de protegerse contra la verdad.

La salud del presidente Ronald Reagan, que ha cumplido ya setenta años, se había prestado a muchas conjeturas, hasta que una bala

Publicado originalmente el 2 de diciembre de 1981.

de nueve milímetros le penetró bajo el alerón izquierdo y se le incrustó muy cerca de la columna vertebral. La apariencia desenvuelta y la sonrisa de propaganda de pasta dentífrica con que salió del hospital hicieron pensar a muchos que Reagan era tan buen vaquero en la vida real como en el cine. Pero las conjeturas no terminaron. Todavía se dice que el presidente de Estados Unidos perdió los ímpetus de su quinta juventud después del atentado, y que su jornada de trabajo se había reducido a no más de tres horas diarias. Sin embargo, quienes lo vimos en Cancún vestido de guayabera y con un maquillaje que parecía más bien un embalsamamiento en vida, no tuvimos la impresión de que desfalleciera en las discusiones intensas ni en sus numerosos compromisos sociales. En todo caso, para disipar los rumores, Reagan se había sometido poco antes a un examen médico a fondo, y había abandonado el hospital con unas ínfulas de gladiador invencible que suscitaron más sospechas que convicción.

Los soviéticos, en cambio, no han modificado en absoluto el hermetismo tradicional del poder ruso. Sus dirigentes mueren a edades bíblicas, sin ningún anuncio previo a la opinión pública. Hace unos tres años, mientras yo estaba en Moscú, murió el gran poeta y héroe de la guerra Constantin Simonov. La noticia era de dominio público 48 horas después de la muerte, y había sido publicada en el mundo entero, menos en la Unión Soviética. Los escritores consternados por aquella pérdida irreparable, que habían velado una noche entera junto al féretro del poeta, no supieron explicarnos por qué una noticia conocida de todos sólo había de divulgarse de un modo oficial casi setenta horas después de ocurrida. Nadie tenía una respuesta segura, de modo que nos conformamos con la más poética: mientras no se divulgara la noticia oficial, era como si en cierto modo Constantin Simonov continuara vivo.

La apoteosis de este hermetismo parece ser la enfermedad del presidente del Consejo de Estado y secretario general del Partido Comunista soviético, Leónidas Breznev. A los 75 años, este enigma humano aparece y desaparece de la escena pública, a veces en un estado de deterioro físico impresionante, y poco después con la potencia y los arrestos de un toro de lidia. Nunca, que se recuerde, ha habido una sola información sobre su estado de salud, que, sin duda, se ha de convertir en uno de los grandes misterios de nuestro tiempo.

Hace unos años, sin embargo, los servicios de inteligencia de Francia lograron abrir una brecha en el enigma, y estuvieron a punto de provocar un incidente diplomático de mucha gravedad. Se cuenta, en efecto, que en el curso de una visita oficial a París, el presidente Breznev olvidó desaguar el inodoro en su residencia de honor. Los servicios de inteligencia recogieron sus insignes materias fecales, las sometieron a un análisis de laboratorio y lograron establecer qué medicamentos estaba tomando el visitante ilustre. Por los medicamentos, desde luego, los científicos franceses dedujeron el carácter de las dolencias. Sin embargo, las consideraciones de orden político prevalecieron por encima de cualquier cosa, y el veredicto de los médicos permanece todavía en las tinieblas de las razones de Estado.

En Francia, el precedente más espectacular de un presidente enfermo es el de Georges Pompidou. Murió en la cama del poder, en medio del rumor mundial de que estaba enfermo de gravedad desde hacía muchos años, mientras sus servicios de prensa lo negaban sin parpadear. Quienes conocen a Mitterrand, sin embargo, saben que no es por este precedente, ni por ningún otro, que decidió por iniciativa propia publicar un boletín semestral de su estado de salud. Para él es un asunto de principios. Es muy propio de su carácter jugar con las cartas sobre la mesa y exigir de los otros la misma conducta. Lo que tal vez no había pensado es que una torcedura dorsal durante una partida de tenis pudiera tener algún interés para la opinión pública. Por otra parte, el lumbago no afecta tanto a la salud como a la dignidad, y el saber disimularlo y soportarlo con una sonrisa no sólo hace parte de la buena educación, sino que es hasta cierto punto uno de los compromisos del poder.

El lumbago del presidente Mitterrand era conocido desde finales del verano pasado por sus amigos más cercanos y sus colaboradores inmediatos. Hace unos dos meses, en México, cuando tuve oportunidad de estar cerca de él en distintas ocasiones, nadie advirtió su sufrimiento a pesar de la actividad febril que se impuso durante la visita oficial. El único momento en que debió interrumpir un programa fue cuando recorría los laberintos arqueológicos del templo mayor. En Cancún, a pesar de las jornadas agotadoras y la intensa vida social, no hizo ningún gesto que permitiera vislumbrar lo que parece ser el síntoma más notable de su dolencia: el mal humor. Lo que no pareció prever el presidente Mitterrand es que su estoicismo y su discreción iban a sustentar una suposición pública de la cual lo menos que puede decirse es que es apresurada, y no exenta además de una cierta dosis de perversión política.

«CÓMO SUFRIMOS LAS FLORES»

Para que vuelva a entrar la buena suerte en una casa desollada por la desgracia no hay nada más eficaz que un ramo luminoso de flores amarillas. Es incluso un conjuro invencible contra las nubes oscuras que suelen perturbar en ciertos días inciertos el oficio misterioso de escribir. Cuando los dedos se nos enredan en la tecla equivocada, cuando no conseguimos que los personajes respiren con su aliento propio en el ámbito de la novela, cuando uno no encuentra la palabra compasiva que los ayude a morir sin dolor, es porque algo falta en el aire del cuarto en que se escribe. Y lo que falta casi siempre es una flor.

De modo que no es por superstición caribe, sino por una experiencia acendrada y fructífera, que nunca me aventuro a escribir sin que haya en el vaso de mi escritorio una rosa amarilla. «Las flores son gente», le oí decir alguna vez a un niño de siete años, que sin duda sabía muy bien lo que quiso decir. En efecto, la conducta humana de las flores y de las plantas se encuentra establecida en la poesía de siempre, tanto en la buena como en la mala. Es difícil acordarse de las hermanas Brontë sin evocar en algunas de sus páginas más bellas una tibia fragancia de mimosa al atardecer. *Mimosa púdica*, porque sus hojas se retraen, incluso sin que nadie las toque, con el simple paso de una sombra. Nadie tendrá derecho a creerse poeta si no comparte con las hermanas Brontë la idea de que el olor de las mimosas es un ánima en pena que vuelve al mundo de los vivos en busca de una persona amada. En alguno de mis libros me apropié de esa idea con todo el derecho de mi grande admiración. «El jazmín es una flor que sale», escribí, sin que nadie hasta ahora me lo haya reprochado. De la novela más hermosa de William Faulkner conservo el recuerdo persistente de una guía de glicinas junto a la ventana en una tarde de verano ardiente. Siempre he vuelto con el mismo estupor a la leyenda del conde Drácula, tal como Bram Stoker la

Publicado originalmente el 9 de diciembre de 1981.

implantó para siempre en la literatura; pero lo que más me sigue impresionando de esa obra maestra pavorosa no es la condición sobrenatural del vampiro, su facultad de desembarcar en el puerto de Londres convertido en perro, sino la zozobra que causaban en su ánimo las flores del acónito.

Uno de los asombros de mi infancia era el ceremonial con que mi abuela sacaba las flores de nuestro cuarto antes de dormir. Alguien me explicó más tarde que ella tenía razón: el anhídrido carbónico que exhalan las flores puede ser perjudicial para quienes duermen con ellas en un cuarto cerrado. Pero la verdad es que mi abuela tenía relaciones con los misterios mucho más reveladoras que las de los científicos, y lo que siempre me dijo fue que las rosas en los dormitorios suscitan sueños indeseables que nos persiguen hasta la muerte.

Hablamos de estas cosas la otra noche en una reunión de amigos, y uno de ellos hizo una disertación sabia y fascinante sobre el alma de las plantas. Todo empezó cuando alguien se refirió a *El día de los trífidos*, de John Wydham, que es una de las novelas más terroríficas que recuerde. Un trífido –al contrario de lo que muchos creímos al leer la novela– no es una planta asesina, capaz de desarrollar sus tentáculos voraces y exterminar en pocas horas el género humano. No; es un modesto adjetivo para uso de botánicos, que califica algo que está hendido o abierto en tres partes. Sólo que el autor de la novela logró infundirle una significación que hoy ha pasado a ser un símbolo de la amenaza tremenda que representa para los mortales el reino vegetal.

Nuestro amigo, y yo creo que con razón, cree todo lo contrario. «Dentro de las casas las plantas llegan a formar parte del núcleo familiar», nos dijo aquella noche. «Gozan y sufren con nosotros, se alarman ante las amenazas verbales y pueden morir de terror ante una agresión real, contra la cual carecen de defensas.» Los animales, sobre todo los perros domésticos, las ratas y ciertos insectos perniciosos, son para ellas un tormento perpetuo. Esto es posible establecerlo sin ninguna duda mediante el uso de un galvanómetro, que es un instrumento para comprobar la existencia, medir la intensidad y determinar el sentido de una corriente eléctrica mediante la desviación que ésta produce en una aguja magnética. Conectado a una planta, el galvanómetro revela sus reacciones y aun sus sentimientos más íntimos.

Alguna vez se hizo un experimento que hoy es célebre en el mundo entero. Un científico destruyó un filodendro en presencia de otras plantas. Participaron también cuatro estudiantes que desconocían los planes del agresor y los propósitos del experimento. Más tarde se comprobó, mediante el galvanómetro, que las plantas se estremecían de horror frente al victimario y no frente a los testigos, y que inclusive reaccionaban de un modo distinto ante el cuchillo con que fue destruido el primer filodendro.

«Las plantas», continuaba el amigo, «reaccionan ante la felicidad y el placer.» Colocada en una habitación donde una pareja humana hace el amor, una planta vivirá los mismos estados de ánimo de los amantes. El galvanómetro, exacerbado, registrará vibraciones febriles que sólo podrían definirse como un orgasmo.

El centro nervioso de las plantas —concluyó el informante— se localiza en los tejidos de las raíces, los cuales se ensanchan y contraen como los músculos del corazón humano. Además, tienen memoria: son capaces de acumular impresiones y retenerlas por largos períodos de tiempo. Uno puede preguntarse, en consecuencia, qué recuerdos históricos podría almacenar una secoya, ese árbol fabuloso que llega a crecer hasta 150 metros y puede vivir hasta 3.000 años del tiempo humano.

Por otra parte, hay plantas a las cuales se les ha inyectado una fuerte dosis de alcohol, y el resultado se ha visto en su comportamiento: una embriaguez triste. Al día siguiente, el galvanómetro ha revelado síntomas semejantes a los que sentimos los seres humanos por los excesos de una parranda. También parece demostrado que los sonidos armónicos influyen en el crecimiento de algunas plantas. Los autores que prefieren en su estado más primario son Johann Sebastian Bach y, en general, los más barrocos. Pero es posible refinarles el gusto, hasta lograr que experimenten un éxtasis real con Bartók o Schoenberg. En cambio, parece que las plantas de hoy detestan el acid rock, y que sus estridencias hacen disminuir el tamaño de sus hojas.

Me pregunto, después de estas revelaciones que supongo bien fundadas, cómo será el sufrimiento de los bonsais, esos árboles normales que los japoneses convierten en enanos a viva fuerza. Me pregunto qué sentirán las rosas de cultivo industrial, como hay tantas en Colombia, a las cuales se les ha eliminado el aroma. Nuestro amigo botánico no pudo explicarnos el motivo de esta mutilación de la fragancia, pero hay quienes dicen que es una exigencia de los importadores norteamericanos, cuyos clientes adoran las rosas, pero detestan su perfume. El profesor René Dumont, en uno de sus libros de protesta sobre la destrucción del medio ambiente, ha revelado un drama fantástico: cada edición dominical del *New York Times* consume una cantidad de papel fabricada con 200 hectáreas de bosque.

Sin embargo, no todos los que estábamos en aquella reunión parecíamos tan sensibles al sufrimiento de las plantas y el genocidio de los bosques. Hace un instante, además, acabo de tener una prueba imprevista de cómo vemos los hombres estos temas insólitos. En efecto, un amigo, que considero inteligente y serio, me ha llamado por teléfono para preguntarme cuál era el tema de mi nota de esta semana. «Estoy escribiendo sobre el sufrimiento de las plantas y las flores», le contesté. Mi amigo, con una alarma cierta, exclamó:

–¡Ah, carajo! ¿No te estarás volviendo marica?

LA PESTE

El primero de mayo pasado, el niño Jaime Baquero García, de ocho años, murió en Torrejón de Ardoz, España, de una enfermedad que sorprendió a los médicos porque no se parecía a ninguna conocida hasta entonces. Al cabo de dos semanas, cinco personas más habían muerto por la misma causa sin nombre, y otras 2.500 habían sido hospitalizadas en una franja de unos 300 kilómetros entre Madrid y León, por la carretera nacional VI.

La enfermedad empezaba con fiebres intensas, insuficiencia respiratoria y dolores musculares, como la gripe vulgar, pero en la pantalla radiográfica se advertían edemas pulmonares, y aun cerebrales en ciertos casos. Más tarde aparecían trastornos vasculares, un deterioro orgánico general, y por último una parálisis progresiva que causaba la muerte al alcanzar los centros respiratorios. En la mayoría de los casos, los pacientes reaccionaban bien a los tratamientos tentativos, pero muchos de los que eran dados de alta volvían al hospital algún tiempo después, y algunos murieron en la recaída. Es decir, la recuperación no era definitiva, sino un alivio temporal cuyo desenlace era imprevisible. Los médicos privados, con muy buen sentido, la distinguieron con un nombre que más bien era una hipótesis de trabajo: neumonía atípica. Lo único que se sabía con seguridad es que todos los casos ocurrían en un área territorial bien establecida, que la enfermedad no era contagiosa, y que todas las víctimas, sin excepción, eran muy pobres.

Ante la inminencia de un pánico social, y contra el criterio de otros médicos que habían observado el fenómeno con un escepticismo científico mucho más serio, el gobierno asumió el concepto del Centro Oficial de Microbiología, según el cual la causa del flagelo era un microplasma. El señor Jesús Sancho Roff, ministro de Trabajo, Salubridad y Seguridad Social, lo dijo a través de la televisión apenas veinte días después de la muerte del primer niño, y lo

Publicado originalmente en diciembre de 1981.

dijo en una forma que no podía ser más tranquilizadora, ni menos responsable, por supuesto. «Ya hemos localizado al bichito», dijo el ministro; «sabemos su nombre, pero todavía no su apellido, y es tan pequeño que si se cae aquí, en el suelo, se rompe.» Hoy, siete meses después de esa declaración folclórica, el flagelo incógnito ha causado 217 muertos, y más de 10.000 personas han sido hospitalizadas, sin que todavía haya sido posible determinar a ciencia cierta si la naturaleza de su enfermedad es de veras la que suponen los médicos, ni se sabe cuál es su remedio definitivo ni su pronóstico certero. En la Edad Media, cuando los ministros no podían explicar nada por televisión, un flagelo social con semejante poder devastador tenía un nombre genérico y simple: la peste.

Una de las primeras pestes que se recuerdan fue la de Atenas, que ocurrió en los años 430 antes de Cristo, y debió ser la que inspiró a Sófocles el drama de Edipo rey. La peste de Londres, en 1664, produjo más de 100.000 muertos, que era casi la cuarta parte de la población urbana. Daniel Defoe, que fue testigo de esa inmensa desgracia a la edad de cinco años, había de reconstruirla mucho más tarde, con la forma de un reportaje falso, en una novela magistral. Sin embargo, la peste más terrible de la historia humana fue la de 1346 que se llamó la peste negra, que causó 25 millones de muertes en Europa y 23 millones en África en sólo siete años. Es decir: casi la cuarta parte de la humanidad conocida. La última peste de nuestro tiempo, que la padeció la población de Argel hasta 1944, es muy probable que hubiera inspirado a Albert Camus su novela inolvidable.

Hace unos años, cuando vivía el general Franco, las autoridades de la base militar norteamericana de Torrejón —situada dentro de la zona de la peste actual— tuvieron que enfrentarse al problema imaginable de los sisones, especie de avutardas de paso veloz y vuelo lento, que se metían en las turbinas de los aviones de guerra. Los españoles, en el apogeo de su desarrollo industrial, les propusieron a los norteamericanos una hermosa solución olvidada: la cetrería. En efecto, soltaron halcones amaestrados para que destruyeran a los sisones. Sin embargo, aquella resurrección de un arte tan noble y tan antiguo les pareció demasiado cara a los norteamericanos, y optaron por una más barata, más radical y, sobre todo, más propia de nuestro tiempo: irrigaron la región con un tóxico que envenenó a los sisones, y otra vez los aviones de la muerte pudieron volar sin el peligro de los pájaros.

Quince años después, costó mucho trabajo convencer a los asustados habitantes de Torrejón de que el origen de las muertes incomprensibles no venía de aquel tóxico. Más aún: pensaron con muy buen juicio que otros animales lo habían ingerido sin morir, y lo estaban transmitiendo a los humanos, y que había vuelto a resurgir en las legumbres y fluir sin color, sin olor ni sabor en el agua de los ríos. Fue así como organizaron batidas contra todo ser vivo de la región que hubiera podido ser el portador de la muerte, y tal vez no hubieran terminado jamás en sus ímpetus de ángeles exterminadores, si un médico inteligente y tenaz no hubiera establecido que la causa probable de la peste estaba en el aceite de colza. Un aceite desnaturalizado para usos industriales, que algún químico chapucero trató de naturalizar de nuevo, y al parecer inventó el agente de una nueva peste de nuestro tiempo, del cual la ciencia todavía lo ignora todo, salvo que es mortal.

No es extraño, pues, que hubiera sufrido una impresión tan fuerte durante mi visita a España de la semana pasada, al encontrarla en un estado de tensión ante el veneno de colza, y que es apenas comparable a la tensión de su incertidumbre política. Las coincidencias con las pestes medievales, revueltas con los elementos más sofisticados del capitalismo industrial y el frenesí del consumo, no pueden ser sino una tentación casi irresistible para un novelista atento a los misterios humanos de su tiempo.

Uno de mis mejores recuerdos de periodista es la forma en que el gobierno revolucionario de Cuba se enteró, con varios meses de anticipación, de cómo y dónde se estaban adiestrando las tropas que habían de desembarcar en la bahía de Cochinos. La primera noticia se conoció en la oficina central de Prensa Latina, en La Habana, donde yo trabajaba en diciembre de 1960, y se debió a una casualidad casi inverosímil. Jorge Ricardo Masetti, el director general, cuya obsesión dominante era hacer de Prensa Latina una agencia mejor que todas las demás, tanto capitalistas como comunistas, había instalado una sala especial de teletipos sólo para captar y luego analizar en junta de redacción el material diario de los servicios de prensa del mundo entero. Dedicaba muchas horas a escudriñar los larguísimos rollos de noticias que se acumulaban sin cesar en su mesa de trabajo, evaluaba el torrente de información tantas veces repetido por tantos criterios e intereses contrapuestos en los despachos de las distintas agencias y, por último, los comparaba con nuestros propios servicios. Una noche, nunca se supo cómo, se encontró con un rollo que no era de noticias sino del tráfico comercial de la Tropical Cable, filial de la All American Cable en Guatemala. En medio de los mensajes personales había uno muy largo y denso, y escrito en una clave intrincada. Rodolfo Walsh, quien además de ser muy buen periodista había publicado varios libros de cuentos policíacos excelentes, se empeñó en descifrar aquel cable con la ayuda de unos manuales de criptografía que compró en alguna librería de viejo de La Habana. Lo consiguió al cabo de muchas noches insomnes, y lo que encontró dentro no sólo fue emocionante como noticia, sino un informe providencial para el gobierno revolucionario. El cable estaba dirigido a Washington por un funcionario de la CIA adscrito al personal de la embajada de Estados Unidos en Guatemala, y era un informe minucioso de los preparativos de un desembarco armado en

Publicado originalmente el 16 de diciembre de 1981.

Cuba por cuenta del gobierno norteamericano. Se revelaba, inclusive, el lugar donde iban a prepararse los reclutas: la hacienda de Retalhuleu, un antiguo cafetal en el norte de Guatemala.

Un hombre con el temperamento de Masetti no podía dormir tranquilo si no iba más allá de aquel descubrimiento accidental. Como revolucionario y como periodista congénito se empeñó en infiltrar un enviado especial en la hacienda de Retalhuleu. Durante muchas noches en claro, mientras estábamos reunidos en su oficina, tuve la impresión de que no pensaba en otra cosa. Por fin, y tal vez cuando menos lo pensaba, concibió la idea magistral. La concibió de pronto, viendo a Rodolfo Walsh que se acercaba por el estrecho vestíbulo de las oficinas con su andadura un poco rígida y sus pasos cortos y rápidos. Tenía los ojos claros y risueños detrás de los cristales de miope con monturas gruesas de carey, tenía una calvicie incipiente con mechones flotantes y pálidos y su piel era dura y con viejas grietas solares, como la piel de un cazador en reposo. Aquella noche, como casi siempre en La Habana, llevaba un pantalón de paño muy oscuro y una camisa blanca, sin corbata, con las mangas enrolladas hasta los codos. Masetti me preguntó: «¿De qué tiene cara Rodolfo?». No tuve que pensar la respuesta porque era demasiado evidente. «De pastor protestante», contesté. Masetti replicó radiante: «Exacto, pero de pastor protestante que vende biblias en Guatemala». Había llegado, por fin, al final de sus intensas elucubraciones de los últimos días.

Como descendiente directo de irlandeses, Rodolfo Walsh era además un bilingüe perfecto. De modo que el plan de Masetti tenía muy pocas posibilidades de fracasar. Se trataba de que Rodolfo Walsh viajara al día siguiente a Panamá, y desde allí pasara a Nicaragua y Guatemala con un vestido negro y un cuello blanco volteado, predicando los desastres del apocalipsis que conocía de memoria y vendiendo biblias de puerta en puerta, hasta encontrar el lugar exacto del campo de instrucción. Si lograba hacerse a la confianza de un recluta habría podido escribir un reportaje excepcional. Todo el plan fracasó porque Rodolfo Walsh fue detenido en Panamá por un error de información del gobierno panameño. Su identidad quedó entonces tan bien establecida que no se atrevió a insistir en su farsa de vendedor de biblias.

Masetti no se resignó nunca a la idea de que las agencias yanquis tuvieran corresponsales propios en Retalhuleu mientras que Prensa Latina debía conformarse con seguir descifrando los cables secretos. Poco antes del desembarco, él y yo viajábamos a Lima desde México y tuvimos que hacer una escala imprevista para cambiar de avión en Guatemala. En el sofocante y sucio aeropuerto de la Aurora, tomando cerveza helada bajo los oxidados ventiladores de aspas de aquellos tiempos, atormentado por el zumbido de las moscas y los efluvios de frituras rancias de la cocina, Masetti no tuvo un instante de sosiego. Estaba empeñado en que alquiláramos un coche, nos escapáramos del aeropuerto y nos fuéramos sin más vueltas a escribir el reportaje grande de Retalhuleu. Ya entonces le conocía bastante para saber que era un hombre de inspiraciones brillantes e impulsos audaces, pero que, al mismo tiempo, era muy sensible a la crítica razonable. Aquella vez, como en algunas otras, logré disuadirle. «Está bien, che», me dijo, convencido a la fuerza. «Ya me volviste a joder con tu sentido común.» Y luego, respirando por la herida, me dijo por milésima vez:

–Eres un liberalito tranquilo.

En todo caso, como el avión demoraba, le propuse una aventura de consolación que él aceptó encantado. Escribimos a cuatro manos un relato pormenorizado con base en las tantas verdades que conocíamos por los mensajes cifrados, pero haciendo creer que era una información obtenida por nosotros sobre el terreno al cabo de un viaje clandestino por el país. Masetti escribía muerto de risa, enriqueciendo la realidad con detalles fantásticos que iba inventando al calor de la escritura. Un soldado indio, descalzo y escuálido, pero con un casco alemán y un fusil de la guerra mundial, cabeceaba junto al buzón de correos, sin apartar de nosotros su mirada abismal. Más allá, en un parquecito de palmeras tristes, había un fotógrafo de cámara de cajón y manga negra, de aquellos que sacaban retratos instantáneos con un paisaje idílico de lagos y cisnes en el telón de fondo. Cuando terminamos de escribir el relato agregamos unas cuantas diatribas personales que nos salieron del alma, firmamos con nuestros nombres reales y nuestros títulos de prensa, y luego nos hicimos tomar unas fotos testimoniales, pero no con el fondo de cisnes, sino frente al volcán acezante e inconfundible que dominaba el horizonte al atardecer. Una copia de esa foto existe: la tiene la viuda de Masetti en La Habana. Al final metimos los papeles y la foto en un sobre dirigido al señor general Miguel Ydígoras Fuentes, presidente de la República de Guatemala, y en una fracción de segundo en que el soldado de guardia se dejó vencer por la modorra de la siesta echamos la carta al buzón. Alguien había dicho en público por esos días que el general Ydígoras Fuentes era un anciano inservible, y él había aparecido en la televisión vestido de atleta a los 69 años, y había hecho maromas en la barra y levantado pesas, y hasta revelado algunas hazañas íntimas de su virilidad para demostrarles a sus televidentes que todavía era un militar entero. En nuestra carta, por supuesto, no faltó una felicitación especial por su ridiculez exquisita.

Masetti estaba radiante. Yo lo estaba menos, y cada vez menos, porque el aire se estaba saturando de un vapor húmedo y helado y unos nubarrones nocturnos habían empezado a concentrarse sobre el volcán. Entonces me pregunté espantado qué sería de nosotros si se desataba una tormenta imprevista y se cancelaba el vuelo hasta el día siguiente, y el general Ydígoras Fuentes recibía la carta con nuestros retratos antes de que nosotros hubiéramos salido de Guatemala. Masetti se indignó con mi imaginación diabólica. Pero dos horas después, volando hacia Panamá, y a salvo ya de los riesgos de aquella travesura pueril, terminó por admitir que los liberalitos tranquilos teníamos a veces una vida más larga, porque tomábamos en cuenta hasta los fenómenos menos previsibles de la naturaleza. Al cabo de veintiún años, lo único que me inquieta de aquel día inolvidable es no haber sabido nunca si el general Ydígoras Fuentes recibió nuestra carta al día siguiente, como lo habíamos previsto durante el éxtasis metafísico.

EL CAMPO, ESE HORRIBLE LUGAR
DONDE LOS POLLOS SE PASEAN CRUDOS

En una reciente encuesta entre niños de las grandes ciudades de Europa les preguntaron cómo se llama el hombre que lleva las cartas a la casa, cómo se llama el que lleva la leche, el que lleva el periódico y el pan, el que recoge la basura y el que arregla los daños menores de la luz y el agua. La respuesta de los niños fue casi unánime: el portero.

No tienen por qué contestar otra cosa, pues ya sabemos que en estas grandes concentraciones urbanas, donde el nacimiento de una flor es como un milagro de la creación, todo lo que entra en los apartamentos debe pasar por el conducto ordinario e ineludible, y además providencial, del portero. Lo que de niños nos enseñaron a conocer como la naturaleza, que era en realidad todo lo que nos rodeaba en el pueblo, ha terminado por parecer un programa fantástico en la televisión. No es extraño, pues, que un niño que vive en el piso dieciséis, que sólo sale de la casa para ir a la escuela en autobús, que pasa las vacaciones de invierno en una estación de nieve y pasa el verano en una playa urbanizada, ignore que en una época había un hombre de uniforme azul que llevaba las cartas en bicicleta, y que había otro de delantal blanco que no sólo llevaba la leche, sino que era tan puntual que servía al mismo tiempo de despertador. Todos terminaban por formar parte de la familia, se demoraban en la cocina tomando el café y comentando con la gente del servicio los secretos del vecindario, hasta que un día oíamos decir a la hora del almuerzo, con toda naturalidad: «Petra está encinta del cartero». La única inocencia que nos permitíamos los niños de entonces era creer que el hijo que iba a tener Petra no podía ser sino un cartero recién nacido.

Los vientos de la civilización no han logrado exterminar por completo en España a uno de los personajes propios de su vida y su literatura: el sereno. Todavía quedan por ahí algunos de esos viejos

Publicado originalmente el 23 de diciembre de 1981.

jubilados para quienes no había secretos en su calle, pues nada ocurría en ella que ellos no supieran. El sereno era el responsable de la seguridad de su sector y tenía consigo el mazo de llaves de todas las casas. Nadie que regresaba tarde llevaba sus propias llaves, sino que se hacía abrir la puerta por el sereno. Siempre estaba al alcance: bastaba con buscarlo en la taberna de la esquina, donde solía pasar la noche conversando con los otros serenos del barrio, o bastaba con batir palmas para que él acudiera de inmediato. Me pregunto qué pensarán los niños de hoy en las grandes ciudades de España si a alguien se le ocurre contarles cómo era el señor sereno que nos abría la puerta. No le creerían, sin duda, como tampoco formará parte de sus nostalgias de viejos el personaje del afilador de cuchillos y tijeras, que pasaba muy de vez en vez, como los eclipses, y dejaba impregnado el aire de la calle con la música de su caramillo.

De todos esos personajes de nuestra infancia, cada vez menos visibles y evidentes para los niños de hoy, el único que estaba señalado por un aura fatídica era el pobre hombre que llevaba los telegramas. Tal vez los propios mensajeros habían contribuido a formarse esa imagen siniestra, por la manera apremiante con que tocaban a la puerta, hacían sonar un silbato, que siempre parecía de emergencia, y gritaban: «¡Telegrama!». Mucho antes, cuando el mundo nos pertenecía por completo, esa función anunciadora estaba reservada a los presagios. Pero los telegramas, desde su invención, se convirtieron en mensajeros de la muerte. Antes de que alguien tuviera tiempo de abrir la puerta ya había que asistir a la abuela, que había caído en un estado de sopor; aullaban los perros en el patio y las gallinas se subían a pleno sol a dormir en las perchas, con el sentido del tiempo trastornado por el desastre. Uno mismo escrutaba el rostro impenetrable del mensajero cuando entregaba el telegrama, pues parecía imposible que no conociera el texto de nuestra desgracia, y le dábamos las gracias con un hilo de voz, con el corazón desbocado y lamentando en el fondo del alma que no existiera ya la costumbre medieval de ahorcar al portador de malas noticias. Con el tiempo, aquel horror de los telegramas fue derrotado por la burla de su lentitud. Entonces, alguien que iba a viajar le mandó a la amada el siguiente telegrama: «Cuando éste llegue estaré en tus brazos».

Hasta el médico familiar, cuya sola presencia en la casa hacía bajar la fiebre, ha sido reemplazado en las ciudades por una divinidad desconocida cuyo corazón nos desconoce. Alguien contaba hace poco el caso de un enfermo grave a quien los diferentes especialistas de una clínica privada le ordenaron seis análisis distintos. El enfermo murió esa noche, pero veinticuatro horas después, cuando ya había sido enterrado, los análisis revelaron que estaba en perfecto estado de salud. Estos episodios tremendos de la civilización, que por desgracia se cuentan como chistes crueles, sólo son comprensibles en un mundo donde ya hay niños que les preguntan a sus padres si las vacas ponen huevos y si los espaguetis nacen en los árboles.

La televisión no alcanza a resolver esas dudas. Por eso las escuelas francesas obligan a un curso especial que consiste en llevar a los niños para que vivan en el campo durante un mes, al aire libre y con los ojos bien abiertos, de modo que conozcan la otra mitad del mundo que la mitad civilizada no les permite ver. Me imagino que a ellos les ocurre lo mismo que nos ocurrió a los niños rurales la primera vez que nos llevaron a la ciudad. Me imagino que ven una gallina poniendo un huevo con el mismo temor reverencial con que nosotros conocimos el cine; que ven dos perros enmarconados en la calle con la misma emoción con que nosotros veíamos a los bomberos apagando un incendio; que ven pasar los burros de carne y hueso, y los oyen rebuznar de verdad, y les arrancan pelos de las ancas con la misma ilusión de aventura con que nosotros veíamos aterrizar los primeros aviones.

Mi amigo Alejandro Santos Rubino, a quien le llevo por delante casi 42 años de vida, y que acaba de hacer su curso sobre la naturaleza en el oriente de Francia, me ha contado su experiencia con el mismo deslumbramiento con que debieron contar sus viajes los navegantes antiguos. Pero su relato, a 10.000 kilómetros de nuestra patria común, me hizo caer en la cuenta de lo lejos que estábamos de ella también en el tiempo. En efecto, al equipo de Alejandro lo llevaron a ver cómo se corta un árbol. Pero el leñador ya no es de aquellos que se pasaba un día completo picando el tronco con un hacha, como un pájaro carpintero, sino que cortaba el árbol con una precisión científica y sirviéndose de una sierra eléctrica. Vio ordeñar una vaca, pero no como ya lo había visto hacer en las siete colinas de colores de Boyacá, a puro pulso, sino mediante un sistema de ordeño eléctrico cuyos tubos estériles conducían la leche hasta las cámaras de pasteurización. Es decir, que en lós países más industrializados ya casi resulta imposible encontrar un sitio donde los niños urbanos se formen una imagen real de la hermosa y triste barbarie del subdesarrollo. Mis hijos, en cambio, recuerdan como uno de los instantes de su vida la tarde en que vieron un sapo vivo y verdadero por primera vez, en el pueblo caribe donde fueron a visitar a sus abuelos. Fue tanta su emoción que cargaron con un tarro de pintura y una brocha gorda que encontraron a mano y pintaron de amarillo a cuantos sapos encontraron en el pueblo.

POLONIA: VERDADES QUE DUELEN

Una de las impresiones irreparables de mi vida fue mi primera y única visita a Varsovia en el otoño de 1955. No habían pasado todavía diez años desde el final de la segunda guerra, y sus estragos enormes eran demasiado visibles no sólo en la devastación de la ciudad, sino en el espíritu de sus habitantes. Una muchedumbre densa, desharrapada, triste, se deslizaba sin rumbo por las calles escuetas con un rumor de creciente de río, y había grupos atónitos que pasaban horas enteras contemplando las vitrinas de los almacenes del Estado, donde se vendían cosas nuevas que parecían viejas, pero que en todo caso no se podían comprar por sus precios irreales. Había muy pocos automóviles, y los tranvías decrépitos pasaban dando tumbos por las calles desiertas. En algunas esquinas había camiones del Estado con altavoces descomunales que tocaban música popular a todo volumen, y en especial canciones latinoamericanas. Pero esa alegría oficial, impuesta por decreto, no se reflejaba en el ánimo de la gente. Uno se daba cuenta al primer golpe de vista de que la vida era dura, de que los sobrevivientes del cataclismo bélico habían padecido sufrimientos difíciles de imaginar por quienes no los padecimos, y que había una situación de pobreza y amargura que el socialismo no podía remediar con música de camiones en las esquinas.

Hubiera sido frívolo, por decir lo menos, tratar de formarse un juicio sobre la verdad de Polonia a partir de una visión superficial de la calle, porque todo lo que saltaba a la vista podía atribuirse a los horrores de la guerra. Pero tres elementos que parecían de fondo me llamaron la atención al cabo de pocas horas. El primero era la impopularidad de los gobernantes, y en especial entre la juventud. La universidad era un barril de pólvora que podía estallar con la chispa mínima de un error, y las críticas al sistema eran destapadas e implacables.

Publicado originalmente el 30 de diciembre de 1981.

La otra cosa que me llamó la atención, aunque ya fuera demasiado sabida, era el inmenso poder político y espiritual de la Iglesia católica, sustentados ambos en el sentimiento religioso de la población. Las monjas y los sacerdotes, con hábitos demasiado ostensibles, tenían una participación activa en todas las manifestaciones de la vida pública. En la amplia y vacía avenida de Marszalkowa, que era la arteria vital de la ciudad, me sorprendió un Cristo coronado de bombillos eléctricos, a cuyos pies ardían lámparas de aceite encendidas por fieles que rezaban de rodillas en plena calle. Una unidad de la moneda polaca, muy pequeña y brillante, había sido retirada de la circulación porque los traficantes del mercado negro las convertían en medallas de la Virgen para venderlas por el triple de su valor nominal. La tercera cosa que me llamó la atención fue el antisovietismo de los polacos. Un sentimiento arraigado e histórico y que a mí me pareció irremediable.

Al cabo de dos semanas de conversaciones, encuentros casuales y averiguaciones callejeras, no encontré un solo polaco que estuviera satisfecho con su gobierno. Pero ellos mismos me parecieron perdidos en un laberinto de confusiones. Los intelectuales –que en Polonia son los más intelectuales del mundo– estaban atascados en definir matices doctrinarios mientras la situación económica y política adquiría proporciones de catástrofe. Muchos reconocían la necesidad del socialismo para reconstruir su país con una sociedad más justa, pero negaban de plano la competencia de los equipos en el poder. Acusaban a éstos de no tomar en cuenta la realidad del país, pero los mismos que formulaban la acusación fomentaban las huelgas y hacían manifestaciones y encuentros callejeros con la policía para pedir cosas que las condiciones económicas no permitían.

La mayoría de mis informadores consideraban que aquello no era la dictadura del proletariado, como decía la versión oficial, sino el dominio de un grupo del partido comunista que trataba de implantar a toda costa y al pie de la letra el esquema soviético. Los obreros estaban en las mejores condiciones posibles, pero carecían de conciencia política. No entendían por qué el gobierno les contaba que eran ellos quienes estaban en el poder, y tenían que trabajar como burros para comprar un par de zapatos que les costaba el sueldo de un mes. No entendían, y nadie se lo explicaba, por qué los obreros de Occidente, explotados por el capitalismo, tenían mejores condiciones de vida y derecho de huelga. Las respuestas no eran difíciles, pero el partido comunista no hacía un trabajo ideológico eficaz donde más debía hacerlo. En cambio, la Iglesia católica lo hacía sin descanso, en el confesonario y en el púlpito, en la fábrica y en la casa, con el inmenso poder de penetración de los sacramentos. La desvinculación de los dirigentes con la población había de terminar, sin remedio, por ser infranqueable. Así no había socialismo posible: desde los ministerios hasta las cocinas domésticas había un inconformismo justo e incontenible y un embrollo burocrático que sólo un régimen popular auténtico hubiera podido desenredar.

Mi impresión global de aquella visita inolvidable fue que la vida de Polonia estaba muy lejos de ser el socialismo idealizado en el colegio a mis veinte años. Era, por el contrario, una realidad cruda y amarga, cuya tensión interna había de estallar, tarde o temprano, si no se corregía a tiempo. Es decir: si no se hacía una revolución propia dentro de las condiciones reales del país. Cuando regresé a Colombia, pocos meses después, escribí todo esto en un artículo cuyo título parece de hoy: «Con los ojos abiertos sobre Polonia en ebullición». La publicación, desde luego, me valió los reproches de los dogmáticos de aquel tiempo –algunos de los cuales están hoy sentados en las sillas del poder y las finanzas–, y no faltó alguno más original que me acusó de estar a sueldo del Departamento de Estado de Estados Unidos. Ahora, transcurridos veinticuatro años, me veo obligado a señalar con mucha pena que era yo quien tenía la razón, aunque sea una razón indeseable, y que casi todo lo que revelaba mi artículo de periodista novato era el germen primario y la única explicación de fondo del callejón en que se encuentra en este momento atascada la Polonia de hoy.

Un callejón sin salida, por supuesto. Desde que empezó la crisis, hace dieciséis meses, y aun contra el criterio de mis amigos más sabios, he sido un optimista terco. Confiaba en la inteligencia casi legendaria y en el sentido de responsabilidad histórica de los polacos de ambos bandos en conflicto, conscientes sin duda de que tenían en sus manos no sólo la suerte de su patria, sino la de toda la humanidad. Pero siempre existió el riesgo de que los acontecimientos desbordaran a sus propios protagonistas, y algo de esto parece haber ocurrido. Así las cosas, el optimismo sería una forma casi irracional de la temeridad.

No sé, por otra parte, si en América Latina somos conscientes de cuánto nos concierne, y de qué modo tan directo, el drama de Polonia. En efecto, se sabe por rumores públicos, por suposiciones bien fundadas y por conversaciones confidenciales, que cualquier tentativa de la Unión Soviética contra Polonia sería seguida por una réplica inmediata de Estados Unidos en América Central. Y en primer término contra Cuba y Nicaragua, desde luego, dos experiencias difíciles, pero diferentes y ejemplares, que no merecen este destino de rehenes de las equivocaciones ajenas.

Son verdades que duelen, pero hay que decirlas. Lo contrario sería dejarlas en manos de quienes menos las necesitan, que son los antisoviéticos y los anticomunistas profesionales de siempre, y los reaccionarios de siempre, que ahora están saliendo juntos y revueltos por las calles del mundo entero a derramar por Polonia sus lagrimones de cocodrilo.

LA REALIDAD MANIPULADA

Una cosa son las tremendas condiciones históricas de Polonia –que fueron tema de esta columna la semana anterior– y otra cosa bien distinta, indigna e indignante, ha sido el manejo político de la información occidental. Una infamia mayor, no sólo contra la nación polaca, que está tratando de enderezar su destino con tantas dificultades y sacrificios, sino también contra la opinión pública de las democracias capitalistas, que en estas tres semanas ha estado a merced de una histeria informativa con muy pocos precedentes, y con muy pocas excepciones.

Apenas ahora se empieza a saber a duras penas qué sucedió de veras en Polonia, pero ya nadie podrá borrar el escándalo de una prensa hablada, televisada y escrita, que no parecía desear nada más que un cataclismo universal. Este frenesí estuvo en muchos casos contra la corriente de los propios gobiernos occidentales, que por una vez dieron muestras de una prudencia afortunada. No es para menos. Después de la crisis de octubre de 1962, que el historiador inglés Hugh Thomas consideró como la más grave de la comedia humana, esta de Polonia no necesitaba sino un paso en falso, por mínimo que fuera, para acabar con lo poco que ya le va quedando a este mundo.

La culpa, por supuesto, está bien repartida. Si el gobierno polaco no hubiera atrancado sus puertas por dentro, como en efecto lo hizo, no hubiera sido tan fácil distorsionar la verdad. Esto prueba una vez más que lo único más peligroso que una mala información es la falta absoluta de información. Pero ha bastado con que se abran las primeras grietas en la cortina de hierro del general Wojciech Witold Jaruzelski para que se compruebe hasta qué punto puede ser diabólica una realidad manipulada.

Yo estaba en París este lúgubre domingo de invierno, cuando un amigo me llamó por teléfono para decirme, sin aviso previo, que las

Publicado originalmente el 6 de enero de 1982.

tropas soviéticas habían invadido Polonia. Lo primero que me vino a la mente' fue la comprobación de una casualidad: también en las dos ocasiones anteriores en que algo semejante había ocurrido yo me encontraba en París, en circunstancias siempre diferentes. La primera vez, en 1956, cuando era un corresponsal varado, fue el drama de Hungría. Mi reacción, pienso ahora, fue la correcta: me eché a la calle dispuesto a viajar a Viena de cualquier modo para meterme de contrabando en Budapest, como lo estaban haciendo tantos otros periodistas del mundo, y escribir en caliente el reportaje de mi vida. La segunda vez, en septiembre de 1968, encendí medio dormido el receptor de radio de la mesa de noche, como obedeciendo a un presagio, y escuché la noticia: las tropas del Pacto de Varsovia estaban entrando en Checoslovaquia. Mi reacción, pienso ahora, fue la correcta: escribí una nota de repudio por la interrupción brutal de una tentativa de liberalización que merecía una suerte mejor. Esta vez, pienso ahora, también mi reacción fue la correcta: sentí un terror irresistible por el destino de tantos hombres y mujeres inocentes que habían esperado ese domingo para ser felices, sin pensar que era un día señalado para la muerte. En realidad, desde que empezó esta crisis, yo pensaba que una intervención soviética no era posible, a menos que Polonia pretendiera abandonar el Pacto de Varsovia, y pensaba además que aquella determinación inconcebible sería el pretexto que Estados Unidos deseaba para lanzarse a una aventura loca en el Caribe. Aquel domingo aciago –de acuerdo con la noticia que un amigo me acababa de dar por teléfono– lo imposible había ocurrido. De modo que me levanté pensando en el Caribe, aquel mundo remoto y amado donde todavía seguía siendo la noche anterior, una noche de sábado caliente y bulliciosa, como todas las nuestras.

Las informaciones eran entonces tan confusas, y dirigidas con tanta perversidad, que se necesitaron varias horas para darse cuenta de que la intervención soviética no había ocurrido. Pero tanto en la radio como en la televisión quedaba flotando la duda de que hubiera soldados soviéticos disfrazados con uniformes polacos y confundidos con la tropa local. Muy pronto se estableció como un hecho indiscutible que el número de detenidos en los allanamientos nocturnos era de 50.000, que todos habían sido enviados a campos de concentración del mar Báltico, a treinta grados bajo cero y con dos metros de nieve en este invierno implacable, y que los dirigentes del sindicato Solidaridad eran víctimas de torturas infames. Fue muy poco lo que no se dijo sobre la suerte de Lech Walesa: que había sido enviado a Siberia, que el gobierno militar lo mantenía como rehén para mantener la resistencia popular, que había sufrido un infarto cardíaco y se temía que hubiera muerto. Un infarto en el que nadie creía, por supuesto, y en el que nadie hubiera creído, tal vez con razón. Se dijo también que el arzobispo primado de Polonia, monseñor Josef Glemp, había sido arrestado; que un dirigente de Solidaridad se había suicidado, y que Tadeusz Mazowiecki, jefe de redacción del semanario de Solidaridad y uno de los principales consejeros de Walesa, había muerto en prisión como consecuencia de las torturas. Sobre este último, uno de sus antiguos compañeros, residente en París, escribió en *L'Express* de hace dos semanas una elegía tan conmovedora como precipitada. Hoy se sabe que todas estas noticias no sólo eran falsas, sino algo peor: inventadas.

Me conmovió de un modo especial el embajador de Polonia en Francia, que se prestó a una entrevista de televisión en directo tan pronto como tuvo noticias de primera mano sobre la situación de su país. Su imagen fue la de un hombre brillante, de una serenidad a toda prueba y de una buena educación sin un solo resquicio, que respondió en un francés perfecto a las preguntas más impertinentes que he escuchado jamás. «¿Qué diferencia haría usted entre el general Jaruzelski y el general Pinochet?», fue una de ellas. Siempre he creído que no hay grosería más detestable que la que abusa de la buena educación del adversario. Éste fue el caso. Por fortuna, el resultado fue al revés: muchos televidentes que empezaron el programa con una posición contraria a la del embajador terminaron de parte suya. Ese mismo día, en una entrevista por radio, a un viajero escandinavo recién llegado de Polonia le preguntaron si había oído decir que a los prisioneros polacos los estaban deportando a Siberia. El viajero, sorprendido, contestó que no, que no lo había oído decir. Pero desde aquel día quedó flotando en el aire la impresión de que algo de cierto había en la pregunta si el periodista se había atrevido a hacerla.

Con todo, la comprobación más amarga que me quedará de esta experiencia es que la opinión pública europea, que es capaz de llegar a estos extremos frenéticos por la suerte de un país europeo, apenas si se conmueve en estos tiempos por la suerte de nuestros países remotos. El escándalo de la ley marcial en Polonia es apenas una prueba entre otras tantas. En realidad, esa medida de excepción, que corresponde al estado de sitio, está vigente en Colombia desde hace treinta años casi continuos, y a su sombra se han consumado más arbitrariedades, y se han aplicado torturas más atroces, y cometido más crímenes oficiales que todos los que sin duda quedarán de la ley marcial de Polonia. Sin embargo, aun los europeos mejor informados se atreven todavía a celebrar que Colombia sea no sólo una de las últimas democracias de América Latina, sino la más antigua.

Hace unos días, la señora Danielle Mitterrand, la esposa del presidente de Francia, estaba consternada en privado por un documental que había visto sobre la masacre de El Salvador. La señora Mitterrand es presidenta del comité francés de solidaridad con ese país desdichado, del cual muy pocos franceses saben dónde queda, y en el cual se cometieron más de 15.000 crímenes oficiales en el año que acaba de pasar. Es decir, un promedio de cuarenta muertos por día. Me consta que los dirigentes de la Internacional Socialista, con el poder o sin él, tienen un gran interés por la suerte de América Latina. Me consta, por incontables conversaciones privadas, que el presidente Mitterrand comparte la consternación de su esposa por el estado de postración de nuestros países. Pero no creo que ni él ni ella, ni la Internacional Socialista en pleno, lograran movilizar a la opinión pública europea en favor de una causa nuestra, y menos en la forma casi epiléptica en que lo han conseguido en favor de Polonia los desenfrenados medios de información de la Europa occidental. Hace un mes se hicieron en París dos manifestaciones. Una de los artistas argentinos por más de 10.000 desaparecidos en su país, y otra por la masacre sin término de El Salvador. En ambas sólo se vieron los latinoamericanos residentes en Francia, sus escasos amigos europeos y nuestros eternos compadres españoles. «Es natural», me dijo sin inmutarse un periodista francés: «Hace ya mucho tiempo que la América Latina dejó de ser noticia en Europa.»

ESPAÑA: LA NOSTALGIA DE LA NOSTALGIA

Hace veinte años, en México, fui a ver varias veces la película *El último cuplé*, cautivado por la nostalgia de las canciones que tanto le había oído cantar a mi abuela. La semana pasada, en Barcelona, fui con una pandilla de amigos a ver el espectáculo vivo de Sara Montiel, pero ya no por escuchar otra vez las canciones de la abuela, sino cautivado por la nostalgia de aquellos tiempos de México. Cuando las cantaba mi abuela, a mis seis años, las canciones me parecían tristes. Cuando las volví a escuchar en la película, treinta años más tarde, me parecieron mucho más tristes. Ahora, en Barcelona, me parecieron tan tristes que apenas eran soportables para un nostálgico irremediable como yo. Al salir del teatro, la noche era diáfana y tibia y había en el aire una fragancia de rosas de mar, mientras el resto de Europa naufragaba en la nieve. Me sentí conmovido en aquella ciudad hermosa, lunática e indescifrable, donde he dejado un reguero de tantos años de mi vida y de la vida de mis hijos, y lo que entonces padecí no fue la nostalgia de siempre, sino un sentimiento más hondo y desgarrador: la nostalgia de la nostalgia.

Para mi generación, la que andaba por los quince años cuando terminó la guerra civil española, esta desazón de las nostalgias superpuestas tiene sus raíces en España. A nosotros nos correspondió vivir, en un momento en que todos los recuerdos son eternos, lo que nosotros llamamos la segunda conquista de América. Me refiero al desembarco masivo de los republicanos derrotados, que no iban armados con la cruz y la espada como la primera vez, sino con una fuerza del espíritu que nos cambió la vida. Muchos llegaron convencidos de que era un exilio momentáneo. Se decía hasta hace poco, y más en serio de lo que pudiera parecer, que muchos de los que llegaron a México no quisieron moverse del puerto de Veracruz, y ni siquiera deshacer las maletas, para no perder su lugar en los primeros barcos de regreso. En el café de la Parroquia, que es un

Publicado originalmente el 13 de enero de 1982.

salón enorme de azulejos con ventiladores de aspas y mesas de mármol sobre las cuales escriben las cuentas los camareros, como si fuera en Cádiz, la guerra continuaba a gritos. En Buenos Aires, en Bogotá, en Ciudad de México, en La Habana, aparecieron de pronto restaurantes populares que parecían llevados enteros de Madrid o Sevilla, con sus jamones colgados, sus carteles de corridas de toros y sus enormes paellas improvisadas con los ingredientes locales. Los exiliados se demoraban después de que los otros clientes se habían ido, casi al amanecer, y volvían a contarse los unos a los otros, una vez y otra vez, el cuento sin término de la batalla del Ebro o el episodio magnificado del Alcázar de Toledo. En cierta ocasión, cuando viajaba de Veracruz a Cartagena de Indias en un barco español, fui testigo de un instante que me pareció una síntesis perfecta de todo el drama del exilio. Un refugiado había subido al barco para tomarse un brandy en la cantina. El camarero, que al parecer lo conocía desde otros viajes, le preguntó si quería el brandy con agua. El refugiado dijo que no, pero el camarero lo convenció, porque se inclinó hacia él por encima del mostrador, y le dijo con una voz de cómplice: «Es todavía agua de España». En medio de tantas verdades diferentes y confundidas, no sé si los refugiados españoles en América Latina fueron conscientes del viento de renovación con que nos cambiaron tantas cosas esenciales de la vida: las universidades, las librerías, el periodismo y, sobre todo, nuestras revenidas concepciones políticas. De cómo nos enseñaron a amar para siempre a una España menos obligatoria, y por lo mismo más humana, que aquella otra España de aceite de ricino que los clérigos brutos de la escuela primaria nos habían hecho tragar a la fuerza.

En cierto modo, yo también fui un exiliado español. Desde la escuela, influido por los maestros republicanos, me hice el propósito de no pisar tierra española mientras el general Franco estuviera vivo. Fue una determinación tan drástica, que en 1955 hice una escala técnica en el aeropuerto de Madrid y ni siquiera me bajé del avión, a pesar de la lucidez con que J. M. Caballero Bonald había tratado de explicarme en Bogotá que la España eterna era tan cojonuda que continuaba siéndolo a pesar del general Franco. Sólo a los 42 años de mi edad —hace ahora once— tuve bastante uso de razón para darme cuenta de que Caballero Bonald la tenía toda, porque, a pesar de mi resistencia pasiva y anónima, España continuaba en el tiempo y el general Franco seguía sin la menor disposición de morir para complacerme. De modo que llegué a Barcelona en el otoño de 1967, con toda mi familia y con el ánimo de quedarme ocho meses que me sobraban de una novela, y me quedé siete años. Más aún: de algún modo difícil de explicar, todavía no me he ido por completo, ni creo que me vaya nunca.

Yo no era consciente de todo esto hasta la semana pasada, cuando salí del teatro con mis amigos de Barcelona y descubrí de pronto la nostalgia de la nostalgia. Comprendí, como tantos otros de mi generación, que había padecido la nostalgia de España antes de conocerla, no sólo por las evocaciones implacables de los republicanos errantes, sino por la poesía grande que ellos mismos me enseñaron. Julio Cortázar dice en uno de sus libros que después de conocer a Viena no seguía recordándola como la había visto en la realidad, sino como la imaginaba antes de conocerla. A mí me ha ocurrido lo mismo con muchos lugares del mundo, pero no con España. Pues su descubrimiento fue una experiencia platónica: la encontré igual, calle por calle, tarde por tarde, nube por nube, a la España que ya había conocido en su literatura, de modo que conocerla en la realidad no fue más que recordarla.

Encontré que, en efecto, como ya había dicho don Antonio Machado, los campos de Soria eran áridos y fríos, con sierras calvas y cerros cenicientos, donde la primavera dejaba entre la hierba un rastro perfumado de margaritas blancas. Reconocí los pueblos de Andalucía, que parecen dibujados a pluma, y sentí al atardecer los cencerros de los corderos y el olor del tomillo estrangulado por el tropel del rebaño. Vi pájaros que conocía sólo leídos, como las cornejas y los tordos, y árboles que hasta entonces creía imaginarios, como los chopos a la orilla de los ríos, y escuché voces de niños distantes que sólo conocía de oídas en los campos de Moguer, y comprendí el drama de la historia de Castilla en una sola noche de enero en El Escorial, donde la soledad y el frío sólo podían ser comparables a los de la muerte. En Granada fui a buscar la calle de Elvira, para ver si era cierto que allí vivían las manolas, como lo había escrito García Lorca. No las encontré, pero, en cambio, tuve la fortuna de ver la Alhambra como hubieran querido verla siempre los califas: bajo un aguacero torrencial. No pude reprimir un estremecimiento recóndito cuando apareció en la ventanilla del tren una de las ciudades más bellas del mundo: Córdoba, lejana y sola, detrás de cuyos muros había dicho el poeta que acechaba la muerte. Una noche, mientras cenábamos en el desolado comedor del hotel Atlántico, en Cádiz, tuve de pronto la impresión maravillosa de que el edificio había zarpado hacia las Américas. En Burgos, alguien extendió el índice hacia una casa de paredes sombrías y me dijo: «Ahí vive el verdugo».

Sentía una gran nostalgia de aquellas hermosas nostalgias esa noche de la semana pasada en que salí del teatro con mis amigos de

Barcelona. Las Ramblas estaban más concurridas y delirantes que nunca, todavía con las enormes estrellas de luces de colores de la Navidad. En medio de la muchedumbre bulliciosa, de los gringos despistados y las suecas suculentas y casi desnudas en enero, estaban los exiliados de América Latina con sus ventorrillos públicos de baratijas, con sus niños envueltos en trapos, sobreviviendo como pueden mientras llega también para ellos el barco del regreso. Son quizá 250.000 en toda España, y no son muchos los que tienen la suerte de que los quieran tanto en España como queríamos nosotros a los republicanos errantes que nos enseñaron a vivir la nostalgia de la nostalgia.

LOS 166 DÍAS DE FELIZA

La escultora colombiana Feliza Bursztyn, exiliada en Francia, se murió de tristeza a las 10.15 de la noche del pasado viernes 8 de enero, en un restaurante de París. El diario *El Tiempo*, de Bogotá, dio la noticia en primera página en su edición del domingo. Y explicó a sus lectores, en tres líneas, por qué la escultora no estaba en Colombia: «Feliza había viajado hacía dos meses a París en compañía de su esposo, y antes había estado varias semanas en México». Nada más. Pero al día siguiente apareció una nota editorial firmada con unas iniciales que coinciden con las del director del periódico, Hernando Santos, y en la cual se hacían dos preguntas sobre Feliza Bursztyn: «¿Por qué tuvo que irse? ¿Por qué fue víctima de un exilio incomprensible al cual hubiera podido escapar con dos sencillas palabras?». Pero la nota no dice cuáles fueron esas palabras mágicas que acaso hubieran prolongado la vida.

De méritos tan grandes como sus carcajadas, la amiga más querida de sus amigos de todas partes, que no sólo hubiera dicho dos palabras simples, sino cuantas fueran necesarias para volver al único país donde siempre quiso vivir. Si alguien le hubiera hecho la caridad de decírselas a tiempo, tal vez hubiera podido cumplir su deseo y ejercer su derecho de morirse en su cama de Bogotá, rodeada de sus poetas locos, y no tirada por los suelos en un restaurante tapizado de espejos, ante la tenacidad estéril de seis médicos bomberos que trataban de despertarla y el espanto de su esposo y cuatro amigos que la sabíamos muerta para siempre desde el primer instante.

Nadie sabe mejor que mi familia y yo cómo fue la vida de Feliza Bursztyn, minuto a minuto, en los 166 días de su exilio mortal. En nuestra casa de México, donde vivió casi tres meses desde que salió de Bogotá bajo la protección diplomática de la embajada mexicana, hasta cuando pudo viajar a París, no sólo tuvimos tiempo de sobra para hablar de su drama, sino que sólo pudimos hablar de eso,

Publicado originalmente el 20 de enero de 1982.

porque Feliza quedó en una especie de estupor de disco rayado que
no le permitía hablar de otra cosa. Infinidad de veces, guiada por mi
curiosidad invencible de periodista y escritor, me contó hasta los
detalles más ínfimos de su mal recuerdo; llenamos juntos las grietas
vacías, tratamos de entender lo incomprensible, ansiosos de tocar
fondo en un misterio que no parecía tenerlo. En París, adonde llegamos el pasado octubre con muy pocos días de diferencia, nos seguimos viendo con frecuencia. De modo que considero como un derecho, e inclusive como un deber de sanidad social, que sea yo quien
trate de dar respuesta pública a las dos preguntas de H. S., aunque
sólo sea para que sus lectores no sucumban también en la peste del
olvido.

Feliza Bursztyn tuvo que escapar de Colombia —como hubiera
podido hacerlo el protagonista de *El proceso*, de Franz Kafka— para
no ser encarcelada por un delito que nunca le fue revelado. El viernes 24 de julio de 1981 una patrulla de militares al mando de un
teniente se presentó a su casa de Bogotá a las cuatro de la madrugada. Todos vestían de civil, con ruanas largas, debajo de las cuales llevaban escondidas las metralletas, y estaban autorizados por una orden
de allanamiento de un juez militar. Su comportamiento fue correcto, amable inclusive, y la requisa que hicieron de la casa duró casi
cuatro horas, pero fue más ritual que minuciosa. Feliza y su esposo,
Pablo Leyva, tuvieron la impresión de que eran unos muchachos
inexpertos que no sabían lo que buscaban ni tenían demasiado interés en encontrarlo. Lo único que registraron a fondo fue la cama
matrimonial, hasta el extremo de que la desarmaron y la volvieron
a armar. «Tal vez buscaban mis polvos perdidos», comentó más tarde
Feliza con su humor bárbaro. Otra cosa que les llamó la atención
fue una caja de fotografías que Feliza había llevado de La Habana,
pocos días antes, adonde había viajado para asistir a una exposición
de sus obras en la Casa de las Américas. Eran las fotos de una exposición colectiva de fotógrafos colombianos que se había realizado en
La Habana el año anterior, también bajo el patrocinio de la embajada de Colombia en Cuba, y con asistencia de sus funcionarios. La
Casa de las Américas le había pedido a Feliza el favor que las de-
volviera a sus autores, cuyos nombres y direcciones estaban escritos
al dorso de cada foto. Los soldados les echaron una ojeada superficial a casi un centenar y pusieron aparte tres, que se llevaron. Feliza,
que ni siquiera había tenido tiempo de abrir el paquete, no pudo ver
muy bien qué fotos eran, pero le pareció que alguna la había visto
publicada en la prensa de Colombia. También se llevaron una pistola Beretta inservible que un amigo le había regalado a Feliza en 1964,

en una época en que vivía sola en Bogotá, pues todavía no se había casado con Pablo Leyva. «No me atreví ni a tocarla nunca», me dijo Feliza, «por temor de sacarme un ojo.» Fue todo cuanto se llevaron. Es cierto que Feliza no encontró después dos cadenas y tres anillos que había puesto en su mesa de noche antes de dormirse, y que eran las únicas cosas de oro, pero también las que costaban menos en su paraíso de chatarra. Pero siempre insistió, con su buena fe inquebrantable, que no podía suponer algo que no había visto.

Terminada la requisa, Feliza fue llevada, sin su esposo, a las caballerizas de la Brigada de Institutos Militares. Permaneció sentada, sin comer ni beber, durante las once horas del interrogatorio. Le vendaron los ojos y le pegaron en el pecho una banda adhesiva con su número de presidiaria: 5. Ese parche, con ese número, está todavía pegado en la pared de la cocina en su casa de Bogotá. Siempre insistió en que la trataron con mucha corrección, que le pidieron excusas por tener que vendarla, y que ninguna de las incontables preguntas le permitió vislumbrar de qué la acusaban. Se lo preguntó a uno de sus interrogadores invisibles, y éste le dio una respuesta deslumbrante.

—Lo vamos a saber ahora por lo que usted nos diga.

Es sorprendente que hubiera resistido aquella prueba con tanta fortaleza, porque Feliza tenía una limitación pulmonar muy seria, debido a las sustancias tóxicas con que trabajaba, y además una lesión de la columna vertebral de la que no se recuperó nunca. Pero no perdió el sentido del humor en ningún momento de aquellas once horas desgraciadas de nuestra historia patria.

Le preguntaron si conocía a algún escritor, y contestó que sí: a Hernando Valencia Goelkel. Le preguntaron si no conocía a otros, y contestó que sí, pero que no los mencionaba porque eran muy malos escritores. Le preguntaron si no temía que la violaran, y contestó que no, porque toda mujer casada está acostumbrada a que la violen todas las noches. Sin embargo, los distintos interrogadores que nunca pudo ver coincidieron en poner en duda su nacionalidad colombiana. Nunca, en las horas interminables de su exilio, Feliza pareció olvidar que alguien en su propio país le hiciera esa ofensa. «Soy más colombiana que el presidente de la República», solía decir en sus últimos días. Más aún: mucho antes de que tuviera que abandonar a Colombia, una revista les preguntó a varios artistas colombianos en qué ciudad del mundo querían vivir, y Feliza fue la única que contestó: «En Bogotá».

Dos días después del interrogatorio, cuando ya se consideraba a salvo de toda sospecha, Feliza fue citada por un juez militar, que la acusó de tener en su casa un arma sin licencia. El juez le mostró la disposición según la cual aquel delito tenía prevista una pena de cinco años de cárcel. Le hizo firmar una notificación, la citó para dos días más tarde y le advirtió que no podía moverse de Bogotá. Dos días después, con todo el dolor de su alma, se asiló en la sede de la embajada de México.

No es comprensible, pues, que alguien se pregunte ahora por qué se fue Feliza de Colombia. El mismo Hernando Santos, que fue uno de sus amigos más queridos, tuvo la entereza de llamar por teléfono al ministro de la Defensa, general Camacho Leyva, para interceder en favor de ella, cuando todavía estaba detenida. El general le contestó que no podía hacer nada, porque había contra Feliza una denuncia concreta. Pocos días después, sin embargo, cuando todavía Feliza estaba asilada en la embajada de México, la Cancillería colombiana dijo, en un comunicado oficial, que no había ningún cargo contra ella, que podía viajar sin salvoconducto a donde quisiera y volver a Colombia con toda libertad. Pero otros días más tarde, el redactor de asuntos militares de *El Espectador*, de Bogotá, publicó una declaración muy explícita de un alto oficial de las fuerzas armadas de Colombia, que nunca se identificó, pero que tampoco ha sido desmentido por nadie. Este militar sin nombre afirmaba tener pruebas de que Feliza Bursztyn era correo entre los dirigentes cubanos y el M-19, pero que se la había tratado con la mayor consideración por ser mujer y artista. Otras gestiones que amigos de Feliza han hecho después ante autoridades militares han recibido la misma respuesta. Es alarmante, pero ya se sabe: en Colombia, los militares guardan secretos que las autoridades civiles no conocen.

Feliza no estaba en París por placer. Su propósito original era viajar a Estados Unidos, donde viven sus tres hijas, su hermana y su madre, todas ellas de nacionalidad norteamericana. Pero el consulado de Estados Unidos en México, después de consultarlo con el de Bogotá, le negó la visa. Amigos de Feliza le consiguieron entonces, con el Ministerio de Cultura de Francia, una beca de duración indefinida, con un estudio para que siguiera haciendo sus chatarras, y tarjeta de la Seguridad Social para que se vigilara mejor su mala salud. En París la encontró su esposo apenas diez días antes de su muerte, cuando vino de Bogotá a pasar con ella el último año nuevo de su vida.

La mujer que Pablo Leyva encontró en París no era la misma que había despedido en Bogotá. Estaba atónita y distante, y su risa explosiva y deslenguada se había apagado para siempre. Sin embargo, un examen médico muy completo había establecido que no tenía nada más que un agotamiento general, que es el nombre científico de la tristeza. El viernes 8 de enero, a nuestro regreso de Barcelona, Mercedes y yo los invitamos a cenar, junto con Enrique Santos Calderón y su esposa, María Teresa. Era una noche glacial de este invierno feroz y triste, y había rastros de nieve congelada en la calle, pero todos quisimos irnos caminando hasta un restaurante cercano. Feliza, sentada a mi izquierda, no había acabado de leer la carta para ordenar la cena, cuando inclinó la cabeza sobre la mesa, muy despacio, sin un suspiro, sin una palabra ni una expresión de dolor, y murió en el instante. Se murió sin saber siquiera por qué, ni qué era lo que había hecho para morirse así, ni cuáles eran las dos palabras sencillas que hubiera podido decir para no haberse muerto tan lejos de su casa.

«CUENTOS DE CAMINOS»

Hace muchos años estaba esperando un taxi en una avenida central de México, a pleno día, cuando vi acercarse uno que no pensé detener, porque había una persona sentada junto al conductor. Sin embargo, cuando estuvo más cerca comprendí que era una ilusión óptica: el taxi estaba libre.

Minutos después le conté al conductor lo que había visto, y él me dijo con una naturalidad absoluta que no era ni mucho menos una alucinación mía. «Siempre ocurre lo mismo, sobre todo de noche», me dijo. «A veces paso horas enteras dando vueltas por la ciudad sin que nadie me detenga, porque siempre ven una persona en el asiento de al lado.» En ese asiento confortable y peligroso que en algunos países se llama «el puesto del muerto», porque es el más afectado en los accidentes, y que nunca merecía tanto su nombre como en aquel caso del taxi.

Cuando le conté el episodio a Luis Buñuel, me dijo, con un grande entusiasmo: «Eso puede ser el principio de algo muy bueno». Siempre he pensado que tenía razón. Pues el episodio no es en sí mismo un cuento completo, pero es, sin duda, un magnífico punto de partida para un relato escrito o cinematográfico. Con un inconveniente grave, por supuesto, y es que todo lo que ocurra después tendría que ser mejor. Tal vez por eso no lo he usado nunca.

Lo que me interesa ahora, sin embargo, y al cabo de tantos años, es que alguien me lo ha vuelto a contar como si acabara de sucederle a él mismo en Londres. Es curioso, además, que hubiera sido allí, porque los taxis londinenses son distintos a los del resto del mundo. Parecen unas carrozas mortuorias, con cortinillas de encajes y alfombras moradas, con mullidos asientos de cuero y taburetes suplementarios hasta para siete personas, y un silencio interior que tiene algo del olvido funerario. Pero en el lugar del muerto, que no está a la derecha, sino a la izquierda del chofer, no hay una silla para otro

Publicado originalmente el 27 de enero de 1982.

pasajero, sino un espacio destinado al equipaje. El amigo que me lo contó en Londres me aseguró, sin embargo, que fue en ese lugar donde vio a la persona inexistente, pero que el chofer le había dicho —al contrario de lo que dijo el de México— que tal vez había sido una alucinación. Ahora bien: ayer le conté todo esto a un amigo de París, y éste se quedó convencido de que yo le estaba tomando el pelo, pues dice que fue a él a quien le ocurrió el episodio. Además, según me dijo, le sucedió de un modo más grave, pues le refirió al chofer del taxi cómo era la persona que había visto a su lado, le describió la forma de su sombrero y el color de su corbatín de lazo, y el chofer lo reconoció como el espectro de un hermano suyo que había sido muerto por los nazis durante los años de la ocupación alemana de Francia.

No creo que ninguno de estos amigos mienta, como no le mentí yo a Luis Buñuel, sino que me interesa señalar el hecho de que hay cuentos que se repiten en el mundo entero, siempre del mismo modo, y sin que nadie pueda nunca establecer a ciencia cierta si son verdades o fantasías, ni descifrar jamás su misterio. De todos ellos, tal vez el más antiguo y recurrente lo oí por primera vez en México.

Es el eterno cuento de la familia a la cual se le muere la abuelita durante las vacaciones en la playa. Pocas diligencias son tan difíciles y costosas y requieren tantos trámites y papeleos legales como trasladar un cadáver de un estado a otro. Alguien me contaba en Colombia que tuvo que sentar a su muerto entre dos vivos, en el asiento posterior de su automóvil, e inclusive le puso en la boca un tabaco encendido en el momento de pasar los controles de carretera, para burlar las incontables barreras del traslado legal. De modo que la familia de México enrolló a la abuela muerta en una alfombra, la amarraron con cuerdas y la pusieron bien atada en la baca del techo del automóvil. En una parada del camino, mientras la familia almorzaba, el automóvil fue robado con el cadáver de la abuelita encima, y nunca más se encontró ningún rastro. La explicación que se daba a la desaparición era que los ladrones tal vez habían enterrado el cadáver en despoblado y habían desmantelado el coche para quitarse, literalmente, el muerto de encima.

Durante una época, este cuento se repetía en México por todas partes, y siempre con nombres distintos. Pero las distintas versiones tenían algo en común: el que la contaba decía siempre ser amigo de los protagonistas. Algunos, además, daban sus nombres y direcciones. Pasados tantos años, he vuelto a escuchar este cuento en los lugares más distantes del mundo, inclusive en Vietnam, donde me lo repitió un intérprete como si le hubiera ocurrido a un amigo suyo en los años de la guerra. En todos los casos las circunstancias son las mismas, y si uno insiste, le dan los nombres y la dirección de los protagonistas.

Un tercer cuento recurrente lo conocí hace menos tiempo que los otros, y quienes tienen la paciencia de leer esta columna todas las semanas tal vez lo recuerden. Es la historia escalofriante de cuatro muchachos franceses que en el verano pasado recogieron a una mujer vestida de blanco en la carretera de Montpellier. De pronto, la mujer señaló hacia el frente con un índice aterrorizado, y gritó: «¡Cuidado!, esa curva es peligrosa». Y desapareció en el instante. El caso lo conocí publicado en diversos periódicos de Francia, y me impresionó tanto que escribí una nota sobre él. Me parecía asombroso que las autoridades de Francia no le hubieran prestado atención a un acontecimiento de tanta belleza literaria, y que además lo hubieran archivado por no encontrarle una explicación racional. Sin embargo, un amigo periodista me contó hace unos días en París que la razón de la indiferencia oficial era otra: en Francia, esa historia se repite y se cuenta desde hace muchos años, incluso desde mucho antes de la invención del automóvil, cuando los fantasmas errantes de los caminos nocturnos pedían el favor de ser llevados en las diligencias. Esto me hizo recordar que, en efecto, también entre los cuentos de la conquista del oeste de Estados Unidos se repetía la leyenda del viajero solitario que viajaba toda la noche en la carreta de pasajeros, junto con el viejo banquero, el juez novato y la bella muchacha del norte, acompañada por su gobernanta, y al día siguiente amanecía sólo su lugar vacío. Pero lo que más me ha sorprendido es descubrir que el cuento de la dama de blanco, tal como lo tomé de la prensa francesa, y tal como yo lo conté en esta columna, estaba ya contado por el más prolífico de todos nosotros, que es Manolo Vázquez Montalbán, en uno de los pocos libros suyos que no he leído: *La soledad del manager*. Conocí la coincidencia por la fotocopia que me mandó un amigo, que además ya conocía el cuento de tiempo atrás y por fuentes distintas.

El problema de derechos con Vázquez Montalbán no me preocupa: ambos tenemos el mismo agente literario de todos *els altres catalans*, y ya se encargará éste de repartir los derechos del cuento como a bien corresponda. Lo que me preocupa es la otra casualidad de que este cuento recurrente —el tercero que descubro— sea también un episodio de carretera. Siempre había conocido una expresión que ahora no he podido encontrar en tantos y tantos diccionarios inútiles como tengo en mi biblioteca, y es una expresión que de seguro tiene algo que ver con estas historias: «Son cuentos de caminos». Lo malo es que esta expresión quiere decir que son cuentos de mentiras, y estos tres que me persiguen son, sin duda, verdades completas que se repiten sin cesar en distintos lugares y con distintos protagonistas, para que nadie olvide que también la literatura tiene sus ánimas en pena.

OTRA VEZ DEL AVIÓN A LA MULA... ¡QUÉ DICHA!

De modo que hice las maletas, encomendé mi alma a media botella de whisky y me subí en el Concorde. Llevaba casi dos horas en una sala de espera del aeropuerto Charles de Gaulle, de París, que casi en todo sentido parece una estación espacial, y no había dejado de mirar un solo instante, a través de los cristales panorámicos, aquel esbelto pájaro en reposo, con sus inmensas alas extendidas, y no hacía más que preguntarme, entre cada sorbo de whisky puro, por qué había llegado a ser tan cobarde que no tenía valor ni siquiera para desistir de la aventura. Al lado del Concorde pasaban los otros aviones de estirpe más modesta sin que nadie dijera adiós desde las ventanas, sin nadie que llorara de desolación en el muelle como ocurría cuando zarpaban los barcos de otros tiempos, sin dejarnos siquiera el consuelo de sus bramidos de adioses, y el corazón se me encogía cada vez más al comprobar que el avión más veloz y más caro era el más pequeño de todos, y sus ventanas eran apenas tan grandes como la palma de la mano, y su envergadura era menor que la de los primeros aviones de hélice que asombraron al mundo. Meterse en aquel cohete, a dos veces la velocidad del sonido, sólo por llegar a Nueva York tres horas antes que un avión convencional, era una temeridad senil. Sin embargo, ahí estaba yo, entre los ejecutivos impasibles y las radiantes putas de lujo, sintiendo que ni la vida dura ni la vida muelle habían cambiado nada dentro de mí desde aquel ardiente mediodía de quién sabe cuándo en que mi abuelo me subió por primera vez en el tren de Aracataca. Era lo mismo: también ahora iba en brazos del miedo, que es el único abuelo que me quedó desde que se murieron los de carne y hueso.

Un amigo colombiano me había explicado el Concorde con una frase fulminante: «Es igual al DC-3, pero a toda mierda». No tengo que agregar ni quitar una letra a esta definición. Su longitud es casi cuatro veces mayor, pero la altura del techo, la estrechez del corredor

Publicado originalmente el 3 de febrero de 1982.

central, el tamaño de los asientos son los mismos de aquellos aviones primitivos en que atravesábamos selvas y saltábamos montañas con la irresponsabilidad feliz de la juventud.

No había, pues, ninguna razón para tener ahora más miedo que entonces, salvo por la diferencia poética de que las vacas de antaño dejaban de comer para ver pasar los aviones por encima de los potreros y, en cambio, el Concorde navega por un cielo solitario que ya no es de este mundo. Salvo por eso, todo lo demás parece igual. Inclusive por lo que me pareció lo más importante: la ambientación interior. Uno espera encontrarse dentro de una cápsula sideral, con una estética diferente a la de los otros aviones mortales; pero, en cambio, es la misma de los hoteles de provincia donde uno pasaba las noches llorando de soledad. «Por este precio», dijo una señora que regresaba de los servicios sanitarios, «bien podían tener un Picasso colgado en cada Concorde.» Me sorprendió por la lucidez con que había logrado expresar una idea que me hubiera hecho falta para explicar mi desazón.

Una de las pérdidas que más me han dolido y desconcertado en la vida fue la de una noche completa en un vuelo de Los Ángeles a Tokio. No la volví a encontrar jamás, y cada vez que la recuerdo me pregunto qué hubiera hecho con ella, si no sería esa la noche más feliz que me estaba destinada, y que se me perdió para siempre por no quedarme quieto en mi casa. En efecto, salimos de Los Ángeles un domingo a las dos de la tarde y llegamos a Tokio a las dos de la tarde del lunes, después de volar once horas a pleno día. Lo primero que noté al desembarcar era que faltaba en mi vida la noche del domingo, no sólo con sus horas contadas, con su cielo y sus estrellas, sino también con su sueño. Esa noche, en el inmenso hotel de Tokio, donde lo despiertan a uno con computadoras ocultas que cantan como pájaros, yo no me preocupaba por tantas y tantas maravillas de la ciencia, sino que me sentía agobiado por la zozobra de estar tratando de dormir en una noche que no era la mía.

En el Concorde, la confusión del tiempo es más amarga, porque uno sale de París a las once de la mañana y llega a Nueva York a las ocho de la mañana del mismo día. Los más avanzados en estos misterios de la ciencia habíamos terminado por aceptar la confusión convencional de que uno saliera de París a las doce del día y llegara a Nueva York a las dos de la tarde, después de haber volado siete horas. Pero desayunar una vez en París y volver a desayunar otra vez en Nueva York el mismo día a la misma hora es una usurpación inadmisible de los misterios reservados a la poesía. Sin embargo, esas curiosidades físicas que todos aceptamos como normales, pero que yo no he logrado entender nunca por más que mis amigos sabios me las explican con números y dibujos, dejan de ser poéticas cuando uno sabe cuál es el riesgo a que uno se somete para hacerlas posibles. En realidad, este avión supersónico, que es en sí mismo una proeza de la inteligencia humana, vuela a 2.200 kilómetros por hora, o sea, más de seis veces más que su bisabuelo de hélice. Para conseguir esa velocidad de vértigo tiene que elevarse a casi veinte kilómetros de altura, donde ya no hay más aire, donde la temperatura invernal es de 66 grados bajo cero y la presión atmosférica es unas veinte veces menor que la del mar. Para que uno pueda disfrutar del servicio exquisito, tomar todo el vino de champaña que se desea y extasiarse con los mejores quesos del mundo en semejantes condiciones es necesario que el ámbito de la nave se mantenga igual que al nivel del mar. Es decir, que entre la presión exterior y la interior hay una diferencia tan grande, que cien viajeros felices van dentro de una bomba dos veces más veloz que el sonido, y que una simple fisura invisible bastaría para convertirlos a todos en glorioso polvo de estrellas. Sería no sólo la forma más moderna de morir, sino tal vez la única garantizada de morir para siempre en cuerpo y alma.

Por fortuna, en la única revista que encontré a bordo había un artículo consolador, sobre la posibilidad inminente de que el dirigible, el manso y venerable dinosaurio de la aeronáutica, sea resucitado con fines comerciales, cuarenta años después de que el gigantesco *Hindenburg* fuera consumido por las llamas en New Jersey, con un saldo de 36 muertos. El *Hindenburg* había hecho 144 vuelos a través del Atlántico, y su única falla fue la causa de su desastre: estaba inflado con oxígeno, que es un gas inflamable. El nuevo dirigible, en cambio, estará inflado con helio, y hay ya una versión británica que entrará en servicio entre Londres y París dentro de cinco años, con una carga útil de dos toneladas y una velocidad de 115 kilómetros por hora. Pero habrá otra versión norteamericana capaz de llevar setecientos pasajeros a través del Atlántico, con dormitorios, corredores de lujo, salas de fiesta y espacios de recreo, pero a no más de treinta metros sobre el nivel del mar. Algo así como un barco que volara a una velocidad humana de quinientos kilómetros por hora, sin prisas ni sobresaltos, para que sea otra vez verdad el placer de viajar. Fue muy difícil y muy doloroso pasar de la mula al avión, pero ahora vamos bien en el viaje de regreso. Otra vez del avión a la mula.

GRAHAM GREENE: LA RULETA RUSA DE LA LITERATURA

No bien había acabado de leer el último libro de Graham Greene
—*Vías de escape*, que es el segundo volumen de sus memorias desco-
sidas— cuando estalló en Francia el escándalo de su libro siguiente,
Yo acuso, que es, al parecer, un reportaje sigiloso sobre la corrupción
en la muy corrupta y muy hermosa ciudad de Niza. El alcalde de
ésta —como cualquier canciller colombiano en una ocasión reciente—
se apresuró a declarar que el gran escritor inglés, cuyos libros figu-
ran entre los más vendidos del mundo, sólo estaba buscando publicidad para aumentar sus ventas. Graham Greene, que es refractario a
toda clase de declaraciones dramáticas, reiteró para la prensa las denuncias de su libro, y le puso un grave punto final al asunto: «Prefiero que me maten de un tiro a morirme de viejo en mi cama». La
declaración, más que suya, parece de alguno de sus personajes. Hay
que confiar en que no sean muchos los franceses que compartan la
opinión apresurada del alcalde de Niza, que ninguno de ellos le dis-
pute a Graham Greene el derecho indeseable de morirse de muerte
natural, y que todos terminen por entender que los escritores son
una plaga imprevisible, incapaces de callar lo que a su juicio se debe
decir.

Graham Greene, a sus 78 años bien vividos, no podía menos que
hacer lo que ha hecho durante toda su vida: escribir contra toda injusticia. Desde hace tiempo vive en Antibes, una ciudad marina a
treinta kilómetros de Niza, muy cerca de donde vivió y murió Pablo Picasso, y donde han ido a morir por sus propios pies muchos
artistas grandes de nuestro tiempo. Los elefantes tienen un sitio común
para morir, y hasta allí llegan con el último aliento. En ese sentido
se ha dicho muchas veces que la Costa Azul es uno de los más
espléndidos cementerios de elefantes del mundo, y Antibes es uno
de sus recodos más tranquilos y hermosos. Sin duda, quienes sabían
que Graham Greene se había instalado allí desde hace muchos años

Publicado originalmente el 10 de febrero de 1982.

no habían podido eludir la metáfora de los elefantes. Es cierto que casi cada año ha publicado un libro, y que hace apenas cuatro escribió la que para mi gusto es una de sus obras maestras: *El factor humano*. Pero no hacía declaraciones, no se le veía en sitios públicos, salvo en algún bar escondido cuyo propietario lo conoce y lo recibe siempre con un cóctel diabólico inventado por él mismo, y había logrado inclusive la dicha de que ya no le incluyeran cada año en la lista de candidatos al Premio Nobel. «No me lo darán nunca porque no me consideran un escritor serio», me dijo alguna vez, y es posible que tenga razón. Lo que nadie podía imaginarse es que aquel inglés colorado y de aspecto un poco distraído, que no se había convertido en un personaje típico de la región, como tantos otros artistas en reposo, seguía viendo más de lo que parecía, y escrutaba con su atención implacable las entrañas más oscuras y peligrosas de la ciudad.

A mí no me ha sorprendido. Primero, porque creo tener una cierta idea de cómo son los escritores por dentro. Aun en sus instantes más pasivos, cuando están tirados boca arriba en la playa, trabajan como burros. El mismo Graham Greene lo ha dicho: «Escribir es una forma de terapia: a veces me pregunto cómo se las arreglan los que no escriben, o los que no pintan o no componen música, para escapar de la locura, de la melancolía, del terror pánico inherente a la condición humana». Rilke dijo lo mismo de otro modo: «Si usted cree que es capaz de vivir sin escribir, no escriba».

El libro sobre la corrupción de Niza no me ha sorprendido, en segundo término, porque Graham Greene ha estado en incontables lugares de este mundo –como periodista, como espía, como corresponsal de guerra, como turista simple– y todos ellos han aparecido más temprano o más tarde incorporados en la esencia de sus libros. Graham Greene lo reconoce en sus memorias, aunque se pregunta a su vez, como todo escritor, hasta qué punto era consciente de la búsqueda o el aprovechamiento de sus fuentes de inspiración. «Yo no las buscaba», dice: «tropezaba con ellas.» Era muy difícil, por supuesto, que no tropezara con los bajos fondos de Niza.

No me ha sorprendido, en último término, porque, de un modo consciente o inconsciente, Graham Greene fue siempre a buscar sus fuentes de inspiración en lugares distantes y arriesgados. En cierta ocasión, siendo muy joven, jugó a la ruleta rusa. El episodio está contado sin dramatismos en el primer volumen de sus memorias, que llegan hasta cuando cumplió veintisiete años. Desde antes se había hablado de eso con cierta frecuencia, como una extravagancia de la juventud. Pero, si se piensa con más cuidado, Graham Greene no ha dejado casi nunca de jugar a la ruleta rusa: la mortal ruleta rusa de la literatura con los pies sobre la tierra. El último episodio es, sin duda, este libro sobre la cara oculta de Niza, que tantos lectores de Graham Greene estamos ansiosos de conocer.

Es difícil encontrar en este siglo un escritor que sea víctima de tantos juicios apresurados como lo es Graham Greene. El más grave de ellos es la tendencia a que se le considere como un simple escritor de novelas de misterio, y que, aun si así fuera, se olvide con tanta facilidad que muchas novelas de misterio circulan por los cielos más altos de la literatura. Pero el más injusto de esos juicios es el de que Graham Greene no se interesa por la política. Nada más falso. «A partir de 1933», dice él mismo, «la política fue ocupando un lugar mayor en mis novelas.» De su permanencia en Vietnam como enviado del *Times* de Londres para escribir sobre la guerra de independencia contra los franceses nos quedó su novela *El americano impasible*. El más distraído de los lectores debería darse cuenta de que esa novela no es sólo una representación literaria de aquel drama humano, sino una visión profética de la intervención posterior de Estados Unidos en la vida privada de Vietnam.

En este aspecto, Graham Greene nos concierne a los latinoamericanos, inclusive por sus libros menos serios. En *El poder y la gloria* dejó plasmada una visión fragmentaria, pero muy conmovedora, de toda una época de México. *Comediantes* es una exploración en el infierno de Haití bajo la tiranía vitalicia del doctor Duvalier. *Nuestro hombre en La Habana* es una mirada fugaz, pero de una ironía amarga, sobre el burdel turístico del general Fulgencio Batista. *El cónsul honorario* fue una de las pocas noticias que la literatura nos ha dado sobre el despotismo oscuro del general Stroessner en el Paraguay. Por todo esto alguna vez le pregunté si no se consideraba un escritor de América Latina. No me contestó, pero se quedó mirándome con una especie de estupor muy británico que nunca he logrado descifrar.

Cuando se firmaron los tratados del canal de Panamá, hace cuatro años, Graham Greene y yo viajamos a Washington como invitados personales del general Omar Torrijos. Para ambos fue una buena ocasión de entrar en Estados Unidos sin formalismo de inmigración: ambos tenemos limitado nuestro ingreso por motivos que ni el uno ni el otro conocemos muy bien. Nunca olvidaré el alborozo de niño travieso de Graham Greene cuando desembarcamos en la base militar de Andrews, cerca de Washington, con nuestros pasaportes oficiales panameños válidos por esa sola vez, y entre las músicas marciales y el saludo de rigor de veintiún cañonazos. «Estas cosas sólo le pasan a uno», pensaba yo, muerto de risa, mientras la periodista Amparo Pérez trataba de sacarme una declaración trascendental. Graham Greene, en cambio, se inclinó hacia mí, menos serio que yo, y me dijo en francés: «Estas cosas sólo le pasan a Estados Unidos». Esa misma noche, un grupo de amigos tuvo que ocuparse de él, porque quería asistir a una recepción oficial en la Casa Blanca sólo para mentarle la madre al general Pinochet. No es extraño, pues, que un hombre así haya escrito ese libro sobre la cara oculta de Niza, que sus lectores asiduos y devotos del mundo entero estamos ansiosos de leer.

Hace poco, al despertar en mi cama de México, leí en un periódico que yo había dictado una conferencia literaria el día anterior en Las Palmas de Gran Canaria, al otro lado del océano, y el acucioso corresponsal no sólo había hecho un recuento pormenorizado del acto, sino también una síntesis muy sugestiva de mi exposición. Pero lo más halagador para mí fue que los temas de la reseña eran mucho más inteligentes de lo que se me hubiera podido ocurrir, y la forma en que estaban expuestos era mucho más brillante de lo que yo hubiera sido capaz. Sólo había una falla: yo no había estado en Las Palmas ni el día anterior ni en los veintidós años precedentes, y nunca había dictado una conferencia sobre ningún tema en ninguna parte del mundo.

Sucede a menudo que se anuncia mi presencia en lugares donde no estoy. He dicho por todos los medios que no participo en actos públicos, ni pontifico en la cátedra, ni me exhibo en televisión, ni asisto a promociones de mis libros, ni me presto para ninguna iniciativa que pueda convertirme en un espectáculo. No lo hago por modestia, sino por algo peor: por timidez. Y no me cuesta ningún trabajo, porque lo más importante que aprendí a hacer después de los cuarenta años fue a decir que no cuando es no. Sin embargo, nunca falta un promotor abusivo que anuncia por la prensa, o en las invitaciones privadas, que estaré el martes próximo, a las seis de la tarde, en algún acto del cual no tengo noticia. A la hora de la verdad, el promotor se excusa ante la concurrencia por el incumplimiento del escritor que prometió venir y no vino, agrega unas gotas de mala leche sobre los hijos de los telegrafistas a quienes se les sube la fama a la cabeza, y termina por conquistarse la benevolencia del público para hacer con él lo que le da la gana. Al principio de esta desdichada vida de artista, aquel truco malvado había empezado a causarme erosiones en el hígado. Pero me he consolado un poco

Publicado originalmente el 17 de febrero de 1982.

leyendo las memorias de Graham Greene, quien se queja de lo mismo en su divertido capítulo final, y me ha hecho comprender que no hay remedio, que la culpa no es de nadie, porque existe otro yo que anda suelto por el mundo, sin control de ninguna índole, haciendo todo lo que uno debiera hacer y no se atreve.

En ese sentido, lo más curioso que me ha ocurrido no fue la conferencia inventada de Canarias, sino el mal rato que pasé hace unos años con Air France, a propósito de una carta que nunca escribí. En realidad, Air France había recibido una protesta altisonante y colérica, firmada por mí, en la cual yo me quejaba del mal trato de que había sido víctima en el vuelo regular de esa compañía entre Madrid y París, y en una fecha precisa. Después de una investigación rigurosa, la empresa había impuesto a la azafata las sanciones del caso, y el departamento de relaciones públicas me mandó a Barcelona una carta de excusas, muy amable y compungida, que me dejó perplejo, porque en realidad yo no había estado nunca en ese vuelo. Más aún: siempre vuelo tan asustado que ni siquiera me doy cuenta de cómo me tratan, y todas mis energías las consagro a sostener mi silla con las manos para ayudar a que el avión se sostenga en el aire, o a tratar de que los niños no corran por los pasillos por temor de que desfonden el piso. El único incidente indeseable que recuerdo fue en un vuelo desde Nueva York en un avión tan sobrecargado y opresivo que costaba trabajo respirar. En pleno vuelo, la azafata le dio a cada pasajero una rosa roja. Yo estaba tan asustado que le abrí mi corazón. «En vez de darnos una rosa», le dije, «sería mejor que nos dieran cinco centímetros más de espacio para las rodillas.» La hermosa muchacha, que era de la estirpe brava de los conquistadores, me contestó impávida: «Si no le gusta, bájese». No se me ocurrió, por supuesto, escribir ninguna carta de protesta a una compañía de cuyo nombre no quiero acordarme, sino que me fui comiendo la rosa, pétalo por pétalo, masticando sin prisa sus fragancias medicinales contra la ansiedad, hasta que recobré el aliento. De modo que cuando recibí la carta de la compañía francesa me sentí tan avergonzado por algo que no había hecho, que fui en persona a sus oficinas para aclarar las cosas, y allí me mostraron la carta de protesta. No hubiera podido repudiarla, no sólo por su estilo, sino porque a mí mismo me hubiera costado trabajo descubrir que la firma era falsa.

El hombre que escribió esa carta es, sin duda, el mismo que dictó la conferencia de Canarias, y el que hace tantas cosas de las cuales apenas si tengo noticias por casualidad. Muchas veces, cuando llego a una casa de amigos, busco mis libros en la biblioteca con aire distraído, y les escribo una dedicatoria sin que ellos se den cuenta. Pero más de dos veces me ha ocurrido encontrar que los libros estaban ya dedicados, con mi propia letra, con la misma tinta negra que uso siempre y el mismo estilo fugaz, y firmados con un autógrafo al cual lo único que le faltaba para ser mío es que yo lo hubiera escrito.

Igual sorpresa me he llevado al leer en periódicos improbables alguna entrevista mía que yo no concedí jamás, pero que no podría reprobar con honestidad, porque corresponde línea por línea a mi pensamiento. Más aún: la mejor entrevista mía que se ha publicado hasta hoy, la que expresaba mejor y de un modo más lúcido los recovecos más intrincados de mi vida, no sólo en literatura, sino también en política, en mis gustos personales y en los alborozos e incertidumbres de mi corazón, fue publicada hace unos dos años en una revista marginal de Caracas, y era inventada hasta el último aliento. Me causó una gran alegría, no sólo por ser tan certera, sino porque estaba firmada con su nombre completo por una mujer que yo no conocía, pero que debía amarme mucho para conocerme tanto aunque sólo fuera a través de mi otro yo.

Algo semejante me ocurre con gentes entusiastas y cariñosas que me encuentro por el mundo entero. Siempre es alguien que estuvo conmigo en un lugar donde yo no estuve nunca, y que conserva un recuerdo grato de aquel encuentro. O que es muy amigo de algún miembro de mi familia, al cual no conoce en realidad, porque el otro yo parece tener tantos parientes como yo mismo, aunque tampoco ellos son los verdaderos, sino que son los dobles de los parientes míos.

En México me encuentro con frecuencia con alguien que me cuenta las pachangas babilónicas que suele hacer con mi hermano Humberto en Acapulco. La última vez que lo vi me agradeció el favor que le hice a través de él, y no me quedó más remedio que decirle que de nada, hombre, ni más faltaba, porque nunca he tenido corazón para confesarle que no tengo ningún hermano que se llame Humberto ni viva en Acapulco.

Hace unos tres años acababa de almorzar en mi casa de México cuando llamaron a la puerta, y uno de mis hijos, muerto de risa, me dijo: «Padre, ahí te buscas tú mismo». Salté del asiento, pensando con una emoción incontenible: «Por fin, ahí está». Pero no era el otro, sino el joven arquitecto mexicano Gabriel García Márquez, un hombre reposado y pulcro, que sobrelleva con un grande estoicismo la desgracia de figurar en el directorio telefónico. Había tenido la gentileza de averiguar mi dirección para llevarme la correspondencia que se había acumulado durante años en su oficina.

Hacía poco, alguien que estaba de paso en México buscó nuestro teléfono en el directorio, y le contestaron que estábamos en la clínica, porque la señora acababa de tener una niña. ¡Qué más hubiera querido yo! El hecho es que la esposa del arquitecto debió de recibir un ramo de rosas espléndidas, y además muy merecidas, para celebrar el feliz advenimiento de la hija con que soñé toda la vida y que no tuve nunca.

No. Tampoco el joven arquitecto era mi otro yo, sino alguien mucho más respetable: un homónimo. El otro yo, en cambio, no me encontrará jamás, porque no sabe dónde vivo, ni cómo soy, ni podría concebir que seamos tan distintos. Seguirá disfrutando de su existencia imaginaria, deslumbrante y ajena, con su yate propio, su avión privado y sus palacios imperiales donde baña con champaña a sus amantes doradas y derrota a trompadas a sus príncipes rivales. Seguirá alimentándose de mi leyenda, rico hasta más no poder, joven y bello para siempre y feliz hasta la última lágrima, mientras yo sigo envejeciendo sin remordimientos frente a mi máquina de escribir, ajeno a sus delirios y desafueros, y buscando todas las noches a mis amigos de toda la vida para tomarnos los tragos de siempre y añorar sin consuelo el olor de la guayaba. Porque lo más injusto es eso: que el otro es el que goza de la fama, pero yo soy el que se jode viviendo.

LAS ESPOSAS FELICES SE SUICIDAN A LAS SEIS

A veces me entretengo en el supermercado observando a las amas de casa que vacilan frente a los estantes mientras deciden qué comprar, las veo vagar con su carrito por los laberintos de artículos expuestos a su curiosidad, y siempre me pregunto, al final del examen, cuál de ellas es la que se va a suicidar ese día a las seis de la tarde. Esta mala costumbre me viene de un estudio médico del cual me habló hace algunos años una buena amiga, y según el cual las mujeres más felices de las democracias occidentales, al cabo de una vida fecunda de matriarcas evangélicas, después de haber ayudado a sus maridos a salir del pantano y de formar a sus hijos con pulso duro y corazón tierno, terminan por suicidarse cuando todas las dificultades parecían superadas y deberían navegar en las ciénagas apacibles de su otoño. La mayoría de ellas, según las estadísticas, se suicidan al atardecer.

Se ha escrito desde siempre sobre la condición de la mujer, sobre el misterio de su naturaleza, y es difícil saber cuáles han sido los juicios más certeros. Recuerdo uno feroz, a cuyo autor no quiero denunciar aquí porque es alguien a quien admiro mucho y temo librarlo a las furias de las lectoras eventuales de esta nota. Dice así la frase: «Las mujeres no desean más que el calor de un hogar y el amparo de un techo. Viven en el temor de la catástrofe y ninguna seguridad es bastante segura para ellas y a sus ojos el porvenir no es sólo inseguro, sino catastrófico. Para luchar por adelantado contra esos males desconocidos no hay engaño al que no recurran, no hay rapacidad de la que no se sirvan, y no hay ningún placer ni ilusión que no combatan. Si la civilización hubiera estado en manos de las mujeres, seguiríamos viviendo en las cuevas de los montes, y la inventiva de los hombres habría cesado con la conquista del fuego. Todo lo que piden a la caverna, más allá del abrigo, es que sea un grado más ostentosa que la del vecino. Todo lo que piden para la

Publicado originalmente el 24 de febrero de 1982.

seguridad de los hijos es que estén seguros en una cueva semejante a la suya». Por los tiempos en que conocí esta frase, declaré en una entrevista: «Todos los hombres son impotentes». Muchos amigos y, sobre todo, algunos que no lo eran, no pudieron reprimir sus ímpetus machistas y me replicaron con denuestos públicos y privados que podrán resumirse en uno solo: «El ladrón juzga por su condición». Pienso ahora que, tanto en la frase sobre las mujeres como en la mía sobre los hombres, lo único reprochable es la exageración. No hay duda: todos los hombres somos impotentes cuando menos lo esperamos y, sobre todo, cuando menos lo queremos, porque nos han enseñado que las mujeres esperan de nosotros mucho más de lo que somos capaces, y ese fantasma, a la hora de la verdad, inhibe a los humildes y conturba a los arrogantes. En la frase sobre las mujeres, que en realidad fue atribuida a las del imperio romano, falta señalar el horror de esa condición que en nuestros tiempos conduce a tantas amas de casa a tomarse el frasco de somníferos, uno detrás del otro, y mejor si es con un vaso de alcohol, a las seis de la tarde.

No hay nada más difícil, más estéril y empobrecedor que la logística de la casa. Una de las cosas que más me intrigan, y que más admiro en este mundo, es cómo hacen las mujeres para que nunca falte el papel en los baños. Calcular por metros enrollados una necesidad cotidiana que es la más íntima, la menos previsible y la más inveterada de cada miembro de la familia, requiere no sólo un instinto especial, sino un talento administrativo digno de mejor causa. Si no las admirara tanto por tantos motivos, como creo haberlo establecido en mis libros, me bastaría con esa virtud para admirar tanto a las mujeres. Creo que muy pocos hombres serían capaces de mantener el orden de la casa con tanta naturalidad y eficacia, y yo no lo haría por ningún dinero ni ninguna razón de este mundo.

En esa logística doméstica está el lado oculto de la historia que no suelen ver los historiadores. Para no ir muy lejos, he creído siempre que las guerras civiles de Colombia en el siglo pasado no hubieran sido posibles sin la disponibilidad de las mujeres para quedarse sosteniendo el mundo en la casa. Los hombres se echaban la escopeta al hombro, sin más vueltas, y se iban a la aventura. No tomaban ninguna providencia para la vida de la familia mientras ellos estuvieran ausentes, y menos ante la posibilidad de su muerte. Mi abuela me contaba que mi abuelo, siendo muy joven, se fue con las tropas del general Rafael Uribe Uribe, y no volvió a saber de él durante casi un año. Una madrugada tocaron a la ventana de su dormitorio, y una voz que nunca identificó, le dijo: «Tranquilina, si quieres ver a Nicolás, asómate ahora mismo». Ella, que entonces era joven y muy bella, abrió la ventana en el instante y sólo alcanzó a ver el polvo de la cabalgata que acababa de pasar y en la cual, en efecto, iba el marido, que ni siquiera alcanzó a distinguir. Mujeres como ella criaban solas a sus hijos, los hacían hombres para otras mujeres que serían también heroínas invisibles de otras guerras futuras, y hacían mujeres a las hijas para otros maridos guerreros que ni siquiera estaban escritos en las líneas de sus manos, y sostenían la casa en hombros hasta que el hombre volvía. Cómo lo hicieron, con qué ideales y con qué recursos, es algo que no se encuentra en nuestros textos de historia escritos por los hombres. En realidad, en toda la historia de la polvorienta y mojigata Academia de la Historia de Colombia sólo ha habido una mujer. Está allí desde hace apenas poco más de un año, y tengo motivos para creer que vive intimidada por la gazmoñería de sus compañeros de gloria.

La explicación de que las mujeres sometidas a su condición actual de amas de casa terminen por suicidarse a las seis de la tarde, no es tan misteriosa como podría parecer. Ellas, que en otros tiempos fueron bellas, se habían casado muy jóvenes con hombres emprendedores y capaces que apenas empezaban su carrera. Eran laboriosas, tenaces, leales, y empeñaron lo mejor de ellas mismas en sacar adelante al marido con una mano, mientras que con la otra criaban a los hijos con una devoción que ni ellas mismas apreciaron como un milagro de cada día. «Llevaban», como tantas veces he oído decir a mi madre, «todo el peso de la casa encima.» Tal como lo hacían sus abuelas en otras tantas guerras olvidadas. Sin embargo, aquel heroísmo secreto, por agotador e ingrato que fuera, era para ellas una justificación de sus vidas. Lo fue menos muchos años después, cuando el marido que acabaron de criar logró una posición profesional y empezó a cosechar solo los frutos del esfuerzo común, y lo fue mucho menos cuando los hijos acabaron de crecer y se fueron de la casa. Aquél fue el principio de un gran vacío, que no era todavía irremediable porque dejaba una grieta de alivio en el trabajo más aburrido del mundo: los oficios de la casa, con los cuales las perfectas casadas solitarias sobrellevaban las horas de la mañana. Todavía no comían solas si el marido llamaba en el último momento para decir que no lo esperaran a almorzar: algunas amigas en iguales condiciones estaban ansiosas de acompañarlas. No obstante, después de la siesta estéril, de la peluquería obsesiva, de las novelas de televisión o los telefonemas interminables, sólo quedaba en el porvenir el abismo de las seis de la tarde. A esa hora, o bien se conseguían un amante de entrada por salida, de aquellos que ni siquiera tienen tiempo de quitarse los zapatos, o se tomaban de un golpe todo el frasco de somníferos. Muchas, las que habían sido más dignas, hacían ambas cosas.

El comentario de los amigos sería siempre el mismo: «¡Qué raro!, si tenía todo para ser feliz». Mi impresión personal es que estas esposas felices sólo lo fueron, en realidad, cuando tenían muy poco para serlo.

EL FANTASMA PARA EL PROGRESO

El 13 de marzo de 1961, el presidente de Estados Unidos John F. Kennedy reunió en la Casa Blanca a los miembros del Congreso y al cuerpo diplomático de los países latinoamericanos, para lanzar un proyecto espectacular. Lo llamó la Alianza para el Progreso, y estaba compuesto por diez puntos con los que pretendía resolver en un decenio el insondable drama económico, político y social de América Latina, y no tanto por compasión ante las injusticias seculares, sino como un emplasto de emergencia para cerrarle el paso a los vientos nuevos de la revolución cubana. La magnitud desmesurada del proyecto estaba resumida en aquel discurso del presidente Kennedy, cuyo aliento profético no tenía nada que envidiarle al Antiguo Testamento. Yo estaba aquel día histórico en la Casa Blanca, en mi condición de corresponsal de prensa, y no podía menos que evocarlo una vez más y con más fuerza el pasado 24 de febrero, mientras escuchaba el discurso del presidente Ronald Reagan frente a la OEA.

La similitud de los dos discursos, con veinte años de distancia, es sorprendente, sobre todo, como una confirmación —una vez más— de que la historia sólo se repite en comedia. El presidente Kennedy, joven y apuesto, leyó su discurso con un acento juvenil de Harvard University. El presidente Ronald Reagan, masticando su acento de ranchero de California, recitó sin un solo tropiezo las diez páginas apretadas que llevaba en la mano, y a las cuales no dirigió ni una simple mirada de modestia. Pero no fue una improvisación. Varias horas antes, una copia del discurso había sido repartida a las agencias de prensa, con la autorización de transmitir a sus clientes del mundo entero los párrafos más resonantes. La correspondencia exacta entre las líneas escritas y las líneas dichas por el presidente Reagan no podía explicarse sino como una proeza en su interminable carrera de actor: se había aprendido de memoria un discurso de casi 3.000 pa-

Publicado originalmente el 3 de marzo de 1982.

labras. Y no sólo lo repitió sin tropiezos, sino que supo darle el énfasis y hasta las tonalidades de duda de una improvisación. No lo digo como un reproche, sino todo lo contrario: como el reconocimiento a un anciano casi embalsamado, que, sin embargo, ha sido capaz de semejante aventura escolar.

La Alianza para el Progreso fue concebida como un programa de desarrollo social para oponerlo a la expansión soviética en América Latina. Kennedy lo dijo: «En este momento de máxima oportunidad enfrentamos las mismas fuerzas que han hecho peligrar a América a través de su historia, las fuerzas foráneas que una vez más pretenden imponer a los pueblos del nuevo mundo el despotismo del viejo mundo». Reagan, como si el tiempo sólo hubiera pasado para él, dijo: «Si no actuamos con urgencia y de un modo decisivo en defensa de la libertad, nuevas Cubas van a nacer de las ruinas de los conflictos que hay ahora mismo en la zona, vamos a encontrarnos con regímenes incompetentes, totalitarios, exportadores de la subversión y unidos militarmente a la Unión Soviética». Ambos presidentes se dieron golpes de pecho por los errores de sus antepasados. «Permítanme ante todo admitir», dijo Kennedy, «que nosotros, los norteamericanos, no siempre hemos comprendido el significado de esta misión común.» Reagan, por su parte, dijo: «No siempre hemos cumplido con estos ideales; hemos sido políticamente débiles en algunos momentos de nuestra historia, hemos sido económicamente atrasados, injustos socialmente o incapaces de resolver nuestros problemas a través de medios pacíficos». Kennedy, conmovido por la lucha de los países de América Latina, nos llamó «compatriotas». Reagan fue más lejos: «En estos momentos, mi país está dispuesto a ser algo más que un simple buen vecino, es decir, a ser amigo y un hermano». Kennedy citó a Juárez: «La democracia es el destino de la humanidad futura». Pero no citó también otra frase suya que también hubiera sido adecuada: «El respeto al derecho ajeno es la paz». Reagan, con un cinismo admirable, citó a José Martí: «La humanidad está compuesta por dos tipos de hombres, los que aman y crean, así como los que odian y destruyen». Pero no citó otra frase que hubiera definido mejor la ideología de José Martí, en relación con Estados Unidos: «Viví en el monstruo y le conozco sus entrañas, y mi honda es la de David». Para concluir, Kennedy expresó su esperanza de que Cuba «se reincorpore a la sociedad de hombres libres, uniéndose a nosotros en un esfuerzo común». Reagan, pensando, sin duda, en Cuba y tal vez en Nicaragua, dijo que algunos países latinoamericanos se han aislado de sus vecinos latinoamericanos y de su propia herencia. «Que vuelvan a las tradiciones y los ideales comunes de este hemisferio», dijo, «y todos los recibiremos con beneplácito.»

Son demasiadas similitudes para ser casuales. Y son además descorazonadoras. En efecto, la Alianza para el Progreso pretendió fundarse en una inversión de quinientos millones de dólares para el desarrollo social en todos los países de América Latina. El proyecto del presidente Reagan prevé una inversión de 350 millones de dólares para los países de América Central y el Caribe. Una suma risible en el mundo de hoy, que apenas si alcanzaría para aliviar la mortalidad infantil en Haití, que es una de las más dramáticas del mundo. El hecho de que el proyecto esté dirigido al sector continental que Estados Unidos considera como el más amenazado por la expansión soviética, es también muy significativo: los gorilas del Cono Sur pueden dormir tranquilos.

Lo único sorprendente en el discurso del presidente Reagan fue el cambio de su estilo bravucón de vaquero viejo por otro de apariencia evangélica que ha sido todavía más parecido al del presidente Kennedy. Pero esto no tiene por qué tranquilizarnos, sino todo lo contrario. Apenas un mes después de proclamada la Alianza para el Progreso, el propio presidente Kennedy autorizó el desembarco de la bahía de Cochinos, y apenas un año más tarde impuso a Cuba el bloqueo que cinco presidentes sucesivos han mantenido sin el menor asomo de piedad ni lucidez. Sin embargo, para vislumbrar lo que será el proyecto del presidente Reagan dentro de veinte años, basta con releer el párrafo angular del discurso del presidente Kennedy. «Si tenemos éxito», dice el párrafo, «si nuestros esfuerzos son audaces y resueltos, el cierre de este decenio» (es decir, a finales de los años sesenta) «marcará el comienzo de una nueva era en la experiencia americana. Los niveles de vida de toda la familia americana se elevarán, la educación básica estará disponible para todos, el hambre será una experiencia olvidada, la necesidad de una ayuda exterior masiva habrá pasado, la mayoría de las naciones entrarían en un período de desarrollo con sus propios recursos, y aunque todavía quedará mucho por hacer, todas las repúblicas americanas serán dueñas de su propia esperanza y su progreso.» Hay que creer que si el presidente Kennedy, en vez de ser víctima de una bala infame, hubiera sobrevivido a aquella década de sus ensueños —que terminó hace once años—, tal vez habría tenido bastante sentido del humor para morirse de risa de su propio horóscopo de delirio.

CRÓNICA DE MI MUERTE ANUNCIADA

El más conocido abogado defensor de presos políticos fue asesinado la semana pasada en las calles de Bogotá, con cinco tiros de pistola en la cabeza; horas después, el grupo clandestino MAS se atribuyó el crimen y dio a los medios de información una lista de sus próximas víctimas. Aunque esta lista no se había publicado completa hasta el sábado pasado, se sabe que en ella figuraban tres personas conocidas. Una era la periodista María Gimena Duzán, que días antes había sido secuestrada y conducida al centro de operaciones de las guerrillas del M-19, para que hiciera un reportaje forzoso, que, sin embargo, ningún periodista verdadero hubiera rehusado. Otro nombre en la lista era el del doctor Alfredo Vázquez Carrizoza, embajador en Londres bajo el gobierno de Alfonso López Michelsen y actual presidente del Comité de Derechos Humanos de Colombia. El tercero de la lista —modestia aparte— era yo.

El grupo MAS —según ellos mismos lo habían hecho saber— se suponía constituido para luchar contra los secuestros en Colombia, de allí su nombre: Muerte a Secuestradores. Desde el primer momento, sus métodos revelaron un alto nivel técnico, un poder sorprendente y una libertad de acción difícil de explicarse, como no fuera por la complicidad o la complacencia de las autoridades. Se decía que sus miembros eran militares en retiro, financiados por las mafias de traficantes de drogas, algunos de cuyas familias habían sido víctimas de secuestros costosos. Las autoridades colombianas guardaron siempre un silencio misterioso frente a las actividades intrépidas y arrogantes del MAS, y el ministro de la Defensa, general Luis Carlos Camacho Leyva, las definió para la prensa con una frase terminante: son pleitos de mafiosos.

Sin embargo, el asesinato de un penalista y la amenaza a tres personas que nunca han tenido nada que ver con secuestros ni han tenido negocio con las mafias, parecen poner las cosas en su verdadero

Publicado originalmente el 17 de marzo de 1982.

lugar. El MAS es un grupo organizado para combatir acciones políticas con métodos ilegales y para matar a los opositores del sistema. De modo que no parece desacertada la vieja suposición callejera de que en verdad son comandos del servicio de inteligencia militar, armados con los métodos represivos de Argentina, Uruguay y Chile. En realidad, ahora se sabe que los escuadrones de la muerte de esos países estaban formados por militares de carrera que escogían por vocación o convicción la siniestra especialidad del exterminio físico. Muchos de ellos, al parecer, una vez terminada la tarea en sus propios países, están ofreciendo sus servicios en los ajenos. Están concentrándose en Honduras para dirigir acciones contra Nicaragua. Están asesorando oficios de terror de muerte a la Junta de Gobierno de El Salvador. Están yendo, inclusive, más allá de nuestro continente: hasta África del Sur, donde uno de los gobiernos más represores del mundo tiene en muy grande aprecio su salvaje nivel profesional. De paso, sus países de origen resuelven de ese modo el problema de no saber qué hacer con ellos, una vez que su propia eficacia los ha vuelto inservibles.

En Colombia, a pesar de las negativas sistemáticas del gobierno, era evidente la existencia de estos organismos de horror. En julio de 1980 se conoció en los periódicos de Bogotá una carta muy reveladora, que, sin embargo, ningún periódico publicó. Estaba escrita de su puño y letra por un antiguo miembro de un escuadrón de la muerte, que era teniente del ejército, y firmada por sus compañeros: dos sargentos y dos cabos, que decían haber formado parte del Batallón de Inteligencia y Contrainteligencia, más conocido como Batallón Charry Solano.

En la carta se contaba, con tanta minuciosidad que ni el más imaginativo de los autores de ficción hubiera podido inventarlo, un historial espeluznante. Contaban que a mediados de 1978 se organizó un grupo denominado Triple A, cuyo nombre y cuya función eran los mismos que los de su homólogo argentino. Había un escuadrón de propaganda, cuya única misión era la de pintar consignas en los muros de la calle usando una motocicleta del batallón. «En caso de ser descubiertos», decía la carta, «podíamos dejarnos capturar sin decir nada y posteriormente se coordinaría nuestra libertad.» Otro de los grupos, según la carta, fue el que puso aquel año las bombas en tres periódicos de Bogotá: *Alternativa*, *El Bogotano* y *Voz Proletaria*. Aunque no lo decían en la carta, es de presumir, en buena lógica, que fueron también ellos quienes pusieron una carga explosiva en la casa del periodista Enrique Santos Calderón, director de *Alternativa*, cuya esposa, María Teresa, estuvo a punto de perder la vida a causa de la explosión. Las actividades de este grupo, que en aquella ocasión no prosperaron, eran similares a las que hoy está llevando a cabo el MAS, de un modo más sistemático y alarmante.

La carta contaba con sus nombres propios, quiénes habían sido los autores del crimen de Manuel Martínez Quiroz, un dirigente guerrillero que «fue asesinado dentro de una camioneta, después de que se le extrajo toda la información bajo tortura». Algunos episodios llamaban más la atención por su refinamiento espantoso: «A la doctora López le dieron una navaja para que se matara, y ante el desespero de las torturas, ella se cortó las venas a la altura de las muñecas. A Augusto Sánchez le fue alcanzada una cuchilla de afeitar y éste intentó suicidarse, al intentar cortarse la vena aorta, pero como no cortaba suficiente no pudo haber llevado a cabo el hecho. A Iván Moreno Ospina le dejaron una cuchilla en el asiento y éste intentó cortarse las venas a la altura de los brazos». En alguna parte, la carta decía: «Sobre estos crímenes podemos atestiguar en cualquier momento, y sobre los métodos utilizados». Pero hasta donde yo sé, sólo un periódico publicó un fragmento de ella, a pesar que su primer destinatario era el actual presidente de la República.

Tenemos, pues, un escuadrón de la muerte en Colombia, de cuya voluntad depende ahora nuestro destino. Contra el criterio de muchos amigos incrédulos, he decidido tomar esta amenaza con toda la seriedad que ella merece. He declarado y reiterado muchas veces mi repudio por el terrorismo, venga de donde viniere, y cualquiera sea su finalidad, porque lo considero un método de lucha ilegítimo e indigno. Sería poco menos que una ironía que fuera víctima de él. Siempre soñé como lo que soñó también un gran escritor de nuestro tiempo: morir a manos de un marido celoso. Pero, al parecer, éste será otro de mis tantos sueños frustrados.

No hay en este mundo una gloria más fácil que la de asesinarme, no tengo ningún arma de defensa distinta de la máquina de escribir, y a estas alturas no estoy dispuesto a cambiar de vida sólo para vivir unos años de sobra. Lo único triste sería ser víctima del gobierno más chapucero que ha tenido mi país en toda su historia, y por un atentado que no sería ni siquiera un crimen político, sino un simple acto administrativo cometido por miembros de las fuerzas armadas de Colombia, cuyo comandante supremo y primer responsable es el presidente de la República.

EE.UU.: POLÍTICA DE SUPOSICIONES

«Estamos metidos en el peor desastre diplomático en mucho tiempo», ha dicho el senador Cristopher Dodd, miembro del Comité de Relaciones Exteriores de Estados Unidos: «Hemos logrado que todo el mundo nos considere como los villanos de la novela: la derecha de El Salvador cree que somos una pandilla de comunistas, y la izquierda piensa que somos fascistas». Casi toda la prensa de Estados Unidos y Europa, aun la que no se ha distinguido nunca como la más liberal, parecía compartir esa consternación del senador este fin de semana. En realidad, lo que más teme la opinión pública de Estados Unidos es que la América Central se convierta para ellos en un nuevo Vietnam. Pero lo que tal vez no se había pensado es que también podía convertirse en un Vietnam político, del cual les fuera tan difícil salir airosos como de un Vietnam militar. Sin embargo, es eso lo que ha ocurrido. Estados Unidos está ya con el agua al cuello y a punto de hundirse sin remedio en un Vietnam político en América Central. Es eso, con otras palabras, lo que sin duda ha querido decir el senador Dodd. Y no es una buena noticia, porque una potencia desesperada puede tratar de salvarse de un atolladero político por el peor camino de la solución militar. En ésas estamos.

Las dos áreas de fuego, en este momento, son Nicaragua y El Salvador. Pero un análisis tranquilo permite establecer que el problema más grave no es la situación real de esos países, sino el embrollo en que los propios Estados Unidos se han enredado tratando de desenredarlo de la mala manera. Paso a paso, desde el día mismo de su posesión, el presidente Ronald Reagan ha venido construyendo para América Central y el Caribe una política sin escalera de emergencia, y a la hora de la verdad se ha encontrado con que ninguna de sus alternativas son viables, a menos que tuviera para alguna de ellas una buena puerta de escape que le permitiera salvar, en última instancia, aunque fuera la cara. El error principal estuvo en el pun-

Publicado originalmente el 24 de marzo de 1982.

to de partida. Su análisis de la realidad se sustentó en la suposición falsa de que los conflictos en América Central y el Caribe no son el resultado de las condiciones históricas de la región, sino un capítulo más en una vasta conspiración soviética. Tratando de demostrarlo a la fuerza, aun contra las evidencias más nítidas, el gobierno de Estados Unidos ha terminado en un pozo de arenas movedizas, al cual pudiera arrastrarnos a todos si alguien no le hace y nos hace a todos el favor de tirarle una tabla de salvación.

Esto ha sido más notable en las últimas semanas. El gobierno del presidente Reagan ha apelado a todos sus recursos mágicos para tratar de demostrar que Nicaragua es el conducto de abastecimiento de armas para las guerrillas de El Salvador, que los instructores de éstas son soviéticos y cubanos, y que sus operaciones se dirigen por telegrafía sin hilos desde territorio nicaragüense. Tratando de probarlo, el gobierno del presidente Reagan hizo desde aviones y satélites espías un mapa fotográfico completo de la superficie de Nicaragua. La revelación, sin duda, debía de ser definitiva. Casi como un símbolo, el protagonista escogido para aquel golpe de publicidad fue John Hugues, el mismo experto en interpretación fotográfica que reveló en 1962, y por el mismo método, la existencia de cohetes soviéticos en Cuba. Sin embargo, una vez más la historia se repitió en comedia. Hugues logró demostrar, con toda honestidad, que en Nicaragua se están construyendo varios aeropuertos estratégicos con técnicas que parecen de ingeniería soviética, que hay algunos tanques y cañones antiaéreos de fabricación soviética —cosa que ya se sabía de sobra desde que Estados Unidos se negó a vender armas a Nicaragua— y que el ejército nicaragüense es tan numeroso y bien entrenado como tantas veces se ha dicho sin necesidad de semejante espectáculo. Se descubrió también que en el aeropuerto internacional de Managua se están echando las bases de algo que parecen ser unos hangares. A partir de esa suposición gráfica, Hugues concibió la suposición para los Migs 21 soviéticos, que «suponemos les serán enviados a los sandinistas en algún momento de este año». Esto, sumado a la suposición de que quinientos pilotos de guerra nicaragüenses se están entrenando en Bulgaria, le permitió suponer a Hugues que Nicaragua «va a tener tal vez la mejor fuerza aérea de América Central». Después de que el propio secretario de Estado de Estados Unidos, Alexander Haig, mostró en público unas fotos que no eran de lo que él decía, después de que trató de probar que un barco cargado de armas llegó desde Cuba hasta el golfo de Fonseca sin pasar de un océano a otro, y después de que un revolucionario nicaragüense de 19 años se retractó en Washington de lo que le habían hecho decir bajo tortura en El Salvador, el folletín fotográfico de John Hugues terminó de convencer a la prensa de Estados Unidos de que las revelaciones de su gobierno carecen por completo de seriedad.

Hace una semana, en el *Spor Lacustre*, de Managua, Sergio Ramírez –miembro de la Junta de Reconstrucción Nacional– nos habló a un grupo de invitados sobre lo que podría llamarse la política de suposiciones de la Administración Reagan. «Hace casi tres años estamos andando con pluralismo de partidos, con economía mixta y con libertad de expresión», nos decía Sergio Ramírez. Sus datos eran terminantes. En Nicaragua hay once partidos políticos en plena actividad. Seis están contra el gobierno, y actúan en consecuencia, y cinco están aliados en el Frente Patriótico Revolucionario, dentro del cual está el Frente Sandinista de Liberación Nacional. El 60 % de la economía nacional está en manos de particulares; el 35 % está en poder del Estado, y el 5 % está en empresas mixtas. En el país hay veinticinco emisoras de radio de propiedad privada, quince estatales, y una sola –Radio Sandino– es de los sandinistas. Hay tres diarios: *La Prensa* y *El Nuevo Diario*, que son de propiedad privada, y *Barricada*, que es del Frente Sandinista. En Radio Sandino participé en un programa de una libertad a toda prueba: «Línea directa». El programa consiste en que el invitado contesta en directo durante una hora a todas las llamadas que se le hagan de la calle, y que entran sin filtro en el estudio de la emisora. La mayoría de las llamadas, como yo lo había solicitado, eran sobre temas literarios, y en especial preguntas sobre mis libros. Pero muchos oyentes, que se identificaron como todos, no hicieron ninguna pregunta, sino que aprovecharon la llamada para decir por radio todo lo que quisieron contra el gobierno. «Sin embargo», concluyó Sergio Ramírez, «Estados Unidos no ha dado ni un minuto de tregua en casi tres años, con el argumento de que no vamos a ser pluralistas, de que no vamos a tener economía mixta ni va a haber libertad de expresión.»

No: la política de suposiciones no tiene fondo ni término, y es, sin duda, la que ha metido al gobierno de Estados Unidos en el Vietnam político en que se encuentra empantanado, y del cual podría sacarlo quizá la presión de su propia opinión pública. Aunque es difícil, porque la política de suposiciones está fundada en un sistema de pensamiento que se alimenta de sí mismo. Hace poco, en efecto, después de una conversación con una personalidad extranjera de alto nivel, el general Alexander Haig comentó en privado: «Es un hombre tan convincente, que me habría convencido de todo lo que dijo si yo no hubiera sabido de antemano que no era verdad».

Bangkok es la ciudad más fea del mundo. Es inmensa, caótica, infernal, y todavía mucho más fea de lo que yo digo. Muchas veces, en una pesadilla recurrente, me he visto otra vez perdido en aquel playón polvoriento de casas miserables y pagodas espléndidas, y siempre he vuelto a padecer el horror de que nunca más encontraré la manera de salir. Me he vuelto a sentir aturdido por el fragor apocalíptico de los automóviles locos que circulan como quieren por el lado izquierdo, al modo inglés, y por los petardos de las motocicletas y el alboroto de los triciclos chinos y los palanquines de tracción humana. Son tres millones de personas apelotonadas a la orilla de un delta vasto e inmóvil, con un eterno vapor de podredumbre, y con el color de ciénaga revuelta de los grandes ríos asiáticos. En uno solo de sus barrios, el Sam Peng, viven 150.000 chinos por kilómetro cuadrado.

Como todo el mundo lo sabe, Bangkok es la capital de Tailandia, que antes fue el reino fabuloso de Siam. De modo que es apenas natural que su imagen estuviera vinculada a los sueños de mi infancia y a mis lecturas juveniles. *Ana y el rey de Siam* fue una de las películas inolvidables de mi generación. Hace unos años, esa imagen volvió a ponerse de moda con la película *Emmanuelle*, que era prohibida para menores, cuando en realidad debía serlo también para mayores, y no por su crudeza, sino por su poder de mistificación de la vida real. Entonces se organizaron caravanas de turistas cautivos, que soñaban con hacer el amor en barcas idílicas que navegaban por canales de flores. Lo que encontraron no eran más que los barrios lacustres de casi todas las ciudades fluviales del Tercer Mundo, con mercados flotantes de comidas típicas más o menos venenosas, y artículos de artesanía popular. El lugar más apropiado para hacer el amor, como siempre, son los hoteles americanos con aire puro y sábanas limpias.

Publicado originalmente el 31 de marzo de 1982.

Aparte del espectáculo histórico y turístico del palacio real, el único consuelo que uno encuentra en Bangkok son tres estatuas de Buda que bien valen un viaje, y la manía nacional de los masajes. Hay un buda de oro macizo, un buda de esmeralda integral, y un buda acostado que mide sesenta metros y pesa ochenta toneladas. Este último es casi tan grande como el santuario donde yace, que fue construido en torno suyo, y que tendría que ser demolido para poder sacarlo. Aparte de eso, los servicios del turismo inventan toda clase de distracciones para ganar tiempo, pero lo único sensato que se puede hacer es escapar cuanto antes de aquella ciudad insoportable. No es difícil, además, pues una de las rarezas incomprensibles de Bangkok es que las veintiuna compañías de aviación más importantes del mundo prestan allí un servicio continuo. *Puerta del Asia* la llaman en los afiches de publicidad. Podría ser tan falso como su nombre primitivo: «La ciudad de los ángeles». Pero no lo es. Una noche, desesperado por aquel espanto de ciudad, pregunté en el aeropuerto cuándo había un avión para Nueva Delhi, y había siete jumbos de grandes compañías europeas en las próximas horas.

Los salones de masajes, tan populares en Tailandia, no son en muchos casos sino una forma disimulada de la prostitución, y por lo mismo son tan deprimentes como en cualquier otra parte del mundo. Pero hay una tradición del masaje medicinal que es una institución patriótica. Al parecer, los tailandeses lo consideran como una medicina buena para todo, como la acupuntura para los chinos, y lo practican entre sí a toda hora y en todas partes. En los mercados, mientras esperan a los clientes, los mercaderes se hacen masajes entre sí. Los novios en los parques y en los cines se dan masajes recíprocos con una inocencia real. Los anuncios de masajes están por todas partes, ocupan páginas enteras en los periódicos, y lo primero que ofrecen en los hoteles, antes que la comida o la bebida o los programas turísticos, es un masaje en la habitación. En algunos casos se puede escoger en un álbum de fotografías a la persona, hombre o mujer, que se prefiere para el masaje. Lo asombroso, por supuesto, es que en muchos casos se trata en realidad de masajes. En los hospitales de medicina tradicional que existen en casi todas las ciudades del Lejano Oriente, las enfermeras y enfermeros se meten en las camas con sus pacientes exhaustos, y tratan de reanimarlos con masajes heroicos. Yo vi a una enfermera corpulenta trabada de piernas con un enfermo que parecía moribundo, y al cual trataba de reanimar con un masaje tan dramático que no se lo daba con las manos, sino con los talones.

El consuelo final es comprar ropa. Igual que en Hong Kong, uno puede llamar a un sastre que le toma las medidas antes del almuerzo, le hace la primera prueba dos horas después, y a las cuatro de la tarde lleva un vestido terminado e impecable de 120 dólares. En una pequeña tienda de camisas me hicieron media docena de seda legítima, sobre medidas, y en dos horas, mientras tanto, el propietario de la fábrica nos hizo una revelación: allí, detrás del mostrador, estaban confeccionando las colecciones que los grandes modistas europeos iban a lanzar en la próxima estación. No le faltaba sentido del humor: cuando me entregó las camisas, me preguntó de qué marca las quería. Y no era broma, pues allí tenían las etiquetas de los grandes nombres de la alta costura europea. Su cálculo era que un traje de noche para mujer comprado en las tiendas del Faubourg de Saint Honoré, en París, costaba doscientas veces más de lo que su confección había costado en Bangkok.

Al contrario de Hong Kong, donde hay un aire de misterio internacional que hace creíbles todas las fábulas de espionaje, en Bangkok se tiene la impresión de que nada puede ocurrir que no esté en la superficie de la vida. No creo que John Le Carré pueda concebir ninguna buena historia que ocurra en Bangkok, ni *La casa noble*, que tantos lectores está ganando en el mundo, podría haber ocurrido allí.

Poco antes de llegar a Bangkok había estado en Hong Kong, y lo primero que me había llamado la atención en los viejos hoteles ingleses, ahora reformados y embellecidos, era que los automóviles de servicio público eran Rolls Royce resplandecientes. Tengo en el mundo muchos amigos que tienen yates y aviones privados, pero no tengo ninguno con Rolls Royce. De modo que no pude resistir la tentación de conocer la ciudad en uno de aquellos transatlánticos de tierra firme, olorosos por dentro a cuero de animal vivo, aunque fuera para contárselo alguna vez a mis lectores. Fue, en efecto, como uno se imagina que es un vehículo espacial. Pero al cabo de una hora, cuando se disponía a subir por la carretera de circunvalación para que viéramos la panorámica de la ciudad desde la colina más alta, el automóvil empezó a corcovear, se resistió a seguir, y entregó su alma al Señor con toda mansedumbre. El conductor no sabía qué hacer con su vergüenza. Yo traté de tranquilizarle con el argumento cierto de que me interesaba más el cuento de un Rolls Royce que no lograba subir una colina, que el cuento obvio de que la había subido como un bólido. Además, era mejor para mis memorias, si algún día las escribo, porque casi treinta años antes me había ocurrido lo mismo en la isla de Capri, pero no en un Rolls Royce, sino en un coche tirado por un caballo viejo y escuálido que se había muerto en la pendiente. El conductor de Hong Kong me dio entonces un dato más: su automóvil no tenía de Rolls Royce sino la carrocería. El motor y todo lo demás era trasplantado de un automóvil norteamericano. La semana siguiente, escandalizado con la fealdad de Bangkok, comprendí que allí no podía ocurrir –como no ocurrió– nada parecido al cuento del Rolls Royce. Ni el del pobre caballo de Capri.

«PEGGY, DAME UN BESO»

En un largo muro blanco, frente a mi casa de México, amaneció el viernes pasado un letrero enorme: *Peggy, dame un beso.* Está pintado con un soplete de tinta indeleble, de esos que se usan para la guerra política de las paredes, y se le nota el pulso tenso e intenso de los letreros clandestinos escritos con el alma en un hilo en el sigilo de la madrugada, mientras los cómplices vigilan las esquinas para dar el aviso oportuno. Sin embargo está fuera de las áreas urbanas donde suelen librarse aquellas guerras de sombras, y adonde no llegan ni siquiera los desahogos murales de la cercana ciudad universitaria. Pero es bastante grande como para que Peggy lo vea al pasar, sin ninguna duda, por muy distraída que vaya y por muy indiferente que sea, y bastante desolado como para tocar su corazón de piedra.

Cuando lo descubrí, acababa de leer los periódicos, que en estos tiempos es como tomarse un frasco de aceite de ricino en ayunas. Había pensado, leyendo las noticias de Guatemala, que tal vez nos ha surgido en América Latina lo único que nos faltaba para colmo de peras en olmo: un *ayatolah.* Había pensado que la noche de Polonia era cada vez más oscura, que el gobierno de Reagan necesitaba cada vez más una tabla de salvación para salir airoso del pantano en que se ha metido, y que nunca, en fin, desde que tengo memoria, había visto tan incierto el destino de mi país. Había intentado, como todas las mañanas al despertar, formarme una visión panorámica del mundo a través de la prensa, y en todas partes había hallado memoria amarga de todo, y no sólo de mí mismo, como don Juan Tenorio en otros tiempos menos tormentosos. De modo que sentí un soplo de consuelo al descubrir que aún quedaba alguien tan cerca de mi casa, cuyo único problema en este mundo era que Peggy le diera un beso.

El semanario italiano *L'Espresso* publicó hace poco un artículo sobre la suposición de que el sexo está pasando de moda, y de que

Publicado originalmente el 7 de abril de 1982.

el amor a la antigua regresa por sus fueros. Revelaba el resultado de encuestas, según las cuales más hombres y mujeres cometen cada vez menos el acto sexual, y que inclusive hay parejas que siguen siendo felices cuando ya no lo hacen. Se atribuía esta disminución al frenesí sexual de los años sesenta, en el cual, al parecer, la humanidad se había gastado casi todas sus reservas eróticas. Y hay estadísticas para demostrarlo: el 30 % de las muchachas y el 55 % de los muchachos habían tenido experiencias sexuales a los quince años durante el apogeo de los sesenta, mientras que sólo el 4 % de las muchachas y el 13 % de los muchachos de quince años reconocieron haberlas tenido a finales del decenio.

No creo, sin embargo, que estas estadísticas sirvan para demostrar que nos estamos cansando del sexo, sino que le estamos dando en nuestras vidas la proporción que le corresponde en justicia, mientras que devolvemos al amor otros ingredientes que le habíamos quitado. A lo largo de mi vida, he asistido a un proceso de liberación sexual en dos países donde parecía menos probable: Colombia y España. En este último país, que era una inmensa casa de Bernarda Alba desde el Cantábrico hasta el Mediterráneo, se empezaron a notar las tremendas presiones sociales contra el cinturón de castidad desde mucho antes de la muerte del general Franco. Hace apenas unos quince años, cuando la necesidad fue más fuerte que la moral y se abrieron las puertas al turismo europeo, los guardias civiles espantaban de las playas a las valquirias que escapaban de las nieves del norte vestidas apenas con bikinis lineales. «Zorras», decían escandalizadas las buenas madres de familia que las veían desde la ventana. En los hoteles, aun en los más modernos y caros, estaban prohibidas las visitas en los cuartos, y más aún si eran del mismo sexo. Para mí, el primer síntoma de que algo estaba cambiando de pronto en aquella sociedad medieval fue la clausura por falta de clientela del famoso hotel de paso de la ciudad: el *meublé* de Pedralbes. Era un palacio decadente, con un cuarto chino donde todo era como en China, y un cuarto persa donde todo era como en Persia, con cortinas de peluche como en todos los burdeles del mundo y espejos de cuerpo entero hasta en los techos, tal vez para que los clientes tuvieran la impresión de que les daban por el mismo dinero la misma felicidad muchas veces repetida. Mis hijos, cuya escuela primaria era contigua a aquel paraíso secreto, no tenían mejor diversión durante el recreo que subirse en la pared divisoria para aguaitar lo que pasaba del otro lado. Lo más divertido que pasaba, en realidad, era que los recamareros serviciales tapaban las placas de los automóviles que entraban, para que los otros clientes no pudieran ver de quién eran, en la vana ilusión de guardar secretos en una ciudad pequeña y murmuradora donde las cosas se sabían aun antes de que sucedieran.

Todo aquello me recordaba a la Bogotá de los años cuarenta, cuando llegué por primera vez desde la costa caribe, a los trece años de edad y ya con la virginidad perdida, como era de buen uso en mi tierra. Mi madre, como todas, me había puesto en guardia contra los dos peligros más graves que nos acechaban en aquellas alturas: la pulmonía y el matrimonio a la fuerza. En realidad, acostumbrados a desvestirnos en cualquier parte a treinta grados a la sombra, los caribes (y no los caribeños, como se dice ahora, no sé por qué) andábamos a merced de los vientos cruzados de los Andes, y muchos se morían de pulmonía de un modo tan fulminante y triste como los turistas bogotanos se ahogaban en el mar. Para esto nos aconsejaban desvestirnos siempre con las puertas cerradas, y salir del cine con la boca tapada con un pañuelo, como todavía se hace en Bogotá, no sé con qué fundamento científico.

El otro peligro era el matrimonio a la fuerza. En efecto, acostumbrados desde niños a gatear en cuartos ajenos, o acostumbrados a que nuestras propias tías nos gatearan en los nuestros, los costeños en Bogotá seguíamos creyendo que podía hacerse lo mismo con igual impunidad, y al final nos encontrábamos casi siempre con la embarazosa situación de estar embarazados. Era, además, la menos terrorífica de las opciones, porque estábamos en los tiempos de gloria de las enfermedades de Venus. En los tranvías, en los vespasianos públicos, en todas partes, había letreros para recordarlo: «Si no le temes a Dios, témele a la sífilis». De modo que el único remedio contra la soledad eran los bailes de los sábados, con cuotas de a dos pesos, y en los cuales se veía a fondo el único lado permitido del amor: los boleros apretados, las citas al día siguiente a la salida de la misa, las cartas perfumadas, los cines furtivos, las lágrimas en la almohada solitaria, la poesía.

Todo eso se había ido en los años sesenta, barrido por el ventarrón del sexo puro. No me pareció mal. Al contrario: siempre he creído que uno nace con sus polvos contados, y que los que no se usan a tiempo se pierden para siempre. Pero es mejor el sexo con todo lo demás, que es el amor completo. Eso es sin duda lo que viene ahora, a juzgar por los anuncios del corazón. Las novelas de amor han vuelto a ser las más vendidas. Los novios vuelven a besarse en la calle. Hace unos días, mi hijo de dieciocho años le pidió a su madre que lo enseñara a bailar el bolero, porque el bolero ha vuelto, cantado y bailado, y en las ciudades de América Latina y España se están abriendo discotecas de penumbra para vivirlo de nuevo. Siempre he creído que el amor salvará de la destrucción al género humano, y estos signos que parecen regresivos son todo lo contrario: luces de esperanza. Por eso deseo con ansiedad que Peggy lea el letrero que alguien ha escrito para ella frente a mi casa.

Por favor, Peggy, dale un beso.

CON LAS MALVINAS O SIN ELLAS

El primer jueves fueron catorce. El siguiente fueron veinte y el tercero fueron treinta y dos. Aparecían en la plaza de Mayo, de Buenos Aires, todos los jueves a las once de la mañana, y durante varias horas clamaban por la aparición de sus hijos secuestrados por los servicios de seguridad de las fuerzas armadas. El gobierno, sorprendido por aquella novedosa y temible forma de acción, las llamó «las locas de la plaza de Mayo». Hoy son más de 2.500 en toda Argentina, y en el mundo entero se las conoce y se las respeta con el nombre de las «madres de la plaza de Mayo».

El movimiento empezó sin un plan definido hace ahora cinco años exactos: el 30 de abril de 1977. Al principio, la dictadura no sabía cómo proceder, pero a medida que los clamores aumentaban y se fortalecían, fue aumentando también el torniquete de la represión. Las madres eran dispersadas a punta de bayoneta, las encerraban en calabozos inmundos, llenos de orines, de ratas y cucarachas, pero el jueves siguiente había en la plaza un número cada vez mayor. En cierta ocasión detuvieron a cuarenta y las hicieron pasar la noche en una celda donde había un joven muerto. El 8 de diciembre de 1977, tres madres fueron arrestadas durante una manifestación, junto con dos monjas francesas que las acompañaban, y nunca más se supo de ellas. Entonces las «madres de la plaza de Mayo» comprendieron que habían pasado del punto sin retorno, y que sólo una organización interna muy sólida y un buen respaldo mundial podían ponerlas a salvo del salvajismo de la dictadura. Fue así como se creó la Asociación de Madres de la Plaza de Mayo —el 14 de mayo de 1978– que todos los jueves, a las once de la mañana, con lluvia o con sol, se manifiesta a gritos pero sin violencia frente al palacio de gobierno, y que ahora tiene una delegación que viaja por el mundo entero en busca de solidaridad. Nunca, desde su primer día, han sido recibidas por ningún funcionario de alto nivel.

Publicado originalmente el 14 de abril de 1982.

En Argentina hay unos 4.000 presos políticos identificados y localizados, de 15.000 que han pasado por las cárceles desde 1972. Se calcula que en los combates hubo unos 5.000 muertos de las guerrillas. A fines de 1980, *Selecciones del Reader's Digest* calculó que las bajas oficiales ascendían a 650. No obstante, el gobierno argentino no dio nunca la cifra de bajas suyas en el combate de Tucumán con el ERP, que fue largo y sangriento. El general Harguindey, hablando alguna vez en términos generales, calculó que las bajas oficiales habían sido de más de un millar. El almirante Emilio Massera, que había sido miembro de la junta militar en la época más bárbara de la represión, calculó a la topa tolondra que las pérdidas totales habían sido de unos 50.000 muertos.

El número de desaparecidos es mucho más difícil de calcular, porque no se sabe a ciencia cierta cuáles están vivos y cuáles están muertos. Hay informaciones muy respetables según las cuales es imposible encontrar los cadáveres, porque la fuerza aérea argentina tenía helicópteros especiales para tirarlos a presión en el mar, de modo que nunca más salieran a flote. De acuerdo con el dato más reciente de Amnistía Internacional el número de desaparecidos asciende a 15.000. Pero las «madres de la plaza de Mayo» tienen sus estadísticas y sus métodos propios. «Nosotras llamamos desaparecidos a toda persona cuyo destino se desconoce», ha dicho su presidenta, Hebe Pastor de Bonafini, en una entrevista muy esclarecedora que publicó en diciembre pasado la revista *Testimonio Latinoamericano*. Su sistema es lógico. Si una madre denuncia que a su hijo lo mataron a la puerta de su casa, y que otro desapareció, se cuenta a éste como desaparecido, pero no al muerto. Sin embargo, si una madre ha visto que a su hijo lo mataron en la puerta de su casa, pero nunca le entregaron el cadáver, entonces no lo cuenta como muerto, sino como desaparecido. Otra cosa es que las «madres de la plaza de Mayo» no consideran que esté muerto ningún desaparecido, mientras no se demuestre sin lugar a dudas. La razón, según ellas, es también muy lógica: «Hay madres cuyos hijos desaparecieron, y por esas cosas extrañas que uno no se explica reaparecieron más tarde en cárceles comunes». En cuatro casos, por lo menos, las madres que fueron a visitar a los hijos reaparecidos apenas si lograron reconocerlos. «No dejaban ir al baño a los muchachos durante dos o tres días», ha contado una de ellas, «para que tuvieran que hacer sus necesidades encima, de modo que cuando las madres les veían los encontraban todos sucios y malolientes, encontraban seres que no eran sus hijos». Los desaparecidos no eran sólo guerrilleros activos. «El 99 % de los desaparecidos no cayó en combate ni en ninguna guerra», dice una madre. «Se llevaron a muchos médicos, a muchos psiquiatras, a muchos periodistas que denunciaron la represión.» Unos trescientos abogados fueron muertos o encarcelados, y hasta hoy no se tiene noticias de ellos. Con las denuncias recibidas de fuentes directas, las «madres de la plaza de Mayo» calculan que el número de desaparecidos es de 30.000. Esta es la cifra que maneja en la actualidad la prensa mundial, y por ella tienen que responder las autoridades argentinas, mientras no se pruebe otra distinta.

Entre estos desaparecidos hay muchos niños de pocos años. Algunos desaparecieron junto con sus padres, otros se quedaron solos porque sus padres desaparecieron y otros fueron dados a luz en la cárcel. En 1979 se encontraron dos niños en una playa de Valparaíso, en Chile. Al parecer, habían sido llevados en automóviles por militares argentinos, y fueron dados en adopción a un matrimonio chileno. Una visitadora social que se interesó en el caso logró identificarlos como los hijos de un matrimonio uruguayo que había desaparecido en Argentina. Se dice que hay militares argentinos que han adoptado niños de desaparecidos. Algunas comadronas que han atendido partos en las cárceles han pasado la información a la familia de la madre presa. Veintitrés de estos casos fueron denunciados por la Fundación Habeas para defensa de los derechos humanos –de la cual soy presidente– en un congreso que celebró la Unicef en México con motivo del Año Internacional del Niño. Hasta el día de hoy, Habeas no ha recibido ninguna respuesta.

La situación de los desaparecidos es tal vez la más dolorosa y grave de las realidades argentinas que el general Leopoldo Galtieri ha tratado de borrar de una sola plumada con la ocupación militar de las islas Malvinas. Estamos de acuerdo: las Malvinas son argentinas. En ese sentido, el general Galtieri no ha hecho más que poner las cosas en su puesto. Pero lo ha hecho con un acto legítimo cuya finalidad es torcida. La corona inglesa, por su parte, al mandar una flota de cuarenta barcos de guerra con un príncipe a bordo, no ha hecho más que tratar de reparar la humillación con el ridículo. Es un acto de capa y espada que sólo se le podía ocurrir a un imperio polvoriento. Pero cualesquiera sean los resultados de esta guerra de naftalina, el general Galtieri no conseguirá impedir que el próximo jueves, a las once de la mañana, esté en la plaza de Mayo la manifestación de siempre con las madres de siempre, cuyo quinto aniversario se cumple dentro de pocos días. Estarán, como siempre, frente a la dictadura más sangrienta de este siglo en América Latina, pidiéndole las cuentas que la dictadura tendrá que rendir, tarde o temprano, y con las Malvinas o sin ellas.

UNA TONTERÍA DE ANTHONY QUINN

«*Cien años de soledad* sería ideal para un serial de cincuenta horas de televisión, pero García Márquez no quiere venderlo», ha declarado a una revista española el actor Anthony Quinn. Y agregó: «Yo le ofrecí un millón de dólares y no quiso, porque García Márquez es comunista, y no quiere que se sepa que ha recibido un millón de dólares. Porque luego vino, después de la cena, y me dijo aparte: "¿Cómo se te ocurre ofrecerme ese dinero en público? Otra vez me lo ofreces sin que haya ningún testigo"».

Lo único malo que tiene esta declaración, aparte de su infantilismo, es que no es cierta. La realidad, como siempre, es más interesante, y sólo por eso quiero contar el cuento tal como sucedió en una de mis tantas llegadas a México, hace unos cinco años. Los periodistas del aeropuerto, que de tanto vernos han terminado por ser mis amigos, me dijeron que Anthony Quinn había dicho la noche anterior por la televisión mexicana que estaba dispuesto a darme un millón de dólares por los derechos para el cine de *Cien años de soledad*. Yo les dije a los periodistas, y ellos lo publicaron por todas partes al día siguiente, que aceptaba venderlos con la condición de que no fuera uno, sino dos millones: uno, para mí, y otro, para la revolución en América Latina. Esa misma semana, y antes de verse conmigo, Anthony Quinn replicó en la televisión: «Yo le doy el millón de dólares para él, pero el otro que se lo consiga en otra parte». La respuesta me pareció tan certera y divertida, que acepté la amable invitación de unos amigos comunes para comer con Anthony Quinn. Fue una cena muy grata. Anthony Quinn, a los 62 años, conservaba todavía una vitalidad atropellada, y me pareció simpático y afectuoso, y un poco obsesionado con la velocidad del tiempo. Se habló de todo, pero no dijo una palabra sobre su oferta de la televisión, y eso me produjo un gran alivio. Fue la primera y la última vez que le vi.

Publicado originalmente el 21 de abril de 1982.

Lo que Anthony Quinn no supo nunca es que, cuando él hizo su oferta en la televisión, hacía mucho tiempo que un consorcio de productores de Estados Unidos y Europa había ofrecido dos millones de dólares por los derechos para el cine de *Cien años de soledad*. La impresión que les quedó a muchos amigos míos fue que el gran actor metido a productor había ofrecido lo que ofreció sólo por darse ínfulas de que andaba tirando a manos llenas un millón de dólares. No era la primera vez que me ocurría. A finales de los años sesenta, en Barcelona, un editor de leontina y cigarro habano apareció en la televisión con dos millones de pesetas en efectivo –que entonces eran unos 70.000 dólares–, y dijo, abanicándose con los billetes, que ése era el anticipo que me ofrecía por mi próximo libro. Esa noche, por supuesto, se ganó gratis el derecho a no publicar ni el próximo ni ninguno de mis libros.

Los ingleses consideran que es de muy mala educación hablar en público de los hijos, de las enfermedades y del dinero. Pero como no soy inglés, a Dios gracias, sino de la calle Mayor de Aracataca, tengo otros pudores mucho menos frívolos. Me gusta hablar de mis hijos porque son iguales a su madre: bien plantados, inteligentes y serios. Me gusta hablar de mi úlcera duodenal, que sólo se me alivia cuando escribo, porque los amigos no son sólo para compartir la buena vida, sino también para joderse con uno. Me gusta decir cuánto dinero gano y cuánto pago por las cosas, porque sólo yo sé el trabajo que me cuesta ganármelo, y me parece injusto que no se sepa. La única excepción a esta norma es que nunca hablo de dinero con los editores y los productores de cine, porque tengo un agente literario que habla por mí mejor que yo; primero, porque es mujer, y después, porque es catalana. Muchos editores la detestan por la ferocidad con que defiende los centavos de los escritores, sobre todo de los jóvenes y más necesitados, y el día que no la detesten empezaré a sospechar que se pasó al bando contrario.

Mi experiencia con los productores de cine, a partir de *Cien años de soledad*, es una de las más sorprendentes de mi vida. En general, no hablan más que de dinero, pero a la hora de la verdad todos son como Anthony Quinn: no se les ve por ningún lado. Son grandilocuentes, inseguros e imprevisibles. Mercedes, mi esposa, les tiene terror, porque llegan a la primera cita con proyectos espaciales, arrasan con el bar y la despensa, hablan con el mundo entero desde nuestro teléfono sin preguntar cuánto te debo, y nunca más se vuelve a saber de ellos. El italiano Paolo Bini, esposo de la bella Rossana Schiaffino, vino hace unos tres años a nuestra casa de Cuernavaca porque quería producir un cuento mío dirigido por Ruy Guerra.

A éste le mandó su billete de avión a Río de Janeiro, y todos hablamos del proyecto durante un domingo entero. Esa misma semana apareció en la revista *Variety*, de Los Ángeles –donde sólo anuncian los productores más afortunados–, un anuncio de página entera sobre la película que íbamos a hacer, como si ya estuviera hecha. Bini se fue con una copia del cuento en inglés, para proponerle a Robert de Niro que hiciera el papel estelar, y prometió ponerse en contacto con nuestros agentes para comprar los derechos de mi cuento y establecer los honorarios de Ruy Guerra. Ésa fue la última vez que le vimos. La única noticia que tuve de él desde entonces fue cuando le dijo a algunos amigos de Roma que nos había anticipado a Ruy Guerra y a mí una buena cantidad de dólares para que trabajáramos en el guión y que nosotros nos la habíamos robado.

Billy Friedkin –el director y productor de *El exorcista* y de *French Connection*– es un hombre distinto, por fortuna, pero con las rarezas de todos los productores grandes. Friedkin vino a México hace varios años con la idea de hacer en cine *El otoño del patriarca*. Es un hombre muy joven, impecable, que había ganado una fortuna con sus películas, y el dinero que le sobró después de comprar un avión privado quería donarlo para escuelas públicas en Israel y Bolivia. Tenía ideas tan atractivas para llevar mi novela al cine, que logró convencerme. Hablando de todo, me contó que el autor de *El exorcista*, que es una novela de segunda, había recibido una suma modesta por los derechos del libro, pero en cambio aceptó una participación en los beneficios de la película, y se ganó diecisiete millones de dólares. Yo entendí que aquélla era una sugerencia elegante, y se lo dije a mi agente. De modo que cuando Friedkin habló con ella sobre los derechos del libro ella le dijo que aceptábamos las mismas condiciones que el autor de *El exorcista*. Friedkin me llamó por teléfono, y con la misma elegancia con que hacía todo, desistió del proyecto. Nunca más supe de él, salvo por los periódicos, cuando se casó en París con Jeanne Moreau y, poco después, cuando se divorciaron.

El único que en realidad no me habló nunca de dinero parece ser el único que en realidad lo tiene: Francis Ford Coppola, el director de *El padrino*. Cuando Coppola hizo *Apocalypse Now*, en Manila, el director de fotografía le habló muchas veces de su ilusión de hacer en cine *Cien años de soledad*. En el verano de 1979, Coppola y yo coincidimos en el Festival de Cine de Moscú, y él me invitó a cenar, pocos días después, en un ruidoso e inmenso restaurante de Leningrado. Hablamos un poco de sus películas y de mis libros, y me contó lo que su fotógrafo le había dicho sobre *Cien años de soledad*, pero en ningún momento planteó la posibilidad de hacerlo en cine. Lo único que de veras le interesó fue cuando supo que mi hijo mayor había hecho un curso de alta cocina en París. Coppola, que es un gran comedor y un cocinero de primer orden, se dejó arrastrar por la inspiración súbita de meterse con mi hijo en la cocina del restaurante para preparar la comida que íbamos a comernos. Fue una noche inolvidable.

Con todo, mi reticencia de que se hagan en cine *Cien años de soledad*, y en general cualquiera de mis libros publicados, no se funda en la extravagancia de los productores. Se debe a mi deseo de que la comunicación con mis lectores sea directa, mediante las letras que yo escribo para ellos, de modo que ellos se imaginen a los personajes como quieran, y no con la cara prestada de un actor en la pantalla. Anthony Quinn, con todo y su millón de dólares, no será nunca para mí ni para mis lectores el coronel Aureliano Buendía. El único que podría hacer ese papel, sin pagar ni un centavo, es el jurista colombiano y gran amigo mío Mario Latorre Rueda. Por lo demás, he visto muchas películas buenas hechas sobre novelas malas, pero nunca he visto una buena película hecha sobre una buena novela.

«EL PEZ ES ROJO»

El pez es rojo, que acabo de leer con un cierto retraso, es uno de esos libros que parecen una novela fantástica por el rigor con que están apegados a la realidad. Sus autores son dos norteamericanos muy conocidos por sus investigaciones periodísticas. Warren Hinckle se ganó hace dos años el Premio George Polk por sus revelaciones sobre actividades de la CIA dentro de Estados Unidos que son ilegales. William Turner es un antiguo funcionario del FBI que ha escrito, entre otros libros, un reportaje exhaustivo sobre el asesinato de Robert Kennedy. Ambos se dedicaron durante varios años a rastrear la documentación y los protagonistas de la guerra secreta que la CIA ha adelantado contra Cuba desde 1959. Sin excluir, por supuesto, el medio centenar de atentados contra Fidel Castro. El resultado es este libro fascinante, que va más allá de cualquier obra de ficción. *El pez es rojo* era el nombre cifrado del desembarco de la bahía de Cochinos.

Leyendo este libro asombroso, uno se pregunta cómo es posible que los gobiernos sucesivos de un país como Estados Unidos, cuyo genio creador es una de las maravillas de este siglo, hayan podido perseverar durante veinte años en semejante sistema de disparates y despilfarrar sin medida tanta imaginación inútil y tanto dinero estéril. Es apenas concebible que sus servicios de inteligencia hayan sido tan poco inteligentes como para establecer un contubernio entrañable con lo peor de su sociedad: las mafias del juego, los traficantes de narcóticos, los delincuentes de la índole más siniestra, y todo eso en nombre de los intereses más altos de la democracia. Uno de sus socios eminentes, Howard Hughes —el excéntrico multimillonario muerto del miedo a los microbios hace pocos años—, trató de que, a cambio de sus servicios, Estados Unidos prolongara la guerra en Vietnam, sólo para vender más helicópteros de sus fábricas, e intentó sobornar al presidente Lyndon B. Johnson con 100.000 dólares en

Publicado originalmente el 28 de abril de 1982.

efectivo para que suspendiera los ensayos nucleares en Nevada, porque creía que le hacían daño para la salud. Dentro de este ámbito de locura, la CIA tenía barcos artillados que navegaban sin tropiezos con banderas inocentes, y que cambiaban de color y de forma con tanta frecuencia que era imposible identificarlos. En cierto momento, esa flota de la guerra secreta, que era autónoma e impune, y no tenía nada que ver con las fuerzas armadas del país, fue la más numerosa y destructiva de todo el Caribe. Cuánto ha costado y sigue costando esta aventura manicomial es algo que el talento magistral de los autores de este libro no ha podido establecer. Tal vez porque nadie lo sabe a ciencia cierta.

Cuesta trabajo creer que en el origen de todo esto estuviera nada menos que el creador del agente secreto 007. Así fue. En la primavera de 1960 —según cuentan Hinckle y Turner—, el senador John F. Kennedy, que poco después sería el nuevo presidente de Estados Unidos, ofreció un almuerzo a su autor favorito, Ian Fleming. El senador le preguntó al escritor qué se le ocurriría a James Bond si se le encomendara la tarea de eliminar a Fidel Castro. Fleming contestó, sin pestañear, que había tres cosas importantes para los cubanos: el dinero, la religión y el sexo. A partir de esa premisa, imaginó tres proyectos. El primero era arrojar sobre Cuba una cantidad fabulosa de dinero falsificado, como una cortesía de Estados Unidos. El segundo era arreglárselas para que apareciera en el cielo cubano una cruz luminosa, como un anuncio de la vuelta inminente de Cristo a la Tierra, para exterminar el comunismo. El tercero era lanzar panfletos sobre Cuba, firmados por la Unión Soviética, para advertir a la población que las pruebas atómicas de Estados Unidos habían contaminado de radiactividad las barbas de los revolucionarios, y que esto los volvería impotentes. Fleming suponía que después de esta advertencia los revolucionarios se afeitarían la barba, incluso Fidel Castro. Y concluyó: «Sin barbas no hay revolución».

John Pearson, el biógrafo de Fleming, escribió más tarde que todo lo dicho en aquel almuerzo histórico era una broma del novelista, para dar a entender que lo único eficaz para derrotar a Fidel Castro era tratar de ponerlo en ridículo. Pero la CIA lo tomó al pie de la letra, y el único proyecto que no tuvo en cuenta fue el de los dineros falsificados, porque no les pareció original. En efecto, había sido estudiado por los nazis para desorganizar la economía de Inglaterra. La idea de obligar a los revolucionarios a afeitarse no habría sido eficaz, pues éstos se afeitaron por propia iniciativa poco tiempo después, y la revolución siguió su curso. Antes de eso, los laboratorios de la CIA habían inventado unos polvos que si se echaban en los zapatos harían caer todos los pelos del cuerpo, pero no encontraron a nadie que los echara dentro de las botas de Fidel Castro.

La gran mayoría de los proyectos, incluido el desembarco en la bahía de Cochinos, que era el más ambicioso, terminaron en el fracaso. Pero algunos dirigidos a destruir la economía fueron certeros. «Aviones del centro de armamento naval de Lake China», dice el libro, «sobrevolaron Cuba, regando cristales en las nubes, que precipitaron lluvias torrenciales sobre áreas no agrícolas y dejaron áridos los cañaverales.»

Más conocida fue la acción de un grupo terrorista que, en marzo de 1970, recibió de la CIA un frasco con el virus de la fiebre porcina africana, para que lo introdujeran en Cuba. Seis semanas más tarde, la isla sufrió el primer brote de fiebre porcina africana del hemisferio occidental, y unos 300.000 cerdos tuvieron que ser sacrificados.

Los fracasos menos explicables, por supuesto, han sido los de los atentados a Fidel Castro. En realidad, Castro tiene una vida cotidiana imprevisible, sus servicios de seguridad son muy difíciles de penetrar, y la contrainteligencia cubana está considerada como una de las más eficaces del mundo. Pero eso no es suficiente para explicar el fracaso de más de cincuenta atentados preparados por la CIA con sus recursos más sabios. Hay que pensar que existe un elemento diferente que escapa a las computadoras de la CIA, y que tal vez no sea del todo ajeno a la magia del Caribe.

Cuando el presidente Kennedy mandó a Cuba al abogado neoyorquino James Donovan, en 1963, para que negociara la liberación de un grupo de prisioneros norteamericanos, la CIA preparó, sin que Donovan lo supiera, un regalo especial para Fidel Castro. Era un equipo de pesca submarina en cuyos tanques de oxígeno habían puesto bacilos de tuberculosis. El propio Donovan no sabe por qué, pero el equipo no le pareció digno de un jefe de Estado, y lo cambió por otro que él mismo compró en Nueva York. «De todos modos», ha dicho un agente de la contrainteligencia cubana, «nosotros hubiéramos revisado el equipo.»

Los fracasos más sorprendentes fueron los de los tres atentados que la CIA preparó contra Fidel Castro durante su larga visita a Chile, en 1971. En el primero, Castro iba a ser asesinado, durante una conferencia de prensa, con una ametralladora escondida dentro de una cámara de televisión. «Era algo similar al asesinato de Kennedy», dijo el hombre de la CIA que dirigió el atentado, «porque la persona que iba a matar a Castro estaba provista de documentos que le harían aparecer como un agente desertor de los servicios cubanos en Moscú.» Pero, a la hora de la verdad, a uno de los asesinos le dio un ataque de apendicitis, y el otro no se atrevió a disparar solo. El otro atentado estaba previsto durante la visita de Fidel Castro a las minas de Antofagasta, en el norte de Chile. Un automóvil descompuesto en el camino obligó a detener la caravana oficial. Dentro de ese automóvil había cuatrocientas toneladas de dinamita conectadas a un detonador eléctrico. Pero, por razones todavía inexplicadas, la dinamita no estalló. El tercer intento debía ser un disparo desde otro avión en tierra, cuando Fidel Castro hiciera escala en Lima; pero un cambio en la posición de los dos aviones determinó que el piloto de la CIA se negara a disparar, por temor de no poder escapar a tiempo. Un cuarto atentado, también frustrado, fue el que intentó una bella agente de la CIA que tuvo acceso a Fidel Castro y estaba dispuesta a echarle en la bebida unas cápsulas de veneno. Pero las había introducido en Cuba dentro de un frasco de *cold cream*, y cuando quiso utilizarlas no las encontró: se habían disuelto.

Hay tres casos que no cuentan los autores de *El pez es rojo*. Uno de ellos fue cuando electrificaron con alto voltaje los micrófonos de la tribuna donde iba a hablar Fidel Castro. La seguridad cubana lo descubrió a tiempo, y su explicación fue la más simple: «Ya habíamos pensado que a alguien se le iba a ocurrir alguna vez». El otro atentado que nunca ocurrió fue el que debía intentar un empleado de la cafetería del hotel Habana Libre, a quien la CIA le había dado unas cápsulas inodoras, incoloras e insípidas, y cuyo efecto mortal era bastante retardado, para que el criminal pudiera escapar. Se trataba de echarlas en el batido de frutas que Fidel Castro solía tomarse cuando llegaba a la cafetería en la madrugada. El agente esperó más de seis meses y, cuando por fin apareció Fidel Castro, las cápsulas ya habían perdido su efecto. La CIA las cambió por otras de actividad indefinida si se conservaban en congelación. El agente las puso sobre el congelador, y cuando Fidel Castro volvió, al cabo de cuatro meses, le preparó el batido de frutas de siempre; pero a última hora no pudo romper el hielo que había cubierto la cápsula.

Con todo, el mayor peligro en que se ha visto Fidel Castro, y que tampoco está citado en este libro fantástico —no fue un atentado—, fue después de la derrota de la invasión de la bahía de Cochinos, cuando regresaba del frente en un jeep descubierto. Dos supervivientes de la derrota, que se habían escondido detrás de unos arbustos, le vieron pasar a menos de diez metros, y uno de ellos le tuvo en su mira por breves segundos. Pero no se atrevió a disparar.

UN PAYASO PINTADO DETRÁS DE UNA PUERTA

Hace más de treinta años, la pintora Cecilia Porras pintó un payaso de tamaño natural en el revés de la puerta de una cantina del barrio de Getsemaní, muy cerca de la calle tormentosa de la Media Luna, en Cartagena de Indias; lo pintó con la brocha gorda y los barnices de colores de los albañiles que estaban reparando la casa, y al final hizo algo que pocas veces hacía con sus cuadros: firmó. Desde entonces, la casa donde estaba la cantina ha cambiado muchas veces: la he visto convertida en pensión de estudiantes con oscuros aposentos divididos con tabiques de cartón, la he visto convertida en fonda de chinos, en salón de belleza, en depósito de víveres, en oficina de una empresa de autobuses y, por último, en agencia funeraria. Sin embargo, desde la primera vez en que volví a Cartagena al cabo de casi diez años, la puerta había sido sustituida. La busqué en cada viaje, a sabiendas de que las puertas de esa ciudad misteriosa no se acaban nunca, sino que cambian de lugar, y hace poco la volví a encontrar instalada como en su propia casa en un burdel de pobres del barrio de Torices, donde fui con varios de mis hermanos a rescatar nuestras nostalgias de los malos tiempos. En el revés de la puerta estaba el payaso pintado. Como era apenas natural, la compramos como si fuera un puro capricho de borrachos, la desmontamos del quicio y la mandamos a casa de nuestros padres en una camioneta de alquiler que nunca llegó. Pero no me preocupé demasiado. Sé que la puerta intacta está por ahí, empotrada en algún quicio ocasional, y que el día menos pensado volveré a encontrarla. Y otra vez a comprarla.

Eso es lo que más me ha fascinado siempre de Cartagena: el raro destino de sus casas y de sus cosas. Todas parecen tener vida propia, tanto más cuanto más muertas parecen, y van cambiando de forma y de utilidad en el tiempo, mudándose de sitio y de oficio mientras sus dueños pasan de largo por la vida sin demasiado ruido.

Publicado originalmente el 5 de mayo de 1982.

Es una magia de origen. Nadie se ha sorprendido nunca de que la casa más bella de la ciudad haya sido el tremendo palacio de torturas de la Inquisición, que las cárceles tenebrosas de la colonia estén ahora convertidas en alegres bazares de artesanía, y que haya un restaurante de pescado en la que fuera la mansión de lujo del marqués de Valdehoyos. De modo que hay que considerar como la cosa más natural del mundo que el Museo de Arte Moderno –al cabo de innumerables peripecias de la casa y de los cuadros– haya encontrado por fin su sitio en las antiguas bodegas coloniales del puerto.

Por la época en que Cecilia Porras pintó el payaso detrás de la puerta, tuve una relación de casualidad, pero muy asidua y grata, con ese edificio en abandono. Yo daba mis primeros pasos de periodista en *El Universal*, que acababa de fundarse a muy pocas cuadras de allí, y lo primero que aprendí del oficio fue la mala costumbre de vivir al revés: durmiendo de día y trabajando de noche. En la madrugada, cuando se paraba el rumor de llovizna de los teletipos, me iba con los linotipistas a las bodegas del puerto, cuyo celador insomne era el único amigo dispuesto a recibirnos a esa hora. Allí permanecíamos hasta el amanecer, tomando aquel ron de caña que parecía de fósforo vivo, y escuchando las historias fantásticas del celador.

Desde el lugar en que nos sentábamos a conversar veíamos el muelle de los Pegasos, con sus veleros de mala muerte, que iban resucitando a medida que aumentaba la madrugada. Nunca podré olvidar en el resto de mi vida aquellos amaneceres irreales de mi juventud; siempre recordaré qué tristes nos quedábamos cuando las goletas se iban, me acordaré del loro que adivinaba el porvenir en la casa de camas alquiladas de Matilde Arenales, de las jaibas que se salían caminando de los platos de sopa que servían en las fondas de maricas del mercado, del viento de tiburones, los tambores remotos, la luz amarga de los primeros días de abril, mientras el celador nos contaba sin cansancio la historia de la casa. Pues ése era su tema único: la historia de la casa. Golpeando las paredes con el puño, detectaba puertas tapiadas, arcadas con columnas y capiteles escondidos, como si aquélla no fuera una sola casa, sino un sistema de muchas casas superpuestas a través de los años. Más tarde, había de darme cuenta de que sus historias eran falsas, pero no me sentí defraudado, sino todo lo contrario, porque sus fábulas eran mejores que la realidad. Fue él quien me habló de una esclava fascinante por la cual un rico de la época había pagado su peso en oro, y había tenido que matarla para librarse de su hechizo. «Está enterrada aquí», decía, golpeando un vacío en el muro. Me contó que, durante el sitio de

Vernón, los habitantes de la ciudad habían capturado una patrulla de
ingleses que trataban de infiltrarse por el lado de tierra, y fueron
descuartizados, asados y devorados por los soldados de la plaza. Fue
él quien me habló por primera vez de Blacamán, mitad mago, mitad bandido, que fue llevado a Cartagena, nadie supo de dónde, para
embalsamar a un virrey que murió ahogado en un aljibe mientras
estaba de paso por la ciudad. Blacamán lo había embalsamado tan
bien, que el virrey muerto siguió gobernando mejor que cuando estaba vivo, y así se supo mantener el orden entre los esclavos alzados
y los blancos codiciosos, hasta que llegó el nuevo virrey e impuso el
orden a sangre y fuego.

Ya por esa época, algunos de los cuadros que habían de estar colgados en esos muros estaban a punto de ser pintados. Cecilia Porras
pintaba en la terraza de su casa de Manga, mirando hacia un patio
sombreado por los palos de mango y matas de guineo, pero los cuadros que pintaba no estaban inspirados en el patio, sino en otros rincones de la ciudad, con una luz distinta que ella misma inventaba.
Pocos años después conocí a Enrique Grau, a la salida de un cine, en
Bogotá, y durante mucho tiempo no hicimos otra cosa que contarnos los argumentos completos de las películas que ya habíamos visto,
hasta que descubrimos por casualidad que era él quien había ilustrado el primer cuento que yo publiqué en mi vida, y que ése era además el primer cuento que él había ilustrado en la suya. Grau vivía
en un apartamento por cuyas ventanas posteriores se veía el cementerio, y donde hacíamos unas fiestas ruidosas en cuyos silencios
casuales escuchábamos el rumor de los muertos pudriéndose en el
patio. Eduardo Ramírez Villamizar, en cambio, quien me hizo el
gran favor de ilustrar un folleto de publicidad que yo había escrito
por necesidad, vivía en una casa de La Perseverancia mucho antes de
que vivir en La Perseverancia estuviera de moda, y era una casa
grande y desnuda sin más muebles que un catre de penitente y un
caballete de pintar. Alejandro Obregón, a quien yo había conocido
antes en Barranquilla, en el burdel poblado de tortugas y alcaravanes
de Pilar Ternera, iba por esos días a Bogotá. Una tarde me dijo que
iría a dormir a mi cuarto, y como el timbre estaba descompuesto, le
dije que me despertara con una piedrecita en el vidrio de la ventana. Obregón tiró un ladrillo que encontró en una construcción vecina, y yo desperté cubierto con una granizada de vidrio. Pero él
entró sin ningún comentario, me ayudó a sacar un colchón que
guardaba debajo de mi cama para los peregrinos trasnochados, y se
tendió a dormir en el suelo, sin más cobijas que la bufanda de seda
italiana que llevaba en el cuello, y con los brazos cruzados sobre el

pecho como las estatuas yacentes de las viejas catedrales. Se despertó muy temprano y, con sus intensos ojos de agua fijos en el cielo raso, dijo:

—Eritreno. ¿Qué significa eritreno?

—No sé —le dije—, pero algún día encontraré dónde ponerla.

Necesité más de veinte años para encontrar un sitio donde colgar esa palabra enigmática en una de mis novelas más recientes. Casi tanto tiempo como el que necesitaron los cuadros del Museo de Arte Moderno de Cartagena para encontrar un muro donde quedar colgados para siempre. Ahora lo tienen. Sin embargo, aún sigue faltando un cuadro: un payaso pintado detrás de una puerta.

Carlo di Lucca, heredero y presidente de un vasto imperio industrial, no sólo era uno de los hombres más influyentes de Italia a los 36 años, sino tal vez el más elegante y simpático. Las fiestas mundanas de Roma o de Milán no tenían sentido sin él. Además de ser un conversador brillante en cinco idiomas perfectos, tocaba el piano, la guitarra y el saxofón como un profesional, cantaba y bailaba como si fuera su oficio, y era un piloto experto, un deportista múltiple, un prestidigitador asombroso y un imitador increíble de los personajes de moda. A pesar de las agobiadoras solicitudes que lo asediaban, tanto en su trabajo como en la vida social, su matrimonio era armónico y estable. Su esposa, bella y distinguida, parecía feliz. Tenían un hijo único, Piero, de ocho años.

La personalidad de este hombre fascinante había suscitado una inquietud secreta en el corazón de Silvio Peñalver, un emigrante latinoamericano, tímido y muy capaz, que en pocos años había logrado una buena posición en alguna de las empresas menores de Carlo di Lucca. Para Peñalver, su patrón era el paradigma del hombre feliz, y esa comprobación le resultaba intolerable por razones de orden moral que él mismo no hubiera podido explicar. Le molestaba sobre todo su doble personalidad: la del trabajo, donde era mezquino y autoritario, y la de su vida pública, donde su encanto era tan deslumbrante que no parecía natural. En el curso de una fiesta de los ejecutivos de la empresa, a la cual Peñalver fue invitado con su esposa por primera vez, éste concibió el mal pensamiento de que a Carlo di Lucca le estaba haciendo falta una desgracia, aunque sólo fuera para conocer los límites de la felicidad. Sin embargo, fue una idea fugaz que no dejó ninguna huella en su corazón.

Peñalver estaba encendiendo la motocicleta para volver a su casa, un domingo de primavera, cuando el pequeño Di Lucca apareció

Publicado originalmente el 12 de mayo de 1982.

por entre los setos. Había estado jugando solo en el inmenso jardín de su casa, y como ocurría con frecuencia, había logrado burlar a su gobernanta y al resto de la servidumbre que se ocupaba de vigilarlo sin sosiego. Fascinado por la motocicleta nueva, el niño le pidió a Peñalver que lo llevara a dar una vuelta, y él decidió complacerlo. Antes de arrancar, le hizo poner el casco protector que llevaba siempre para su hijo, y le dio algunas indicaciones de seguridad. El niño, acostumbrado al eterno rigor de su casa, las cumplió encantado. Se trataba de una sola vuelta, por supuesto, pero el niño insistió en una segunda, y luego en una tercera, cada vez más lejos de la casa. De pronto, Peñalver tomó conciencia de que en aquel instante tenía en sus manos la insoportable felicidad de Carlo di Lucca. Fue una inspiración súbita y embriagante. Entonces dio una vuelta completa, sin ningún plan preconcebido, y apretó a fondo el acelerador y se alejó de la casa. El pequeño Piero cantaba de júbilo.

La primera llamada telefónica la hizo Peñalver desde una cafetería, tapando la bocina con un pañuelo, como lo había visto hacer en el cine. El mayordomo que contestó al teléfono le informó lo que ya sabía: Carlo di Lucca había salido una hora antes para el aeropuerto, y su esposa estaba en Holanda. En pocas palabras, Peñalver le explicó entonces al mayordomo que hablaba en nombre de un inexistente movimiento de liberación proletaria, que el hijo único de Carlo di Lucca estaba en su poder, y que sólo sería liberado mediante el cumplimiento de dos condiciones inapelables: la entrega de cincuenta millones de dólares en efectivo, y la introducción de una serie de reformas de fondo que les dieran una mayor participación a los obreros en el imperio industrial de Carlo di Lucca. La voz era seria y terminante, y el plazo feroz para salvar la vida del pequeño Piero apenas si daba tiempo para pensar: veinticuatro horas. Carlo di Lucca recibió la noticia cuando el avión de Nueva York acababa de decolar, y lo hizo volver a Roma de inmediato.

Así empezó la jornada más terrible en la vida de aquel hombre acostumbrado a los paraísos artificiales del poder. Para su hijo, en cambio, había de ser un domingo distinto.

En realidad, Peñalver sabía hacerse querer de los niños, en especial de su hijo, y además conocía muy bien todos los lugares de diversión infantil de la ciudad. No hubo uno al que no llevara al pequeño Piero, que de pronto se sintió liberado de las normas rígidas y los convencionalismos estrechos de sus vigilantes. Vio una película de bandidos, comió helados y dulces hasta la saciedad, aprendió a remar en el lago del parque, caminó descalzo y hasta se revolcó en el barro, y se subió en todos los aparatos de la ciudad mecánica.

Nunca, desde su nacimiento, había experimentado un sentimiento igual de libertad.

Al anochecer, Peñalver llegó a su apartamento de Parioli con el pequeño Piero, que aún no parecía fatigado de ser tan feliz. Su mujer y su hijo, que había vivido también un buen domingo, lo esperaban para cenar. Peñalver explicó la presencia de Piero de la manera más simple: el niño había querido dormir con ellos, pues sus padres no estarían en Roma aquella noche, y había sido tanta su insistencia que el propio Carlo di Lucca le había dado el permiso antes de viajar a Nueva York.

Fue una cena muy divertida. El niño de Peñalver y el dichoso Piero se entendían muy bien, y éste pudo comer por primera vez lo que quería y rechazar lo que no le gustaba, y violar todas las leyes de la urbanidad en la mesa sin que nadie lo reprendiera. Peñalver tranquilizó a su esposa: todo era una broma. Le parecía inmoral que Carlo di Lucca fuera tan feliz, y quería darle al menos un domingo de angustia. Ángela le hizo ver que, de todos modos, aquella broma pesada le costaría el puesto. Peñalver contaba con la complicidad de Piero para que no lo descubrieran, pero en todo caso estaba dispuesto a regresar a su país, donde empezaban a cambiar las condiciones políticas que los habían obligado a emigrar. Ángela, que era seria y lúcida, comprendió que a esas alturas no tenía más camino que compartir la suerte de su esposo. El noticiero de televisión acababa de aliviarla: no se dijo una palabra sobre el caso. Terminaron de acuerdo: al día siguiente, muy temprano, el niño volvería a su casa sano y salvo.

Carlo di Lucca no durmió un solo instante. La discusión con sus socios fue larga y difícil, pero al amanecer estaban a punto de llegar a un acuerdo. Las maletas de dinero venidas de diversas fuentes se habían ido acumulando en el despacho, y los cincuenta millones estaban siendo preparados para la entrega. A las siete de la mañana, cuando sólo se esperaba la llamada final para establecer los pormenores del rescate, los sorprendió la noticia de que Piero había vuelto.

En efecto, Peñalver lo había llevado en la motocicleta hasta el parque cercano. Y allí lo había despedido con indicaciones precisas de volver a su casa sin rodeos. El niño se alejó sin mucho entusiasmo, un poco triste de que la gran aventura de su vida hubiera terminado. Ni él ni su amable secuestrador se habían dado cuenta de que dos agentes de los muchos que vigilaban el sector —uno disfrazado de lechero y otro disfrazado de barrendero público— los habían descubierto.

Carlo di Lucca, estragado por la tensión y la vigilia, salió corriendo a recibir a su hijo. Casi al instante se detuvo frente a ellos el coche policial donde llevaban preso a Peñalver. Carlo di Lucca comprendió entonces la verdad, y descargó contra su subalterno toda la furia reprimida durante casi veinte horas de ansiedad. El niño, todavía en brazos de su padre, tuvo un instante de incertidumbre. Pero cuando la patrulla arrancó con su escándalo de luces y sirenas, se soltó de su padre y corrió detrás del coche policial, llorando a gritos, para impedir que se llevaran a la cárcel al falso papá que le había regalado su único domingo feliz.

LA VAINA DE LOS DICCIONARIOS

Uno de los placeres de la vida es encontrar las imbecilidades de los
diccionarios. Para mí, en especial, constituyen una cierta forma de
venganza contra el destino, porque mi abuelo el coronel me enseñó
desde muy niño que los diccionarios no sólo lo sabían todo, sino
que además no se equivocaban nunca. El suyo, que era un mamo-
treto muy viejo y ya a punto de desencuadernarse, tenía pintado en
el lomo un Atlas corpulento con la bola del mundo sobre los hombros. «Esto quiere decir que el diccionario tiene que cargar con el
mundo entero», me decía mi abuelo, a quien, sin duda, no se le
ocurrió nunca buscar la nota sobre Atlas en el propio diccionario.
De haberlo hecho, se habría dado cuenta de que ese dibujo era un
error muy grave. Atlas, en efecto, era uno de los titanes de la mitología griega que provocó una guerra contra los dioses, por lo cual
lo condenó Zeus a sostener el firmamento sobre sus espaldas. El firmamento, por supuesto, y no el mundo, como estaba dibujado en el
lomo del diccionario, porque ni el propio Zeus sabía en sus tiempos
que la Tierra era redonda como una naranja.

En todo caso, el hábito de mi abuelo de consultar para todo el
diccionario se me quedó a mí para siempre, y debieron pasar muchos años antes de que descubriera con mi propia alma que no sólo
los diccionarios no lo saben todo, sino que además cometen equivocaciones, casi siempre muy divertidas. Con el tiempo he terminado
por confiar más en mi instinto del idioma, tal como se oye en la calle, y en las leyes infalibles del sentido común. De todos modos,
consulto siempre el diccionario, pero no antes de escribir, sino después, para comprobar si estamos de acuerdo.

El otro día, después de decidir, por mi cuenta y riesgo, que se
puede decir pitoniso cuando el vidente es un hombre, descubrí que
ningún diccionario incluye la palabra, aunque ninguno la prohíbe.
El de la Real Academia la define así: «Sacerdotisa de Apolo que

daba los oráculos en el templo de Delfos, sentada en el trípode».
Una pizca de sentido común permitía pensar que la palabra no existe en masculino, porque eran mujeres quienes hacían en el templo
de Delfos el hermoso oficio de adivinas, pero que nada se oponía a
que se les llamara pitonisos si hubieran sido hombres, como los hay
tantos en nuestro tiempo y, sobre todo, en nuestros medios de la
prensa.

En cambio, hay errores imperdonables en los diccionarios. El
más escandaloso de ellos me parece el de la inolvidable María Moliner, en su *Diccionario de uso del español*, cuando define la palabra día:
«Espacio de tiempo que tarda el Sol en dar una vuelta completa
alrededor de la Tierra». En primer término, siempre me ha resultado incómodo que se diga espacio de tiempo. No; o es espacio o es
tiempo, porque, aunque sean magnitudes conjugadas, son dos cosas
bien distintas. Pero lo que ahora me interesa no es eso, sino la barbaridad de que sea el Sol el que da la vuelta completa alrededor de
la Tierra, y no ésta sobre sí misma, como nos enseñaron en la escuela. El error, al parecer, tiene su origen en el diccionario de la
Real Academia Española, que define el día de este modo: «Tiempo
que el Sol emplea en dar, aparentemente, una vuelta a la Tierra».
La precaución del *aparentemente* no resuelve el enigma, porque no
queda claro si los reales académicos quisieron decir que la cosa parece así, aunque en realidad no lo sea, o si quisieron decir que ellos
no lo saben a ciencia cierta. De todos modos, el modesto *Petit La-
rousse*, que no se da ínfulas de nada, trae una definición diáfana:
«Tiempo que tarda la Tierra en dar la vuelta sobre sí misma».

A veces, los diccionarios se dan cuenta de que han hecho el ridículo, y lo corrigen en una edición posterior. Eso le ocurrió al de
la Real Academia con la famosa e inefable definición de perro:
«Mamífero doméstico de la familia de los cánidos, de tamaño, forma y pelaje muy diversos, según las razas, pero siempre con la cola
de menor longitud que las patas posteriores, una de las cuales
levanta el macho para orinar». Se prestó a tantas burlas esta precisión
excesiva —y entre ellas una muy feroz e inteligente de Guillermo
Cabrera Infante en su novela *Tres tristes tigres*–, que en las ediciones
más recientes del diccionario de la Real Academia ya los perros no
levantan la pata posterior para orinar, aunque sigan haciéndolo en
la vida real.

Otra cosa que me inquietó siempre del diccionario de la Real
Academia es la definición de los colores. Amarillo: «Del color semejante al del oro, el limón, la flor de la retama, etcétera». A mi modo
de ver las cosas desde la América Latina, el oro era dorado, no

conocía las flores de la retama, y el limón no era amarillo, sino verde. Desde antes me había llamado la atención el romance de García Lorca: «En la mitad del camino cortó limones redondos, y los fue tirando al agua hasta que la puso de oro».

Necesité muchos años para viajar a Europa y darme cuenta de que el diccionario tenía razón, porque, en realidad, los limones europeos son amarillos.

Sin embargo, me parece justo decir que, en medio de tantos tropiezos, hay un gran escritor escondido en la Real Academia, y es el que ha escrito las definiciones de las plantas. Todas son excelentes, de una andadura elegante, pero, en especial, la de una planta con la cual tengo un pleito pendiente desde la infancia, porque me la daban en ayunas como vermífugo. Me refiero al paico, pazote o epazote, que viene definido así en el diccionario de la Real Academia: «Planta herbácea anual, de la familia de las quenopodiáceas, cuyo tallo, asurcado y muy ramoso, levanta hasta un metro de altura; tiene las hojas lanceoladas, algo dentadas y de color verde oscuro; las flores, aglomeradas en racimos laxos y sencillos, y las semillas nítidas y de margen obtusa. Toda la planta despide un olor aromático, y se toman en infusión, a manera de té, las flores y las hojas. Oriunda de América, se ha extendido mucho por el Mediodía y el centro de Europa, donde se encuentra como si fuese espontánea entre los escombros de los edificios».

Hay, por supuesto, una dimensión de las palabras que los diccionarios no pueden establecer, y es la de su significado subjetivo. Hace algunos meses, mi amigo Argos, en su columna inclemente de *El Espectador*, de Bogotá, se preguntaba qué diferencia hay entre un barco y un buque. El diccionario de la Academia describe el buque de este modo: «Barco con cubierta que, por su tamaño, solidez y fuerza, es adecuado para navegaciones o empresas marítimas de importancia». Esto permite preguntarse, en primer término, qué empresas marítimas puede acometer un buque sin tener que navegar, puesto que las dos funciones las establece el diccionario como diferentes. Y permite pensar, en segundo término, que un buque no sirve para empresas fluviales, porque sólo se dice que sirve para empresas marítimas. Pero lo importante está dicho, y es que un barco es un buque. Sin embargo, para mí hay una diferencia subjetiva que me obliga a utilizar ambas palabras con un sentido diferente. En casa de los abuelos, los barcos eran sólo los de mar, como los que transportaban el banano desde Santa Marta hasta Nueva Orleans. En cambio, los buques eran los del río Magdalena, con dos chimeneas, alimentados con leña e impulsados con una rueda de madera en la popa. Para ambos, de todos modos, había un nombre genérico: vapor.

Otra cosa que se preguntaba Argos, el otro día, era el significado exacto del verbo perecer, a propósito de un herido a puñaladas que, según se dijo, pereció unas horas después en el hospital. A Argos no le parecía correcto, pero no sabía por qué, y a mí tampoco me parece, y yo tampoco sé por qué. Hay un instinto del idioma que indica, sin lugar a dudas, que los enfermos de los hospitales no perecen, sino que se mueren, cualquiera sea el motivo, a menos que les caiga el techo encima. En cambio, una persona puede haber perecido en una catástrofe aérea, si fue ésa la causa de su muerte, aunque ésta haya ocurrido, en realidad, varios días después en el hospital. Casi me atrevería a decir que el acto de perecer puede no ser simultáneo con el de morir, aunque el uno tiene que ser consecuencia del otro. Pero, por fortuna, yo no soy diccionario para atreverme a decir tanto.

JURADO EN CANNES

Los casi 2.000 fotógrafos que vinieron al XXXV Festival de Cannes no corrían demasiado este año detrás de las aspirantes a actrices que se desnudaban para ellos en la playa. No; las arenas de la Costa Azul están ahora tapizadas con una alfombra exquisita de pechugas desnudas, las más caras del mundo, de modo que ya nadie se toma el trabajo de mirar otra vez porque una adolescente con agallas se desnude cuan larga sea en mitad de la calle. A nadie en este festival parece importarle nada distinto de lo que ocurre en la noche perpetua de las veinte salas de cine, donde se pasan 42 películas diferentes cada día durante dos semanas agotadoras. Antes, las estrellas de cada delegación subían al escenario después de la proyección, protegidas por la certidumbre de que nadie se atrevería a rechiflar en público a una mujer hermosa. Tal vez el último grito de esa feria mundana fue el que lanzó Johnny Weissmuller al llegar a Cannes hace unos tres años, para recoger las cenizas de su propio Tarzán juvenil, antes de hundirse para siempre en las tinieblas de un hospital psiquiátrico. Todo eso se lo llevó el viento. Este año, los únicos que subieron al escenario en la noche inaugural fueron un grupo de hombres, maduros, tímidos y feos, a quienes muy pocos podían reconocer por su cara, pues siempre habían estado detrás de las cámaras. Con la excepción de uno, que había estado delante y detrás al mismo tiempo: el francés Jacques Tati. Eran directores de los más grandes, a quienes el festival rindió este año un homenaje merecido: el norteamericano Joseph Losey, el austríaco Billy Wilder, el húngaro Miklos Jancsó, el japonés Akira Kurosawa, el italiano Michelangelo Antonioni. Cada uno improvisó un discurso breve, inteligente y con sentido del humor, y esto pareció dar en su conjunto el tono nuevo del festival. Antonioni contó que un admirador le había parado en la calle para decirle: «Sus películas me han hecho crecer». Aquélla fue para él una revelación alarmante, según

Publicado originalmente el 26 de mayo de 1982.

dijo, porque el hombre medía apenas un metro con cuarenta centímetros.

Esta derrota del *star system*, que durante tantos años fue el atractivo mayor del festival, podría ser el signo de los nuevos tiempos. «Ahora todo el mundo sabe que el actor es sólo un medio de expresión del director», ha dicho un crítico. Geraldine Chaplin, que es una mujer inteligente y culta, y además una excelente actriz, piensa que esa concepción del oficio de actor será muy positiva, y en especial para el actor mismo. «Así nos veremos obligados a aprenderlo todo», me dijo al término de una reunión del jurado, del cual ambos formamos parte. «Terminaremos por escribir y dirigir nuestras propias películas.» Por lo que a ella se refiere, no lo recogerá del suelo: su padre hacía hasta la música.

En el mismo tono, es también natural que el festival se haya inaugurado con la proyección de uno de los mastodontes colosales de la historia del cine: *Intolerancia*, de D. W. Griffith. Fue una noche memorable, consagrada a un precursor de los más grandes, que en 1916 se gastó dos millones de dólares de los de entonces para hacer este espectáculo fabuloso que parece más actual que muchas de las películas más ambiciosas de hoy. Su duración inicial debía ser de ocho horas, y sólo para transportar todos los días los caballos y los elefantes hasta la Babilonia reconstruida en los estudios hubo que hacer un ferrocarril especial. Con todo, su estreno en el Liberty Theater de Nueva York tuvo la misma suerte desdichada de la *Consagración de la primavera*, de Stravinski, y fue el primer fracaso histórico del cine.

Lillian Gish, la madre recurrente que mece la cuna al principio de cada episodio de *Intolerancia*, tiene ahora 85 años. La periodista francesa Catherine Laporte la entrevistó hace algunos días en su apartamento de Nueva York, y la encontró como estratificada en sus recuerdos. «Tiene el cutis todavía fresco», escribió, «los labios pintados en forma de corazón, los cabellos teñidos de un rojo extraño y los bucles a la antigua, y toda ella exhala un encanto pasado de moda, pero desmentido por una impertinencia muy contemporánea.» De modo que Lillian Gish hubiera podido venir en persona a recoger sus últimas nostalgias. Sin embargo, esos laureles le fueron reservados al pianista Stanley Kilburn, de 87 años, que acompañó la película con su música original, como era de uso corriente en los tiempos del cine mudo. La acompañó durante tres horas en la sesión vespertina, y durante otras tres horas en la sesión nocturna, sin un respiro, sin un instante de decaimiento, como lo hizo tantas veces desde la primera vez, a los dieciséis años.

Otra personalidad cuyo nombre quedará vinculado para siempre
a este festival –cualquiera sea su suerte en las calificaciones finales–
es el cineasta turco Yilmaz Guney, que escribió y dirigió desde la
cárcel de Kaisiri, en Turquía, la película que más había impresionado al público del festival hasta este fin de semana: *Yol*. Los negati-
vos pasaron a Suiza y luego a Francia por conductos clandestinos,
y allí se terminó el trabajo de laboratorio. Yilmaz Guney, que a los
45 años cumplía su tercera condena por delitos políticos, se fugó de
la cárcel en octubre pasado y desapareció sin dejar rastro. Hasta la
noche del estreno de su película en Cannes, cuando reapareció sin
aviso previo bajo los proyectores del Palacio del Cine, con esmoquin y bufanda de seda y el cabello pintado de rubio como el más
mundano y tranquilo de los espectadores. En ese momento había
una manifestación de turcos encapuchados en la puerta del teatro,
gritando consignas intraducibles contra la represión militar en su
país. Dos días después, el gobierno turco pidió a Francia la extradición de Yilmaz Guney.

No obstante, la personalidad que se robó la atención de las cámaras desde la noche inaugural fue el ministro de Cultura de Francia,
Jack Lang, un antiguo actor de teatro de 42 años con una cabellera
alborotada de loco feliz. Ahora que es ministro, y aun dentro del esmoquin alquilado, Jack Lang sigue siendo igual a sí mismo: informal
y simpático, de una creatividad incontenible y un dinamismo sin tregua, y con una voluntad de renovación que de algún modo ha empezado a proyectarse en el estilo de este festival lastrado todavía por
demasiados convencionalismos.

Desde su creación, el esmoquin es obligatorio para las sesiones
nocturnas. La razón no me cabe en la cabeza. Hasta donde yo sé, esa
chaqueta sin gracia se llama como se llama porque era la que se ponían los caballeros ingleses para fumar en un cuarto aparte, de modo
que el humo del tabaco no contaminara el resto de la casa ni sus
buenos vestidos. Es decir, todo lo contrario de lo que se hace ahora.
Yo no me lo he puesto nunca, ni pienso ponérmelo, y no por prejuicio ni porque me parezca feo. Mis motivos son más profundos.
Yo tenía unos ocho años en Aracataca cuando vi en un periódico
la foto impresionante del cadáver embalsamado de Enrique Olaya
Herrera, un antiguo presidente de Colombia que había muerto en
Roma, y había sido expuesto en cámara ardiente durante varios días
en el Capitolio nacional. Lo que más me impresionó fue que estaba
vestido de etiqueta. Desde entonces pensé que esa clase de ropa sólo
se usaba para exhibir a los muertos ilustres, y aquella idea se me
convirtió para siempre en una superstición. De modo que he sido

tal vez el primer jurado de Cannes que ha roto la mala costumbre del esmoquin.

Jack Lang, tal vez sin proponérselo, me dio una buena mano en esta rebelión solitaria: la noche de gala fue al Palacio del Cine en esmoquin porque, en realidad, a un ministro no le quedaba más remedio. Pero la noche siguiente, después de un rápido viaje de ida y vuelta a París, y con el pretexto de que no había tenido tiempo de cambiarse, reapareció con los mismos pantalones de vaquero y el *pullover* rojo con que lo conocí hace años en México cuando era el director del Festival de Teatro de Nancy. Aquel gesto, sin duda muy bien calculado, fue como un ventarrón de naturalidad en un festival que se ha distinguido siempre por su rigor de selección, pero también por su conservatismo polvoriento.

INFIDENCIAS DE UN JURADO EN CANNES

Esta vez no se oyeron las rechiflas habituales cuando el presidente del jurado, el italiano Giorgio Strehler, reveló el veredicto unánime en la ceremonia de clausura del XXXV Festival de Cannes. Aquel público compacto de especialistas y aficionados del mundo entero parecía aprobar de ese modo el acuerdo nada fácil de dos mujeres y ocho hombres de nacionalidad diversa, cuya reunión decisiva, la noche anterior, había durado nueve horas sin una sola pausa. Yo era uno de ellos.

Esta vez, como tantas otras, la Palma de Oro fue repartida entre dos películas. Nunca me ha gustado esta solución, que siempre parece de compromiso y que en todo caso es intermedia. Creo que una profundización en el juicio de las películas empatadas tiene que conducir sin remedio al hallazgo de valores que harían prevalecer a una sobre la otra. La solución, por supuesto, habría sido resolver la duda mediante una votación. Pero desde el principio nos habíamos impuesto el compromiso de no apelar a ese recurso sino en casos extremos. Más aún: cuando el jurado es de número par, como era nuestro caso, el presidente tiene derecho a resolver los empates electorales con un voto suplementario; pero Giorgio Strehler renunció desde el principio a ese privilegio para sentirse en situación igual al resto del jurado. Esto nos obligó a analizar cada película con argumentos razonados. El director del festival, Robert Favre le Bret, comentó, cuando lo supo, que en 35 años no había visto un jurado que hablara tanto.

Mi candidato para la Palma de Oro sin compartir fue siempre *Missing*, del griego naturalizado francés Costa-Gavras, que revela, a través de un caso particular, toda la tragedia humana del golpe militar de Chile y denuncia la complicidad de grandes funcionarios de Estados Unidos. Me pareció que la única limitación de esta película era su escritura clásica, dentro del ámbito de un concurso donde uno tiene derecho a esperar invenciones renovadoras. La actuación de

Publicado originalmente el 2 de junio de 1982.

Jack Lemmon, como había de reconocerlo el veredicto, no sólo parecía la mejor del festival, sino también la mejor de su carrera. Sin embargo, en el curso del debate se fue definiendo con claridad que todos los jurados estaban de acuerdo en que *Missing* era una de las dos mejores películas, pero no todos pensábamos que fuera bastante buena para obtener sola la Palma de Oro.

La otra película favorita era *Yol*, del turco Yilmaz Guney, que en noventa minutos de proyección intensa hace sentir en las entrañas cuán terrible es el drama simple de estar vivo en la Turquía de hoy. Al contrario de *Missing*, que tuvo al servicio de su buena causa la inmensa maquinaria de producción de Hollywood, la película turca había sido escrita y planeada en la cárcel, hasta en sus detalles más ínfimos, por un preso político, y desde allí dirigida por interpuesta persona, hasta el punto de que no parece posible determinar con justicia quién es en última instancia su autor verdadero. Con todas las cosas raras que han ocurrido en la historia del cine, no creo que haya ocurrido antes nada tan raro como esto.

Desde el principio del debate quedó claro que *Yol* era la película que había impresionado más a fondo a todos los jurados, pero no existía la misma unanimidad en cuanto al premio que se le debía otorgar. Para mí era muy claro que parecía hecha sobre medida para el premio especial del jurado, por una verdad que puede parecer paradójica: aunque es imposible precisar quién es su autor verdadero, la película tiene una respiración personal que es uno de sus encantos mayores. En todo caso, los reglamentos del festival, que Giorgio Strehler se había aprendido de memoria como si fuera un guión de teatro, establecen de un modo expreso que la Palma de Oro y el premio especial del jurado no son un premio primero y un premio segundo, sino que son dos premios paralelos del mismo nivel. Con todo, más bien por razones prácticas, la fórmula del *ex aequo* se impuso. Pero sólo en la ceremonia final fuimos conscientes los jurados de que habíamos hecho por la armonía interna del Tercer Mundo algo más que repartir un premio. Habíamos conseguido el milagro de que un griego y un turco se subieran a un escenario para besarse de felicidad ante los ojos del mundo entero.

El resto de los premios fueron fáciles. El homenaje a la obra conjunta de Michelangelo Antonioni, tomando como punto de referencia *Identificazione de una donna*, que es su película más refinada a los setenta años de su edad, era un acuerdo que todos llevábamos resuelto en el corazón. Mi única duda general sigue siendo que el premio especial del jurado, atribuido a *La notte di San Lorenzo*, de los hermanos Taviani, y el premio de la *mise-en-scène*, atribuido a *Fitz-*

carraldo, del alemán Werner Herzog, quedaron en posiciones intercambiadas.

En efecto, desde que vi *La notte di San Lorenzo* me sentí estremecido por su fluidez y deslumbrado por la luz de diamante de la Toscana, pero me quedé al final con la zozobra de no saber a ciencia cierta lo que sus autores admirables me habían querido decir. No estuve solo en ese abismo: otros dos jurados pidieron verla de nuevo. Pero al cabo de un día de reflexión decidí no asistir a la repetición privada, porque mis relaciones con el arte han sido siempre de amor a primera vista, y no recuerdo ninguna obra de ningún género que me haya impresionado más la segunda vez que la primera. En la discusión final, por supuesto, mis reservas no fueron un obstáculo para la unanimidad.

El premio especial del jurado me parecía más adecuado para *Fitzcarraldo*, no porque fuera mejor o peor, sino porque tiene ese aliento misterioso, indefinible y devastador que permite identificar de inmediato una auténtica obra de arte: la inspiración. Es esa magia oculta lo que le permite a la película alcanzar con el mismo impulso las alturas más sublimes de la locura y los abismos más insondables de la chapucería poética. Esto no lo tenía ninguna otra película del festival, y es un milagro cada vez más raro en las artes contemporáneas. Si Herzog no despuntó en el primer lugar, a mi juicio, fue porque se disolvió en el conformismo de un final a la manera de Rossini, en lugar del apocalipsis wagneriano que todos esperábamos para quedar completos después de tantos desafueros de la imaginación.

Muchos espectadores creyeron ver en el delirio de *Fitzcarraldo* algo como un saqueo a mano desarmada en las novelas contemporáneas de la América Latina. Sobre todo, en el episodio de un personaje que se vale de un bloque de hielo para impresionar a los indios de la Amazonia, y la imagen de un barco encallado en medio de la selva. Por eso fui muy claro en el seno del jurado en el sentido de que estaba dispuesto a renunciar a lo que pareciera mío, porque además no eran imágenes esenciales, sino todo lo contrario: simples tropiezos de la película. Sin embargo, cuando Werner Herzog me llamó por teléfono al día siguiente para decirme, con una amabilidad muy suya, que le gustaría hacer algo conmigo, no pude resistir la mala educación de contestar: «No se preocupe, Herzog: ya lo hemos hecho». Le doy excusas públicas.

ROMA EN VERANO

He vuelto a Roma, al cabo de una muy larga ausencia, y la he vuelto a encontrar como siempre: más bella, y más sucia, y más loca que la vez anterior. El verano estalló de pronto la semana pasada, con ese calor que parece de vidrio líquido, y la moda femenina, que este año dejó las puertas abiertas a toda clase de desafueros de formas y colores, convirtió a la ciudad eterna en la más moderna y juvenil del mundo.

Creo que fue Julio Cortázar quien observó en alguno de sus libros que después de conocer una ciudad seguía recordándola para siempre, no como era en la realidad, sino como se la imaginaba antes de conocerla. Esto me parece cierto, salvo con Roma, pues es la única ciudad que siempre me imaginé tal como fue cuando la conocí, que es como sigue siendo siempre. Tal vez la única de la que puedo decir que la recordaba sin conocerla.

Estuve aquí por primera vez en el verano de 1955 –hace ahora la módica suma de veintisiete años–, como enviado especial de *El Espectador*, de Bogotá, a los funerales de un Papa que aún no había muerto, pero que tenía hipo desde hacía varios meses. Un médico amigo me había dicho en Colombia que ése es un síntoma del cáncer de esófago, y que si no se conseguía controlarlo era una causa segura de muerte por deshidratación. Yo conocía un antecedente literario: el hermoso cuento de Somerset Maugham de un inmigrante inglés a quien le sorprendió el hipo en un transatlántico de lujo que navegaba por el océano Índico, y al cabo de pocos días de esfuerzos estériles su cadáver fue arrojado a las aguas envuelto en la bandera británica. El Papa, como se sabe, no corrió la misma suerte. En cambio, fui yo quien estuvo a punto de morir el mismo día de mi llegada a Roma, un alucinante domingo de julio en que había, como siempre, una huelga de todo, e Italia parecía, como siempre, al borde del desastre. «Esto es igual que Aracataca», me dije, abru-

Publicado originalmente el 9 de junio de 1982.

mado por el calor y el polvo, mientras recorría la estación solitaria buscando en vano un alma caritativa que me ayudara a cargar las maletas. De pronto, un esquirol de los que nunca faltan, aun en las mejores familias, no sólo me ayudó a cargarlas por cincuenta liras de aquellos tiempos, sino que se ofreció para conseguirme un hotel en la cercana Via Nazionale.

Era un edificio muy viejo y reconstruido con materiales varios, en cada uno de cuyos pisos había un hotel diferente. Sus ventanas estaban tan cerca de las ruinas del Coliseo, que no sólo se veían los miles y miles de gatos adormilados por el calor en las graderías, sino que se percibía su olor intenso de orines fermentados. Mi buen acompañante, que se ganaba una comisión por llevar clientes a los hoteles, me recomendó el del tercer piso, porque era el único que tenía las tres comidas incluidas en el precio. Además, la recepcionista era una mujer gorda y floral, con una cálida voz de soprano, y parecía muy sensible a la idea de que un caribe de veintitrés años hubiera atravesado el océano para conocerla. Eran las cinco de la tarde, y en el vestíbulo había diecisiete ingleses sentados, todos hombres y todos con pantalones cortos, y todos cabeceando de sueño. Al primer golpe de vista me parecieron iguales, como si fuera uno solo dieciséis veces repetido en una galería de espejos; pero lo que más me llamó la atención fueron sus rodillas óseas y rosadas. Siempre había querido mucho a los ingleses, hasta este año funesto de las Malvinas, en que una imbecilidad de su gobierno me los sacó del corazón sin remedio. Sin embargo, no sé qué rara facultad oculta del Caribe me sopló al oído que aquella sucesión de rodillas rosadas era un mensaje aciago. Entonces le dije a mi acompañante que me llevara a otro hotel donde no hubiera tantos ingleses sentados en el vestíbulo, y él me llevó sin preguntarme nada al del piso siguiente. Esa noche, los diecisiete ingleses y todos los huéspedes del hotel del tercer piso se envenenaron con la cena.

Así empezó para mí aquel verano inolvidable. Por la mañana, la ciudad estaba casi vacía, porque muchos romanos se iban a la playa. Roma era todavía una ciudad con muy pocos automóviles, y el único lujo que podían pagarse los deslumbrantes automovilistas de hoy eran unas Vespas rudimentarias que se metían por todas partes y atropellaban a los transeúntes aun sobre los andenes. Al contrario de lo que hacíamos en el trópico, que abríamos puertas y ventanas para que entrara el fresco de la calle, los romanos cerraban las casas con persianas herméticas. Así lo hacían desde los tiempos del Imperio, y así lo siguen haciendo, con toda la razón, porque así impiden que se meta en las casas el aire ardiente de la calle. Después de un almuerzo ligero a base de pasta —esa comida prodigiosa que cambia de sabor con sólo cambiar de forma— hacían una siesta lisa y densa que se parecía demasiado a la muerte. A esa hora no había un alma en la calle, el sol se quedaba inmóvil en el centro del cielo, y el silencio era tan intenso que no parecía posible. Pero un poco después de las seis de la tarde, todas las ventanas se abrían de golpe para convocar el aire fresco que empezaba a moverse, y una muchedumbre jubilosa se echaba a las calles en medio de los petardos de las Vespas, los gritos de los vendedores de sandías y las canciones de amor entre las flores de las terrazas, y sin ningún otro objetivo que el de vivir. Hoy todo sigue igual. Los italianos, en efecto, descubrieron desde hace mucho tiempo que no hay más que una vida, y esa certidumbre los ha vuelto refractarios a la crueldad.

Los únicos seres despiertos a las tres de la tarde en aquel verano de hace veintiséis años eran las putitas tristes de la Villa Borghese, que hacían de día lo que todas las otras hacen de noche, inclusive trasnocharse. El tenor Rafael Rivero Silva y yo vivíamos en dos cuartos contiguos de una pensión cercana, cuyo único defecto era estar a la vuelta del jardín zoológico, de modo que uno despertaba a medianoche asustado por el rugido de los leones. Después del almuerzo, mientras Roma dormía, nos íbamos en una Vespa prestada a ver las putitas vestidas de organza azul, de popelina rosada, de lino verde, y a veces encontrábamos alguna que nos invitaba a comer helados. Una tarde no fui. Me quedé dormido después del almuerzo, y de pronto oí unos toquecitos muy tímidos en la puerta del cuarto. Abrí medio dormido, y vi en la penumbra del corredor una imagen de delirio. Era una muchacha desnuda, muy bella, acabada de bañar y perfumar, y con todo el cuerpo empolvado.

«Buona sera», me dijo con una voz muy dulce. «Mi manda il tenore.»

Al contrario de lo que sucedió con el personaje de Somerset Maugham, el Papa se recuperó en mitad del verano y volvió a las audiencias públicas. Yo asistí a una de ellas en el patio de Castelgandolfo, que era su residencia de verano. Lo vi muy cerca, con un hábito inmaculado y unas manos parasitarias que parecían restregadas con lejía, y en aquel instante me di cuenta de que yo tenía que buscar un tema más fructífero e inmediato que el de su muerte. Hice bien, porque cuando el Papa murió, tres años después, yo no estaba ya en este mundo, sino en el otro: en Caracas. Pero la imagen de aquella muchacha en sus puros y hermosos cueros a las tres de la tarde se me quedó para siempre en la memoria, como uno de los tantos milagros que sólo son posibles en el sopor de Roma en verano.

LO QUE NO ADIVINÓ EL ORÁCULO

El jueves nos fuimos a consultar al oráculo. A las siete de la mañana tomamos en Atenas un autobús refrigerado y tres horas después estábamos en Delfos, la patria del oráculo, la ciudad sagrada de Apolo que fue en su tiempo el ombligo del mundo. El autobús iba lleno de gringos domesticados que seguían con mucho juicio en folletos de colores las explicaciones que el guía griego nos hacía en su inglés casi imaginario.

En realidad, el idioma universal no es el inglés, sino el inglés mal hablado. Si uno lo habla apenas bien no encontrará quien entienda lo que uno dice. En las largas pausas sin información nos dejábamos narcotizar por la música universal, que no es la de Mozart, como dicen los entendidos, sino esa música infinita de malos entendidos que suena sin misericordia en todos los ascensores del mundo.

El viaje fue lento, cauteloso, pues los choferes griegos tienen instrucciones de tomar su oficio con calma para no asustar a las señoras jubiladas que vienen de Nevada, de Maryland, de Kentucky, acompañadas por viejos maridos que a veces no son suyos, sino prestados a escondidas para jugar al amor otoñal después de consultar el oráculo. Viajamos despacio a través de trigales soleados y olivos milenarios, y después por desfiladeros pavorosos donde volaban unos pájaros enormes y oscuros que en épocas mejores fueron las águilas de Zeus. A un cierto momento, el guía se atrevió a decir: «A la derecha pueden ver una torre del siglo XV». Lo dijo con una cierta vergüenza, y con razón, pues en un país donde uno se encuentra de pronto comiendo con una cuchara del siglo VII antes de Cristo, un pedazo de torre como aquélla no tiene más interés que una estación de gasolina. Sin embargo, los guías cumplen con su deber, porque los turistas esperan que se les diga todo por el dinero que pagan, y de todos modos, si no se lo dicen lo preguntan. Por eso, cada vez que llego por

Publicado originalmente el 16 de junio de 1982.

primera vez a una ciudad, me inscribo en un programa turístico y salgo de eso de una vez por todas. A partir de entonces, sé que todo lo que vea lo tengo que descubrir por mis propios medios, puesto que ya conozco todo lo conocido. Más aún: en ciudad de México, después de vivir allí veinte años, me inscribí en una caravana sólo por la curiosidad de saber cómo le enseñan la ciudad a los turistas, y me quedé sorprendido de cuántas cosas habían pasado inadvertidas para mis ojos de residente.

Sin embargo, debo reconocer que me interesa más la leyenda que la realidad histórica, y, por consiguiente, en Grecia me interesa más Homero que Herodoto. En mi visita al oráculo, por lo mismo, me interesaban más las fuentes del drama de Edipo que la historia de tantos tiranos que encontraron en aquel lugar su desgracia o su fortuna. La emoción empezó en el transcurso del viaje, cuando dijo el guía: «En este lugar, según la leyenda, Edipo mató al rey Layo, su padre». Pero fue ésa la única mención que se hizo en todo el viaje. Al parecer, el drama de Edipo se considera aquí como ficción pura, tanto como las aventuras de Ulises o la desgracia de Medea. En cambio, no sé por qué extraña trasposición, los personajes de la mitología han sido aceptados en los dominios de la vida real.

A uno le hablan de Prometeo encadenado y expuesto a la ferocidad de las aves de rapiña en la cima de una montaña, y le cuentan que Apolo luchó contra la serpiente Pitón hasta que logró suplantarla, y le explican el mundo a través de los dioses innumerables y las diosas traviesas como si fueran más reales que los hombres y las mujeres de Sófocles. En cambio, las mejores verdades, las más humanas, se ocultan por pudor. Del Partenón, que se sostiene apenas como si fuera hecho de cáscaras de huevo, se nos dice que fue el gran templo de Atenas, que en el siglo XIII fue convertido en santuario católico por los cruzados y en mezquita turca dos siglos después, pero se nos oculta en cambio el que fuera su destino más humano: residencia ocasional de las cortesanas de algún rey de Macedonia en el siglo IV antes de Cristo. Asimismo, del oráculo se nos cuenta que las pitonisas debían pasar de los cincuenta años, que debían ser feas y vulgares y que «desde el momento en que se consagraban al servicio del dios debían abandonar a sus maridos y a sus hijos». Pero no se nos dice la razón, y es que al principio eran las vírgenes más jóvenes y hermosas del país, cuyos encantos terminaban por ablandar al más incorruptible de los peregrinos.

De modo que cuando llegamos a la cumbre del santuario de Delfos ya el guía nos había contado todo, pero no nos había dado ningún elemento nuevo sobre el drama de Edipo, que a fin de cuentas era lo único que me interesaba del oráculo. Cuentan que la pitonisa, antes de profetizar, se purificaba en las aguas de la cercana fuente de Castalia y masticaba hojas de laurel y aspiraba vapores de incienso y mirra, hasta el punto de que apenas si era dueña de sí misma cuando debía responder a las preguntas que le hacían los viajeros llegados de todo el mundo conocido, y que bien podían ser reyes o mendigos. Cuentan que sus respuestas eran alaridos y contorsiones incomprensibles que los sacerdotes descifraban a su manera. De modo que era imposible conocer el sentido exacto de la adivinación, y como todas las adivinaciones, sólo podía entenderse a fondo después de que se cumplía. La más célebre, sin duda, fue la que recibió el rey Creso, famoso por sus riquezas sin cuento, cuando quiso saber si convenía hacer la guerra contra los persas, cuyo reino estaba al otro lado del río Halys. El oráculo contestó: «Si Creso atraviesa el río, destruirá un gran reino». Creso lo hizo y fue derrotado, con lo cual se cumplió la predicción, pues destruyó su propio reino, que era uno de los más poderosos de su tiempo. En cambio, al contrario de lo que ocurría en la realidad, la predicción que recibió Edipo, rey de Tebas, fue directa y explícita: la peste sería conjurada el día en que se descubriera quién había sido el asesino de Layo, el rey anterior. Edipo lo descubrió, como se sabe, y descubrió al mismo tiempo su propia identidad y su propio destino. Y así nació para siempre la única estructura literaria de una perfección absoluta: el investigador que descubre que él mismo es el asesino.

Lo más impresionante del santuario de Delfos, sin duda, es el lugar donde fue construido. Uno estaría dispuesto a creer que, en efecto, era el ombligo del mundo si no se conocieran los altos de Machu Picchu, en los Andes, donde se tiene de veras la impresión de haber cambiado de planeta. Uno estaría dispuesto a postrarse de admiración ante estas construcciones de piedra y de sueño si no se conociera el ámbito mágico de Uxmal y Chichen Itza, en Yucatán, donde todavía parece sentirse la respiración de los seres que lo vivieron. Pero la comparación no es justa, porque los centros ceremoniales de México están casi intactos, y en cambio los monumentos de Grecia son apenas los restos de un saqueo histórico despiadado.

En realidad, aquí se viene a conocer los lugares y a imaginar, a través de tantas lecturas atrasadas y del inglés aproximado de los guías, cómo eran los monumentos antes de que pasaran por aquí las hordas imperiales de los países que hoy se sienten civilizados. Perdida en la constelación de las Cícladas hay una isla minúscula −Milos− de la cual nadie se acordaría al pasar si no fuera porque allí fue encontrada la Venus sin brazos que es el atractivo mayor –junto con la Gioconda– del Museo del Louvre.

En el Museo de Delfos, por puro milagro, queda la estatua de un auriga fundido en un bronce que todavía parece vivo, y que para mi gusto es una de las obras más asombrosas de las artes de todos los tiempos. Pero el resto no son sino los escombros que quedaron después del saqueo. Porque lo mejor de todo este mundo –salvo los lugares, que por fortuna no se pueden llevar– no está donde los dioses lo pusieron, sino en el Museo Británico, en Londres, o en el Louvre, en París. A pesar de la sabiduría y el poder adivinatorio de este oráculo de miércoles que ya no se acuerda de Edipo.

LA DURA VIDA DEL TURISTA

Tan pronto como subimos a bordo, una voz untuosa de mujer bien servida ordenó en cuatro idiomas por los altavoces que los visitantes bajaran a tierra porque el barco se disponía a partir, y no había acabado de decirlo cuando el barco partió sin ningún otro anuncio. Al cabo de tantos años sin navegaciones ni regresos sentí revivir una emoción casi olvidada viendo borrarse en las brumas de junio las casas apelotonadas y descoloridas del Pireo, y permanecí allí con el ánimo dispuesto para no pensar en nada durante varios días. Aquellos fueron los únicos cinco minutos de descanso a fondo en todo el viaje. Apenas habíamos salido del puerto cuando la misma voz de mujer, con un énfasis más perentorio que el anterior, ordenó a todos los pasajeros reunirnos en cubierta para una maniobra simulada de salvamento. De modo que acudimos en masa con los chalecos salvavidas colgados del cuello y mirándonos unos a otros con nuestras caras de imbéciles, mientras la sirena del barco lanzaba bramidos de naufragio y el primer oficial impartía instrucciones precisas y alarmantes que ninguno de los quinientos turistas de aquella catástrofe de mentira escuchaba con la atención debida. Cinco minutos después todo había terminado. Pero sólo por breves minutos, pues no bien nos habíamos quitado los salvavidas cuando ya nos estaban convocando al salón principal para una conferencia sobre las incontables islas del mar Egeo que íbamos a conocer en los próximos días. Así empezó una semana frenética, que si nos sirvió para algo fue para darnos cuenta en carne propia de que no hay oficio más ingrato y agotador que el de turista de cuerpo entero. Esto es más grave en Grecia que en ninguna otra parte. En efecto, no sé por qué tuve siempre la idea de que los griegos tenían algo del carácter desordenado y expansivo de los italianos. Y no es así: son locos para el lado contrario. Desde el capitán del barco hasta el muchacho que carga las maletas tienen un sentido de la autoridad que se parece mucho al

Publicado originalmente el 23 de junio de 1982.

autoritarismo, y son rigurosos y puntuales, pero de un modo distinto del de los ingleses, por fortuna, hasta el extremo de que uno tiene la impresión de ser prisionero de un organismo de relojería. Estas virtudes son ideales para el turista cuadrado, al cual hay que indicarle todo. Pero quienes tenemos la pretensión de hacer las cosas de un modo distinto, tropezamos sin remedio con las talanqueras del orden.

Eso fue lo que nos sucedió cuando dos matrimonios amigos resolvimos salirnos del programa y nos quedamos tres días en la isla de Mikonos. Bajamos con ocho maletas en un lugar donde los viajeros sensatos no llevan sino un traje de baño y un cepillo de dientes. Los guardias de la aduana local, que tal vez no habían visto nunca un equipaje semejante, se empecinaron en hacernos una requisa a fondo. En vano les explicamos que ya la requisa había sido hecha en el Pireo, cuando entramos al país. La hicieron otra vez. Una semana más tarde, en el puerto de Heraklion, en la isla de Creta, estuvieron a punto de repetirla por tercera vez. «Ya nos la han hecho dos veces», le dije al guardia, un griego de ojos soñadores y barba tupida. «¿Y cómo sé que es cierto?», me preguntó él. «Porque yo le doy mi palabra», le dije. El hombre me dio una palmada en la espalda y nos dejó pasar con una sonrisa de las grandes.

Fue nuestra única victoria en diez días. De resto, nos costaba trabajo comer porque no estábamos incorporados a ningún grupo, porque llegábamos al desayuno cuando faltaba media hora para cerrar el comedor o porque queríamos el pescado al horno y no asado a la plancha como estaba previsto. Teníamos la impresión de que había una sola manera de hacer las cosas, cualesquiera que fueran, y que hacerlas de un modo distinto era como romper el orden del universo. No recuerdo una mirada de mayor asombro que la del oficial de guardia del barco que me encontró escrutando el mar a las doce de la noche cuando ya todas las actividades estaban terminadas y todos los viajeros en sus camarotes, como se les había recomendado, porque al día siguiente había que levantarse a las seis de la mañana para la primera excursión en la isla de Rodas. Pero cuando hicimos el esfuerzo por marcar el paso de todos, nos encontramos en un mundo ajeno, un mundo feroz y vertiginoso, del cual no nos será fácil reponernos con otras dos semanas de vacaciones en quién sabe qué playa olvidada. Las excursiones desde Mikonos a la isla de Delos salen todos los días a las nueve de la mañana y regresan a la una de la tarde. Es decir, que en tres horas hay que reconstruir con la imaginación casi la cuarta parte de la historia de la humanidad. El resultado final es que lo único que se recuerda a ciencia cierta no es la forma y el lugar en que nació Apolo, ni la decisión de que los nacimientos y las muertes sólo podían ocurrir en la isla de enfrente porque nadie podía nacer ni morir en Delos. No: lo único que se recuerda es la hilera de excusados públicos y colectivos, donde los ciudadanos notables se sentaban a dar del cuerpo mientras dilucidaban asuntos de la más grande importancia. Sólo por casualidad cae uno en la cuenta, varios meses después, de que aquellos excusados de visita no eran en Delos, sino en Éfeso, donde habíamos estado con la misma prisa cinco días más tarde. El espacio y el tiempo terminan por unificarse en la memoria, sometida a una prueba excesiva. ¿Quién fue primero, Píndaro o Cleopatra?

Lo peor para mí es que por andar corriendo detrás de tantas piedras viejas uno termina por no conocer la vida real de los lugares que visita. Grecia está tan viva como en los tiempos de Pericles, pero las agencias de viajes sólo siguen mostrando aquéllos y no los fascinantes tiempos de hoy. En todas las islas hay calles enteras de almacenes donde sólo se venden pieles de animales finos en pleno verano y joyas magníficas, muchas de las cuales son reproducciones excelentes de las muy antiguas que están en los museos. Eso fue algo que me sorprendió también en Nueva Delhi, la capital de India, donde las únicas colas interminables que se ven en las calles son las de las matronas frente a las joyerías. Mujeres impasibles, asediadas por hordas de mendigos leprosos. Recuerdo que entré en un hotel de lujo muerto de hambre al cabo de un largo viaje desde Tailandia. Y el alma se me instaló en su almario cuando sentí el exquisito olor de carne asada que flotaba en el aire. Sólo después descubrí que aquella fragancia apetitosa era la de los muertos incinerados al aire libre en el río cercano. En las islas griegas, por el contrario, uno se pregunta dónde han escondido la miseria: no hay un mendigo ni un perro en la calle. Rodas sigue siendo una ciudad hermosa. Es difícil entender que san Juan Evangelista haya podido concebir los horrores del Apocalipsis en la isla de Patmos, cuyas colinas tibias y cuyos mares internos no se pueden parecer a nada más que al paraíso perdido. En un bar de Mikonos, donde tal vez no lleguen los productores de cine, está el ser humano más bello del mundo sirviéndoles cerveza helada y pulpos fritos a los turistas. Pero no es fácil descubrir estas cosas, porque en los programas de los turistas no está incluida la vida de hoy. La que nuestros remotos descendientes conocerán dentro de 3.000 años, cuando los barcos de la América Latina colosal los lleven a conocer las ruinas de Nueva York y los guías les describan un lugar de Manhattan donde no habrá nada, pero donde les dirán que estuvo en otro tiempo el Empire State o la estación de gasolina de la calle Cuarenta y Cinco.

ESTÁ DE MODA SER DELGADO

Sí, está de moda en casi todo el mundo, y aun en el tercero, donde a tantos seres humanos les cuesta tanto trabajo comer para sobrevivir. Hace unos años, los artículos más leídos en periódicos y revistas eran los relacionados con el cáncer. Ahora lo son los que hablan de la dieta, entendida ésta como las restricciones alimenticias para adelgazar y no como un «régimen que se manda observar a los enfermos o convalecientes en el comer y el beber», según la inefable descripción del diccionario de la Academia. Los libros sobre esta materia son cada día más numerosos y solicitados. En las reuniones sociales, más que la política y los signos del zodíaco, las conversaciones sobre métodos para recobrar la línea son casi obsesivas. Siempre hay alguien que pretende haber encontrado una dieta ideal —e irreal, por supuesto—, que permite adelgazar como una gacela sin ningún sacrificio. Se reparten copias entre los amigos. Se cuentan puntos de calorías, se habla de comida antes de comer, y cuando se llega a la mesa se tiene tanta hambre que hay un acuerdo unánime: «Hoy no hago dieta, empiezo el lunes». Hay quienes no sólo cuentan puntos sino que pesan los alimentos en la mesa con un granatario de farmacéutico. La duración de las conversaciones telefónicas aumenta porque hay un tiempo suplementario destinado a hablar de la dieta. A veces lo invitan a uno a comer, y el anfitrión es tan discreto que decide cocinar para que nadie engorde y se termina comiendo peor que en el hospital.

Al cabo de tantos años de estar viviendo dentro de esta logia de dietistas puedo sacar algunas conclusiones generales. La más curiosa, desde luego, es la de que los hombres son mucho más obsesivos que las mujeres por la conservación de su línea, sobre todo después de cierta edad. Parece ser que las mujeres renuncian más temprano. Recuerdo una amiga esclava de su silueta que, en medio de la pachanga ruidosa y multitudinaria de sus treinta años, me dijo: «El sueño

Publicado originalmente el 30 de junio de 1982.

de mi vida es cumplir los sesenta para poder comerme todo lo que me dé la gana». Es probable que cuando los cumpla –y el día está lejano– se sienta atravesando una segunda juventud, y entonces sea más intensa que ahora su ansiedad por mantener el peso. Pienso, en cambio, que los hombres tenemos el sentimiento contrario, y que, a medida que nuestra vida avanza, tenemos una mayor preocupación por no parecer más feos de lo que Dios nos hizo. La mejor solución, desde luego, es ser rico en la India, donde el tamaño de la panza se considera en proporción directa con la respetabilidad.

Los fabricantes de alimentos y los propietarios de restaurantes empiezan a preocuparse. Acabo de comprobar que en Italia hay una campaña publicitaria para convencer a los clientes de que el plato nacional, o sea, las pastas en todas sus formas, tiene la virtud mágica de no engordar si se las come solas. En todo caso, durante muchos años se dijo, y nunca fue desmentido, que la cantante de ópera de peso completo, María Callas, que en sus mocedades pesaba casi cien kilos, recobró su figura corporal para siempre con una dieta drástica de espagueti. La creencia de que las pastas no engordan si se comen solas está muy generalizada en Italia. Sobre todo entre la gente de cine, que es la que más tiene que cuidar su apariencia para vivir. Sin embargo, Mónica Vitti es una de las mujeres más bellas y esbeltas que conozco, y la he visto comerse dos platos de espagueti a la putanesca y un conejo entero con berenjenas y, enseguida, dos kilos de helado de crema, mientras veía en la televisión una película de pandilleros. Nunca he podido saber, y siempre he olvidado preguntárselo, si la cara de complacencia infinita con que miraba la pantalla era por el placer del helado después de haber comido tanto, o por la felicidad con que los bandidos ametrallaban a los policías.

Como es natural, ante la obsesión de la dieta ha surgido la obsesión contraria: tratar de demostrar los peligros de la dieta. Hace poco, una amiga se encontró con un amigo que parecía haber envejecido treinta años durante los seis meses en que se habían dejado de ver. «Pero qué te ha pasado», exclamó ella, convencida de que aquel pobre hombre era víctima de una enfermedad fatal. Pero el amigo le contestó: «Es que hice la dieta de los carbohidratos». Este método de adelgazamiento, que se volvió muy popular hace años, con el prestigio real o falso de haber sido creado para los pilotos de la fuerza aérea del Canadá, tiene ahora, en efecto, la rara reputación de ser muy eficaz, no sólo para adelgazar, sino también para envejecer sin necesidad de vivir demasiados años. Según la Academia de Ciencias de Estados Unidos –citada por una agencia de prensa–, ciertas dietas pueden provocar diversos tipos de cáncer, como casi todo, al fin y al cabo, pues si uno cree lo que lee, aun en revistas especializadas y serias, se termina por pensar que lo que produce el cáncer es el hecho simple de estar vivo. Pero los datos de la Academia de Ciencias de Estados Unidos son precisos y alarmantes: las dietas podrían ser la causa del 40 % de los casos de cáncer en los hombres y del 60 % en las mujeres.

Menos mal que otro artículo sobre el mismo asunto, publicado hace poco en el *New York Times*, dice la misma cosa, pero vista por el lado positivo: una dieta acertada puede prolongar la vida más allá de los límites imaginables: «Si la respuesta del ser humano a la restricción de los alimentos fuera similar a la de los animales de laboratorio», dice el artículo, «la duración máxima de la vida podría extenderse hasta 140 años, y el promedio actual de vida podría aumentar a más de 120».

Nada me gusta más en este mundo que comer. Tengo la inmensa suerte de que ningún problema me quita el hambre, sino todo lo contrario, me la estimula. Hasta el punto de que en una mala época puedo estar comiendo sin pausas durante todo el día. Además, quedo encerrado, entonces, en un triángulo vicioso: cuando no me está saliendo bien lo que escribo, caigo en cierta desmoralización que me produce un hambre insaciable, y de tanto comer para tratar de saciarla termino por engordar sin ningún control, y esta gordura me produce un estado de desmoralización que me impide escribir bien.

De modo que tengo razones científicas, inclusive profesionales, para preocuparme por las dietas. Pero no creo mucho en ellas, porque me parece que todo lo que entra por la boca engorda, así como me parece que todo lo que sale de ella envilece. Es un mal destino: haber pasado la mitad de la vida sin comer porque no tenía con qué, y tener que pasar igual la otra mitad, sólo por no engordar.

EL AMARGO ENCANTO
DE LA MÁQUINA DE ESCRIBIR

Los escritores que escriben a mano, y que son más de lo que uno se imagina, defienden su sistema con el argumento de que la comunicación entre el pensamiento y la escritura es mucho más íntima, porque el hilo continuo y silencioso de la tinta hace las veces de una arteria inagotable. Los que escribimos a máquina no podemos ocultar por completo cierto sentimiento de superioridad técnica, y no entendemos cómo fue posible que en alguna época de la humanidad se haya escrito de otro modo. Ambos argumentos, desde luego, son de orden subjetivo. La verdad es que cada quien escribe como puede, pues lo más difícil de este oficio azaroso no es el manejo de sus instrumentos, sino el acierto con que se ponga una letra después de la otra.

Se ha hecho mucha literatura barata sobre las diferencias entre un texto escrito a mano y otro escrito a máquina. Lo único cierto, sin embargo, es que la diferencia se nota al leerlos, aunque no creo que nadie pueda explicarlo. Alejo Carpentier, que era escritor a máquina, me contó alguna vez que en el curso de la escritura tropezaba con párrafos de una dificultad especial, que sólo lograba resolver escribiéndolos a mano. También esto es tan comprensible como inexplicable, y sólo podrá admitirse como uno más de los tantos misterios del arte de escribir. En general, yo pienso que los escritores iniciados en el periodismo conservan para siempre la adicción a la máquina de escribir, mientras quienes no lo fueron permanecen fieles a la buena costumbre escolar de escribir despacio y con buena letra. Los franceses, en general, pertenecen a ese género. Hasta los periodistas: hace poco, en Cancún, me llamó la atención encontrar al director del *Nouvel Observateur*, Jean Daniel, escribiendo a mano su nota editorial con una caligrafía perfecta. El famoso café Flore, de París, llegó a ser uno de los más conocidos de su

Publicado originalmente el 7 de julio de 1982.

tiempo porque allí iba Jean-Paul Sartre todas las tardes a escribir las obras que todos esperábamos con ansiedad en el mundo entero. Se sentaba muchas horas con su cuaderno de escolar y su estilógrafo rupestre, que muy poco tenía que envidiar a la pluma de ganso de Voltaire, y tal vez no era consciente de que el café se iba llenando poco a poco de los turistas de todas partes que habían atravesado los océanos sólo por venir a verle escribir. Sin embargo, no había necesidad de verlo para saber que era una obra escrita a mano.

En cambio, es difícil imaginar a un norteamericano que no escriba a máquina. Hemingway, hasta donde lo sabemos por sus confesiones y las infidencias de sus biógrafos, usaba los dos sistemas –como Carpentier–, y ambos del modo más extraño: de pie. En su casa de La Habana se había hecho construir un facistol especial en el que escribía con lápices de escuela primaria, a los cuales sacaba punta a cada instante con una navaja de afeitar. Su letra era redonda y clara, un poco dibujada, y de su oficio original de periodista le había quedado la costumbre de no contar por páginas el rendimiento de su trabajo, sino por el número de palabras. A su lado, en una mesa tan alta como el facistol, tenía una máquina de escribir portátil y, al parecer, en un estado más bien deplorable, de la cual se servía cuando dejaba de escribir a mano. Lo que no se ha podido establecer es cuándo y por qué usaba a veces un sistema y a veces el otro. En cuanto a la rara costumbre de escribir de pie, él mismo da una explicación muy suya, pero que no parece satisfactoria: «Las cosas importantes se hacen de pie», dijo, «como boxear». Hay el rumor de que sufría de alguna dolencia sin importancia pero que le impedía permanecer sentado durante mucho tiempo. En todo caso, lo envidiable no era sólo que pudiera escribir lo mismo a mano o a máquina, sino que pudiera hacerlo en cualquier parte y, al parecer, en cualquier circunstancia. Se sabe que alguna vez, en el curso de un combate, se fue a la retaguardia a escribir un despacho de prensa sentado en el suelo y con el cuaderno apoyado en las rodillas. En su hermoso libro *París era una fiesta*, nos contó una radiante tarde de otoño en que estuvo en la librería de Sylvia Beach esperando a que llegara James Joyce, y nos contó cómo caminó después hasta la brasería Lip y cómo permaneció allí escribiendo en una mesa apartada hasta que se hizo de noche, y el local se llenó, y ya no le fue posible escribir más.

No es frecuente que los escritores que escriben a máquina lo hagan con todas las reglas de la mecanografía, que es algo tan difícil como tocar bien el piano. El único que yo he conocido capaz de escribir con todos los dedos y sin mirar el teclado era el inolvi-

dable Eduardo Zalamea Borda, en la redacción de *El Espectador*, en
Bogotá, quien, además, podía contestar preguntas sin alterar el ritmo de su digitación virtuosa. El extremo contrario es el de Carlos
Fuentes, que escribe sólo con el índice de la mano derecha. Cuando fumaba, escribía con una mano y sostenía el cigarrillo con la
otra, pero ahora que no fuma no se sabe a ciencia cierta qué hace
con la mano sobrante. Uno se pregunta asombrado cómo su dedo
índice pudo sobrevivir indemne a las casi 2.000 páginas de su novela *Terra nostra*.

En general, los escritores a máquina lo hacemos con los dos índi-
ces, y algunos buscando la letra en el teclado, igual que las gallinas
escarban el patio buscando las lombrices ocultas. Sus originales suelen estar plagados de enmiendas y tachaduras, y en un tiempo fueron
el horror de los linotipistas, que tantos y tan útiles secretos del oficio
nos enseñaron en la juventud, y que hoy han sido reemplazados por
las hermosas mecanógrafas de la composición fotográfica, que ojalá
nos enseñaran también otros tantos y apetitosos secretos de la vejez.
Algunos originales eran tan difíciles de descifrar, que muchos escritores tenían que ser encomendados siempre a un linotipista de cabecera que conociera a fondo sus jeroglíficos personales. Yo era uno de
aquellos escritores, pero no por lo intrincado de mis originales, sino
por mis desastres ortográficos, de los cuales no estoy a salvo todavía
en estos tiempos de gloria.

Lo peor es que cuando uno se vuelve mecanógrafo esencial ya
resulta imposible escribir de otro modo, y la escritura mecánica termina por ser nuestra verdadera caligrafía. Hasta el punto de que
hace falta la ciencia para interpretar el carácter de un escritor por las
alternativas de la presión que ejerce sobre el teclado. En mis tiempos de reportero juvenil escribía a cualquier hora y en cualquiera de
las máquinas paleolíticas de la redacción de los periódicos, y en las
cuartillas de un metro que cortaban del papel sobrante en la rotati-
va. La mitad de mi primera novela la escribí en ese papel en las madrugadas ardientes y olorosas a miel de imprenta del periódico *El
Universal*, de Cartagena, pero luego la continué en el dorso de unos
boletines de aduana que estaban impresos en un papel áspero y de
mucho cuerpo. Ése fue el primer error: desde entonces, sólo puedo
escribir en un papel como ése: blanco, áspero y de 36 gramos. Después tuve la desdicha de conocer una máquina eléctrica que no sólo
era más fluida, sino que parecía ayudarme a pensar, y ya no pude
usar nunca más una máquina convencional. El tiempo agravó las
cosas: ahora sólo puedo escribir en máquina eléctrica, siempre de
la misma marca, con el tipo de la misma medida, y sin un solo

tropiezo, porque hasta el mínimo error de mecanografía me duele en el alma como un error de creación. No es raro, pues, que el único cuadro que tengo frente al escritorio donde escribo sea el afiche de una máquina de escribir destrozada por un camión en medio de la carretera. ¡Qué dicha!

LA NOCHE CALIENTE DE AMSTERDAM

Había sido el viernes más cálido de Amsterdam y tal vez de toda Holanda en 152 años, y no era necesario sentirlo para saberlo. «Todo el mundo se ha vuelto loco», me dijo el servicial empleado del hotel, empapado de sudor dentro de su librea de paño. «Hasta las computadoras», dijo. Las hermosas valquirias venidas del norte en su éxodo migratorio anual hacia las playas del Mediterráneo andaban descalzas por las calles con las sandalias al hombro, se sentaban durante horas en el borde de los canales para refrescarse los pies, y una de ellas se quitó la poca ropa que llevaba encima y se echó al agua en pelotas, cuan larga y dorada era, en medio de los aplausos de los turistas adormilados bajo los toldos de las cafeterías. No la rescató la policía, sino una ambulancia de la Seguridad Social, preocupada de lo que pudiera ocurrirle a quien bebiera aquellas aguas apacibles y venenosas. Había músicas improvisadas en las calles, había predicadores y maromeros, había profetas y atracadores, y había un mimo que se ganó en tres horas más que un conductor de taxi en una semana, imitando el modo de andar de los transeúntes. Era una parodia tan admirable que no se podía creer.

Alguna vez he dicho que lo que más me gusta de Amsterdam es lo mucho que se parece a Curaçao, y aquella tarde eran idénticas. Había un aire lunático, un olor de frutas podridas y pájaros muertos que revolvía las nostalgias del Caribe. Las únicas que no disfrutaron del calor fueron las pobres muchachas que se exhiben en las vitrinas, y que en tiempos normales constituyen el atractivo mayor de los visitantes. Se quedaron en sus casitas de muñecas haciendo el chocolate en sus cocinitas de muñecas, tomándoselo con pan ensopado en las camitas de muñecas donde no vino nadie a hacer el amor. A nadie se le hubiera ocurrido venir, por supuesto, si todo el que quiso hacer el amor aquella tarde babilónica lo hizo sin que le molestara nadie en los recodos otoñales de los parques, agonizando de un amor

Publicado originalmente el 14 de julio de 1982.

verdadero y gratis entre los tulipanes y las avestruces. Fue algo tan insólito, que los universitarios se echaron a las calles en una jubilosa manifestación de protesta contra todo, menos contra el calor, y quemaron automóviles y se pelearon a palos con la policía, y pidieron a gritos lo que no parecía posible que fuera pedido por alguien en Holanda: «Afuera la reina». En medio de tantos desafueros, un globo amarillo y enorme con letreros pintados daba vueltas en el cielo, y sus tripulantes, impávidos en la canasta colgante, le decían adiós con la mano a la muchedumbre como si de veras se fueran para siempre. «Volar es natural en los holandeses», pensé. El globo siguió dando vueltas en el cielo hasta que se hizo de noche como a las once, y sólo quedó la luz malva del verano nórdico, y todavía no encontraban a alguien en el hotel que fuera capaz de componer las computadoras enloquecidas por el calor.

Yo lo supe a esa hora, cuando subí cargando las maletas hasta el cuarto que me asignaron en el quinto piso, y encontré una pareja del mismo sexo —aunque nunca supe de cuál— retozando en la cama. Protesté en recepción, no porque la pareja fuera del mismo sexo, sino porque me hubieran dado una habitación donde ya se estaba haciendo el amor. Entonces el empleado, inconmovible, ensopado en sudor, me dijo que todos los horrores eran posibles aquella noche, porque todo en el hotel estaba a merced de las computadoras. En efecto, teníamos una tarjeta perforada en la misma argolla de la llave, y con ella era posible hacer todo: cerrar la puerta con un sistema de seguridad sin moverse de la cama y abrirla del mismo modo, controlar once canales de televisión desde cualquier lugar de la habitación, programar el despertador electrónico, abrir y cerrar las puertas del ascensor cuando ya estaban bloqueadas para quienes no fueran clientes del hotel, y pedir comidas o bebidas a la habitación mediante un código de señales luminosas. Yo conocía otros hoteles automatizados, pero el único que lo estaba hasta ese punto era el New Otani, de Tokio. Uno programaba el teléfono para que lo despertara a una hora, y al día siguiente, a esa hora precisa, el cuarto se llenaba de trinos de pájaros que parecían cantar de veras entre los árboles, pájaros que por la voz se podían imaginar de todos los tamaños y colores, y que cantaban con tanta pasión que no invitaban a levantarse, sino, al contrario, a permanecer en estado de gracia en medio de aquel bosque imaginario. Lo único que no se había querido preguntar era cómo acallarlos, y durante todo mi primer día en Tokio fui perseguido por la algarabía de tantos pájaros despertadores.

En Amsterdam, en efecto, todos los horrores fueron posibles aquella noche. La consternación del empleado era justa: la única manera de saber qué habitaciones estaban disponibles y cuáles estaban ocupadas era yendo de piso en piso con una llave maestra, y abriéndolas todas. A partir de la medianoche, los ascensores no obedecieron a las órdenes de las tarjetas, y subían y bajaban sin orden ni concierto y no se detenían si no se les dirigía desde dentro. Cuando lo necesitábamos, el empleado llamaba a la recepción y alguien subía a buscarlo en el piso en que se hubiera parado por su propia voluntad, y lo llevaba como si fuera de cabestro hasta donde lo esperábamos. Yo, que les tengo a los ascensores tanto miedo como a los aviones, y por las mismas causas, pregunté qué pasaba si nos quedábamos encerrados. «No lo sabemos», me dijo el empleado. «Es la primera vez que esto nos sucede, de modo que dependemos de la voluntad de Dios nuestro señor.» Pero no era eso lo único que nos faltaba. Un maletero del hotel nos había dicho que la habitación 507 estaba libre con seguridad, porque él había sacado el equipaje una hora antes y había acompañado al huésped hasta el taxi. Pero fue imposible abrir la puerta de ese cuarto, porque algún defecto de computación la había bloqueado por dentro con el dispositivo de seguridad. Había que esperar hasta que el cerebro mágico recobrara la lucidez del invierno para que todo volviera a funcionar como en la vida real. Y no ocurrió hasta el día siguiente. Mientras tanto, la solución que aceptamos, al cabo de cuatro horas de aventuras, no fue la más afortunada.

Esa tarde habían alquilado el dormitorio de una suite para alguien que no la quiso ocupar completa. Así que nos acomodaron en la sala contigua: el sofá se convirtió en una cama, y subieron otra del depósito. Apenas si cabían las dos en el cuarto, pero después de tantas subidas y bajadas heroicas no había tiempo para remilgos. Sólo al día siguiente, cuando el cielo se volvió a nublar y fue posible reanudar la comunicación con el resto del mundo, supimos que Gloria y Álvaro Castaño, que viajaban con nosotros, habían pasado también la mitad de la noche subiendo y bajando escaleras, buscándonos de cuarto en cuarto para cenar, hasta que fueron conscientes de su ilusión y se fueron a dormir extenuados.

Fue una noche de espantos. Junto a la mesa de la cama había un tablero de comandos desde donde se controlaban sin moverse todos los servicios de la habitación. Apenas empezábamos a dormirnos cuando la televisión se encendió sola, y vimos a Marlene Dietrich en sus mejores años, con sus piernas mitológicas y su voz de ceniza, y a pesar del cansancio no pudimos resistir la tentación de verla hasta el final. Pero la canción no había terminado cuando el receptor de televisión se apagó solo, y quedó sonando una señal de alarma y una ráfaga de luz intermitente, roja e intensa como un mensaje jupiterino imposible de descifrar. Nos resignábamos a convivir con aquella luz, cuando se apagó por su propia cuenta, y el teléfono empezó a sonar de urgencia. Pero lo único que oíamos al otro lado de la línea era siempre la misma voz que decía: *Sorry*. Poco antes del amanecer se encendieron todas las luces, y el cuarto se llenó con la miel derretida del piano de Richard Clyderman que nos disparó a todo volumen su repertorio completo de música enlatada. A esa hora, el tablero nos mandaba señales insondables de galaxias remotas y naufragios intemporales. En vano traté de desconectarlo de la corriente eléctrica, porque no estaba integrado al sistema central con el enchufe simple de los mortales, sino con un cable maestro imposible de cortar sin correr el riesgo de que funcionara algún mecanismo de autodestrucción que nos arrastrara a todos a la muerte. El calor era todavía insoportable, como en el Caribe a las dos de la tarde, pero abrir la ventana era peor que tenerla cerrada. Afuera seguía el estrépito de las motocicletas, el despelote de la música, el trueno interminable de una ciudad apacible y hermosa que recobraba de pronto su verdadera vocación luciferina. De modo que dormimos como pudimos hasta las ocho de la mañana, cuando alguien abrió la puerta con la llave maestra, sin tocar, y una fila de camareros solemnes introdujeron en la habitación un suculento desayuno para once personas. Un momento después, llamado de urgencia, el empleado que había sido nuestro ángel guardián de la noche anterior llegó al último límite de la paciencia. Aquel desayuno para once personas no podía ser más que otro error de las computadoras, que ya se suponían normalizadas por el cambio de clima. Como movido por un soplo revelador, el hombre se volvió hacia mí con una expresión afligida, y me dijo: «Espero que usted no vaya a escribir esto, por favor». Juro que hasta ese instante no se me había ocurrido. «Por supuesto que no», le dije. «Ni más faltaba.»

LOS POBRES TRADUCTORES BUENOS

Alguien ha dicho que traducir es la mejor manera de leer. Pienso también que es la más difícil, la más ingrata y la peor pagada. *Traduttore, traditore*, dice el tan conocido refrán italiano, dando por supuesto que quien nos traduce nos traiciona. Maurice-Edgar Coindreau, uno de los traductores más inteligentes y serviciales de Francia, hizo en sus memorias habladas algunas revelaciones de cocina que permiten pensar lo contrario. «El traductor es el mono del novelista», dijo, parafraseando a Mauriac, y queriendo decir que el traductor debe hacer los mismos gestos y asumir las mismas posturas del escritor, le gusten o no. Sus traducciones al francés de los novelistas norteamericanos, que eran jóvenes y desconocidos en su tiempo —William Faulkner, John Dos Passos, Ernest Hemingway, John Steinbeck—, no sólo son recreaciones magistrales, sino que introdujeron en Francia a una generación histórica, cuya influencia entre sus contemporáneos europeos —incluidos Sartre y Camus— es más que evidente. De modo que Coindreau no fue un traidor, sino todo lo contrario: un cómplice genial. Como lo han sido los grandes traductores de todos los tiempos, cuyos aportes personales a la obra traducida suelen pasar inadvertidos, mientras se suelen magnificar sus defectos.

Cuando se lee a un autor en una lengua que no es la de uno se siente un deseo casi natural de traducirlo. Es comprensible, porque uno de los placeres de la lectura —como de la música— es la posibilidad de compartirla con los amigos. Tal vez esto explica que Marcel Proust se murió sin cumplir uno de sus deseos recurrentes, que era traducir del inglés a alguien tan extraño a él mismo como lo era John Ruskin. Dos de los escritores que me hubiera gustado traducir por el solo gozo de hacerlo son André Malraux y Antoine de Saint-Exupéry, los cuales, por cierto, no disfrutan de la más alta estimación de sus compatriotas actuales. Pero nunca he ido más allá

Publicado originalmente el 21 de julio de 1982.

del deseo. En cambio, desde hace mucho traduzco gota a gota los *Cantos* de Giacomo Leopardi, pero lo hago a escondidas y en mis pocas horas sueltas, y con la plena conciencia de que no será ése el camino que nos lleve a la gloria ni a Leopardi ni a mí. Lo hago sólo como uno de esos pasatiempos de baños que los padres jesuitas llamaban placeres solitarios. Pero la sola tentativa me ha bastado para darme cuenta de qué difícil es, y qué abnegado, tratar de disputarles la sopa a los traductores profesionales.

Es poco probable que un escritor quede satisfecho con la traducción de una obra suya. En cada palabra, en cada frase, en cada énfasis de una novela hay casi siempre una segunda intención secreta que sólo el autor conoce. Por eso es sin duda deseable que el propio escritor participe en la traducción hasta donde le sea posible. Una experiencia notable en ese sentido es la excepcional traducción de *Ulysses*, de James Joyce, al francés. El primer borrador básico lo hizo completo y solo August Morell, quien trabajó luego hasta la versión final con Valery Larbaud y el propio James Joyce. El resultado es una obra maestra, apenas superada —según testimonios sabios— por la que hizo Antonio Houaiss al portugués de Brasil. La única traducción que existe en castellano, en cambio, es casi inexistente. Pero su historia le sirve de excusa. La hizo para sí mismo, sólo por distraerse, el argentino J. Salas Subirat, que en la vida real era un experto en seguros de vida. El editor Santiago Rueda, de Buenos Aires, la descubrió en mala hora, y la publicó a fines de los años cuarenta. Por cierto, que a Salas Subirat lo conocí pocos años después en Caracas trepado en el escritorio anónimo de una compañía de seguros y pasamos una tarde estupenda hablando de novelistas ingleses, que él conocía casi de memoria. La última vez que lo vi parece un sueño: estaba bailando, ya bastante mayor y más solo que nunca, en la rueda loca de los carnavales de Barranquilla. Fue una aparición tan extraña que no me decidí a saludarlo.

Otras traducciones históricas son las que hicieron al francés Gustav Jean-Aubry y Phillipe Neel de las novelas de Joseph Conrad. Este gran escritor de todos los tiempos —que en realidad se llamaba Jozef Teodor Konrad Korzeniowski— había nacido en Polonia, y su padre era precisamente un traductor de escritores ingleses y, entre otros, de Shakespeare. La lengua de base de Conrad era el polaco, pero desde muy niño aprendió el francés y el inglés, y llegó a ser escritor en ambos idiomas. Hoy lo consideramos, con razón o sin ella, como uno de los maestros de la lengua inglesa. Se cuenta que les hizo la vida invisible a sus traductores franceses tratando de imponerles su propia perfección, pero nunca se decidió a traducirse a sí mismo. Es curioso, pero no se conocen muchos escritores bilingües que lo hagan. El caso más cercano a nosotros es el de Jorge Semprún, que escribe lo mismo en castellano o en francés, pero siempre por separado. Nunca se traduce a sí mismo. Más raro aún es el irlandés Samuel Beckett, Premio Nobel de Literatura, que escribe dos veces la misma obra en dos idiomas, pero su autor insiste en que la una no es la traducción de la otra, sino que son dos obras distintas en dos idiomas diferentes.

Hace unos años, en el ardiente verano de Pantelaria, tuve una enigmática experiencia de traductor. El conde Enrico Cicogna, que fue mi traductor al italiano hasta su muerte, estaba traduciendo en aquellas vacaciones la novela *Paradiso*, del cubano José Lezama Lima. Soy un admirador devoto de su poesía, lo fui también de su rara personalidad, aunque tuve pocas ocasiones de verlo, y en aquel tiempo quería conocer mejor su novela hermética. De modo que ayudé un poco a Cicogna, más que en la traducción, en la dura empresa de descifrar la prosa. Entonces comprendí que, en efecto, traducir es la manera más profunda de leer. Entre otras cosas, encontramos una frase cuyo sujeto cambiaba de género y de número varias veces en menos de diez líneas, hasta el punto de que al final no era posible saber quién era, ni cuándo era, ni dónde estaba. Conociendo a Lezama Lima, era posible que aquel desorden fuera deliberado, pero sólo él hubiera podido decirlo, y nunca pudimos preguntárselo. La pregunta que se hacía Cicogna era si el traductor tenía que respetar en italiano aquellos disparates de concordancia o si debía vertirlos con rigor académico. Mi opinión era que debía conservarlos, de modo que la obra pasara al otro idioma tal como era, no sólo con sus virtudes, sino también con sus defectos. Era un deber de lealtad con el lector en el otro idioma.

Para mí no hay curiosidad más aburrida que la de leer las traducciones de mis libros en los tres idiomas en que me sería posible hacerlo. No me reconozco a mí mismo, sino en castellano. Pero he leído alguno de los libros traducidos al inglés por Gregory Rabassa y debo reconocer que encontré algunos pasajes que me gustaban más que en castellano. La impresión que dan las traducciones de Rabassa es que se aprende el libro de memoria en castellano y luego lo vuelve a escribir completo en inglés: su fidelidad es más compleja que la literalidad simple. Nunca hace una explicación en pie de página, que es el recurso menos válido y por desgracia el más socorrido en los malos traductores. En este sentido, el ejemplo más notable es el del traductor brasileño de uno de mis libros, que le hizo a la palabra *astromelia* una explicación en pie de página: «Flor imaginaria inventada por García Márquez». Lo peor es que después leí no sé dónde que las astromelias no sólo existen, como todo el mundo lo sabe, en el Caribe, sino que su nombre es portugués.

Y DE LA GUAYABA, ¿QUÉ?

Me molesta hablar una vez más de mi posible regreso a Colombia habiendo tantos temas de actualidad más atractivos, pero la amable invitación que me hizo el presidente electo para que asista al acto de su posesión, el 7 de agosto, lo ha convertido –aunque él no quisiera ni yo tampoco– en un asunto de interés público. No hubiera querido hablar de esto antes de haber recibido en mi casa la invitación oficial, y de haberla contestado en los términos que merece un mensaje tan cordial y humano de un viejo y muy apreciado amigo a quien la vida le ha echado encima el gran honor y la inmensa desgracia personal de ser presidente de la República dentro de un sistema que no tiene remedio. Sin embargo, aún no ha llegado a mis manos 72 horas después de que se hizo público sin advertírmelo, lo cual hace pensar que una de las primeras cosas que debe hacer el nuevo mandatario será agilizar los sistemas telegráficos de Colombia. Así las cosas, he decidido ocuparme de la invitación no recibida para salirle al paso a la especulación creciente de que mi silencio era un desaire personal al presidente electo.

Muchos colombianos, yo entre ellos, no estamos de acuerdo con las ideas políticas de Belisario Betancur, pero eso no habría sido un obstáculo para que aceptara su invitación. Cuando algunos amigos comunes me preguntaron hace varias semanas cuál sería mi actitud en caso de recibirla, les contesté que mi respuesta sería negativa por razones que no tenían nada que ver con las ideas políticas del presidente electo, y mucho menos con su persona. Más aún: les pedí que se lo hicieran saber a él para que todo el trámite se quedara sin publicidad.

A pesar de eso, la invitación ha sido hecha, y creo entender la razón: si Belisario invitaba a un numeroso grupo de escritores y artistas y no me invitaba también a mí, hubiera podido interpretarse como una discriminación. De modo que, además de la invitación,

Publicado originalmente el 28 de julio de 1982.

tengo que agradecerle a Belisario el haber afrontado el riesgo de la negativa con la mayor elegancia.

Mis razones son más profundas. El 26 de marzo del año pasado, cuando Mercedes y yo salimos de Colombia bajo la protección diplomática de México, la reacción de las autoridades más altas no sólo fue frívola, sino de una vulgaridad inadmisible. El presidente Turbay Ayala repitió una vez más el socorrido argumento de que yo trataba de sumarme a la campaña de descrédito internacional contra Colombia, sin preguntarse siquiera si ese descrédito no estaba ya mejor sustentado por sus propios actos de gobierno. El canciller Lemos Simonds, a quien siempre tuve como un hombre inteligente, cometió la tontería de decir que el mío era un acto de publicidad para mi libro inminente. El ministro de la Defensa dijo que el único que me perseguía en Colombia era un agente de la policía que deseaba un autógrafo. Al cabo de 697.203 minutos de haber salido de Colombia, no conozco ningún acto ni ninguna declaración que permitan atribuir a esos tres funcionarios ni siquiera la atención de rendir tributo a la buena fe.

Sin embargo, en los meses siguientes a mi salida se han conocido algunos hechos que exigían por lo menos una explicación del gobierno, no tanto por mí como por respeto a la opinión pública. Varios miembros del Movimiento Diecinueve de Abril (M-19), de cuya seriedad no tengo ninguna duda, han declarado desde la cárcel que la justicia militar trataba de establecer cuáles eran mis vínculos con ellos, sobre todo en relación con el desembarco del año pasado en el sur del país. A algunos se les preguntó en concreto cuál había sido mi participación en el supuesto entrenamiento que les impartieron en Cuba y varios fueron torturados para arrancarles una declaración que sirviera para acusarme. Uno por lo menos declaró para la prensa, y se publicó en forma destacada, que había firmado bajo tortura un pliego de cargos falsos contra mí.

Nada de esto era nuevo. Desde mucho antes del desembarco, Álvaro Fayad –a quien conocí y admiré como un hombre serio e inteligente desde que no era todavía uno de los dirigentes mayores del M-19– declaró, y se publicó en la prensa, que en el interrogatorio insaciable y brutal que le hicieron trataban de hacerle revelar a la fuerza una supuesta complicidad mía con su movimiento. No obstante, estas publicaciones no han merecido ninguna atención del gobierno. A pesar de tantas revelaciones públicas, el ministro de la Defensa volvió a repetir la semana pasada, en una entrevista de prensa, el chiste bobo del policía que deseaba un autógrafo. La buena fe del ministro de la Defensa quedaría establecida si pudiera probar que no conoce las actas de los interrogatorios, que hace pocas semanas debieron pasar de la justicia militar a la civil cuando se levantó el estado de sitio. Pero aun así su situación sería muy grave, porque revelaría un grado inquietante de compartimentación entre los altos mandos de las fuerzas armadas.

Hay más. Poco después de mi salida de Colombia, cuando se inauguró el nuevo aeropuerto de Barranquilla, algunos miembros de su comitiva presidencial les contaron a amigos míos –sin duda para que me lo hicieran saber– que quien me previno de lo que se intentaba contra mí fue el propio presidente Turbay Ayala, a través de emisarios no oficiales y sin que yo supiera cuál era el origen de la información. Según ellos, el presidente lo hacía porque no quería que se intentara nada contra mí, pero no podía impedirlo, y decidió hacérmelo saber de trasmano para que me pusiera a salvo mientras se calmaban los ánimos. Lo que el presidente no se imaginaba –según sus infidentes– fue que yo no me escurriría del país en silencio, sino que lo haría con el escándalo deliberado y necesario de la protección diplomática. No le presté ningún crédito a esa versión, que me pareció una página más en esa gran novela de realismo mágico que es la vida real de Colombia. Pero unos meses después, uno de los ministros del presidente Turbay Ayala la repitió como si fuera con pleno conocimiento de causa y en presencia de varias personalidades intelectuales y políticas. Conozco el nombre del ministro y los de cada una de esas personalidades, así como la fecha y el lugar de la reunión, pero tengo el derecho profesional de no revelar las fuentes. Además lo que importa en realidad de esta versión no es saber si es falsa o cierta, sino lo mucho que revela sobre la crisis de credibilidad de este gobierno, que al fin se acaba para alivio de todos.

Regresar a Colombia en estas condiciones –a pesar de mi nostalgia, ya casi irresistible– sería admitir como ciertas las negativas sistemáticas de las más altas autoridades, y no tanto en mi caso personal como en el de las numerosas víctimas de la represión oficial. Sería olvidar de una sola plumada por qué Feliza Bursztyn no podrá estar el 7 de agosto en la primera fila de invitados del nuevo presidente, como fue su deseo hasta el instante de morir. Sería contribuir al perdón y el olvido de un gobierno que se los ha negado sin grandeza a una oposición armada que parece dispuesta a desarmarse por el bien de todos. En efecto, con la misma frescura con que lo ha hecho conmigo, este gobierno se ha conformado con negar que ha cometido toda clase de abusos: allanamientos, torturas, desapariciones de presos y, en general, todas las formas de un terrorismo oficial tan abominable como el de sus contrarios. Regresar mientras no existan condiciones para que estas cosas se aclaren –como si nada hubiera pasado y sólo por adornar la fiesta de un buen amigo mío– sería conceder la razón a las peores razones del peor gobierno que ha tenido mi país en toda su historia.

LA VEJEZ JUVENIL DE LUIS BUÑUEL

La magnífica autobiografía de Luis Buñuel, que acaba de publicarse, empieza con un capítulo deslumbrante sobre la facultad humana que más nos condiciona e inquieta: la memoria. Cuenta don Luis que su madre la perdió por completo los últimos diez años de su vida y que leía una misma revista muchas veces con el mismo deleite porque siempre le parecía nueva. «Llegó a no reconocer a sus hijos, a no saber quiénes éramos, ni quién era ella», dice. «Yo entraba, le daba un beso, me sentaba un rato a su lado, y luego salía y volvía a entrar.» Ella lo recibía con la misma sonrisa y le invitaba a sentarse como si lo viera por primera vez y sin recordar cómo se llamaba.

Lo que no dice don Luis, y que tal vez nadie sabe a ciencia cierta, es si su madre era consciente de su desgracia. A lo mejor no lo era: quizá su vida volvía a empezar cada minuto y terminaba en el siguiente, con una coincidencia fugaz y sin dolor de la que habían desaparecido no sólo los malos recuerdos, sino también los buenos, que en última instancia son los peores porque son la semilla de la nostalgia. Sin embargo, no es este enigma lo que más me ha impresionado de este libro excelente, sino la fuerza con que me ha puesto a pensar por primera vez en algo que suele estar siempre muy lejos de nuestras preocupaciones: la certidumbre de la vejez. En su momento leí con una gran admiración el libro de Simone de Beauvoir sobre este tema —que es tal vez el más minucioso y documentado que se ha escrito—, pero en ninguna de sus páginas me produjo esta impresión de desastre biológico de que habla Luis Buñuel. Según él, a los setenta años empezó por no recordar los nombres propios con tanta facilidad como antes. Más tarde empezó a olvidar dónde había dejado el encendedor, dónde puso las llaves, cómo era la melodía que oyó una tarde de lluvias en Biarritz. Esto le preocupa —ahora que tiene 82 años— porque le parece el principio de un proceso que terminará por arrastrarlo al limbo del olvido en que vivió su madre en los últimos años. «Hay que

Publicado originalmente el 4 de agosto de 1982.

haber empezado a perder la memoria, aunque sólo sea a retazos, para darse cuenta de que esta memoria es lo que constituye nuestra vida», dice. Por fortuna, su propio libro demuestra que el drama de Luis Buñuel no es la pérdida de la memoria, sino el miedo de perderla.

En realidad, éste es un libro de recuerdos, y tener facultades para haberlos reconstruido en forma tan vívida es una proeza que niega de plano cualquier amenaza de la amnesia senil. Hace poco le dije a un amigo que me disponía a escribir mis memorias, y aquél me replicó que todavía no estaba en edad para eso. «Es que quiero empezar cuando todavía me acuerdo de todo», le dije. «La mayoría de las memorias se escriben cuando ya su autor no se acuerda de nada.» Pero éste no es el caso de Luis Buñuel. La precisión de sus evocaciones, de la vida medieval de Calanda, de la Ciudad Universitaria de Madrid —que tanto influyó en su generación—, de la época del surrealismo y, en general, de tantos momentos estelares de este siglo demuestran que hay todavía en este anciano invencible un germen de juventud que nunca se extinguirá. Es verdad, como él lo dice, que perdió hace mucho tiempo el oído, lo cual le privó del placer incomparable de la música. Tiene que leer a duras penas con una lupa y un rayo de luz especial porque está perdiendo la vista, y dice haber perdido también el apetito sexual. Su última película, *Ese oscuro objeto del deseo*, la hizo hace cinco años, y él considera que será la última. Es decir, es cierto que está enfermo, aburrido por la falta de oficio, con la sensación de que sus amigos lo han abandonado y pensando en la muerte cada vez con más frecuencia e intensidad. Pero un hombre que es capaz de analizar su propia vida en la forma en que él lo ha hecho y dejar un testimonio como éste de su mundo y su tiempo no es sin duda el viejo decrépito que él mismo cree ser.

Uno se consuela pensando que la vejez no es más que un estado de ánimo. Cuando vemos pasar a un anciano que no puede con su alma tenemos la tendencia a creer que esos son infortunios que sólo les ocurren a los otros. Se piensa, y ojalá con razón, que nuestra voluntad no tendrá fuerzas para oponerse a la muerte, pero sí para cerrarle el paso a la vejez. Hace unos años encontré en la sala de espera de un aeropuerto de Colombia a un condiscípulo de mi edad que parecía tener el doble. Un rápido examen permitía descubrir que su vejez prematura no era tanto un hecho biológico, como pura y simple negligencia suya. No pude contenerme. Le dije, entre otras muchas cosas, que su mal estado no era culpa de Dios, sino suya, y que yo tenía derecho a reprochárselo porque su deterioro no sólo le envejecía a él, sino a toda nuestra generación. Hace poco le pedí a un amigo que viniera a México. «Allí no», me contestó en el acto, «porque hace veinte años que no voy a México y no quiero ver mi vejez en la cara de mis amigos.» Me di cuenta inmediatamente que él tenía la misma norma que yo: no facilitarle nada a la vejez. Mi padre, que ahora tiene 81 años, tiene una vitalidad y un aspecto excepcionales, y sus hijos sabemos que su secreto contra la vejez es muy simple: no piensa en ella.

Hay excepciones, por supuesto, buenas y malas, y lo mejor en este asunto es no pensar sino en las excepciones buenas. Miguel Barnet, el escritor cubano, escribió la biografía de un antiguo esclavo. En el momento de la entrevista, Barnet pudo comprobar que, en efecto, el anciano tenía los 104 años que decía tener, y su memoria era tan buena que parecía un archivo viviente de la historia de su país. Por otra parte, el doctor Grave E. Bird –citado por Simone de Beauvoir– hizo un estudio de cuatrocientas personas mayores de cien años, y sus resultados son consoladores. «La mayoría de ellos», concluye el estudio, «tenían planes precisos para el porvenir, se interesaban por los asuntos públicos, manifestaban entusiasmos juveniles, tenían un apetito sólido y un sentido del humor muy agudo, y una resistencia extraordinaria. Eran optimistas y no manifestaban miedo a la muerte.» En cuanto a la actividad sexual de los viejos, hay evidencia de que hacia los noventa años se inicia en ese aspecto una segunda adolescencia. La única condición parece ser que se haya sido activo toda la vida anterior. Nada enfría más que la frialdad. Tengo un amigo de 85 años a quien alguien le acusó de ser un viejo verde porque le gustan las muchachas de catorce años. Su respuesta fue aplastante: también a los muchachos de catorce años les gustan y nadie les llama viejos verdes.

El problema es que la sociedad, fingiendo veneración y respeto, termina por volvernos viejos a la fuerza. «Con la india más vieja se prueba la flecha», dice un proverbio guajiro. Hace algún tiempo, cuando yo le propuse a un productor que hiciera en cine *El coronel no tiene quien le escriba*, me contestó de plano: «Los viejos no se venden». En Francia –que en 1970 tenía el promedio de viejos más alto del mundo– se ha conseguido la jubilación a los sesenta años. Es un escándalo. La mejor prueba de la injusticia de esa decisión es que no hay seres más agresivos en este mundo que los ancianos franceses: se disputan los taxis a golpes de paraguas con los jóvenes, se saltan los turnos en las colas a codazos y son capaces de una procacidad devastadora en una disputa callejera. Yo me había preguntado siempre si esos viejos saben que son viejos. No lo sé. Sólo sé que la semana pasada un hombre de 54 años, que se siente en la plenitud de su vida, le dio a un niñito de cinco años un billete de cien pesos. El niñito, feliz, corrió a mostrárselo a su padre y le dijo: «Me lo dio aquel viejito que está allá». El viejito que estaba allá, por supuesto, era yo.

TAMBIÉN EL HUMANITARISMO TIENE SU LÍMITE

El 6 de diciembre de 1980, a las doce del día, la embarcación *El Socorro*, con bandera de Bahamas, fue capturada frente a Cayo Confites, en aguas territoriales cubanas, por unidades de la Marina de guerra. Se encontraron a bordo 514 pacas de marihuana, que pesaban 19.506 kilogramos. Su capitán era el norteamericano Vincent Salvatore Simone, y la tripulación estaba compuesta en su totalidad por doce colombianos. Todos fueron condenados a diez años de cárcel por violación de las aguas territoriales y tráfico de droga.

Más tarde, el 20 de abril de 1981, la embarcación *Liliana*, con bandera de Honduras, fue capturada en la punta de Maisi, en el extremo oriental de Cuba, y se encontraron a bordo 56.000 libras de marihuana. La tripulación estaba compuesta por nueve colombianos. Ocho de ellos fueron condenados a ocho años de cárcel por tráfico de droga y a otros cuatro años por entrada ilegal al país. Pero uno de ellos, por razones que las autoridades cubanas no han explicado, fue condenado a diez años por tráfico de droga y a sólo dos años por entrada ilegal al país.

Ésta era la situación en noviembre del año pasado, cuando vine a La Habana con un paquete de cartas de las familias de los colombianos presos, en las que me pedían hacer algo para obtener su liberación. Todas coincidían en un punto: el Ministerio de Relaciones Exteriores de Colombia se negaba a toda gestión, por considerar que se trataba de delincuentes comunes. Las familias se dirigían a mí porque un año antes había conseguido el indulto para otros diez colombianos presos en iguales circunstancias, y también lo había hecho por las súplicas de las familias. Al igual que aquella vez, en ésta le hice la solicitud informal del indulto al presidente Fidel Castro en persona, y él la presentó al Consejo de Estado.

Sin embargo, cuatro meses después, cuando volví a Cuba para enterarme del estado en que se encontraban las gestiones, me infor-

Publicado originalmente el 11 de agosto de 1982.

maron de que ya los colombianos presos no eran veintiuno, sino treinta. En efecto, el 13 de febrero de este año, a las 3.20 de la tarde, la Marina de guerra cubana había capturado, a tres millas de las costas de la provincia de Holguín, a un yate de placer con bandera colombiana cuyo nombre mundano no le sirvió de nada: *Lucky Star*, es decir, la estrella de la buena suerte. La embarcación llevaba a bordo cuatrocientas pacas de marihuana, con un peso de 8.300 kilogramos. Su capitán era un ecuatoriano radicado en Colombia. A pesar de que no tenía ninguna solicitud para la liberación de estos nuevos presos pedí el favor de que fueran incluidos en la lista anterior (inclusive el ecuatoriano), y las autoridades cubanas lo hicieron aun antes de que se celebrara el juicio correspondiente.

En esta ocasión, una casualidad que merece ser contada agregó un nuevo nombre a la lista. Mis amigos del periódico *El Heraldo*, de Barranquilla, me habían pedido averiguar si no estaban presos en Cuba un piloto comercial y su hijo de dieciocho años, que unos meses antes habían salido de La Guajira en una avioneta de un motor, con rumbo a Florida, y no se había vuelto a saber nada de ellos. La esposa del piloto había hecho toda clase de averiguaciones inútiles en el trayecto: Jamaica, Haití, Bahamas y aun Estados Unidos. La última posibilidad que la esposa vislumbraba, como una lucecita de esperanza, era que estuvieran presos en Cuba. Esta historia conmovió tanto a Fidel Castro, que ordenó una investigación a fondo. Pero fue inútil. «Lo siento mucho», me dijo entonces Fidel Castro, «porque nada nos hubiera complacido más que haberle dado una respuesta favorable a esa pobre mujer.» Pero, en cambio, la investigación reveló que, además de los treinta marineros, estaba preso un piloto colombiano que no era el que yo buscaba.

En efecto, a principios de abril de 1981, este hombre volaba desde la costa caribe de Colombia hacia la de Florida, como copiloto de un Cessna de dos motores, pilotado por el norteamericano Allen Jackson, quien había sido aviador de guerra en Vietnam. A las nueve de la noche, cuando sobrevolaron Haití, el colombiano, que era el que conocía la ruta, le indicó al piloto que pusiera rumbo norte hasta ver nuevas luces que, sin duda, serían las de las Bahamas, y que allí hiciera el ángulo hacia el oeste para llegar a Florida. Luego se durmió, y cuando el norteamericano lo despertó, casi a las diez de la noche, estaba perdido. Tenían gasolina sólo para una hora, y no les quedaba otro recurso que aterrizar donde pudieran. «Por fortuna», ha dicho el colombiano, «había una luna como de mediodía, y eso fue lo que nos salvó». Después de buscar una pista sin encontrarla, aterrizaron en un tramo recto de carretera, sin saber siquiera en qué país estaban.

La historia es casi fantástica para los cubanos, porque muy pocos países del mundo tienen una barrera espacial como la de Cuba: su cielo se considera poco menos que invulnerable. Sin embargo, estos aviadores perdidos sobrevolaron la Sierra Maestra y pasaron sobre la base norteamericana de Guantánamo, y nunca fueron detectados. Más aún: esa noche acamparon cerca del avión, que no fue descubierto hasta el día siguiente, cuando ellos mismos se presentaron en un puesto de policía que les fue indicado por un grupo de campesinos. En el avión se encontraron 1.404 libras de pastillas de dilaudid y de ruaalude lemon 174, que son estupefacientes de lujo. Las autoridades cubanas aceptaron que se incluyera también este piloto en la lista de los indultados, que de este modo fueron 31. Sin embargo, la semana pasada, cuando vine una vez más a Cuba para agilizar la salida de los presos, me encontré que en mayo había sido capturado un nuevo barco con dieciséis colombianos más. Uno de ellos se encontró en un estado de salud tan deteriorado, que desde entonces está sometido a cuidados intensivos en un hospital.

En realidad, este drama es infinito. El canal de los Vientos, que es el estrecho que separa Cuba de Haití, así como la costa norte cubana, son zonas de navegación muy difícil, y sólo expertos logran sortear sus riesgos incontables. No obstante, ésa es la ruta obligada de los barcos cargados de droga que vienen de Colombia hacia Estados Unidos. Por otra parte, las embarcaciones están a duras penas en condiciones de navegar, y sólo gentes muy necesitadas e inexpertas se atreven a embarcarse en semejante aventura.

Ninguna de las naves capturadas venía con destino a Cuba, uno de los pocos países del mundo que las Naciones Unidas han declarado limpios de drogadicción. Los cubanos los capturan no sólo porque violan sus aguas territoriales, sino porque han suscrito tratados internacionales contra el tráfico de droga, que se han esmerado en cumplir aun en las circunstancias más arduas. Estados Unidos, por pura sevicia política, inventa contra Cuba toda clase de infundios en relación con el tráfico de droga, pero ellos saben muy bien que la barrera establecida por los cubanos es la más difícil de franquear por los traficantes que se dirigen a Estados Unidos. Los grandes tiburones del tráfico no viajan en estos barcos perdularios, que vienen casi siempre al mando de aventureros gringos de tercera categoría. También éstos son condenados a penas muy duras. Pero no las cumplen por mucho tiempo. Cada vez que un norteamericano influyente viene a Cuba –y vienen muchos más de los que uno supone– se lleva de regreso un lote de compatriotas liberados, para usarlos como trofeos. Los presentan como víctimas del infierno comunista, pero la mayoría son, en realidad, agentes de la CIA o traficantes de droga. En enero de este año se llevaron ocho.

Los colombianos, en cambio, no tienen ni quien les escriba. Pero escriben. «Lo único que nos ofrece nuestra querida Colombia es traficar marihuana», me dice uno de los presos en una carta que me mandó a México hace poco. «Para nosotros, no es más que un trabajo fuera de la ley, pero que nos permite no morirnos de hambre.» Es difícil no pensar que este hombre tiene toda la razón. Sin embargo, por lo que a mí se refiere, esta carta y todas las que ya siento venir se quedarán sin respuesta, pues no estoy dispuesto a interceder ni por los dieciséis presos más recientes ni por ninguno de los que, sin duda, serán capturados después. La razón es muy simple: a este paso, por puro humanitarismo fácil, tanto yo como las autoridades cubanas terminaremos por convertirnos en servidores involuntarios, pero eficaces, de los verdaderos traficantes. Las familias de estos presos, en todo caso, tienen ahora oportunidad de apelar al nuevo gobierno de Colombia, que acaso tenga mejor corazón que el que acaba de irse para bien de todos.

FRASES DE LA VIDA

Hay en México un dicho tan hermoso como enigmático: «El que come y canta, loco se levanta». Siempre he tenido la curiosidad de averiguar su origen, aunque pienso que es una deformación poética de otro dicho español: «Al que come y canta, un sentido le falta». Comer cantando fue uno de los sueños de la infancia, sueño prohibido como tantos otros, pues siempre se nos dijo que cantar en la mesa espantaba a los duendes.

Estamos llenos de frases como ésa desde el nacimiento. La costumbre de vivir con ellas, de regir por ellas nuestra conducta, se atribuye con demasiada facilidad a los abuelos, cuando en realidad es la vida toda la que nos llena con ellas en cada instante. Los viejos refranes españoles fueron los primeros y tal vez los que más temprano nos despertaron el sentido de la poesía. «Culos conocidos de lejos se dan silbos», decía uno de ellos, por desgracia caído en desuso. Las madres que no querían ver a las hijas coqueteando en la calle, viendo bailes ajenos por las ventanas, tenían una manera de disuadirlas: «El buen paño, en el arca vende». Para referirse a alguien muy atolondrado, decían: «Es de las que confunden el culo con las témporas». Durante muchos años lo oí decir, y nunca tuve el cuidado de averiguar qué eran las témporas, hasta hace muy poco tiempo, cuando una necesidad del oficio me obligó a preguntárselo al diccionario, y éste me contestó: «Tiempo de ayuno que prescribe la Iglesia en las cuatro estaciones». La absoluta falta de relación entre los dos términos del refrán me hizo apreciarlo más de lo que ya lo apreciaba desde niño.

No todos los refranes españoles estaban al alcance de los niños, por supuesto. Había uno bárbaro, pero indiscutible: «Más pueden dos tetas que dos carretas». Los valencianos decían lo mismo, pero más a la valenciana: «Tira més un pèl de figa que una maroma de barco». Para hablar de una mujer que nunca se puso ropa fina, decían: «A la

Publicado originalmente el 18 de agosto de 1982.

que nunca llevó bragas, las costuras le hacen llagas». Sin embargo, casi tan deslumbrantes como los refranes son las desfiguraciones que se suelen hacer de los más corrientes. «Ojos que no ven, pisan mierda», dice uno. Y otro, muy chulo, de Madrid: «Mal de muchos, *apidemia*». Y otro, de un realismo inclemente: «Al que a buen árbol se arrima, si no lo ve nadie, orina». Un escritor original tituló su libro reciente: *Cría ojos*. Con todo, en materia de desfiguraciones, ningunas me parecieron tan divertidas como las que hacíamos en la juventud a las obras de misericordia: visitar al desnudo, enterrar a los enfermos, dar posada a los muertos, dar de comer a los agonizantes. Las posibilidades eran enormes, y cada quien las arreglaba como mejor convenía a sus gustos.

En México —tierra de grandes dichos— hay uno estupendo para referirnos a algo que nos deja indiferentes: «Eso me hace lo que el viento a Juárez». El origen, de acuerdo con un informador bien ilustrado, fue un huracán, en Oaxaca, que arrasó con toda la población pero dejó intacta, en la plaza Mayor, la estatua de don Benito Juárez. El más certero de los refranes que recuerdo es del departamento colombiano de Antioquía: «Habla más que un perdido cuando lo encuentran».

Ramón Gómez de la Serna —a quien algún día habrá que reivindicar como uno de los grandes escritores de la lengua castellana— nos llenó la adolescencia de frases hermosas con sus mejores greguerías. «El melón sabe a fresca y limpia madrugada», dijo. Y otra: «La flauta canta por la nariz». Y otra, entre miles: «Puso a secar tantos guantes que parecía haber recibido una ovación de aplausos». Por la época en que las greguerías estaban de moda, una generación entera de poetas colombianos dio en la flor de jugar con las palabras. Arturo Camacho Ramírez, cuya inteligencia del idioma asombraba a todos, hizo conversiones de frases corrientes: a la vida social la convirtió en la socia vidal, a santa Teresita la convirtió en tanta cerecita, y al contraalmirante Piedrahíta, en el piedralmirante Contrahíta. Con esa técnica hizo la mejor: «Se botan forrones». Al poeta Jorge Rojas se le atribuyen dos frases de aquella cosecha. Una: «La tortuga es una totuma llena de lentitud». Y otra: «La sirena no abre las piernas porque se quedó escamada». La cartilla de leer había marcado la pauta con aquellas frases, que no pretendían nada más que aliteraciones útiles y, sin embargo, se asomaban a los límites de la poesía: «Otilia lava la tina», «La mula va al molino», «El adivino se dedica a la bebida».

Ya de adultos, la publicidad nos ha ido llenando la vida de frases y consignas que terminan por ser refranes de nuestro tiempo. Durante muchos años, una empresa aérea se anunció con una frase que se incorporó a nuestro hablar cotidiano: «Volar es natural en los holandeses». La imagen del holandés errante, que tal vez nace dentro de nosotros, se hizo más vívida con esa frase, que al fin y al cabo no decía una verdad. Durante la segunda guerra mundial, una famosa marca de cigarrillos se vio obligada a quitarle a su cajetilla el color verde tradicional. «El verde se fue a la guerra», fue su explicación publicitaria. Se decía de un modo corriente que tarde o temprano su radio sería un Phillips, o que había un Ford en su futuro, aunque en el fondo del corazón todos sabíamos que no era cierto. En Colombia, en cambio, una compañía de seguros contra incendios concibió la frase de la verdad: «El fósforo tiene cabeza, pero no tiene corazón».

Los sabores, los sonidos y los olores nos han obligado siempre a forzar el idioma para describirlos. Hace muchos años, en una alcoba ajena, oí durante toda la noche a un cordero amarrado en el patio que lanzaba un balido idéntico y de una regularidad inclemente. La dueña de la alcoba, deslumbrada por la simetría de aquel lamento, dijo en la oscuridad: «Parece un faro». Hace menos años, en París, la bella y voluntariosa Tachia Quintana improvisó una tisana con cuantas hierbas secas encontró en los armarios, y cuando la probé creí encontrar la descripción exacta: «Sabe a procesión». Por eso entendí tan bien al Che Guevara cuando probó la primera bebida que se hizo en Cuba para sustituir la Coca-Cola, y dijo, sin vacilar: «Sabe a cucaracha». Nadie, que se sepa, se ha comido una cucaracha, pero es difícil que alguien no entienda a qué sabía aquel refresco nuevo. Cuántas veces hemos tomado un café que sabe a ventana, un pan que sabe a baúl, un arroz que sabe a depósito, una sopa que sabe a rincón. Un amigo probó en un restaurante de París unos espléndidos riñones al jerez, y dijo, suspirando: «Sabe a mujer». En un ardiente verano de Roma probé una vez un helado que no me dejó la menor duda: sabía a Mozart.

EL MAR DE MIS CUENTOS PERDIDOS

Durante muchos años quise escribir el cuento del hombre que se extraviaba para siempre en los sueños. El hombre soñaba que estaba durmiendo en un cuarto igual a aquel en que dormía en la realidad, y también en ese segundo sueño soñaba que estaba durmiendo, y soñando el mismo sueño en un tercer cuarto igual a los dos anteriores. En aquel instante sonaba el despertador en la mesa de noche de la realidad, y el dormido empezaba a despertar. Para lograrlo, por supuesto, tenía que despertar del tercer sueño al segundo, pero lo hizo con tanta cautela, que cuando despertó en el dormitorio de la realidad había dejado de sonar el despertador. Entonces, despierto por completo, tuvo el instante de duda de su perdición: el dormitorio era tan parecido a los otros de los sueños superpuestos, que no pudo encontrar ningún motivo para no poner en duda que también aquél era un sueño soñado. Para su gran infortunio, cometió por eso el error de dormirse otra vez, ansioso de explorar el cuarto del segundo sueño para ver si allí encontraba un indicio más cierto de la realidad, y como no lo encontró, se durmió a su vez dentro del sueño segundo para buscar la realidad en el tercero, y luego en el cuarto y en el quinto. De allí —ya con los primeros latidos de terror— empezó a despertar de nuevo hacia atrás, del quinto sueño al cuarto, y del cuarto al tercero, y del tercero al segundo, y en su impulso desatinado perdió la cuenta de los sueños superpuestos y pasó de largo por la realidad. De modo que siguió despertando hacia atrás, en los sueños de otros cuartos que ya no estaban delante, sino detrás de la realidad. Perdido en la galería sin término de cuartos iguales, se quedó dormido para siempre, paseándose de un extremo al otro de los sueños incontables sin encontrar la puerta de salida a la vida real, y la muerte fue su alivio en un cuarto de número inconcebible que jamás se pudo establecer a ciencia cierta.

Publicado originalmente el 25 de agosto de 1982.

Durante mucho tiempo pensé que no había escrito este cuento de horror porque su parentesco con Jorge Luis Borges era demasiado evidente, pero además inferior a todos sus cuentos. Sin embargo, ahora que lo recuerdo y lo escribo, he caído en la cuenta de que el cuarto en que lo hago —con la máquina de escribir frente a una ventana por donde se mete sin permiso todo el mar Caribe— es un cuarto igual al que siempre quise para el sueño del cuento: cuadrado justo y de paredes lisas y sin color, con una sola puerta y una sola ventana, y ningún otro mueble distinto de la cama simple y la mesa de noche con un despertador que había de repetirse sin respiro en cada uno de los cuartos soñados, pero que había que soñar en el cuarto real. Ahora que lo veo en la realidad me he dado cuenta de que no era de Borges este cuento, sino de la estirpe más antigua y sobrecogedora de Franz Kafka. En todo caso, nunca lo escribí, y tal vez ése sea su mérito mayor.

No es el único que se quedó sin escribir, ni fue tampoco una excepción en el mundo de la literatura. La vida de los escritores está llena de las obras que nunca escribieron, y que tal vez en muchos casos hubieran sido mejores que las que se escribieron. Pero lo curioso es que ese reguero casi interminable de historias concebidas jamás nacidas constituyen para los escritores una parte invisible e importante de su obra: la parte que nunca verán en sus obras completas.

También durante muchos años, y en una época posterior a la del cuento del hombre que se perdió en los sueños, soñé con escribir un cuento del cual sólo tenía el título: *El ahogado que nos traía caracoles.* Recuerdo que se lo dije a Álvaro Cepeda Samudio en una fragosa noche de la casa de amores de Pilar Ternera, y él me dijo: «Ese título es tan bueno que ya ni siquiera hay que escribir el cuento». Casi cuarenta años después me sorprendo de comprobar cuán certera fue aquella réplica. En efecto, la imagen del hombre inmenso y empapado que debía de llegar en la noche con un puñado de caracoles para los niños se quedó para siempre en el desván de los cuentos sin escribir.

En cambio, perdí mucho tiempo tratando de escribir una vez y otra vez el cuento del hombre que descomponía las máquinas. En cierto modo, éste era una nueva variación del asunto que más me ha obsesionado de un modo ineludible: las pestes. El hombre había llegado caminando a un pueblo de artesanos y había preguntado por alguien a un hombre que laboraba con un tractor. Sin remedio: el tractor no volvió a funcionar. Lo mismo ocurrió a la máquina de coser de la costurera a quien hizo la misma pregunta poco después, y a todas las máquinas de oficios diversos con cuyos propietarios tuvo algo que ver. Hice muchas versiones antes de que el ángel de la guarda, que tan mal se ocupa de los escritores tercos, me convenció de que no insistiera más, por la razón más simple del mundo: era un cuento muy malo.

Siempre creí, en cambio, que era muy bueno otro de los que tampoco pude escribir. Me refiero al que concebí en una enloquecedora tarde de tramontana en Cadaqués, el pueblo más hermoso y mejor conservado de la Costa Brava. Al cabo de tres días de aquel viento inclemente tuve de pronto la revelación deslumbrante de que jamás volvería a ese pueblo porque había de costarme la vida. El personaje de mi cuento debía padecer la misma obsesión durante muchos años, hasta que una noche de fiesta se la reveló a un grupo de amigos en Barcelona. Los amigos, con buena intención de aplicarle a su miedo una cura de burro, lo metieron a la fuerza en un automóvil y se lo llevaron esa misma noche a Cadaqués. El hombre hizo el viaje paralizado por la superstición, y cuando, por fin, vio las luces del pueblo desde la última curva de la montaña, logró zafarse de los amigos y se desbarrancó por un precipicio, incapaz de soportar el terror del regreso.

En ese estado se quedó para siempre el cuento de la muchacha que buscó durante muchos años al desconocido que la violó en un parque, hasta que ella misma descubrió que sólo quería encontrarlo porque no podía vivir sin él. Y el cuento de los niños que conspiraron para matar al rey y al fin lo consiguieron con un caramelo envenenado, y el cuento de los niños que mataron al compañero que lo sabía todo porque no podían soportar que supiera tanto. Hubo uno que terminé: el del hombre que se metió en una armadura de acero para asustar a sus amigos en una fiesta y nunca más pudo salir de ella, de modo que siguió viviendo en ella durante muchos años y se murió dentro de ella de una buena vejez. Estaba a punto de publicarlo cuando lo leyó un amigo providencial, y me hizo caer en la cuenta de que las armaduras de los guerreros no eran una pieza integral —como yo lo creía hasta entonces—, sino que se iban poniendo sobre el cuerpo pieza por pieza, como los trajes de luces de los toreros. De modo que, como tantos otros, también este cuento naufragó para siempre, y con toda justicia, en el mar de los cuentos perdidos.

TERRORISMO CIENTÍFICO

Hace veinte años estaba de moda el colesterol. Las dietas eran rigurosas e insípidas, y no estaban destinadas, como ahora, a regular el peso, sino a impedir que el silencioso asesino se acumulara en la sangre. Alguna revista de divulgación científica publicó la versión de que la berenjena era el preventivo más eficaz del colesterol: sus precios se dispararon hasta un punto en que era como comer pepitas de oro. En las visitas no se hablaba de otra cosa. Como ocurre en cada época con cada enfermedad de moda —tal como ocurre hoy con el cáncer—, la sola palabra adquirió semejante potencia de superstición, que nadie se atrevía a mencionarla. Alguien, hablando de sus males, decía apenas: «Eso». Y ya se sabía que eso no podía ser sino eso: el colesterol.

En esos tiempos sufrí en secreto por un amigo, pues un médico infidente me reveló su diagnóstico: «Tiene el colesterol más alto de la ciudad». No era sorprendente: el plato favorito de mi amigo eran las orejas de cerdo, y la trompa y las patas, que según su médico eran colesterol puro y simple. «No vivirá cinco años», me dijo su médico. Me lo dijo hace treinta años, y el amigo sigue comiendo orejas de cerdo con una facilidad que se le sale por las suyas, y en cambio, su médico acucioso empezó a convertirse en polvo hace tanto tiempo, que ya no recuerdo cuánto. En todo caso, fue mucho antes de que el colesterol se hundiera en el olvido.

Evocando la otra noche, con algunos amigos, las épocas gloriosas del colesterol, terminamos por preguntarnos qué sucede con las enfermedades de moda, que de pronto dejan de serlo sin ningún motivo aparente o, al menos, sin ninguna explicación científica. El terror de nuestra infancia era una de ellas: la amigdalitis. Hay toda una generación de castrados de amígdalas, a los que nunca se les dijo cuáles fueron los motivos reales por los que fueron reducidos a tan inexplicable condición. Yo recuerdo que de niño solía prestar demasiada atención a las conversaciones cifradas de los adultos, y cuando

Publicado originalmente el 1 de septiembre de 1982.

era sorprendido disimulaba mi atención prohibida con un parpadeo muy poco convincente. Me llevaron con el oculista, que al cabo de un examen minucioso ordenó extirparme las amígdalas. Cosa que no se hizo, por fortuna, porque un médico menos dramático aconsejó sustituir la operación por tres cucharadas diarias de jarabe de rábano yodado, y yo tuve el cuidado de ir disminuyendo poco a poco mis parpadeos de susto, hasta que los médicos estuvieron de acuerdo en que mis amígdalas habían sanado por completo.

Más tarde, fue la apendicectomía. Al contrario de la gran mayoría de las enfermedades de moda, la apendicitis ingresó a la vida social con un aura de distinción que la hacía imprescindible. Una mujer no podía permitirse la temeridad de aspirar al reinado de la belleza si en el borde de su traje de baño no se alcanzaba a ver –como una medalla de guerra– la cicatriz de su apendicitis de honor. Uno se hacía operar para no ser menos que el vecino. En el internado donde estudié, éramos despertados varias veces en la noche por el trajín de los enfermeros que se llevaban en la oscuridad a los que iban a ser operados de apendicitis. Los síntomas eran fáciles: un dolor lancinante en la ingle derecha, una especie de adormecimiento en la misma pierna y náuseas con vómitos en los casos más severos. Cuando alguien no se sentía bastante preparado para el examen de álgebra, sólo tenía que fingir esos síntomas e irse al hospital con la seguridad de regresar con una excusa válida cosida en la ingle. Era cierto: los novios se regalaban el apéndice dentro de un frasco de formol, y en la ceremonia de la ruptura había que devolverlo, junto con las cartas de amor.

He estado recordando estos tiempos a propósito de la publicación que hizo el otro día una revista norteamericana sobre el flagelo de moda en Estados Unidos: el herpes. Según ese informe horripilante, veinte millones de norteamericanos han contraído esta enfermedad que no mata, pero que tampoco muere. Se supone, dice la revista, que el agente transmisor se atrinchera en el sistema nervioso, donde está en condiciones de sobrevivir a cualquier ofensiva médica, y vuelve a aparecer a flor de piel cuando encuentra alguna grieta en las condiciones del individuo. Sobre todo, las condiciones emocionales. Esta lepra moderna, que apenas se manifiesta por una úlcera minúscula y recurrente en cualquier sitio del cuerpo, sobre todo en los labios, tiene su manifestación más alarmante y devastadora en los órganos genitales. Es decir: hay que considerarla como una enfermedad venérea. Contagiosa e incurable, por supuesto. Y para colmo de todo, no tiene ni siquiera posibilidad de ser resuelta con recursos quirúrgicos, como era el caso de las amígdalas y el apéndice. La revista lo dice de un modo más simple: «El herpes no te mata, pero tampoco puedes matarlo».

La primera noticia que tuve de este nuevo enemigo público del amor me la dio Carlos Fuentes hace como un año, con un estilo muy suyo: «Maestro: ya no se puede tirar sino en la casa». Pero aun esa notificación inquietante parecía un caramelo para niños al lado de los informes terroríficos de la revista norteamericana. Toda la vida íntima de Estados Unidos –según el informe– está siendo conmovida en sus cimientos. Las relaciones personales sufren modificaciones que pueden ser de fondo. El ritmo social se altera, o el sentimiento que se vislumbra en un porvenir inmediato no es sino uno: el pánico. Una mujer que había padecido el herpes leyó en una revista que éste tenía una relación directa con el cáncer cervical: trató de degollarse con un cuchillo, para que el hijo que tenía en el vientre no viniera al mundo con el estigma de moda. Los noviazgos se interrumpen por la misma causa. Cada día, un número creciente de nuevas víctimas se incorpora a las huestes de los pestíferos. Como consecuencia, un fantasma recorre al país más poderoso del mundo: el fantasma de la impotencia sexual.

De todos modos, lo que parece más desastroso no es el herpes, sino la manera alegre con que se está manejando su información. Lo menos que puede decirse es que no es una manera higiénica: la noticia del flagelo puede tener una gravedad social mucho más perjudicial que la del flagelo mismo. Como en la Edad Media. Cuando el miedo a la peste causaba tantos estragos sociológicos y morales, y tanto desorden social, que se consideraban como otra peste distinta y más temible.

No es posible no preguntarse, habiendo tantos precedentes, si no estamos otra vez en presencia de una nueva campaña de terrorismo científico, cuya finalidad es condicionarnos para quién sabe qué tremenda operación comercial. Hace unos años se proclamó con la misma resonancia que correr era por fin la versión contemporánea de la fuente maravillosa de salud. Primero, todo Estados Unidos, después el mundo entero, se pusieron a correr hasta el delirio. El propio presidente Carter no fue una excepción. Las vitrinas del mundo se llenaron de los folletos para aprender a correr, los zapatos para correr, las camisetas y las botas, y los alimentos adecuados para seguir corriendo hasta la vida eterna. Ahora, saturado el mercado, los mismos que proclamaron sus virtudes están advirtiendo al mundo de los tremendos riesgos de correr. Algo semejante ocurrió con la noticia, ahora desmentida, de que la sacarina producía cáncer, y la de los tampax que causaban trastornos circulatorios, y de tantos otros productos que ahora son absueltos de sus culpas supuestas. Y, como ocurre ahora, cuando la cafeína está siendo objeto de toda clase de injurias, en el momento mismo que las grandes marcas de refrescos se disponen a lanzar al mercado nuevas versiones descafeinadas. Con estos antecedentes, nada de raro tiene que el frankenstein tenebroso del herpes sólo pretenda condiciones para alguna innovación radiante en los hábitos —ya prehistóricos— de las artes del amor. Optimista que es uno.

EL DESTINO DE LOS EMBALSAMADOS

Como uno de los chismes periódicos que divulgan las agencias de prensa, ha surgido ahora la versión de que el cuerpo de Lenin que se exhibe en la plaza Roja de Moscú es, en realidad, una estatua de cera. Se dice que un sobrino de Stalin llamado Budu Svakadze reveló el secreto en un libro que el KGB no permitió publicar en 1952, pero que una copia del manuscrito logró llegar a Israel por correos clandestinos, y desde allí ha sido difundida al mundo por el *Jerusalem Post*. Todo esto es tan difícil de comprobar, que tal vez el método más útil sea tomarse el trabajo de viajar a Moscú, hacer la cola de tres horas bajo las nieves de enero y entrar en el glacial y denso edificio de mármoles incandescentes para tratar de averiguar con ojos propios qué puede haber de cierto en este folletín trasnochado.

Yo lo hice en las dos únicas ocasiones en que he estado en la Unión Soviética –en 1957 y en 1979–, y en ambas tuve la impresión de que el cuerpo de Lenin estaba hecho de su materia natural, aunque es fácil entender que un visitante distraído, o demasiado incrédulo, se sienta inclinado a pensar que es una estatua de cera. La primera vez, el cuerpo de Lenin yacía en su urna de cristal, a la derecha del cuerpo de Stalin, que todavía entonces se consideraba digno de aquella gloria de formaldehído. Lenin había muerto 33 años antes, y Stalin apenas cuatro, y la diferencia se notaba. Este último parecía irradiar un aura de vida, y su bigote histórico de tigre montuno apenas si ocultaba una sonrisa indescifrable. Lo que más me llamó la atención –como ya lo dije en los reportajes que publiqué en aquella ocasión– fueron sus manos delgadas y sensibles, que parecían de mujer. De ningún modo se parecía al personaje sin corazón que Nikita Jruschov había denunciado con una diatriba implacable en el vigésimo congreso de su partido. Poco después, el cuerpo sería sacado de su templo glorioso y mandado a dormir un sueño sin testigos, y tal vez más justo, entre los muertos numerosos de los patios del

Publicado originalmente el 15 de septiembre de 1982.

Kremlin. Muy cerca de la tumba de John Reed, el único norteamericano que alimenta las rosas de aquel jardín quimérico.

El cuerpo de Lenin era menos impresionante, porque estaba menos conservado. En efecto, 33 años son muchos, aun para los muertos, y también en ellos se notan, a través del tiempo, los artificios del embalsamamiento. Al lado de la cabeza de Stalin, enorme y maciza, la de Lenin parecía tan frágil como si fuera de vidrio, y su semblante oriental parecía llegarnos de muy lejos. Tal vez buena parte de esa degradación había sido heredada de sus dos últimos años de vida, que para Lenin habían sido de sufrimientos. En 1922 había sido operado para sacarle una bala que le quedó en el cuello del atentado de agosto de 1918, y el brazo izquierdo le quedó sin vida. El año siguiente sufrió varias recaídas, perdió el habla, se redujo a la nada su fabulosa capacidad de trabajo, y el 21 de enero de 1922 murió devastado por la arterioesclerosis cerebral. Su cerebro, extraído para embalsamar el cuerpo, tenía la consistencia árida de una piedra. La inutilidad del brazo izquierdo se notaba aun después de embalsamado, y la erosión general del cadáver, que ya era evidente la primera vez que lo vi, lo era mucho más la segunda, cuando ya habían transcurrido 55 años de la muerte. Pero en ningún caso me pareció una estatua de cera, entre otras cosas, porque la cera no tiene la buena virtud de envejecer.

En realidad, lo que más me estremeció en las dos ocasiones en que vi la momia de Lenin fue la impresión ineludible de que el cuerpo no se conservaba completo bajo las sábanas de la urna, sino que lo habían cortado por la cintura para facilitar la conservación. Hasta el pecho, en efecto, el relieve del cuerpo era convincente, pero luego se confundía con la superficie del mesón donde estaba acostado, y se dejaba la puerta abierta a cualquier aventura de la imaginación. No era fácil soportar la idea de que la muchedumbre que desfilaba por el mausoleo le estaba rindiendo tributo a un héroe partido por la mitad, cuya parte inferior se había podrido y convertido en polvo en algún basurero distinto.

En todo caso, estas suposiciones son posibles por la mala costumbre de conservar cadáveres para ser adorados por la muchedumbre. Nada se parece menos a la imagen que se tiene de un hombre o una mujer memorables que sus desperdicios mortales arreglados como para una fiesta funeraria. Los motivos de los egipcios eran perdonables, porque creían que mientras se conservara el cuerpo se conservaría también el espíritu, y en ningún caso embalsamaban a sus faraones para la exhibición pública. Los católicos, al revés, piensan que la conservación casual del cuerpo es un indicio de santidad, y lo exponen en sus templos para deleite de sus fieles. Pero es difícil encontrar una justificación doctrinaria para la costumbre creciente de los regímenes comunistas, que parecen confundir el culto de los héroes con el culto de sus momias. Es el caso en Bulgaria, donde se conserva el cuerpo de Dimitrov, y el caso de China, donde se conserva el cuerpo de Mao, y el caso de Vietnam, donde se conserva el cuerpo de Ho Chi Min. No se necesita ser un visionario para suponer que Kim Il Sum, el presidente de Corea del Norte, que desconoce por completo el dulce encanto de la modestia, debe estar ya ansioso por someter su cuerpo glorioso a los buenos oficios de sus embalsamadores.

Por fortuna, Cuba sentó un precedente ejemplar para este lado del mundo con las manos del Che Guevara, que fueron cortadas por la CIA para una identificación a fondo por las huellas digitales. Un antiguo funcionario del gobierno boliviano que desertó de su cargo las llevó después a La Habana, y no faltó quien sugiriera la idea de conservarlas para el culto público. Fidel Castro, que tiene la buena costumbre de llevar estos problemas hasta la última instancia, lo consultó con las muchedumbres al final de un discurso en un acto de masas. La respuesta, que era la que Fidel Castro esperaba, fue unánime y rotunda: nones.

Hay en América Latina otros antecedentes que no son tan consoladores. El general Antonio López de Santa Anna, que gobernó a México varias veces desde 1833, perdió la pierna derecha en la guerra contra los invasores franceses y la hizo enterrar en la catedral, bajo palio de obispo y con todos los honores militares y religiosos, en unos funerales babilónicos presididos por él mismo. Más tarde, el general Álvaro Obregón perdió el brazo izquierdo por una bala de cañón que le disparó Pancho Villa en la batalla de Celaya, y su mano se conserva todavía en la ciudad de México, achicharrada por el formol, en un monumento público, que por razones inescrutables se ha convertido en un sitio de peregrinación de los jóvenes enamorados. El caso más extraño de nuestro tiempo es el del cadáver de Evita Perón, que desapareció de Buenos Aires después de embalsamado y reapareció muchos años después en Italia, bajo la responsabilidad del Vaticano. El hombre que la embalsamó era un catalán grandilocuente que montó guardia en la antesala de la enferma durante las largas semanas de su agonía, pues debía proceder al embalsamamiento en el instante mismo de la muerte para una conservación más convincente y duradera. Mientras esperaba, les hacía ver a los visitantes ilustres el álbum de fotos de sus trabajos más notables. Y entre ellos, su obra maestra: un niño de Montevideo que había muerto a los siete años, y cuyos padres lo hicieron embalsamar sentado en una sillita y vestido de marinero. Todos los años, durante muchos, sus hermanos le celebraron el cumpleaños con los que fueron sus amigos, hasta que todos crecieron, y se casaron y tuvieron otros hijos para embalsamar, y el pobre niño embalsamado, en su sillita de madera y con su vestido de marinero, quedó a merced de las polillas y el olvido en un ropero del dormitorio.

EL AVIÓN DE LA BELLA DURMIENTE

Era bella, elástica, con una piel tierna del color del pan y los ojos de almendras verdes, y tenía el cabello liso y negro y largo hasta la espalda, y un aura de antigüedad oriental que lo mismo podía ser de Bolivia que de Filipinas. Estaba vestida con un gusto sutil: una chaqueta de lince, una blusa de seda de flores muy tenues, unos pantalones de lino crudo, y unos zapatos lineales del color de las buganvillas. «Ésta es la mujer más bella que he visto en mi vida», pensé, cuando la vi en la cola para abordar el avión de Nueva York en el aeropuerto Charles de Gaulle, de París. Le cedí el paso, y cuando llegué al asiento que me habían asignado en la tarjeta de embarque, la encontré instalándose en el asiento vecino. Casi sin aliento alcancé a preguntarme de cuál de los dos sería la mala suerte de aquella casualidad aterradora.

Se instaló como si fuera para vivir muchos años, poniendo cada cosa en su lugar en un orden perfecto, hasta que su espacio personal quedó tan bien dispuesto como una casa ideal donde todo estaba al alcance de la mano. Mientras lo hacía, el oficial de servicio nos ofreció champaña de bienvenida. Ella no quiso, y trató de explicar algo en un francés rudimentario. El oficial le habló entonces en inglés, y ella se lo agradeció con una sonrisa estelar, y le pidió un vaso de agua, con la súplica de que no la despertaran para nada durante el vuelo. Después abrió sobre las rodillas un neceser grande y cuadrado, con esquinas de cobre como los baúles de viaje de las abuelas, y se tomó dos pastillas doradas que sacó de un estuche donde llevaba muchas de diversos colores. Hacía todo de un modo metódico y parsimonioso, como si no hubiera nada que no estuviera previsto para ella desde su nacimiento.

Por último, puso la almohadita en el rincón de la ventanilla, se cubrió con la manta hasta la cintura sin quitarse los zapatos, y se acomodó de medio lado en la silla, casi en estado fetal, y durmió sin una

Publicado originalmente el 22 de septiembre de 1982.

sola pausa, sin un suspiro, sin un cambio mínimo de posición, durante las siete horas pavorosas y los doce minutos de sobra que duró el vuelo hasta Nueva York.

Siempre he creído que no hay nada más hermoso en la naturaleza que una mujer hermosa. De modo que me fue imposible escapar ni un instante al hechizo de aquella criatura fabulosa que dormía a mi lado. Era un sueño tan estable, que en cierto momento tuve la inquietud de que las pastillas que se había tomado no eran para dormir, sino para morir. La contemplé muchas veces centímetro a centímetro, y la única señal de vida que pude percibir fueron las sombras de los sueños que pasaban por su frente como las nubes en el agua. Tenía en el cuello una cadena tan fina que era casi invisible sobre su piel de oro, tenía las orejas perfectas sin perforaciones para los aretes, y tenía un anillo liso en la mano izquierda. Como no parecía tener más de veintidós años, me consolé con la idea de que no fuera un anillo de matrimonio, sino el de un noviazgo efímero y feliz. No llevaba ningún perfume: su piel exhalaba un hálito tenue que no podía ser otro que el olor natural de su belleza. «Tú por tu sueño y por el mar las naves», pensé, a 20.000 pies de altura sobre el océano Atlántico, tratando de recordar en orden el soneto inolvidable de Gerardo Diego. «Saber que duermes tú, cierta, segura, cauce fiel de abandono, línea pura, tan cerca de mis brazos maniatados.» Mi realidad se parecía de tal modo a la del soneto, que al cabo de media hora lo había reconstruido de memoria hasta el final: «Qué pavorosa esclavitud de isleño, yo insomne, loco, en los acantilados, las naves por el mar, tú por tu sueño». Sin embargo, al cabo de cinco horas de vuelo había contemplado tanto a la bella durmiente, y con tanta ansiedad sin destino, que comprendí de pronto que mi estado de gracia no era el del soneto de Gerardo Diego, sino el de otra obra maestra de la literatura contemporánea, *La casa de las bellas durmientes*, del japonés Yasunari Kawabata.

Descubrí esta hermosa novela por un camino largo y distinto, pero que de todos modos concluye con la bella dormida del avión. Hace varios años, en París, el escritor Alain Jouffroy me llamó por teléfono para decirme que quería presentarme a unos escritores japoneses que estaban en su casa. Lo único que yo conocía entonces de la literatura japonesa, aparte de los tristes haikais del bachillerato, eran algunos cuentos de Junichiro Tanizaki que habían sido traducidos al castellano. En realidad, lo único que sabía a ciencia cierta de los escritores japoneses era que todos, tarde o temprano, terminarían por suicidarse. Había oído hablar de Kawabata por primera vez cuando le concedieron el Premio Nobel en 1968, y entonces traté de leer algo suyo, pero me quedé dormido. Poco después se destripó con un sable ritual, tal como lo había hecho en 1946 otro novelista notable, Osamu Dazai, después de varias tentativas frustradas. Dos años antes que Kawabata, y también después de varias tentativas frustradas, el novelista Yukio Mishima, que es tal vez el más conocido en Occidente, se había hecho el harakiri completo después de dirigir una arenga patriótica a los soldados de la guardia imperial. De modo que cuando Alain Jouffroy me llamó por teléfono, lo primero que me vino a la memoria fue el culto a la muerte de los escritores japoneses. «Voy con mucho gusto», le dije a Alain, «pero con la condición de que no se suiciden.» No se suicidaron, en efecto, sino que pasamos una noche encantadora, en la cual lo mejor que aprendí fue que todos estaban locos. Ellos estuvieron de acuerdo. «Por eso queríamos conocerte», me dijeron. Al final, me dejaron convencido de que para los lectores japoneses no hay ninguna duda de que yo soy un escritor japonés.

Tratando de entender lo que quisieron decirme fui al día siguiente a una librería especializada de París y compré todos los libros de los autores disponibles: Shusako Endo, Kenzaburo Oe, Yasushi Inoue, Akutagwa Ryunosuke, Masuji Ibusi, Osamu Dazai, además de los obvios de Kawabata y Mishima. Durante casi un año no leí otra cosa, y ahora yo también estoy convencido: las novelas japonesas tienen algo en común con las mías. Algo que no podría explicar, que no sentí en la vida del país durante mi única visita al Japón, pero que a mí me parece más que evidente.

Sin embargo, la única que me hubiera gustado escribir es *La casa de las bellas durmientes*, de Kawabata, que cuenta la historia de una rara mansión de los suburbios de Kyoto donde los ancianos burgueses pagaban sumas enormes para disfrutar de la forma más refinada del último amor: pasar la noche contemplando a las muchachas más bellas de la ciudad, que yacían desnudas y narcotizadas en la misma cama. No podían despertarlas, ni tocarlas siquiera, aunque tampoco lo intentaban, porque la satisfacción más pura de aquel placer senil era que podían soñar a su lado.

Viví esa experiencia junto a la bella durmiente del avión de Nueva York, pero no me alegro. Al contrario: lo único que deseaba en la última hora del vuelo era que el oficial la despertara para que yo pudiera recobrar mi libertad, y tal vez hasta mi juventud. Pero no fue así. Se despertó sola cuando ya la nave estaba en tierra, se arregló y se levantó sin mirarme, y fue la primera que salió del avión y se perdió para siempre en la muchedumbre. Yo seguí en el mismo vuelo hasta México, pastoreando las primeras nostalgias de su belleza junto al asiento todavía tibio por su sueño, sin poder quitarme de la cabeza lo que me habían dicho de mis libros los escritores locos de París. Antes de aterrizar, cuando me dieron la ficha de inmigración, la llené con un sentimiento de amargura. Profesión: escritor japonés. Edad: 92 años.

BEGUIN Y SHARON,
PREMIOS «NOBEL DE LA MUERTE»

Lo más increíble de todo es que Menájem Beguin sea Premio Nobel de la Paz. Pero lo es sin remedio –aunque ahora cueste trabajo creerlo– desde que le fue concedido en 1978, al mismo tiempo que a Anuar el Sadat, entonces presidente de Egipto, por haber suscrito un acuerdo de paz separada en Camp David. Aquella determinación espectacular le costó a Sadat el repudio inmediato de la comunidad árabe, y más tarde le costó la vida. A Beguin, en cambio, le ha permitido la ejecución metódica de un proyecto estratégico que aún no ha culminado. Pero que hace pocos días propició la masacre bárbara de más de un millar de refugiados palestinos en un campamento de Beirut. Si existiera el Premio Nobel de la Muerte, este año lo tendrían asegurado sin rivales el mismo Menájem Beguin y su asesino profesional Ariel Sharon.

En efecto, vistos ahora, los acuerdos de Camp David no tendrían para Beguin otra finalidad que la de cubrirse las espaldas para exterminar, primero, a la Organización para la Liberación de Palestina (OLP), y establecer luego nuevos asentamientos israelíes en Samaria y Judea. Para quienes tenemos una edad que nos permite recordar las consignas de los nazis, estos dos propósitos de Beguin suscitan reminiscencias espantosas: la teoría del espacio vital, con la que Hitler se propuso extender su imperio a medio mundo, y lo que él mismo llamó la *solución final* del problema judío, que condujo a los campos de exterminio a más de seis millones de seres humanos inocentes.

La ampliación del espacio vital del Estado de Israel y la solución final del problema palestino –tal como las concibe hoy el Premio Nobel de la Paz de 1978– se iniciaron, en la noche del 5 de junio pasado, con la invasión de Líbano por fuerzas militares israelíes especializadas en la ciencia de la demolición y el exterminio. Menájem

Publicado originalmente el 29 de septiembre de 1982.

Beguin trató de justificar esta expedición sangrienta con dos argumentos falsos. El primero fue la tentativa de asesinato del embajador de Israel en Londres, Shlomo Argov, a finales de mayo. El segundo fue el supuesto bombardeo de Galilea por la OLP, refugiada en Líbano. Beguin acusó del atentado de Londres a la resistencia palestina y amenazó con represalias inmediatas. Pero Scotland Yard reveló más tarde que los verdaderos autores habían sido miembros de la organización disidente de Abou Nidal, que en los meses anteriores había asesinado inclusive a varios dirigentes de la OLP. En cuanto al segundo argumento, se comprobó muy pronto que los palestinos sólo dispararon dos o tres veces contra Galilea y causaron un muerto. Los disparos fueron hechos como represalia por los bombardeos de Israel contra los campos de refugiados palestinos, que dieron muerte a varios centenares de civiles.

En realidad, la guerra sin corazón desatada por Beguin con base en aquellos dos pretextos no era nada nuevo para los lectores del semanario israelí *Haclam Haze*, que la había anunciado con todos sus pormenores desde septiembre de 1981. Es decir, nueve meses antes. Contra el refrán según el cual una guerra avisada no mata a nadie, las tropas israelíes –que se consideran entre las más eficaces del mundo– mataron en las primeras dos semanas a casi 30.000 civiles palestinos y libaneses y convirtieron en escombros a media ciudad. Sus pérdidas en el mismo período no habían pasado de trescientas.

Ahora la estrategia de Beguin es muy clara. Al destruir a la OLP ha tratado de eliminar al único interlocutor palestino que parecía capaz de negociar una paz fundada sobre la base de la instalación de un Estado palestino independiente en Cisjordania y Gaza, que el propio Beguin ha proclamado como territorios ancestrales del pueblo judío. Ese acuerdo estaba al alcance de la mano desde el 4 de julio pasado, cuando Yasir Arafat, presidente de la OLP, aceptó el principio de un reconocimiento recíproco de los pueblos de Israel y Palestina, en una entrevista publicada por *Le Monde*, de París, en aquella fecha. Pero Beguin ignoró esa declaración, que entorpecía sus proyectos expansionistas ya en pleno desarrollo, y prosiguió con el establecimiento de un cinturón de seguridad en torno de Israel. Un cambio de gobierno en Siria podría ser el paso inmediato, con la extensión consiguiente de una guerra desigual y sin cuartel, cuyas consecuencias finales son imprevisibles.

Yo estaba en París en junio pasado, cuando las tropas de Israel invadieron Líbano. Por casualidad estaba también el año anterior, cuando el general Jaruzelski implantó el poder militar en Polonia contra la voluntad evidente de la mayoría del pueblo polaco. Y también por casualidad me encontraba allí cuando las tropas argentinas desembarcaron en las islas Malvinas. Las reacciones de los medios de comunicación ante esos tres acontecimientos, así como las de los intelectuales y la de la opinión pública en general, fueron para mí una lección inquietante. La crisis de Polonia produjo en Europa una especie de conmoción social. Yo tuve la buena ocasión de agregar mi firma a la de los muy escogidos y muy notables intelectuales y artistas que suscribieron la invitación para un homenaje al heroísmo del pueblo polaco, que se celebró en el teatro de la Ópera de París, patrocinado por el Ministerio de Cultura de Francia. Sin embargo, algunos anticomunistas profesionales me acusaron en público de que mi protesta no fuera tan histórica como la de ellos. En aquel clima pasional, toda actitud que no fuera maniqueísta se consideraba ambigua.

En cambio, cuando las tropas de Israel invadieron y ensangrentaron Líbano, el silencio fue casi unánime aun entre los más exaltados Jeremías de Polonia, a pesar de que ni el número de muertos ni el tamaño de los estragos admitían ninguna posibilidad de comparación entre la tragedia de los dos países. Más aún: por esas mismas fechas, los argentinos habían recuperado las islas Malvinas, y el Consejo de Seguridad de las Naciones Unidas no esperó 48 horas para ordenar el retiro de las tropas ni la Comunidad Económica Europea lo pensó demasiado para imponer sanciones comerciales a Argentina. En cambio, ni ese mismo organismo ni ningún otro de su envergadura ordenó el retiro de las tropas israelíes de Líbano en aquella ocasión. El gobierno del presidente Reagan, por supuesto, fue el cómplice más servicial de la pandilla sionista. Por último, la prudencia casi inconcebible de la Unión Soviética, y la fragmentación fraternal del mundo árabe acabaron de completar las condiciones propicias para el mesanismo demente de Beguin y la barbarie guerrera del general Sharon. Tengo muchos amigos, cuyas voces fuertes podrían escucharse en medio mundo, que hubieran querido y sin duda siguen queriendo expresar su indignación por este festival de sangre, pero algunos de ellos confiesan en voz baja que no se atreven por temor de ser señalados de antisemitas. No sé si serán conscientes de que están cediendo –al precio de su alma– ante un chantaje inadmisible.

La verdad es que nadie ha estado tan sólo como el pueblo judío y el pueblo palestino en medio de tanto horror. Desde el principio de la invasión a Líbano empezaron en Tel Aviv y otras ciudades las manifestaciones populares de protesta que aún no han terminado, y que en el pasado fin de semana habían alcanzado una fuerza emocionante. Eran más de 400.000 israelíes proclamando en las calles que aquella guerra sucia no es la suya porque está muy lejos de ser la de su dios, que durante tantos y tantos siglos se había complacido con la convivencia de palestinos y judíos bajo el mismo cielo. En un país de tres millones de habitantes, una manifestación de 400.000 personas equivaldría en términos proporcionales a una de casi treinta millones en Washington.

Es con esa protesta interna con la que me siento identificado cada vez que conozco las noticias de las hostilidades de los Beguines y los Sharones en Líbano, y en cualquier parte del mundo, y a ella quiero sumar mi voz de escritor solitario por el gran cariño y la admiración inmensa que siento por un pueblo que no conocí en los periódicos de hoy sino en la lectura asombrada de la Biblia. No le temo al chantaje del antisemitismo, como no le he temido nunca al chantaje del anticomunismo profesional, que andan juntos y a veces revueltos, y siempre haciendo estragos semejantes en este mundo desdichado.

SE NECESITA UN ESCRITOR

Me preguntan con frecuencia qué es lo que me hace más falta en la vida, y siempre contesto la verdad: «Un escritor». El chiste no es tan bobo como parece. Si alguna vez me encontrara con el compromiso ineludible de escribir un cuento de quince cuartillas para esta noche, acudiría a mis incontables notas atrasadas y estoy seguro de que llegaría a tiempo a la imprenta. Tal vez sería un cuento muy malo, pero el compromiso quedaría cumplido, que al fin y al cabo es lo único que he querido decir con este ejemplo de pesadilla. En cambio, no sería capaz de escribir un telegrama de felicitación ni una carta de pésame sin reventarme el hígado durante una semana. Para estos deberes indeseables, como para tantos otros de la vida social, la mayoría de los escritores que conozco quisieron apelar a los buenos oficios de otros escritores. Una buena prueba del sentido casi bárbaro del honor profesional lo es sin duda esta nota que escribo todas las semanas, y que por estos días de octubre va a cumplir sus primeros dos años de soledad. Sólo una vez ha faltado en este rincón, y no fue por culpa mía: por una falla de última hora en los sistemas de transmisión. La escribo todos los viernes, desde las nueve de la mañana hasta las tres de la tarde, con la misma voluntad, la misma conciencia, la misma alegría y muchas veces con la misma inspiración con que tendría que escribir una obra maestra. Cuando no tengo el tema bien definido me acuesto mal la noche del jueves, pero la experiencia me ha enseñado que el drama se resolverá por sí solo durante el sueño y que empezará a fluir por la mañana, desde el instante en que me siente ante la máquina de escribir. Sin embargo, casi siempre tengo varios temas pensados con anticipación, y poco a poco voy recogiendo y ordenando los datos de distintas fuentes y comprobándolos con mucho rigor, pues tengo la impresión de que los lectores no son tan indulgentes con mis metidas de pata como tal vez lo serían con el otro escritor que me hace falta.

Publicado originalmente el 6 de octubre de 1982.

Mi primer propósito con estas notas es que cada semana les enseñen algo a los lectores comunes y corrientes, que son los que me interesan, aunque esas enseñanzas les parezcan obvias y tal vez pueriles a los sabios doctores que todo lo saben. El otro propósito —el más difícil— es que siempre estén tan bien escritas como yo sea capaz de hacerlo sin la ayuda del otro, pues siempre he creído que la buena escritura es la única felicidad que se basta de sí misma.

Esta servidumbre me la impuse porque sentía que entre una novela y otra me quedaba mucho tiempo sin escribir, y poco a poco —como los peloteros— iba perdiendo la calentura del brazo. Más tarde, esa decisión artesanal se convirtió en un compromiso con los lectores, y hoy es un laberinto de espejos del cual no consigo salir. A no ser que encontrara, por supuesto, al escritor providencial que saliera por mí. Pero me temo que ya sea demasiado tarde, pues las tres únicas veces en que tomé la determinación de no escribir más estas notas me lo impidió, con su autoritarismo implacable, el pequeño argentino que también yo llevo dentro.

La primera vez que lo decidí fue cuando traté de escribir la primera, después de más de veinte años de no hacerlo, y necesité una semana de galeote para terminarla. La segunda vez fue hace más de un año, cuando pasaba unos días de descanso con el general Omar Torrijos en la base militar de Farallón, y estaba el día tan diáfano y tan pacífico el océano que daban más ganas de navegar que de escribir. «Le mando un telegrama al director diciendo que hoy no hay nota, y ya está», pensé, con un suspiro de alivio. Pero no pude almorzar por el peso de la mala conciencia y, a las seis de la tarde, me encerré en el cuarto, escribí en una hora y media lo primero que se me ocurrió y le entregué la nota a un edecán del general Torrijos para que la enviara por télex a Bogotá, con el ruego de que la mandaran desde allí a Madrid y a México. Sólo al día siguiente supe que el general Torrijos había tenido que ordenar el envío en un avión militar hasta el aeropuerto de Panamá, y desde allí, en helicóptero, al palacio presidencial, desde donde me hicieron el favor de distribuir el texto por algún canal oficial.

La última vez, hace ahora seis meses, cuando descubrí al despertar que ya tenía madura en el corazón la novela de amor que tanto había anhelado escribir desde hacía tantos años, y que no tenía otra alternativa que no escribirla nunca o sumergirme en ella de inmediato y de tiempo completo. Sin embargo, a la hora de la verdad, no tuve suficientes riñones para renunciar a mi cautiverio semanal, y por primera vez estoy haciendo algo que siempre me pareció imposible: escribo la novela todos los días, letra por letra, con la misma paciencia, y ojalá con la misma suerte, con que picotean las gallinas en los patios, y oyendo cada día más cerca los pasos temibles de animal grande del próximo viernes. Pero aquí estamos otra vez, como siempre, y ojalá para siempre.

Ya sospechaba yo que no escaparía jamás de esta jaula desde la tarde en que empecé a escribir esta nota en mi casa de Bogotá y la terminé al día siguiente bajo la protección diplomática de la embajada de México; lo seguí sospechando en la oficina de Telégrafos de la isla de Creta, un viernes del pasado julio, cuando logré entenderme con el empleado de turno para que transmitiera el texto en castellano. Lo seguí sospechando en Montreal, cuando tuve que comprar una máquina de escribir de emergencia porque el voltaje de la mía no era el mismo del hotel. Acabé de sospecharlo para siempre hace apenas dos meses, en Cuba, cuando tuve que cambiar dos veces las máquinas de escribir porque se negaban a entenderse conmigo. Por último, me llevaron una electrónica de costumbres tan avanzadas que terminé escribiendo de mi puño y letra y en un cuaderno de hojas cuadriculadas, como en los tiempos remotos y felices de la escuela primaria de Aracataca. Cada vez que me ocurría uno de estos percances apelaba con más ansiedad a mis deseos de tener alguien que se hiciera cargo de mi buena suerte: un escritor.

Con todo, nunca he sentido esa necesidad de un modo tan intenso como un día de hace muchos años en que llegué a la casa de Luis Alcoriza, en México, para trabajar con él en el guión de una película. Lo encontré consternado a las diez de la mañana, porque su cocinera le había pedido el favor de escribirle una carta para el director de la Seguridad Social. Alcoriza, que es un escritor excelente, con una práctica cotidiana de cajero de banco, que había sido el escritor más inteligente de los primeros guiones para Luis Buñuel y, más tarde, para sus propias películas, había pensado que la carta sería un asunto de media hora. Pero lo encontré, loco de furia, en medio de un montón de papeles rotos, en los cuales no había mucho más que todas las variaciones concebidas de la fórmula inicial: por medio de la presente, tengo el gusto de dirigirme a usted para... Traté de ayudarlo, y tres horas después seguíamos haciendo borradores y rompiendo papel, ya medio borrachos de ginebra con vermouth y atiborrados de chorizos españoles, pero sin haber podido ir más allá de las primeras letras convencionales. Nunca olvidaré la cara de misericordia de la buena cocinera cuando volvió por su carta a las tres de la tarde y le dijimos sin pudor que no habíamos podido escribirla. «Pero si es muy fácil», nos dijo, con toda su humildad. «Mire usted.» Entonces empezó a improvisar la carta con tanta precisión y tanto dominio que Luis Alcoriza se vio en apuros para copiarla en la máquina con la misma fluidez con que ella la dictaba. Aquel día —como todavía hoy— me quedé pensando que tal vez aquella mujer, que envejecía sin gloria en el limbo de la cocina, era el escritor secreto que me hacía falta en la vida para ser un hombre feliz.

EL CUENTO DESPUÉS DEL CUENTO

Clotilde Armenta, que es un personaje de mi novela más reciente, exclamó de pronto en alguna parte del libro: «¡Dios mío, qué solas estamos las mujeres en el mundo!». Rossana Rossanda, que es uno de los seres humanos más inteligentes que conozco, me preguntó en una entrevista de prensa cómo había llegado yo a esa conclusión. «¿Desde cuándo lo sabes?», fue su pregunta concreta. Ningún periodista me había puesto a pensar tanto sobre el comportamiento de alguno de mis personajes. Sobre todo, ninguno, como Rossana Rossanda en esa ocasión, me había obligado a pensar tan en serio sobre el papel de las mujeres en mis libros —y tal vez en mi vida—, que es algo de lo cual muchos críticos han hablado no sólo más de lo que deben, sino inclusive más de lo que saben.

El personaje de Clotilde Armenta, que no existió en la realidad, fue inventado por mí, de cuerpo entero, porque me hacía falta como contrapeso a Pura Vicario, la madre de la protagonista principal. El carácter de Clotilde Armenta lo fui construyendo a medida que lo escribía, de acuerdo con los meandros imprevistos del drama. Siempre tuve la intuición de que el crimen de la realidad no se pudo impedir porque en la vida real no existió una mujer como ella, y en algún momento tuve la tentación de que, en efecto, lo impidiera en el libro. Sin embargo, a cada paso me daba cuenta de que lo único que ella podía hacer para impedirlo era solicitar la ayuda de otros, y casi siempre esos otros eran hombres. Era una realidad, no sólo dentro de la ficción, sino dentro de las condiciones sociales del pueblo. En la culminación del drama, yo mismo descubrí, no sin cierto deslumbramiento, que era allí donde radicaba la impotencia de Clotilde Armenta para impedir el crimen. Entonces fue cuando exclamó: «¡Dios mío, qué solas estamos las mujeres en el mundo!». No lo dije yo. Lo dijo ella, aunque sea algo difícil de entender por alguien que no sea escritor. Sin embargo, creo que ella y yo lo descu-

Publicado originalmente el 13 de octubre de 1982.

brimos al mismo tiempo, y que al descubrirlo nos dimos cuenta de que lo sabíamos desde hacía mucho tiempo pero no lográbamos explicárnoslo. Fue eso lo que le contesté a Rossana Rossanda para una entrevista que publicó hace pocos meses en su periódico, *Il Manifesto*, de Roma.

Uno de los primeros lectores del libro me dijo: «Esto no es más que un sucio asunto de mujeres». Otro me señaló que era un drama de jóvenes, pues, en realidad, ninguno de sus protagonistas era mayor de veinticinco años, y este lector creyó entender que el libro era una prueba de que fueron los prejuicios de los adultos los que determinaron la tragedia. En todo caso, mi convicción es que la participación de las mujeres fue decisiva en el drama, y esto corresponde a mi convicción de que el machismo es un producto cultural de las sociedades matriarcales. La persona que comandaba el drama desde las sombras era Pura Vicario, la madre de Ángela —cosa que no ocurrió, por cierto, en la realidad—, y no creo que lo hiciera por vocación, sino porque pensaba que la familia no sería capaz de sobrevivir al repudio social si sus hijos no lavaban la afrenta. Ángela Vicario descubrió esa verdad mucho más tarde, en el hotel del puerto de Riohacha, cuando volvió a ver al esposo que la había repudiado y descubrió que lo amaba por encima de todo, y comprendió que la madre era la única responsable de la desgracia. Entonces la vio tal como era: «Una pobre mujer consagrada al culto de sus defectos». En todo caso, a mi modo de ver, lo que revela mejor la injusticia y la miseria de aquella sociedad es que la mujer más libre del pueblo, y en realidad la única libre, era María Alejandrina Cervantes, la puta grande.

Otro aspecto que le interesaba mucho a Rossana Rossanda era el ingrediente de la fatalidad en el drama. En realidad nunca me interesó la fatalidad como factor determinante. Lo que se parece a la fatalidad en la *Crónica de una muerte anunciada* no es más que un elemento mecánico de la narración. Tal como en el *Edipo rey*, de Sófocles —aunque parezca extraño en una tragedia griega—, cuya esencia no es la fatalidad de los hechos sino el drama del hombre en la búsqueda de su identidad y su destino.

En mi novela, mi trabajo mayor fue descubrir y revelar la serie casi infinita de coincidencias minúsculas y encadenadas que dentro de una sociedad como la nuestra hicieron posible aquel crimen absurdo. Todo era evitable, y fue la conducta social, y no el *fatum*, lo que impidió evitarlo. Rossana Rossanda no sólo estaba de acuerdo, sino que tal vez descifró la clave más inquietante. «Éste no es el drama de la fatalidad», me dijo, «sino el drama de la responsabilidad.»

Más aún: el drama de la responsabilidad colectiva. Yo creo, incluso, que la novela termina por desprestigiar el mito de la fatalidad, puesto que trata de desmontarlo en sus piezas primarias y demuestra que somos nosotros los únicos dueños de nuestro destino.

Todo esto me parece más evidente cada vez que evoco el día aciago en que ocurrieron los hechos en la realidad. Yo no fui testigo presencial, pero conocía muy bien el lugar y conocía muy bien a los protagonistas, que al fin y al cabo eran todos los habitantes del pueblo. Recuerdo que cuando conocí la noticia y sus pormenores, mi primera reacción fue de rabia, pues, por más que le daba vueltas y más vueltas, todo me parecía evitable. A partir de entonces, todos los testigos con quienes he seguido hablando se siguen preguntando cómo fue que ellos mismos no pudieron impedirlo, y en todos he encontrado tanta ansiedad por justificar sus actos de aquel día que he creído reconocer en esa ansiedad un cierto sentimiento de culpa. Yo creo que lo que los paralizó fue la creencia, consciente o inconsciente, de que aquel crimen ritual era un acto socialmente legítimo.

Las circunstancias en que Bayardo San Román volvió con la esposa repudiada no fueron tampoco las mismas que en el libro. Debo reconocer que en este caso la realidad fue más aleccionadora. Todo fue, al parecer, un rumor que circuló casi veinte años después entre los testigos. Según ese rumor, el marido había hecho toda clase de gestiones para volver con la esposa repudiada, y fue ella quien no quiso aceptarlo. Sin duda, el tiempo no había pasado con igual velocidad ni con igual intensidad para ella y para su marido. Pero lo que entonces me interesaba era que aquella tentativa de reconciliación –tal vez inventada por los propios testigos– se divulgó de inmediato entre los sobrevivientes, y éstos divulgaron el rumor como si fuera un hecho cumplido que los viejos esposos habían vuelto a reunirse y vivirían felices para siempre. Tal vez sentían que todos necesitábamos de esa reunificación, porque era como el final de la culpa colectiva, como si el desastre de que todos éramos culpables pudiera no sólo ser reparado sino borrado para siempre de la memoria social. Lo malo para todos es que siempre aparece un aguafiestas desperdigado cuya única función en el mundo es recordar lo que los otros olvidan.

OBREGÓN O LA VOCACIÓN DESAFORADA*

Hace muchos años, un amigo le pidió a Alejandro Obregón que lo ayudara a buscar el cuerpo del patrón de su bote, que se había ahogado al atardecer, mientras pescaban sábalos de veinte libras en la ciénaga grande. Ambos recorrieron durante toda la noche aquel inmenso paraíso de aguas marchitas, explorando sus recodos menos pensados con luces de cazadores, siguiendo la deriva de los objetos flotantes, que suelen conducir a los pozos donde se quedan a dormir los ahogados. De pronto, Obregón lo vio: estaba sumergido hasta la coronilla, casi sentado dentro del agua, y lo único que flotaba en la superficie eran las hebras errantes de su cabellera. «Parecía una medusa», me dijo Obregón. Agarró el mazo de pelos con las dos manos y, con su fuerza descomunal de pintor de toros y tempestades, sacó al ahogado entero, con los ojos abiertos, enorme, chorreando lodo de anémonas y mantarrayas, y lo tiró como un sábalo muerto en el fondo del bote.

Este episodio, que Obregón me vuelve a contar porque yo se lo pido cada vez que nos emborrachamos a muerte –y que además me dio la idea para un cuento de ahogados–, es tal vez el instante de su vida que más se parece a su arte. Así pinta, en efecto, como pescando ahogados en la oscuridad. Su pintura con horizontes de truenos sale chorreando minotauros de lidia, cóndores patrióticos, chivos arrechos, barracudas berracas. En medio de la fauna tormentosa de su mitología personal anda una mujer coronada de guirnaldas florentinas, la misma de siempre y de nunca, que merodea por sus cuadros con las claves cambiadas, pues en realidad es la criatura imposible por la que este romántico de cemento armado se quisiera morir. Porque él lo es como lo somos todos los románticos, y como hay que serlo: sin ningún pudor.

Publicado originalmente el 20 de octubre de 1982.
* Esta es la nota de presentación del catálogo de la exposición que el pintor colombiano Alejandro Obregón inauguró esta semana en el Metropolitan Museum and Art Center de Coral Gable, Fla. (EE.UU.).

La primera vez que vi a esa mujer fue el mismo día en que conocí a Obregón, hace ahora 32 años, en su taller de la calle de San Blas, en Barranquilla. Eran dos aposentos grandes y escuetos por cuyas ventanas despernancadas subía el fragor babilónico de la ciudad. En un rincón distinto, entre los últimos bodegones picassianos y las primeras águilas de su corazón, estaba ella con sus lotos colgados, verde y triste, sosteniéndose el alma con la mano. Obregón, que acababa de regresar de París y andaba como atarantado por el olor de la guayaba, era ya idéntico a este autorretrato suyo que me mira desde el muro mientras escribo, y que él trató de matar una noche de locos con cinco tiros de grueso calibre. Sin embargo, lo que más me impresionó cuando lo conocí no fueron esos ojos diáfanos de corsario que hacían suspirar a los maricas del mercado, sino sus manos grandes y bastas, con las cuales lo vimos tumbar media docena de marineros suecos en una pelea de burdel. Son manos de castellano viejo, tierno y bárbaro a la vez, como don Rodrigo Díaz de Vivar, que cebaba sus halcones de presa con las palomas de la mujer amada.

Esas manos son el instrumento perfecto de una vocación desaforada que no le ha dado un instante de paz. Obregón pinta desde antes de tener uso de razón, a toda hora, sea donde sea, con lo que tenga a mano. Una noche, por los tiempos del ahogado, habíamos ido a beber gordolobo en una cantina de vaporinos todavía a medio hacer. Las mesas estaban amontonadas en los rincones, entre sacos de cemento y bultos de cal, y los mesones de carpintería para hacer las puertas. Obregón estuvo un largo rato como en el aire, trastornado por el tufo de la trementina, hasta que se trepó en una mesa con un tarro de pintura, y de un solo trazo maestro pintó a brocha gorda en la pared limpia un unicornio verde. No fue fácil convencer al propietario de que aquel brochazo único costaba mucho más que la misma casa. Pero lo conseguimos. La cantina sin nombre siguió llamándose El Unicornio desde aquella noche, y fue atracción de turistas gringos y cachacos pendejos hasta que se la llevaron al carajo los vientos inexorables que se llevan al tiempo.

En otra ocasión, Obregón se fracturó las dos piernas en un accidente de tránsito, y durante las dos semanas de hospital esculpió sus animales totémicos en el yeso de la entablilladura con un bisturí que le prestó la enfermera. Pero la obra maestra no fue la suya, sino la que tuvo que hacer el cirujano para quitarle el yeso de las dos piernas esculpidas, que ahora están en una colección particular en Estados Unidos. Un periodista que lo visitó en su casa le preguntó con fastidio qué le pasaba a su perrita de aguas que no tenía un instante de sosiego, y Obregón le contestó: «Es que está nerviosa porque ya sabe que la voy a pintar». La pintó, por supuesto, como pinta todo lo que encuentra en todo lo que encuentra a su paso, porque piensa que todo lo que existe en el mundo se hizo para ser pintado. En su casa de virrey de Cartagena de Indias, donde todo el mar Caribe se mete por una sola ventana, uno encuentra su vida cotidiana y además otra vida pintada por todas partes: en las lámparas, en la tapa del inodoro, en la luna de los espejos, en la caja de cartón de la nevera. Muchas cosas que en otros artistas son defectos son en él virtudes legítimas, como el sentimentalismo, como los símbolos, como los arrebatos líricos, como el fervor patriótico. Hasta algunos de sus fracasos quedan vivos, como esa cabeza de mujer que se quemó en el horno de fundición, pero que Obregón conserva todavía en el mejor sitio de su casa, con medio lado carcomido y una diadema de reina en la frente. No es posible pensar que aquel fracaso no fue querido y calculado cuando uno descubre en ese rostro sin ojos la tristeza inconsolable de la mujer que nunca llegó.

A veces, cuando hay amigos en casa, Obregón se mete en la cocina. Es un gusto verlo ordenando en el mesón las mojarras azules, la trompa de cerdo con un clavel en la nariz, el costillar de ternera todavía con la huella del corazón, los plátanos verdes de Arjona, la yuca de San Jacinto, el ñame de Turbaco. Es un gusto ver cómo prepara todo, cómo lo corta y lo distribuye según sus formas y colores, y cómo lo pone a hervir a grandes aguas con el mismo ángel con que pinta. «Es como echar todo el paisaje dentro de la olla», dice. Luego, a medida que hierve, va probando el caldo con un cucharón de palo y vaciándole dentro botellas y botellas y botellas de ron de Tres Esquinas, de modo que éste termina por sustituir en la olla el agua que se evapora. Al final, uno comprende por qué ha habido que esperar tanto con semejante ceremonial de sumo pontífice, y es que aquel sancocho de la edad de piedra que Obregón sirve en hojas de bijao no es un asunto de cocina, sino pintura para comer. Todo lo hace así, como pinta, porque no sabe hacer nada de otro modo. No es que sólo viva para pintar. No: es que sólo vive cuando pinta. Siempre descalzo, con una camiseta de algodón que en otro tiempo debió servirle para limpiar pinceles y unos pantalones recortados por él mismo con un cuchillo de carnicero, y con un rigor de albañil que ya hubiera querido Dios para sus curas.

Ernest Miller Hemingway llegó por primera vez a La Habana en abril de 1928, a bordo del vapor francés *Orita*, que lo llevó de Le Havre a Cayo Hueso en una travesía de dos semanas. Lo acompañaba su segunda esposa, Pauline Pfeiffer, con quien se había casado apenas diez meses antes, y ni él ni ella debían tener por aquella ciudad del Caribe un interés mayor que el de una escala tropical de dos días después del vasto océano y el bravo invierno de Francia. Hemingway tenía treinta años, había sido corresponsal de prensa en Europa y chofer de ambulancias en la primera guerra mundial, y había publicado, con un cierto éxito, su primera novela. Pero todavía estaba lejos de ser un escritor famoso, y seguía necesitando un oficio secundario para comer y no tenía una casa estable en ninguna parte del mundo. Pauline, en cambio, era lo que entonces se llamaba una mujer de sociedad. Sobrina de un magnate norteamericano de los cosméticos, que la mimaba como a una nieta, lo tenía todo en la vida, inclusive la belleza estelar y el humor incierto de la esposa de Francis Macomber. Pero aquél no era su mejor abril. Estaba encinta y aburrida del mar, y el único deseo de ambos era llegar cuanto antes a Cayo Hueso, donde iban a instalarse para que Hemingway terminara su segunda novela: *Adiós a las armas*.

De esas 48 horas de Hemingway en La Habana no quedó ninguna huella en su obra. Es verdad que en sus artículos de prensa él solía hacer revelaciones muy inteligentes sobre los lugares que visitaba y la gente que conocía, pero entonces se había impuesto un receso como periodista para consagrarse por completo a escribir novelas. Sin embargo, seis años después escribió su primer artículo de reincidente, y era sobre un tema cubano. A partir de entonces escribió una media docena sobre su estancia en Cuba, pero en ninguno de ellos hizo revelaciones útiles para la reconstitución de su vida

Publicado originalmente el 27 de octubre de 1982.

privada, pues se referían de un modo general a su pasión dominante en aquella época: la pesca mayor. «Esta pesca», escribió en 1956, «era en otro tiempo lo que nos llevaba a Cuba.» La frase permite pensar que en el momento de escribirla, cuando ya Hemingway llevaba veinte años viviendo en La Habana, los motivos de su residencia eran más hondos o al menos más variados que el placer simple de pescar.

Cerca del bar El Floridita está el hotel Ambos Mundos, donde Hemingway alquilaba una habitación cada vez que se quedaba a dormir en tierra, y terminó por hacer de ella un sitio permanente para escribir cuando regresó de la guerra civil española. Años después, en su entrevista histórica con George Plimpton, dijo: «El hotel Ambos Mundos era un buen sitio para escribir». Cuando uno piensa en la meticulosidad con que Hemingway escogía los lugares para escribir, su preferencia por aquel hotel sólo podría tener una explicación: sin proponérselo, tal vez sin saberlo, estaba sucumbiendo a otros encantos de Cuba, distintos y más difíciles de descifrar que los grandes peces de septiembre y más importantes para su alma en pena que las cuatro paredes de su cuarto. Sin embargo, cualquier mujer que debiera esperar a que él terminara su jornada de escritor para volver a ser su esposa no podía soportar aquel cuarto sin vida. La bella Pauline Pfeiffer lo había abandonado en sus momentos más duros. Pero Martha Gellhorn, con quien Hemingway se casó poco después, encontró la solución inteligente, que fue buscar una casa donde su marido pudiera escribir a gusto, y al mismo tiempo, hacerla feliz. Fue así como encontró en los anuncios clasificados de los periódicos el hermoso refugio campestre de Finca Vigía, a pocas leguas de La Habana, que alquiló primero por cien dólares mensuales, y que Hemingway compró más tarde por 18.000 al contado. A muchos escritores que tienen casas en distintos lugares del mundo les suelen preguntar cuáles consideran como su residencia principal, y casi todos contestan que es aquella donde tienen sus libros. En Finca Vigía, Hemingway tenía 9.000 y, además, cuatro perros y 34 gatos.

Vivió en La Habana veintidós años en total. Allí pasó casi la mitad de su vida útil de escritor, y escribió sus obras mayores: parte de *Tener o no tener*, *Por quién doblan las campanas*, *Al otro lado del río y entre los árboles*, *París era una fiesta* e *Islas en el golfo* y, además, hizo incontables tentativas de la rara novela proustiana sobre el aire, la tierra y el agua, que siempre quiso escribir. Sin embargo, son esos los años menos conocidos de su vida, no sólo porque fueron los más íntimos, sino porque sus biógrafos han coincidido en pasar sobre ellos con una fugacidad sospechosa.

Cómo era ese Hemingway secreto fue la pregunta que se hizo el joven periodista cubano Norberto Fuentes, en junio de 1961, cuando su jefe de redacción lo mandó a Finca Vigía para que escribiera un artículo sobre el hombre que la semana anterior se había volado la cabeza con un tiro de rifle en el paladar. Lo único que Norberto Fuentes sabía de Hemingway en aquel momento era lo poco que su padre le había contado una tarde en que lo encontraron por casualidad en el ascensor de un hotel. En alguna ocasión –cuando no tenía más de diez años– lo vio pasar en el asiento posterior de un largo Plymount negro, y tuvo la impresión fantástica de que lo llevaban a enterrar sentado en la carroza fúnebre más conocida en las cantinas de la ciudad. A partir de aquellas vivencias fugaces, Norberto Fuentes se empeñó en la tarea colosal de averiguar cómo era el Hemingway de Cuba, que algunos de sus biógrafos póstumos parecían interesados no sólo en ocultar, sino también en tergiversar. Necesitó veinte años de pesquisas meticulosas, de entrevistas arduas, de reconstituciones que parecían imposibles, hasta rescatarlo de la memoria de los cubanos sin nombre que de veras compartieron su ansiedad cotidiana: su médico personal, los tripulantes de sus botes de pesca, sus compinches de las peleas de gallos, los cocineros y sirvientes de cantinas, los bebedores de ron en las noches de parranda de San Francisco de Paula. Permaneció meses enteros escudriñando los rescoldos de su vida en Finca Vigía, y logró descubrir los rastros de su corazón en las cartas que nunca puso en el correo, en los borradores arrepentidos, en las notas a medio escribir, en su magnífico diario de navegación, donde resplandece toda la luz de su estilo. Estableció por percepción propia que Hemingway había estado dentro del alma de Cuba mucho más de lo que suponían los cubanos de su tiempo, y que muy pocos escritores han dejado tantas huellas digitales que delaten su paso por los sitios menos pensados de la isla. El resultado final es este reportaje encarnizado y clarificador de casi setecientas páginas que acabo de leer en sus originales, y que nos devuelve al Hemingway vivo y un poco pueril que muchos creíamos vislumbrar apenas entre las líneas de sus cuentos magistrales. El Hemingway nuestro: un hombre azorado por la incertidumbre y la brevedad de la vida, que nunca tuvo más de un invitado en su mesa y que logró descifrar como pocos en la historia humana los misterios prácticos del oficio más solitario del mundo.

LA CÁNDIDA ERÉNDIRA Y SU ABUELA IRENE PAPAS

Hace muchos años, en una noche de parranda de un remoto pueblo del Caribe, conocí a una niña de once años que era prostituida por una matrona que bien hubiera podido ser su abuela. Andaba en un burdel ambulante que iba de pueblo en pueblo, siguiendo el itinerario de las fiestas patronales y llevando consigo su propia carpa, su propia banda de músicos y sus propios puestos de alcoholes y comidas. Yo tenía entonces unos dieciséis años y era consciente de que tarde o temprano sería escritor. La niña era uno de los seres más escuálidos que recuerde, y su actitud no tenía nada que ver con su oficio. Casi podía decirse que no tenía la menor idea de lo que estaba haciendo, sino que parecía repetir una lección aprendida de memoria. Su estancia en el pueblo fue sólo de tres días, pero la memoria que dejó duró mucho tiempo. Al parecer, había sido seducida a la edad de diez años por un tendero libidinoso, que le dio un plátano maduro a cambio de su virginidad, y la vieja matrona que la disfrutaba ejercía sobre ella un dominio inclemente mediante el terror.

Nunca olvidé aquel episodio, y a medida que pasaba la vida se iba definiendo en mi memoria la certidumbre de que la matrona era su abuela. Cuando escribí *Cien años de soledad* me pareció que aquel recuerdo era adecuado para la iniciación sexual del adolescente que más tarde había de convertirse en el coronel Aureliano Buendía, y así lo utilicé. En el momento de escribirlo se me ocurrió algo que era fundamental: por qué la abuela explotaba a la nieta. Y entonces supe que lo hacía para pagarse el valor de la casa que se había incendiado por culpa de un descuido de la niña.

Aunque no era la primera vez que me sucedía, me llamó la atención que aquella imagen siguiera persiguiéndome, a pesar de que ya la había utilizado. Sin embargo, no lograba sentirla como una novela, sino como un drama en imagen. Era más cine que literatura. De

Publicado originalmente el 3 de noviembre de 1982.

modo que lo escribí en forma de guión y sólo muchos años después decidí someterla a un segundo tratamiento novelizado.

El nombre de la niña, que se me ocurrió a última hora, lo había conocido en México y es un nombre tarasco: Eréndira. En cambio, nunca se me ocurrió un nombre convincente para la abuela, como no se me había ocurrido tampoco para el coronel que no tenía quien le escribiera, ni para el viejo patriarca de más de doscientos años que a veces se oía llamar Nicanor y a veces Zacarías. Parece tonto, pero está muy lejos de serlo: si el nombre no se ajusta al personaje no se lo cree nadie, y hay muchas novelas en este mundo, inclusive novelas buenas, que se desbarrancan en el olvido porque los personajes tienen nombres equivocados. Algún día, con más tiempo, quisiera hacer algunas reflexiones y contar experiencias propias en relación con los nombres de los personajes. Juan Rulfo —cuyos personajes tienen los nombres más hermosos y sorprendentes de nuestra literatura— me dijo alguna vez que él los encuentra en las lápidas de los cementerios, mezclando nombres de unos muertos con los apellidos de los otros, hasta lograr sus combinaciones incomparables: Fulgor Sedano, Matilde Arcángel, Toribio Altrete y tantos otros. Es algo tan importante que la actriz griega Irene Papas se resistía a aceptar el papel de la abuela en la película mientras yo no le pusiera un nombre. «Si no tiene un nombre no lograré sentir que soy yo», me dijo. Pero yo también fui sincero: si no sabía el nombre no podía ponerle uno cualquiera, porque corría el riesgo de que se nos volviera un personaje distinto. Irene Papas decidió entonces ponerle al personaje un nombre secreto, sólo para ella, para poder evocarlo y meterse con facilidad dentro de su pellejo. Me prometió no decirlo nunca, y si alguna vez lo dice, espero no conocerlo.

El primer tratamiento del guión cinematográfico fue escrito hace catorce años. Durante todo ese tiempo, las diferentes tentativas de realización se habían frustrado por motivos diversos. Pero todas las condiciones que siempre parecían dispersas empezaron a integrarse hace unos dos años, hasta convertirse en una aventura compacta, capaz de instalar en la realidad un sueño muy antiguo. Ruy Guerra, el director brasileño nacido en Mozambique, había esperado varios años con una paciencia de portugués hasta que el sueño estuviera completo. Ahora me parece una vivencia irreal aquella noche de quién sabe cuándo en Barcelona, cuando él y yo nos pusimos de acuerdo en que él sería el director de la película. Estábamos en una sala tan grande, él sentado en un extremo y yo sentado en el otro, que yo tenía que atravesarla cada vez que iba a servirle un trago. De modo que lo resolvimos con un sentido práctico digno de dos poetas: destapamos una botella para cada uno, y sólo cuando acabamos de tomarla dimos por terminada la conversación. Entonces eran las siete de la mañana y apenas podíamos caminar, pero ambos teníamos la convicción de que tarde o temprano haríamos la película. Desde entonces, casi como si fuera un rito memorable, Ruy Guerra y yo conservamos la costumbre de encontrarnos en cualquier parte del mundo y en los momentos menos pensados, pero siempre que nos sentamos a beber lo hacemos cada uno de su botella propia.

Yo no estaba muy seguro de que el personaje de la abuela le fuera bien a Irene Papas, que es una de las más grandes actrices de nuestro tiempo. Siempre me había imaginado a la abuela como está escrita: con una gordura inmensa y unos enormes ojos diáfanos y unos setenta años de edad. Hice todo lo posible por convencer a Simone Signoret de que tenía el tipo perfecto y que con un poco de trapos más y un maquillaje adecuado podía ganar lo que le faltaba. Pero no fue posible, y en las diversas ocasiones en que lo discutimos su argumento fue siempre el mismo y muy respetable: al cabo de una carrera larga y brillante, Simone Signoret había conseguido imponer la misma imagen que tiene en la realidad, y no quería malograrla con la encarnación de un personaje desalmado. En cambio, cuando conocí a Irene Papas en un hotel de Roma me impresionó con la fuerza devastadora de un huracán y me sedujo de inmediato su corazón de griega desmandada, pero me pareció demasiado joven y esbelta para representar a la abuela. Ruy Guerra me pidió un poco de confianza. «De acuerdo», le dije, «ya lo veremos en la pantalla.» Por el resto del reparto no hubo problema. Durante muchos años le repetí a Ruy Guerra mi convicción de que Brasil era un país lleno de Eréndiras por todas partes, y allí encontró él a Claudia Ohana, que es una réplica embellecida del original. Ulises, el adolescente holandés de la historia, apareció como hecho sobre medida con sus resplandores angélicos en una escuela de danza clásica en Alemania Federal. Todo el resto era fácil.

La semana pasada, mientras mis amigos del mundo entero celebraban mi fiesta nobiliaria, todo el interés de mi alma estaba concentrado en una hacienda en ruinas, a setenta kilómetros de San Luis Potosí, en México, donde se acaba de iniciar la filmación de la historia escrita hace catorce años. Es una empresa babélica: un autor colombiano, un director brasileño nacido en Mozambique, una actriz griega y otra brasileña, y el resto alemanes, franceses y mexicanos, en una producción franco-alemano-mexicana. Cada quien habla como puede en la lengua que puede, pero todo el mundo se entiende a través de la historia. Para mí, sin embargo, la mayor alegría me la proporcionó el tener que admitir, una vez más, que la realidad termina por imponerse a la fuerza sobre cualquier tentativa mistificadora de la imaginación. En efecto, cuando vi a Irene Papas metida en su pellejo de abuela, confirmé lo que había pensado en Roma: era demasiado joven y esbelta para el personaje inventado por mí. Pero en cambio me bastó ese mismo golpe de vista para descubrir –no sin cierta vergüenza de mí mismo– que era idéntica a aquella abuela desalmada de la realidad que conocí hace tantos años en una noche de parranda del Caribe.

USA: MEJOR CERRADO QUE ENTREABIERTO

Hace unos dieciocho años acompañé a Mercedes y a nuestros dos hijos a la ciudad fronteriza de Nuevo Laredo, donde hay un puente de hierro que tiene una pata en México y otra en Estados Unidos. Los tres pasaron al otro lado con el objetivo de solicitar una visa de reingreso a México, pues las suyas estaban vencidas. La mía también lo estaba, por supuesto, pero yo no podía acompañarlos al otro lado, porque Estados Unidos me negó inclusive un permiso simple de tres horas para atravesar el puente. El paso de gente en ambos sentidos era constante y numeroso. Como en casi todas las fronteras del mundo, hay muchos que viven en un lado y trabajan en el otro, de modo que son conocidos de los funcionarios de ambos lados y ni siquiera les exigen una identificación. Pero los controles de inmigración y aduana en los dos extremos del puente eran severos con los desconocidos, y mucho más con los no mexicanos; de modo que ni siquiera intenté convencer a nadie. Me senté en un escaño de madera que estaba frente al lado mexicano del puente y me dispuse a leer un paquete de revistas en ambos idiomas, mientras mi familia regresaba de aquel raro viaje al exterior. Tardaron menos de lo que todos suponíamos, pero antes de regresar ocurrió algo que sin duda no podré pasar por alto en mis memorias: ocurrió que Mercedes quería traerme un suéter de regalo, pero no se decidía a escoger el color. De modo que se paró frente a la puerta de una tienda del otro mundo y desde allí me mostraba los suéteres disponibles, hasta que le indiqué por señas cuál prefería. Tengo este episodio muy bien anotado, no sólo por ser tan insólito y divertido, sino porque me parece un buen ejemplo de los extremos de ridiculez a que puede conducirlo a uno la estupidez ajena.

Ésa fue la primera vez que Estados Unidos me negó una visa y, desde entonces, cada viaje mío a ese país —con permisos provisionales y condicionados— ha dado origen a incidentes extraños. Para

Publicado originalmente el 10 de noviembre de 1982.

empezar, nunca he podido saber por qué fui declarado inaceptable para entrar en Estados Unidos. En 1959, cuando solicité en Bogotá mi primera visa para trabajar como corresponsal de la agencia cubana de noticias en Nueva York, me dieron de inmediato una tarjeta de residente; disfruté de ella durante casi un año, hasta que abandoné la agencia y me vine a México. Un funcionario de la embajada de Estados Unidos en este país me encontró sin dificultad y me pidió devolver las tarjetas de residentes de toda la familia. Me sorprendió la eficacia con que encontraron mi dirección, así como había de sorprenderme después que no la encontraran nunca para devolverme los dólares excedentes de la última liquidación de impuestos que había hecho en Nueva York.

Durante más de diez años fueron inútiles todos mis esfuerzos para que me concedieran la visa o, al menos, para que alguien me explicara cuál era el motivo de mi inelegibilidad. Un amigo que creyó descifrar un código secreto de la embajada donde trabajaba me dijo el motivo: actos terroristas en Camerún. No me sorprendió, porque estoy acostumbrado a esta clase de disparates, sobre todo teniendo conciencia de que siempre he sido un enemigo público del terrorismo y que nunca en mi vida he estado en Camerún. Sin embargo, la razón oficial, que distintos consulados me han repetido muchas veces a lo largo de tantos años, ha sido siempre la misma. Se me atribuye el cargo frívolo de pertenecer, o haber pertenecido, a un partido comunista o a alguna organización afiliada. Podría ser cierto, y no tendría nada de que arrepentirme; pero el caso es que no lo es. Nunca he pertenecido a ningún partido de ninguna clase.

La primera vez que aceptaron darme una visa de una semana, y circunscrita a la isla de Manhattan, fue en 1971, cuando la Universidad de Columbia, de Nueva York, me ofreció el grado de doctor *honoris causa* en Letras. Mi alegría de volver a Nueva York se ensombreció mucho por otro incidente tan divertido como lamentable. El Departamento de Estado, temiendo que las autoridades de inmigración del aeropuerto de Nueva York hicieran algo indebido que pudiera repercutir en la prensa, mandó desde Washington un funcionario, que debía recibirme a las ocho de la noche en el aeropuerto, acompañarme luego al hotel y regresar de inmediato en el avión más próximo, para estar al día siguiente en su oficina. Sólo que mi avión no iba desde Francfort, sino desde Barranquilla (Colombia), y no llegó a las ocho de la noche, sino a las cuatro de la madrugada.

Encontré al pobre hombre muerto de hambre y de sueño, después de haber leído casi tres veces durante la espera una versión inglesa de *El coronel no tiene quien le escriba*. Lo había conseguido al menos para saber quién era y qué había escrito el hombre que iba a recibir en el aeropuerto. Al amanecer, cuando me dejó en el hotel, quise ponerle un autógrafo en el libro, pero él me confesó avergonzado que era de una biblioteca circulante y no se podía escribir nada en sus páginas. Salió disparado, tratando de alcanzar un avión del alba que le permitiera estar a tiempo en su oficina, y me dejó con el mal sabor de haberle estropeado una noche completa a un pobre empleado público, mal pagado y sin ningún sentido del humor, y que no tenía nada que ver con la imbecilidad de los burócratas que no se atrevían a concederme la visa completa ni a negármela completa.

Una de las cosas que me gustan menos de los gringos es su conciencia de pecadores. Viven enredados con ella. Y donde más se nota es, por cierto, en este problema que ellos mismos se han creado con sus visas a escritores y artistas latinoamericanos. Tengo incontables amigos cuya entrada les ha sido prohibida en Estados Unidos. Invitado perpetuo de las universidades y otros organismos culturales norteamericanos, Julio Cortázar tiene que someterse a toda clase de vueltas cada vez que quiere cumplir un compromiso en ese país. Sin embargo, el único cargo que pueden hacerle –aparte del de ser un escritor que piensa con su propia cabeza– es que siempre ha sido partidario de la revolución cubana y ahora lo es del proceso de Nicaragua. Carlos Fuentes, cuyas ideas políticas las proclama él mismo cada vez que puede, aun dentro de Estados Unidos, es un inelegible a quien le conceden un permiso provisional muy limitado. Son muchos los escritores, artistas y profesores de América Latina que no son víctimas del mismo sistema de discriminación. Es decir, se nos permite la entrada a Estados Unidos cuando vamos a prestar algún servicio. Si no, nos la niegan con el argumento revenido de los vínculos comunistas.

En ese sentido, los casos de la crítica de arte argentina Marta Traba y del profesor y crítico uruguayo Ángel Rama constituyen un escándalo muy especial. Al cabo de varios años de excelentes servicios en la Universidad de Maryland, se les ha notificado sin más vueltas que deben abandonar el país. A Ángel Rama se le ofrece la opción más humillante: apelar como defensor y prometer mediante declaración jurada que renuncia a su pretendida vocación comunista. A Marta Traba le niegan inclusive esa opción.

Todo esto me parece no sólo estúpido, sino además inconsecuente: si nos impiden la entrada a nosotros, sería racional que se la impidieran también a nuestros libros, pues si los talentos ocultos del

Departamento de Justicia lo pensaran dos veces se darían cuenta de algo que ya Hitler había descubierto, y es que los libros son más peligrosos que quienes los escriben. El hecho de que esto no les importe a los gobiernos de Estados Unidos permite pensar que la prohibición de ingreso no es un acto defensivo de la sociedad norteamericana, como sus gobernantes dicen, sino que es un simple castigo imperial contra sus críticos.

LA PENUMBRA DEL ESCRITOR DE CINE

En Fregene, una localidad marina cerca de Roma, murió hace poco mi muy querido amigo Franco Solinas, uno de los escritores de cine mejor calificados de nuestro tiempo. Creo que no alcanzó a terminar su último guión, que había trabajado junto con el director Costa Gavras, sobre el tema actual y apasionante del pueblo sin tierra de Palestina. Varios directores de renombre mundial debieron de quedarse esperando su turno, pues solían aceptar las esperas largas e imprevisibles a que les obligaban los numerosos compromisos de Franco Solinas. Éste era, de todos modos, un caso raro en su medio: no aceptó nunca trabajar en más de un guión al mismo tiempo, y a ése consagraba toda su energía, su paciencia infinita y su autocrítica implacable, durante un tiempo que era imposible calcular de antemano. Un año de trabajo diario era su promedio para cada guión. Su obra maestra fue, sin duda, *La batalla de Argel*, que escribió para el director Gillo Pontecorvo, para quien escribió también *La queimada*. Para Costa Gavras escribió *Estado de sitio*, y para Joseph Losey escribió *Mr. Klein*. La lista de sus películas no es muy larga, pero es de muy alta calidad. Para mi gusto, fue uno de los profesionales más rigurosos en uno de los oficios más difíciles y menos servidos, y también de los más ingratos. Prueba de esto último es que la noticia de la muerte de Franco Solinas ha pasado casi inadvertida, aun para las publicaciones especializadas, y muy pocos amigos personales y admiradores sabemos a ciencia cierta lo que hemos perdido.

Ésta es, en todo caso, una oportunidad para reflexionar sobre el destino de penumbra de los escritores de cine. Nadie sabe quiénes son, a menos que sean conocidos como escritores de otra cosa, y en este caso, hasta ellos mismos tienen la tendencia a pensar que su trabajo para el cine es secundario. Un recurso para comer. Las revistas de cine se fijan, por encima de todo, en el director —no sin razón—

Publicado originalmente el 17 de noviembre de 1982.

y pocas veces recuerdan que, antes de llegar a la pantalla, toda película tiene que haber pasado por la prueba de fuego de la letra escrita. De modo que son los escritores, y no los directores, quienes suministran la base literaria que sustenta la película. Lo cual, por cierto, no está bien ni para la literatura ni para el cine.

Después de la segunda guerra mundial, los escritores de cine vivieron su cuarto de hora con la aparición en primer plano del guionista Cesare Zavattini, un italiano imaginativo y con un corazón de alcachofa, que le infundió al cine de su época un soplo de humanidad sin precedentes. El director que realizó sus mejores argumentos fue Vittorio de Sica, su gran amigo, y estaban tan identificados que no era fácil saber dónde terminaba uno y dónde empezaba el otro. Fueron ellos las dos estrellas mayores del neorrealismo, en cuyo cielo había otras tan radiantes como Roberto Rossellini. Juntos hicieron *Ladrones de bicicletas*, *Milagro en Milán*, *Umberto D* y otras inolvidables. Se hablaba entonces de las películas de Zavattini como se habla de las películas de Bertolucci: como si aquél fuera el director. En la práctica, fueron muy pocas las películas italianas de aquellos tiempos cuyos guiones no pasaron por el rastrillo purificador de Zavattini, quien aparecía siempre en el último lugar de los créditos sólo porque éstos eran dados por orden alfabético. Su fecundidad era tal que, dicen quienes lo conocían en ese tiempo, tenía un archivador enorme atiborrado de argumentos sintéticos en tarjetas. Los productores, siempre escasos de temas, acudían a él desesperados. En alguna ocasión, uno de ellos le pidió con urgencia una historia de amor, y Zavattini le preguntó muy en serio: «¿La quiere sin perrito o con perrito?». Toda una generación fanática del cine se fue a estudiar en el Centro Experimental de Cinematografía, en Roma, con la esperanza de que fuera Zavattini quien lo enseñara.

Fue una excepción. En realidad, el destino del escritor de cine está en la gloria secreta de la penumbra, y sólo el que se resigne a ese exilio interior tiene alguna posibilidad de sobrevivir sin amargura. Ningún trabajo exige una mayor humildad. Más aún: debe considerarse como un factor transitorio en la creación de la película y es una prueba viviente de la condición subalterna del arte del cine. Mientras éste necesite de un escritor, o sea, del auxilio de un arte vecino, no logrará volar con sus propias alas. Ése es uno de sus límites. El otro, y todavía más grave, por supuesto, es su compromiso industrial. El propio director termina por darse cuenta, tarde o temprano, de que tampoco es mucho lo que él puede hacer dentro del estrecho margen de creación que le dejan las cuentas del productor, por un lado, y los fantasmas prestados del escritor, por el otro. Es un milagro que todavía pueda tener la impresión de que ha logrado expresarse a fondo dentro de ese callejón enrarecido. Por eso me asombra tanto, y me alegra tanto, cada vez que encuentro una película capaz de hacerme llorar, que es lo que uno va buscando en el fondo de su alma cuando se apagan las luces de la sala.

En estos días de tantas entrevistas, una pregunta que se ha repetido sin tregua es la de mis relaciones con el cine. Mi única respuesta ha sido la misma de siempre: son las de un matrimonio mal avenido. Es decir, no puedo vivir sin el cine ni con el cine, y, a juzgar por la cantidad de ofertas que recibo de los productores, también al cine le ocurre lo mismo conmigo. Desde muy niño, cuando el coronel Nicolás Márquez me llevaba en Aracataca a ver las películas de Tom Mix, surgió en mí la curiosidad por el cine. Empecé, como todos los niños de entonces, por exigir que me llevaran detrás de la pantalla para descubrir cómo eran los intestinos de la creación. Mi confusión fue muy grande cuando no vi nada más que las mismas imágenes al revés, pues me produjo una impresión de círculo vicioso de la cual no pude restablecerme en mucho tiempo. Cuando por fin descubrí cómo era el misterio, me atormentó la idea de que el cine era un medio de expresión más completo que la literatura, y esa certidumbre no me dejó dormir tranquilo en mucho tiempo. Por eso fui uno de los tantos que viajaron a Roma con la ilusión de aprender la magia secreta de Zavattini, y también uno de los que apenas lograron verlo a distancia. Ya para entonces había ganado en Colombia una batalla para el cine. Cuando llegué a *El Espectador* de Bogotá, en 1954, la única crítica de cine posible en el país era la complaciente. De no ser así, los exhibidores amenazaban con suspender los anuncios de las películas, que son para la prensa una apreciable fuente de ingresos. Con el respaldo de los directores del periódico, que asumieron el riesgo, escribí entonces la primera columna regular de crítica de cine durante un año. Los exhibidores, que al principio asimilaron mis notas desfavorables como si fuera aceite de ricino, terminaron por admitir la conveniencia de contar con un público bien orientado.

Fue también por la ilusión de hacer cine que vine a México hace más de veinte años. Aun después de haber escrito guiones que luego no reconocía en la pantalla, seguía convencido de que el cine sería la válvula de liberación de mis fantasmas. Tardé mucho tiempo para convencerme de que no. Una mañana de octubre de 1965, cansado de verme y no encontrarme, me senté frente a la máquina de escribir, como todos los días, pero esa vez no volví a levantarme sino al cabo de dieciocho meses, con los originales terminados de *Cien años de soledad*. En aquella travesía del desierto comprendí que no había un acto más espléndido de libertad individual que sentarme a inventar el mundo frente a una máquina de escribir.

EL LUJO DE LA MUERTE

He dicho muchas veces que no tengo corazón para enterrar a los amigos. Sin embargo, el pasado 2 de noviembre, día de todos los muertos, quise acompañar a la esposa de uno muy querido que sería incinerado en el improbable panteón de las Lomas. El cuerpo había pasado la noche en el motel funerario que tiene la agencia Gayoso en la avenida Félix Cuevas de la ciudad de México, la cual había hecho los trámites de la incineración y el transporte final hasta el horno crematorio. La cita era a las once de la mañana y todos suponíamos que sería un acto más bien técnico, sin ceremonias de ninguna clase, que no podía durar más de dos horas. Pero al llegar al panteón nos hicieron ver que había otros cadáveres en turno, y que el de nuestro amigo tenía que esperar por lo menos hasta las cinco de la tarde. En la lúgubre sala de espera, helada, sin una flor y sin un escaño miserable donde sentarse, estaban alineados contra la pared en posición vertical los ataúdes usados de los que habían tenido la precaución de morirse más temprano. Aquellos ataúdes habían sido vendidos por las agencias funerarias y habían servido para la velación y el transporte, pero era obvio que los deudos que los habían pagado a precio de oro no tenían nada que hacer con ellos, de modo que alguien se encargaría de venderlos otra vez para otros muertos futuros. El chofer de la carroza que había llevado el cuerpo de nuestro amigo, dijo: «¿Por qué no vuelven mañana y tratan de ser los primeros?». Esa sola pregunta, formulada por alguien que sin duda conocía mejor que nosotros estos dramas de la burocracia fúnebre, nos hizo vislumbrar de pronto cuál era la clase de día que nos esperaba.

Ana María Pecanins se hizo cargo de la situación, y ha relatado aquella experiencia en una carta a la prensa que no debía pasar inadvertida, porque es apenas un botón de muestra del desamparo en que se encuentran los sobrevivientes frente a las agencias funerarias después de que los servicios han sido pagados. Hace unos meses,

Publicado originalmente el 24 de noviembre de 1982.

también Fernando Benítez contó en un periódico cómo habían sido tratados por Gayoso los parientes de un escritor que no tenían dinero para pagar la cuenta de los funerales, una cuenta tal vez mayor que la suma total de derechos de autor percibidos en toda su vida por el amigo muerto. La revista del Instituto Nacional del Consumidor también se ha ocupado en varias ocasiones del precio desmesurado de la muerte en México, pero su prédica, como tantas otras sobre otros temas mortales, se ha perdido para siempre en el desierto. Es como si las agencias funerarias en el mundo entero gozaran de un fuero especial que las pusiera a salvo de cualquier sanción por sus abusos.

Ana María Pecanins ha contado que el único funcionario que encontró en el crematorio le dio una explicación tan realista que más parecía de un panadero: «El horno está ocupado», le dijo, «el horneador está dentro y no terminará de hornear en tres horas». No hubo más información. Ana María llamó entonces a la agencia Gayoso pensando obtener un auxilio suplementario después de haber pagado los servicios completos, y un empleado que dijo llamarse Ricardo López le informó que la responsabilidad de la empresa termina en el momento en que el cadáver sale de la casa funeraria. Punto: colgó el teléfono. Ana María, con su temeridad catalana, volvió a marcar el mismo número, y esa vez le contestó otro funcionario, quien le explicó con la voz colorida de los comerciantes de la muerte que nada podía hacerse para apresurar la incineración. «Por desgracia», dijo, sin saber acaso que estaba inventando un proverbio desolador, «la suerte es de los que llegan primero.» No hubo, en efecto, nada que hacer. El servicio, el apoyo y la comprensión contratados por los vendedores de la muerte que prometen hasta la entrada al cielo con trompetas angélicas, habían cesado para siempre.

Aquél había sido un drama más, y de los menos graves, de cuantos ocurren a cada minuto en el mundo por la voracidad y el corazón de piedra de las agencias funerarias. En México, donde el negocio de la muerte es uno de los más despiadados y de los más fructíferos, los abusos suelen invadir los territorios más esquivos de la literatura fantástica. «El servicio dura apenas diez o quince minutos máximo», dice el folleto de propaganda de una agencia funeraria. «No es deprimente, puede ir uno hasta de día de campo. Es muy bonito. No es un panteón tradicional, es muy moderno, está alfombrado, tiene luz, vitrales, aire acondicionado y cuenta con filtros de ventilación dentro de las criptas.»

El Instituto del Consumidor ha calculado que existen en México 195 agencias funerarias con registro legal, y 110 que actúan de un modo casi clandestino. Sobre todo estas últimas, que se rigen más bien por las leyes de la oferta y la demanda coyunturales que por una tarifa establecida, participan en una pavorosa rebatiña de cadáveres en las puertas y corredores de los hospitales. Pero aun en las funerarias de los ricos, los agentes vendedores carecen de una norma precisa para establecer los precios del servicio. Se guían más bien por el aspecto y el estado del cliente en el momento de cerrar el negocio. El precio del ataúd determina el valor de todo el servicio, y no es posible combinar un ataúd caro con un servicio modesto, o al contrario. Al fin y al cabo, la muerte no es más que un viaje, por muy eterno que sea, y las agencias no han encontrado una razón para no organizar sus servicios como las excursiones turísticas en las que todo va incluido, hasta las posibilidades del amor ocasional. El negocio es fabuloso: en 1976, sólo las funerarias legales de México se ganaron 175 millones de pesos, equivalentes a un 76 % de utilidades en relación con sus costos de operación.

La concepción nos viene de Estados Unidos y es muy simple: el lujo de la muerte es de primera necesidad. El norteamericano medio no tiene en ningún momento un nivel de vida más alto que el nivel de su muerte. Ni nunca es más bello que en el ataúd: sus propios parientes se asombran de cuánto les favorece el embalsamamiento, con cuánta ternura sonríen y cuán comprensivos y amorosos parecen con la cabeza apoyada en las almohadas de la muerte, y tal vez se duelan en secreto de que no se hubiera inventado la posibilidad de embalsamar en vida a los seres difíciles. Pero es una ilusión que cuesta caro, y detrás de ella prospera uno de los comercios más descorazonados y sucios del mundo. Hace muchos años, en un libro fascinante sobre el tráfico funerario en Estados Unidos, leí una anécdota de horror. Una viuda de clase media había invertido sus últimos ahorros para darle a su marido muerto unos funerales más lujosos que el de sus posibilidades reales. Todo parecía acordado, cuando un funcionario de la agencia mortuoria le llamó por teléfono para decirle que el cadáver era más alto de lo previsto en el contrato, y que ella debía pagar en consecuencia una suma suplementaria. La viuda no tenía un centavo más. Entonces el funcionario, con la voz melodiosa de los de su oficio, le dio la solución. «En ese caso», dijo, «le suplico darnos la autorización para serrucharle los pies al cadáver.» La pobre viuda, por supuesto, encontró donde pudo el dinero que no tenía, sólo para que la agencia le hiciera la caridad de enterrar entero a su marido.

BUENO, HABLEMOS DE MÚSICA

En una de esas encuestas que proliferan a diario me han preguntado, como tantas veces, cuál es la música que me llevaría, si sólo pudiera llevarme un disco a una isla desierta. No he dudado un instante la respuesta: las *Suites para chelo solo*, de Juan Sebastián Bach. Y si sólo pudiera llevarme una de ellas, escogería la número uno. Conozco distintas versiones, y entre ellas, por supuesto, la de Pau Casals. Además de su valor histórico es una versión excelente, pero la grabación es tan antigua que es mucho lo que se pierde de su excelencia. En realidad, la versión que más me conmueve es la de Maurice Gendron, y por consiguiente sería ésta la que me llevaría a la isla desierta, junto con un libro único: una buena antología de la poesía española del Siglo de Oro.

Este tema me ofrece la oportunidad de contestar a otra pregunta que los periodistas me hacen con frecuencia sobre mis relaciones con la música. Les contesto siempre la verdad: la música me ha gustado más que la literatura, hasta el punto de que no logro escribir con música de fondo porque le presto más atención a ésta que a lo que estoy escribiendo. Sin embargo, nunca voy mucho más lejos en mis explicaciones, entre otras cosas porque tengo la impresión de que mi vocación musical es tan entrañable que forma parte de mi vida privada. Por lo mismo, cuando estoy solo con mis amigos muy íntimos no hay nada que me guste más que hablar de música. Jomi García Ascot, que es uno de estos amigos, publicó un libro excelente sobre sus experiencias de melómano empedernido, y allí incluyó una frase que me oyó decir alguna vez: «Lo único mejor que la música es hablar de música». Sigo creyendo que es verdad.

Lo raro es que cuando uno dice que le gusta la música se piensa casi siempre en la música que por pura pereza mental se ha dado en llamar música clásica. También se la llama música culta, lo que no resuelve el problema, pues pienso que la música popular también es

Publicado originalmente el 1 de diciembre de 1982.

culta, aunque de una cultura distinta. Aun la simple música comercial, que no siempre es tan mala como suelen decir los sabios de salón, tiene derecho a llamarse culta, aunque no sea el producto de la misma cultura de Mozart. Al fin y al cabo, los grandes maestros de todos los tiempos saben que el manantial más rico de su inspiración es la música popular. La foto más conmovedora en la vasta y hermosa iconografía de Béla Bartók es una en que aparece recogiendo una canción de labios de una campesina con una grabadora de cilindro, que nada tenía que envidiar a la primera que construyó Edison, y en la cual quedaron grabadas para la historia las preciosas líneas del *Corderito de María*.

Todo esto para mí es más simple: música es todo lo que suena, y el trabajo de establecer si es buena o mala es posterior. Tengo más discos que libros, pero muchos amigos, sobre todo los más intelectuales, se sorprenden de que la lista en orden alfabético no termine con Vivaldi. Su estupor es más intenso cuando descubren que lo que viene después es una colección de música del Caribe –que es, de todas, sin excepción, la que más me interesa. Desde las canciones ya históricas de Rafael Hernández y el trío Matamoros, los tamboritos de Panamá, los polos de la isla de Margarita, en Venezuela, o los merengues de Santo Domingo. Y, por supuesto, la que más ha tenido que ver con mi vida y con mis libros: los cantos vallenatos de la costa del Caribe de Colombia, de los cuales habría que hablar un día de estos en una nota distinta. Jamaica y la Martinica tienen una música grande, y fue Daniel Santos quien divulgó algunas canciones que estuvieron de moda hace muchos años sin que casi nadie supiera que eran de Curaçao con letra en papiamento. Debo decir, sin embargo, que la canción más bella que escuché jamás en esa región alucinada fue la que cantaba una niña indígena de unos nueve años en las islas San Blas de Panamá. La niña cantaba con una hermosa voz primitiva, acompañándose con una sola maraca, mientras se mecía a grandes bandazos en la misma hamaca donde dormía un niño de pocos meses. Me quedé como extasiado, flotando en la magia de la canción y lamentando con el alma no haber llevado conmigo una grabadora. Nuestro guía local nos dijo –sin pretender ningún juego de palabras– que era una canción de cuna de los indios cunas. Fue tanta mi impresión que al día siguiente le conté mi emoción al general Omar Torrijos para que me facilitara el regreso a las islas con una grabadora, pero él me disuadió con su raro y demoledor sentido común. «No vuelvas más», me dijo, «que esas cosas suceden una sola vez en la vida.» No volví, por supuesto, pero la certidumbre de que nunca más volveré a escuchar aquella canción es una de las muy pocas amarguras de mi vida.

Tengo versiones inencontrables en ningún lugar del Caribe, que, sin embargo, las he encontrado donde menos podía imaginarse: en los mercados de discos latinos de la calle Catorce de Nueva York. Tengo discos de salsa, desde luego, pero con la conciencia de que no es una música nueva, sino la continuación exiliada y sofisticada para bien de la música tradicional de Cuba. Como lo dijo hace pocos días en una entrevista Dámaso Pérez Prado, el inmortal, que es uno de mis ídolos más antiguos y tenaces, como debe constar en los archivos de los periódicos en que escribí mis primeras notas. Me alegra comprobar, por otra parte, que mi pasión por la música del Caribe está bien correspondida. Hace unos años recibí en Barcelona un telegrama de alguien que solicitaba mi ayuda para escribir sus memorias y que se firmaba con el seudónimo de *El Inquieto Anacobero*. Un seudónimo cuyo titular es conocido de todo el Caribe: Daniel Santos, el jefe. Más tarde me llamó por teléfono desde Nueva York mi amigo Rubén Blades para decirme que quería cantar algunos de mis cuentos, y yo le contesté que encantado, inclusive por la curiosidad de saber qué clase de transposición endiablada podía quedar de semejante aventura. Lo digo sin ironía: nada me hubiera gustado en este mundo como haber podido escribir la historia hermosa y terrible de Pedro Navajas. Por último, en el reciente aluvión telefónico que estremeció mi casa de México, una de las primeras llamadas fue la del otro gigante de la canción, Nelson Ned. Hace pocos años perdí la amistad de algunos escritores sin sentido del humor porque declaré en una entrevista –pensándolo de veras– que uno de los más grandes poetas actuales de la lengua castellana era mi amigo Armando Manzanero.

Hablar de música sin hablar de los boleros es como hablar de nada. Pero también eso es motivo para una nota distinta, y tal vez interminable. En este género, Colombia tiene un mérito que sólo Chile le disputa, y es haberse mantenido fiel al bolero a través de todas las modas, y con una pasión que sin duda nos enaltece. Por eso debemos sentirnos justificados con la noticia cierta de que el bolero ha vuelto, que los hijos les están pidiendo con urgencia a sus padres que les enseñen a bailarlo para no ser menos que los otros en las fiestas del sábado, y que las viejas voces de otros tiempos regresan al corazón en los homenajes más que justos que se rinden en estos días a la memoria inmemorial de Toña la Negra. Sin embargo, y sin ninguna duda, mi respuesta a la pregunta de siempre fue muy bien pensada y sincera: el disco que me llevaría a una isla desierta es la *Suite número uno para chelo solo*, de Juan Sebastián Bach. Terco que es uno.

LA LITERATURA SIN DOLOR

Hace poco incurrí en la frivolidad de decirle a un grupo de estudiantes que la literatura universal se aprende en una tarde. Una muchacha del grupo –fanática de las bellas letras y autora de versos clandestinos– me concretó de inmediato: «¿Cuándo podemos venir para que nos enseñe?». De modo que vinieron el viernes siguiente a las tres de la tarde y hablamos de literatura hasta las seis, pero no pudimos pasar del romanticismo alemán, porque también ellos incurrieron en la frivolidad de irse para una boda. Les dije, por supuesto, que una de las condiciones para aprender toda la literatura en una tarde era no aceptar al mismo tiempo una invitación para una boda, pues para casarse y ser felices hay mucho más tiempo disponible que para conocer la poesía. Todo había empezado y continuado y terminado en broma, pero al final yo quedé con la misma impresión que ellos: si bien no habíamos aprendido la literatura en tres horas, por lo menos nos habíamos formado una noción bastante aceptable sin necesidad de leer a Jean-Paul Sartre.

Cuando uno escucha un disco o lee un libro que le deslumbra, el impulso natural es buscar a quién contárselo. Esto me sucedió cuando descubrí por casualidad el *Quinteto para cuarteto de cuerdas y piano*, de Béla Bartók, que entonces no era muy conocido, y me volvió a suceder cuando escuché en la radio del automóvil el muy bello y raro *Concierto gregoriano para violín y orquesta*, de Ottorino Respighi. Ambos eran muy difíciles de encontrar, y mis amigos melómanos más cercanos no tenían noticias de ellos, de modo que recorrí medio mundo tratando de conseguirlos para escucharlos con alguien. Algo similar me está sucediendo desde hace muchos años con la novela *Pedro Páramo*, de Juan Rulfo, de la cual creo haber agotado ya una edición entera sólo por tener siempre ejemplares disponibles para que se los lleven los amigos. La única condición es que nos volvamos a encontrar lo más pronto posible para hablar de aquel libro entrañable.

Publicado originalmente el 8 de diciembre de 1982.

Por supuesto, lo primero que les expliqué a mis buenos estudiantes de literatura fue la idea, tal vez demasiado personal y simplista, que tengo de su enseñanza. En efecto, siempre he creído que un buen curso de literatura no debe ser más que una guía de los buenos libros que se deben leer. Cada época no tiene tantos libros esenciales como dicen los maestros que se complacen en aterrorizar a sus alumnos, y de todos ellos se puede hablar en una tarde, siempre que no se tenga un compromiso ineludible para una boda. Leer estos libros esenciales con placer y con juicio es ya un asunto distinto para muchas tardes de la vida, pero si los alumnos tienen la suerte de poder hacerlo terminarán por saber tanto de literatura como el más sabio de sus maestros. El paso siguiente es algo más temible: la especialización. Y un paso más adelante es lo más detestable que se puede hacer en este mundo: la erudición. Pero si lo que desean los alumnos es lucirse en las visitas, no tienen que pasar por ninguno de esos tres purgatorios, sino comprar los dos tomos de una obra providencial que se llama *Mil libros*. La escribieron Luis Nueda y don Antonio Espina, allá por 1940, y allí están resumidos por orden alfabético más de un millar de libros básicos de la literatura universal, con su argumento y su interpretación, y con noticias impresionantes de sus autores y su época. Son muchos más libros, desde luego, de los que harían falta para el curso de una tarde, pero tienen sobre éstos la ventaja de que no hay que leerlos. Ni tampoco hay que avergonzarse: yo tengo estos dos tomos salvadores en la mesa donde escribo, los tengo desde hace muchos años, y me han sacado de graves apuros en el paraíso de los intelectuales, y por tenerlos y conocerlos puedo asegurar que también los tienen y los usan muchos de los pontífices de las fiestas sociales y las columnas de periódicos.

Por fortuna, los libros de la vida no son tantos. Hace poco, la revista *Pluma*, de Bogotá, le preguntó a un grupo de escritores cuáles habían sido los libros más significativos para ellos. Sólo decían citarse cinco, sin incluir a los de lectura obvia, como la Biblia, *La Odisea* o *El Quijote*. Mi lista final fue ésta: *Las mil y una noches*; *Edipo rey*, de Sófocles; *Moby-Dick*, de Melville; *Floresta de la lírica española*, que es una antología de don José María Blecua que se lee como una novela policíaca, y un *Diccionario de la lengua castellana* que no sea, desde luego, el de la Real Academia. La lista es discutible, por supuesto, como todas las listas, y ofrece tema para hablar muchas horas, pero mis razones son simples y sinceras: si sólo hubiera leído esos cinco libros —además de los obvios, desde luego—, con ellos me habría bastado para escribir lo que he escrito. Es decir, es una lista de carácter profesional. Sin embargo, no llegué a *Moby-Dick* por un camino fácil. Al principio había puesto en su lugar a *El conde de Montecristo*, de Alejandro Dumas, que, a mi juicio, es una novela perfecta, pero sólo por razones estructurales, y este aspecto ya estaba más que satisfecho por *Edipo rey*. Más tarde pensé en *Guerra y paz*, de Tolstoi, que, en mi opinión, es la mejor novela que se ha escrito en la historia del género, pero en realidad lo es tanto que me pareció justo omitirla como uno de los libros obvios. *Moby-Dick*, en cambio, cuya estructura anárquica es uno de los más bellos desastres de la literatura, me infundió un aliento mítico que sin duda me habría hecho falta para escribir.

En todo caso, tanto el curso de literatura en una tarde como la encuesta de los cinco libros conducen a pensar, una vez más, en tantas obras inolvidables que las nuevas generaciones han olvidado. Tres de ellas, hace poco más de veinte años, eran de primera línea: *La montaña mágica*, de Thomas Mann; *La historia de San Michele*, de Axel Munthe, y *El gran Meaulnes*, de Alain Fournier. Me pregunto cuántos estudiantes de literatura de hoy, aun los más acuciosos, se han tomado siquiera el trabajo de preguntarse qué puede haber dentro de estos tres libros marginados. Uno tiene la impresión de que tuvieron un destino hermoso, pero momentáneo, como algunos de Eça de Queiroz y de Anatole France, y como *Contrapunto*, de Aldous Huxley, que fue una especie de sarampión de nuestros años azules; o como *El hombrecillo de los gansos*, de Jacobo Wassermann, que tal vez le deba más a la nostalgia que a la poesía; o como *Los monederos falsos*, de André Gide, que acaso fueran más falsos de lo que pensó su propio autor. Sólo hay un caso sorprendente en este asilo de libros jubilados, y es el de Hermann Hesse, que fue una especie de explosión deslumbrante cuando le concedieron el Premio Nobel en 1946, y luego se precipitó en el olvido. Pero en estos últimos años sus libros han sido rescatados con tanta fuerza como antaño por una generación que tal vez encuentra en ellos una metafísica que coincide con sus propias dudas.

Claro que todo esto no es preocupante sino como enigma de salón. La verdad es que no debe haber libros obligatorios, libros de penitencia, y que el método saludable es renunciar a la lectura en la página en que se vuelva insoportable. Sin embargo, para los masoquistas que prefieran seguir adelante a pesar de todo hay una fórmula certera: poner los libros ilegibles en el retrete. Tal vez con varios años de buena digestión puedan llegar al término feliz de *El paraíso perdido*, de Milton.

CENA DE PAZ EN HARPSUND

Al cabo de casi dos horas de viaje nocturno por una carretera glacial, llegamos a la residencia campestre de Harpsund, donde el primer ministro de Suecia, Olof Palme, nos había invitado a cenar aquella tranquila noche del 9 de diciembre. Mercedes y yo estábamos preparados para descubrir entre la bruma un castillo medieval de aquellos de los cuentos de Andersen, y nos encontramos en cambio con una casa muy sencilla y limpia junto a un lago dormido en el hielo, y en medio de un prado apacible donde había otras casas iguales para los invitados. Aquel conjunto es la residencia campestre de los primeros ministros de Suecia.

Todos los actos que aquella semana agotadora se llevaron a cabo en Estocolmo terminaban por convertirse en homenajes públicos a la América Latina. Algunos espíritus puros de Colombia abrigaban el temor provinciano de que nuestra delegación cultural fuera a hacer el ridículo en la muy civilizada Escandinavia. Lo que hizo, en cambio, no fue sólo una labor excelente de afirmación cultural, sino una demostración emocionante de que nuestra identidad es ya bastante específica como para ser exportada sin reservas. La propia reina Silvia, que está en la vida real con los pies sobre la tierra, me habló de su pesar por no haber tenido ocasión ni tiempo para aprender a bailar la cumbia con el conjunto de nuestra delegación cultural. Me dijo que la había bailado una vez y deseaba descifrar a fondo el secreto de esa danza tan nuestra, cuya elegancia natural dejó en Suecia un rastro de dignidad y buen gusto. Tal vez nuestro único mérito haya sido ése: haber tenido el decoro de mostrarnos tal como somos, y no como quisiéramos que los otros creyeran que somos.

La cena en la casa campestre del primer ministro Olof Palme no fue una excepción: también aquella reunión, que había sido despojada de todo carácter oficial y se ofrecía como un encuentro entre dos viejos amigos, terminó convertida por la misma dinámica de los

Publicado originalmente el 22 de diciembre de 1982.

hechos en un homenaje a la América Latina. Era un grupo muy
reducido de amigos comunes. Allí estaba la señora Danielle Mitterrand, la esposa del presidente de Francia, que no oculta su satisfacción de ser el alma del comité francés de solidaridad con El Salvador. Estaban Regis Debray y Pierre Schori, francés el primero y
sueco el segundo, pero ambos vinculados de modo muy estrecho
a la América Latina. Había un grupo muy escogido de escritores
suecos, entre ellos el presidente del Pen Club Internacional, y nuestro muy querido Sven Linqvist, autor de un estudio muy serio y
muy bien divulgado sobre las relaciones entre la propiedad de la
tierra y el poder político en América Latina. Estaba, por último, el
antiguo primer ministro turco, Bulen Ecevit, un hombre de brazo
fuerte y corazón generoso, que cumplió varios meses de cárcel después de ser derrocado, y que hasta la semana pasada carecía de permiso para salir de su país. Olof Palme le invitó a esta cena íntima,
pero no como político, sino como poeta, que, según él mismo ha
dicho, es su vocación dominante. En su breve y amable brindis de
aquella noche, el primer ministro sueco lo contó con su sentido del
humor habitual: «Me alegro mucho en mi fuero interno de que las
autoridades turcas entendieran que son tan inocentes nuestras extra-
vagancias de esta noche, que le dieron a nuestro amigo Bulen Ecevit el permiso para venir».

La sensibilidad de Olof Palme por los sufrimientos de América
Latina –que es común a la mayoría de los suecos que conozco– está,
por cierto, en el origen de nuestra amistad. Nos presentó François
Mitterrand hace muchos años en su casa de la calle de Bievre, en
París, dentro del paréntesis de alguna de sus tantas derrotas anteriores. Había allí personalidades políticas y literarias de todas partes, de
modo que la conversación era suculenta y al mismo tiempo muy
divertida. De pronto, sin que nada especial hubiera ocurrido, Olof
Palme me hizo llegar el mensaje de que deseaba salir a tomarse una
cerveza con los latinoamericanos. Fuimos a La Coupole, como era
de rigor a la medianoche, y durante más de dos horas estuvo Olof
Palme interrogándonos sobre la situación de nuestros países, con una
versación y un interés que nos dejó sorprendidos. Ninguno de nosotros advirtió a un matrimonio de adultos tranquilos que seguía la
conversación con un gran interés desde una mesa vecina. Al final,
cuando Olof Palme se empeñó en pagar la cuenta, la mujer de la
otra mesa le preguntó en sueco si había pagado con dinero suyo o
con dinero del estado sueco. Palme se sentó entonces a la mesa de
sus compatriotas desconocidos y les dio toda clase de explicaciones.
En realidad había pagado con dinero suyo, pero consideraba de

todos modos que habría sido legítimo pagar con dinero del estado, porque le parecía que aquella reunión informativa sobre América Latina era un acto oficial importante del primer ministro sueco.

En la cena de su casa campestre logró también cautivarnos con sus recuerdos de nuestros países remotos. Evocó una conversación que sostuvo con Pablo Neruda en su casa de Isla Negra, en 1969, un año antes de la victoria electoral de Salvador Allende. «Hablamos toda la noche frente al fuego», dijo, «rodeados de los soberbios mascarones de proa que habían navegado por todos los mares del mundo. Hablamos, y Neruda era inagotable en sus reflexiones sobre la dictadura como fenómeno omnipresente de la historia latinoamericana, inagotable como el movimiento incesante de la resaca del Pacífico que aquella noche subía hasta la casa.» Su brindis puso sobre la mesa el tema de América Latina, y allí estuvo hasta la hora tardía en que nos levantamos para dormir.

Al término de la velada, el primer ministro me pidió que hiciera para sus invitados una síntesis de la situación de América Central en este momento. Yo llevaba tres días sin dormir, abrumado por las solicitudes insaciables de aquel jubileo mortal, pero la petición del primer ministro me pareció tan importante que me metí en un análisis minucioso de casi dos horas, hasta que Pierre Schori, muerto de risa, me interrumpió para decirme: «No sigas, Gabriel; ya estamos convencidos». Fue así como surgió la idea del llamado a los seis presidentes de América Central para que hagan un esfuerzo inmediato en favor de la paz en la región. El sentido de ese llamado, que correspondía al de mi exposición, era que nunca había estado la América Central tan cerca de una guerra generalizada, pero que tampoco —tal vez por eso mismo— nunca habían sido más propicias las condiciones para una solución negociada.

DESDE PARÍS, CON AMOR

Vine a París por primera vez una helada noche de diciembre de 1955. Llegué en el tren de Roma a una estación adornada con luces de Navidad, y lo primero que me llamó la atención fueron las parejas de enamorados que se besaban por todas partes. En el tren, en el metro, en los cafés, en los ascensores, la primera generación después de la guerra se lanzaba con todas sus energías al consumo público del amor, que era todavía el único placer barato después del desastre. Se besaban en plena calle, sin preocuparse de no estorbar a los peatones, que se apartaban sin mirarlos ni hacerles caso, como ocurre con esos perros callejeros de nuestros pueblos que se quedan colgados los unos de las otras, haciendo cachorros en mitad de la plaza. Aquellos besos de intemperie no eran frecuentes en Roma –que era la primera ciudad europea donde yo había vivido–, ni tampoco, por supuesto, en la brumosa y pudibunda Bogotá de aquellos tiempos, donde era difícil besarse aun en los dormitorios.

Eran los tiempos oscuros de la guerra de Argelia. Al fondo de las músicas nostálgicas de los acordeones en las esquinas, más allá del olor callejero de las castañas asadas en los braseros, la represión era un fantasma insaciable. De pronto, la policía bloqueaba la salida de un café o de uno de los bares de árabes del Boulevard Saint Michel y se llevaban a golpes a todo el que no tenía cara de cristiano. Uno de ellos, sin remedio, era yo. No valían explicaciones: no sólo la cara, sino también el acento con que hablábamos el francés, eran motivos de perdición. La primera vez que me metieron en la jaula de los argelinos, en la comisaría de Saint Germain-des-Près, me sentí humillado. Era un prejuicio latinoamericano: la cárcel era entonces una vergüenza, porque de niños no teníamos una distinción muy clara entre las razones políticas y las comunes, y nuestros adultos conservadores se encargaban de inculcarnos y mantenernos la confusión. Mi situación era todavía más peligrosa, porque, si bien

Publicado originalmente el 29 de diciembre de 1982.

los policías me arrastraban porque me creían argelino, éstos descon-
fiaban de mí dentro de la jaula cuando se daban cuenta de que, a
pesar de mi cara de vendedor de telas a domicilio, no entendía ni la
jota de sus algarabías. Sin embargo, tanto ellos como yo seguimos
siendo visitantes tan asiduos de las comisarías nocturnas que terminamos por entendernos. Una noche, uno de ellos me dijo que para
ser preso inocente era mejor serlo culpable, y me puso a trabajar
para el Frente de Liberación Nacional de Argelia. Era el médico
Amed Tebbal, que por aquellos tiempos fue uno de mis grandes
amigos de París, pero que murió de una muerte distinta de la guerra después de la independencia de su país. Veinticinco años después, cuando fui invitado a las fiestas de aquel aniversario en Argel,
declaré a un periodista algo que pareció difícil de creer: la revolución argelina es la única por la cual he estado preso.

Sin embargo, el París de entonces no era sólo el de la guerra de
Argelia. Era también el del exilio más generalizado que ha tenido
Latinoamérica en mucho tiempo. En efecto, Juan Domingo Perón
–que entonces no era el mismo de los años siguientes– estaba en el
poder en Argentina, el general Odría estaba en Perú, el general
Rojas Pinilla estaba en Colombia, el general Pérez Jiménez estaba
en Venezuela, el general Anastasio Somoza estaba en Nicaragua, el
general Rafael Leónidas Trujillo estaba en Santo Domingo, el general Fulgencio Batista estaba en Cuba. Éramos tantos los fugitivos de
tantos patriarcas simultáneos, que el poeta Nicolás Guillén se asoma-
ba todas las madrugadas a su balcón del hotel Grand Saint Michel, en
la calle Cujas, y gritaba en castellano las noticias de Latinoamérica
que acababa de leer en los periódicos. Una madrugada gritó: «Se
cayó el hombre». El que se había caído era sólo uno, por supuesto,
pero todos nos despertamos ilusionados con la idea de que el caído
fuera el de nuestro propio país.

Cuando llegué a París yo no era más que un caribe crudo. Lo
que más le agradezco a esta ciudad, con la cual tengo tantos pleitos
viejos, y tantos amores todavía más viejos, es que me hubiera dado
una perspectiva nueva y resuelta de Latinoamérica. La visión de conjunto, que no teníamos en ninguno de nuestros países, se volvía
muy clara aquí en torno a una mesa de café, y uno terminaba por
darse cuenta de que, a pesar de ser de distintos países, todos éramos
tripulantes de un mismo barco. Era posible hacer un viaje por todo
el continente y encontrarse con sus escritores, con sus artistas, con
sus políticos en desgracia o en ciernes, con sólo hacer un recorrido
por los cafetines populosos de Saint Germain-des-Près. Algunos no
llegaban, como me ocurrió con Julio Cortázar –a quien ya admiraba

desde entonces por sus hermosos cuentos de *Bestiario*–, y a quien esperé durante casi un año en el Old Navy, donde alguien me había dicho que solía ir. Unos quince años después le encontré, por fin, también en París, y era todavía como lo imaginaba desde mucho antes: el hombre más alto del mundo, que nunca se decidió a envejecer. La copia fiel de aquel latinoamericano inolvidable que, en uno de sus cuentos, gustaba de ir en los amaneceres brumosos a ver las ejecuciones en la guillotina.

Las canciones de Brassens se respiraban con el aire. La bella Tachia Quintana, una vasca temeraria a quien los latinoamericanos de todas partes habíamos convertido en una exiliada de las nuestras, realizaba el milagro de hacer una suculenta paella para diez en un reverbero de alcohol. Paul Coulaud, otro de nuestros franceses conversos, había encontrado un nombre para aquella vida: *la misère dorée*: la miseria dorada. Yo no había tenido una conciencia muy clara de mi situación hasta una noche en que me encontré de pronto por los lados del jardín de Luxemburgo sin haber comido ni una castaña durante todo el día y sin lugar donde dormir. Estuve merodeando largas horas por los bulevares, con la esperanza de que pasara la patrulla que se llevaba a los árabes para que me llevara a mí también a dormir a una jaula cálida, pero por más que la busqué no pude encontrarla. Al amanecer, cuando los palacios del Sena empezaron a perfilarse entre la niebla espesa, me dirigí hacia la Cité con pasos largos y decididos y con una cara de obrero honrado que acababa de levantarse para ir a su fábrica. Cuando atravesaba el puente de Saint Michel sentí que no estaba solo entre la niebla, porque alcancé a percibir los pasos nítidos de alguien que se acercaba en sentido contrario. Lo vi perfilarse en la niebla, por la misma acera y con el mismo ritmo que yo, y vi muy cerca su chaqueta escocesa de cuadros rojos y negros, y en el instante en que nos cruzamos en medio del puente vi su cabello alborotado, su bigote de turco, su semblante triste de hambres atrasadas y mal dormir, y vi sus ojos anegados de lágrimas. Se me heló el corazón, porque aquel hombre parecía ser yo mismo que ya venía de regreso.

Ése es mi recuerdo más intenso de aquellos tiempos, y lo he evocado con más fuerza que nunca ahora que he vuelto a París de regreso de Estocolmo. La ciudad no ha cambiado desde entonces. En 1968, cuando me trajo la curiosidad de ver qué había pasado después de la maravillosa explosión de mayo, encontré que los enamorados no se besaban en público, y habían repuesto los adoquines en las calles, y habían borrado los letreros más bellos que se escribieron jamás en las paredes: *La imaginación, al poder*; *Debajo del pavimento está la playa*;

Amaos los unos encima de los otros. Ayer, después de recorrer los sitios que alguna vez fueron míos, sólo pude percibir una novedad: unos hombres del municipio vestidos de verde, que recorren las calles en motocicletas verdes y llevan unas manos mecánicas de exploradores siderales para recoger en la calle la caca que un millón de perros cautivos expulsan cada veinticuatro horas en la ciudad más bella del mundo.

FELIPE

Estuve dos veces la semana pasada con Felipe González y Carmen, su esposa, en el hogar tranquilo que tienen dentro del palacio de la Moncloa. La casa es lo menos hogareño que uno se pueda imaginar, y más parece un escenario de teatro para una pieza de don Jacinto Benavente —a quien Dios tenga en su Santo Reino— que un lugar para vivir. Pero los González lo han logrado hasta ahora —y espero que por mucho tiempo—, y no tanto por el decorado como por el modo de ser, naturales, dentro del aire enrarecido del poder. Aun para mí, que me considero, a mucha honra, como el ser humano más refractario a la formalidad, aquellas dos visitas largas y sosegadas fueron una lección inolvidable.

Todavía no me acostumbro a la idea de que mis amigos lleguen a ser presidentes, ni he podido superar el prejuicio de que me impresionen las casas de gobierno. Estas últimas tienen un olor propio, una especie de hálito sobrenatural que tal vez sus habitantes sean los últimos en percibir, y que a mí me causan una incertidumbre que apenas si logro dominar. Por eso, aunque Felipe González y yo hemos entrado sin corbata a algunos lugares donde otros se sentirían inhibidos aun con el esmoquin, yo me sentí obligado a ponérmela, no tanto por un homenaje a aquellos santos lugares como para que no pareciera que estaba usurpando el derecho de ser informal donde esto no fuera de buena educación. Llegué con Mercedes y nuestro hijo menor en un mediodía radiante y dulce de este raro invierno de Madrid, y Felipe había salido con alguno de sus asesores a dar una vuelta por el parque apacible dentro del cual se encuentra el palacio de la Moncloa. Cuando lo vi venir por entre los árboles con un suéter azul de mangas largas, que le daba más bien un aire de universitario que de presidente, me sentí demasiado vestido para la ocasión. Menos mal que él llevaba también una corbata. La primera que le veía alrededor del cuello desde aquella noche fugaz de

Publicado originalmente el 5 de enero de 1983.

hace ocho años en que nos conocimos en un populoso cuarto de hotel en Bogotá.

Habíamos ido con Enrique Santos Calderón y Antonio Caballero a hacerle una entrevista para la revista *Alternativa*, que era la oveja descarriada de la prensa nacional, pero Felipe no se asustó de nuestra mala reputación política, sino que de alguna manera distinta, pero muy inteligente, terminó de acuerdo con nosotros sin necesidad de decirlo. La verdad, sin embargo, fue que de algún modo tanto él como nosotros comprendíamos que aquella entrevista no era más que un pretexto y quedamos de acuerdo en encontrarnos al día siguiente para conversar sin testigos ni magnetofones. Lo hicimos por iniciativa del propio Felipe, en un ambiente al mismo tiempo acogedor e insospechable: entre los estantes de una librería, donde los clientes, absortos, apenas si se apercibían de nuestra presencia. Me pareció que aquella manera de estar casi invisible, pero sin necesidad de esconderse, era para Felipe un hábito cotidiano de la clandestinidad, en la cual había vivido tantos años en los malos tiempos de España. Sin embargo, donde en realidad nos hicimos amigos fue en otras épocas diferentes, en las distintas casas que tenía el general Omar Torrijos en Panamá. Uno llegaba casi sin anuncio previo a la antigua base militar de Farallón, donde reventaban sin tregua las olas indómitas del Pacífico, o llegaba al paraíso cautivo de la isla de Contadora, y se encontraba siempre con alguien que tenía algo que decir sobre el destino de la América Latina, y en especial sobre la América Central. Sobre todo tres personas que habían de ser claves en la batalla sorda y difícil por la recuperación del canal de Panamá: Carlos Andrés Pérez, Alfonso López Michelsen y el propio Omar Torrijos. Entre ellos, el joven Felipe González, que andaba por los treinta y pocos años cuando ya los otros tres eran presidentes, parecía sólo un discípulo privilegiado que se movía en la cátedra con tanta versación y tanto interés como sus maestros. Su carrera hacia la victoria ha sido tan fulminante que todo esto parece ocurrido hace muchos años, con una distancia histórica que ya ofrece hasta una cierta perspectiva para el análisis. Tal vez ésa fue la razón por la cual, cuando vi a Felipe González paseando por el parque de la Moncloa, me costó trabajo acostumbrarme a la idea de que nuestro amigo de vacaciones en Farallón y Contadora se había convertido en presidente del gobierno en su país con apenas cuarenta años mal contados. En realidad, para mí seguía siendo uno más de los muchos sobrevivientes de aquel avión del general Torrijos en que todos andábamos por todos lados a toda hora. Por entre soles y tempestades, sin pensar tal vez que era un avión señalado por la muerte.

Como ocurrió en nuestra primera tarde en Bogotá, como ocurrió tantas veces en Panamá, en México y aun en las islas San Blas, adonde hicimos alguna vez un viaje de regreso por el tiempo, Felipe González y yo ocupamos las casi diez horas de las dos visitas hablando de América Latina. En mí no es raro, pues es una, y tal vez la más dominante, de mis tres obsesiones. Lo raro es que no lo sea tampoco en un hombre como Felipe González, que tiene tanto que hacer y tanto que pensar para gobernar como es debido un país tan difícil. «Es el más grande especialista que conozco en el tema de América Central», les dije a los periodistas españoles, que todo lo querían saber. No era una exageración. En realidad, no conozco a nadie que no sea latinoamericano y que se interese tanto por nuestra suerte, consciente tal vez de que, de algún modo, la suerte de España y la nuestra podrían ser complementarias. En la primera visita de una tarde de domingo completa me llamó la atención algo que es insólito en un presidente: durante cinco horas nadie lo hizo pasar al teléfono, ni se vio al eterno ayudante de siempre que le hiciera un papelito con un recado urgente. Todo el tiempo era para los amigos con quienes estaba, que es, para mi modo de ver las cosas, la mejor prueba del respeto a la amistad. En la segunda visita ocurrió algo todavía más significativo. Los González nos invitaron a ver una película en una sala improvisada de la Moncloa y cada siete minutos había una interrupción para cambiar el rollo en el proyector. Felipe y yo aprovechamos aquellos tres minutos de intermedio para seguir nuestro diálogo sobre América Latina. La muy grave situación en América Central, por supuesto, era un tema específico. Cuando salimos de allí a la una de la madrugada y con la certidumbre de que habíamos perdido el último avión de las Américas, yo iba impulsado por la idea real de que no sólo había aprendido mucho sobre quiénes somos y para dónde vamos los latinoamericanos y los españoles, sino también por la convicción de que nuestros caminos siguen estando cruzados, que muchos de sus trechos hay que hacerlos juntos, y que Felipe puede ser un hombre decisivo no sólo para España, sino también para nuestro destino común. No pude pensar en otra cosa en las diez horas siguientes, mientras volaba a través de los cielos solitarios y estrellados de Cristóbal Colón, y tratando de escribir por primera vez a 10.000 pies de altura esta nota de mis tormentos semanales.

LAS VEINTE HORAS DE GRAHAM GREENE
EN LA HABANA

Graham Greene ha hecho en La Habana una escala de veinte horas, a la cual le han dado toda clase de interpretaciones los corresponsales locales de la prensa extranjera. No era para menos: llegó en un avión ejecutivo del gobierno de Nicaragua acompañado por José de Jesús Martínez, un poeta y profesor de matemáticas panameño que fue uno de los hombres más cercanos al general Omar Torrijos, y fueron recibidos en el aeropuerto por funcionarios del protocolo dentro de la mayor discreción, de modo que ningún periodista se enteró de esa visita sino después de que había terminado. Fueron conducidos a una casa de visitantes distinguidos reservada, en general, para los jefes de estado de países amigos, y pusieron a su disposición un solemne Mercedes Benz negro de los que sólo se usaron durante la sexta reunión cumbre de los países No Alineados, hace cuatro años. No lo necesitaban, en realidad, pues no salieron de la casa, donde los visitaron algunos viejos amigos cubanos, que se enteraron de la noticia porque el mismo escritor la hizo saber. El pintor René Portocarrero, que es su amigo desde los tiempos en que Graham Greene pasó por aquí para estudiar el ambiente de *Nuestro hombre en La Habana*, recibió el recado demasiado tarde y cuando llegó a la visita el escritor ya se había marchado por donde vino. Apenas si comió una vez en aquellas veinte horas, picando un poco de todo como un pajarito mojado, pero se tomó en la mesa una botella de buen vino tinto español y durante su estancia fugaz se consumieron en la casa seis botellas de whisky. Cuando se fue, nos dejó la rara impresión de que ni él mismo supo a qué vino, como sólo podía ocurrirle a uno de esos personajes de sus novelas, atormentados por la incertidumbre de Dios.

Pasé por su casa dos horas después de la llegada, porque me hizo llamar por teléfono tan pronto como supo que estaba en la ciudad,

Publicado originalmente el 19 de enero de 1983.

y esto me produjo una muy grande alegría, no sólo por la antigua e inagotable admiración que le tengo como escritor y como ser humano, sino porque habían pasado muchos años desde la última vez en que nos vimos. Había sido —como él mismo lo recordaba— cuando ambos viajamos a Washington en la delegación panameña a la firma de los tratados del canal. Algunos periódicos especularon entonces que la invitación había sido una maniobra de Torrijos para adornar su delegación con los nombres de dos escritores famosos que nada tenían que ver con aquella fiesta. En realidad, ambos habíamos tenido que ver con las negociaciones del tratado mucho más de lo que suponía la prensa, pero no fue ni por aquello ni por esto por lo que el general Torrijos nos invitó a acompañarlo a Washington, sino porque no pudo resistir a la tentación de hacerle una burla cordial a su amigo el presidente Jimmy Carter. El caso es que a Graham Greene y a mí —como a tantos otros escritores y artistas de este mundo— se nos tiene prohibida la entrada a Estados Unidos desde hace muchos años por razones que ni los propios presidentes han podido explicar nunca, y el general Torrijos se había empeñado en resolvernos el problema. Les planteó el asunto a muchos de los funcionarios de alto rango que lo visitaron por aquellos tiempos, y por último lo llevó hasta el propio presidente Carter, quien le manifestó su sorpresa y prometió resolverlo a la mayor brevedad, pero se le acabó el tiempo de su poder antes de dar una respuesta. Cuando estaba integrando la delegación para ir a Washington, a Torrijos se le ocurrió la idea de meternos de contrabando en Estados Unidos a Graham Greene y a mí. Era una obsesión: poco antes, le había propuesto a Greene que se disfrazara de coronel de la Guardia Nacional y fuera a Washington en misión especial ante el presidente Carter, sólo por hacerle a éste una de sus bromas habituales. Pero Graham Greene, que es más serio de lo que pudiera parecer por algunos de sus libros, no quiso prestar su cuerpo glorioso para un episodio que, sin duda, hubiera sido uno de los más divertidos para sus memorias. Sin embargo, cuando el general Torrijos nos propuso asistir a la ceremonia de los tratados con nuestras identidades propias pero con pasaportes oficiales panameños e integrados a la delegación de ese país, ambos aceptamos con un cierto regocijo infantil. De modo que llegamos juntos a la base militar de Andrews. Ambos con pantalones de vaqueros y camisas de mezclilla en medio de una delegación de caribes vestidos de negro y aturdidos por el estampido de veintiún cañonazos de júbilo y las notas marciales del himno norteamericano, que parecían formar parte de la burla. Consciente de la carga literaria del momento, Graham Greene me dijo al oído cuando bajábamos por la escalerilla del avión: «Dios mío, qué cosas las que le suceden a Estados Unidos». El propio Carter no pudo menos que reír con sus dientes luminosos de anuncio de televisión cuando el general Torrijos le contó su travesura.

Al cabo de tantos años me encontré con un Graham Greene rejuvenecido, cuya lucidez sigue siendo su virtud más sorprendente e inalterable. Hablamos, como siempre, un poco de todo. Pero lo que más me llamó la atención fue el sentido del humor con que evocaba los cuatro juicios que debe enfrentar esta semana en distintos tribunales de Francia, como consecuencia del folleto acusatorio que publicó contra la mafia de Niza. Para muchos conocedores de los bajos fondos de la Costa Azul, las revelaciones de Greene no decían nada nuevo. Pero los amigos del escritor temimos por su vida. Él no se inmutó, sino que siguió adelante con su denuncia. «Para morir de un cáncer en la próstata», dijo, «prefiero morir de un tiro en la cabeza.» Yo dije entonces, no recuerdo dónde, que Graham Greene estaba jugando a la ruleta literaria, como jugó en su juventud con un Smith y Wesson calibre 32, según lo había contado en sus memorias. Él recordó esta declaración mía durante la visita y la tomó como punto de partida para contarnos los pormenores de sus cuatro procesos judiciales.

Hacia la una de la madrugada pasó a visitarlo Fidel Castro. Se conocieron al principio de la revolución, muy al principio, cuando Graham Greene asistió a la filmación de *Nuestro hombre en La Habana*. Se volvieron a ver varias veces, en los viajes periódicos de Graham Greene pero, al parecer, no se habían visto en los dos últimos, porque esta vez, cuando se dieron la mano, Graham Greene dijo: «No nos veíamos desde hace dieciséis años». Ambos me parecieron un poco intimidados y no les fue fácil empezar la conversación. Por eso le pregunté a Graham Greene qué había de cierto en el episodio de la ruleta rusa que él ha contado en sus memorias. Sus ojos azules, los más diáfanos que conozco, se iluminaron con los recuerdos. «Eso fue a los diecinueve años», dijo, «cuando me enamoré de la institutriz de mi hermana.» Contó que, en efecto, había jugado entonces al juego solitario de la ruleta rusa con un viejo revólver de un hermano mayor, y en cuatro ocasiones diferentes. Entre las dos primeras hubo una semana de intervalo, pero las dos últimas fueron sucesivas y con pocos minutos de diferencia. Fidel Castro, que no podía pasar por alto un dato como ése sin agotar hasta las últimas precisiones, le preguntó para cuántos proyectiles era el tambor del revólver. «Para seis», le contestó Graham Greene. Entonces, Fidel Castro cerró los ojos y empezó a murmurar cifras de multiplicación.

Por último, miró al escritor con una expresión de asombro y le dijo: «De acuerdo con el cálculo de las probabilidades, usted tendría que estar muerto». Graham Greene sonrió con la placidez con que lo hacen todos los escritores cuando se sienten viviendo un episodio de sus propios libros, y dijo: «Menos mal que siempre fui pésimo en matemáticas». Tal vez porque se hablaba de la muerte, Fidel Castro se fijó de pronto en el semblante juvenil y saludable del escritor, y le preguntó qué ejercicios hacía. Era una pregunta que no podía faltar, porque Fidel Castro considera la cultura física como una de las claves de la vida. Hace varias horas de ejercicios todos los días, con las mismas proporciones descomunales de todo lo que emprende, y les aconseja un régimen semejante a sus amigos. Sus condiciones físicas son excepcionales para un hombre de 56 años y a ellas atribuye su buena salud mental. Por eso se sorprendió tanto cuando Graham Greene le contestó que nunca había hecho ningún ejercicio en toda su vida, y, sin embargo, se sentía muy lúcido y sin ningún trastorno de salud a los 79 años. Además, reveló que no tenía ningún régimen de alimentación especial, que dormía entre siete y ocho horas diarias, cosa que también era sorprendente en un anciano de costumbres sedentarias, y además se bebía, a veces, hasta una botella de whisky al día y un litro de vino con cada comida, sin haber padecido nunca la servidumbre del alcoholismo.

Por un instante, Fidel Castro pareció poner en duda la eficacia de su régimen de salud. Pero muy pronto comprendió que Graham Greene era una excepción admirable, pero nada más que una excepción. Cuando nos despedimos, ya me estaba inquietando la certidumbre de que aquel encuentro, tarde o temprano, iba a ser evocado en el libro de memorias de alguno de nosotros tres, o quizá de los tres.

REGRESO A MÉXICO

Alguna vez dije en una entrevista: «De la ciudad de México, donde hay tantos amigos que quiero, no me va quedando más que el recuerdo de una tarde increíble en que estaba lloviendo con sol por entre los árboles del bosque de Chapultepec, y me quedé tan fascinado con aquel prodigio que se me trastornó la orientación y me puse a dar vueltas en la lluvia, sin encontrar por dónde salir».

Diez años después de esa declaración he vuelto a buscar aquel bosque encantado y lo encontré podrido por la contaminación del aire y con la apariencia de que nunca más ha vuelto a llover entre sus árboles marchitos. Esta experiencia me reveló de pronto cuánta vida mía y de los míos se ha quedado en esta ciudad luciferina, que hoy es una de las más extensas y pobladas del mundo, y cuánto hemos cambiado juntos, la ciudad y nosotros, desde que llegamos sin nombre y sin un clavo en el bolsillo, el 2 de julio de 1961, a la polvorienta estación del ferrocarril central.

La fecha no se me olvidará nunca, aunque no estuviera en un sello de un pasaporte inservible, porque al día siguiente muy temprano un amigo me despertó por el teléfono y me dijo que Hemingway había muerto. En efecto, se había desbaratado la cabeza con un tiro de fusil en el paladar, y esa barbaridad se quedó para siempre en mi memoria como el principio de una nueva época. Mercedes y yo, que teníamos dos años de casados, y Rodrigo, que todavía no tenía uno de nacido, habíamos vivido los meses anteriores en un cuarto de hotel en Manhattan. Yo trabajaba como corresponsal en la agencia cubana de noticias de Nueva York, y no había conocido hasta entonces un lugar más idóneo para morir asesinado. Era una oficina sórdida y solitaria en un viejo edificio de Rockefeller Center, con un cuarto de teletipos y una sala de redacción con una ventana única que daba a un patio abismal, siempre triste y oloroso a hollín helado, de cuyo fondo subía a toda hora el estruendo de las ratas dispu-

Publicado originalmente el 26 de enero de 1983.

tándose las sobras en los tarros de basura. Cuando aquel lugar se hizo insoportable, metimos a Rodrigo en una canasta y nos subimos en el primer autobús que salió para el sur. Todo nuestro capital en el mundo eran trescientos dólares, y otros cien que Plinio Apuleyo Mendoza nos mandó desde Bogotá al consulado colombiano en Nueva Orleans. No dejaba de ser una bella locura: tratábamos de llegar a Colombia a través de los algodonales y los pueblos de negros de Estados Unidos, llevando como única guía mi memoria reciente de las novelas de William Faulkner.

Como experiencia literaria, todo aquello era fascinante, pero en la vida real –aun siendo tan jóvenes– era un disparate. Fueron catorce días de autobús por carreteras marginales, ardientes y tristes, comiendo en fondas de mala muerte y durmiendo en hoteles de peores compañías. En los grandes almacenes de las ciudades del sur conocimos por primera vez la ignominia de la discriminación: había dos máquinas públicas para beber agua, una para blancos y otra para negros, con el letrero marcado en cada una. En Alabama pasamos una noche entera buscando un cuarto de hotel, y en todos nos dijeron que no había lugar, hasta que algún portero nocturno descubrió por casualidad que no éramos mexicanos. Sin embargo, como siempre, lo que más nos fatigaba no eran las jornadas interminables bajo el calor ardiente de junio ni las malas noches en los hoteles de paso, sino la mala comida. Cansados de hamburguesas de cartón molido y de leche malteada, terminamos por compartir con el niño las compotas en conservas. Al término de aquella travesía heroica habíamos logrado confrontar una vez más la realidad y la ficción. Los partenones inmaculados en medio de los campos de algodón, los granjeros haciendo la siesta sentados bajo el alero fresco de las ventas de caminos, las barracas de los negros sobreviviendo en la miseria, los herederos blancos del tío Gavin Stevens, que pasaban para la misa dominical con sus mujeres lánguidas vestidas de muselina: la vida terrible del condado de Yocknapatapha había desfilado ante nuestros ojos desde la ventanilla de un autobús, y era tan cierta y humana como en las novelas del viejo maestro.

Sin embargo, toda la emoción de aquella vivencia se fue al carajo cuando llegamos a la frontera de México, al sucio y polvoriento Laredo que ya nos era familiar por tantas películas de contrabandistas. Lo primero que hicimos fue entrar en una fonda para comer caliente. Nos sirvieron para empezar, a manera de sopa, un arroz amarillo y tierno, preparado de un modo distinto que en el Caribe. «Bendito sea Dios», exclamó Mercedes al probarlo. «Me quedaría aquí para siempre aunque sólo fuera para seguir comiendo este arroz.» Nunca se hubiera podido imaginar hasta qué punto su deseo de quedarse sería cumplido. Y no por aquel plato de arroz frito, sin embargo, porque el destino había de jugarnos una broma muy divertida: el arroz que comemos en casa lo hacemos traer de Colombia, casi de contrabando, en las maletas de los amigos que vienen, porque hemos aprendido a sobrevivir sin las comidas de nuestra infancia, menos sin ese arroz patriótico cuyos granos nevados se pueden contar uno por uno en el plato.

Llegamos a la ciudad de México en un atardecer malva, con los últimos veinte dólares y sin nada en el porvenir. Sólo teníamos aquí cuatro amigos. Uno era el poeta Álvaro Mutis, que ya había pasado las verdes en México, pero que todavía no había encontrado las maduras. El otro era Luis Vicens, un catalán de los grandes que se había venido poco antes de Colombia, fascinado por la vida cultural de México. El otro era el escultor Rodrigo Arenas Betancur, que estaba sembrando cabezas monumentales a todo lo ancho de este país interminable. El cuarto era el escritor Juan García Ponce, a quien había conocido en Colombia como jurado de un concurso de pintura, pero apenas si nos recordábamos el uno del otro, por el estado de densidad etílica en que ambos nos encontrábamos la noche en que nos vimos por primera vez. Fue él quien me llamó por teléfono tan pronto como supo de mi llegada, y me gritó con su verba florida: «El cabrón de Hemingway se partió la madre de un escopetazo». Ése fue el momento exacto —y no las seis de la tarde del día anterior— en que llegué de veras a la ciudad de México, sin saber muy bien por qué, ni cómo, ni hasta cuándo. De eso hace ahora veintiún años y todavía no lo sé, pero aquí estamos. Como lo dije en una memorable ocasión reciente, aquí he escrito mis libros, aquí he criado a mis hijos, aquí he sembrado mis árboles.

He revivido este pasado —enrarecido por la nostalgia, es cierto— ahora que he vuelto a México como tantas y tantas veces, y por primera vez me he encontrado en una ciudad distinta. En el bosque de Chapultepec no quedan ni siquiera los enamorados de antaño, y nadie parece creer en el sol radiante de enero, porque en verdad es raro en estos tiempos. Nunca, desde nunca, había encontrado tanta incertidumbre en el corazón de los amigos. ¿Será posible?

SÍ: YA VIENE EL LOBO

Muchos amigos de Nicaragua, inclusive algunos que están bien informados, piensan que las voces de alarma que los dirigentes sandinistas hacen oír cada cierto tiempo en el mundo entero no corresponden a una amenaza real en sus fronteras, sino que son como los gritos de diversión con que el pastor de la fábula anunciaba que ya venía el lobo. Sin embargo, la amenaza desde territorio de Honduras no sólo es verdadera y constante, sino que cuenta cada vez con mayores recursos, y si no ha llegado hasta sus últimas consecuencias es porque distintos sectores del gobierno de los Estados Unidos no han logrado ponerse de acuerdo para una decisión final.

Hace unos meses, un oficial del ejército argentino, mandado por su gobierno como maestro de represión en Honduras, desertó de su empleo por la actitud de los Estados Unidos en la guerra de las Malvinas, y reveló a la prensa todos sus secretos. Su confesión espontánea no dejaba ninguna duda de que la Agencia Central de Inteligencia de los Estados Unidos estaba madurando una agresión en grande contra el gobierno de Nicaragua, y que contaba no sólo con antiguos militares de Somoza y mercenarios del mundo entero, sino también con asistentes oficiales argentinos, chilenos e israelíes. Este convencido de última hora permitía pensar también que había sido una patraña del gobierno argentino el retiro anunciado de sus maestros de represión en América Central, también por la actitud de los Estados Unidos en la guerra de las Malvinas. La patraña acabó de confirmarse con visos de burla sangrienta hace algunas semanas, cuando, otra vez, el gobierno militar argentino volvió a anunciar que retiraba a los asesores que ya se suponían retirados y que, al parecer, todavía hoy continúan dando clases en sus escuelas siniestras, lo cual es —ahora sí— una versión moderna de la fábula del lobo, pero al revés.

Todo esto recuerda, a quienes tenemos una buena memoria de periodistas, las vísperas del desembarco en bahía de Cochinos, en

Publicado originalmente el 2 de febrero de 1983.

abril de 1961. En esa ocasión, como todo el mundo sabe ahora, se llegó hasta el extremo de pintar las insignias de la aviación cubana en el fuselaje de aviones de guerra de los Estados Unidos, los cuales bombardearon la base de San Antonio de los Baños, en Cuba, con el propósito –cumplido a medias– de destruir a los pocos aviones cubanos que podían enfrentarse al desembarco. Los aviones disfrazados regresaron a la Florida, y sus pilotos, que en realidad eran exiliados cubanos, se presentaron ante la prensa como desertores de la aviación revolucionaria que había bombardeado su propia base con aviones robados. Es cierto que las condiciones de América Central y el Caribe y del mundo no son las mismas de hace veinte años, pero también es verdad que el gobierno de Ronald Reagan actúa como si no lo supiera. De modo que los nicaragüenses tienen razón, una vez más, en gritar tan fuerte como puedan, y está bien que sus amigos los ayudemos a gritar tan fuerte como podamos, porque es verdad que ya viene el lobo, el lobo, y que viene pisando con pasos de animal tan grande que hasta el pastor más ingenuo se daría cuenta de que no viene solo.

Las maniobras conjuntas que 1.700 soldados norteamericanos y 4.000 hondureños comenzaron ayer en las fronteras de Honduras con Nicaragua no contribuyen, ni mucho menos, a la paz, que ya más de medio mundo está deseando para América Central, ni son un paso para la solución pacífica negociada que tantos gobiernos de buena voluntad están tratando de conseguir, ni revelan en sus protagonistas ningún ánimo real de poner término a la sangría constante que padece esa desdichada cintura de las Américas. No comparto los temores de quienes piensan que semejante despliegue militar es apenas una pantalla para encubrir una invasión masiva de Nicaragua. No es así como suelen suceder las cosas. Pero estoy de acuerdo con quienes piensan que son decisivas para mejorar las condiciones profesionales del ejército de Honduras, que está demasiado bien adiestrado para la represión interna, pero no para una guerra internacional. Las maniobras permitirían también a los Estados Unidos introducir equipo bélico mejor y mayor, y dejarlo en Honduras después del retiro de sus tropas, y no sólo al servicio de las fuerzas armadas hondureñas, sino también de los somocistas y sus pandillas de mercenarios. En cambio, comparto el temor de quienes piensan que el peligro real no está en las maniobras de tierra que se llevarán a cabo en la Mosquitia, sobre el mar Caribe, sino en las maniobras navales que se llevarán al mismo tiempo en el golfo de Fonseca, sobre el océano Pacífico. Este hermoso lugar, cuyas costas son compartidas por Honduras, El Salvador y Nicaragua, es la esquina caliente de América Central. Los tres países limítrofes montan allí una guardia constante y tensa, y cualquier provocación que en cualquier otro lugar sería resuelta con una protesta diplomática formal, podría ser allí el principio de una deflagración irreparable. A los Estados Unidos les gusta servirse dos veces del mismo plato, aun de los más amargos, y no sería asombroso que intentaran en el golfo de Fonseca una provocación semejante a la del golfo de Tomkin, en el mar de China, que les sirvió de pretexto para intervenir en Vietnam. Las analogías, por desgracia, son cada día más inquietantes.

A los numerosos periodistas que vinieron a mi casa de México el 21 de octubre pasado, a las seis de la mañana, les expresé mis temores de una invasión inminente a Nicaragua desde el territorio de Honduras, y les dije que había que hacer lo imposible por evitarla. No hablaba por decir algo resonante en la mañana del Nobel, no; el proyecto de invasión a Nicaragua desde Honduras lo había preparado la CIA bajo los auspicios del anterior secretario de Estado de los Estados Unidos, Alexander Haig, y su sucesor, George Shultz, lo había encontrado servido cuando tomó posesión del cargo. Era lo mismo que le había ocurrido al presidente John F. Kennedy, en 1961, cuando encontró servido en su mesa el proyecto de invasión a Cuba preparado por su antecesor, el general Eisenhower. Dos personalidades que hablaron por separado con George Shultz a principios de octubre lo encontraron preocupado por lo que pudiera ocurrir entre Honduras y Nicaragua, y aunque no les habló del proyecto, les dio seguridades de que él se oponía a cualquier acción de guerra en esa línea de alta tensión, y que haría lo que estuviera a su alcance por impedirla. Yo tenía versiones directas de esas conversaciones cuando dije lo que dije a los periodistas, y hoy creo y celebro que tal vez el señor Shultz haya logrado impedir el desastre a principios de diciembre, que era la época prevista. Lo que no sabemos hoy es si aquello no fue más que un aplazamiento.

Todo esto obliga a una movilización más activa, eficaz y coherente para el logro de una solución pacífica global al drama de América Central. La oposición armada de El Salvador ve aumentar cada día sus posibilidades de una victoria total sobre un ejército cada día más dividido y desmoralizado. Cinco mil rebeldes, la mitad de ellos con armas automáticas sofisticadas, avanzan frente a un ejército regular que abandona sus posiciones y sus armas sin combatir. Los propios Estados Unidos deben ser conscientes de que aquélla es para ellos y sus socios salvadoreños una guerra que no van a ganar, pero que puede prolongarse con una crueldad que ninguno de los dos bandos merece. Los cancilleres de México, Colombia, Venezuela y Panamá se reunieron hace poco en la isla de Contadora, tratando de encontrar la fórmula de paz necesaria para América Central, pero la reunión, al parecer, fue más bien un torneo de buena voluntad que de sentido práctico. No tengo ninguna duda de la sinceridad y la antigüedad de los buenos deseos de México, y me consta que la paz en esta región, como en su propio país, es uno de los propósitos más entrañables del presidente de Colombia. Sin embargo, en lugar de prestar atención al clamor, ya casi mundial, por un acuerdo político que los propios gobernantes nicaragüenses y los mismos rebeldes salvadoreños están dispuestos a patrocinar, el gobierno del presidente Reagan prefiere seguir mostrando sus dientes de lobo, de lobo, de lobo, en unas maniobras de ya viene el lobo que, en el menos grave de los casos, habría que repudiar de todos modos como una impertinencia estúpida.

ESTÁ BIEN, HABLEMOS DE LITERATURA

Jorge Luis Borges dijo en una vieja entrevista que el problema de los jóvenes escritores de entonces era que en el momento de escribir pensaban en el éxito o el fracaso. En cambio, cuando él estaba en sus comienzos sólo pensaba en escribir para sí mismo. «Cuando publiqué mi primer libro», contaba, «en 1923, hice imprimir trescientos ejemplares y los distribuí entre mis amigos, salvo cien ejemplares, que llevé a la revista *Nosotros*». Uno de los directores de la publicación, Alfredo Bianchi, miró aterrado a Borges y le dijo: «¿Pero usted quiere que yo venda todos esos libros?». «Claro que no», le contestó Borges, «a pesar de haberlos escrito no estoy completamente loco». Por cierto, que el autor de la entrevista, Alex J. Zisman, que entonces era un estudiante peruano en Londres, contó al margen que Borges le había sugerido a Bianchi que metiera copias del libro en los bolsillos de los sobretodos que dejaran colgados en el ropero de sus oficinas, y así consiguieron que se publicaran algunas notas críticas.

Pensando en este episodio recordé otro tal vez demasiado conocido, de cuando la esposa del ya famoso escritor norteamericano Sherwood Anderson encontró al joven William Faulkner escribiendo a lápiz con el papel apoyado en una vieja carretilla. «¿Qué escribe?», le preguntó ella. Faulkner, sin levantar la cabeza, le contestó: «Una novela». La señora Anderson sólo acertó a exclamar: «¡Dios mío!». Sin embargo, unos días después Sherwood Anderson le mandó decir al joven Faulkner que estaba dispuesto a llevarle su novela a un editor, con la única condición de no tener que leerla. El libro debió ser *Soldier's Pay*, que se publicó en 1926 —o sea, tres años después del primer libro de Borges—, y Faulkner había publicado cuatro más antes de que se le considerara como un autor conocido, cuyos libros fueran aceptados por los editores sin demasiadas vueltas. El propio Faulkner declaró alguna vez que después de esos primeros

Publicado originalmente el 9 de febrero de 1983.

cinco libros se vio forzado a escribir una novela sensacionalista, ya que los anteriores no le habían producido bastante dinero para alimentar a su familia. Ese libro forzoso fue *Santuario*, y vale la pena señalarlo, porque esto indica muy bien cuál era la idea que tenía Faulkner de una novela sensacionalista.

Me he acordado de estos episodios en los orígenes de los grandes escritores en el curso de una conversación de casi cuatro horas que sostuve ayer con Ron Sheppard, uno de los redactores literarios de la revista *Time*, que está preparando un estudio sobre la literatura de América Latina. Dos cosas me dejaron muy complacido de esa entrevista. La primera es que Sheppard sólo me habló y sólo me hizo hablar de literatura, y demostró, sin el menor asomo de pedantería, que sabe muy bien lo que es. La segunda es que había leído con mucha atención todos mis libros y había estudiado muy bien, no sólo por separado, sino también en su orden y en su conjunto, y además se había tomado el trabajo arduo de leer numerosas entrevistas mías para no recaer en las mismas preguntas de siempre. Este último punto no me interesó tanto por halagar mi vanidad –cosa que, de todos modos, no se puede ni se debe descartar cuando se habla con cualquier escritor, aun con los que parecen más modestos–, sino porque me permitió explicar mejor, con mi experiencia propia, mis concepciones personales del oficio de escribir. Todo escritor entrevistado descubre de inmediato –por cualquier descuido ínfimo– si su entrevistador no ha leído un libro del cual le está hablando, y desde ese instante, y acaso sin que el otro lo advierta, lo coloca en situación de desventaja. En cambio, conservo un recuerdo muy grato de un periodista español, muy joven, que me hizo una entrevista minuciosa sobre mi vida creyendo que yo era el autor de la canción de las mariposas amarillas, que por aquella época sonaba por todas partes, pero que no tenía la menor idea de que aquella música había tenido origen en un libro y que, además, era yo quien lo había escrito.

Sheppard no hizo ninguna pregunta concreta, ni utilizó una grabadora, sino que cada cierto tiempo tomaba notas muy breves en un cuaderno de escolar, ni le importó qué premios me habían dado antes o ahora, ni trató de saber cuál era el compromiso del escritor, ni cuántos libros había vendido, ni cuánto dinero me había ganado. No voy a hacer una síntesis de nuestra conversación, porque todo cuanto en ella se habló le pertenece ahora a él y no a mí. Pero no he podido resistir a la tentación de señalar el hecho como un acontecimiento alentador en el río revuelto de mi vida privada de hoy, donde no hago casi nada más que contestar varias veces al día las mismas preguntas con las mismas respuestas de siempre. Y peor aún: las mismas preguntas, que cada día tienen menos que ver con mi oficio de escritor. Sheppard, en cambio, y con la misma naturalidad con que respiraba, se movía sin tropiezos con los misterios más densos de la creación literaria, y cuando se despidió me dejó ensopado en la nostalgia de los tiempos en que la vida era más simple y uno disfrutaba del placer de perder horas y horas hablando de nada más que de literatura.

Sin embargo, nada de lo que hablamos se me fijó de un modo más intenso que la frase de Borges: «Ahora, los escritores piensan en el fracaso y en el éxito». De un modo o de otro, les he dicho lo mismo a tantos escritores jóvenes que encuentro por esos mundos. No a todos, por fortuna, los he visto tratando de terminar una novela a la topa tolondra para llegar a tiempo a un concurso. Los he visto precipitándose en abismos de desmoralización por una crítica adversa, o por el rechazo de sus originales en una casa editorial. Alguna vez le oí decir a Mario Vargas Llosa una frase que me desconcertó de entrada: «En el momento de sentarse a escribir, todo escritor decide si va a ser un buen escritor o un mal escritor». Sin embargo, varios años después llegó a mi casa de México un muchacho de veintitrés años que había publicado su primera novela seis meses antes y que aquella noche se sentía triunfante porque acababa de entregar al editor su segunda novela. Le expresé mi perplejidad por la prisa que llevaba en su prematura carrera, y él me contestó, con un cinismo que todavía quiero recordar como involuntario: «Es que tú tienes que pensar mucho antes de escribir porque todo el mundo está pendiente de lo que escribes. En cambio, yo puedo escribir muy rápido, porque muy poca gente me lee». Entonces entendí, como una revelación deslumbrante, la frase de Vargas Llosa: aquel muchacho había decidido ser un mal escritor, como, en efecto, lo fue hasta que consiguió un buen empleo en una empresa de automóviles usados, y no volvió a perder el tiempo escribiendo. En cambio —pienso ahora—, tal vez su destino sería otro si antes de aprender a escribir hubiera aprendido a hablar de literatura. Por estos días hay una frase de moda: «Queremos menos hechos y más palabras». Es una frase, por supuesto, cargada de una muy grande perfidia política. Pero sirve también para los escritores.

Hace unos meses le dije a Jomi García Ascot que lo único mejor que la música era hablar de música, y anoche estuve a punto de decirle lo mismo sobre la literatura. Pero luego lo pensé con más cuidado. En realidad, lo único mejor que hablar de literatura es hacerla bien.

MEMORIAS DE UN FUMADOR RETIRADO

En una época casi irreal en que todo el mundo era joven, el crítico mexicano de cine Emilio García Riera se quedó dormido en un cuarto de hotel mientras leía fumando en la cama. El cigarrillo resbaló de sus labios al mismo tiempo que resbaló el libro de sus manos, y cuando despertó estaba a punto de morir asfixiado en un cuarto lleno de humo y sobre un colchón en llamas. No fue posible convencer al administrador del hotel de que había sido un accidente común, que debía estar previsto en los contratos de seguro, como los vasos que se quiebran y las alfombras que se estropean porque se deja abierta la llave de la bañera, y que, por consiguiente, no era justo que trataran de cargar el precio del colchón quemado en la cuenta de un crítico de cine cuyo único lujo burgués era fumar dormido. No hubo nada que hacer: el hotel cobró el colchón a precio de colchón nuevo.

He recordado este percance de juventud leyendo un artículo sobre los peligros de fumar, y entre los cuales no se menciona el cáncer como uno de los más temibles. No: parece que en Estados Unidos, donde el temor al fuego es una especie de obsesión patriótica, el vicio de fumar ocasiona más incendios que cualquier otra causa. Inclusive más que cocinar. «Se calcula», dice el artículo distribuido por el servicio de noticias del *New York Times*, «que no menos de 2.500 personas mueren cada año en incendios provocados por cigarrillos, y unas 25.000 resultan lesionadas en incendios del mismo origen, en los que se registran pérdidas de más de trescientos millones de dólares.» Es probable, además, que estos desastres ocurran en lugares donde no está prohibido fumar, lo cual puede dar una idea de cuál sería el tamaño de los estragos si no existiera ningún límite al albedrío de los fumadores.

Algún piloto me explicó una vez por qué en los aviones está prohibido fumar sólo durante el despegue y el aterrizaje, y no recuerdo

Publicado originalmente el 16 de febrero de 1983.

su explicación, quizá porque no me pareció muy convincente. Sin embargo, cada vez que veo a alguien fumando durante un vuelo tengo la impresión ineludible de que está cometiendo una imprudencia y que nos está sometiendo a todos los pasajeros a un riesgo adicional, además de los muchos a que nos somete por sí sola la navegación aérea. A un vecino de asiento que me preguntó el otro día sobre el océano Atlántico si me molestaba que fumara, le contesté que no, siempre que él tuviera la amabilidad de fumarse su cigarrillo apagado. Quería decirle que el humo no me estorbaba para nada, pero que no podía soportar la tensión de ver una brasa ardiendo dentro de un ámbito artificial sometido a una presión de mil metros a 15.000 pies de altura, y disparado a una velocidad de novecientos kilómetros por hora.

Hasta hace unos cinco años no estaba prohibido fumar en los retretes de los aviones. Ahora no sólo hay letreros alarmantes que lo impiden, sino que en las instrucciones verbales que se imparten a través de los altavoces se subraya con un énfasis sospechoso, y a veces sin ningún motivo aparente, que está prohibido fumar en los lavabos. Hay indicaciones muy creíbles de que esa prohibición fue el resultado de un accidente atroz que ocurrió hace unos seis años en un aeropuerto de París, cuando un avión gigante de una empresa latinoamericana se precipitó a tierra a pocos metros de la pista. La investigación del accidente, que yo sepa, no fue nunca divulgada, pero hay versiones muy serias de que los pasajeros murieron asfixiados por el humo de las materias plásticas incendiadas en un lavabo. Al parecer, un pasajero había dejado allí un cigarrillo encendido.

Es fácil imaginar por qué me siento tan a gusto contando estos horrores. Sucede que soy un fumador retirado, y no de los menores. Hace poco le oí decir a un amigo que prefiere ser un borracho conocido que un alcohólico anónimo. Yo había dicho otra cosa menos inteligente, pero tal vez más sincera en ese momento: «Prefiero morirme antes que dejar de fumar». Sin embargo, antes de dos años había dejado. De eso hace ahora catorce años, y había fumado desde la edad de dieciocho, y a un ritmo que no le conozco a muchos fumadores empedernidos. En el momento en que me detuve, me fumaba cuatro cajetillas de tabaco negro en catorce horas: ochenta cigarrillos. Alguien había calculado que de esas catorce horas útiles en la vida malgastaba cuatro horas completas en el acto simple de sacar el cigarrillo, buscar los fósforos y encenderlo. Fumaba en exceso, pero no era un adicto catastrófico: nunca me quedé dormido fumando, ni quemé un sillón o una alfombra en una visita, ni fumé desnudo, pero caminando con los zapatos puestos —que es una de las cosas de peor suerte que se pueden hacer en la vida–, ni olvidé un cigarrillo encendido en ninguna parte, y mucho menos, por supuesto, en el lavabo de un avión. No estoy tratando de hacer proselitismo, aunque suelo hacerlo y me gusta, como a todos los conversos. Al contrario, debo decir que en mis largos y dichosos años de fumador no tuve nunca un acceso de tos, ni ningún trastorno del corazón, ni ninguno de los males mayores y menores que se atribuyen a los grandes fumadores. En cambio, cuando dejé de fumar contraje una bronquitis crónica que me costó mucho trabajo superar. Más aún, no dejé de fumar por ningún motivo especial, y nunca me sentí ni mejor ni peor, ni se me agrió el carácter ni aumenté de peso, y todo siguió como si nunca hubiera fumado en mi vida. O mejor aún: como si aún siguiera fumando.

Durante muchos años repetí un chiste flojo: «La única manera de dejar de fumar es no fumar más». Mi mayor sorpresa en este mundo es que cuando dejé de fumar comprendí que aquél no era un chiste flojo, sino la pura verdad. Pero la forma en que ocurrió merece recordarse, por si estas líneas llegan ante los ojos de alguien que quisiera dejar de fumar y no ha podido. Sucedió en Barcelona, una noche en que salimos a cenar con el médico Luis Feduchi y su esposa, Leticia, y él andaba feliz porque había dejado el cigarrillo hacía un mes. Admirado de su fuerza de voluntad, le pregunté cómo lo había conseguido, y me lo explicó con argumentos tan convincentes, que al final aplasté la colilla de mi cigarrillo en el cenicero, y fue el último que me fumé en la vida. Dos semanas después el doctor Luis Feduchi volvió a fumar, primero en una pipa apagada, después en una pipa encendida, y después en dos, en tres y en cuatro pipas diferentes, y ahora en una preciosa colección de cuarenta pipas de todas las clases. A veces, para descansar de tantas pipas, fuma tabacos puros de todas las marcas, sabores y tamaños. Su explicación es válida: nunca me dijo que había dejado de fumar, sino que había dejado el cigarrillo.

Todas estas experiencias –que tal vez no sean más que las ráfagas de envidia que a veces deben sentir los curas que colgaron los hábitos– me permiten pensar que, a fin de cuentas, tal vez sea lo mismo fumar que no fumar. Pero que quienes dirigen las campañas contra el tabaquismo no debían ser los médicos y psicólogos –que, después de todo, no han logrado convencer a muchos–, sino que debía de ser una de las tantas y fructíferas atribuciones de los bomberos.

Un joven de Checoslovaquia abandonó su país con el ánimo de hacer fortuna. Al cabo de veinticinco años, casado y rico, volvió a su pueblo natal, donde su madre y su hermana tenían un hotel. Sólo por hacerles una broma, el viajero dejó a su esposa en otro hotel del poblado y tomó una habitación en el hotel de la madre y la hermana, quienes no lo reconocieron después de tantos años de separación. Su propósito, al parecer, era identificarse al día siguiente durante el desayuno. Pero a medianoche, mientras dormía, la madre y la hermana lo asesinaron para robarle el dinero.

Éste es el nudo de *El malentendido*, la conocida obra de teatro de Albert Camus, inspirada en una de esas historias sin origen cierto que la tradición oral transmite —con muy ligeras modificaciones—, no sólo en el espacio, sino también en el tiempo. Roger Quillot, autor de las notas con que el drama de Camus fue publicado en la edición de La Pléyade, dice que la historia se encuentra con muchas variantes en numerosos países y que desde la Edad Media aparecía en la tradición oral o en la prensa. «M. Paul Benicaou me señaló en particular una vieja canción de Nivernais, *El soldado muerto por su madre*», escribe Roger Quillot. «De igual modo, en *Mon Portrait*, de Louis Claude de Saint-Martin, se refiere esta historia como un caso policíaco que habría ocurrido en Tours en junio de 1796. Por último, el escritor latinoamericano Domingo Sarmiento asegura que la misma leyenda es muy conocida en Chile, y una acción idéntica es el tema de la tragedia titulada *El 24 de febrero*, de Zacarías Werner.»

No sé si existía, aunque debería existir, una antología de esas historias que se repiten por todo el mundo y de las cuales —quienes las cuentan— aseguran haber sido testigos presenciales. O bien los narradores mienten, cosa que es muy probable, o bien es cierto que las historias ocurren una y otra vez a través de distintas culturas y de épocas diversas. Una de ellas, de las cuales se ha hablado otras veces

Publicado originalmente el 23 de febrero de 1983.

en esta columna, es la del automovilista que recoge en la carretera a una mujer solitaria, que desaparece de su asiento vecino en el transcurso del viaje. Hay un dato constante: en todas las versiones de los distintos países: en el sitio donde la mujer es recogida ha habido un accidente atroz en el que ha muerto una mujer vestida del mismo modo. La última vez que escribí sobre esto recibí numerosas cartas en las que me decían que el mismo caso había ocurrido en lugares diversos, y en algunas se daban hasta los nombres de los protagonistas. Alguien me mandó la fotocopia de varias páginas de un libro de mi amigo el escritor catalán Manolo Vázquez Montalbán, que había sido publicado mucho antes de que la prensa francesa publicara la historia ocurrida en el verano anterior. Vuelvo al tema ahora porque un amigo de México, cuya palabra no se puede poner en duda, me cuenta que vivió la misma historia un día de la semana pasada, a pleno sol, cuando regresaba desde Taxco a la ciudad de México por una autopista tan concurrida que uno se pregunta a veces cómo es que no se han instalado semáforos en algunas esquinas.

Sin embargo, la más extraña, horrorosa y complicada de estas historias recurrentes se supone que ocurrió en algún lugar de Afganistán hace muchos años. Es la de un hombre que se encontró por casualidad en un mercado con una mujer que le pareció la más bella del mundo. De acuerdo con las costumbres locales, no trató de seducir a la hermosa con los sanos recursos occidentales, sino que concertó la boda con sus padres. La muchacha aceptó por obediencia, pero le puso al marido la condición no sólo de dormir en habitaciones separadas, sino también la de no tener ningún tipo de relaciones sexuales, salvo en las escasas ocasiones en que ella lo dispusiera. El marido se sometió a semejantes normas contra natura hasta una noche en que descubrió que su esposa solía escapar de la casa mientras él dormía para visitar un amante secreto, que mantenía desde antes de su matrimonio en una cabaña no muy distante de la suya. Entonces el marido la siguió armado con su espada, esperó a que ella saliera de la casa ajena para volver a la suya y decapitó al amante con un tajo certero. Luego limpió la espada con tanto cuidado que cuando la esposa la examinó —sospechando quién podía ser el autor del crimen— no encontró ningún rastro que le permitiera culpar al marido. Éste, por su parte, coronó por fin su ambición de dormir y folgar con la mujer más bella del mundo, la cual terminó por ser feliz con él y le dio tres hijos. Muchos años después, cuando pasaron por casualidad frente a la cabaña del amante muerto, la mujer no pudo disimular su nerviosismo y le pidió al marido que se alejaran de allí lo más pronto posible. Entonces el marido cometió la imprudencia que lo delató. «En aquel tiempo no tenías tanta prisa», dijo. La mujer no hizo ningún gesto revelador, pero aquella noche, cuando el marido regresó a su casa, encontró a los tres hijos decapitados con la misma espada con que él había decapitado a su rival y nunca más en su vida volvió a tener la menor noticia de la mujer más bella del mundo.

La historia, con toda clase de variaciones, se repite con frecuencia por todas partes; pero el último que la contó fue un profesor universitario que aseguró haber estado en Afganistán y haber conocido al protagonista. Y añadió un dato terminante: el hombre tenía una cicatriz en la espalda, causada por su propia mujer con la espada insaciable cuando trató de decapitarlo también a él. Esto convertiría en contemporánea una historia que se suponía muy antigua, de los tiempos en que las espadas se anticipaban a las armas de fuego en los crímenes pasionales, y cuando no era posible concebir una historia con un final feliz, de esos que hoy se consideran como un desastre literario.

Leí *Las mil y una noches* cuando apenas empezaba a tener uso de razón, y tal vez sea una más de las razones por las cuales las sigo apreciando como mi libro inolvidable. Ahora bien: cada vez que oigo contar la historia del amante decapitado creo emociones dormidas en aquellas lecturas brumosas de mi infancia, pero no logro encontrar la historia en las distintas versiones que tengo de los relatos fantásticos de Scherezade. Tropiezo siempre, en cambio, con otra parecida y tremenda: la historia de la mujer que en su casa sólo comía granitos de arroz, uno por uno y pinchándolos siempre con un alfiler, hasta que su marido descubrió que no comía porque de noche escapaba de la casa para irse a comer muertos en el cementerio. Y tropiezo con otra de las más hermosas que he leído jamás: la historia del pescador que le pide a su vecino un plomo para su red, con la promesa de que le dará a cambio el primer pescado de la jornada. Cumple su promesa, y cuando la mujer del vecino destripa el pez para prepararlo, le encuentra en el estómago un diamante del tamaño de una avellana. Encuentro éstas y muchas historias de maravillas, pero no logro encontrar el origen de la otra terrible de la mujer más bella del mundo que decapitó a sus tres hijos porque el marido había decapitado al amante. ¿Habrá un lector benévolo que me ayude a encontrarlo?

¿PARA QUÉ SIRVEN LOS ESCRITORES?

La forma airada, burlona o despectiva en que la prensa de Estados Unidos ha reaccionado ante la reunión de casi quinientos intelectuales en la venerable Universidad de la Sorbona, en París, es más bien un síntoma de que su trascendencia fue mucho más significativa de lo que pudiera parecer a simple vista. Los argumentos contrarios, en general, son los más frívolos. Los que se repiten con más ahínco es que los invitados viajan gratis en primera clase, y algunos, sobre todo los norteamericanos, lo hicieron en el Concorde, lo cual sólo quiere decir que fueron menos cómodos pero llegaron primero; que todos fueron alojados en el hotel Plaza Athenée, que es el vividero predilecto de los grandes burgueses del mundo cuando visitan París, y que se hartaron de manjares exquisitos en los restaurantes más refinados. No hay nada falso en estas informaciones, pero tampoco nada raro.

Tanto el mundo de este lado como el del otro lado se han complacido siempre en complacer a los intelectuales, mientras éstos no levanten la mano contra sus gobiernos soberanos, y nunca he visto nada de reprochable en que escritores, artistas y científicos disfruten de la buena vida que los burgueses han tomado para ellos solos. Si el gobierno francés los invitó, está muy bien que lo haya hecho con todos los honores, y habría estado muy mal que lo hubiera hecho de otro modo.

El otro reproche que hace la prensa de Estados Unidos es menos frívolo. Dice que Francia está tratando de recuperar un liderazgo cultural que perdió hace mucho tiempo, y que para lograrlo está dispuesta a gastarse hasta el dinero que no tiene y en un momento en que el país —como casi todos los del mundo, aun los más desarrollados— atraviesa el desierto de la crisis y se enfrenta al fantasma del desempleo. La prensa de Estados Unidos, aun la que suele ser la más serena, ha aprovechado la ocasión para decir que Francia no es

Publicado originalmente el 2 de marzo de 1983.

ya la de los grandes días de gloria de su himno nacional, que hace más de veinticinco años que sus novelistas no escriben una gran novela, ni sus poetas cantan con la misma voz de otros tiempos, ni sus músicos hacen lo mejor, ni sus pintores hacen ni siquiera lo menos peor. Son exageraciones de verdades que, sin duda, los franceses conocen mejor que nadie, pero ellas sirven más bien para celebrar que para reprochar las buenas intenciones de un buen ministro de la Cultura que trata de recuperar el paraíso perdido. Está en su derecho, y si algo podemos hacer los amigos de Francia por ayudarlo no hay en eso nada de reprochable, sino todo lo contrario.

Yo tengo y he tenido siempre grandes reservas, en general, por los congresos de escritores y artistas. Sobre todo, en los últimos tiempos, en que se han puesto de moda hasta un punto en que cualquier intelectual más o menos solicitado podría pasar el año entero viajando por el mundo entero y de ese modo malgastar su tiempo entero sin tener que hacer nada más fructífero. Según cálculos a primera vista, y según se desprende de las cartas y telegramas de invitación que han pasado por mis manos en los últimos días, en este año que apenas comienza habrá 63 congresos, encuentros, reuniones o seminarios masivos de escritores.

Los costos sumados alcanzarían sin duda para resolver la situación de muchos escritores de menores recursos que de veras quisieran tener mejores condiciones para escribir. Pero no es eso lo que importa tanto, como el hecho demostrable por la experiencia de años anteriores de que los congresos de escritores no sirven para nada. Otra cosa son, sin duda, los congresos científicos, en los cuales se discuten e intercambian secretos útiles para el género humano. Pero los escritores no tenemos secretos que intercambiar, ni su divulgación —en caso de que los tuviéramos— serviría para nada. Durante muchos años me negué a asistir a congresos de escritores, pero cada vez me ha costado más trabajo decir que no, por razones que casi nunca tienen algo que ver con la literatura. El resultado ha sido siempre el mismo: me he aburrido a más no poder, asediado por preguntas cuyas respuestas todo el mundo conoce, o abrumado por las discusiones que los profesionales de los congresos sostienen sin tregua, aunque sólo sea para que el congreso exista. En general, no hay muchas probabilidades de que de veras todos los asistentes tengan alguna posibilidad de participación. No quiere decir, por supuesto, que los escritores y artistas sean borregos fáciles de pastorear. Al contrario, a muy pocos escritores les gustan los congresos, y si asisten a ellos es por razones que no tienen nada que ver con el congreso mismo. La mayoría —sobre todo, los que escriben bien— se aburren a muerte durante los debates y sólo desean que se levante la sesión para que vuelva a empezar la vida. La verdad es que unos asisten para poder viajar, otros asisten por conocer lugares que de otro modo no podrían visitar, o por volver a encontrarse con amigos que de otro modo no podrían ver. Este último motivo es tal vez el más perdonable de todos, y el único, en definitiva, que me ha movido para asistir alguna vez a una reunión multitudinaria de intelectuales.

La otra noche se discutía en una fiesta cuáles son los métodos que debían utilizar los estados benévolos para promover la creación artística. Mi respuesta fue bien simple: lo único que el estado puede hacer es asegurarles a los escritores las proteínas necesarias desde que nacen, y luego asegurarles las condiciones para que puedan hacer su oficio sin sobresaltos y con una independencia absoluta. Es una verdad de perogrullo, pero, por desgracia, no hay otra.

Por fortuna, lo que Jack Lang ha intentado en la Sorbona no es que los escritores escriban mejor, sino algo original aun en el París de Francia, donde tantas cosas originales se han inventado. Ha tratado de que los artistas y los economistas se pusieran de acuerdo sobre lo que se puede hacer desde los predios de la cultura para enfrentar la crisis de este mundo. Un invitado norteamericano, quién sabe si en serio o en burla, dijo que los economistas han enredado de tal modo la economía que tal vez sean los artistas los únicos capaces de desenredarla. No sé por qué esta frase me hizo recordar el hermoso y sabio discurso de Saint-John Perse cuando recibió el Premio Nobel, y en el cual demostró que los métodos de la poesía podían ser de una enorme utilidad para la investigación científica. Me acordé también de otro episodio menos significativo, pero que, de todos modos, venía muy al caso. Hace varios años, un profesor de sociología de una universidad de Estados Unidos hizo una encuesta entre los novelistas latinoamericanos para averiguar cuál era el método que éstos utilizaban en sus esfuerzos por descifrar la realidad de sus países. El profesor consideraba que la metodología de los sociólogos había fracasado en esa tentativa, y pensaba que el método de los novelistas podía ser útil también para los sociólogos.

Nunca conocí los resultados de su encuesta, pero sé que la respuesta de varios escritores fue la misma: lo único que hacían para tratar de interpretar la realidad era observar la vida, otra verdad de perogrullo que, sin duda, era la que el ministro de la Cultura de Francia, que es un hombre inteligente y febril, esperaba escuchar también como resultado final del encuentro en la Sorbona. Al fin y al cabo, también a la cultura, como a tantas cosas de nuestro tiempo, le está haciendo falta una buena dosis de sentido común.

EL PAPA, EN EL INFIERNO

Hay muchos motivos para preguntarse qué raros designios de Dios inspiraron al papa Juan Pablo II la determinación imprevisible de visitar el infierno de la América Central. El mismo día de su llegada a Costa Rica estaban enterrando en Managua a diecisiete milicianos adolescentes abatidos por bandas somocistas que penetraron en Nicaragua desde Honduras, y seis guatemaltecos eran fusilados por el régimen militar de su país, presidido por un general fanático que oficia como sumo sacerdote de una secta religiosa que nada tiene que ver con la Iglesia católica. En menos de un año, el régimen de este ayatollah de grueso calibre ha dado muerte a más de 10.000 indígenas en el genocidio más barato de estos tiempos en nuestra América. Sin embargo, tal vez el país donde resultaba más embarazosa la visita del Papa era la república de El Salvador, en cuya capital fue asesinado, en un acto con muy pocos antecedentes, nada menos que el arzobispo primado, en el altar mayor de la catedral y en el instante de la elevación.

Sin embargo, hasta donde se supo, no fue con El Salvador ni con Guatemala con los que el Vaticano tuvo más reticencias para la visita del pontífice máximo, sino con un tercer país, Nicaragua, cuyo gobierno no ha matado a nadie y cuenta además con la colaboración activa de sacerdotes católicos en niveles muy elevados y con el apoyo del clero popular. No obstante, la presencia de sacerdotes en el gobierno fue el principal obstáculo en las largas negociaciones secretas que precedieron y que, por fin, hicieron posible la visita. La decisión original del Vaticano era que el Papa fuera a Nicaragua sólo como jefe espiritual de la Iglesia, sin tomar en cuenta para nada a las autoridades terrenales. Éstas, con toda razón, se permitieron recordar a los emisarios papales que Juan Pablo II es el gobernante máximo de un estado con el cual Nicaragua no sólo mantiene relaciones diplomáticas, sino que son relaciones muy buenas. No lo son tanto,

Publicado originalmente el 9 de marzo de 1983.

en cambio, las del gobierno con la jerarquía eclesiástica, a causa de la colaboración de sacerdotes católicos en el proceso de transformación social que se lleva a cabo en el país. La pretensión del Vaticano, ya en el último caso, era que estos sacerdotes no estuvieran presentes a la llegada del Papa, para que no tuviera que saludarles, y con ellos al que ocupa nada menos que el cargo de canciller –el padre Manuel d'Escoto. No había ningún problema, porque aquél debería estar en la Conferencia de los Países No Alineados de Nueva Delhi; pero los otros estarían allí, con todo su derecho. El Vaticano terminó por aceptar que la visita fuera oficial y no sólo pastoral, y el gobierno de Nicaragua fue tan estricto en sus reglas de cortesía que el comandante Daniel Ortega, miembro de la Junta de Gobierno, había de llegar tarde a la reunión de Nueva Delhi para estar presente en Managua a la llegada del sumo pontífice.

Estos tejemanejes, cuyos propósitos políticos eran inocultables, obligaban a preguntarse, por consiguiente, qué raros designios de Dios determinaron esta visita. No parecía probable que los informadores del Papa fueran los mejores con respecto al drama de América Central, una tierra tan distante del Vaticano que uno tiene derecho a preguntarse si Juan Pablo II sabía con precisión sobre cuál de los dos océanos estaban las costas de Honduras; una pregunta que, por lo demás, estoy seguro de que muy pocos lectores de esta nota se atreverían a contestar sin vacilación. Cuando el papa Juan Pablo II me hizo el inmenso favor de recibirme en audiencia privada pocos meses después de su elección, me causó una agradable y muy grata impresión, de la cual he hablado mucho en cuantas oportunidades he tenido. Había una rara contradicción entre su fortaleza física, su estructura de atleta, y el calor humano, casi tierno, de sus buenas maneras. Pero algo más me llamó la atención, y fue un cierto y comprensible condicionamiento mental que le impedía entender una situación de cualquier parte del mundo si no la relacionaba con la de Europa del Este. De acuerdo con el propósito de mi visita, le hice una exposición muy breve sobre la situación de los presos y desaparecidos políticos en Argentina, que era lo que causaba mayor preocupación en aquel momento. El Papa expresó su estupor con una frase: «¡Qué horror, es como en la Europa del Este!». Fui capaz de imaginarme en aquel momento que el Papa no lo creía tanto como lo decía, sino que alguien le había preparado para mi visita y tal vez le había dicho, con el simplismo de ciertos curas de las viejas novelas españolas, que yo era un comunista de los que comen niños crudos, algo que nunca he sido ni seré, entre otras cosas porque no corresponde ni a mi concepción de la vida ni a mi formación ni a

mi carácter. Pero era probable que alguien se lo hubiera dicho al
Papa para que no fuera a decirme algo comprometedor, y al buen
pastor no se le había ocurrido nada más que defenderse con un lati-
guillo invencible: «Es como en la Europa del Este». Yo no estaba
ahí, por supuesto, para entablar con el sumo pontífice una polémica
sobre las analogías y diferencias entre el Oriente y el Occidente,
sino para tratar de que nos ayudara a encontrar a los desaparecidos
del sur, pero tuve razones para preocuparme por lo que iba a hacer
en su inminente visita a México, que sería la primera de su mandato. La jerarquía eclesiástica de México no se distingue por su men-
talidad progresista, y no era probable que sus informaciones le per-
mitieran al Papa colocarse del lado de los justos en su primer viaje a
las Américas. Así debió ser, en efecto, porque hubo una diferencia
muy grande entre el discurso que pronunció el Papa con motivo de
su llegada y el que pronunció después de que vio con sus propios
ojos la miseria de ciertas provincias mexicanas, cuya gravedad, dígase lo que se diga, no admitía comparaciones con la pobreza indudable de algunos sectores de la Europa del Este. Cuando visitó más
tarde Brasil, en cambio, Juan Pablo II se comportó como un buen
pastor, bien informado sobre la realidad gracias a la sensibilidad social de los obispos brasileños influyentes, entre ellos don Helder
Cámara, desde luego, y don Paulo Evaristo Arns, cardenal de São
Paulo. Ojalá que el mismo Dios que ha querido que el Papa baje
ahora a los infiernos de América Central haya querido que sus informadores hayan sido más justos.

EL ACUERDO DE BABEL

La semana anterior, cuando se preparaba la última conferencia cumbre de los países No Alineados, la ciudad de Nueva Delhi tenía un aspecto candoroso de novia nueva, en sus amplias avenidas acabadas de barrer y las aceras pintadas de blanco. Los leprosos, los encantadores de serpientes, los levitadores que en plena calle tendían una sábana en el aire y se acostaban sobre ella, todo lo que hacía de este país y de esta ciudad el territorio más fascinante y misterioso del mundo había desaparecido para alegrar la vista de más de un millar de visitantes, y entre ellos unos sesenta jefes de estado y de gobierno. Nada hacía pensar en la posibilidad de que algo sangriento pudiera ocurrir, hasta que un comando iraquí armado como para una guerra desembarcó en el aeropuerto dispuesto a enfrentarse a tiros con otro comando iraní que había llegado el día anterior. La cosa no era tan simple de resolver: los iraquíes habían aterrizado en un avión especial sin permiso de las autoridades indias y se habían introducido en la ciudad por la fuerza. En cualquier momento, aquellos dos grupos feroces hubieran podido trabarse a tiros en el recinto mismo de la conferencia, y tal vez terminar de una vez por todas con el movimiento multinacional más heterogéneo y difícil de cuantos existen en el mundo, pero que es también, sin duda, el que tiene un porvenir más promisorio.

Ni Irak ni Irán parecían dispuestos a impedir la desgracia. Ante el llamado a la paz que hizo la señora Indira Gandhi en el discurso inaugural, un delegado iraní declaró: «Nos complace mucho esta iniciativa, pero la solución de este conflicto no está aquí, sino en el campo de batalla». Otros dos fantasmas que pesaban sobre la conferencia desde antes de su inauguración eran los de la intervención soviética en Afganistán, que es un país No Alineado, y la vieja disputa sobre quién debía sentarse en el sillón de Kampuchea. En su informe final sobre sus tres años como presidente de los No Alineados,

Publicado originalmente el 16 de marzo de 1983.

Fidel Castro se refirió al primero de estos problemas con mucha precisión, y reveló cuáles habían sido sus esfuerzos de mediador para lograr una solución a la presencia soviética en Afganistán. Es un conflicto que Fidel Castro lleva sin duda muy cerca del corazón, pues surgió apenas unos meses después de que él asumió la presidencia del movimiento y no dejó de pesar un solo instante sobre sus hombros. Esta séptima conferencia, por supuesto, tampoco había de resolverlo, y es uno de los más incómodos de cuantos arrastra la señora Gandhi desde el principio de su mandato. En cuanto al problema de Kampuchea, el propio canciller de Vietnam, que es un viejo amigo con un sentido del humor inagotable, me dijo: «Hemos hecho todo lo que nos correspondía para que el asunto de Kampuchea no fuera un obstáculo». La solución, por supuesto, era la más fácil: dejar el sillón vacío mientras la vida resuelve el problema.

Cuando uno piensa en la serenidad, la paciencia y la madurez que deben hacer falta para ser un jefe de estado, hay razones para preguntarse cómo es que en medio de tantas responsabilidades enormes haya hombres de poder que parecen dispuestos a jugárselas todas por cualquier tropiezo baladí. Fue ése el caso del muy admirable Yasir Arafat, que amenazó con retirarse de la conferencia si no le adjudicaban el turno que quería para pronunciar su discurso. Poco antes, un gobernante africano hizo la misma amenaza si no se permitía la entrada al país de un miembro de su delegación que estaba detenido en el aeropuerto porque no se había vacunado contra la fiebre amarilla. El hecho merece una precisión: cuando uno viene a la India, los médicos occidentales hacen toda clase de advertencias alarmantes —y aun alarmistas— contra las enfermedades terroríficas que amenazan a los extranjeros, y obligan a aplicarse cinco vacunas distintas y a hacerse tratamientos preventivos contra el tifus y la malaria. A los indios no les preocupan demasiado estos miedos occidentales, pero en cambio son de una severidad intransigente con las vacunas de la fiebre amarilla, porque ésta no existe en la India y los varios millones de monos sagrados que hay aquí se contagian con facilidad. Es un temor tan arraigado en los indios protectores de sus deidades, que el delegado africano —a pesar de las amenazas de su presidente— tuvo que regresar a su país.

Lo asombroso es que las cosas marchen. En realidad, el movimiento de los No Alineados, ahora que ha redondeado su centenar de miembros con el ingreso de Colombia, es la representación moderna de la torre de Babel. No tanto por la diversidad de sus idiomas, que son muchos, desde luego. Una imagen triste de lo que ha sido el colonialismo se refleja muy bien en el hecho de que los delegados de cien países del Tercer Mundo tienen que entenderse entre sí en tres idiomas europeos: inglés, francés y español.

Los únicos que escapan a este yugo son los árabes, que se suben en la tribuna de los oradores con sus chilabas de santos y a cantar sus penas en la misma lengua en que han cantado sus antepasados desde hace casi tantos milenios cuantos tiene el mundo. Los indios, en cambio, cuyos billetes de banco tienen la denominación escrita en catorce lenguas nacionales, se ven obligados a hablar inglés no sólo en la tribuna, sino aun en sus relaciones privadas. Sin embargo, no es por la confusión de las lenguas que se define la dimensión babélica de este movimiento, sino por la multiplicidad de sus ideas y sus posiciones políticas. Aquí hay desde el extremo más reaccionario hasta el más progresista. Quienes se opusieron en Colombia al ingreso del país en los No Alineados utilizaron el argumento, más bobo que simplista, de que íbamos a caer en los brazos de la Unión Soviética, porque esos eran los designios de Fidel Castro. Lo que tendrían que aclarar ahora es algo que sabían de sobra y nunca dijeron: en el instante en que la delegación colombiana ingresó en el recinto de los nuevos miembros hacía ya veinticuatro horas que Fidel Castro no era el presidente del movimiento. Y no lo era –pienso yo– con un gran alivio, mientras la señora Gandhi, con una serenidad y una dulzura que no parecen faltarle aun en los instantes de la peor incertidumbre, se preparaba para pastorear por los próximos tres años a los dirigentes de las dos terceras partes del mundo.

Pues al cabo de cinco días de discursos interminables, repetitivos y en su mayoría soporíferos, las divergencias espinosas del primer día se habían canalizado por lo único que estos cien países tienen en común: la urgencia de ponerse de acuerdo a cualquier precio en la lucha contra la desgracia sin término del subdesarrollo y la explotación extranjera. Lo único, digo, pero que es más que suficiente para sufrir estos cinco días agotadores. Quienes tuvimos la paciencia de vivirlos de cerca –a veces con un sentimiento de incertidumbre, a veces de desaliento, pero también a veces con una gran esperanza– no olvidaremos con facilidad la experiencia de haber visto cómo surgió un acuerdo final, útil y alentador, en medio de la algarabía de Babel.

¡MANOS ARRIBA!

Muy pocos años antes de su muerte, el general Omar Torrijos visitó varias haciendas ganaderas de la costa caribe de Colombia, y a todo el que quiso oírlo le manifestó su entusiasmo por el desarrollo técnico de aquella industria, en especial en cuanto a la genética. Tanto fue así, que repitió la visita más de una vez, e inclusive estuvo pensando en servirse de la experiencia colombiana para tratar de implantarla en Panamá. Sin embargo, lo que más le impresionó, en definitiva, fueron los ejércitos privados con que los magnates de la industria ganadera protegían sus bienes y sus vidas de las incursiones frecuentes de guerrilleros o bandoleros comunes. Torrijos, que, como todo el mundo lo sabe, era un hombre de armas, se sorprendió no sólo de la cantidad de gente armada y con instrucción militar que lo escoltaba durante sus visitas, sino de la cantidad y la clase de las armas de guerra que llevaban consigo. Al mismo tiempo, el gobierno de Colombia se sentía obligado a protegerlo con una escolta suplementaria que no lo perdía de vista un solo instante. «Uno termina por no saber si lo están protegiendo o si lo están vigilando», dijo alguna vez, con su humor de siempre, y decidió renunciar, contra su propio corazón, a uno de los pocos placeres solitarios de su edad adulta, que era el de viajar de incógnito por los países vecinos, y en especial por Colombia. Ese placer, entre paréntesis, estuvo a punto de crear un problema grave entre Panamá y Brasil, cuando los servicios de seguridad de este último país descubrieron que el general Torrijos –que ya no era jefe de gobierno, pero que seguía siendo comandante de la Guardia Nacional– estaba viajando por tierras brasileñas con el nombre cambiado. Por muy inocente y explicable que sea la intención, es un delito grande entrar en Brasil e inscribirse en los hoteles con la identidad tergiversada, y sólo la comprensión del gobierno impidió que la travesura del general Torrijos se convirtiera en un incidente grave, y se mantuvo en secreto. Hasta este momento, supongo.

Publicado originalmente el 23 de marzo de 1983.

Pero lo que viene a cuento es que el general Torrijos, ante las tropas privadas de los terratenientes colombianos, tuvo la impresión de estar viviendo de nuevo una experiencia que ya había vivido años antes en El Salvador, donde hizo un curso de especialización en sus primeros años de militar. «Así empezó todo», decía, refiriéndose a las bandas de criminales a sueldo que sembraban el terror en El Salvador, y que habían llegado al extremo inimaginable de acribillar a un arzobispo en el altar mayor de la catedral y en el instante de la elevación. Pensaba que los integrantes de los ejércitos privados —muchos de los cuales, sin duda, no eran más que campesinos sin empleo— terminarían por derivar hacia la práctica de una delincuencia común incontrolable. Era tan sincera su alarma, que consideró como un deber transmitírsela al presidente de Colombia, doctor Turbay Ayala, y a su ministro de la Defensa, el general Camacho Leyva, en el curso de una entrevista informal que sostuvo con ellos en Cartagena. El doctor Turbay lo escuchó con el paternalismo sarraceno con que solía tratar al general Torrijos, pero el general Camacho Leyva no disimuló su contrariedad por lo que le pareció una impertinencia, y cambió de tema con una frase terminante.

«No se equivoque, general», le dijo al general Torrijos. «En este país hay paz social.»

También aquel incidente quedó como un secreto entre muy pocos. Hasta este momento, en que me ha parecido oportuno evocarlo ante la imprevisible carrera de armamento civil y militar que está padeciendo Colombia. De una manera u otra, con mayor o menor intensidad, con diferentes motivos y razones distintas, mi país ha vivido una guerra interna con cuentagotas desde el primer instante de su ser natural. De modo que ha sido siempre un país de gente armada, y me temo que sea ésa su naturaleza real, por debajo del manto de legalismo con que tratamos de convencer al mundo, e inclusive a nosotros mismos. No parece probable que en ningún otro país haya tanta gente armada, ni con tantos ánimos para usar sus armas. Al término de la guerra de los mil días, en 1903, los coroneles menores de edad y de ambos bandos volvieron a los colegios con sus armas al cinto, y no faltaban los que se trenzaban a tiros por un pleito de trompos durante el recreo. Mi abuelo, que era un coronel revolucionario con vocación pacífica, durmió siempre con el revólver debajo de la almohada, y para mí era algo cotidiano desde que tengo recuerdos que todo el que entrara en la casa y saliera de ella llevara sus armas a la vista en los tiempos intrépidos de Aracataca. Supongo que lo único que ha cambiado desde entonces es que ahora se llevan un poco más escondidas.

Sin embargo, creo que nunca como en los últimos tiempos ha habido una aceleración más inquietante del armamentismo nacional. Si no recuerdo mal, fue el mismo general Camacho Leyva quien recomendó hace unos tres años a los civiles pacíficos que aprendiéramos a defendernos de agresiones que las autoridades no estaban en condiciones de prevenir o contrarrestar. Aquello fue como el anuncio de lanzamiento de una nueva marca de pomada con virtudes afrodisíacas, pues el propio Instituto de Industrias Militares abrió a los civiles su tienda bien surtida de armas para matar. Las exigencias no eran más difíciles que las necesarias para obtener un pasaporte, y según mis datos de hace dos años, se vendían con sus licencias respectivas hasta doscientas armas diarias. Esto quiere decir que desde el anuncio de promoción del general Camacho Leyva deben haberse vendido unas 200.000 armas cortas con licencia, y sólo en el almacén de Bogotá. Más aún: en condiciones especiales, una persona puede adquirir un arma corta y otra larga, de modo que el cálculo puede ser insuficiente. Y todo esto sin contar las armas sin licencia, que son las que más se venden.

Un amigo mal pensado me decía hace algunos años en Bogotá que bastaba poner de acuerdo a todos los porteros y celadores del país para tener un pie de fuerza civil tan numeroso como el militar. No es fácil saber cuántos hay, pero es, sin duda, uno de los oficios más solicitados y tal vez de los mejor pagados en estos malos tiempos. Más que espacio y comodidad, en los edificios de apartamentos de las ciudades de Colombia se está vendiendo seguridad armada. La suma es astronómica: entre los militares, guerrilleros urbanos y rurales, los terroristas, los traficantes de droga y de todo lo demás, los contrabandistas de toda índole, los atracadores comunes, los asesinos a sueldo, los celadores y guardaespaldas, y los ya incontables civiles de buena índole con licencia para no dejarse matar, tal vez los escritores somos de los pocos colombianos que ya no tenemos más armas que la máquina de escribir. Es un arsenal de proporciones incalculables, cuyas posibilidades de destrucción ponen la carne de gallina. El tema es bien conocido, desde luego, pero no me ha parecido nada superfluo pedir a los lectores aunque sean cinco minutos de reflexión sobre el pronóstico sobrecogedor del general Omar Torrijos, un militar de buen corazón que tenía más de visionario que de guerrero.

«ALSINO Y EL CÓNDOR»

Entre las muchas que se disputaban este año la postulación para el Oscar a la mejor película extranjera, cuatro llegaron con márgenes muy estrechos a la decisión final: la turca *Yol*, que compartió con *Missing*, de Costa-Gavras, la Palma de Oro en el Festival de Cannes del año pasado; *Fitzcarraldo*, alemana, de Werner Herzog, que ganó en el mismo certamen el premio a la mejor dirección; *La noche de San Lorenzo*, Italia, de los hermanos Taviani, que en el mismo certamen se llevó el premio especial del jurado, y *Alsino y el cóndor*, de Nicaragua, dirigida por el chileno exiliado en México Miguel Littin, que andaba dando sus primeros pasos por el mundo. Las tres primeras las conocía muy bien, porque me correspondió discutir sobre ellas en mi condición de jurado en Cannes, y todas son de una calidad tan alta que en un momento determinado se disputaban el primer lugar para la Palma de Oro. En cambio, tenía muy buenas referencias de amigos que habían visto *Alsino y el cóndor* en privado, pero no había tenido oportunidad de verla. Acabo de verla ahora, sorprendido por la noticia de que fue escogida en Los Ángeles como candidata al premio de la mejor película extranjera, en medio de competidores tan bien calificados. Es muy buena.

Sin embargo, tal vez su excelencia no es su mérito mayor, sino el hecho de que lo sea a pesar de las condiciones casi inverosímiles en que fue realizada. Al principio no había ni argumento ni plata, pero el Instituto del Cine de Nicaragua quería que Miguel Littin hiciera una película para ellos, y Miguel Littin quería hacerla, sobre una idea antigua y no muy promisoria, inspirada en un cuento del escritor chileno Pedro Prado, sobre un niño del campo que se tiraba de los árboles porque quería volar.

Era un buen ejemplo de la obsesión lírica de Miguel Littin, que es el aspecto más vulnerable de sus películas, pero al cual se rinde siempre como a una amante ilusoria, a pesar de las duras críticas de

Publicado originalmente el 30 de marzo de 1983.

los críticos y de las aún más duras y secretas de los amigos que lo queremos. Por fortuna, no hay maestra más cabeza dura que la realidad. Recorriendo los campos de Nicaragua en busca de ambiente para su niño volador, en busca de árboles para que volara, en busca de justificaciones sociales para que fuera creíble la aventura de su Ícaro tropical, Miguel Littin descubrió en la memoria colectiva los recuerdos nunca contados de la guerra de liberación de Nicaragua, y se encontró de pronto —tal vez sin saberlo— con una película distinta, pero mucho más verídica y conmovedora que la que buscaba. No hay en esto nada nuevo ni raro: así ha sido el arte desde siempre.

Las circunstancias en que fue realizada podrían servir de argumento para otra película. El gobierno de Nicaragua participaba con toda clase de recursos —civiles y militares, materiales y morales—, pero sumando todo lo que se pudo conseguir en efectivo no se alcanzaban a reunir más de 60.000 dólares, que era mucho menos de lo que iba a cobrar un actor norteamericano, indispensable para el drama. Cuba contribuyó con equipo técnico, e inclusive con uno de sus directores de fotografía mejor calificados —Jorge Herrera, de 56 años—, que había asentado su prestigio con *Lucía* y *La primera carga al machete*. México contribuyó con tres actores y otros se ofrecieron como voluntarios. Nicaragua hizo la contribución más sustancial con tropas armadas, carros de combate y la única tanqueta de que disponían, y con un helicóptero que estaba destinado a ser una de las estrellas de la película. Su gloria duró muy poco: al cabo de una semana de rodaje sufrió un accidente mortal, con catorce personas a bordo, mientras hacía labores de rescate en una zona de inundaciones, y hubo que rehacer todo lo hecho hasta entonces.

El otro helicóptero, marca Bell, que es el único de que dispone el gobierno nicaragüense, cumplió su misión artística hasta el final, pero con veleidades que ningún productor le habría permitido a su estrella mejor cotizada. Cuando menos se pensaba, tenía que desplazarse a las zonas de conflicto de la frontera con Honduras, y el rodaje quedaba en suspenso hasta que el helicóptero volvía a quedar disponible. Nada, en general, permitía hacer planes definitivos. Las propias tropas de la película tenían que movilizarse cuando menos se pensaba para la defensa de las fronteras, y cuando volvían llevaban caras nuevas, armas distintas y, a veces, hasta un ánimo distinto, y había que rehacer muchas escenas para no incurrir en contradicciones visuales. En alguna ocasión, al volver de un combate protagonizaron una escena con proyectiles reales, sin que el director se diera cuenta porque se habían acabado los cartuchos de fogueo. En otra ocasión los habitantes del pueblo quisieron incendiar una tanqueta —como lo hacían durante la guerra– porque gracias a ella los somocistas de la película habían ganado un combate, de acuerdo con el guión. Un actor nicaragüense hizo con tanta propiedad el papel de sargento de Somoza que despertó las sospechas de la población, pensando que tal vez era un antiguo miembro de la Guardia Nacional infiltrado en la película. Un mal día, mientras filmaba a bordo del helicóptero, el fotógrafo Jorge Herrera se apretó las sienes con las dos manos y se quedó inmóvil con una mirada de deslumbramiento. «Era como si estuviera viendo algo que sólo él podía ver», dice Miguel Littin. Había muerto de una congestión cerebral fulminante.

El resultado de tantas contrariedades e incertidumbres fue esta película, donde el niño que quería volar no es más que un elemento circunstancial. Lo interpretó Alan Esquivel, el hijo de un trabajador de la construcción, que no sabía leer a los trece años y se aprendía los diálogos que un asistente le leía en voz alta. Es, sin duda, un actor nato, y el propio Miguel Littin dice que al cabo de pocos días le bastaba con hacerle las mismas indicaciones que a un profesional. Sin embargo, a mi modo de ver, muy personal, el verdadero drama de esta película ejemplar, el que de veras convence y conmueve, es el del capitán Frank, un instructor norteamericano interpretado de un modo magistral por su compatriota Dean Stockel. No es un actor muy conocido en la actualidad, pero los que tengan buena memoria para los nombres del cine recordarán sin duda que es el mismo que a los diez años hizo el papel de *El niño del caballo verde*, de Joseph Losey. Stockel no sólo aceptó hacer la película por una cantidad irrisoria, sino que soportó con estoicismo y buen humor los contratiempos innumerables y resistió con seriedad a las presiones políticas que se le hicieron de distintos frentes. No hay duda de que es un hombre inteligente que sabía muy bien lo que estaba haciendo.

En realidad, el capitán Frank, que se pasea a lo largo de toda la película más solo que nadie en su helicóptero solitario, no lo hace por dinero, ni por espíritu de aventura, sino por la convicción de que su tarea, y aun su sacrificio, es un tributo al triunfo de la justicia y la verdad. Es ésa la dimensión más patética de los equivocados. Y en el caso del capitán Frank lo es todavía más porque es un ejemplar perfecto, lúcido y humano, de la tercera generación que Estados Unidos manda a morir en sus guerras sucias posteriores a la última guerra mundial. Toda una cosecha invaluable de muchachos como éste fue mandada al matadero de Corea, otra al de Vietnam, y una tercera, ahora, al infierno de América Central, donde el gobierno del señor Reagan está demostrando, una vez más, que el

país más poderoso y fascinante del mundo es refractario a las lecciones terribles de su propia historia. No es posible que Dean Stockel
no sea consciente de que el capitán humano y un poco mesiánico
que encarna se ha dejado meter en una trampa sin salida, donde
lo menos grave que le ocurre es que nadie lo quiere. Estoy seguro
de que lo sabe, y ése es el gran servicio que le ha prestado a su país,
al ponerlo frente a este espejo revelador de su extraño e inmerecido
destino.

LAS MALVINAS, UN AÑO DESPUÉS

Un soldado argentino que regresaba de las islas Malvinas al término de la guerra llamó a su madre por teléfono desde el Regimiento I de Palermo, en Buenos Aires, y le pidió autorización para llevar a casa a un compañero mutilado cuya familia vivía en otro lugar. Se trataba –según dijo– de un recluta de 19 años que había perdido una pierna y un brazo en la guerra y que además estaba ciego. La madre, feliz del retorno de su hijo con vida, contestó horrorizada que no sería capaz de soportar la visión del mutilado y se negó a aceptarlo en su casa. Entonces el hijo cortó la comunicación y se pegó un tiro: el supuesto compañero era él mismo, que se había valido de aquella patraña para averiguar cuál sería el estado de ánimo de su madre al verlo llegar despedazado.

Ésta es apenas una más de las muchas historias terribles que durante estos últimos doce meses han circulado como rumores en Argentina, que no han sido publicadas en la prensa porque la censura militar lo ha impedido y que andan por el mundo entero en cartas privadas recibidas por los exiliados. Hace algún tiempo conocí en México una de esas cartas y no había tenido corazón para reproducir algunas de sus informaciones terroríficas. Sin embargo, revistas inglesas y norteamericanas celebraron este 2 de abril el primer aniversario de la aplastante victoria británica, y me parece injusto que en la misma ocasión no se oiga una voz indignada de la América Latina que muestre algunos de los aspectos inhumanos e irritantes del otro lado de la medalla: la derrota argentina. La historia del joven inválido que se suicidó ante la idea de ser repudiado por su madre es apenas un episodio del drama oculto de aquella guerra absurda.

Ahora se sabe que numerosos reclutas de 19 años, que fueron enviados contra su voluntad y sin entrenamiento a enfrentarse con los profesionales ingleses en las Malvinas, llevaban zapatos de tenis y muy escasa protección contra el frío, que en algunos momentos era de

Publicado originalmente el 6 de abril de 1983.

30 grados bajo cero. A muchos tuvieron que arrancarles la piel gangrenada junto con los zapatos y 92 tuvieron que ser castrados por congelamiento de los testículos, después de que fueron obligados a permanecer sentados en las trincheras. Sólo en el sitio de Santa Lucía, 500 muchachos se quedaron ciegos por falta de anteojos protectores contra el deslumbramiento de la nieve.

Con motivo de la visita del Papa a Argentina, los ingleses devolvieron 1.000 prisioneros. Cincuenta de ellos tuvieron que ser operados de las desgarraduras anales que les causaron las violaciones de los ingleses que los capturaron en la localidad de Darwin. La totalidad debió ser internada en hospitales especiales de rehabilitación para que sus padres no se enteraran del estado en que llegaron: su peso promedio era de 40 o 50 kilos, muchos padecían de anemia, otros tenían brazos y piernas cuyo único remedio era la amputación y un grupo se quedó interno con trastornos psíquicos graves.

«Los chicos eran drogados por los oficiales antes de mandarlos al combate», dice una de las cartas de un testigo. «Los drogaban primero a través del chocolate y luego con inyecciones, para que no sintieran hambre y se mantuvieran lo más despiertos posible.» Con todo, el frío a que fueron sometidos era tan intenso que muchos murieron dormidos. Tal vez fueron los más afortunados, porque otros murieron de hambre tratando de extraer la pasta de carne que se petrificaba dentro de las latas.

En este sentido, mucho es lo que se sabe sobre la barbarie de la logística alimenticia que los militares argentinos practicaron en las Malvinas. Las prioridades estaban invertidas: los soldados de primera línea apenas si alcanzaban a recibir unas sardinas cristalizadas por el hielo, los de la línea media recibían una ración mejor y, en cambio, los de la retaguardia tenían a veces la posibilidad de comer caliente.

Frente a condiciones tan deplorables e inhumanas, el enemigo inglés disponía de toda clase de recursos modernos para la guerra en el círculo polar. Mientras las armas de los argentinos se estropeaban por el frío, los ingleses llevaban un fusil tan sofisticado que podía alcanzar un blanco móvil a 200 metros de distancia y disponía de una mira infrarroja de la más alta precisión. Tenían además trajes térmicos y algunos usaban chalecos antibalas que debieron de ocasionarles trastornos mentales a los pobres reclutas argentinos, pues los veían caer fulminados por el impacto de una ráfaga de metralleta y, poco después, los veían levantarse sanos y salvos y listos para proseguir el combate. Las tropas inglesas estaban una semana en el frente y luego una semana a bordo del *Camberra*, donde se les concedía un descanso verdadero con toda clase de diversiones urbanas en uno de los parajes más remotos y desolados de la Tierra.

Sin embargo, en medio de tanto despliegue técnico, el recuerdo más terrible que conservan los sobrevivientes argentinos es el salvajismo del batallón de gurkhas, los legendarios y feroces decapitadores nepaleses que precedieron a las tropas inglesas en la batalla de Puerto Argentino. «Avanzaban gritando y degollando», ha escrito un testigo de aquella carnicería despiadada. «La velocidad con que decapitaban a nuestros pobres chicos con sus cimitarras de asesinos era de uno cada siete segundos. Por una rara costumbre, la cabeza cortada la sostenían por los pelos y le cortaban las orejas.» Los gurkhas afrontaban al enemigo con una determinación tan ciega, que de setecientos que desembarcaron sólo sobrevivieron setenta. «Estas bestias estaban tan cebadas», concluye el testigo, «que una vez terminada la batalla de Puerto Argentino, siguieron matando a los propios ingleses, hasta que éstos tuvieron que esposar a los últimos para someterlos.»

Hace un año, como la inmensa mayoría de los latinoamericanos, expresé mi solidaridad con Argentina en sus propósitos de recuperación de las islas Malvinas, pero fui muy explícito en el sentido de que esa solidaridad no podía entenderse como un olvido de la barbarie de sus gobernantes. Muchos argentinos, e inclusive algunos amigos personales, no entendieron bien esta distinción. Confío, sin embargo, en que el recuerdo de los hechos inconcebibles de aquella guerra chapucera nos ayude a entendernos mejor. Por eso me ha parecido que no era superfluo preguntar otra vez y mil veces más –junto a las madres de la plaza de Mayo– dónde están los 8.000, los 10.000, los 15.000 desaparecidos de la década anterior.

REGRESO A LA GUAYABA

El gerente de *El Espectador*, de Bogotá, donde yo era reportero de planta a los veinticuatro años, se acercó aquella brumosa tarde de julio a mi escritorio y me preguntó qué pensaba hacer el viernes siguiente. «Nada especial», le contesté. «¿Por qué?» Y él me contestó con una naturalidad pavorosa: «Es para que te vayas a cubrir la conferencia de los cuatro grandes». Esa conferencia entre el general Eisenhower, Harold MacMillan, Nikita Jruschov y Edgar Faure tendría lugar en Ginebra una semana después, y yo no tenía ninguna documentación para viajar. Ni siquiera la cédula de identidad, pues los periodistas solíamos identificarnos con nuestro carné profesional, que en aquellos tiempos tenía más crédito que los documentos oficiales. Sin embargo, con ese optimismo irracional que siempre he considerado como mi mejor virtud, acepté la misión: viajaría a Ginebra 72 horas después.

Hasta entonces había conocido casi todo el país en mi inolvidable oficio de enviado especial a donde quiera que ocurría algo raro, pero nunca se me había pasado por la cabeza la idea de viajar al exterior. No entendía por qué había de hacerlo ni nunca había pensado, en mi miseria secreta de los años anteriores, que alguna vez tuviera recursos para hacerlo. Sólo aquel día supe cuántas cosas difíciles eran necesarias para obtener un pasaporte: certificado de nacimiento, certificado de policía, constancia de paz y salvo con el fisco, libreta del servicio militar, tres clases de vacunas distintas y dos cartas de recomendación de personas con solvencia moral reconocida. Terminé anhelante el artículo que estaba escribiendo, fui a la oficina del poeta Álvaro Mutis, que desde entonces era un viajero empedernido, y le puse mi problema sobre la mesa. Álvaro se burló de mi ofuscación y llamó de urgencia a un gestor de una agencia de viajes, que en 48 horas me convirtió en el viajero mejor documentado del país. Su método era tan simple que no parecía

Publicado originalmente el 13 de abril de 1983.

posible: me hizo tomar veinticuatro fotografías de pasaporte, me hizo firmar unas seis hojas de papel sellado y me pidió para gastos y servicios una suma de dinero que hoy parece ridícula, pero que era más que mi sueldo mensual: mil pesos. Al día siguiente me llevó de la mano por todas las oficinas pertinentes, subiendo por escaleras excusadas, entrando por puertas prohibidas, abriéndonos paso por entre muchedumbres que esperaban su turno frente a ventanillas inmóviles, y por todas partes me hacían firmar papeles, me tomaban huellas digitales, y pocas horas después me entregaban mis documentos legítimos. Lo único para lo cual no tuve que prestar mi cuerpo fue para las vacunas. Muchos años más tarde me contaron, no sé con cuánto fundamento, que la agencia de viajes disponía de los buenos servicios de un bobo amaestrado que se hacía vacunar tantas veces cuantas fueran necesarias, y con nombres distintos, para que los clientes no tuvieran que padecer el martirio de la vacuna.

Lo menos creíble de la historia es que después de aquel milagro vertiginoso perdí en Bogotá el avión de Europa porque los compañeros del periódico me hicieron la noche anterior una fiesta de despedida tan borrascosa que llegué tarde al aeropuerto. Por fortuna, la vida era tan fácil en aquellos tiempos que el avión se había descompuesto en la primera escala de Barranquilla y logré alcanzarlo en otro avión nacional que salió de Bogotá tres horas más tarde. El vuelo de Bogotá a París duró 42 horas, con escalas en Nassau, Azores, Lisboa y Madrid, y en dos ocasiones hubo que cambiarle una hélice a aquel Super Constellation que parecía a punto de desbaratarse por entre las turbulencias.

Llegué a Ginebra, en tren, el 17 de julio de 1955, con el propósito de regresar a Colombia unas semanas después, pero, en realidad, no he vuelto todavía ni una sola vez con una certidumbre de regreso. Desde que tengo recursos para hacerlo, he estado yendo todos los años por lo menos una vez, y casi siempre por un lapso de tres meses. Me encuentro con mis amigos de siempre, cada vez más viejos, más amigos, más serenos, y lo único que nos hace diferentes es que las nostalgias de ellos son nostalgias del tiempo, y en cambio las mías lo son no sólo del tiempo, sino también de ellos. En todos esos viajes, la única vez en que tuve un incidente ingrato fue hace dos años, cuando salí del país bajo la protección diplomática de México, y fue una experiencia tan amarga que he decidido expulsarla para siempre de la memoria. Pero su huella es irreparable, porque han sido éstos los únicos dos años que he consagrado a reflexionar, casi en cada minuto, en cada sitio del mundo, cuál ha sido la razón verdadera por la cual no vivo en Colombia, y no he podido encontrarla.

Una noche, a la salida de un teatro de Barcelona, una mujer madura y hermosa se interpuso en mi camino y me dijo: «Usted no existe». Su observación lúcida, y tal vez demasiado obvia, me infundió un estado de consuelo, porque no fue dicha con amargura, sino con una inmensa compasión. Fue lo contrario de lo que siento cuando me oigo decir, casi todos los días, que soy un hombre feliz porque no hago más que viajar. La verdad es que nada me aburre tanto, quizá porque nada es más estéril. Tal vez lo único que me queda de haber visto el mundo entero son algunos instantes de Barcelona, que logro recuperar siempre cada vez que vuelvo, y una pasión irresistible por la disparatada y entrañable ciudad de México, que es la única que me disputa la nostalgia de mi país, y donde tengo amigos del alma cuya discreción y cuyo temor de importunar se confunden a veces con el olvido.

Aparte de eso, no recuerdo sino cuartos de hoteles y restaurantes raros. A estos últimos, si no son de veras muy buenos, prefiero un pollo hervido en casa con una pareja de buenos amigos. Los hoteles, en cambio, cuanto más impersonales, me resultan preferibles. Hoteles nuevos, iguales en todas partes del mundo, desde Buenos Aires hasta Helsinki, con las mismas teclas con dibujos para indicar los servicios, los mismos colores en el tapiz de las paredes, los mismos muebles de plástico, el mismo aire oloroso a medicina refrigerada. Un estilo universal que permite viajar sin viajar, porque es como estar en cualquier parte. El único riesgo es cuando suena el teléfono, al amanecer, y uno despierta sin saber en qué ciudad del mundo está durmiendo. Un amigo viajador, hace muchos años, me dio la fórmula para conjurar este percance: todas las noches, antes de acostarme, pongo en el teléfono un cartón con el nombre de la ciudad en que me encuentro.

Sin embargo, sea como sea, y sea donde sea, son muy pocos los instantes en que no me pregunto cuál ha sido la razón verdadera por la cual no regresé a Colombia en el mismo avión que me llevó a París hace veintidós años. Por todo el mundo me persiguen las noticias de mi país, por todo el mundo persigo a mis amigos de mi país sólo para hablar siempre de lo mismo, y muchas veces he despertado en la soledad de la noche perturbado por el fantasma ecuatorial del olor de la guayaba. Muchas veces, en los insomnios de estos dos últimos años, he llegado a pensar que tal vez el error de mi vida esté más atrás de donde lo he buscado. Tal vez no tenga origen en haber aceptado y prolongado tanto aquel viaje a Ginebra, sino en haberme ido del pueblo ardiente y polvoriento donde mis padres me aseguran que nací, y en el cual sueño que estoy –inocente, anónimo y feliz– casi todas las noches. En ese caso no sería tal vez el mismo que soy, pero acaso hubiera sido alguien mucho mejor: un personaje simple de las novelas que nunca hubiera escrito. Aquí estoy, pues, con la ilusión de que no sea ya demasiado tarde para remendar el pasado.

AMÉRICA CENTRAL, ¿AHORA SÍ?

El 21 de octubre del año pasado, cuando los periodistas me sacaron de la cama a las seis de la mañana en mi casa de México, yo tenía un tema en la punta de la lengua: la invasión inminente de Nicaragua desde el territorio de Honduras. Lo denuncié a los cuatro vientos. Con base en informaciones que no admitían la menor duda y que no provenían de fuentes nicaragüenses, sino norteamericanas. Más aún: había otra persona más preocupada que yo por la inminencia de esa agresión, y era el propio secretario de Estado de Estados Unidos, George Shultz, quien así se lo había confesado en conversaciones privadas a varios latinoamericanos prominentes que le habían visitado en los días anteriores. El proyecto había sido elaborado por la CIA cuando el secretario de Estado era el general Alexander Haig, y su sucesor lo había recibido como una papa caliente que no sabía dónde poner, y al final, según parece ahora, no le quedó otro recurso que comérsela.

Lo más inquietante de las cosas que suceden en América Central es que casi todas son de dominio público y, sin embargo, se manejan como si en realidad fueran infundios puros. Apenas unas horas después de que se publicó y transmitió por todas partes mi denuncia de la invasión inminente, el semanario *Newsweek* y el diario *The New York Times* publicaron el plan hasta en sus detalles más ínfimos, y aun con fotografías a todo color de los preparativos que se adelantaban en el territorio de Honduras y muy cerca de la frontera con Nicaragua. Pero 24 horas antes, el presidente de Honduras, Roberto Suazo Córdova, me había mandado una carta muy respetuosa en la cual rechazaba mis declaraciones y me invitaba a viajar a su país para que comprobara su falsedad sobre el terreno. La carta terminaba con una frase que cobra en estos momentos una significación especial: «Honduras no levantará jamás un arma contra sus vecinos».

Publicado originalmente el 20 de abril de 1983.

Mi determinación de no atender aquella invitación amable y tendenciosa no la tomé solo. Amigos más versados que yo en los misterios de la región y, sobre todo, algunos periodistas extranjeros que se la saben de memoria, me convencieron con toda razón de que iba a meterme en la trampa de no encontrar nada en una frontera extensa y difícil, y a regresar a México con el compromiso ético de divulgarlo también a los cuatro vientos como un acto de contrición ineludible. Había, además, una rara unanimidad en cuanto a los poderes reales del presidente Suazo Córdova. Nadie creía, y menos ahora, que en realidad dispusiera de alguna facultad de decisión, pues había sido impuesto en unas elecciones de fantasía sólo para improvisar en Honduras una apariencia democrática que le hacía falta a Estados Unidos para utilizarla como base de sus intervenciones encubiertas en todo el ámbito de América Central, aun en contra de los deseos y el criterio de su propio secretario de Estado. El poder real lo tiene y lo ejerce con puño de hierro el ministro de Defensa, mientras que el presidente sin presidencia entretiene sus ocios mandándoles telegramas para despistar a los favorecidos por la Fundación Nobel. Habría que recordar que este triste papel de portaaviones de tierra firme le ha sido impuesto a Honduras en otras ocasiones de rememoración ingrata. La más escandalosa de ellas, por supuesto, fue haber servido de trampolín a la expedición del general Castillo Armas, en 1954, que derrocó al último gobierno democrático que ha existido en Guatemala. De modo que, cuando el presidente Suazo Córdova se comprometió en su respuesta a que Honduras no levantaría un arma contra sus vecinos, había motivos de sobra para no creerle. Una vez más, por desgracia, la historia ha dado la razón a los incrédulos.

El porvenir de esa guerra es imprevisible. Alguien que tenía por qué saberlo en Estados Unidos tuvo hace poco la frescura de declarar que la invasión desde Honduras no tiene la finalidad de derrocar al gobierno de Nicaragua, sino la de mantenerlo en un estado de emergencia permanente. Esto, que podría parecer una tontería, tiene todas las apariencias de ser verdad. En primer término, porque el gobierno sandinista está bien asentado en el poder, con un apoyo popular más sustancial y decidido de lo que dicen muchos de sus adversarios; y en término segundo, porque los invasores parecen más dispuestos a sobrevivir para cobrar su sueldo que a morir por unas ideas que no tienen. Hay quienes piensan, por otra parte, que la costosa teoría de hostigar sin ganar sólo pretende por parte del gobierno de Ronald Reagan convertir la invasión de Nicaragua en una carta de negociación global; es decir, pararla a cambio de condiciones favorables por parte de los guerrilleros de El Salvador. Lo cual probaría, una vez más, que el principal problema de los gobiernos de Estados Unidos es su concepción simplista de América Latina.

Lo malo es que cualquier tentativa de solución negociada en América Central será inútil sin el acuerdo de Estados Unidos, y éste ha encontrado siempre la manera de escamotear todas las que se han propuesto. Ha habido hasta ahora unas 18. Ha habido una iniciativa conjunta de Francia y de México. Ha habido gestiones concretas del primer ministro de Suecia, Olof Palme, y del presidente del gobierno español, Felipe González. Ha habido una búsqueda constante por parte de la Internacional Socialista. El gobierno de Estados Unidos ha sido hostil a todas ellas, con un argumento que es como para morirse de risa: dice que son injerencias europeas en los asuntos continentales. Y lo dice nada menos que un gobierno que apoyó a una potencia extracontinental −el Reino Unido− en su guerra con Argentina. Dentro de aquella misma definición intervencionista habría que incluir la accidentada y todavía incomprensible visita del papa Juan Pablo II a América Central, cuyos únicos resultados reales fueron los de radicalizar las posiciones extremas de los protagonistas.

En todo caso, las iniciativas de acuerdos negociados no son de ahora ni todas son europeas. Al contrario, en El Salvador la propusieron los propios movimientos armados cuando su situación militar no era buena, y la han retirado con mayor énfasis ahora que esa situación es tan favorable que tienen al gobierno arrinconado, a pesar de la ayuda constante de Estados Unidos y a pesar de la sabiduría de sus asesores militares. Nicaragua, en todo momento, se ha manifestado dispuesta a aceptar toda propuesta de diálogo. México, por su parte, primero con Francia y luego con Venezuela, ha sido siempre un abanderado del acuerdo pacífico. Este clamor de paz, que viene ya de hace mucho tiempo y que crece cada día, empieza a convertirse en un ser difícil de eludir para el gobierno de Estados Unidos, que siente crecer ese clamor dentro de su propio ámbito, en el Congreso y en la prensa, en las universidades y en la calle. Es increíble que a tan pocos años de la derrota de Vietnam, el país más poderoso y próspero del mundo tenga que encarar otra vez a los mismos fantasmas.

El presidente Felipe González, tal vez el hombre que más se ha interesado por la suerte de América Central en estos años, decía hace pocos meses en privado que la falla mayor de las iniciativas de negociación era que todas estaban fundadas en la mejor voluntad, pero que ninguna proponía un método concreto para su aplicación. Ése es el mérito de la irrupción intempestiva y original del presidente Belisario Betancur en el ajedrez confuso de América Central, del cual Colombia estaba ausente o jugando con fichas infames después de su participación excelente en la recuperación del canal de Panamá. Cualesquiera sean los resultados de esa irrupción, el solo hecho de haberla emprendido, y además con una fórmula práctica, es algo alentador. Confío, además, en que será fructífero.

CON AMOR, DESDE EL MEJOR OFICIO DEL MUNDO

Un día de la semana pasada llegamos Mercedes y yo a almorzar en casa de unos amigos, y los encontramos en un estado de ofuscación que sólo logramos disipar cuando ellos nos aclararon el motivo: esa mañana habían oído por radio que yo estaba en Popayán, y en vista de eso no sólo se apresuraron a desinvitar a los otros invitados, sino que se sintieron un poco agraviados porque no les hubiéramos dado ninguna clase de excusas oportunas. La verdad es que también nosotros habíamos oído por radio aquella mañana la noticia falsa de mi viaje a Popayán, y más bien me causó un cierto alivio. «Qué maravilla», me dije; «puesto que ya estoy en Popayán, no tendré que levantarme tan temprano para ir a Popayán.» El viaje, que en realidad tenía previsto para dos días con el fin de calcular con ojos propios la magnitud de la tragedia, no tuvo que cumplirse en la realidad porque ya estaba cumplido en la ficción de las noticias habladas. Además, los amigos entendieron, y todos terminaron en la cocina improvisando un almuerzo en familia.

No era aquél nuestro primer almuerzo conflictivo desde que regresamos a Colombia la semana anterior. El otro lo había sido mucho más, porque era en el palacio presidencial, por invitación de los muy ilustres dueños de casa. No se trataba de un acto oficial, sino de una comida privada en la residencia ocasional de un viejo amigo y con asistencia de sólo tres parejas de amigos comunes que habíamos escogido de común acuerdo. La tensión empezó para mí aquella mañana cuando leí un despacho de una agencia nacional de noticias, según el cual yo asistía al mediodía a una reunión solemne de intelectuales y artistas de los más distinguidos y convocados en el palacio de Nariño por el presidente Betancur para condecorarme con la Cruz de Boyacá. Para mí, aquella noticia era un disgusto por partida doble. En primer término, el presidente no me había dicho nada sobre medallas colgadas en la solapa, a pesar de que varias ve-

Publicado originalmente el 27 de abril de 1983.

ces habíamos hablado por teléfono en los días recientes para acordar ciertos pormenores de la invitación. En término segundo, tengo ideas muy personales —aunque también muy bien compartidas— sobre los usos y los abusos con que algunos de nuestros presidentes han ejercido la facultad suprema de conceder la Cruz de Boyacá. Hace varios años, un presidente amigo me consultó sus deseos de concedérmela, y por fortuna entendió sin resentimientos mi negativa, que se fundaba en mi creencia de que no es digno aceptar honores —por muy altos y nobles que sean— cuando hay que sobrellevarlos en malas compañías. La prodigalidad presidencial con la Cruz de Boyacá llegó a extremos de circo en los días finales del gobierno anterior, cuando las últimas que quedaban en las gavetas del poder fueron repartidas como caramelos entre los amigos personales y políticos del presidente, que se iba para bien de la patria. Así las cosas, fui con mucho gusto al almuerzo privado del presidente Betancur con la sensación desapacible de haber sido víctima por lo menos de un malentendido, pero con la determinación firme de decirle que no a la Cruz de Boyacá con los mismos argumentos —ahora mejor sustentados— de la ocasión anterior. Por fortuna, al término de un almuerzo muy grato y de una sobremesa que se prolongó hasta las seis de la tarde no hubo en ningún momento la menor tentativa de condecoración sobre seguro, y mucho menos a mansalva. Otra vez, como tantas anteriores, todo se explicaba por el hecho simple de que era una noticia inventada.

Pero aquella comprobación no me servía de consuelo. Al contrario: no tenía aún una semana de haber llegado a Colombia y aquélla era sólo una más de las noticias falsas sobre mí que se publicaban a diario. Apenas el viernes anterior había logrado desbaratar a tiempo la tentativa más escandalosa, cuando Enrique Santos Calderón tuvo la buena estrella de consultarme por teléfono algún detalle de una entrevista que alguien decía haberme hecho a mi llegada a Bogotá y que *El Tiempo* se disponía a publicar el domingo siguiente. La entrevista era falsa desde el principio hasta el fin, y sus jóvenes autores habían logrado burlar con su sangre fría nada envidiable la buena fe de Enrique Santos Calderón. Éste me puso en contacto con uno de sus autores, a quien la sangre no se le calentó ni un grado cuando me confirmó por teléfono que, en efecto, había inventado la entrevista con la complicidad de un compañero cuando se dio cuenta de que era imposible conseguir que yo le concediera una entrevista auténtica. «Es muy grave que sea inventada», le dije yo, «pero es peor aún que sea tan mala.» En realidad, la mejor entrevista conmigo que se ha publicado entre las incontables que me han hecho fue una inventada en Caracas. Pero en vez de protestar felicité a su autor, porque era una síntesis perfecta de casi todo lo que yo había declarado para la prensa en los últimos 15 años, y todo organizado y mejorado de tan buena manera y con tanta precisión y tanta inteligencia que ya hubiera querido yo mismo hacerla igual. No era éste el caso de la entrevista apócrifa de Bogotá, que no pasaba de ser una burla chapucera de la ética profesional.

Estas erosiones del oficio, por supuesto, no son apreciables sólo en Colombia. La barbaridad más indigna la cometió una revista española hace varios años, cuando me atribuyó una declaración según la cual el sueño de mi vida era figurar en la enciclopedia soviética. Me alarmó la atribución; primero, porque era falsa; segundo, porque nunca he sabido a ciencia cierta si figuro o no en la enciclopedia soviética; tercero, porque me importa un bledo si figuro o no en cualquier enciclopedia de cualquier parte, y pienso que si en cualquiera de ellas se omite un dato que interesa a los lectores el perjuicio es mayor para la propia enciclopedia que para el dato omitido; y cuarto, porque la falsedad tenía el propósito definido de ocasionarme un perjuicio personal y –lo que es mucho más grave– engañar a los lectores. No hice ninguna rectificación porque tengo por norma no hacerlas, y ya estoy demasiado curtido en esta guerra para empezar a hacerlas. Además, con todo lo que quiero a España pensé que las infamias de su prensa eran un asunto de los españoles. Pero que estas cosas ocurran en mi país –al cual no le he hecho nunca ningún daño consciente y, en cambio, he consagrado casi todos los minutos de mi vida a tratar de hacerle bien– es algo que no puedo pasar por alto sin el temor de ocasionarle un perjuicio por omisión.

En dos semanas aquí he encontrado motivos de sobra para preguntarme con alarma hacia dónde va este periodismo apresurado y sin control ético. Hay casos en que las faltas son de voluntarismo puro; como el de un periodista de Medellín, a quien le declaré frente a una grabadora que pensaba solicitar la conmutación de la pena del colombiano condenado a muerte en Estados Unidos, y la noticia se transformó en que yo encabezaría un movimiento nacional para que no lo maten. Fui muy explícito en que los únicos argumentos a que podía apelarse eran de carácter humanitario, que el país donde se comete el delito tiene derecho a juzgar al delincuente de cualquier nacionalidad y que en el caso del compatriota condenado a muerte en Estados Unidos no tenía ninguna utilidad el tratado de extradición –a todas luces inconveniente– que ese país acordó con el gobierno anterior de Colombia. Sin embargo, todas esas precisiones se esfumaron en la declaración publicada por la prensa. En otros casos, la falta es por negligencia pura. En Cartagena, este último viernes, unos quince colegas me asediaron a preguntas cuando llegué a la ciudad. Frente a quince grabadoras declaré, y supongo que quedó grabado, que el periódico que quiero fundar «pondría el interés nacional por encima de todo». En una de las transcripciones de la entrevista, que fue hecha con una puntuación mal inspirada en *El otoño del patriarca*, me hicieron decir que mi periódico «estaría por encima de todos los intereses nacionales». Lo cual no es sólo todo lo contrario de lo que dije, sino una barbaridad inadmisible.

Para cualquier hombre público −como yo he terminado por serlo, muy a mi pesar y para mi infortunio−, estos infundios y accidentes de mal manejo son de una gravedad tremenda; pero lo son mucho más para quienes antes que nada nos consideramos periodistas. Para nosotros, más que para las víctimas, estos atentados cada vez más frecuentes y escandalosos a la moral del periodismo nos parecen delitos de la más alta peligrosidad, porque terminarán por dañar y pervertir por completo el mejor oficio del mundo.

LA SUERTE DE NO HACER COLAS

Una de las ventajas del Premio Nobel es que nunca más hay que volver a hacer la cola en ninguna parte. Esto lo había leído hace algunos años en un libro de Edgar Wallace, y desde hace algunos meses he tenido la ocasión de comprobarlo en carne propia. En el mundo urbano de hoy, donde con tanta frecuencia se tiene la impresión de que los individuos no cabemos en la muchedumbre, el privilegio de no hacer la cola es uno de los más apetecibles. Sin embargo, no tengo la impresión de que sea esto lo que suscita más envidias, sino el raro parecido que el Premio Nobel tiene con la lotería. Uno se encuentra con muchas caras por todas partes, y en cada una alcanza a vislumbrar un sentimiento distinto. Pero el que más se repite es el del asombro de encontrarse frente a alguien a quien el destino le puso de pronto en las manos la módica suma de 170.000 dólares. El pudor del dinero me ha parecido siempre un defecto –y no una virtud casi teologal, como lo consideran los ingleses–, de modo que nunca tengo inconveniente en hablar del premio en efectivo para tranquilizar a los interlocutores que se preguntan, sin atreverse a plantearlo, cómo se siente una persona que un 21 de octubre se despertó con semejante suma de dinero imprevista, casi como Gregorio Samsa se despertó una mañana convertido en un gigantesco insecto.

Cuando se trata en público este tema, apenas si logro convencer al auditorio de que el aspecto utilitario del premio no me causó ninguna emoción, por la razón muy simple de que desde hace años estaba convencido de que no era cierto que el Premio Nobel –además del honor y la desdicha– tuviera dentro una gratificación en efectivo. Esa certidumbre me venía desde que Pablo Neruda me contó los secretos de su premio, incluso los pormenores de la entrega de los miserables 42.000 dólares devaluados que fueron la recompensa de aquel año. Poco después, el poeta compró una casa de campo en

Publicado originalmente el 4 de mayo de 1983.

Normandía, que era la antigua caballeriza de un castillo local, por entre cuyas arboledas se deslizaba apenas un río cubierto de lotos. Los domingos invitaba a almorzar a sus amigos, que nos íbamos en tren durante 20 minutos desde la estación de Montparnasse, en París, y lo encontrábamos sentado como un Papa en su cama papal, y muerto de risa como siempre de saber que parecía un Papa y que sus mejores amigos nos moríamos de risa de que lo pareciera.

El aspecto de la mesa donde servían el almuerzo tenía para él tanta importancia como la tenían para Matilde las cosas que sucedían en la cocina, y la arreglaba con tanta meticulosidad como escribía un poema con su hilo interminable de tinta verde. La última vez que estuvimos allí se distrajo en la conversación de los aperitivos, y ya íbamos a sentarnos cuando descubrió en la mesa un error de composición que sólo él podía detectar, y nos hizo volver a la sala mientras lo rehacía todo hasta que la mesa quedara cantando con voz propia. Se decía por todas partes que aquella casa apacible había sido comprada con el Premio Nobel. Pero yo pensaba en el fondo de mi corazón que el origen debía ser distinto, porque estaba convencido de que no era cierto que dieran dinero además del honor y la desdicha.

Estuve tan cerca de Neruda en aquellos días de su premio, que no pude olvidarme de él ni un solo instante cuando me correspondió vivir la misma suerte, hasta el punto de que lo único que se me ocurrió cuando volví al hotel después de la coronación solemne fue llamar por teléfono a Matilde Neruda desde Estocolmo a Santiago de Chile para darle las gracias por lo mucho que ella y el poeta ya muerto me habían ayudado a sobrellevar aquel trance. Pero aún no había terminado de recordarlo, porque al día siguiente me citaron a las 11 de la mañana en la sede de la Fundación Nobel, y en el instante en que colgué el teléfono sufrí el mismo sobresalto que había sentido Neruda, según él mismo me había contado poco antes de viajar a Chile para morir.

También a él lo habían citado solo, y no junto con los otros premiados, como había ocurrido para todas las ceremonias anteriores. De modo que nos fuimos Mercedes y yo, sintiéndonos tan solos de estar solos por primera vez en aquella semana interminable, que a dos amigos que encontramos en la puerta los invitamos a que nos acompañaran en la limusina con ínfulas de carroza fúnebre que teníamos a nuestra disposición para nosotros solos.

La sede de la Fundación Nobel está en un edificio sobrio, que más bien parece el de un banco, y sus ventanas dan sobre un parque nevado que Neruda me había descrito hasta en sus detalles menos visibles. Fuimos recibidos en un ámbito de silencio, a través de pasillos y salones alfombrados en cuyos muros se veían los recuerdos de mis antecesores, y no pude evitar un estremecimiento recóndito ante aquella súbita evidencia del tiempo. «Dentro de cien años», pensé, «el escritor premiado pasará por este salón y no sabrá siquiera quién era yo cuando vea mi retrato colgado en la pared.»

Todo lo demás ocurrió tal como Neruda me lo había contado. Me pidieron firmar el libro de honor de la casa, me entregaron la medalla y el diploma que habían retenido en la ceremonia de la noche anterior para que no tuviera que cargarlos, me pidieron que firmara un formulario impreso por medio del cual cedía a la Fundación Nobel los derechos de autor de mi conferencia y de mi brindis por la poesía –que en los apuros de las últimas horas había improvisado a cuatro manos con el poeta Álvaro Mutis–, y luego firmé ejemplares de mis libros en sueco para los empleados de la fundación, y fui presentado a cada uno de ellos con la informalidad genuina y cómoda de los suecos. Por último me invitaron a pasar al despacho del presidente de la fundación, que nos había impresionado a todos por su elegancia y su simpatía, y fue entonces cuando sentí pasar la misma ráfaga de revelación de que Neruda me había hablado. Se me ocurrió que allá por la cuarta década del siglo se había acabado todo el dinero del legado de Alfred Nobel; pero la fundación, con muy buen sentido, decidió seguir adelante con el premio sin más gratificación que la gloria de merecerlo. Pensaban, también con muy buen sentido, que los escritores y científicos estarían de acuerdo con esta solución providencial y estarían dispuestos a guardar el secreto no sólo hasta la muerte, sino aun después de ella. La revelación me pareció tan lúcida que en el momento en que entré en el despacho del presidente ya había tomado la decisión de aceptar el trato. Sin embargo, la conversación transcurrió en un clima convencional, sin novedad alguna, frente a una ventana única donde empezaba a nevar sin consuelo. Hablamos un poco de todo. Menos, por supuesto, del dinero del premio, porque los suecos son tan discretos que el asunto había sido resuelto desde antes, dentro del mayor sigilo, y sin que yo lo supiera. En todo caso, desde aquella mañana me quedé con la mala idea de que todo había sido una premonición ajena: sólo dentro de cien años habría de convertirse en una realidad irreparable para el premiado de las bellas letras, que abandonaría aquella oficina sostenido apenas por el consuelo de no tener que hacer más colas por el resto de su vida, y en un mundo tan difícil donde tal vez habrá colas interminables para hacer otras colas.

VIENDO LLOVER EN GALICIA

Mi muy viejo amigo, el pintor, poeta y novelista Héctor Rojas Herazo —a quien no veía desde hacía mucho tiempo—, debió sufrir un estremecimiento de compasión cuando me vio en Madrid abrumado por un tumulto de fotógrafos, periodistas y solicitantes de autógrafos, y se acercó para decirme en voz baja: «Recuerda que de vez en cuando debes ser amable contigo mismo». En efecto, fiel a mi determinación de complacer todas las demandas sin tomar en cuenta mi propia fatiga, hacía ya varios meses —quizá varios años— en que no me ofrecía a mí mismo un regalo merecido. De modo que decidí regalarme en la realidad uno de mis sueños más antiguos: conocer Galicia.

Alguien a quien le gusta comer no puede pensar en Galicia sin pensar antes que en cualquier otra cosa en los placeres de su cocina. «La nostalgia empieza por la comida», dijo el Che Guevara, tal vez añorando los asados astronómicos de su tierra argentina, mientras se hablaba de asuntos de guerra en las noches de hombres solos en la Sierra Maestra. También para mí la nostalgia de Galicia había empezado por la comida, antes de que hubiera conocido la tierra. El caso es que mi abuela, en la casa grande de Aracataca, donde conocí mis primeros fantasmas, tenía el exquisito oficio de panadera, y lo practicaba aun cuando ya estaba vieja y a punto de quedarse ciega, hasta que una crecida del río le desbarató el horno y nadie en la casa tuvo ánimos para reconstruirlo. Pero la vocación de la abuela era tan definida, que cuando no pudo hacer panes siguió haciendo jamones. Unos jamones deliciosos, que, sin embargo, no nos gustaban a los niños —porque a los niños no les gustan las novedades de los adultos—, pero el sabor de la primera prueba se me quedó grabado para siempre en la memoria del paladar. No volví a encontrarlo jamás en ninguno de los muchos y diversos jamones que comí después en mis años buenos y en mis años malos, hasta que probé por casualidad

Publicado originalmente el 11 de mayo de 1983.

—40 años después, en Barcelona— una rebanada inocente de lacón. Todo el alborozo, todas las incertidumbres y toda la soledad de la infancia me volvieron de pronto en ese sabor, que era el inconfundible de los lacones de la abuela. De aquella experiencia surgió mi interés de descifrar su ascendencia, y buscando la suya encontré la mía en los verdes frenéticos de mayo hasta el mar y las lluvias feraces y los vientos eternos de los campos de Galicia. Sólo entonces entendí de dónde había sacado la abuela aquella credulidad que le permitía vivir en un mundo sobrenatural donde todo era posible, donde las explicaciones racionales carecían por completo de validez, y entendí de dónde le venía la pasión de cocinar para alimentar a los forasteros y su costumbre de cantar todo el día. «Hay que hacer carne y pescado porque no se sabe qué les gusta a los que vengan a almorzar», solía decir cuando oía el silbato del tren. Murió muy vieja, ciega, y con el sentido de la realidad trastornado por completo, hasta el punto de que hablaba de sus recuerdos más antiguos como si estuvieran ocurriendo en el instante, y conversaba con los muertos que había conocido vivos en su juventud remota. Le contaba estas cosas a un amigo gallego la semana pasada, en Santiago de Compostela, y él me dijo: «Entonces tu abuela era gallega, sin ninguna duda, porque estaba loca». En realidad, todos los gallegos que conozco, y los que vi ahora sin tiempo para conocerlos, me parecen nacidos bajo el signo de Piscis.

No sé de dónde viene la vergüenza de ser turista. A muchos amigos, en pleno frenesí turístico, les he oído decir que no quieren mezclarse con los turistas, sin darse cuenta de que, aunque no se mezclen, ellos son tan turistas como los otros. Yo, cuando voy a conocer algún lugar sin disponer de mucho tiempo para ir más a fondo, asumo sin pudor mi condición de turista. Me gusta inscribirme en esas excursiones rápidas, en las que los guías explican todo lo que se ve por las ventanas del autobús, a la derecha y a la izquierda, señores y señoras, entre otras cosas porque así sé de una vez todo lo que no hay que ver después, cuando salgo solo a conocer el lugar por mis propios medios. Sin embargo, Santiago de Compostela no da tiempo para tantos pormenores: la ciudad se impone de inmediato, completa y para siempre, como si se hubiera nacido en ella. Siempre he creído, y lo sigo creyendo, que no hay en el mundo una plaza más bella que la de Siena. La única que me ha hecho dudar es la de Santiago de Compostela, por su equilibrio y su aire juvenil, que no permite pensar en su edad venerable, sino que parece construida el día anterior por alguien que hubiera perdido el sentido del tiempo. Tal vez esta impresión no tenga su origen en la plaza misma, sino en el hecho de estar –como toda la ciudad, hasta en sus últimos rincones– incorporada hasta el alma a la vida cotidiana de hoy. Es una ciudad viva, tomada por una muchedumbre de estudiantes alegres y bulliciosos, que no le dan ni una sola tregua para envejecer. En los muros intactos, la vegetación se abre paso por entre las grietas, en una lucha implacable por sobrevivir al olvido, y uno se encuentra a cada paso, como la cosa más natural del mundo, con el milagro de las piedras florecidas.

Llovió durante tres días, pero no de un modo inclemente, sino con intempestivos espacios de un sol radiante. Sin embargo, los amigos gallegos no parecían ver esas pausas doradas, sino que a cada instante nos daban excusas por la lluvia. Tal vez ni siquiera ellos eran conscientes de que Galicia sin lluvia hubiera sido una desilusión, porque el suyo es un país mítico –mucho más de lo que los propios gallegos se lo imaginan–, y en los países míticos nunca sale el sol. «Si hubieran venido la semana pasada, habrían encontrado un tiempo estupendo», nos decían, avergonzados. «Este tiempo no corresponde a la estación», insistían, sin acordarse de Valle-Inclán, de Rosalía de Castro, de los poetas gallegos de siempre, en cuyos libros llueve desde el principio de la creación y sopla un viento interminable, que es tal vez el que siembra ese germen lunático que hace distintos y amorosos a tantos gallegos.

Llovía en la ciudad, llovía en los campos intensos, llovía en el paraíso lacustre de la ría de Arosa y en la ría de Vigo, y en su puente, llovía en la plaza, impávida y casi irreal, de Cambados, y hasta en la isla de la Toja, donde hay un hotel de otro mundo y otro tiempo, que parece esperar a que escampe, a que cese el viento y resplandezca el sol para empezar a vivir. Andábamos por entre esta lluvia como por un estado de gracia, comiendo a puñados los únicos mariscos vivos que quedan en este mundo devastado, comiendo unos pescados que siguen siendo peces en el plato y unas ensaladas que seguían creciendo en la mesa, y sabíamos que todo aquello estaba allí por virtud de la lluvia, que nunca acaba de caer. Hace ahora muchos años, en un restaurante de Barcelona, le oí hablar de la comida de Galicia al escritor Álvaro Cunqueiro, y sus descripciones eran tan deslumbrantes que me parecieron delirios de gallego. Desde que tengo memoria les he oído hablar de Galicia a los gallegos de América, y siempre pensé que sus recuerdos estaban deformados por los espejismos de la nostalgia. Hoy me acuerdo de mis 72 horas en Galicia y me pregunto si todo aquello era verdad, o si es que yo mismo he empezado a ser víctima de los mismos desvaríos de mi abuela. Entre gallegos –ya lo sabemos– nunca se sabe.

¿SABE USTED QUIÉN ERA MERCÈ RODOREDA?

La semana pasada pregunté por Mercè Rodoreda en una librería de Barcelona y me dijeron que había muerto hace un mes. La noticia me causó una pena muy grande, primero por la admiración muy justa que siento por sus libros y segundo por el hecho inmerecido de que la noticia de su muerte no se hubiera publicado fuera de España con el despliegue y los honores debidos. Al parecer, pocas personas saben fuera de Cataluña quién era esa mujer invisible que escribía en un catalán espléndido unas novelas hermosas y duras como no se encuentran muchas en las letras actuales. Una de ellas —*La plaza del Diamante*— es, a mi juicio, la más bella que se ha publicado en España después de la guerra civil.

La razón de que se la conozca tan poco, aun dentro de España, no puede atribuirse a que hubiera escrito en una lengua de ámbito reducido, ni a que sus dramas humanos transcurran en un rincón secreto de la muy secreta ciudad de Barcelona, pues sus libros han sido traducidos a más de diez idiomas y en todos ellos han sido objeto de comentarios críticos mucho más entusiastas de los que merecieron en su propio país. «Éste es uno de los libros de alcance universal que haya escrito el amor», escribió en su momento el crítico francés Michel Cournot, refiriéndose a *La plaza del Diamante*. Diana Athill, sobre la versión inglesa, escribió: «Es la mejor novela publicada en España en muchos años». Y un crítico del *Publishers Weekly*, en Estados Unidos, escribió que era una novela extraña y maravillosa. Sin embargo, hace algunos años, y con motivo de alguno de tantos aniversarios, se hizo una encuesta entre escritores españoles de hoy para tratar de establecer, según su criterio, cuáles eran los diez mejores libros escritos en España después de la guerra civil, y no recuerdo que alguno hubiera mencionado a *La plaza del Diamante*. En cambio, muchos citaron con toda justicia *La forja de un rebelde*, de Arturo Barea. Lo curioso es que este libro, cuyos cuatro tomos apreta-

Publicado originalmente el 18 de mayo de 1983.

dos habían sido publicados a fines de la cuarta década de este siglo en Buenos Aires, no había sido todavía publicado en España, y, en cambio, *La plaza del Diamante* llevaba ya veintiséis ediciones en catalán. Yo la leí en castellano por esos tiempos, y mi deslumbramiento fue apenas comparable al que me había causado la primera lectura de *Pedro Páramo*, de Juan Rulfo, aunque los dos libros no tienen en común sino la transparencia de su belleza. A partir de entonces, no sé cuántas veces la he vuelto a leer, y varias de ellas en catalán, con un esfuerzo que dice mucho de mi devoción.

La vida privada de Mercè Rodoreda es uno de los misterios mejor guardados de la muy misteriosa ciudad de Barcelona. No conozco a nadie que la haya conocido bien, que pueda decir a ciencia cierta cómo era, y sus libros sólo permiten vislumbrar una sensibilidad casi excesiva y un amor por sus gentes y por la vida de su vecindario que es quizá lo que les da un alcance universal a sus novelas. Se sabe que pasó la guerra civil en la casa familiar de San Gervasio, y su estado de alma de ese tiempo es evidente en sus libros. Se sabe que después se fue a vivir a Ginebra, y que allí escribió al rescoldo de sus nostalgias. «Cuando empecé a escribir la novela apenas si recordaba cómo era la plaza del Diamante», escribió en uno de sus prólogos, que son muestras ejemplares de su conciencia de novelista. Alguien que no sea otro escritor podría sorprenderse de que la autora hubiera logrado una recreación tan minuciosa y lúcida de sus lugares y sus gentes a partir de una vivencia remota, casi perdida entre las brumas de la infancia. «Sólo recordaba», ha escrito en el prólogo de una edición catalana, «cuando tenía trece o catorce años, que una vez, por la fiesta mayor de Gracia, fui a caminar por las calles con mi padre. En la plaza del Diamante habían levantado una carpa, como en otras plazas, por supuesto, pero la que siempre recordé fue aquélla. Al pasar frente a esa caja de música, yo, a quien mis padres habían prohibido bailar, tenía unos deseos desesperados de hacerlo, y andaba como un ánima en pena por las calles adornadas.» Mercè Rodoreda suponía que fue a causa de esta frustración que muchos años después, en Ginebra, empezó su novela con aquella fiesta popular. En general, esa ansiedad de bailar, que sus padres reprimieron siempre porque no era admisible en una chica decente, ha sido identificada por la propia escritora como la contrariedad original que le dio el impulso para escribir.

Pocos autores han hecho precisiones tan certeras y útiles sobre el proceso subconsciente de la creación literaria como las que hizo Mercè Rodoreda en los prólogos de sus libros. «Una novela es un acto mágico», escribió. Hablando de *El espejo roto* —su novela más larga– hizo otra revelación casi alquímica: «Eladi Farriols, muerto y tendido en una biblioteca de una casa señorial, me resolvió el primer capítulo del modo más inesperado». En otra parte dijo: «Las cosas tienen una gran importancia en la narración. Y la han tenido siempre, mucho antes de que Robbe-Grillet escribiera *Le voyeur*». Conocí esta declaración mucho después de que su autora me hubiera deslumbrado con la sensualidad con que hace ver las cosas en el aire de sus novelas, mucho después de que me hubiera asombrado la luz nueva con que las iluminan sus palabras. Un escritor que todavía sabe cómo se llaman las cosas tiene salvada la mitad del alma, y Mercè Rodoreda lo sabía a placer en su lengua materna. En castellano, en cambio, no todos los escritores lo sabemos, y en algunos se nota más de lo que nosotros mismos creemos.

Creo –si no recuerdo mal– que Mercè Rodoreda es la única escritora (o el único escritor) que he visitado sin conocerla, impulsado por una admiración irresistible. Supe por nuestro editor común, hace unos doce años, que ella estaba en Barcelona por pocos días, y me recibió en un apartamento provisional, amueblado de un modo muy sobrio y con una sola ventana que daba sobre el jardín crepuscular de Monterolas. Me sorprendió su aire distraído que más tarde encontré definido en uno de sus prólogos: «Quizá la más marcada de mis múltiples soledades sea una especie de inocencia que me hace sentirme bien en el mundo en que me ha tocado vivir». Entonces yo sabía que junto a la vocación literaria tenía una vocación paralela, tan dominante como la otra, y era la de cultivar flores. Hablamos de eso, que yo consideraba como otra forma de escribir, y entre rosas y rosas trataba de hablarle de sus libros y ella trataba de hablarme de los míos. Me llamó la atención que de todo lo escrito por mí le interesaba más que nada el gallo del coronel que no tenía quien le escribiera, y a ella le llamó la atención que me gustara tanto la rifa de la cafetera en *La plaza del Diamante*. Tengo hoy un recuerdo entre nieblas de aquel extraño encuentro, que sin duda no fue uno de los recuerdos que ella se llevó a la tumba, pero para mí fue la única vez en que conversé con un creador literario que era una copia viva de sus personajes. Nunca supe por qué, al despedirme en el ascensor, me dijo: «Usted tiene mucho sentido del humor». Nunca más tuve noticias de ella hasta esta semana, en que supe por casualidad, y en mala hora, que le había ocurrido el único percance que podía impedirle seguir escribiendo.

EL REINO UNIDO LOS HA HECHO ASÍ

Las réplicas oficiales del gobierno británico a las denuncias de crueldades y actos salvajes cometidos por los soldados gurkhas en la reconquista de las islas Malvinas han insistido en que esas versiones son infundios puros inspirados en una leyenda negra. Voceros del Ministerio de Defensa del Reino Unido se han empeñado en negar la condición mercenaria de los gurkhas, y el teniente coronel David Morgan –que es el comandante del batallón– ha negado inclusive que sus nepaleses feroces hayan tenido una participación activa en las Malvinas, y ha rechazado con indignación que se les haya calificado de monstruos bestiales. Y con una frase que parece más bien una de aquellas ambigüedades corrosivas de Bernard Shaw, creyó poner término al debate: «Gurkhas are just bloody good soldiers». Lo único que les ha faltado a los voceros británicos y a los partidarios suyos que escriben cartas a los periódicos es decir que los gurkhas no existen.

Sin embargo, la historia reciente tiene demasiadas pruebas sangrientas no sólo de que sí existen, sino de que jugaron un papel tenebroso en la reconquista de las Malvinas. En el libro *Los chicos de la guerra*, del argentino Daniel Kon, publicado hace nueve meses en Buenos Aires por la editorial Galerna, un soldado que regresó de la guerra cuenta: «Los gurkhas parecían completamente drogados. Se mataban entre ellos mismos. Avanzaban gritando, sin apenas protegerse. No era difícil matarlos, pero eran demasiados. Tal vez matabas a uno o dos, pero el siguiente te mataba a ti. Eran como robots: un gurkha pisaba una mina y volaba por el aire, y el que venía detrás no se preocupaba en lo más mínimo: pasaba por la misma zona sin inmutarse, y a lo mejor también volaba. Parecían no tener instinto de supervivencia. Iban barriendo zonas con sus ametralladoras Mag, que pesan más que un fusil. Si al adentrarse en nuestras líneas encontraban alguna lata de ración de nuestras provisiones, las abrían

Publicado originalmente el 25 de mayo de 1983.

por la mitad de un cuchillazo, comían un poco y seguían peleando, siempre gritando. No les interesaba nada, ni siquiera sus propias vidas. Los ingleses que venían detrás de los gurkhas lo tenían muy fácil: encontraban el camino casi despejado». Ocho testigos más cuentan en el mismo libro que vieron cómo un gurkha hacía desnudar a un prisionero argentino y lo hacía caminar por el campo dándole patadas y golpes con un fusil. Dicen que otros gurkhas lo agarraron por fin de los cabellos, lo empujaron hasta que quedó arrodillado en el suelo y le cortaron el cuello. Lo mismo hicieron con cuatro o cinco prisioneros más.

Al leer este relato, alguien me dijo que parecía magnificado por el miedo del narrador. No obstante, la conducta de los gurkhas ha sido descrita en términos mucho más dramáticos por los propios ingleses que han estado al lado de ellos en otras infamias más largas y sangrientas que la de las Malvinas. Al fin y al cabo, los gurkhas han participado con los ingleses en no menos de diez guerras grandes y en numerosas campañas de conquista y represión colonial. La más importante de ellas fue la guerra de independencia de la India, donde un batallón gurkha, al mando del brigadier general Reginal Dyer, disparó sin discriminación, a mansalva y sobre seguro, contra una manifestación pacífica de civiles, y mataron a 379 —entre hombres, mujeres y niños— e hirieron a más de un millar. Este episodio bárbaro se conoce como la «matanza de Amritsar» y está reconstruida con una escalofriante fidelidad en la película *Gandhi*, que barrió con casi todos los Oscar de Hollywood y fue dirigida por un caballero británico: sir Richard Attenborough.

Otro inglés, nada menos que el mariscal de campo vizconde Slim, quien comandó las tropas inglesas en Birmania durante la segunda guerra mundial, ha contado episodios alucinantes sobre la conducta de los gurkhas que peleaban en sus filas. Su libro, *Defeat into victory*, es más que revelador. «En cierta ocasión», cuenta el mariscal Slim, «algunos gurkhas se presentaron ante su general y con gran orgullo abrieron una canasta de la cual sacaron tres cabezas ensangrentadas de japoneses y las pusieron sobre la mesa. Luego, con sus maneras más finas, le ofrecieron al general, para su cena, los pescados frescos que llevaban en la misma canasta.»

No debió ser por casualidad que los ingleses destinaron sus gurkhas más encarnizados para pelear contra los japoneses en Birmania y Malasia durante la segunda guerra mundial. «La 17.ª División, al mando del mayor general D. T. Punch Cowan, compuesta sólo por gurkhas, tenía la misión de emboscar y cazar japoneses», según lo ha escrito un testigo de aquella guerra espantosa que saturó los cines dominicales de nuestra juventud. «El 1 de mayo de 1945, desde aviones norteamericanos, varios comandos de gurkhas fueron lanzados en paracaídas sobre *Elephant Point*, donde estaban las fuerzas japonesas que custodiaban las vías de acceso a Rangún. Los gurkhas debían despejar la ruta para que los aliados entraran en Birmania, pero cuando tomaron tierra ya los japoneses habían evacuado el lugar.» Sólo quedaban treinta, que –según suponía el mariscal Slim– habían dejado allí como observadores. Los gurkhas los hicieron prisioneros y los degollaron a todos.

Sin embargo, tal vez ninguno de los relatos atroces sobre los gurkhas sea más revelador de su carácter que el de la batalla de Imphal-Kohima, en la cual los mercenarios nepaleses exterminaron a un número incalculable de japoneses. Después de la batalla, algunos gurkhas estaban recogiendo cadáveres en sitios inaccesibles para las excavadoras. En eso, un japonés que era levantado por dos gurkhas demostró que no estaba tan muerto como parecía. Un gurkha blandió su cuchillo para acabar con el prisionero, cuando intervino un oficial británico que pasaba por el lugar. «No, Johnny, no lo mates», le dijo. El gurkha, con su cuchillo levantado, miró al oficial entre atónito y dolorido. «Pero, *sahib*», protestó, «no podemos enterrarlo vivo.»

Los oficiales ingleses que han desmentido las atrocidades de los gurkhas en las Malvinas han insistido en que no son mercenarios, sino militares de élite al servicio de la Corona. Peor si es así, porque sería un reconocimiento de que el ejército británico admite como suya una moral que no se compadece con el derecho de gentes. Los gurkhas son guerreros a sueldo al servicio de un ejército extranjero, y esto define, sin más vueltas, su condición de mercenarios. En efecto, los gurkhas son contratados por oficiales británicos que los seleccionan entre los mejores, después de recorrer durante varias semanas las aldeas de cuatro tribus del minúsculo y legendario reino de Nepal, en las estribaciones del Himalaya. Los escogidos, que no son más de 400 al año, ingresan en el ejército británico con un sueldo básico de diez libras esterlinas al mes, más otras tres y 20 chelines para comer. Su interés primordial es ahorrar lo más posible para enviar dinero a sus familiares, cuya pobreza es legendaria.

La ferocidad y la disciplina casi sobrenatural de los gurkhas no son, por supuesto, una condición genética, sino elementos sustanciales de un oficio aprendido. Desde 1815, cuando los oficiales ingleses de la honorable Compañía de las Indias contrataron a los primeros guerreros gurkhas para que los ayudaran a apoderarse de la India, esas malas artes inculcadas no han hecho sino perfeccionarse. Para eso existe el centro de entrenamiento de Hong Kong, donde los hambrientos nepaleses recién contratados, que no conocen la electricidad ni ninguna otra invención de nuestro siglo, son adiestrados como animales en el oficio de matar. En el ejército británico hay un número constante de 10.000 repartidos entre la Hong Kong Field Force, que mantiene el control de la colonia inglesa dentro del territorio chino; la reserva estratégica del Reino Unido, que está estacionada en el Reino Unido como un cuerpo de bomberos colonial para enfrentar cualquier emergencia en cualquier parte del mundo —como en la reciente de las islas Malvinas—, y un destacamento especial que protege al sultanato autoritario de Brunei contra las pretensiones armadas de Indonesia, una verdadera fuerza colonial, anacrónica y vergonzosa, que el Reino Unido mantiene como un rezago indigno de sus tiempos de gloria. De modo que no hay nada extraño en su comportamiento criminal de las Malvinas, ni tienen por qué parecer delirantes los testimonios de tantos supervivientes argentinos. Una fuente tan seria e insospechable como la revista británica *The Economist* ha escrito hace poco que los gurkhas «son guerreros eficaces, pero despiadados»: una vez desenvainado su famoso cuchillo no puede guardarse sin sangre. «Cuando pierden la oportunidad de derramar la ajena», ha escrito otro inglés, «los gurkhas satisfacen la tradición cortándose un dedo.»

ESTOS ASCENSORES DE MIÉRCOLES

En *La vida de Archibaldo de la Cruz* —una película inolvidable de don Luis Buñuel— ocurría el episodio tremendo de una monja que entraba por la puerta de un ascensor cuando el ascensor no estaba en el piso, y la mujer infortunada se precipitaba con un alarido de espanto hasta el abismo del sótano. En algún periódico se publicó hace mucho tiempo la noticia de dos mecánicos de ascensor que trataban de reparar uno trabajando en el fondo del conducto y, de pronto, la caja descendió sin remedio y los aplastó contra el piso. Conozco a la hija de un matrimonio amigo que a los 12 años se quedó encerrada durante dos horas en un ascensor en tinieblas, y nunca más se recuperó del espanto, a pesar de los muchos tratamientos médicos y psicológicos a que fue sometida. La niña —para decirlo del modo menos dramático posible— se volvió loca.

Sin embargo, la historia de ascensores más horrible que he oído contar ocurrió en Caracas hace muchos años. Una familia que vivía en una casa de tres pisos con ascensor se fue a Europa por tres meses, y antes de salir, como lo hacían siempre, desconectaron la electricidad en los controles de la entrada principal. Una criada se había quedado poniendo orden en el piso superior, pero estaba de acuerdo con sus patrones en que bajaría por la escalera, echaría la llave a la puerta de la calle y volvería todas las semanas a hacer la limpieza. Pero en el momento en que los dueños de la casa salían debió recordar algo urgente, y trató de alcanzarlos con el ascensor. La interrupción de la electricidad la sorprendió en mitad de camino, y nadie se enteró, hasta tres meses después, cuando la familia regresó de Europa y encontró los restos putrefactos en el ascensor. Me cuesta mucho trabajo no pensar en ésta y en otras muchas historias horribles cuando tengo que entrar en un ascensor. En alguna época me tranquilizaba mucho viajar en esos ascensores modernos de hoteles caros y edificios de lujo que tienen un teléfono para pedir auxilio.

Publicado originalmente el 1 de junio de 1983.

Pero mi confianza se volvió humo en una ocasión en que alguien que viajaba conmigo descolgó el teléfono para dar aviso de una parada irregular y no logró que le contestara nadie. La explicación que tuvo fue que el personal encargado de atender ese teléfono se había ido a almorzar a la hora en que ocurrió la emergencia –por fortuna– momentánea. Desde entonces tengo la costumbre de averiguar quién oye los timbres de alarma de botones rojos con una campana dibujada que se encuentran en todos los ascensores del mundo, y en la inmensa mayoría de los casos habría que admitir que no sirven sino para darles a los pasajeros una sensación de seguridad sin ningún fundamento. En realidad, muchos de esos timbres no suenan en ninguna parte. No funcionan en la realidad, sino en la imaginación de los viajeros ilusos, pero nadie lo sabe porque nadie ha necesitado de ellos en mucho tiempo. Un mecánico de ascensores de México me decía hace poco que en el servicio regular de mantenimiento es obligatorio establecer el estado de las campanas de alarma, pero no siempre lo hacen, porque los mecánicos están tan familiarizados con sus ascensores que no les alarma que la alarma no funcione. Además –me dijo uno de ellos–, la mayoría de los timbres de emergencia son inútiles, porque casi todos funcionan con electricidad, y son muy pocos los daños de ascensores que ocurren por causas distintas de una falla eléctrica. De modo que la alarma no suena por las mismas razones por las que no funciona el ascensor.

En los edificios de apartamentos, aun en los más caros, la alarma suena en el cuarto del portero, el cual tiene una llave simple con la que abre la puerta del ascensor en un minuto. El problema es que el portero no está siempre en su puerta, aunque su nombre lo indique, y los más eficaces tienen tantas prerrogativas merecidas que salen a descansar con su familia durante los fines de semana. El otro día, en un edificio de apartamentos de Barcelona, descubrí por casualidad que el portero no duerme en su cubil, sino en la casa de su familia, de modo que si alguien se queda encerrado, lo mejor que puede hacer es echarse a dormir enroscado en el piso del ascensor hasta las siete de la mañana, si es que tiene la buena suerte –¿o la mala suerte?– de estar solo en su desgracia, o si su percance no ocurre en pleno invierno y amanece congelado.

En un edificio residencial de París, que cuesta su peso en oro, todos los servicios son tan modernos que han prescindido de la portera, una de las instituciones más antiguas y legendarias de la ciudad. En efecto, las porteras del París de otros tiempos tenían tan buen crédito que la literatura francesa, y no sólo la de Balzac, sino en especial las novelas de criminales y detectives, tenía que recurrir a ellas sin remedio para que los relatos más fantásticos parecieran verdaderos. Un testimonio de una portera sobre alguno de sus inquilinos podía ser definitivo ante una autoridad judicial. Pero cada día que pasa más porteras de París son sustituidas por ingenios electrónicos deshumanizados, mucho más eficaces que sus viejas antecesoras cascarrabias, pero, en todo caso, incapaces de rescatar a un pobre inquilino atrapado en un ascensor. El problema del timbre de alarma en los edificios sin porteras ha sido resuelto instalándolo en el apartamento del responsable del edificio, cuyo cargo es eventual y rotativo, y que, por supuesto, no tiene ninguna obligación de estar en su casa en espera de que alguien se quede encerrado en el ascensor. El hecho final es que la soledad del ascensor es una de las más temibles, sobre todo para quienes padecen de claustrofobia, y saben que podrían soportar cualquier cosa menos un minuto de encierro en un ascensor.

Nuestros abuelos, que eran tan severos, eran mucho más humanos en su concepción de la vida. A ninguno de ellos se le hubiera ocurrido inventar un ascensor como los más usados en estos tiempos, cuya seguridad radica en todo lo contrario de lo que uno quisiera para sentirse seguro. Son sepulcros blindados. En Nueva York, donde en realidad se tiene tanta conciencia de la peligrosidad de los ascensores que se les trata como vehículos arriesgados, lo único que falta es que se enciendan letreros como en los aviones: «Ajuste el cinturón de seguridad». Cuando uno entra en los ascensores multitudinarios de Manhattan, el ascensorista, como un general en una batalla, ordena: «Póngase de frente a la puerta». Lo cual, sin duda, facilitaría la evacuación inmediata. Pero todo eso son los resultados del hermetismo de los ascensores de hoy. Antes, en cambio, los abuelos eran conscientes de que el uso del ascensor, por efímero y rutinario que fuera, era de todos modos un viaje y había que hacerlo con la mayor felicidad que fuera posible. De modo que construían unas obras de arte, no sólo de la técnica, sino también de la ebanistería, con ventanas por todos lados que no sólo servían para respirar, sino para ver el paisaje interior de la casa. Uno no subía con el aliento cortado por el temor de que se fuera la luz, sino que iba viendo la vida: los enamorados del primer piso, que esperaban besándose a que el ascensor regresara; la anciana inválida que fingía bordar frente a la puerta abierta del segundo piso, cuando, en realidad, lo que hacía era disfrutar a su vez con el espectáculo de la vida que subía y bajaba en el ascensor; o el alborozo del niño que nos decía adiós con la mano cuando nos veía pasar de largo por el piso tercero. Todo eso se acabó con los temibles cajones de acero de hoy, cuya única ventaja —porque alguna tenían que tener— es que, en caso de urgencia, como ocurre con más frecuencia de lo que uno cree, los amantes sin techo pueden oprimir el botón del freno para hacer un amor vertical de gallo triste, mientras alguien maldice en algún piso intermedio. Estos modernos ascensores de miércoles que se quedan parados de pronto en cualquier parte, sin permiso de nadie. Menos mal que pueden servir para tanto las cosas que no sirven.

CARTAGENA: UNA COMETA EN LA MUCHEDUMBRE

Había un hombre impasible que trataba de elevar una cometa en medio de la muchedumbre inmensa. Había una mujer vendiendo un armadillo amaestrado que, según decía a voz en cuello, era capaz de hacer las operaciones aritméticas, salvo la división, marcando las respuestas cifradas con golpecitos rítmicos de las patas. Había un cura decrépito que llamaba tanto la atención como un papa porque iba vestido de un modo insólito para estos tiempos: con una sotana negra de aquellas que se usaban en los versos del compatriota Luis Carlos López. Había un transatlántico iluminado que navegaba por las calles en medio de un viento de banderas. Había un presidente conservador que las multitudes aclamaban con la espontaneidad y el entusiasmo que hubieran querido para ellos algunos presidentes liberales, y había un príncipe de quince años saludado por los descendientes de quienes hace apenas 170 expulsaron a tiros a sus antecesores, y el presidente del gobierno socialista español, a quien adoraban como si fuera un rey tropical, a pleno sol y con 34 grados de calor. Había todo eso y mucho más, como para que los visitantes ilustres no olvidaran que estaban en el mundo alucinado y alucinante del Caribe, donde aun las ilusiones más locas terminan por ser ciertas y se conoce el otro lado de la realidad. Era el miércoles 1 de junio de 1983, y la muy noble ciudad de Cartagena de Indias celebraba los primeros 450 años de su vida y sus milagros.

Todos los años, por el 11 de noviembre, la ciudad celebra el aniversario puntual de su independencia proclamada, pero ni aun los más viejos recuerdan una fiesta callejera tan concurrida, entusiasta y pacífica como ésta. Ha sido —y ése fue uno de sus méritos mayores— el orden del júbilo dentro del caos. El buen humor no desfalleció ni un instante. Al final de la agotadora jornada, la explanada del muelle de los Pegasos se oscureció con una muchedumbre de no menos de 100.000 personas en torno a la tarima de los músicos. Arriba, en

Publicado originalmente el 8 de junio de 1983.

la terraza del baluarte de San Ignacio, estaban los invitados especiales encabezados por el príncipe adolescente don Felipe de Borbón y el presidente Felipe González con su séquito numeroso. Alguien dijo entre la muchedumbre de la calle: «Hoy estamos otra vez como hace cuatrocientos años, nosotros aquí abajo, y allá arriba los españoles». Pero era evidente que no se trataba de un sentimiento hostil, sino la expresión de un anhelo general, que era el de ver a todos los partícipes de la fiesta, tanto los de arriba como los de abajo, tanto los nacionales como los invitados de fuera, celebrando el cumpleaños en la misma plaza y bajo el mismo techo de estrellas abigarradas. La noche, una de las más cálidas que se recuerden, era más que propicia.

Sin embargo, como ocurre casi siempre en estos casos, las cosas que no se ven suelen ser, por lo menos, tan importantes como las que están a la vista. La presencia de los cuatro cancilleres del grupo de Contadora –Colombia, México, Panamá y Venezuela– y la oportunidad que sin duda tuvieron de intercambiar ideas con sus colegas del resto de América Latina en los callejones secretos de la fiesta, pudo haber sido providencial para la paz en América Central. Los esfuerzos más recientes trataban de conseguir que los cancilleres de Honduras y Nicaragua se reunieran a discutir sus discrepancias en presencia de los cuatro cancilleres de Contadora reunidos hace unos días en Panamá. No fue posible aun algo que debía parecer tan fácil. En Cartagena, por una de esas coincidencias a las cuales no son del todo ajenas las conveniencias políticas, los cancilleres de Honduras y Nicaragua comulgaron juntos en la misma misa de acción de gracias. El de Nicaragua, que es el padre Miguel d'Escoto, no sólo está acostumbrado a comulgar desde su edad más tierna, y aun en la que tiene ahora, que ya lo es menos, sino que no le han faltado oportunidades de dar él mismo la comunión a otros cancilleres de este mundo. El canciller de Honduras, en cambio, y al menos hasta donde se sabe, no tiene ninguna investidura que le permita hacerlo. Sin embargo, son viejos amigos entre sí, y al margen de los pleitos públicos oficiales hablan de ellos en privado con una naturalidad y un sentido del humor que no permiten entender muy bien por qué les cuesta tanto trabajo a sus países llegar a un acuerdo de buenos vecinos. En una de las recepciones de Cartagena, alguien les dijo a ambos: «Ya que ustedes están comulgados y por consiguiente inspirados en la gracia de Dios, enciérrense ahora mismo en ese cuarto y no vuelvan a salir hasta que no estén de acuerdo». Lo que más duele es que, sin duda, a ambos les hubiera gustado hacerlo, pero se sienten impedidos por las contradicciones internas y los intereses ineludibles de sus propios gobiernos. Hay que confiar, a pesar de todo, en que el ambiente de cordialidad que se impuso en el cumpleaños de Cartagena haya contribuido de algún modo a aflojar la tensión.

En todo caso, las tentativas no fueron inútiles, y de este cumpleaños de Cartagena queda, sin duda, una nueva esperanza de encontrar un remedio para todas esas guerras de América Central, guerras dispersas y sin futuro previsible, a las que bastaría con añadir un grado más de los muchos que ya tienen de confusión para provocar un cataclismo mundial. El presidente Felipe González, que es sin discusión el hombre de aquellos lados del océano que mejor nos conoce, tal vez tuvo ocasión de clarificar aún más sus ideas sobre la situación actual de América Central. Ya en el Encuentro dentro de la Democracia, celebrado en Madrid hace poco más de un mes, se hablaba en voz baja, aunque no muy baja, de la posibilidad de que el presidente Felipe González actuara como mediador entre América Central y Estados Unidos. La necesidad de esta mediación no la niega nadie. Ni siquiera los propios cancilleres de Honduras o El Salvador, que no vacilan en reconocer en privado que es Estados Unidos, y no sus propios gobiernos, lo que constituye un obstáculo para lograr un acuerdo en un conflicto que ya está causando más muertes y desgracias que un temblor de tierra. No parece casual que el presidente del gobierno español haya querido tener estos contactos de Cartagena antes de su próxima visita a Estados Unidos, donde su voz tiene un crédito grande y su autoridad para hablar por nosotros no puede ser disputada. Confiemos en que éstas no sean ilusiones de periodista, sino clarividencias de poeta, y que algún día no lejano se pueda hablar de esta hermosa fiesta de Cartagena como del cumpleaños histórico donde, por fin, se rompió la piñata de la paz en América Central. Todo es posible en una ciudad donde alguien logra elevar una cometa de colores en medio de la muchedumbre.

PORTUGAL, NUEVE AÑOS DESPUÉS

Uno se pregunta qué fue de aquella pléyade de militares portugueses de alto rango que en abril de 1974 hicieron la jubilosa revolución de los claveles. La foto de una muchacha poniendo una flor en el cañón del fusil de un soldado amigo no sólo dio la vuelta al mundo por su belleza, sino que se impuso de inmediato como un símbolo de una vida nueva. Portugal era una fiesta. Nadie dormía, nadie tenía un horario de trabajo fijo, y el tiempo apenas si alcanzaba para celebrar la victoria sobre una de las dictaduras más antiguas y crueles del mundo, y para disfrutar, en plena calle y a voz en cuello, de la libertad recobrada. «Nadie puede entendernos mejor que ustedes», le dijo por aquellos tiempos un miembro del Consejo de la Revolución a un grupo de periodistas latinoamericanos. «Los europeos, aun los más comprensivos, tratan de interpretarnos con una óptica de país desarrollado y no encuentran cómo meternos a la fuerza dentro de sus esquemas.» Por motivos históricos y geográficos, siendo uno de los países más pobres del mundo, pero con una posición estratégica esencial para las potencias occidentales, Portugal estaba obligada a sentarse a la mesa de los países más ricos de la tierra, pero hablando un idioma nuevo que nadie entendía porque a nadie le convenía entenderlo, y con los fondillos remendados y los zapatos rotos, pero con la dignidad que le imponía el haber sido en otro tiempo el dueño casi absoluto de todos los mares.

La presión tremenda de ese drama se reflejaba en todos los aspectos de la vida portuguesa. Todo se había vuelto político. Desde la plaza del Rossío, en el corazón de Lisboa, hasta el rincón más remoto y olvidado de la provincia no había un centímetro de pared, ni un anuncio de carretera, ni el pedestal de una estatua que no tuviera pintado un letrero político. «Unidad sindical», pedían a brocha gorda los comunistas, mientras acusaban a los socialistas de querer dividir la clase obrera para dejarla a merced de la socialdemocracia

Publicado originalmente el 15 de junio de 1983.

europea. «Socialismo sí; pero con libertades», decían sin más explicaciones los socialistas. «Fuera el imperialismo capitalista y el social-imperialismo», decía un partido de extrema izquierda, cuyo radicalismo intransigente se confundía con la línea de candela de la provocación. «Viva Cristo Rey», gritaba la reacción católica. «El voto es el arma del pueblo», decían los liberales. Y los anarquistas, con su ingenio incansable, corregían en la pared de enfrente: «El arma es el voto del pueblo». De día, en medio del desorden alborozado, los militantes del Ejército de Salvación tocaban el trombón en la puerta de los grandes almacenes y fomentaban el pánico con sus diatribas pavorosas contra el alcohol y el sexo. Muy tarde en la noche, cuando el cansancio doblegaba por fin el activismo desaforado, la reacción hacía reventar bombas de alto poder y envenenaba al mundo con el rumor infame de que al hermoso e idílico Portugal de las canciones se lo había llevado el carajo.

En medio del estruendo ensordecedor, había una inteligencia distinta: el Movimiento de las Fuerzas Armadas (FMA), dirigido por una cosecha de oficiales jóvenes, y cuyo poder político, unido a su poder de fuego y a su popularidad inmensa, hacía de ellos mucho más que árbitros simples de la situación. Los pesimistas por la ruina de la economía nacional decían con un gran desprecio: «Portugal no produce sino portugueses». Los dirigentes del FMA replicaban: «La mayor riqueza de un pueblo es su población». La mayoría de ellos eran antiguos universitarios reclutados por la dictadura como carne de cañón para las guerras coloniales. Trabajaban sin horarios, sin pausas, lo mismo en la administración pública que en las campañas de politización de los campesinos. La democracia había empezado por los cuarteles: oficiales y soldados se tuteaban, dormían en el mismo cuarto y comían la misma comida en la misma mesa. Por primera vez en la historia de la humanidad las tropas tenían derecho a desobedecer una orden si sus oficiales no les decían para dónde iban y con qué propósito. La respuesta a todos los niveles era la misma: «Vamos para un socialismo inventado por nosotros mismos, de acuerdo con nuestras condiciones propias, independiente de todo centro internacional de poder y, al mismo tiempo, construido con imaginación y humanidad».

Lo que nos preguntábamos todos los periodistas asombrados que visitábamos Portugal por aquellos días era cómo los militares de una dictadura infame habían llegado a comprender que todo cambio era imposible sin una integración real con el pueblo, y cómo habían tomado conciencia de esa realidad. El proceso, en verdad, fue muy simple. Cuando la guerra se agudizó contra los movimientos de liberación de las colonias africanas –y en especial Angola y Mozambique–, los oficiales de la dictadura, que eran aristócratas de solemnidad, decidieron improvisar una oficialidad de clase media que sirviera de carne de cañón en los dominios sublevados. Para eso abrieron, en primer término, las puertas de la academia militar, donde se formaban los oficiales de carrera, y en segundo término, empezaron a reclutar universitarios para convertirlos en oficiales milicianos con el grado inmediato de subtenientes. De modo que en el curso de pocos años cambió por completo la composición de la clase de los mandos medios. «Fue muy fácil», dijo un oficial a los periodistas, «que nuestra promoción, con una edad promedio de 28 años, sufriera una transformación ideológica en el sentido de las aspiraciones populares.» Otro oficial decía: «Nuestra conciencia se formó en las largas noches de reflexión en los campamentos de África, conversando con los soldados, que en realidad eran universitarios uniformados, y con los prisioneros que capturábamos entre las guerrillas y que nos estremecieron con el ejemplo de su decisión y su claridad». Los nombres de los militares que dirigían el cambio llegaron a ser legendarios: Vasco Gonçalves, Costa Gomes, Melo Antúnes, Otelo de Carvalho, Vasco Lourenço, Correia Jesuino, Rosa Coutinho. Todos pertenecían al Consejo de la Revolución, que era el organismo rector del proceso, y ocupaban cargos claves en el gobierno. Melo Antúnes, un fumador nervioso y sonriente que pasaba casi sin darse cuenta de una conversación política a una discusión sobre literatura, era considerado por sus compañeros como uno de los ideólogos más antiguos y lúcidos del FMA. Sin embargo, tal vez fue Otelo de Carvalho el más carismático de todos y el que pareció más decidido a asumir la responsabilidad política del país para llevar el proceso de cambio hasta sus últimas consecuencias. Qué fue lo que se lo impidió y qué fue lo que se llevó a Portugal por un camino distinto es algo muy difícil y, sobre todo, muy largo de establecer. Pero el hecho es que los promotores y protagonistas mayores de aquella revolución casi poética fueron relegados, si no al olvido, al menos a la penumbra. De ahí que sea tan significativa una noticia a la cual no se le ha dado, ni siquiera en Portugal, la atención que merece. Me refiero a la creación de la Asociación Veinticinco de Abril, integrada no sólo por todos los miembros del Consejo de la Revolución original, sino por más de 1.500 oficiales de las fuerzas armadas. Mil trescientos de ellos son todavía activos, o sea, la cuarta parte de la oficialidad actual. «Se trata de ampliar y profundizar el espíritu democrático de las fuerzas armadas», ha dicho en privado uno de sus fundadores. «No tenemos aspiraciones políticas inmediatas ni queremos intervenir en las condiciones actuales.» Pero la mayoría de ellos están de acuerdo en que las fuerzas de la reacción son cada vez más activas e influyentes en Portugal y que la vigilancia insomne de los héroes en reposo del FMA puede impedir, llegado el caso, que Portugal regrese a su pasado sombrío. Ante las acusaciones de que se trata de una organización subversiva, sus promotores señalan que se ajusta del todo a la Constitución vigente y que es tan legal y pública como otras muchas de diverso carácter que existen en Portugal. Más aún: uno de sus miembros más distinguidos es el propio presidente de la república, el general Eanes. Lo cual demuestra, una vez más, lo que desde hace tantos siglos se sabe, y es que Portugal es un país muy raro. Por decir lo menos.

VALLEDUPAR, LA PARRANDA DEL SIGLO

Un día de 1963, durante el festival de cine de Cartagena, le pedí a Rafael Escalona que me reuniera a los mejores conjuntos de música vallenata para oír todo lo que se había compuesto en los siete años en que yo había estado fuera de Colombia. Escalona, que ya era compadre mío desde unos 12 años antes, me pidió que fuera el domingo siguiente a Aracataca, adonde él llevaría la flor y nata de los compositores e intérpretes de las hornadas más recientes. El acuerdo se llevó a cabo en presencia de la muy querida amiga y periodista sagaz Gloria Pachón –que hoy es la esposa del senador Luis Carlos Galán– y ella publicó la noticia al día siguiente con un título que a todos nos tomó por sorpresa: «Gran festival vallenato el domingo en Aracataca». Todos los fanáticos del vallenato de aquellos tiempos, que no éramos muchos, pero sí suficientes para llenar la plaza del pueblo, nos encontramos el domingo siguiente en Aracataca. El escritor Álvaro Cepeda Samudio llevó tres camiones de cerveza helada, y los repartió gratis entre la muchedumbre. Escalona llegó tarde, como de costumbre, pero también como de costumbre llegó bien, con nadie menos que con Colacho Mendoza, de quien nadie dudaba entonces que iba a ser lo que es hoy: uno de los maestros del acordeón de todos los tiempos. Mientras los esperábamos, el centro de la fiesta fue Armando Zabaleta, quien nos dejó admirados con el modo de cantar su canción más reciente y magnífica: *La garra del águila*. Era un buen comienzo, porque aquella canción era la crónica muy bien contada de la visita que Escalona había hecho poco antes al presidente Guillermo León Valencia en su palacio, y estaba, por consiguiente, en la línea del vallenato clásico que fue creado para contar cantando y no para bailar. Tanto es así, que en el festival de la semana pasada, alguien se disponía a bailar cuando Alejo Durán el Grande estaba en uno de sus grandes momentos, y se interrumpió para decir: «Si me bailas me voy».

Publicado originalmente el 22 de junio de 1983.

Aquella pachanga de Aracataca no fue el primer festival de la música vallenata –como ahora pretenden algunos– ni quienes la promovimos sin saber muy bien lo que hacíamos podemos considerarnos como sus fundadores. Pero tuvimos la buena suerte de que les inspirara a la gente de Valledupar la buena idea de crear los festivales de la leyenda vallenata. Así fue, y en 1967 se llevó a cabo el primero, con todas las de la ley, y en la ciudad de Valledupar, que es la sede natural por derecho propio. El primer rey elegido fue el rey de reyes, Alejo Durán, que de ese modo le dio al certamen su verdadero tamaño histórico. Aunque ya para esa época la música vallenata empezaba a treparse por la cornisa de los Andes tratando de conquistar Bogotá, todavía no lograba conquistar el corazón de muchos fuera de su ámbito original. En Bogotá –por los años cuarenta– se transmitía los domingos un programa de radio con música para bailar que se llamaba *La hora costeña*, y que muy pronto se convirtió en una parranda matinal para los estudiantes caribes. Allí se tocaban el porro y la cumbia, el fandango y el mapale, pero ni un solo vallenato. Y no sólo porque los costeños sabíamos que el vallenato no era para bailar sino para escuchar, sino porque nadie de allá arriba sabía de su existencia y de su pureza. En la costa caribe, en cambio, el programa de más prestigio en esa época era una hora de canto de un hombre de Ciénaga –Guillermo Buitrago– a quien hay que reconocerle, entre otros muchos méritos, el de haber sido el primero que puso la música vallenata en el comercio. Ya Rafael Escalona, con poco más de 15 años, había hecho sus primeras canciones en el Liceo Celedón de Santa Marta, y ya se vislumbraba como uno de los herederos grandes de la tradición gloriosa de Francisco el Hombre, pero apenas si lo conocían sus compañeros de colegio. Además, los creadores e intérpretes vallenatos eran gente del campo, poetas primitivos que apenas si sabían leer y escribir, y que ignoraban por completo las leyes de la música. Tocaban de oído el acordeón, que nadie sabía cuándo ni por dónde les había llegado, y las familias encopetadas de la región consideraban que los cantos vallenatos eran cosas de peones descalzos, y si acaso, muy buenas para entretener borrachos, pero no para entrar con la pata en el suelo en las casas decentes. De modo que el joven Rafael Escalona, cuya familia era nada menos que parienta cercana del obispo Celedón, se escandalizó con la noticia de que el muchacho compusiera canciones de jornaleros. Fue tal el escándalo doméstico, que Escalona no se atrevió nunca a aprender a tocar el acordeón, y hasta el día de hoy compone sus canciones silbadas, y tiene que enseñárselas a algún acordeonista amigo para poder oírlas. Sin embargo, la irrupción de un bachiller en el vallenato tradicional le introdujo un ingrediente culto que ha sido decisivo en su evolución. Pero lo más grande de Escalona es haber medido con mano maestra la dosis exacta de ese ingrediente literario. Una gota de más, sin duda, habría terminado por adulterar y pervertir la música más espontánea y auténtica que se conserva en el país.

De modo que hay una prehistoria del vallenato que sus fanáticos de hoy —que son muchos, aun más allá de nuestras fronteras— apenas si han oído nombrar. Es un mundo cerrado, con un olimpo propio, cuyos dioses viven ya respirando los aires enrarecidos de la leyenda. Francisco Moscote, a quien se recuerda con el buen nombre de Francisco el Hombre porque le ganó al diablo en un duelo de acordeón, está tan implantado en la mitología popular que ahora no se sabe a ciencia cierta si en realidad existió. Pacho Rada, otro de los primitivos grandes, tenía raíces tan bien sembradas en el corazón de su pueblo, que una noche le tomaron preso en la población de Plato, pero el inspector de policía cometió el error de dejarle el acordeón en la cárcel. Pacho Rada, tal vez de puro aburrido, se puso a tocar y a cantar, y el pueblo se despertó escandalizado de que estuviera preso un hombre investido de tanta gloria, y entonces invadieron la cárcel y lo sacaron a la calle. De estos dos precursores se habla como si hubieran muerto sin edad después de haber vivido durante siglos. Uno piensa que tal vez fuera cierto cuando ve a los que todavía quedan vivos, y cuya serenidad y cuya sabiduría hacen pensar que viven en un tiempo distinto del nuestro. Leandro Díaz es una especie de patriarca mítico. A pesar de que es ciego de nacimiento ha vivido desde muy joven de su buen oficio de carpintero, y nunca podré olvidar el día en que Rafael Escalona me llevó a conocerlo en su taller, porque estaba haciendo una mesa con las luces apagadas, y no se oía nada más que el rumor del serrucho y los golpes del martillo en las tinieblas. Más aún: durante la guerra mundial, cuando no fue posible importar más acordeones de Alemania, la tradición no sufrió ni una grieta, porque el ciego Leandro Díaz reparaba los acordeones más antiguos hasta dejarlos como nuevos. La semana pasada, cuando lo oí cantar otra vez después de casi 20 años, y me envolvió con la belleza de *La diosa coronada* —que no sólo es su canción más hermosa sino una nota muy alta de nuestra poesía—, tuve la sensación de haber entrado por primera vez en el ámbito prohibido de la leyenda. Sin embargo, a su lado no era menos mítico Emiliano Zuleta cantando, con su voz estragada por los años y el alcohol de caña, los versos magistrales de *La gota fría*, que para mi gusto es una canción perfecta, y por tanto, un punto de referencia que no pueden perder de vista los creadores de hoy. La lista no se acaba fácil: Chico Bolaño, Toño Salas, Lorenzo Morales y tantos otros. Sin embargo, lo más alentador es que el manantial no se seca: Julito Rojas, el rey elegido este año, no llega todavía a los 30 años.

Fue dentro de ese ámbito místico donde transcurrió el XVI Festival de la Leyenda Vallenata, y fue por eso y por nada más por lo que tuvo la autenticidad y la resonancia que había empezado a perder en años anteriores. Un equipo de la televisión holandesa que registró cada minuto de aquella parranda sin una sola tregua se llevó una impresión de la cual no alcanzarán a reponerse en mucho tiempo. No podían entender que existiera en este mundo de horrores un lugar como aquél, donde las casas no se cerraban nunca, y todo el que quería entraba a comer donde quisiera a cualquier hora del día y de la noche en que tuviera hambre y siempre encontraba una mesa servida, y todo el que tuviera sueño entraba a dormir a cualquier hora donde quisiera y siempre encontraba una hamaca colgada. Y todo eso sin un instante ni un resquicio de silencio: el espacio total estaba saturado de música.

Convencido de que aquél no era un fenómeno local sino una condición propia del país, uno de los técnicos holandeses que se dejaron arrastrar por aquel torbellino anotó en su diario: «Todos los colombianos están locos». Lo cual será, por fortuna, una nota de alivio para la mala imagen que tan bien ganada tenemos por estos días en la prensa extranjera. En síntesis: el XVI Festival de la Leyenda Vallenata ha sido una prueba más –y de las mejores– de que la cultura popular no es tan aburrida, no huele tan mal como lo creen y lo sienten los intelectuales puros. Mal de muchos, consuelo de corronchos.

«¿QUÉ HAY DE MALO EN LA MALA PRENSA?»

Tal vez ningún país está tan preocupado como Colombia por lo que se piensa de él en el exterior. Ninguno está tan susceptible a las noticias y comentarios de prensa que puedan afectar su imagen ante el mundo. Y, sin embargo, son muy pocos los que han dado tanto de que hablar a la prensa extranjera en los últimos años. Tal parece como si los colombianos tuviéramos que sobrellevar el destino de ser exportadores de noticias raras. Buenas y malas, pero muchas de ellas en páginas primeras y aun con fotografías en colores. Pero la misma inquietud, a veces desproporcionada, que nos causan las noticias perjudiciales para nuestra imagen externa se transforma ante las buenas noticias en un impulso irresistible de magnificarlas hasta el ridículo. En cualquiera de los dos extremos somos víctimas de la exageración, ya sea por la vergüenza o por el regocijo. A este paso, ofuscados por las contradicciones de nuestra propia imagen en el espejo del mundo, corremos el riesgo de terminar por no saber a ciencia cierta cómo somos en la realidad.

El balance parece ser desfavorable, sobre todo en el inventario de los últimos meses. Un día cualquiera de esta semana se encontraban en un mismo periódico dos noticias enfrentadas. Una se refería a la actuación de los ciclistas colombianos en la Vuelta a Francia, que, al parecer, era motivo de admiración y entusiasmo para la prensa francesa; la otra decía que un colombiano es el dueño y señor de una isla del Caribe destinada al tráfico y comercialización de la cocaína. Uno termina por preguntarse con la mano en el corazón con cuál de las dos noticias se queda, y termina tal vez por no quedarse con ninguna, deprimido por la evidencia de que las malas noticias derrotan a las buenas.

Así es. Los colombianos, en el exterior, considerábamos como un acto de justicia poética que nuestra mala fama de traficantes de drogas fuera, en cierto modo, compensada por el renglón de expor-

Publicado originalmente el 29 de junio de 1983.

tación más hermoso del mundo: las flores. En diciembre pasado, la nevada ciudad de Estocolmo parecía un jardín de rosas amarillas. Estaban por todas partes y era imposible entrar en alguna sin que uno fuera recibido por una tormenta de rosas que, en realidad, parecían caídas del cielo con una profusión mayor que la de la nieve. La bella y gentil reina de Suecia le hizo a uno de sus invitados la confidencia de que en Estocolmo era imposible encontrar rosas en invierno, de modo que aquellas turbulencias amarillas habían sido importadas de algún remoto país de calores perpetuos donde las rosas florecían sin reposo durante todo el año. La reina, por supuesto, tuvo la discreción de no decir qué país era ése ni cuánto habían costado tantas toneladas de rosas transportadas por avión a través del océano. Pero los colombianos sabíamos, con un justo orgullo patriótico, que eran rosas colombianas. Por eso no fue posible reprimir un estremecimiento de pudor hace pocas semanas, cuando se publicó en el mundo entero la noticia de que se había descubierto un contrabando de cocaína entre un cargamento de flores colombianas. La mala noticia, una vez más, había derrotado a la buena, y era justo suponer que la reina de Suecia, al leer la prensa aquella mañana, tal vez se había preguntado si las rosas amarillas de su fiesta no llevaban oculta también entre sus pétalos la ponzoña intempestiva de otra nieve más constante e insidiosa que la de las noches eternas de los inviernos de Suecia.

No es fácil contrarrestar los éxitos espectaculares de la delincuencia. Se dice que la mafia colombiana ha terminado por derrotar y suplantar a la mafia irlandesa y siciliana en los muelles de Nueva York. No es una condición reciente. En Gran Bretaña se cuenta que hace muchos años vinieron a Colombia dos expertos de Scotland Yard contratados para adiestrar a la policía colombiana en la lucha contra los carteristas callejeros. A su llegada al aeropuerto de Bogotá, los dos expertos fueron despojados de todo cuanto llevaban en los bolsillos y obligados de ese modo a regresar a su país de inmediato con el honor hecho trizas. En otro orden de cosas, un oficial del servicio de contrainteligencia de Venezuela contaba hace muchos años en privado, no sin una cierta admiración, que los espías colombianos son los más duros de exprimir, pues no hay martirio psicológico ni físico que los obligue a revelar sus secretos. Igual comportamiento —decía— lo observaban los delincuentes comunes sometidos a la tortura. La fama del ingenio colombiano para sortear los escollos de la legalidad y burlar los controles policiales se encuentra muy bien sustentada por la realidad en el mundo entero, y se funda, por supuesto, en una facultad reprobable. Pero sería injusto no reconocerlo, aunque sea en lo más secreto de nuestro fuero interno, como el fruto de un talento nacional pervertido por la adversidad social.

Sin embargo –las abuelas lo dicen muy bien–, el mismo Dios que manda la enfermedad manda el remedio. Al lado de un funcionario de la cancillería que fue sorprendido en México con un contrabando de cocaína, al lado de un honorable parlamentario que trataba de venderla en Nueva York, al lado de un condenado a la silla eléctrica por el cargo supuesto de asesinato desalmado y de un malabarista de las finanzas que sorprendió a los bancos de Estados Unidos con un préstamo múltiple de unos 200 millones de dólares, al lado de ellos y de tantos otros que sustentan la mala imagen en los periódicos, hay otros millares de colombianos –y de latinoamericanos de todas partes, por supuesto– que andan por el mundo con la patria a cuestas, sin que nadie se pregunte cómo hacen para vivir sin delinquir. Uno se los encuentra en las buhardillas de Europa o de Estados Unidos, durmiendo a veces debajo de los puentes de medio mundo, trabajando como hormigas arrieras para hacer las buenas noticias de cualquier día sin ayuda de nadie. Son los aprendices.

Cinco mil ochenta y cuatro aprendices de escritores que se han apretado el cinturón hasta el último agujero para terminar su libro sin molestar a nadie, los aprendices de músicos que tocan en el tren subterráneo, no tanto para recoger unas monedas como para gozar con la resonancia de sus voces en los socavones, los aprendices de teatro que levantan su carpa en las esquinas, los aprendices de pintores en quienes nadie ha de fijarse mientras no los descubra un traficante de artes que les compre sus cuadros a 10 para revenderlos a 10.000. Son como minas ocultas en un sendero inocente, que irán estallando poco a poco. Por todas partes y donde menos se espera y de cuyos años amargos y azarosos de aprendizaje no volverá a acordarse nadie –y ellos mismos menos que nadie– cuando lleguen por fin las vacas gordas.

Sin embargo –mientras llegan–, es a ellos a quienes primero desnudan en los aeropuertos, porque los policías no pueden entender que viajen en avión siendo tan pobres, a menos que lleven un tubo de drogas escondido en el trasero. Es a ellos a quienes primero agarran cuando empiezan las redadas, porque no se puede pensar que no se hayan muerto de hambre sin robar, ni se puede pensar que no sean terroristas estando tan peludos y tan pálidos y tan jodidos. También ellos son fruto del mismo talento nacional que alienta a los protagonistas de nuestra mala imagen en el exterior. Sólo que van en sentido contrario y al ritmo imperceptible de la perseverancia y la paciencia, como la tortuga del cuento.

NUEVE AÑOS NO ES NADA

El presidente Belisario Betancur constituyó hace unos días la comisión de once hombres y una sola mujer que debe organizar la participación de Colombia en la celebración de los 500 años de las Américas. La iniciativa podría parecer prematura, puesto que el aniversario anunciado no se cumple el año entrante, sino el 12 de octubre de 1992. Sin embargo, su misma prematurez supuesta revela la importancia que se le atribuye al acontecimiento y da una idea del tamaño de los actos que se quieren preparar para celebrarlo. Tanto es así que un miembro distinguido de la comisión inició la lista de sus sugerencias con una frase que nadie tomó como una broma, sino todo lo contrario, como una muestra de realismo crudo: «Tenemos que darnos prisa, porque ya no faltan sino nueve años».

Ojalá que estos nueve años nos alcancen a los españoles y a los americanos para ponernos de acuerdo sobre el nombre definitivo que debería dársele a la llegada de Cristóbal Colón a nuestras tierras. Los europeos decidieron llamarlo, por su cuenta, el Descubrimiento de América, y nosotros lo aceptamos así, sin reflexionar demasiado sobre la facilidad del término. Germán Arciniegas, hablando más en serio de lo que suele hacerlo en privado, dijo alguna vez que, más que descubrimiento, la llegada de Colón y el tenebroso tiempo de la conquista debía llamarse *el cubrimiento de América*. Tal vez no sea justo tampoco llegar a esos extremos, pero no cabe duda de que la relación de descubridores y descubiertos no es la que mejor conviene a la verdad histórica y al cariño recíproco, y la fiesta de los 500 años es una buena ocasión para ponernos de acuerdo.

Sería gracioso jugar a la ficción de que todo hubiera sido al revés. La civilización maya, que había logrado un desarrollo todavía apreciable y asombroso en sus escombros, tenía conocimientos de astronomía suficientes para orientarse a través de los océanos. Nada hubiera tenido de raro que un grupo de ellos se hubiera aventurado

Publicado originalmente el 6 de julio de 1983.

a explorar el Atlántico hasta mucho más allá de su horizonte y que hubiera llegado al puerto de Cádiz siglos después de fundado por el propio Hércules y en un tiempo en que era uno de los centros más importantes del mundo conocido. Con el mismo derecho con que lo hizo Cristóbal Colón, los navegantes mayas —o los aztecas, o los incas— habrían podido declarar a la península Ibérica como parte de su reino y emprender su conquista e imponer su religión y su lengua, y celebrar dentro de nueve años el medio milenio del descubrimiento de Europa.

Uno no puede menos que pensar en estas fantasías históricas cuando visita —con el fervor merecido con que lo hacemos los latinoamericanos— los lugares de España que tienen una relación más directa con la historia de América. Cuando se ve el puerto de Palos de Moguer, que los maestros de la escuela primaria, con su retórica grandilocuente, nos habían enseñado a imaginar como el muelle original de nuestras vidas, es imposible no asombrarse de que en verdad hayan salido de un atracadero tan pobre las naves destinadas a cambiar la forma y el sentido del mundo. Pero el mismo asombro se siente del otro lado del espejo cuando se contempla la belleza sideral de los altos de Machu Picchu, o el equilibrio mágico de los centros ceremoniales de Yucatán, o la cositeria lírica de la orfebrería quimbaya, o el vasto planeta interior de la Amazonia que se alcanza a ver a simple vista desde la Luna.

Una de las fascinaciones de la infancia era una litografía en la que se representaba el regreso de Colón de su primer viaje. Los Reyes Católicos, que acababan de expulsar de España a los árabes, lo hicieron ir hasta Barcelona, donde lo recibieron con un esplendor que tal vez no fue tanto en la realidad como en la litografía. Lo digo con base en una comprobación curiosa: el gobierno de la ciudad de Barcelona lleva un diario de todo cuanto ocurre en ella desde la Edad Media, y el día en que los Reyes recibieron a Colón no se hizo una anotación especial, sino una mención pasajera entre muchas otras sobre un navegante que regresó de algún viaje y que fue recibido en audiencia por los Reyes. Ese episodio, tal como lo describe la litografía, hace pensar en nuestros antepasados caribes como unos morenos altos y apuestos, cubiertos de plumas y collares y toda clase de adornos de oro, y cargados de frutos extraños de aspecto venenoso y de animales raros que debieron parecer de pesadilla a los testigos de la audiencia. Sin embargo, en lo que conocemos como el diario de Colón —que es apenas la reconstrucción hecha por el padre Las Casas—, nuestros antepasados no están descritos con tanto asombro. Se dice que estaban muy bien hechos, de muy hermosos cuerpos y muy buenas caras, y que tenían los cabellos gruesos y casi como sedas de caballos. La descripción no deja dudas de que tenían el cuerpo pintado y que no eran ni blancos ni negros, sino del color de los nativos de las islas Canarias. Se deduce también de aquella primera visión que los habitantes de la isla de Guananí andaban por la playa como sus madres los parieron, aunque, al parecer, no había ninguna mujer entre los que se acercaron a recibirlos en sus almadías, como tampoco había ninguna en las carabelas que llegaban y como sólo hay una en la comisión colombiana del quinto centenario. Lo cual autoriza a suponer que el machismo proverbial de los latinoamericanos pudo haber llegado en las carabelas (porque ya lo había de sobra en España), pero de todos modos ya era de uso común entre nosotros. Si la memoria no me falla, también era una sola la mujer que figuraba en la litografía, y esto parece ajustarse a la verdad histórica.

En todo caso, la litografía viene a cuento más que nada porque permite preguntarse qué fue de aquellos pobres caribes que Colón llevó consigo en su primer viaje de regreso. Hay quienes ponen en duda que fueran los españoles los primeros en llegar desde Europa a las Américas, pero en cambio es una verdad absoluta que los caribes de la litografía fueron los primeros americanos que llegaron a Europa. Sin embargo, no sólo se ignora hoy quiénes eran, sino que es muy poco lo que se conoce de su destino final. Se sabe que fueron bautizados y que por lo menos uno de ellos se murió de frío en un monasterio de Aragón. Pero inclusive este dato, que cito de memoria y sin la posibilidad inmediata de comprobarlo en la tarde de un viernes apresurado, merecería una mayor atención de los historiadores. Lo menos que podríamos hacer en memoria de los primeros americanos que respiraron el aire de Europa —y con motivo de los 500 años de su viaje involuntario, pero de todos modos histórico— sería establecer a fondo y de una vez por todas qué fue de ellos y de sus tristes huesos. Tal vez nos llevemos todos la sorpresa de descubrir cuántos españoles ilustres descienden de ellos, a tanta honra para nosotros como para ellos. Aunque nunca logremos penetrar el misterio de quiénes eran antes de que llegaran arrastrando sus nostalgias amargas a una Europa dispersa y en ruinas, sin papas, ni chocolate, ni maíz, ni tomates, y que tal vez no hubiera conocido jamás —sin el oro de América— el esplendor del Renacimiento. En efecto, tenemos que darnos prisa —en ambos lados del océano—, pues nueve años son muy pocos para acabar de hacer juntos tantas cosas comunes que se nos han quedado sin terminar.

CONTADORA, CINCO MESES DESPUÉS

Contadora es una isla panameña de 7,5 kilómetros cuadrados, en el océano Pacífico, que se llama como se llama porque en ella se encontraban y se contaban durante la colonia las perlas conseguidas en los mares vecinos. A principios de este siglo se convirtió en un paraíso secreto de millonarios gringos y terminó por ser uno de los sitios más famosos del mundo cuando allí encontró refugio el sha de Irán pocos meses antes de morir. En la actualidad su vieja imagen de guarida de lujo ha sido sustituida por otra que muy bien podría convertirse en el símbolo de una vida mejor para la América Central. Sin embargo, transcurridos cinco meses desde que se reunieron por primera vez en Contadora los cancilleres de Colombia, México, Panamá y Venezuela, no es fácil todavía establecer sin duda alguna si tanto esfuerzo ha valido la pena.

Los logros de Contadora, en realidad, han sido demasiado sutiles para que se les considere como noticia grande en medio de los acontecimientos bárbaros que ocurren a diario en este mundo. Ha sido casi una labor de hormiguitas solitarias asediadas por los intereses contrarios de las partes comprometidas en América Central y por el escepticismo sustentado por tantos tiempos de desengaños. La prensa de Colombia, que debía ser una de las más atentas a las vicisitudes de Contadora, no le ha prestado tanta atención como la de otros países, y ha sido, sin lugar a dudas, la más incrédula. En privado, muchos colombianos de los que hacen opinión apenas si consideran los esfuerzos de Contadora como algo que merezca desperdiciar palabras en un cóctel, y no falta quien considere la participación de Colombia en esa empresa como una aventura de consolación de un presidente idealista que no logró hacer oír la música de la paz en su propio país y resolvió irse con ella para otra parte.

Sin embargo, un examen cuidadoso de las condiciones de América Central en estos cinco meses permitiría pensar, con muy escaso

Publicado originalmente el 13 de julio de 1983.

margen de dudas, que los logros de Contadora no sólo son positivos, sino en cierto modo espectaculares. Y no tanto por lo que se ha conseguido que suceda en América Central, que quizá no es mucho, como por lo que se ha logrado que no suceda. Por lo pronto, Estados Unidos no ha enviado tropas regulares a El Salvador, ni ha invadido a Nicaragua, ni ésta está en guerra con Honduras y Costa Rica, ni el presidente Reagan ha podido hacer lo que le da la gana, frente a las reticencias del Congreso y la inconformidad creciente de su opinión pública. Uno tiene derecho a preguntarse, y los periodistas tenemos por lo menos la obligación de tratar de averiguarlo, hasta qué punto los trabajos de Contadora han tenido que ver con esas verdades históricas.

El pasado 8 de enero, cuando los cancilleres de Colombia, México, Panamá y Venezuela se reunieron por primera vez en Contadora para examinar los factores de conflicto en América Central y el Caribe, no menos de 18 iniciativas de arreglo negociado habían sido echadas al canasto de la basura. El presidente del gobierno español, Felipe González, había dicho en privado que el fracaso de tantas y tan meritorias iniciativas se debía a que, más allá de sus buenos propósitos, ninguna de ellas proponía un método concreto para ponerlas en práctica. En aquel momento era casi de dominio público que Honduras se estaba prestando como base para una invasión de opositores a Nicaragua, entrenados y abastecidos por Estados Unidos, a pesar de que el presidente, Suazo Córdova, se había comprometido ante el mundo a que su país no levantaría nunca un arma contra sus vecinos. Al sur, la democrática y pacífica Costa Rica, que tanto se ha preciado de no tener un ejército de soldados, sino de maestros de escuela, no sólo se hacía la de la vista gorda ante los grupos antisandinistas que actuaban con toda libertad desde su territorio, sino que en algún momento pareció dispuesta a crear unas fuerzas armadas capaces de enfrentarse a Nicaragua. El presidente, Luis Alberto Monge, le dijo por aquellos días a un diplomático que la situación económica de su país era tan desesperada que no encontraba cómo resistir la presión de Estados Unidos para que tuviera una participación beligerante en aquel pleito infernal de vecinos. Nicaragua, por su parte, se armaba para la defensa con un potencial de guerra que la hundía cada vez más en su círculo vicioso: si no se armaba sería destruida, y si se armaba, como lo estaba haciendo —con armas conseguidas donde pudiera y como pudiera—, daba un pretexto más de los muy pocos que necesitaba Estados Unidos para señalarla como un fortín soviético en el Caribe. «Lo que más nos molesta», dijo por esos días un miembro de la Junta de Gobierno de Nicaragua, «es que todo esto nos fuerza a actuar contra nuestros propósitos de economía mixta y pluralismo político.» En El Salvador, mientras la guerrilla ganaba terreno contra unas fuerzas armadas corruptas y desmoralizadas, el gobierno del presidente Reagan sólo esperaba un pretexto –como el de Pearl Harbour, en el Japón, o como el del golfo de Tomkin, en Vietnam– para que el Congreso y la opinión pública admitieran una intervención directa. Ésa era la situación explosiva, el pasado 9 de abril, cuando el presidente de Colombia, Belisario Betancur, tomó la decisión imprevista e imprevisible de visitar en 24 horas a los otros tres presidentes del grupo de Contadora.

Por qué cedió a este impulso intempestivo el presidente Betancur es quizá un secreto que sólo conoce su almohada. Sus consejeros más cercanos no compartían su optimismo, y mucho menos ante el proyecto original, que era el de visitar también a los presidentes de los cinco países en conflicto. Un argumento de peso fue que aquel viaje iba a parecer una mala copia de la gira reciente e infortunada del Papa por América Central. El mismo Betancur ha dicho que su decisión era más que coherente: su preocupación primordial es la paz, y la paz es una sola, dentro y fuera del país. Sea cual fuere el incentivo inmediato de aquella iniciativa, lo cierto es que le dio un vuelco completo a la política exterior servil y chambona del gobierno precedente, convirtió a Colombia en un protagonista real de la historia de estos tiempos y le dio a Contadora un contenido concreto y el método que les había hecho falta a las tentativas anteriores.

Hay razones para pensar, en primer término, que fue la mediación del grupo de Contadora la que consiguió que Costa Rica y Nicaragua llegaran a un acuerdo para la pacificación de sus fronteras. Es cierto que no ha logrado tener una influencia decisiva en la lucha interna de El Salvador, sobre todo porque no ha obtenido de los gobiernos en conflicto la autorización para sostener conversaciones oficiales con los grupos armados. Pero ha logrado lo que hace seis meses no parecía posible ni por un milagro: sentar en una misma mesa –el pasado 20 de abril– a los cinco cancilleres de los países centroamericanos y lograr de ellos, por primera vez en estos tiempos, un acuerdo de diálogo común. Este encuentro permitió asimismo, y también por primera vez, conocer a fondo el criterio de cada uno de esos países en relación con sus conflictos, y de este modo establecer un diagnóstico más certero de la crisis. Ése fue el principio de una serie de intercambios múltiples cuyo contenido y cuyos alcances no son fáciles de precisar, pero que han culminado en estos días con la elaboración de un proyecto secreto que parece haber infundido un nuevo aliento de optimismo a los protagonistas de este drama intenso y difícil. El plan está siendo estudiado en este momento por los gobiernos interesados con un sigilo que no ha dejado ni una grieta para los periodistas más astutos, y algo tiene que ver con su enigma la visita a México del subsecretario de Estado para América Latina, Richard Stone.

Cualquiera que sea el contenido de ese documento, hay motivos para creer que el gobierno de Reagan no ha de tomarlo tan a la ligera como lo ha hecho con otras tantas propuestas de arreglo pacífico en América Central. Más aún: su política para esa región –a pesar de su endurecimiento aparente y sus peligrosos disparates– no es hoy igual a la que era hace cinco meses. En efecto, cuando la oposición demócrata de Estados Unidos trató de capitalizar con propósitos electorales las gestiones de Contadora, el presidente Reagan imprimió a su política en América Central un giro imprevisto. En esa ocasión, por primera vez, aceptó ante el Congreso que un régimen distinto del de su país podía existir en América Central, siempre que no fuera dependiente de la Unión Soviética o Cuba. Nada de esto, desde luego, hubiera sido posible sin la influencia de Contadora sobre la opinión pública de Estados Unidos, que cada día observa con mayor claridad la naturaleza real del problema. Por los mismos motivos, cada día es más difícil que Reagan encuentre el pretexto que le haría falta para justificar una intervención directa en América Central, pues el grupo de Contadora –aun si sólo estuviera cruzado de brazos– es un testigo con demasiada autoridad como para no tomarlo en cuenta antes de lanzarse al disparate final.

¿QUÉ LIBRO ESTÁS LEYENDO?

Hay una pregunta muy frecuente entre escritores: ¿qué estás leyendo? Primero, porque es raro que un escritor le pregunte a otro qué está escribiendo, y segundo, porque se supone que el escritor, por una necesidad propia del oficio, debe estar siempre leyendo algún libro que merece ser recomendado. La respuesta es casi siempre evasiva, porque a partir de una cierta edad uno no sabe muy bien qué libro está leyendo a ciencia cierta, ofuscado un poco por la sensación desoladora de que todo lo que valía la pena ya fue leído en otro tiempo, y las horas que antes se dedicaban a la lectura se nos van ahora en picotear por aquí y por allá, con la esperanza de encontrarse por fin con una nueva e intempestiva revelación.

Se ha dicho mucho —y se ha dicho bien— que el hábito de la lectura se adquiere muy joven o no se adquiere nunca. También se dice, quién sabe con cuánta razón, que es necesario inculcárselo a los niños. Parece más probable que se adquiera por contagio: en general, los hijos de buenos lectores suelen serlo también. De modo que el hábito de leer suele ser de la familia entera. Algo semejante ocurre con el gusto por la música. Sólo que en ambos casos la presión de los adultos puede tener efectos contrarios: la aversión a la lectura y a la música. Alguna vez le oí decir a un gran profesor de música que a los niños no se les debía forzar a aprender el piano con aquellas prácticas cotidianas que de veras parecían sesiones de tortura. Su fórmula era más humana: hay que tener el piano en la casa para que los niños jueguen con él.

Parece que los poetas son los lectores más ávidos y perseverantes. De los novelistas, en cambio, se dice que sólo leen para saber cómo están escritas las novelas de los otros escritores, y descubrir en ellas hasta los tornillos más ocultos del oficio. Algo así como desmontar todas las piezas de un reloj para descubrir cómo está hecho y armarlo de nuevo, de manera que los otros no tengan secretos artesa-

Publicado originalmente el 20 de julio de 1983.

nales que uno no esté en condiciones de aprovechar. Sin embargo, tanto los poetas como los novelistas, como quizá todos los lectores habituales, se encuentran de pronto en una esquina de la vida en que ya no hallan nada nuevo que leer, y optan por lo más frecuente, que es leer de nuevo sus libros favoritos de siempre, rendidos ante la evidencia de que ya no se escriben libros como los de antes. Es entonces cuando surge la pregunta desoladora: ¿qué estás leyendo? Y no es raro que le contesten: nada.

En primer término, como todos los hábitos, también el de la lectura se extingue. Pero tal vez no sea por cansancio ni porque llegue a su término el interés por la literatura. La razón podría ser más simple. En los primeros años, cuando acabamos de contraer el sarampión de la lectura, uno tiene a su disposición para leer, en el orden que quiera y a la hora que pueda, una cantidad incalculable de libros escritos en 10.000 años. Puede empezarse por casualidad: un ejemplar descuadernado de *Las mil y una noches* que se descubre por puro azar, entre muchos trastos viejos y papeles de archivo, dentro de un baúl olvidado. Pero si hubiera que empezar en orden —después de los cuentos infantiles y la media tonelada de historietas gráficas—, el libro más aconsejable sería la Biblia. En nuestros tiempos jóvenes había el inconveniente grave de la versión de Casiodoro de Reina y Cipriano de Varela, cuyo lenguaje era el mismo del viejo *Padrenuestro*, y la partición incansable en versículos numerados que más bien parecían versos mal medidos y peor rimados.

Más tarde, cuando uno lee la inolvidable trilogía de Thomas Mann —*El joven José*, *José y sus hermanos* y *José en Egipto*—, uno se pregunta por qué toda la Biblia no está escrita así, como un relato intenso de doscientos tomos, cuya lectura podría durar toda la vida. Otro obstáculo serio era que nuestros muy católicos abuelos nos inculcaban el pavor metafísico de la que ellos llamaban la Biblia protestante —que es la que se encuentra en la mesa de noche de casi todos los hoteles del mundo, con la intención inequívoca de que el huésped se la robe—, y trataban de meternos a la fuerza por el mal camino de la Biblia católica comentada, en la cual las cosas no debían decir lo que en realidad querían decir, sino otra muy diferente, ordenada por el comentarista marginal, cuyas notas eran más largas que el texto mismo. Era así como el hermoso y cachondo Cantar de los cantares no debía leerse como lo que es, sino como una metáfora lunática del matrimonio de Cristo con la Iglesia. Dentro de ese orden pueril, uno se preguntaba qué diablos quería decir entonces aquel verso apasionado: «Hay miel y leche debajo de tu lengua, hermana».

Sólo para leer los libros indispensables se le iría a uno la mitad de la vida. Pero la otra mitad se le iría en preguntar lo mismo: ¿qué estás leyendo? Y la única respuesta de alguien que ha sido un buen lector tal vez sea siempre la misma: ya no leo, releo. El poeta Álvaro Mutis hace cada cierto tiempo lo que él llama «los festivales Proust», que consisten en una relectura de páginas selectas del gran novelista francés, y hace unos tres años se volvió a despachar, casi sin un respiro, las novelas completas de Balzac. Más vale no hacerle nunca la pregunta consabida, porque se corre el riesgo de ser mandado a releer todo Conrad. En cambio, al viejo maestro catalán don Ramón Vinyes le preguntaba uno qué debía leer, y la respuesta estaba casi siempre condicionada por el estado de su humor, pero cuando éste era el mejor, contestaba sin vacilar: «Lo más seguro en estos tiempos es no leer nada».

El gran peligro de la relectura es la desilusión. Autores que nos deslumbraron en su momento podrían –y casi siempre pueden– resultar insoportables. Es algo como lo que sucede con la novia de colegio, siempre que uno no haya tenido la precaución de casarse con ella y envejecer con ella, intercambiando arrugas y virtudes. Como lector, en mi caso, hay pasiones juveniles que han sobrevivido a todo, y las tres más importantes son Herman Melville, Robert Louis Stevenson y Alejandro Dumas. En cambio el maestro William Faulkner, sin cuyas lecciones escritas tal vez no hubiera aprendido los mejores recursos del oficio, no me parece fácil de leer en estos tiempos. En cierto modo, lo había previsto. Hacia 1949 le solté a don Ramón Vinyes mi temor de que Faulkner no fuera sino un retórico que años después resultara insoportable, y el viejo sabio contestó con una frase que hoy me parece mucho más enigmática que entonces: «No te preocupes, que si Faulkner estuviera aquí, estaría sentado en esta mesa».

Hay, sin duda, un factor contra el hábito de la lectura, y es que los últimos libreros bien orientados y buenos orientadores se murieron hace tiempo, y las librerías son cada vez menos lugares de tertulias vespertinas. Uno tenía su librero personal, como tenía su médico de familia y su cepillo de dientes. Ese librero profesional, que atendía en persona su negocio como el dentista atendía su gabinete, sabía con sólo leer los catálogos qué libros le interesaban a cada uno de sus clientes, y muy pocas veces se equivocaba. De modo que uno llegaba a la tertulia de las seis y encontraba ya reservado un paquete de novedades que alcanzaban para un mes de trasnochos placenteros. Hoy, las librerías son grandes y vistosos mercados de libros de actualidad, fabricados a propósito para vender de un solo golpe y leerlos para pasar el tiempo y tirarlos después en el cajón de la basura. Hasta el placer de la relectura es difícil, porque uno va a la librería a comprar un libro que se conoció hace dos años, y nadie le da razón de él. Así es: si hay un lugar donde se aprecia cuánto ha cambiado el mundo no es en una base de lanzamiento de satélites espaciales, sino en la librería de la esquina. Si es que todavía existe. Con razón, un excelente escritor contemporáneo y activo, a quien le preguntaron por teléfono, la semana pasada, qué libro estaba leyendo, contestó sin pensarlo dos veces: «Ya no leo sino la revista *Time*».

BATEMAN

Estuve con Jaime Bateman —el comandante del Movimiento 19 de Abril— en febrero de este año. Nos vimos fuera del país, con la anuencia del presidente de Colombia y movido por mi propósito impenitente de propiciar la paz interna al amparo de la amnistía amplia y completa promulgada meses antes. Había oído hablar tanto de él a tantos amigos comunes que desde el primer momento sentí como si fuéramos conocidos viejos. Era un caribe típico. Mucho mejor conservado que yo en la propia salsa, pero en ambos eran demasiado evidentes los dos signos congénitos que muy pocos suelen percibir en el alma caribe: una tristeza recóndita, que no logramos ocultar por completo con nuestros aspavientos proverbiales y nuestra música ardiente, y un sistema de reservas cautelosas en las relaciones personales. Muchas horas de aquella larga y colorida conversación se nos fueron en añorar juntos el olor de la guayaba. Pero de lo que más hablamos, por supuesto, fue del asunto central de nuestro encuentro: la política. Su concepción del cambio social dentro de las condiciones propias del país y sin ninguna dependencia de ningún centro internacional de poder me pareció muy convincente.

En cambio, su posición frente a la amnistía me pareció todavía un poco confusa, y daba la impresión de que estaba a merced de distintos vientos contrarios, propios y ajenos, que le impedían encontrar el buen camino en un instante que no sólo era crucial para su vida, sino para la de todos nuestros compatriotas. Esto me pareció una inconsecuencia, porque nadie había luchado tanto como él en los últimos años por conseguir esa amnistía, con una campaña magistral, que logró encender una pasión nacional insaciable desde la clandestinidad, hasta el extremo sin precedentes de que se convirtió en un factor determinante de nuestra política y en el núcleo mismo de las elecciones presidenciales. Luego, cuando el promotor de aquella fiebre providencial se encontró con un sueño realizado,

Publicado originalmente el 27 de julio de 1983.

parecía cogido en su propia trampa y sin un proyecto político definido para amaestrar la fuerza incontenible que él mismo había desencadenado. No sé si era consciente de que su incertidumbre nos afectaba a todos de un modo grave, y sobre todo a quienes habíamos jugado todas nuestras cartas en la lotería de la paz. En todo caso, aquél era el primer tropiezo grande, y quizá irreparable, de un gobierno todavía inexperto pero de muy buena fe, que era el que más arriesgaba con aquella amnistía ilusoria, y merecía, por lo mismo, una suerte menos veleidosa.

Sin embargo, nadie que hubiera hablado con Bateman aquellos días podía poner en duda que su búsqueda de una solución para su incertidumbre era intensa y positiva y que su fe en el porvenir estaba fundada en una sola posibilidad: el diálogo. Ésa parecía ser la clave de su personalidad. Uno no tenía la impresión de estar hablando con un guerrero, sino con un iluminado que estaba dispuesto a apelar a cualquier medio –inclusive la guerra– para lograr un diálogo unificador. Tal vez a eso se debía la leyenda de su ubicuidad, que creció y se ramificó hasta un tamaño mítico en los últimos años. Nadie sabía a ciencia cierta dónde estaba Jaime Bateman, pero la verdad es que estaba en todas partes. Tal vez ningún conspirador tan buscado de un modo tan feroz haya vivido menos en la clandestinidad. Su explicación era, sin duda, un sofisma, pero hasta el último instante le resultó eficaz. «Uno no debe esconderse», decía, «porque corre el riesgo de que lo encuentren.» Su vida social era tan intensa como la de cualquier político legal. En Bogotá se sabía que cenaba en las casas de personas conocidas, que asistía a reuniones de diferentes tendencias, que iba inclusive a lugares públicos de diversión y que sus viajes por el interior y el exterior del país eran constantes. Muchos que tenían noticias de esta vida activa la interpretaban como un gusto infantil por la aventura; nada más equivocado. Bateman era un hombre prudente y astuto, y su movimiento había creado para él un sistema de seguridad que sólo fracasó por culpa de una avioneta de un solo motor con un piloto inexperto perdido en la tormenta. Lo que pasaba era que Bateman no podía estar un solo instante, como se dice, sin hacer política. Su pasión era el diálogo. Y al parecer murió buscando el que tal vez iba a ser decisivo.

Cuando nos vimos, hace ahora cinco meses, no se daba un instante de tregua en la disposición de hacer contactos con el gobierno, ya fuera a través de la comisión de paz o de cualquier otro intermediario calificado. Pero su objetivo más alto era una entrevista personal con el presidente Belisario Betancur. Una entrevista secreta o pública, pero que permitiera llegar a un acuerdo sin intermediarios. Sin embargo, al cabo de varios meses de tentativas frustradas tenía la impresión de que el gobierno y sus representantes, que estaban adelantando conversaciones casi públicas con otros movimientos armados, no tenían ningún interés en dialogar con el suyo. Su creencia tenía un fundamento difícil de discutir. Bateman había esperado en vano durante varios días a que el presidente de la comisión de paz, Otto Morales Benítez, concurriera a una cita concertada con mucha dificultad, y en todo caso muy arriesgada. «Me puso conejo», fue la expresión, muy colombiana, que utilizó Bateman, desahogándose de su rabia por lo que consideraba una burla oficial. Sin embargo, tal vez murió sin saber que la verdad era todavía más extraña: Otto Morales Benítez no supo nunca que tenía concertada esa entrevista ni tenía ninguna noticia de ella casi dos meses después de que debía haberse realizado.

En todo caso, Bateman estaba dispuesto a empezar de nuevo cuantas veces fuera necesario, y trasladarse si era preciso hasta el otro lado del mundo para concertar un acuerdo de paz. No era una metáfora: al término de nuestra entrevista había tomado la decisión de viajar a Nueva Delhi, durante la Conferencia de los Países No Alineados, para tratar de encontrarse con el presidente Betancur. No sé si éste hubiera estado dispuesto a recibirlo en aquéllas o en otras circunstancias, pero la cancelación imprevista de su viaje a la India frustró otra oportunidad muy valiosa de dar un paso grande hacia la paz.

Para mí, el aspecto más serio y tranquilizador de la forma en que los sucesores de Bateman dieron al mundo la noticia oficial de su muerte fue el hecho de que su decisión de diálogo se mantiene intacta. Más aún: la condición original del retiro del ejército de ciertas zonas de guerrilla parece haber sido sustituida por la de una tregua en las hostilidades. Hay mucha distancia entre esas dos propuestas, y quienes deseamos con tanto ahínco la felicidad de nuestras gentes no podemos menos que señalarlo con la esperanza reverdecida.

La pregunta más generalizada, sin embargo, es si el M-19 sigue siendo el mismo, para bien o para mal, sin el atractivo personal de Bateman, sin su autoridad indiscutida, sin su imaginación, sin su amplitud mental y, sobre todo, sin su maravilloso instinto para la publicidad. La misma forma en que el M-19 manejó la noticia de su muerte hace pensar a muchos que el movimiento no logrará sobrevivir a su orfandad. En efecto, uno no entiende por qué no fue la propia dirección la que dio primero que nadie la noticia del accidente, si éste era ya de dominio público desde hacía casi tres meses. La explicación que dio Álvaro Fayad para justificar esa demora es, sin duda, convincente, pero es un argumento emocional que pasó por alto los enormes riesgos políticos de tan largo silencio. Fue éste el que hizo posible que prosperaran tantas versiones perversas. Sobre todo la de que Bateman se había fugado del país con los fondos del movimiento. La versión fue soltada por un noticiero de radio sin precisar la fuente, y repetida por los periódicos más importantes y serios sin ninguna comprobación, pero con un despliegue de primera página que no era, en todo caso, un ejemplo de buena fe. Por esa brecha abierta para siempre no era difícil que se filtraran otros infundios: que Bateman había sido ejecutado por la línea dura, que su avión había sido derribado por un cohete enemigo o destruido por una bomba colocada en la cabina por la propia organización. La circunstancia de que tal vez el cadáver no aparezca nunca en la selva impenetrable de Urabá será también un factor en contra de la verdad. Por lo visto, el destino de Bateman, desde su nacimiento, parecía no ser otro que el de la leyenda. Quiera Dios que sus herederos tengan la grandeza de realizar sus sueños, que no eran otros –aunque parezca increíble– que los de una paz justa y fructífera.

NO SE PREOCUPE: TENGA MIEDO

Decía la otra noche Oriana Fallaci que los grandes miedos de su vida no los ha sentido en la guerra ni en ninguno de los instantes de riesgo que ha afrontado en la práctica de su oficio. No; el miedo mayor lo ha sentido antes de hacer alguna de sus grandes entrevistas, y no por cierto las más ruidosas. Una declaración como aquélla no podía menos que promover una conversación sobre el miedo profesional, que sería el que padece toda persona responsable en el momento en que afronta la realidad de su profesión. Alguien que le tenía miedo al avión –como tantos viajeros de hoy– contó que el más intenso de su vida lo había sentido en la cabina de una enorme nave transatlántica. Lo había invitado el piloto, que en esa forma trataba de demostrarle que la seguridad y el método rutinario de la tripulación eran la vacuna más eficaz contra el terror del vuelo. La demostración fue muy convincente, hasta el instante en que el avión se colocó en la cabecera de la pista y la torre de control dio la orden de despegue. Entonces, tanto el piloto como el copiloto hicieron una pausa instantánea en su trabajo y se santiguaron al unísono. Un buen católico que estaba presente discutió con argumentos muy lúcidos que el acto de santiguarse no era señal de miedo, sino un deber simple de buen creyente. Pero la mayoría de los presentes –que tal vez éramos idólatras anónimos– estuvimos de acuerdo en que también los pilotos de las grandes líneas tienen miedo de volar y su mérito mayor consiste en hacerlo, y hacerlo bien, a pesar del miedo. Siguiendo el tema, le preguntaron a un cirujano si no había sido víctima alguna vez del terror profesional. La pregunta era adecuada, porque tal vez ningún oficio se practica, como la cirugía, en el límite mismo de la vida y la muerte. En este caso era aún más adecuada, porque se le planteaba a un cirujano que había tenido en su mesa de operaciones a más de cuatro jefes de estado. Su respuesta fue un poco sesgada, pero terminante: «Hace poco tuve el honor

Publicado originalmente el 3 de agosto de 1983.

de operar a mi viejo maestro de cirugía». Por ese camino, desde luego, la conversación siguió en torno a todos los miedos imaginables, y la conclusión final fue que todo profesional serio –lo confiese o no– tiene casi el deber de sentir miedo en el momento de las grandes responsabilidades del oficio.

De paso, se contaron muchas anécdotas demostrativas de que no es cierto –como tanto se dice y se repite– que todo miedo es en el fondo el miedo a la muerte. Para muchos hay uno peor, el miedo escénico. Es decir, ese terror de hablar en público, que sólo quienes lo padecemos sin remedio conocemos hasta qué extremo de confusión puede conducir. Aun quienes logran dominarlo están amenazados por acciones imprevistas. Uno de nuestros contertulios, que había sido presidente de la república, contó que en cierta ocasión, mientras improvisaba un discurso, olvidó por completo el nombre del agasajado, que se encontraba al lado suyo; no sólo no pudo recordar el nombre en aquel momento, sino que lo olvidó muchos años después, cuando nos refirió la anécdota.

Por cierto, que entre los miedos de que no se habló habría que incluir en un renglón el miedo de no recordar los nombres de las personas conocidas que uno encuentra de pronto. Sobre todo cuando éstas no son tan compasivas como para sacarnos del apuro, sino todo lo contrario. De allí que los escritores recordemos con tanta gratitud a los lectores bien educados que nos solicitan una dedicatoria en un libro y tienen el buen cuidado de decir su nombre, aun a sabiendas de que el autor tendría que conocerlo. «Hombre», les dice uno con alivio infinito, «¿cómo se te ocurre que no voy a recordar cómo te llamabas?». Pues en realidad no hay un aprieto más aterrador que el de estar con un libro abierto y una pluma en la mano frente a alguien que espera en silencio una dedicatoria cordial. Hay un recurso: preguntar el nombre pero sin levantar la vista. Pero nunca falta alguien que diga: «Cómo vas a saberlo si ni siquiera me has mirado». Y entonces uno levanta la vista y se encuentra con el pavor de ver un rostro conocido cuyo nombre se ha borrado para siempre de la memoria.

El miedo de viajar parece ser uno de los más comunes e intensos. No me refiero al miedo de los riesgos materiales del viaje, sino a esos aletazos metafísicos que lo despiertan a uno más temprano que de costumbre el día previsto para emprender el viaje. El remedio no está siquiera en la experiencia. Al contrario: el miedo aumenta cuanto más se viaja y la desazón indefinida es más atroz cuanto más se acostumbra uno a viajar. Es algo como la incertidumbre de lo desconocido, que sólo cesa al término del viaje. Todo esto se ha vuelto mucho más terrible con los nuevos inventos. Antes, cuando se andaba en medios de transporte a la escala humana, uno sentía que se alejaba de su casa con el alma en su almario. Pero ahora no. Uno atraviesa el Atlántico en doce horas y llega al otro lado con la certidumbre desapacible de haber llegado sólo con el cuerpo, mientras el alma sigue viajando a la zaga por sus propios medios y a una velocidad mucho más racional: a lomos de una mula. A veces pasa uno hasta ocho días como perdido en una bruma de irrealidad, arrastrando el cuerpo vacío que llegó en el avión, hasta el instante feliz en que por fin el alma acaba de llegar.

En aquella conversación, que varias veces alcanzó la tensión del juego de la verdad, no podía faltar quien recordara el miedo más antiguo y oscuro de la especie humana: el miedo del amor. No fueron muchos hombres los que admitieron sentirlo en toda experiencia nueva como si fuera la primera vez, pero los pocos que no le tenían miedo al miedo estuvieron de acuerdo en que no hay miedo más irreprimible y deprimente. «Lo que pasa», dijo una señora al borde de la madurez, «es que cuanto más inteligente se es, más miedo se tiene. De modo que el ser humano más inteligente le tiene miedo a todo.» Y, sobre todo, por supuesto, a la muerte. Sin embargo, la conversación había terminado. Nadie tuvo el valor de confesar su miedo a la muerte. Yo, por mi parte, me conformé con admitir que el sentimiento más nítido que me suscita la idea de mi muerte no es tanto de miedo como de rabia por su tremenda injusticia. Peor aún en un escritor que vive de contar sus experiencias, y que, sin embargo, tiene que vivir resignado al desastre final de no poder contar la más importante y dramática de todas: la experiencia de la muerte. Tal vez Oriana Fallaci no había imaginado nunca hasta qué extremos tan inconsolables iba a llevarnos la espontánea confesión de su miedo.

Se dice que algunos jefes de estado africanos, cuyas esposas no son bellas, tienen derecho a llevar otra mujer que sí lo sea cuando hacen viajes oficiales al exterior. El que lo dijo no está muy seguro de lo que decía, y tal vez no se atrevió a dar más detalles, y mucho menos en relación con lo que todos quisieran saber: hasta dónde llegaban los derechos del jefe de estado con la esposa prestada, si dormían en alcobas distintas o si, por el contrario, la falsa primera dama tenía los mismos derechos y deberes que la legítima hasta las últimas consecuencias de la intimidad. En todo caso, la sospecha de que es cierta esta versión de la esposa postiza parece confirmada por el hecho notable y notorio de que en las recepciones de la alta política internacional las mujeres que más llaman la atención por su belleza, por su elegancia y por su originalidad son esas sílfides de ébano que vuelven invisibles a las primeras damas occidentales y dejan sin aliento a sus maridos, y a veces —por cierto— ni siquiera se toman el trabajo de parecer esposas de nadie.

Puesto que el origen de todo suele encontrarse en la Biblia, es inevitable evocar lo que le pasó a Abraham —o Abram, como se escribe ahora— cuando se fue a Egipto con su esposa Saray, que se tenía por una de las mujeres más bellas de su tiempo. Temeroso de que los egipcios lo mataran para quedarse con ella, Abram la convenció de que se hiciera pasar por su hermana, «a fin de que me vaya bien por causa tuya, y viva yo en gracia a ti». Y en efecto, tan bien les fue a ambos que el faraón se enamoró de Saray «y trató bien a Abram por causa de ella y le dio ovejas, asnos, siervos, siervas, asnas y camellos». Hasta que Jehová —Jahveh, como se escribe ahora— castigó al faraón con grandes plagas por un pecado que ignoraba haber cometido y expulsó de su reino a Abram y Saray. La moraleja de este remoto episodio del Génesis, aplicado a la supuesta costumbre de ciertos monarcas africanos, es que alguien se crea al pie de la letra el

Publicado originalmente el 10 de agosto de 1983.

cuento de la esposa alquilada y se encuentre de pronto con su casa llena de plagas por haber seducido a la legítima.

La primera vez que se oyó hablar sin la menor clemencia de la fealdad histórica de una primera dama fue durante el reinado de Eleanor, la esposa del presidente de Estados Unidos Franklin Delano Roosevelt. El escritor Ricardo Muñoz Suay, que la conoció en su juventud cuando fue a Washington con una misión de la república española durante la guerra civil, dice que no la apreció nunca por su generosidad, pues en cambio de la ayuda urgente que fueron a pedirle lo único sustancioso y memorable que les dio en aquel verano ardiente fue un cartucho de helado a cada uno de los miembros de la misión. Pero dice también que era una mujer tan corpulenta, tan desproporcionada y tan fea, en definitiva, que terminó por parecerle de una hermosura extraña. No era, sin embargo, la más fea de su tiempo, pues el cetro lo llevó hasta su muerte la periodista mundana Elsa Maxwell, y lo llevó con tanto orgullo que no tuvo necesidad de casarse con ningún jefe de estado para que su fealdad mereciera el reconocimiento universal. La grande ironía de su vida fue que cuando era muy niña ganó un concurso de belleza infantil, y alguna vez publicó la foto de ese acontecimiento para que fueran más prudentes los padres que llevan en la cartera los retratos de sus hijos para mostrárselos a todo el mundo.

Nunca fue más indiscreta la prensa de Estados Unidos que con motivo de la visita a ese país del primer ministro soviético Nikita Jruschov y de su esposa Nina. Alguien tuvo el mal gusto de calcular que los dos juntos pesaban casi 200 kilos y que parecían vestidos por sus enemigos con ropas mandadas por correo y tal vez a las direcciones equivocadas. Además de ser descorteses, aquellas observaciones iban cargadas de perversidad política, porque no disimulaban el propósito de marcar un contraste con la apostura de la pareja más pareja que hubo jamás en Estados Unidos y que era la que entonces estaba en el poder: John F. Kennedy y Jacqueline Bouvier. Él llevaba unas chaquetas deportivas impecables, con el estilo perfecto e insoportable de los chicos millonarios de Harvard. Y ella era sin más vueltas la modelo mejor vestida del mundo, y una de las más bellas, sin ninguna duda, con sus ojos de ternera feliz y sus huesos magistrales. En cambio, Nikita Jruschov aparecía en los actos oficiales con el que parecía ser el uniforme de los funcionarios más altos de la Unión Soviética desde la muerte de Stalin: unas chaquetas bolsudas que parecían heredadas de un muerto más grande y unas mangas que apenas si dejaban ver el borde de las uñas. El colmo de su estilo montaraz pareció revelarse cuando se quitó un zapato y golpeó con él su pupitre de jefe de estado en las Naciones Unidas. Pero, en todo caso, el centro de las burlas era ella, la rozagante y discreta Nina, con sus trajes campesinos de grandes flores coloradas y sus zapatos de gansa. Sin embargo, el tiro les salió por la culata a los infelices cronistas sociales de Nueva York, porque, al final de la visita, la primera dama soviética se había ganado el corazón de las amas de casa de Estados Unidos, que, al fin y al cabo —como hasta su nombre lo indica—, son las que mandan en su casa. Tal vez de haber hecho lo que ahora se atribuye a ciertos mandatarios africanos, Nikita habría cometido el grave error de olvidar que también en el amor —al contrario de lo que se dice— muchas veces es mejor lo bueno conocido que lo malo por conocer.

En todo caso, lo que sería más injusto de la supuesta costumbre africana sería que las primeras damas no tuvieran la misma prerrogativa de sus maridos. Es decir: que ellas no pudieran cambiar a su vez a sus feos jefes de estado por un sustituto más presentable cuando tuvieran que salir de su país a mostrarse en las vitrinas sin corazón de la política internacional. Cuántas primeras damas legítimas que ahora creemos alquiladas tienen que soportar y arrastrar por los salones mundanos a tantos maridos verdaderos que parecen ejemplares fugitivos de algún jardín zoológico del horror. Y no sólo africanos, por supuesto, sino de cualquier parte, pues ya es demasiado sabido que a la hora de ser feos los hombres no tenemos límites temporales ni geográficos, así seamos tan jefes de estado como John F. Kennedy o como Idi Amin Dada.

JACK, EL DESMESURADO

Hace unos diez años se presentó en mi casa de México un joven francés que decía ser el director del festival de teatro de Nancy. Sin embargo, su aspecto no era el de un dirigente, sino el de un actor experimental: llevaba pantalones de vaquero, zapatos de tenis, una chaqueta de cuero muy usada, y tenía una hermosa cabeza con el pelo alborotado y los ojos de loco. Su esposa, una mujer pequeña y fuerte como suele imaginarse a las grandes mujeres del evangelio, lo acompañaba no sólo con su presencia silenciosa y sabia, sino con una solidaridad de granito que se le notaba a simple vista sin necesidad de que dijera nada. Tenía que ir a Cuba al día siguiente, en un viaje relámpago de 24 horas, para tratar de convencer a los cubanos de que mandaran un grupo teatral al festival de aquel año. No sé quién les dijo que solicitaran mi ayuda para tratar de saltarse las instancias burocráticas, y por eso habían ido a verme sin más cartas de presentación que su voluntad de irse esa misma noche para La Habana a cualquier precio. Aquel muchacho arrollador se parecía más a un gitano de Cádiz que a un francés, pero su nombre no tenía nada de gitano ni de latino. Se llamaba Jack Lang. Hoy –con 44 años apenas cumplidos– es el ministro de la Cultura de Francia.

Uno no podía menos que evocar esta anécdota durante los días pasados, cuando Jack Lang y Monique, su esposa de siempre, hacían a Colombia una visita medio oficial y medio de descanso, y conversábamos a toda hora de las cosas raras que nos han sucedido a ambos, y que ambos hemos hecho juntos y separados, desde que nos conocimos de aquel modo tan casual en la casa de México. Tratando de reconstruir el espejo roto de la memoria, recordamos nuestro segundo encuentro, que no fue menos imprevisto que el primero. Fue en París, hace ahora unos cuatro años, cuando se hablaba sin muchas ilusiones de la posibilidad de que François Mitterrand fuera una vez más candidato a la presidencia de Francia. Jack Lang supo no

Publicado originalmente el 17 de agosto de 1983.

sé cómo que yo estaba en París, y debo confesar que no recordaba muy bien quién era cuando me llamó por teléfono para invitarme a almorzar en su casa. Pensé que era un gesto tardío de gratitud por aquel viaje a Cuba que ya parecía tan remoto, pero en realidad fue algo más grato y espectacular: François Mitterrand era un amigo muy cercano de Jack Lang —cosa que no supe hasta entonces— y el almuerzo era en honor suyo.

Muchos factores contribuyeron a la victoria de Mitterrand, pero a mí no me cabe duda de que la inteligencia de Jack Lang, su desmesurada imaginación creadora, su capacidad para convertir en realidad los sueños más delirantes y su invencible fuerza de trabajo tuvieron mucho que ver con ella. Tuve la suerte de estar muy cerca de aquel ascenso incontenible del socialismo francés, y fue una experiencia inolvidable.

El día en que François Mitterrand tomó posesión de la presidencia de la república, en una ciudad estremecida por una explosión de rosas rojas y los coros del *Himno de la alegría*, le pregunté a Jack Lang quién iba a ser el ministro de la Cultura, y él me contestó sin la menor malicia en la sonrisa: «Quienquiera que sea, será sin duda amigo nuestro». Al día siguiente, cuando compré los periódicos, me di cuenta de que su respuesta había sido una amable y grata tomadura de pelo: Jack Lang, con su cabello alborotado y sus ojos de loco, había sido nombrado ministro de la Cultura.

No me atrevería a decir que ha sido el mejor ministro de Mitterrand —y tal vez lo ha sido—, pero ha sido sin discusión el que más se ha hecho notar. Aunque sólo hubiera sido por su concepción desmesurada, expansiva y casi astronómica de la cultura. Cuando tuvo que definirla ante la Asamblea Nacional, de la cual solicitó un presupuesto sin precedentes, sólo necesitó cuatro palabras: «La culture c'est la vie». La cultura es la vida, o sea, todo lo que el hombre hace en el universo. La definición dejaba casi en cueros a la que usa la Unesco en sus papeles oficiales: «Cultura es todo lo que el hombre agrega a la naturaleza». Iba inclusive mucho más lejos de la que parece ser una definición más justa: «La cultura es el aprovechamiento social de la inteligencia humana». La definición de Jack Lang tiene sobre todas las otras la enorme ventaja de ser tan desmedida que no hay por dónde atacarla. El resultado de esa concepción descomunal fue que la Asamblea francesa aprobó para el ministerio un presupuesto dos veces mayor que el anterior y el más grande que se le ha asignado jamás a la cultura en Francia: seis mil millones de francos, que en ese momento eran equivalentes a más de mil millones de dólares.

No podía ser menos con un gobierno que empezó por declarar que la cultura era el centro del poder de cambio en la sociedad francesa, y que, por consiguiente, debía considerarse que no había sólo uno, sino 44 ministros de la Cultura. Es decir, todo el gabinete ministerial, encabezado desde luego por el propio presidente de la república. El programa de nacionalizaciones, el proyecto de descentralización, que era una vasta operación destinada a reforzar el poder de los elegidos y, por consiguiente, de los electores, todos los proyectos de transformación económica y social debían ser considerados como un fenómeno cultural de proporciones inmensas.

Jack Lang, dentro de esos criterios, empezó su programa totalizador con una acción directa en los otros organismos del estado. Primero, con los grandes ministerios. Fue así como se firmó un acuerdo con el Ministerio de Justicia destinado a ejercer una acción cultural dentro de las cárceles. Fue así también como se firmó, al término de una discusión muy prolongada, un acuerdo semejante con el Ministerio de la Defensa para llevar la acción cultural hasta el alma misma de las fuerzas armadas, y fue así como se llegó a un acuerdo con el Ministerio de la Salud Pública para que la dinámica de la cultura pudiera penetrar en los hospitales y quizá hasta los suspiros finales de los moribundos.

Sin embargo, para quienes conocemos la arrogancia de los franceses y en especial en los dominios de la cultura, el mérito mayor de Jack Lang es estar tratando de inculcar en su país la noción de diálogo con otras culturas del mundo. Cuatro millones y medio de trabajadores emigrados, además de sus familias, son para Francia un enorme problema psicológico, económico, político y social. «Pero sobre todo», dice Jack Lang, «son un inmenso problema cultural que constituye todo un desafío.» A partir de esa concepción, todas las instituciones de Francia han recibido la orientación de abrirse a todas las culturas del mundo, y no con el deplorable espíritu de caridad colonial que era demasiado evidente en otros tiempos, sino inclusive de un modo que no parecía posible: con una cierta humildad. Si ese sueño de Jack Lang llega a realizarse −como él lo espera en su optimismo sin límites− merecerá de sobra el sobrenombre secreto con que lo distinguimos sus amigos: Jack, el desmesurado.

CASI MEJOR QUE EL AMOR

En una entrevista de prensa que Soledad Mendoza me hizo en Caracas hace algunos años declaré con bastante conocimiento de causa que todos los hombres somos impotentes. Muchos de mis amigos venezolanos –como buenos machistas caribes– me amenazaron con demostrarme en carne propia que ellos no lo son. Un médico a quien le conté esta anécdota la otra noche me dijo: «Adviérteles a tus amigos que no hagan tantos alardes de su masculinidad, porque a lo mejor se descubre que se han hecho implantar una prótesis». Es decir: un mecanismo artificial de erección. Ninguno de los presentes en aquella reunión tan instructiva había oído hablar nunca de este recurso de nuestro tiempo, que parece ser la solución final de la pesadilla más temible y a la vez más secreta de nosotros, los pobres hombres. Sin embargo, parece ser que cada día más amantes frustrados tienen que apelar a este artificio –sobre todo en Estados Unidos– y que los resultados son de lo más consoladores.

Al contrario de lo que ocurre a menudo, el problema de los impotentes reales es que tienen que demostrarlo con hechos y aun con testimonios de sus mujeres para que el cirujano tome la decisión de implantarles la masculinidad postiza. En realidad hay dos clases de impotencia. Una, la más frecuente, es de origen psíquico y no requiere una prótesis mientras el psiquiatra o el psicólogo no hayan agotado todas sus artes. La otra tiene una causa orgánica y en términos generales es irreversible. Es para estos casos para los que la prótesis constituye el único recurso de salvación.

Cuando la causa no es muy evidente, los médicos tienen un sistema que permite establecer la naturaleza del efecto. Una vez hechas todas las pruebas médicas para demostrar que, en efecto, el paciente no tiene erecciones, todavía hay que hacer la prueba de colocar unos electrodos en la base y en la cabeza del órgano sexual y conectarlos durante la noche a un aparato que registra –como un elec-

Publicado originalmente el 24 de agosto de 1983.

trocardiograma– las erecciones que tiene el paciente mientras duerme. Un hombre normal tiene entre seis y ocho erecciones en una buena noche de sueño. Si en dos noches consecutivas se demuestra que el paciente dormido no tiene ninguna erección se puede considerar que su impotencia es orgánica y no psicológica, y sólo entonces está permitida la implantación de la prótesis.

Hay otras causas de impotencia transitoria cuyas consecuencias son remediables. Una de ellas es la prostatitis, cuyos efectos contra la erección desaparecen tan pronto como desaparece la causa. Otra es la arteriosclerosis, por obstrucción de las arterias que irrigan los cuerpos cavernosos del pene. Esta causa es bastante frecuente y se remedia con la construcción de un *bypass* –como el que se hace a los cardíacos– para restituir la circulación normal de la sangre en los órganos genitales. Sin embargo, muchos médicos aseguran que la mayoría de sus pacientes por impotencia son adolescentes drogadictos. La marihuana, la cocaína, la heroína y aun los tranquilizantes que se venden como los más inocuos producen un debilitamiento de la potencia, y no sólo por causas orgánicas sino también porque los efectos psicológicos de la adicción afectan de un modo muy profundo al apetito venéreo. Menos mal –dicen los médicos– que también este tipo de impotencia cesa poco tiempo después de que se interrumpe la adicción. Por último habría que citar una causa de muy alta frecuencia: la fidelidad conyugal absoluta. O, por decirlo de otro modo más feroz: la falta de amor. Está demostrado que hombres normales y bien servidos, que han disfrutado de una larga y fructífera vida matrimonial, son víctimas a partir de cierta edad de los estragos de la rutina. Algunos de ellos deciden acudir al médico, y si éste es un profesional inteligente debería darles un consejo que tal vez fuera providencial: «Pruebe con otra». Es triste, pero cierto. Y no sólo en el caso de los hombres, desde luego.

En síntesis, parece que la causa más común de la impotencia orgánica irreversible es la diabetes avanzada. Esto no quiere decir –aclaran los médicos– que todos los diabéticos sean incapaces de pasar una buena noche sin dormir. Pero entre el 10 % y el 15 % de ellos son propensos a perder el sueño sólo por no poder. Otra causa son los golpes, los accidentes o cualquier enfermedad que lesione de un modo grave los mecanismos del sistema erectivo. En estos casos no hay más recursos que sustituirlo por un mecanismo artificial.

El doctor Neftalí Otero, un urólogo colombiano que es en realidad uno de nuestros buenos cerebros fugados en Chicago, les ha hecho a los lectores de esta columna el favor de explicarme con una versación ejemplar los diferentes sistemas de prótesis genital que se están usando cada vez más en Estados Unidos pero que sin duda están disponibles en nuestros países, tal vez menos urgidos, pero sí más asustados por el fantasma del querer y no poder. El artificio más usual es el de Small Carrion —llamado así, como todos los otros, por el nombre de su inventor— y consiste en colocar por medios quirúrgicos en los cuerpos cavernosos un par de cilindros rígidos de silicón. Es una operación muy sencilla, que demora entre 30 y 45 minutos con anestesia local y después de la cual se puede disfrutar para siempre de una erección invencible. El hombre queda en condiciones de llevar una vida sexual sin sobresaltos y su único problema se reduce a que su sastre encuentre la manera de que la solución no se le note demasiado.

Otro sistema muy usado es el de Jonás, que viene a ser un perfeccionamiento del anterior. La diferencia es que en este caso los cilindros de silicón son flexibles, pero tienen un resorte que se maneja a voluntad y permite plegar y replegar el saxofón del modo más conveniente. Es —para entenderlo mejor— una solución con bisagras que permite guardar el instrumento cuando ya la música se ha acabado, y se evita así el problema de pedirle favores al sastre. No es para reírse: el sistema es serio y tan útil como el original, y tiene sobre éste la ventaja de que funciona a cualquier hora y en cualquier parte, y hasta con susto y sin amor.

El sistema que más se acerca a la perfección, sin embargo, es el llamado de Scott —como el de la emulsión del pescado a cuestas, qué casualidad— y sus diferencias con el de la realidad son apenas perceptibles. Consiste en colocar dentro de los cuerpos cavernosos los mismos cilindros de silicón, pero conectados en este caso, mediante unos tubos conductores, con un recipiente de un líquido estéril. Este recipiente se encuentra debajo de los músculos del abdomen y está conectado, a su vez, con dos válvulas implantadas dentro del testículo derecho. A la hora de la verdad, mientras los mortales comunes y corrientes rogamos a Dios que todo nos salga bien, el depositario de este ingenio magistral sólo tiene que oprimir una de las dos válvulas para inyectar de este modo el líquido estéril en los cuerpos cavernosos. El milagro no sólo es inmediato, sino que uno puede regular la presión a voluntad y según las preferencias de la feliz adversaria. Una vez terminado el episodio se puede empezar otra vez. Y así tantas veces cuantas uno quiera, hasta donde alcance la vida, y al final basta con apretar la otra válvula para que el líquido regrese intacto y sin remordimientos a su recipiente secreto. Nada es más barato en relación con su servicio: el aparato cuesta 4.000 dólares y su implantación quirúrgica cuesta unos 2.500 en Estados Unidos.

Ahora, que si el médico es un buen amigo puede que acepte el pago en cómodas mensualidades y hasta sin cuota inicial. Es increíble. Tal vez —en la larga y tenebrosa historia del hombre— no se ha inventado otra cosa que se parezca más a la verdad.

¿EN QUÉ PAÍS MORIMOS?

El periodista Germán Santamaría nos ha puesto a los colombianos frente a frente con el pavor de nuestro propio fantasma. En dos artículos abrumadores que publicó la semana pasada en *El Tiempo*, los distraídos habitantes de las ciudades hemos comprendido que el infierno no está más allá de la muerte –como nos lo enseñaron en el catecismo–, sino a sólo cuatro horas por carretera de los cumpleaños de corbata negra y los torneos retóricos y las fiestas de bodas medievales de las sabanas de Bogotá. Está en el corazón de Colombia, en un vasto atardecer que conocemos como el Magdalena Medio, donde las tierras son feraces y las aguas generosas, y donde las injusticias son inmensas y seculares. Como síntesis de los horrores que vio y comprobó durante su breve visita, Santamaría ha dicho que la violencia es tan intensa y salvaje en aquel paraíso de pesadilla que éste puede considerarse –según sus palabras– como un Salvador chiquito. Pero su propio testimonio lo contradice por defecto. En realidad, el Magdalena Medio, cuya extensión es de 50.000 kilómetros cuadrados, tiene más de dos veces el tamaño de la República de El Salvador, que sólo mide 21.393. Además, la proporción de asesinatos es también comparable, pues el Magdalena Medio tiene una población que no pasa de 800.000, mientras que El Salvador –que es uno de los países más densos del mundo– tiene un poco más de cuatro millones de habitantes. No: no es un Salvador chiquito, sino otro mucho más grande que el de la América Central, y todavía mucho peor, por ser más confuso y olvidado.

Santamaría ha dicho en sus artículos que por el río Magdalena bajan los cadáveres podridos con los gallinazos encima, y que las autoridades de la ribera han decidido no recogerlos por su bundancia y su mal estado. Ha contado que en la aldea de Santo Domingo fueron exterminados todos los hombres, y que sus viudas, con los niños, pasan las noches en los montes vecinos desveladas por el terror.

Publicado originalmente el 31 de agosto de 1983.

Ha contado que en la vereda de Los Mangos mataron a 13 campesinos sólo porque habían asistido al velorio de dos compañeros suyos asesinados. Desde entonces, nadie se atreve a reclamar sus muertos, y los que tienen suerte son enterrados sin identidad en fosas comunes. Los otros son arrojados al río, para que se los coman los gallinazos mientras les queda algo que comer y para que sus despojos terminen por calcinarse al sol en algún playón olvidado. Hace poco, un campesino que logró escapar de una matanza empezó su relato con una frase que barrió de un solo trazo a muchos años de literatura tremenda: «Los muertos fuimos cinco».

Quiénes son los autores de este genocidio, y con qué propósitos, es algo que no se puede establecer con precisión absoluta ni en los artículos de Santamaría ni en otros muchos testimonios que llegan a las ciudades desde el infierno del Magdalena Medio. Una cosa queda en claro: los autores materiales son bandas armadas de pistoleros a sueldo, que matan a pleno día, unas veces a cara descubierta y otras con la cara pintada, y a quienes todo el mundo conoce pero no se atreve a denunciar. Su método, por desgracia, es inmemorial en la historia de Colombia y nos resulta familiar por su barbarie. Los cadáveres que flotan en las aguas, que yacen sin dueño en las veredas, han sido despellejados a cuchillos y aparecen con los órganos genitales cortados y a veces metidos en la boca, y sin lengua ni orejas.

Son las mismas señas de identidad de aquella otra violencia que asoló al país desde 1948 y que causó una mortandad calculada por la prensa de la época en 450.000 hombres, mujeres y niños en diez años. Que esta tragedia vuelva a salir a flote tan pronto como las condiciones sociales le son propicias, y que lo haga con las mismas formas de su salvajismo primitivo, es algo que hace pensar en quién sabe qué componentes enfermizos e irremediables de nuestra personalidad nacional.

Los testimonios que sustentan el relato de Germán Santamaría son tan apasionados y contradictorios que constituyen en sí mismos una prueba de la complejidad y la virulencia del profundo drama social que se vive en el Magdalena Medio. El personero de Aguachica dice sin más vueltas que las bandas son pagadas por latifundistas para robarles sus tierras a los campesinos pobres. En Puerto Boyacá, un diputado liberal ha dicho que las matanzas actuales son la reacción de los ganaderos contra la explotación y los secuestros a que los han sometido los guerrilleros durante 20 años. Señala, en concreto, a las Fuerzas Armadas Revolucionarias de Colombia (FARC), que han actuado en la región desde hace mucho tiempo, y a las que atribuye toda clase de crímenes y el cobro de más de 500 millones de pesos en rescates. Por su parte, el general Daniel García Echeverry, comandante de la XIV Brigada, con sede en Puerto Berrío, interpreta la situación –de acuerdo con el testimonio de Germán Santamaría– del modo siguiente: «Las FARC obligaron a 156 finqueros para quedarse con las tierras, porque lo que se está viviendo aquí es un enfrentamiento de los partidos liberal y conservador, que están desarmados, contra los comunistas, que están armados por medio de las FARC». Sin embargo, el MOIR, un movimiento legal que ha repudiado la lucha armada y el terrorismo como método de lucha política, ha visto caer asesinados en la región por lo menos a diez de sus dirigentes. Y Santamaría concluye: «Inicialmente todo era como una campaña para eliminar físicamente a la izquierda en el Magdalena Medio. Pero después, sin pararse en contemplaciones de matices ideológicos internacionales, arremetieron contra comunistas y moiristas, y después contra los ladrones de ganado en el campo, y luego contra los rateros del pueblo, y, finalmente, están matando hasta a los homosexuales». Es tal el estado de confusión que, en una reciente visita a Puerto Berrío, el procurador general de la República, Carlos Jiménez Gómez, se encontró reunido con un grupo de ganaderos entre los cuales estaban tres que él mismo había incluido en una lista de organizadores de bandas armadas, y fueron éstos quienes le pidieron cuentas por exponerlos a la vindicta pública. Tal vez, en medio de tantos intereses contrapuestos, haya, en realidad, muchos culpables en todos los lados, y los únicos inocentes sean los pobres campesinos despojados de todo, que llegan huyendo del terror a las ciudades, sin otro destino más seguro que la miseria o la delincuencia común.

En los albores de su mandato, hace ahora casi un año, el presidente, Belisario Betancur, prometió, con más seguridad de la que era prudente, que en su gobierno no se derramaría una gota de sangre. Su buena estrella, que tanto nos ha ayudado a todos, no lo ayudó en el Magdalena Medio. Tan consciente de esto ha sido él mismo que una de sus iniciativas más ambiciosas fue emprender para aquella región martirizada un vasto y costoso programa de rehabilitación a largo plazo, que empezó a caminar en junio pasado. Sin embargo, lo que desde entonces era más urgente, para que hoy no fuera demasiado tarde, era poner término con justicia a esa carnicería luciferina que no le hace honor a nadie, y menos a un gobernante que quiso ser el más pacífico de nuestra historia. No sería justo –por decir lo menos– que al cabo de tantos esfuerzos lograra conseguir la paz en El Salvador –como tanto lo deseamos– y no pudiera lograrla en este otro Salvador interno que nos devora las entrañas.

Por fortuna, todavía el presidente Betancur es el más popular que hemos tenido en este siglo. A pesar del desgaste natural del poder, al cumplirse el primer año del suyo las encuestas demuestran que más del 60 % de la opinión pública sigue creyendo en él, y entre ellos nos contamos muchos que no lo quisimos como candidato ni votamos por él. Ésa es una fuerza volcánica incontenible, y tal vez la única que nos queda para enfrentarnos con buena fortuna al engendro tentacular del Magdalena Medio. El paso inmediato sería entender qué es lo que allí ocurre a ciencia cierta, cuál es la verdad, toda la verdad, e inclusive mucho más que toda ella, y sólo el presidente de la república tiene la autoridad y la información para explicárnosla con una de esas charlas sencillas, de maestro sabio, que tanto alivio nos han causado en otros instantes difíciles de su gobierno. Sólo una conciencia nacional bien formada y mejor dirigida podrá salvarnos del desastre. Sólo el presidente puede y debe forjarla. Pues los rumbos que va tomando el Magdalena Medio —y Dios no lo quiera— amenazan con convertir el tiempo de su gobierno en uno de los más sangrientos de nuestra historia.

ME ALQUILO PARA SOÑAR

Me pregunto qué fue de ella. La conocí en Viena hace 28 años, comiendo salchichas con papas hervidas y bebiendo cerveza de barril en una taberna de estudiantes latinos, y se hubiera dicho que era la única austríaca legítima en la mesa, no sólo por su suculenta pechuga otoñal, sus lánguidas colas de zorros en el cuello del abrigo y el acento de quincallería con que hablaba un castellano primario. Pero no: había nacido en Armenia —la de Colombia— y se había ido a Austria muy joven, entre las dos guerras, a estudiar música y canto. En aquel momento andaba por los 40 años muy mal llevados, pues nunca debía haber sido bella y había empezado a envejecer antes de tiempo. Pero en cambio era uno de los seres humanos más simpáticos que he conocido en mi vida. Y también el más temible.

Viena era entonces —y desde entonces lo fue para siempre— la ciudad de *El tercer hombre*. Una antigua ciudad imperial que la historia había de convertir en una remota capital de provincia, y cuya posición geográfica entre los dos mundos irreconciliables que dejó la segunda guerra mundial había acabado de reducirla a lo que fue Estambul en otro tiempo: el paraíso del mercado negro y el espionaje mundial. Carol Reed y Graham Greene no hubieran podido escoger un ámbito más adecuado para una gran película. Y para una gran novela, por supuesto, que es lo que queda en la casa para siempre después de que se encienden las luces del cine y sus hermosos fantasmas de carne y hueso empiezan a fugarse de la memoria. Pero tampoco hubiera podido imaginarme un ámbito más adecuado para aquella compatriota fugitiva que seguía comiendo en la taberna estudiantil de la esquina sólo por fidelidad a su origen, pues tenía recursos de sobra hasta para comprarla de contado con todos sus comensales dentro. Nunca dijo su verdadero nombre, pues siempre la conocimos con el que la conocieron siempre sus amigos más antiguos de Viena: Frau Roberta.

Publicado originalmente el 7 de septiembre de 1983.

En esa época ya no le quedaba de su vocación de soprano sino la calidad de aceite tibio de la voz y la suntuosidad pectoral, de modo que no había en mí ninguna intención secundaria la noche en que se nos hizo más tarde que de costumbre y la invité a dar un paseo por el Danubio para ver si era en realidad tan azul como en los valses. No lo era, por cierto, como era fácil de imaginar, sino un torrente denso que no alcanzaba a reflejar la hermosa luna de primavera que se mantenía sin pudor en el centro del cielo. Yo, que siempre he sido un nostálgico a la defensiva, comprendí que un remoto viernes como hoy, cuando ya fuera incrédulo y viejo, iba a acordarme de aquella noche como una de las buenas noches de mi vida (y que tal vez lo iba a escribir, como acabo de hacerlo), y traté de convencerme desde entonces de que era una noche turbia e insípida que no merecería el homenaje de una nostalgia. Sin embargo, una vez más, el destino jugó sucio, porque no me mandó aquella noche sólo con el Danubio y con la luna, sino que me puso la trampa de Frau Roberta. Ahora, tratando de acordarme de ella, no he podido impedir el recuerdo de la noche en que ella estaba, y me parece injusto, pero irremediable. Porque fue en ese mismo instante cuando cometí la impertinencia feliz de preguntarle a Frau Roberta cómo había hecho para asimilarse de tal modo a aquel mundo tan distante y tan distinto de los riscos de vientos del Quindio, y ella me contestó con su verdad de un solo golpe: «Me alquilo para soñar».

Era cierto. Muchos años antes, cuando la nieve se hizo más fría por el hambre, no apeló al recurso fácil de pedir un pasaje de regreso al calor de la patria y olvidarse para siempre de *La Bohème* y de *Tann-häuser*, sino que llamó a la primera puerta que le gustó para vivir y pidió trabajo. Le preguntaron qué sabía hacer, y también en ese caso contestó la verdad: «Sé soñar». Aquella frase, que sólo un ama de casa austríaca estaba en condiciones de entender, no sólo cambió el rumbo de una honesta familia católica y pequeñoburguesa ejemplar, sino que fue el principio del bienestar y la fortuna de Frau Roberta.

En realidad, su único compromiso, como lo había propuesto, era soñar. No le costaba ningún trabajo, porque sabía hacerlo muy bien desde niña. Era la tercera de los hijos de un tendero próspero de algún pueblo cercano de Armenia –de cuyo nombre tal vez no quiero acordarme por prudencia– y desde que aprendió a hablar instauró en la casa la buena costumbre de contar los sueños en ayunas, que es la hora en que se conservan más puras sus virtudes premonitorias. Una vez, a los siete años, soñó que uno de sus hermanos era arrastrado por un torrente, y la madre, que todo lo creía, le prohibió al hijo lo que más le gustaba, que era bañarse en la quebrada.

Pero Frau Roberta –que quién sabe cómo se llamaba en aquellos tiempos prehistóricos del viejo Quindío– tenía ya desde entonces un sistema original e intransmisible de interpretar los sueños. «Lo que ese sueño significa», dijo, «no es que se vaya a ahogar, sino que no debe comer nada dulce.» La sola interpretación era una infamia cuando la advertencia era para un niño de siete años que no podía vivir sin sus postres. La madre, que nunca puso en duda la facultad adivinatoria de la hija, hizo respetar la advertencia con mano dura. Pero un mal día el hermano señalado se atragantó con una bola de caramelo que se estaba comiendo a escondidas y no fue posible salvarlo de una muerte atroz por asfixia.

Frau Roberta no pensó nunca que aquella virtud pudiera ser un oficio, hasta que la vida la agarró en Viena por la garganta y la obligó a apreciar las posibilidades comerciales de sus sueños. Fue aceptada en la primera casa en que tocó, sin que esa preferencia obedeciera a ninguna visión de la noche anterior –como sería fácil pensarlo–, porque su facultad tenía un límite, y era que servía para los otros pero nunca para ella misma. Empezó con un sueldo modesto, apenas suficiente para los gastos menudos, pero le dieron en la casa un buen cuarto y las tres comidas. Sobre todo el desayuno, que era el momento en que toda la familia se sentaba a conocer –dicho por ella según los sueños de la noche anterior– el destino inmediato de cada uno de sus miembros: el padre, que era un funcionario importante de la administración de correos; la madre, que era una mujer alegre y apasionada de la música de cámara romántica, y dos niños de 11 y 9 años. Todos eran muy religiosos, y por lo mismo, propensos a la superstición vergonzante. Y todos, hasta los niños, tenían el sentido del humor del padre, que recibió encantado a Frau Roberta en su casa. «Es un placer», le dijo, «conocer a la única persona en este mundo que trabaja durmiendo.»

Se quedó para siempre. Durante muchos años, sobre todo en los más tremendos de la segunda guerra, cuando sus sueños se llenaron de obuses que significaban amenazas de dolencias hepáticas, y de aviones en llamas que significaban domingos apacibles, o carnicerías de trincheras que significaban tesoros escondidos en algún lugar de la casa. Por esa época soñaban tanto los miembros de la familia que ella hizo un esfuerzo sincero para aplicar su método de interpretación a los sueños ajenos. Pero fue inútil: sólo ella sabía soñar. De modo que con el tiempo sólo ella podía determinar a la hora del desayuno lo que cada quien debía hacer aquel día y cómo debía hacerlo, hasta que su voluntad terminó por ser la única admisible en la casa. Su dominio sobre la familia fue total, y aun el suspiro más tenue tenía la raíz en su almohada visionaria. Por los días en que la conocí había muerto el dueño de la casa –liberado para siempre de la esclavitud del correo por una cuantiosa herencia que recibió al final de la guerra– y había tenido la elegancia de favorecer a Frau Roberta en el testamento, con la única condición de que siguiera soñando para la familia hasta donde le alcanzaran los sueños.

Mientras me contaba esta historia maravillosa frente al Danubio espeso no pude reprimir la sospecha de que Frau Roberta era tal vez la estafadora más feliz y original de cuantas habían pasado por el mundo. Y se lo di a entender del modo más delicado. «Lo único que quisiera», le dije, «es saber si es verdad que usted sueña.» Ella me envolvió con una mirada de compasión. «Es verdad», me dijo sonriendo. «Y por eso he venido esta noche, para decirte que anoche tuve un sueño que tiene que ver contigo: debes irte enseguida y no volver a Viena antes de cinco años.» En el primer tren de la madrugada, por supuesto, me fui para Roma. De eso hace 28 años –como ya lo he dicho– y todavía no he vuelto a Viena.

269 MUERTOS

Un disparo fatal contra un avión civil con 269 personas a bordo, en tiempo de paz o de guerra y cualquiera que sea la causa o las razones, es sin más vueltas un asesinato masivo. No hay ideología que lo sustente, ni cálculo político que lo justifique, ni dios que lo perdone. Es un acto inconcebible, si es deliberado, pero lo es mucho más si sucede por equivocación, porque en los cielos más altos del poder y con el grado de desarrollo actual de las ciencias de la guerra, ninguna de esas dos circunstancias permite forjarse muchas ilusiones sobre el destino de la especie humana.

El derribamiento del avión de la Korea Air Lines por un avión de guerra soviético está dentro de ese cuadro. Es un acontecimiento tan irracional, tan inhumano, tan estúpido inclusive desde el punto de vista político, que muchos amigos de la Unión Soviética –cuidadosos de no dar argumentos útiles a sus enemigos– no hubiéramos podido creerlo si el propio gobierno de este país no lo hubiera admitido. Pero el hecho mismo de que lo hubiera admitido con tanto retraso, y sólo cuando las denuncias encarnizadas y farisaicas de Estados Unidos le impidieron seguir demorando la verdad, es algo que agrava mucho más la gravedad. En la carrera armamentista sin control ni medidas en que están enfrentadas las dos grandes potencias, la Unión Soviética le ha regalado a su adversario un arma imprevista y demoledora en un campo en el que ya son maestros viejos y sabios los gobiernos de Estados Unidos: el aprovechamiento, a veces honesto y legítimo, pero muchas veces torcido y manipulado, de la información pública.

Con todo, en medio del festín de buitres del gobierno norteamericano y de los retazos de verdades a medias de la Unión Soviética, la otra mitad del mundo que ve la contienda desde la galería no acaba de acomodar en su puesto todas las fichas del rompecabezas. La historia parece comenzar cuando el Boeing 747 de la KAL, al mando

Publicado originalmente el 14 de septiembre de 1983.

de un capitán surcoreano de 46 años y con 18.000 horas de vuelo, se salió de la ruta que debía seguir desde Anchorage (Alaska) hasta Seúl, la capital de Corea del Sur. Cuándo ocurrió esa desviación, y por qué, es algo que no se ha podido establecer hasta ahora. De acuerdo con el plan de vuelo, la nave debía eludir por el sur el cordón de las islas Kuriles, que señalan el límite del espacio aéreo soviético. En la carta de vuelo de esa ruta hay una línea muy visible con un letrero perentorio: «Todo avión que penetre en este territorio restringido puede ser abatido sin advertencia». La razón es muy conocida. En esa extensa área está la península de Kamchatka, que es el más famoso laboratorio de cohetes de la Unión Soviética, y una avanzada de radares de largo alcance para detectar cualquier ataque procedente de Estados Unidos. También en esa área se encuentra una base importante para un centenar de submarinos nucleares capaces de alcanzar el territorio norteamericano con sus cohetes, y está, por último, el puerto de Vladivostok, donde se concentra casi toda la defensa naval del Pacífico. Es un vasto arsenal de destrucción y defensa que los soviéticos quieren mantener lejos de observadores indiscretos.

Siempre de acuerdo con las escasas informaciones de la agencia Tass, el avión surcoreano fue detectado por los radares desde el momento en que atravesó la línea prohibida, y fue rastreado durante dos horas y media, hasta que fue localizado y más tarde derribado por un cohete. Esta desviación no es ninguna prueba de mala fe. Sucede en ocasiones, y aun a los pilotos más expertos. En 1973, un avión de la Panamerican, cuyo piloto debió cometer un error en la programación de su computadora, salió de Nueva York para París, pero al amanecer, en lugar de la torre Eiffel, lo que vio fue el desierto del Sáhara: se había desviado 1.000 millas del rumbo correcto. En otra ocasión, un 707 de Colombia que volaba de San Juan de Puerto Rico a Madrid, se encontró de pronto con el sol en la cola, cuando debía tenerlo enfrente: la verdad es que había derivado tanto, que había dado la vuelta completa y estaba regresando a Bogotá. La explicación era tan divertida como increíble: el navegante se había guiado por la sofisticada brújula electrónica, que tenía un defecto, y nunca la comparó con la humilde brújula magnética que inventaron los chinos, y que sigue siendo la más confiable. De modo que no era raro, ni mucho menos, que el avión de la KAL creyera estar en un lugar y estar en realidad en otro muy distinto. Lo sospechoso –y que permite poner en duda la inocencia del piloto– es que mientras éste se negaba a responder a los llamados y señales de los soviéticos, hizo varios contactos con los controladores radiales de Tokio. Y no dijo nada sobre la situación difícil en que se encontraba. Peor aún: minutos antes de ser derribado, el control del radar de Japón estableció que el avión perdido estaba a 115 millas al norte de la isla japonesa de Okkaido —o sea, dentro del espacio aéreo soviético— y no a 115 millas al sur, como acababa de decirlo el piloto.

Esto derrota también la hipótesis de que el avión tuviera descompuestos todos sus sistemas de radio. Un 747 tiene cuatro equipos independientes de VHF y dos de HF que le permiten comunicarse nada menos que con 20.000 frecuencias. Pero aun si llegara a ocurrir la casualidad imposible de que en efecto se quedara sin contacto radial, al avión coreano le quedaba todavía el recurso de un código de señales con las alas, con las luces y aun con las ruedas, que los aviones de guerra soviéticos dicen haber utilizado sin obtener respuesta alguna durante más de dos horas.

En este punto, es imposible que un novelista no intente suposiciones que pudieran parecer fantasías. La primera de ellas sería pensar que el avión se había despresurizado, y todos sus ocupantes estaban (ya) muertos en sus asientos mientras el vuelo continuaba en línea recta al mando del piloto automático. Otra suposición podría ser que alguna materia letal había contaminado el oxígeno y envenenado a todos los pasajeros, incluida la tripulación. Pues bien: estas fantasías de ficción científica no son en realidad de un novelista febril sino de un comandante del 747 con más de 30 años de experiencia, que lejos de considerarlas disparatadas, le parecen más que posibles. Hay muchos antecedentes, pero sin duda el más fascinante es el de un avión espía de Estados Unidos —el famoso U2— que salió de su base en Carolina del Norte con destino a Panamá, pero nunca llegó allí sino que fue a estrellarse en un pico de los Andes de Bolivia. La única explicación es que el piloto había tomado su altura de crucero, había puesto los mandos automáticos y había muerto de repente en algún lugar anterior a Panamá.

No había confusión posible con el avión coreano. Dos horas y media son mucho tiempo para identificar y recapacitar, en un país que no sólo es uno de los más refinados del mundo en su industria de guerra, sino que ha demostrado ser además uno de los más cautelosos en sus decisiones políticas: la cola de un 747 tiene 35 pies de altura y sus insignias son visibles para un avión de caza que lo persigue y lo rodea. Por otra parte, todos los pilotos entrenados en la intersección aérea aprenden a reconocer cualquier clase de avión de cualquier marca en cualquier parte del mundo, aunque sólo sea por la silueta.

Nada de esto quiere decir, por supuesto, que la culpa ineludible de la Unión Soviética sea una demostración de la inocencia absoluta de Estados Unidos. El reconocimiento tardío que hizo este país de que había un avión espacial norteamericano en el área donde ocurrió la desgracia, hace indispensable algo más que una explicación, y nadie se ha acordado de pedirla. Los muertos se quedarán muertos, y pasará mucho tiempo antes de que la Unión Soviética se reponga de este percance, pero la búsqueda de todos los culpables no puede terminar en punta. Pues hay dos acontecimientos importantes que todavía no han ocurrido. El primero es que la caja negra del avión abatido –donde, sin duda, reposan otros pedazos de la verdad– no ha sido rescatada aún del fondo de los mares y la Unión Soviética parece haber empeñado todo su poder para conseguirla. El otro hecho que todavía no ha ocurrido en medio del escándalo y el regocijo publicitario de Estados Unidos y sus satélites es que el gobierno de ese país no ha dicho en ningún momento que el avión coreano no tuviera en realidad una misión de espionaje.

AQUEL TABLERO DE LAS NOTICIAS

Desde la tercera década de este siglo, y durante unos diez años, existió en Bogotá un periódico que tal vez no tenía muchos antecedentes en el mundo. Era un tablero como el de las escuelas de la época, donde las noticias de última hora estaban escritas con tiza de escuela, y que era colocado dos veces al día en el balcón de *El Espectador*. Aquel crucero de la avenida de Jiménez de Quesada y la Carrera Séptima —conocido durante muchos años como la mejor esquina de Colombia— era el sitio más concurrido de la ciudad, sobre todo a las horas en que aparecía el tablero de las noticias: las doce del día y las cinco de la tarde. El paso de los tranvías se volvía difícil, si no imposible, por el estorbo de la muchedumbre, que esperaba impaciente.

Además, aquellos lectores callejeros tenían una posibilidad que no tenemos los de ahora, y era la de aplaudir con una ovación cerrada las noticias que les parecían buenas, de rechiflar las que no les satisfacían por completo y de tirar piedras contra el tablero cuando las consideraban contrarias a sus intereses. Era una forma de participación activa e inmediata, mediante la cual *El Espectador* —el vespertino que patrocinaba el tablero— tenía un termómetro más eficaz que cualquier otro para medirle la fiebre a la opinión pública.

Aún no existía la televisión, y había noticieros de radio muy completos, pero a horas fijas, de modo que antes de ir a almorzar o a cenar, uno se quedaba esperando la aparición del tablero para llegar a casa con una visión más completa del mundo. Una tarde se supo —con un murmullo de estupor— que Carlos Gardel había muerto en Medellín, en el choque de dos aviones. Cuando eran noticias muy grandes, como ésa, el tablero se cambiaba varias veces fuera de sus horas previstas, para alimentar con boletines extraordinarios la ansiedad del público. Esto se hacía casi siempre en tiempos de elecciones, y se hizo de un modo ejemplar e inolvidable cuando el vuelo resonante del *Concha Venegas* entre Lima y Bogotá, cuyas peripecias se

Publicado originalmente el 21 de septiembre de 1983.

vieron reflejadas, hora tras hora, en el balcón de las noticias. El 9 de abril de 1948 –a la una de la tarde–, el líder popular Jorge Eliecer Gaitán cayó fulminado por tres balazos certeros. Nunca, en la tormentosa historia del tablero, una noticia tan grande había ocurrido tan cerca de él. Pero no pudo registrarla, porque ya *El Espectador* había cambiado de lugar y se habían modernizado los sistemas y los hábitos informativos, y sólo unos pocos nostálgicos atrasados nos acordábamos de los tiempos en que uno sabía cuándo eran las doce del día o las cinco de la tarde porque veíamos aparecer en el balcón el tablero de las noticias.

Nadie recuerda ahora en *El Espectador* de quién fue la idea original de aquella forma directa y estremecedora de periodismo moderno en una ciudad remota y lúgubre como la Bogotá de entonces. Pero se sabe que el redactor responsable, en términos generales, era un muchacho que apenas andaba por los 20 años y que iba a ser, sin duda, uno de los mejores periodistas de Colombia sin haber ido más allá de la escuela primaria. Hoy –al cumplir 50 años de actividad profesional–, todos sus compatriotas sabemos que se llamaba, y sigue llamándose, José Salgar.

La otra noche, en un homenaje interno del periódico, José Salgar dijo, más en serio que en broma, que con motivo de este aniversario había recibido en vida todos los elogios que suelen hacerse a los muertos. Tal vez no ha oído decir que lo más sorprendente de su vida de periodista no es haber cumplido medio siglo –cosa que le ha sucedido a muchos viejos–, sino al revés: el haber empezado a los 12 años en el mismo periódico, y cuando ya llevaba casi dos buscando trabajo de periodista. En efecto, siempre que volvía de la escuela, por allá, por 1939, José Salgar se demoraba contemplando por la ventana las prensas de pedal donde se imprimía el *Mundo al Día*, un periódico de variedades muy solicitado en su tiempo, cuya sección más leída era ya un periodismo puro. Se llamaba «Lo vi con mis propios ojos», y eran experiencias de los lectores contadas por ellos mismos. Por cada nota enviada y publicada, *Mundo al Día* pagaba cinco centavos, en una época en que casi todo costaba cinco centavos: el diario, una taza de café, lustrarse los zapatos, el viaje en tranvía, una gaseosa, una cajetilla de cigarrillos, la entrada al cine infantil y muchas otras cosas de primera y segunda necesidad. Pues bien, José Salgar, desde los diez años cumplidos, empezó a mandar sus experiencias escritas, no tanto por el interés de los cinco centavos como por el de verlas publicadas, y nunca lo consiguió. Por fortuna, pues de haber sido así habría cumplido el medio siglo de periodista desde hace dos años, lo cual hubiera sido casi un abuso.

Empezó en orden: por lo más bajo. Un amigo de la familia que trabajaba en los talleres de *El Espectador* –donde se imprimía entonces *El Espectador*– lo llevó a trabajar con él en un turno que empezaba a las cuatro de la madrugada. A José Salgar le asignaron la dura tarea de fundir las barras de metal para las linotipias, y su seriedad le llamó la atención a un linotipista estrella –de aquellos que ya no se hacen–, el cual, a su vez, llamaba la atención de sus compañeros por dos virtudes distinguidas: porque se parecía como un hermano gemelo al presidente de la República, don Marco Fidel Suárez, y porque era tan sabio como él en los secretos de la lengua castellana, hasta el punto de que llegó a ser candidato a la Academia de la Lengua. Seis meses después de estar fundiendo plomo de linotipias, José Salgar fue mandado a una escuela de aprendizaje rápido por el jefe de redacción –Alberto Galindo– aunque fuera para aprender las normas elementales de la ortografía, y lo ascendió a mensajero de redacción. A partir de allí hizo toda la carrera por dentro, hasta ser lo que es hoy, subdirector del periódico y su empleado más antiguo. En los tiempos en que empezó a escribir el tablero de noticias le hicieron una foto callejera con un vestido negro de solapas anchas cruzadas y un sombrero de ala inclinada, según la moda del tiempo impuesta por Carlos Gardel. En sus fotos de hoy no se parece a nadie más que a sí mismo.

Cuando yo ingresé en la redacción de *El Espectador* –en 1953–, José Salgar fue el jefe de redacción desalmado que me ordenó como regla de oro del periodismo: «Tuérzale el cuello al cisne». Para un novato de provincia que estaba dispuesto a hacerse matar por la literatura, aquella orden era poco menos que un insulto. Pero tal vez el mérito mayor de José Salgar ha sido el saber dar órdenes sin dolor, porque no las da con cara de jefe, sino de subalterno. No sé si le hice caso o no, pero en vez de sentirme ofendido le agradecí el consejo, y desde entonces –hasta el sol de hoy– nos hicimos cómplices.

Tal vez lo que más nos agradecemos el uno al otro es que mientras trabajamos juntos no dejábamos de hacerlo ni siquiera en las horas de descanso. Recuerdo que no nos separábamos ni un minuto durante aquellas tres semanas históricas en que al papa Pío XII le dio un hipo que no se le quitaba con nada, y José Salgar y yo nos declaramos en guardia permanente, esperando que ocurriera cualquiera de los dos extremos de la noticia: que al Papa se le quitara el hipo o que se muriera. Los domingos nos íbamos en el carro por las carreteras de la sabana, con la radio conectada, para seguir sin pausa el ritmo del hipo del Papa, pero sin alejarnos demasiado, para poder regresar a la redacción tan pronto como se conociera el desenlace. Me acordaba de esos tiempos la noche de la semana pasada en que asistimos a la cena de su jubileo, y creo que hasta entonces no había descubierto que tal vez aquel sentido insomne del oficio le venía a José Salgar de la costumbre incurable del tablero de las noticias.

UN TRATADO PARA TRATARNOS MAL

Entre los males numerosos que nos dejó en Colombia el gobierno que expiró el año pasado para bien de todos, hay dos que pueden seguir ocasionándonos males más amargos, aun más allá de la vida de los autores. Me refiero al tratado de extradición y el tratado de asistencia legal mutua firmados entre Colombia y los Estados Unidos. Los juristas han dicho ya mucho sobre ellos y tienen, sin duda, mucho más que decir, y hay que confiar en que van a decirlo sin reticencias. Pero quienes somos indoctos de solemnidad en la ciencia del Derecho no necesitamos saber si esos dos tratados son inconstitucionales —como parece que lo son— ni si son ilegales e injustos —como tal vez lo sean—, sino que nos basta con estar convencidos de que son indignos. Y lo son, por supuesto, para ambos países, porque cada uno de ellos se compromete por igual a la indignidad de entregarle al otro a sus propios ciudadanos para que los juzgue y los condene según sus leyes.

Lo normal —y digno, desde luego— es todo lo contrario: que cada uno de los países contratantes devuelva al delincuente supuesto a su país de origen para que éste lo juzgue por los actos cometidos en él. Los Estados Unidos, más allá de la indignidad, no tienen muchos motivos para temer por el trato que puede darse en Colombia a sus ciudadanos, aun si son delincuentes extraditables. La experiencia enseña que ha bastado siempre la intervención de un funcionario consular o diplomático norteamericano, para que sus compatriotas sean tratados no sólo con las consideraciones debidas, sino también con las que no se deben. En cambio, hoy basta con ser el titular de un pasaporte colombiano en regla para ser sometido a toda clase de vejámenes en las aduanas de los Estados Unidos, donde todo colombiano, por el solo hecho de serlo, es considerado y tratado como un traficante de droga. El ex presidente conservador Misael Pastrana Borrero nos pintaba hace pocos días en la prensa

Publicado originalmente el 28 de septiembre de 1983.

un cuadro terrorífico del infierno que es para los colombianos de cualquier condición el ingreso en los Estados Unidos. No se refirió –sin duda, porque no lo conoce– al caso de la esposa de otro ex presidente colombiano que fue sometida a graves irrespetos en la aduana de Nueva York, a pesar de que se identificó a tiempo y sin ninguna duda con un pasaporte especial. Pero aún: la reclamación apenas formal que hizo su esposo ante la embajada de los Estados Unidos en Bogotá no mereció ni siquiera una excusa verbal. En realidad, los únicos que estamos a salvo de estos atropellos impávidos somos los que tenemos prohibido el ingreso ordinario a los Estados Unidos desde hace más de 15 años, sólo porque la justicia de ese país –por sus pistolas– nos considera delincuentes políticos. Y tal vez –¿por qué no?– susceptibles de ser reclamados como tales a nuestro propio país dentro de todo lo que es posible por el tratado de extradición. Si así somos recibidos por el hecho simple de ser colombianos, hay motivos de sobra para preguntarnos cuál será el tratamiento de reyes que les espera en las cárceles norteamericanas a los compatriotas extraditados.

Hay muchos antecedentes para darnos cuenta de que en ningún caso estarían dispuestos los Estados Unidos a aceptar un tratamiento igual para sus ciudadanos. Hace unos siete años, la policía de control de drogas de los Estados Unidos decidió que las mujeres colombianas que llegaran al aeropuerto de Miami fueran desnudadas y sometidas a una requisa que no excluía ni a sus partes más íntimas. La medida se puso en práctica de inmediato. Fueron inútiles para impedirla todos los esfuerzos de las autoridades colombianas, hasta que éstas resolvieron darles una bienvenida igual a las viajeras norteamericanas que llegaran a Bogotá. Al cabo de pocas horas, los Estados Unidos revocaron en Miami sus infames prácticas ginecológicas.

Hasta hace unas semanas, gente de mucho peso se expresaba con fluidez y versación contra el tratado de extradición. Pero incluso algunos de sus impugnadores abrieron fuego contra él cuando se encontraron en el mismo lado de la trinchera con los traficantes de drogas. Es decir, que a pesar de ser conscientes de que el tratado es indigno, algunos de sus antiguos opositores prefieren callar, por temor de aparecer aliados de las mafias. Pues no: el tratado es indigno y no lo dignifica el hecho de que los traficantes de droga lo repudien por razones distintas de las buenas, desde luego.

Al contrario. Que el tratado haya iniciado su actuación pública contra las mafias ha servido para hacer más evidentes sus peligros desde el primer instante de su ser natural. En efecto, los traficantes de drogas, cuyos nombres y fotos eran de dominio público desde hace mucho tiempo, vivían libres con sus aviones de príncipes, sus gustos babilónicos y sus parques zoológicos sin que nadie perturbara su impunidad feliz. Tenían visas privilegiadas en Estados Unidos, donde nunca hubieran podido ser lo que son sin la complicidad de autoridades venales y socios con poder y clientes bien colocados. Su mercado estaba allá y no acá, y aquel era su paraíso. No se necesitaría de una perspicacia demasiado aguda para preguntarse por qué la justicia que se hizo durante tanto tiempo la de la vista gorda, se despertó de pronto y con una furia luciferina sólo cuando los traficantes de drogas tuvieron la desastrosa idea de irrumpir con ínfulas vandálicas en la política nacional. Es casi imposible impedir el mal pensamiento de que el tratado de extradición levantó su mandoble no como un instrumento de la justicia, sino como un garrote de persecución política y retaliación personal. En cambio, se presume de que cinco ejecutivos del grupo financiero más poderoso de Colombia están huyendo de nuestra justicia en los Estados Unidos y no se sabe hasta ahora que el gobierno colombiano haya acudido al tratado para que vengan a pagar lo que deben.

Lo que será el manejo político del tratado de extradición no es difícil imaginarlo. Menos de 24 horas después de que un acusado de traficante de droga colombiano se fugara a su país para no ser extraditado a los Estados Unidos, el embajador norteamericano en Bogotá dijo en una rueda de prensa de alto nivel que el fugitivo estaba en Cuba. La verdad que el embajador debía saber que está en Brasil. Pero el interés de Estados Unidos de vincular el tráfico de drogas con la política de izquierda en América Latina no se detiene ante nada, y el tratado de extradición le será muy útil en esa campaña. Éste es apenas el preludio.

En resumen, el acuerdo maldito está inspirado en el mismo espíritu represivo del estatuto de seguridad, la tortura sistemática y la violación sin frenos de los derechos humanos, que tanto contribuyeron a hacer del gobierno pasado uno de los más funestos de nuestra historia. No es fácil imaginarlo como un instrumento del gobierno actual, que tantas cosas ha hecho en sus primeros tres meses, y que si algo mejor puede y debe hacer es librarnos de una vez por todas no sólo de los malos recuerdos sino también de estas malas herencias. Que el presidente Betancur me perdone si imagino mal, pero no puedo imaginarlo a él mandando a un compatriota suyo a Estados Unidos —o a cualquier otro país— para que lo castiguen en otro idioma y con otro dios, aunque sea allá donde haya cometido el peor de los delitos. Como tampoco puedo imaginar al gobierno de los Estados Unidos mandando para Colombia a tantos y tantos norteamericanos que durante tantos años se han pasado a la justicia colombiana por donde han querido. Lo malo es que el presidente Betancur sólo seguirá siéndolo por 39 meses más, mientras que el tratado de extradición pretende haber sido suscrito para siempre.

LAS GLORIAS DEL OLVIDO

Una de las injusticias de la literatura es que no existe una clasificación escalonada de los escritores de acuerdo con su calidad. En música se sabe que hay un paraíso más alto donde están sentados para siempre Johann Sebastian Bach, Mozart, Beethoven, Bartók —y tal vez los Beatles—, pero hay todo un olimpo de compositores de segunda, y aun de tercera categoría, que escuchamos y admiramos a pesar de la certidumbre de que no son eternos. Ocurre lo mismo con los pintores. No hay más que pasearse por los museos del mundo para darse cuenta de que junto a Goya y Velázquez, junto a Leonardo y Botticelli, junto a Rembrandt y Picasso, hay muchos colgados en la antesala de la eternidad que sin duda merecen estar donde están, pero en niveles distintos. En literatura no: o se es un escritor de primera línea o uno no encuentra donde ponerlo, y no sólo en los innumerables compartimientos del corazón, sino ni siquiera en los estantes de la biblioteca. En ese sentido, el criterio más justo es el del mundo del boxeo: hay pesos pesados, pesos welter, pesos medios, pesos mosca, y cada cual disfruta de una gloria universal dentro de sus límites respectivos. En literatura, en cambio, sólo los pesos pesados van al cielo.

Hablábamos de esta injusticia la otra noche con el escritor Pedro Gómez Valderrama, a propósito de un escritor que ambos admiramos sin ningún pudor, a pesar de ser conscientes de que no es uno de los más grandes: Somerset Maugham. El problema es dónde ponerlo. Sus novelas, que le hicieron famoso, sobre todo por sus adaptaciones al cine, no merecen ni un recuerdo piadoso. En cambio, hay un mundo de tesoros ocultos en sus casi 300 cuentos, muchos de los cuales no son más que obras maestras. Curioso: igual cosa ocurre con Hemingway, y sin embargo no nos cabe ninguna duda de que es y tal vez seguirá siendo para siempre una estrella de la primera división. Maugham, al contrario, es un autor que se

Publicado originalmente el 5 de octubre de 1983.

olvida, aunque se sabe de la existencia de grandes lectores, críticos respetables y escritores consagrados que quisieran subirlo a un piso más alto, pero no se atreven. Así como hay muchos que lo siguen leyendo en secreto, y hasta algunos escritores que siguen nutriendo con la lectura la propia obra, y sin embargo lo niegan en público más de tres veces y mucho después de que ha cantado el gallo.

Pensando en el destino injusto de Maugham, no es posible eludir el recuerdo de otros tantos escritores que por un momento nos parecieron grandes porque nos deleitaron como si en efecto lo fueran, y que han sido arrasados por el tiempo. Uno de ellos es Aldous Huxley, a quien sin duda la generación de hoy, en ningún país, no ha oído ni siquiera mencionar. Se sorprenderían al saber que por lo menos durante una década su novela *Contrapunto* estaba considerada como una pieza capital de las letras de este siglo, y que nadie que quiera ser o parecer culto tenía el coraje de admitir que no la había leído. Su predestinación al olvido, sin embargo, tuvo una prueba que parece sobrenatural: Aldous Huxley murió en California el mismo día en que fue asesinado el presidente John F. Kennedy, de modo que la noticia —sin espacio ni tiempo para homenajes póstumos— se quedó traspapelada en el cementerio de las causas perdidas.

Un contendor muy apreciado de Aldous Huxley en el mercado de las vanidades del mundo fue el mamífero más raro de su época: Lin Yutang, un chino norteamericanizado que además de vender como salchichas sus libros numerosos en casi todos los idiomas, hizo un diccionario chino-inglés e inventó una máquina de escribir en chino. Su libro *La importancia de vivir* llegó a considerarse en Occidente como un compendio de la felicidad oriental, y sus ejemplares se volvían polvo en las manos de tanto ser leídos con una especie de avidez atónita. Eran los años de la posguerra, en los cuales irrumpió otro nombre que puso a temblar a los consagrados: Curzio Malaparte, un italiano con una concepción descomunal del arte de escribir, que impuso en el mundo, con el título de uno de sus libros, una palabra alemana de significado devastador: *Kaputt*. Con todo, ese libro que lo consagró en la primera fila no fue el que se leyó con más pasión, sino otro posterior, *La piel*, sin duda uno de los más vendidos de aquellos tiempos. Cuando lo estaba leyendo por primera vez, en una sórdida pensión de estudiantes de Bogotá, tuve en mitad de camino la ráfaga de pavor de no querer morirme antes de saber cómo terminaba. Entre los muchos episodios que hoy parecerían truculentos, sin duda el más impresionante era el de un manatí del acuario de Nápoles que le fue servido en una cena de gala al comandante de las tropas norteamericanas en Italia y que éste había rechazado porque era igual a una niña hervida que llevaba a la mesa en una fuente adornada con algas y coliflores. Hace unos años, buscando otra cosa me encontré de pronto con este recuerdo lancinante de la juventud, y me quedé perplejo preguntándome qué clase de lectores incautos éramos en aquellos tiempos.

Se leía entonces otros libros capaces de estremecernos por motivos que hoy nos resultan misteriosos, y que no nos atrevemos a releer por el temor a romper el encanto. Recuerdo *El hombrecillo de los gansos*, del alemán Jacobo Wassermann –biógrafo incidental de Cristóbal Colón–; recuerdo *Primavera mortal* –del húngaro Lajos Zilahy– y recuerdo por supuesto el libro que conmovió al mundo con una fuerza cuya naturaleza no fue nunca descifrada: *La historia de San Michele*, del médico sueco Axel Munthe. Este último, cuyas virtudes de escritor eran más que evidentes, tuvo la debilidad muy propia del cine de nuestro tiempo de querer exprimir el limón hasta más allá de la cáscara y escribió una segunda parte de su libro capital. En todo caso, ninguno de estos autores se asomó siquiera a la gloria desmesurada de otro de los grandes olvidados de la literatura: Vicente Blasco Ibáñez, que sin duda fue el escritor español más conocido y aclamado del presente siglo en el mundo entero. La recepción popular que se le tributó en Nueva York en 1920 hace todavía menos comprensible la magnitud de su olvido.

Queda todavía por establecer si estos autores borrados de la memoria merecían de veras su suerte. Pero hay otros de los cuales se puede y se debe decir sin vacilación que no la merecían. Es el caso de Anatole France, Premio Nobel de 1921, que ejerció una fascinación justa no sólo en Francia sino en todo el ámbito latino, y del cual son muy pocos los que hoy pueden hablar sin conocimiento de causa. Su caso es más triste aún que el de Alejandro Dumas, porque a éste lo leen todavía algunos franceses desperdigados, aunque un poco a escondidas, como los estudiantes que fuman en el baño. Es el caso del ruso Leónidas Andréiev, que irrumpió en el ámbito de la moda con su novela *Sashka Zhegulov*, y luego desapareció para siempre. Fue una fugacidad injusta, pues si en realidad su novela más famosa no parecía animada por un aliento perdurable, muchos de sus cuentos –sencillos y hermosos– merecen leerse todavía más que las obras de algunos de sus contemporáneos. Es el caso también de Thomas Mann, de quien se encuentran todavía ediciones imprevistas y evocaciones ocasionales, pero que en todo caso parece ya cubierto a medias por las cenizas del olvido. Son comprobaciones tristes pero saludables, sobre todo cuando surgen de conversaciones casuales entre escritores. Es como si de pronto recordáramos –con la voz del pequeño argentino que todos llevamos dentro– que tal vez ya vaya siendo hora de poner nuestras barbas en remojo. Aunque sólo sea por si acaso.

WILLIAM GOLDING, VISTO POR SUS VECINOS

Siempre he tenido una gran curiosidad por la forma en que los seres humanos reciben las noticias que pueden cambiar su vida. Y en el caso de los escritores, por supuesto, me había hecho siempre la pregunta que casi todos los periodistas y los amigos me han hecho desde hace un año: «¿Qué se siente cuando se gana el Premio Nobel?». He dado casi siempre una respuesta distinta, según quién sea el interlocutor, porque la verdad es que no tengo un recuerdo muy definido. Había tantos rumores desde los días precedentes (como los había habido por la misma época en los años anteriores), que cuando recibí la noticia ya no sabía muy bien qué sentía. Contra todas las leyendas, la confirmación irreparable la tuve el 21 de octubre de 1982, en nuestra casa de México, cuando sonó el teléfono a las 6.05 de la mañana. Mercedes contestó medio dormida y me pasó el auricular, diciendo: «Te llaman de Estocolmo». Una voz masculina, en un español perfecto con un leve acento nórdico, y que se identificó como redactor del periódico más importante de Estocolmo, me dijo que la Academia Sueca había dado cinco minutos antes la noticia oficial. No sé muy bien qué dijo después, porque yo estaba en ese instante consternado por el terror, pensando en el discurso que debía pronunciar casi dos meses después en Estocolmo al recibir el premio. Ese terror fue el único sentimiento definido que me acompañó, no solamente durante los días interminables y las noches insomnes en que escribí las 15 páginas más difíciles de mi vida, sino que persistió hasta el instante en que acabé de leerlas en público en el salón de actos de la Academia Sueca. Todo lo que ocurrió después –hasta hoy– fue pura rutina.

Hago esta evocación porque el jueves pasado, cuando conocí la noticia de que a William Golding le habían dado el Premio Nobel de Literatura, volví a preguntarme con toda inocencia: «¿Cómo se sentiría cuando le dieron la noticia?». Estuve todo el día leyendo cables de

Publicado originalmente el 12 de octubre de 1983.

agencias de prensa para ver si alguno lo decía, pero las informaciones carecían de esos detalles humanos que no parecen importantes pero que son en realidad los que nos conmueven. Por la tarde, sin embargo, ocurrió una de esas cosas increíbles que no pueden llamarse casualidades, porque son mucho más que eso, y que los escritores no nos atrevemos a contar por el temor de que nadie las crea.

Ocurrió que a las cinco de la tarde del jueves, como estaba previsto desde hacía una semana, vino a mi casa Andrew Graham-Yool, un periodista de *The Guardian*, de Londres, para hablar de amigos comunes y hacer tal vez una entrevista. Hablamos del tema del día, desde luego, que era su compatriota William Golding. Sabíamos de él todo lo que puede aprenderse en los libros, y yo le había seguido la pista muy de cerca desde que leí en Barcelona la versión castellana de *El señor de las moscas*. Más tarde se publicaron *El dios Escorpión* y *La oscuridad visible*, pero me parece que Golding estaba publicado en castellano desde mucho antes. De modo que el nuevo Premio Nobel no era tan desconocido en nuestra lengua como se había dicho en las primeras horas. Además, según me lo confirmó Graham-Yool, en el Reino Unido es un escritor muy leído y premiado. Sin embargo, mientras conversábamos yo no lograba apartar de la mente la pregunta de cómo habría transcurrido su día en Broadchalke, el pueblecito de unas 600 personas donde vive, cerca de Salisbury, Inglaterra. Fue entonces cuando ocurrió lo increíble. «Yo tengo una tía que es vecina suya en ese pueblo», me dijo Graham-Yool con toda naturalidad. «Si quiere, la llamamos por teléfono.» Sacó del bolsillo su libreta de direcciones y dos minutos después la señora Betty Graham-Yool oyó sonar el timbre a las once de la noche y tuvo que salir chorreando agua de la bañera para contestarle a un sobrino que le dijo desde 10.000 kilómetros de distancia: «Estoy aquí con el Premio Nobel de Literatura del año pasado, que quiere saber algunas cosas sobre el premio Nobel de este año». La tía, muy británica, no dio ninguna muestra de asombro, sino que pidió por favor un minuto, mientras se secaba.

La curiosidad fue satisfecha. Al contrario de los escritores de las Américas, que conocemos la noticia al amanecer, los europeos la conocen a la una de la tarde, que es la hora en que el sobrio Lars Gyllensten, secretario de la Academia Sueca, hace el anuncio oficial. De modo que William Golding no fue despertado por nadie, sino que se enteró de su buena nueva como cualquier vecino: oyendo por radio las noticias del mediodía.

Visto por la señora Betty Graham-Yool, el nuevo Premio Nobel se parece de un modo sorprendente a la imagen que un lector podría haberse formado por sus libros. Es un hombre de barba y cabellos blancos, que vive con su esposa Ann y sus dos hijos –un varón y una mujer–, pero que a sus 72 años no puede considerarse como un viejo, porque lleva una vida muy activa. Su segunda vocación es la música, pero no sólo para oírla, sino para ejecutarla en cualquiera de estos instrumentos: el violín, la viola, el piano o el oboe. Su tercera vocación es la navegación, como ya deben de haberlo imaginado sus lectores y como resulta natural en alguien que admira tanto a otro gran escritor de alta mar: Herman Melville. Su cuarta vocación es la egiptología. Sin embargo, hace poco se descubrió una quinta vocación, que es la de jinete. Se ha comprado un caballo y en las tardes de buen tiempo se le ve galopar por los campos vecinos con tanta propiedad como si lo hubiera hecho toda la vida.

Alguien con quien había hablado antes de conversar por teléfono con la señora Graham-Yool me había dicho con razón que era fácil inventar la vida de un escritor inglés de 72 años que vive en el campo. «Seguro que tiene un perro y que los domingos trabaja en el jardín», me dijo. Golding –que se levanta a escribir a las cinco de la mañana y que, además, tiene que sacar tiempo para sus otras cuatro vocaciones– no es aficionado a las flores, pero, en cambio, su esposa cultiva unas orquídeas que son la admiración de la aldea. La señora Graham-Yool reiteró que el jardín de los Golding es uno de los más bellos de Inglaterra. Dijo, por último, que le gusta ver al nuevo Premio Nobel cabalgando con su magnífica estampa de vikingo, y se apresuró a aclarar que no es un hombre insociable, sino que se mantiene un poco al margen de sus vecinos, más bien por timidez.

En todo caso, la jornada del jueves transcurrió en Broadchalke como otra cualquiera. Nadie perturbó la paz virgiliana de Eble Thatch, la cabaña con techo de palma donde los Golding recibían llamadas telefónicas y telegramas del mundo entero. No en vano ellos y los otros 600 habitantes son ingleses y saben que un Premio Nobel no cae del cielo todos los días, pero que, en todo caso, no es algo tan importante como para perturbar la vida privada de un buen vecino. Sin duda aterrorizado también por el discurso que debe pronunciar en Estocolmo dentro de 60 días interminables.

PASTERNAK, VEINTIDÓS AÑOS DESPUÉS

Se ha publicado por estos días la noticia de que en Moscú se celebró un acto que puede considerarse como un homenaje casi oficial al escritor Boris Pasternak, Premio Nobel de Literatura de 1958 y quien dos años después murió en una especie de exilio interior. El acto consistió en la lectura de algunos poemas suyos ante unas 500 personas, y la agencia de prensa europea que dio la noticia precisó que había sido anunciado en los periódicos y en carteles murales, y que la mayoría de los asistentes eran jóvenes.

La noticia —al contrario de muchas otras de índole semejante que las agencias occidentales nos mandan de aquellos mundos— merecía la atención que le fue prestada, pero faltó advertir que este aparente deshielo en torno del gran poeta y novelista no es nada nuevo en la Unión Soviética, y que hace mucho tiempo que su nombre y su obra no son tan misteriosos ni conflictivos como en efecto lo fueron alguna vez. Hace ya varios años que un gran poeta de la generación penúltima —Andrei Voznesenski— publicó algunos de los poemas póstumos de Pasternak en una revista literaria, que como todas las de la Unión Soviética, por supuesto, era una revista oficial, y escribió para ellos una presentación en la cual se hablaba de sus virtudes sin la menor reticencia. También en esa ocasión las agencias de prensa occidentales registraron el hecho como algo extraordinario, y también como si fuera el primero después del escándalo de su Premio Nobel.

Lo que valdría la pena sería establecer de una vez por todas qué fue lo que sucedió en realidad con Pasternak en la Unión Soviética. Su padre, el pintor Leónidas Osipovich Pasternak, muerto en el Reino Unido poco después de la segunda guerra mundial, pintó algunos retratos oficiales que no están a salvo por completo de la retórica política de su tiempo, y que todavía se exhiben en museos de Moscú y Leningrado. El mismo Boris fue conocido desde muy joven como

Publicado originalmente el 19 de octubre de 1983.

un poeta de gran inspiración y talento, y desde 1922 se situó en la primera línea con su libro *Mi hermana la vida*, seguido por numerosos poemas líricos y de contenido social. Al parecer, sus problemas empezaron hacia 1935, bajo la noche oscura de Stalin, y nada se volvió a saber de él en Occidente hasta 1957, cuando el editor italiano Giangiacomo Feltrinelli sacó de contrabando los originales de la novela *El doctor Zhivago* y la publicó en Italia y luego en el mundo entero. La novela, a pesar de algunos tramos excelentes, no es ni mucho menos lo mejor de ese poeta inmenso que fue Pasternak, así como las novelas de Pär Lagerkvist –Premio Nobel de 1951– sólo sirvieron para ocultar al gran lírico sueco que había detrás de ellas. En todo caso, entre las muchas desgracias en la vida de Pasternak no fue la menor el que sólo fuera conocido en Occidente por *El doctor Zhivago*, un libro que la mayoría conoce sin haberlo leído, gracias a la película que hizo David Lean, y la cual a su vez no se recuerda tanto por lo que en ella se contaba como por la almibarada canción de supermercado que le hizo Maurice Jarre sobre medida. Las circunstancias rocambolescas de la publicación, el desdichado incidente de su Premio Nobel, su muerte que se consideró prematura a los 70 años y la comercialidad descomunal de la película fueron los ingredientes que hicieron famoso a Pasternak en el mundo entero por las peores razones, sin que el mundo entero conociera nunca las razones verdaderas de su grandeza ni de su infortunio.

He estado dos veces en la Unión Soviética. La primera fue hace 26 años, cuando el Festival de la Juventud. Nadie hablaba entonces de Boris Pasternak, ni allá ni en ninguna parte, pero un año después –con motivo del Premio Nobel– en todo el mundo se hablaba de él, salvo en la Unión Soviética. No era para menos: el poeta estaba condenado entonces por el cargo fácil de desviacionista, la Unión de Escritores lo había expulsado con escándalo y los libros glorificados en otros tiempos habían sido prohibidos. A Pasternak no le impidieron viajar a Estocolmo para recibir el Premio Nobel –como tanto se ha dicho sin fundamento y como había de ocurrir más tarde con Solzhenitsin–, pero él se vio obligado a rechazarlo, según sus propias palabras, «por la significación que se le ha dado a este honor en la sociedad en que vivo».

Sin embargo, la segunda vez en que fui a la Unión Soviética, hace cuatro años y como invitado al Festival de Cine de Moscú, creo que no hubo una conversación con escritores y artistas en la que no se evocara el nombre de Pasternak, siempre sin escondrijos y con la admiración más entusiasta. Pero nadie podía decir en realidad qué era lo que había pasado antes para que fuera repudiado, ni qué había pasado después para que dejara de serlo. Entre los muchos chismes que se contaban sobre eso, varias veces oí decir que Jruschov –bajo cuyo reinado ocurrió el escándalo– había sido informado del peor modo por sus consejeros cuando no había leído *El doctor Zhivago*, y que cuando lo leyó, varios años después, expresó en privado su contrariedad y su arrepentimiento, pero ya Pasternak había muerto.

Entre los entusiastas del gran poeta encontré otros dos grandes de las generaciones posteriores: Yevgueny Yevtushenko y Andrei Voznesenski. Este último guarda con fervor poemas manuscritos y recuerdos imborrables de sus encuentros con Pasternak, conoce de memoria gran parte de sus versos y fue uno de los adelantados de la reparación pública. Yevtushenko, por su parte, tuvo la buena idea de invitarme a una peregrinación emocionante que se quedó para siempre en mi memoria como si hubiera sido ayer: me llevó a conocer la tumba de Pasternak.

Como tal vez se sabe, el poeta murió en la aldea de Peredelkino, a 30 kilómetros de Moscú, donde hay una colonia de escritores, y con un enorme y sombrío pabellón para los escritores retirados que se pasean entre las brumas del verano, solitarios o en parejas silenciosas, por las alamedas crepusculares. Muy cerca de ese pabellón, y a pocos pasos de la casa donde Pasternak vivió sus últimos años de soledad y donde murió en silencio, empieza el cementerio de la aldea, que quizá es uno de los más humanos del mundo. Son varias filas de tumbas escalonadas en una colina apacible; y en cada una de ellas, detrás de un marco de cristal, hay una fotografía del muerto y una pintura que ilustra sin metáforas la causa de la muerte. Hay una matrona rozagante, de esas que, sin duda, eran capaces de tumbar un caballo agarrándolo por las orejas, y junto a su retrato está pintado el rayo que la mató durante una tormenta. Hay un retrato del médico de la aldea que murió por un paro del corazón, pintado ahora en la tumba con un realismo conmovedor; el retrato de la niña paralítica con su silla de ruedas eternizada a todo color; todos los muertos del tranquilo recodo de Peredelkino glorificados junto a la razón de su muerte.

En la vertiente posterior de la colina, dentro de un cerco de cemento y en un espacio casi tan grande como el que debió tener su dormitorio de vivo, estaba el túmulo de Pasternak. No recuerdo si había grabados en la piedra, como en todas partes, el nombre y las fechas, pero recuerdo muy bien que era la única tumba que no tenía el retrato de su habitante ni tenía pintada la causa de su muerte, tal vez porque no hubo ningún artista en el pueblo que supiera cómo pintar la tristeza. Era un instante de una intensidad difícil de describir, y no supe qué hacer de inmediato ni encontré nada que decir ante la austeridad casi medieval de aquella tumba y la densidad del sitio y el rumor siempre nocturno —aun a pleno día— del viento entre los árboles. De pronto, obedeciendo a una orden del alma, arranqué del suelo un manojo de arbustos silvestres con unas cuantas florecitas de monte y lo puse frente a la tumba. Poco después, de regreso a Moscú, Yevtushenko me dijo: «Lo que más me impresionó es el respeto tremendo que le tienes a la muerte».

No había podido olvidar aquella frase a finales de esa semana, cuando un consejero para la Cultura del Comité Central de la URSS aceptó contestarme una muy larga serie de preguntas sobre la situación de los disidentes en aquel país. Fue una entrevista cordial, pero con instantes difíciles y nada clarificadores, cuyas notas conservo para cualquiera de estos días. Lo que ahora me interesa recordar es que antes de empezar —y sin el menor ánimo de provocación, sino con el deseo de imprimir desde el principio a nuestro encuentro el sello de la sinceridad más pura— le dije a mi interlocutor: «El viernes llevé flores a la tumba de Pasternak». Él me miró con una especie de melancolía muy antigua, y me dijo: «Ya lo sé, y me parece muy bien».

BISHOP

Tal vez no se ha hablado bastante de las analogías entre la muerte de Salvador Allende y la de Maurice Bishop. Más grave aún esta última, si resulta ser cierta —como parece serlo— la versión de que Bishop estaba desarmado y con las manos en alto en señal de rendición cuando fue asesinado por unidades militares, mientras que Allende se había enfrentado al ejército con una ametralladora que sabía manejar muy bien. No disminuye esta semejanza el hecho de que el uno había sido abatido por una fuerza de derecha y el otro por una fuerza que se proclama de izquierda. El día en que se justifique con cualquier argumento que las fuerzas del progreso se sirvan de los mismos métodos infames de la reacción, será ésa la hora —para decirlo en buen romance— de que nos vayamos todos al carajo. La declaración oficial del gobierno cubano, en la cual es más que evidente el estilo personal de puño y letra de Fidel Castro, lo dice de un modo más rotundo: «Ninguna doctrina, ningún principio o posición proclamada revolucionaria y ninguna división interna justifica procedimientos atroces como la eliminación física de Bishop y el grupo destacado de honestos dirigentes muertos en el día de ayer».

Nadie más ajeno a esos métodos que el mismo Maurice Bishop, que en 1979 conquistó el poder con una acción sustentada más bien por la presión popular que por las armas. En cambio, los protagonistas principales del drama que culminó con la muerte de Bishop son famosos por su vocación de violencia. Bernard Coard, su rival más visible y promotor del golpe mortal, ha sido siempre temible por la crueldad de sus decisiones, y no tiene ni mucho menos la inmensa popularidad de Bishop. El general Hudson Austin, el hombre que asumió el mando del país después de la muerte del primer ministro, es un matón del peor estilo, y la represión feroz que ha implantado desde sus orígenes es un mal anuncio del porvenir de esta isla de 110.000 habitantes que, dígase lo que se quiera, no tendrá el

Publicado originalmente el 26 de octubre de 1983.

poder suficiente para perturbar a nadie. Estados Unidos, que en los últimos años había hecho más que lo posible para poner término al proceso pacífico que Bishop impulsaba con su popularidad inmensa, no podía soñar con dos aliados más serviles —aunque involuntarios e inconscientes— que estos bandoleros en mala hora extraviados en la política.

Las vidas de estos tres hombres siguieron un curso paralelo hasta el viernes pasado. Bernard Coard y Maurice Bishop obtuvieron casi al mismo tiempo sus diplomas de abogados, el primero en Estados Unidos y el segundo en Londres. Abrieron juntos su primera oficina en Saint George, la capital de la isla. Juntos emprendieron y terminaron la lucha contra el régimen de sir Eric Gairy poco después de que Granada se independizara del Reino Unido, que la mantuvo bajo régimen colonial continuo desde 1876.

Sir Eric Gairy era un primer ministro un poco lunático que repartía su tiempo entre reprimir a sus opositores y especular en público —hasta en las Naciones Unidas— sobre el misterio de los platillos voladores. Sus sicarios mataron a tiros al padre de Bishop, en cuyo honor fue bautizado el cuartel de Fort Rupert, donde Bishop fue muerto. Bernard Coard y Maurice Bishop seguían juntos en marzo de 1979, cuando derribaron a Gairy e iniciaron el proceso revolucionario del cual Coard era viceprimer ministro, y que ahora pretendía continuar solo y a su mala manera. El general Hudson Austin, por su parte, fue el comandante de las fuerzas armadas desde el principio de la revolución, y su irrupción en el poder supremo parece ser un remedio de última hora para reprimir la rabia popular por la muerte de Bishop. Sin su protección, no parece posible que Coard pueda siquiera asomarse a la calle.

Bishop se había enfrentado a las numerosas tentativas de desestabilización que el gobierno de Estados Unidos había promovido contra el suyo, con una determinación admirable y un gran valor. Pero, en privado, se refería a ellas con un sentido del humor que era uno de los rasgos apreciables de su personalidad. Lo vi por última vez en marzo pasado, cuando fuimos en el mismo avión a la cumbre de los No Alineados en Nueva Delhi. Aunque lo conocía desde mucho antes, en esa ocasión tuve oportunidad de conocerlo mejor. Me sorprendió su capacidad de concentración: durante nueve horas continuas, casi sin parpadear, sin comer ni beber, leyó hasta el final un libro de 400 páginas sobre el desastre económico del Tercer Mundo, y subrayó párrafos y llenó un cuaderno de notas, cambiando apenas de posición en el asiento. Por fortuna, le sobraron horas para conversar en aquel vuelo interminable, y habló de su isla con una pasión que resultaba conmovedora cuando uno recordaba que es un territorio de 311 kilómetros cuadrados en un rincón perdido del Atlántico, y que no produce nada más que nuez moscada. El tema central de aquellos días era la foto aérea del aeropuerto que los cubanos están construyendo en Granada, y que el presidente Reagan había mostrado a la prensa como una prueba de que los soviéticos estaban instalando una base militar en el Caribe. La pretendida revelación de Reagan fue la repetición en comedia del drama que protagonizó el presidente Kennedy cuando mostró las fotos de las instalaciones de cohetes en Cuba en 1962. La realidad es más sencilla. Granada –descubierta por Cristóbal Colón en su tercer viaje– tiene una enorme riqueza potencial en el aprovechamiento turístico de sus playas doradas y sus paraísos secretos. Pero hasta ahora no había tenido recursos para construir un aeropuerto capaz de recibir aviones grandes. Cuba, mediante un acuerdo civil, emprendió hace casi dos años la construcción de una pista de 3.000 metros, que permitirá a Granada explotar a fondo sus recursos turísticos. El gobierno de Estados Unidos insiste, sin embargo, en que la verdadera finalidad de la obra es estratégica, pues una pista de ese tamaño permite las operaciones de los más modernos aviones de guerra de la Unión Soviética. Bishop evocaba muerto de risa este argumento, y no podía menos que hacerlo cuando se recordaba el espectáculo del presidente Reagan mostrando los depósitos de materiales de construcción en las fotos ampliadas como si fueran silos de armas mortíferas.

Lo que más impresionaba en la personalidad de Bishop era su simpatía, capaz de proyectarse en la muchedumbre. Es difícil encontrar otro hombre más elegante en la tribuna, no sólo por su estampa de casi dos metros y por su gracia caribe, sino por su inglés impecable, cultivado en la salsa propia de las universidades inglesas, y por la fluidez y la magia de sus palabras. Como ocurrió con Salvador Allende, había que matarlo para sustituirlo en el poder, pero nadie podrá sustituirlo en la memoria de su pueblo. Tenía 39 años. No había otro igual a él en Granada, ni en muchas leguas a la redonda. De modo que el drama de la isla sin él apenas ha comenzado.

Por 1956, la editorial Gallimard de París patrocinó una ruidosa campaña de prensa para vender el libro de versos de una niña de siete años llamada Minou Drouet, a quien se quería colocar de una vez por todas como un genio de las letras. Entre las muchas publicaciones de propaganda que se hicieron entonces hubo una encuesta entre los escritores y artistas más famosos del momento, los cuales se prestaron al juego editorial con frases más o menos convencionales. Pero Jean Cocteau le puso término al asunto con una sola frase mortal: «Tous les enfants sont des poetes, sauf Minou Drouet». Dicho en buen cristiano: «Todos los niños son poetas, menos Minou Drouet».

Me acordaba esta semana de aquel episodio mientras leía –en mi condición de jurado– los casi 200 cuentos finales del millar escritos por niños colombianos para un concurso de literatura infantil. No todos los concursantes tenían el aliento poético, pero los pocos que no lo tenían no era por culpa de ellos, sino por la de los adultos. Quiero decir: de todos nosotros, los padres, los maestros, los escritores, que les hemos transmitido a los niños una noción de la literatura que tal vez sea buena para nosotros y hecha por nosotros, pero que sin duda no tiene nada que ver con la magia de los niños. Tuve esta sensación nítida hace muchos años, cuando hice mi primera y última tentativa de escribir un ejercicio de cuento para niños. No era un tema improvisado. Desde hacía tiempo me daba vueltas en la cabeza la idea de un ángel decrépito que se cayera por la lluvia y que terminara sus días en un gallinero, picoteado por las gallinas y reducido a una triste condición de juguete de los niños. Puesto que la historia no me parecía creíble para los adultos que hace tanto tiempo dejaron de creer en los ángeles, pensé que sería buena para engañar a los niños. La escribí pensando en ellos, pero no como hablan los niños, sino con la entonación bobalicona y con el lenguaje de débil mental con que los adultos les hablamos a los hijos cuando

Publicado originalmente el 2 de noviembre de 1983.

empiezan a descubrir el mundo. Una vez terminado se lo mostré a los míos —que entonces tenían ocho y seis años de edad—, y ellos lo leyeron sólo una vez con mucha atención y me lo devolvieron, diciendo: «Tú crees que los niños somos pendejos». Yo no lo creía, en realidad, pero entendí lo que querían decirme, de modo que volví a escribir el cuento completo con todos mis convencionalismos de persona mayor y sólo conservé el título original: *Un señor muy viejo con unas alas enormes.* Por cierto, que mis hijos, creyéndome ofendido, aprovecharon el día de mi cumpleaños para hacerme un desagravio con una frase que conservo como un ejemplo de lo que es en verdad el talento puro de los niños para la poesía. «Papá», me dijeron a coro, «nosotros queremos que cuando tú seas niño seas como nosotros y que tengas un papá como tú.»

Había muchos cuentos hermosos en el concurso, pero los malos eran aquellos en que se notaba la mano perturbadora de los adultos. Para empezar estaban los peores, que son aquellos en que los niños quieren imitar los cuentos infantiles de la literatura universal, contados por los adultos, con reyes malvados y princesas encantadas, y hadas madrinas y madrastras infames. No hay duda de que los niños repiten esas historias por una de dos razones: o porque suponen que eso es literatura —tal como los adultos se lo han enseñado— o porque suponen, con razón, que los adultos somos tan cretinos que creemos que los niños creen que eso es la literatura, y escriben así para engañarnos, aunque son conscientes de escribir sobre un mundo falso y ajeno a ellos por completo. Lo recuerdo muy bien. Los niños de mi edad, en Aracataca, escuchábamos con una especie de éxtasis celestial los relatos de aventuras sexuales de los compañeros más avanzados —muchas de ellas inventadas, sin duda—, de modo que cuando escuchábamos después los cuentos para niños que nos contaban los adultos era como comer después del almuerzo. Nos contaban con aquel énfasis hipócrita que la cucarachita Martínez se sentaba al atardecer a la puerta de su casa, empolvada con almidón, con los labios pintados de carmín y con un traje de volantes y un lazo de organza en la cabeza, esperando a que pasara el ratoncito Pérez para preguntarle: «Ratoncito Pérez, ¿te quieres casar conmigo?». Y los niños, que ya habíamos visto tantas cucarachitas Martínez sentadas en la puerta de sus casas al otro lado del puente, pensábamos en lo profundo del alma: «Mira qué puta». Y así seguíamos hasta el final, siguiendo en la mente el cuento paralelo como sin duda era en la realidad, mientras los adultos trataban de hacernos creer que lo único que cucarachita Martínez quería del ratoncito Pérez era que la ayudara a revolver la olla en el fogón. Mamola.

No hay razones para no creer que los niños de hoy no hagan lo mismo, pero los adultos siguen sumidos en el mismo limbo de inocencia de nuestros tiempos. El mal resultado, por supuesto, es que cuando se les pide que escriban un cuento para la escuela lo escriben con la misma hipocresía de los adultos, para que le guste a la maestra, y cuando se les pide que escriban para un concurso lo hacen para que le guste al jurado. Sólo que en este caso el jurado único de última instancia era uno a quien no se le ha olvidado cómo era cuando era niño, y no le gustó la mayoría de los cuentos que sólo fueron escritos para que le gustaran.

No sé si fueron los padres o si fueron los maestros, o a veces unos y a veces otros, o a veces ambos, quienes decidieron meter la mano a última hora para tratar de arreglar los cuentos, y lo que consiguieron fue acabar de estropearlos. En efecto, es inverosímil que un niño de ocho años escriba un cuento de más de cinco cuartillas a máquina sobre la guerra espacial y lo haga sin un solo error de ortografía y apenas con las fallas de sintaxis propias de los buenos papás que ayudan a los hijos a hacer sus tareas. Faltan en esos cuentos la originalidad, la locura, la imaginación irracional, que hacen fascinantes y sabios a los niños. Una prueba del grado de esterilidad a que los han arrastrado los adultos es que una gran cantidad de cuentos que hubieran sido buenos por su inventiva y su hermosa irrealidad han sido resueltos por los niños con el recurso de los sueños. Todo lo que no se atreven a contar como real, porque saben que los adultos lo rechazarían, los niños lo cuentan como si hubiera ocurrido en un sueño. De modo que los cuentos soñados son numerosos en este concurso, en el cual, por fortuna, hubo muchos concursantes que se atrevieron a escribir como quisieron, sin consultarlo con los adultos, y hubo muchos adultos certeros que les permitieron hacerlo. Ellos serán los premiados.

Creo que fue Marshall McLuhan —y tal vez en su libro *El medio es el mensaje*— quien se atrevió a decir que la infancia es una invención del siglo XVII. Antes, las etapas de la vida eran sólo la adolescencia, la madurez y la ancianidad, y a los niños se les consideraba, con un criterio un poco bárbaro, como seres humanos chiquitos sin personalidad propia. En cierto modo, a pesar de que ya en las Naciones Unidas se toman en cuenta hasta los derechos humanos del niño, hay todavía muchos adultos que siguen pensando como antes del siglo XVII. Entre ellos están los que les corrigen los cuentos a los niños, que es algo tan cruel y tan grave como cortarles las alas.

Es comprensible, en consecuencia, que entre los cuentos mejores del concurso estuvieran aquellos cuyos protagonistas son animales.

Éstos son más del 80 %. La impresión que queda de su lectura es que los niños nuestros tienen con los animales que les rodea la comunicación real que no tienen con los adultos. No comprenden a mamá, pero en cambio comprenden muy bien los motivos del lobo, sobre todo si el lobo les sirve de instrumento para expresarse sin riesgos. Con todo, es una liberación a medias, porque los niños escritores terminan de todos modos por decir que era un conejo bueno porque le gustaba la escuela, cuando todos los que tenemos memoria recordamos que la escuela no le gustaba ni a los malos. No le gustaba a nadie, y con toda la razón. Los niños mienten, por supuesto, como siempre se ha dicho, pero no como siempre se ha dicho, sino porque los adultos les vamos enseñando a medida que los criamos. Es sólo cuando no nos hacen caso cuando son poetas verdaderos. Como no lo fue Minou Drouet, y como sí lo fue la niña colombiana de siete años que escribió este prodigio de ternura: «Cuando yo sea grande quiero ser un gran médico en un gran hospital de Nueva York, y cuando los enfermos se mueran me voy a morir con ellos».

TEODORO

Teodoro Petkoff –candidato socialista a la presidencia de Venezuela– estaba preso en el cuartel San Carlos, de Caracas, a principios de 1962, mientras la llamarada de la guerrilla se extendía por todo el país. Había sido capturado en el curso de una operación urbana y recluido en una celda de alta seguridad, de la cual parecía imposible fugarse. Apenas había cumplido los 30 años, pero ya era un dirigente destacado del partido comunista y tenía un pasado brillante como resistente universitario contra la dictadura militar de Marcos Pérez Jiménez. Desde el instante mismo en que fue capturado tuvo un objetivo único que no le dio un instante de tregua en los largos meses de reclusión. Ese objetivo, que se consideraba poco menos que fantástico, era fugarse de una cárcel militar de la cual nadie había logrado escapar hasta entonces.

Lo consiguió en pocos meses con un plan deslumbrante. Un sábado de visitas, una amiga suya le llevó escondidas varias cápsulas llenas de sangre fresca de vaca. Cuando quedó solo en la celda, Teodoro empezó a quejarse de un malestar cuyos síntomas precisos no le dejaban ninguna duda al médico de la prisión: una úlcera gástrica. Era una dolencia fingida, por supuesto, pero el proyecto era tan meticuloso que Teodoro se había aprendido de memoria hasta las manifestaciones más sutiles de la enfermedad que le convenía aparentar. El médico le recomendó reposo –que no era nada difícil en una celda de alta seguridad– y le prescribió un tratamiento severo. Pero aquella noche, Teodoro se tragó las cápsulas, despertó a la prisión con sus gritos y los guardias que acudieron lo encontraron postrado por una crisis de vómitos de sangre. Lo trasladaron al hospital militar, donde las medidas de seguridad no eran tan rigurosas, y antes del amanecer se descolgó por la ventana del séptimo piso con ayuda de una cuerda que alguien le hizo llegar.

Publicado originalmente el 9 de noviembre de 1983.

Fue una fuga tan espectacular que cuatro años después, cuando Teodoro fue capturado de nuevo en el estado de Falcón, donde operaba la guerrilla comandada por Douglas Bravo, la prisión del cuartel San Carlos le pareció poca cosa al gobierno para mantenerlo a buen recaudo. De modo que lo mandaron a una colonia marítima donde la fuga era imposible: la isla de Tacarigua. Otra vez la obsesión de Teodoro en cada instante de su reclusión siguió siendo la misma: evadirse.

La primera tentativa, cuya audacia revela muy bien cuáles eran las condiciones de la cárcel, fracasó por una filtración. El rescate debía intentarlo una célula guerrillera durante el traslado de Teodoro de un lugar a otro de la isla, acompañado por un solo guardián; pero éste lo llevó con la pistola apoyada en la nuca y con la advertencia de que le volaría el cráneo de un balazo si alguien intentaba interceptar el vehículo. Teodoro comprendió entonces, como lo había comprendido la primera vez, que su única esperanza era hacerse cambiar de prisión. Le costó tiempo y trabajo, pero lo consiguió.

Tuvo la suerte de que lo trasladaran otra vez al cuartel San Carlos, donde estaban recluidos no menos de 50 compañeros suyos. Cuando llegó, ya el plan de fuga estaba adelantado. Un túnel que el partido comunista había empezado a construir desde hacía casi 20 meses llegaba por esos días a su término feliz. A lo largo de muchos años, incluidos los interminables y feroces de la dictadura de Juan Vicente Gómez, los políticos presos habían iniciado la construcción de túneles desde las celdas hacia la calle, y todos habían sido descubiertos cuando ya era imposible ocultar la tierra de las excavaciones. El más reciente lo había intentado el general Castro León –un conspirador nostálgico de los tiempos de Pérez Jiménez– y por un error de cálculo no desembocó en la calle, sino en las cocinas del cuartel. El nuevo túnel había resuelto el problema de la tierra, excavando desde una casa al lado del cuartel, con una calle de por medio. Todo estaba tan bien planeado que en cierta ocasión empezó a ceder el pavimento con el peso de los vehículos y los inquilinos de la casa consiguieron que las autoridades del cuartel prohibieran el tránsito por aquella calle.

Tres dirigentes se fugaron en febrero de 1967 y en el día más propicio del año: la noche de carnaval. En una fecha así la vigilancia era menos intensa y la búsqueda casi imposible en una ciudad sumergida en el frenesí de la parranda. Era imposible identificar a nadie porque medio mundo andaba disfrazado. Teodoro fue uno de los tres. Pero mientras la mayoría de sus compañeros parecían empeñados en proseguir una guerra que era un error militar evidente, él salió de la cárcel convencido de que era además un error político en el cual no parecía sensato persistir.

Estos dos episodios de la vida de Teodoro Petkoff me llamaron la atención de un modo muy especial desde que alguien me los contó hace ya muchos años, porque dan una clave reveladora de su personalidad. Es un político audaz, de una energía que se le siente hasta en un apretón de manos, pero todos sus actos están comandados por el sentido común. Cuando abandonó la lucha armada, esto requería mucha más valentía que continuar en ella. Teodoro, con lo mejor de su partido de entonces, asumió el riesgo con un proyecto en el cual no se sabe si admirar más la visión o la paciencia: diez años para formar un movimiento nuevo y otros diez para imponerlo. Los diez primeros han transcurrido y el movimiento está implantado. Pase lo que pase en las elecciones venezolanas de diciembre, el partido de Petkoff quedará establecido como una tercera fuerza con posibilidades inminentes de convertirse en la segunda y entrar en la recta final hacia el poder.

Yo mismo, que lo conozco desde hace tantos años y que he seguido de tan cerca su trayectoria espectacular, me sorprendo de que haya llegado a este punto en un tiempo tan breve. Pero me sorprende más que lo haya conseguido sin dejar de ser el hombre humano que ha sido siempre, capaz al mismo tiempo de fugarse de la cárcel como un héroe de cine, de bailar como un muchacho la música de moda hasta el amanecer, o de pasar una noche entera –y a veces sin tomarse un trago– hablando de literatura. «Soy un apasionado lector de novelas –ha dicho en una entrevista–. Son mundos en los que me sumerjo con facilidad.» Y no se trata de un lector cualquiera, sino de uno que ha hecho la proeza de leer dos veces *La montaña mágica*, de Thomas Mann, lo cual es casi un dato decisivo de la personalidad. «La primera vez la leí por compromiso –ha dicho–, pero la segunda vez la agarré así, hojeándola, y de pronto me encontré leyéndola con un inmenso placer.» No es raro para quienes lo conocemos bien: su poesía favorita son los *Veinte poemas de amor y una canción desesperada* de Pablo Neruda. Se sabe que tiene cuentos clandestinos y que rompió los originales de una novela que había escrito en la cárcel.

He señalado este afecto a la literatura por el puro gusto de señalar una afinidad. Pero la verdad es que a Teodoro le interesa todo con la misma pasión –desde la filosofía escolástica hasta el béisbol–, y a esto se debe quizá el que se le note tan poco el paso de los años. Ahora anda por el medio siglo, pero muchas veces, oyéndolo hablar, uno piensa que cambia de edad –desde la adolescencia hasta la madurez– según el tema y la ocasión. Sólo hay dos cosas que le causan miedo, que son las matemáticas y la tribuna de los discursos, pero en ambos casos lo domina muy bien. En cambio no le tiene miedo al tiempo, y eso es tal vez lo que mejor define su vida: le alcanzará para todo.

EL FRENESÍ DEL VIERNES

En el mundo occidental se hace cada vez más frenética la furia del viernes. Es un estado de exaltación perpetua desde el instante en que uno despierta por la mañana y recuerda que el gran día ha llegado por fin: viernes. En los días precedentes, aquella ansiedad se había sentido crecer con pasos de animal grande, pues nada importaba tanto en el futuro como ese día penúltimo cuya noche parece hecha a propósito para todos los excesos. Los amores más inciertos, los rencores más enconados, las ilusiones menos verosímiles, los sueños sin porvenir, todo lo que tiene que ver con la esencia de la vida se encontraba en suspenso durante toda la semana, en espera de que llegara ese viernes providencial que todo lo decide y lo resuelve.

Durante muchos años, la noche predestinada fue la del sábado. El mundo cristiano estallaba de júbilo desde el atardecer, con una sensación de fuegos artificiales en el alma, y la emoción se iba acentuando hasta el instante en que empezaba el baile, empezaba la boda, empezaba la fiesta de cumpleaños aplazada para aquella noche –la más larga e intensa– y la vida se convertía en una enloquecida cresta oceánica que nos mecía durante varias horas al borde de un precipicio feliz y cuya resaca nos arrojaba como náufragos sin esperanza en el vasto playón del domingo. En los tiempos de estudiantes, pasar del sábado al domingo era como cambiar de edades geológicas. El último día era un castigo, agravado más todavía por la memoria aún caliente de la noche anterior, por las ráfagas de música que se nos habían quedado enredadas en el corazón, por un perfume furtivo, por la huella de una voz, por todo lo que fue un instante y tal vez no volvería a serlo jamás. Hasta el otro sábado.

Mientras tanto, el drama era sobrellevar el domingo. Los solteros querían ser casados para no despertar solos en la cama dentro del vasto silencio de la ciudad. Se levantaban temprano. No para ir a la santa misa, como les habían enseñado de niños, sino para esperar en

Publicado originalmente el 16 de noviembre de 1983.

la puerta de la iglesia algún amor perdido que les resolvería la soledad de aquel día funerario que iba como desangrándose, arrastrando
una cola interminable de lagarto senil. «Los domingos huelen a carne cruda», escribí en alguna parte, evocando mis tardes dominicales
de estudiante de provincia en una capital provinciana como era la
Bogotá de aquellos tiempos. Había que levantarse temprano porque
todos lo hacían antes de que empezaran los mejores programas y
uno se quedara rezagado, sin nada que hacer en unas calles que se
desaguaban de su vida matinal y quedaban sumergidas en el sopor
más desolado de cuantos existían entonces, que era el de las tardes
de los domingos. Si esto ocurría no quedaba otro consuelo que tratar de robarle a los otros las migajas de su dicha. Se paraba uno en
la puerta de los cines a esperar que terminara la película para que los
otros salieran, para ver si uno tenía la suerte de encontrar un conocido, aunque fuera el más remoto. Y en el peor de los casos, para
disfrutar del placer incomprensible de disfrutar de la película por las
caras con que salían los que tuvieron la suerte de entrar. Era una
manera efímera y prestada de consolarse: lo más parecido posible a
la solidaridad. De modo que no era raro que alguien se casara de
pronto, sin pensarlo dos veces, sólo para tener con quien compartir
la ciénaga de soledad de los domingos.

Aquella aversión al domingo nos impedía a muchos entender la
aversión contra el lunes. Mientras los que tenían domingos resueltos
iniciaban la semana arrastrando una ristra de recuerdos que no les
dejaban vivir, los solitarios dominicales la iniciábamos con la sensación jubilosa de que el mundo se había poblado de nuevo, de que la
vida empezaba otra vez por un camino mejor, de que el porvenir
era fácil y fructífero porque había mucho tiempo para componer
desde hoy hasta el sábado próximo cualquier entuerto del pasado.
Sin embargo, algún amigo, uno de aquellos empedernidos nostálgicos del sábado que pasaban seis días esperando el siguiente, decía:
«Semana que llega a lunes, se la llevó el carajo». Era su verdad.

La mía era el terror a los martes, tal vez porque desde muy niño
nos habían enseñado que era un mal día para casarse y para embarcarse. Más tarde fue peor. Una locución muy española, ahora caída
en desuso, era *dar con la del martes.* De acuerdo con el diccionario,
esta locución significaba nada menos que morir, y en casos menos
graves sufrir perjuicios y contrariedades o recibir elogios engañosos.
Era una connotación tan siniestra como era despreciativo decir de
algo que no era nada del otro jueves. El martes 13 —ya lo sabemos—
era el día aciago, el día de los marcianos. El consagrado al dios de la
guerra. En cambio, el miércoles —que resuena como un caracol— es

el día de la buena suerte, propicio para comprar y vender, para vivir y amar y hacerse tomar un buen retrato.

El viernes –día de la crucifixión de Cristo– tuvo un mal prestigio, semejante al del martes, hasta que se impuso la semana inglesa. Entonces el viernes adquirió una categoría de víspera, no se sabe si para su fortuna o para su desgracia. En todo caso, si uno tiene la suerte de despertar sin acordarse de que es viernes, no pasarán dos minutos antes de que el timbre del teléfono se encargue de recordárselo. Alguien llama para invitar, y es una invitación más entre las muchas que han venido acumulándose en el curso de la semana. Es un día tenso e intenso, marcado con los signos de la incertidumbre de no saber cuál escoger entre tantas opciones, tantas bodas, tantas cenas para tantos, tantos cumpleaños, tantas oportunidades indeseables de salir fotografiados en las páginas sociales de las revistas mundanas. A veces, en viernes de dolores, uno decide no decidir sino ir de una fiesta a la otra durante toda la noche, hasta que la luz del alba lo sorprende en un estado de frustración feliz, dándose golpes de pecho por sus culpas pasadas, y prometiéndose no recaer en ellas nunca más. Hasta el próximo viernes, por supuesto.

Hay automóviles que salen defectuosos de la fábrica, y hay fabricantes honestos que los cambian por otro nuevo a quien ha tenido la mala suerte de comprarlo. La gente del oficio conoce esos carros por su nombre: carros de viernes. Se les llama así porque fueron ensamblados el día penúltimo de la semana, cuando los obreros se sienten fatigados de los días anteriores y no ven la hora de terminar la jornada para lanzarse al frenesí del viernes. En su ansiedad no aciertan a apretar bien un tornillo, ni a ajustar una pieza, ni a medir con precisión los niveles. Todo queda a medio hacer hasta el lunes, cuando la rutina empezará de nuevo y se iniciarán otra vez las pendientes de los días que nos deslicen y nos arrojen –como en un inmenso caldero de aceite hirviendo– en el desorden del viernes. No hay salvación. Un sábado le pregunté a un amigo fiestero en qué playa había estado que tenía la piel tostada, y me contestó en serio: «No estoy quemado por el sol, sino por los flashes de los fotógrafos de la fiesta de anoche».

¿QUÉ PASÓ AL FIN EN GRANADA?

A medida que pasa el tiempo y retorna la tranquilidad a la isla de Granada, va quedando claro que la invasión por tropas norteamericanas no fue tanto una operación militar como una maniobra enorme de manipulación informativa. Para empezar, el balance de víctimas no corresponde al escándalo: 18 norteamericanos muertos y 91 heridos −muchos de ellos en incidentes y accidentes confusos, y no en combates− y un número nunca establecido de granadinos muertos, 20 de ellos en el curso de un bombardeo a un hospital de enfermos mentales.

Este último episodio es el más oscuro de todos. Según se dijo al principio, el lugar había sido bombardeado con cohetes aéreos porque dos cubanos resistían desde el interior. Sin embargo, una versión de la prensa norteamericana dice ahora que el bombardeo se debió a que el hospital estaba a sólo 150 metros de Fort Frederick, y éste era defendido con ahínco por los últimos soldados del ejército granadino. El hospital −dice la versión− no tenía ninguna señal que permitiera identificarlo. Más aún: la periodista colombiana Laura Restrepo, enviada por la revista *Semana* con los primeros grupos de periodistas que entraron en la isla, se sorprendió de que aquel sanatorio para enfermos mentales −contra el esplendor de su nombre− no fuera más que un grupo de chozas de paja. La revista *Time*, por su parte, le reprocha al Pentágono que no hubiera dicho nada de esa matanza −accidental o no− mientras no fue denunciada por un periodista canadiense. El Pentágono se defendió diciendo que cuando los infantes de Marina ocuparon el hospital los muertos habían sido ya sepultados, y no encontraron ninguna razón para sospechar que el bombardeo de los días precedentes hubiera causado alguna víctima. Pero quienquiera que haya estudiado con cierto cuidado las informaciones de Granada, sobre todo en los primeros días, debe tener motivos para creer que el silencio del Pentágono

Publicado originalmente el 23 de noviembre de 1983.

en relación con los muertos del hospital psiquiátrico pudo no haber sido casual.

En realidad, toda la información de los primeros días —manejada de un modo exclusivo por el gobierno de Estados Unidos, y casi siempre por el presidente Reagan en persona— ha empezado a desmoronarse. Ahora se entiende cómo fue posible que más de 6.000 hombres de Estados Unidos, bien entrenados y con todos los recursos de la guerra moderna, no hubieran podido someter en dos semanas a uno de los ejércitos más reducidos y pobres del mundo, en una isla de 110.000 habitantes desmoralizados que no tenían ni modo ni ganas de resistir. La explicación es simple: no hubo tal resistencia. En primer término porque los granadinos, aún no repuestos del asesinato de su líder más querido —Maurice Bishop—, no debieron de ver a los infantes de Marina como sus enemigos, sino al contrario, como los enemigos de sus enemigos. En segundo término, porque los asesinos de Bishop, repudiados por su pueblo y por la mayoría de su ejército, se metieron debajo de la cama a los primeros tiros. En efecto, parece ser que Bernard Coard —el viceprimer ministro de Bishop que lo derribó a traición— se había escondido con su esposa después de abandonar el poder que había usurpado pocos días antes. Por su parte, el general Hudson Austin —responsable inmediato del asesinato de Bishop— abandonó el ejército a su suerte, y al parecer andaba ofreciendo hasta 3.500 dólares a quien le hiciera el favor de llevarlo en una lancha a Guyana. Los focos de resistencia que quedaron después de la desbandada podían ser reducidos en pocas horas por unas fuerzas de desembarco preparadas para operaciones mucho más gloriosas.

La verdad parece ser que el gobierno de Reagan necesitaba inventar aquella resistencia para justificar la invasión con el supuesto de la militarización masiva de la isla por los cubanos y los soviéticos. Durante más de una semana, las tropas de ocupación se movieron a sus anchas por la isla, sin que ningún periodista de ningún país pudiera entrar para contradecir las versiones oficiales del gobierno de Estados Unidos. Sin embargo, fuera de todo control, aun este último no hizo más que contradecirse a sí mismo.

La primera contradicción enorme fue el motivo de la invasión. De acuerdo con un comunicado inicial, el desembarco tenía como único objetivo proteger la vida de unos 600 estudiantes norteamericanos que estudiaban en la muy prestigiosa facultad de Medicina de Granada. Sin embargo, hasta ahora no se ha demostrado que estuvieran en peligro y, en cambio, sí es probable que estuvieran contentos en el lugar. Nada les había impedido abandonarlo por su propia voluntad,

y los pocos que hicieron la payasada de besar la tierra norteamericana cuando volvieron a ella —como suelen hacerlo los papas de
ahora dondequiera que llegan— parecían olvidarse de que para lograrlo no era necesaria la intervención brutal de 6.000 hombres,
armados como para una guerra mundial. Sin embargo, el presidente Reagan también olvidó demasiado pronto su pretexto original y
no tuvo ningún inconveniente en decir que el desembarco había
sido necesario porque Granada se había convertido en una fortaleza
militar del comunismo internacional. Lo triste es que las supuestas
pruebas de esa afirmación —anunciadas a grandes voces por el gobierno de Estados Unidos— no han logrado convencer sino a los ya
convencidos, algunos de ellos, por cierto, muy respetables por motivos distintos.

El cuento de la ocupación cubana fue tal vez el menos consistente. Los primeros periodistas extranjeros que llegaron a Granada
no pudieron disimular su desilusión frente al aeropuerto que estaban
construyendo los técnicos y obreros de Cuba. El gobierno de Estados Unidos había hecho creer que era un aeropuerto construido
para las naves de guerra soviéticas y no para aviones comerciales que
llevaran turistas pacíficos, inclusive norteamericanos, que son los
más fructíferos. El argumento se fundaba en que la pista iba a tener
3.000 metros de larga, y esta cifra parecía impresionante para quienes no saben que cualquier aeropuerto moderno donde operan los
grandes aviones civiles debe tener esas medidas, sobre todo si se prevé un desarrollo futuro de su capacidad. La misma revista *Time*, con
una pretensión de objetividad, hace esta consideración retorcida: «Es
verdad que la nueva pista no está construida con las estructuras de
protección y los equipos de apoyo que son usuales en los aeropuertos militares, pero podría ser usada por aviones militares pesados
como punto de abastecimiento para los cubanos en ruta hacia África, o para los soviéticos que transporten armas hacia América Central». Es decir: como cualquier aeropuerto corriente y común.

Lo más confuso de todo fue el manejo que hizo el gobierno de
Estados Unidos de la información sobre los cubanos en Granada.
Desde el principio se dijo que eran unos 600 hombres, entre obreros del aeropuerto, maestros, médicos y asistentes militares. Estos
últimos —según un supuesto documento revelado por la Secretaría
de Estado de Estados Unidos— eran sólo 27 con carácter permanente y unos 12 eventuales. Se ha dicho, sin embargo, que debía de haber muchos más que se hacían pasar por trabajadores civiles, porque
todos demostraron tener un buen entrenamiento militar. Hasta los
mismos que lo dijeron sabían, sin duda, que todo cubano mayor de

14 años, hombre o mujer, tiene suficiente formación militar para defender a su país en caso de una invasión extranjera. Las llamadas tropas territoriales, que son milicias civiles, cuentan en la actualidad con 500.000 hombres y mujeres y está previsto que en breve serán el doble.

En todo caso, al principio de la invasión Estados Unidos dijo que los 600 cubanos habían caído prisioneros sin resistir. Después –cuando Cuba pidió su repatriación– se dijo que la mayoría estaba resistiendo en las colinas. Por último, sin ninguna explicación, aparecieron 27 muertos, 57 heridos, y el resto en un campamento de prisioneros donde sólo se permitió el ingreso a dos periodistas: un reportero y un camarógrafo dominicanos, que trataron de convencer a algunos prisioneros, con toda clase de promesas, para que se asilaran en Estados Unidos y confirmaran una versión que éstos tenían preparada sobre las actividades de Cuba en la isla. Eran, por supuesto, agentes de la CIA, que se llevaron la sorpresa de no encontrar a ningún cubano dispuesto a vender su alma al diablo. De haberlo encontrado, la vasta operación de manipulación informativa habría culminado con un acto espectacular. Por no haber sido así, la triste invasión de Granada pasará como uno de los capítulos menos honorables de la historia de Estados Unidos.

NÁUFRAGOS DEL ESPACIO

El hecho tiene una trascendencia enorme: dos náufragos del espacio, los primeros en la historia de la humanidad, han sido rescatados sanos y salvos, después de dos meses de una incertidumbre que parecía en realidad una lenta, intensa y silenciosa agonía. El salvamento providencial, como estaban las cosas, parecía poco menos que imposible. Sin embargo, el azaroso naufragio cósmico no mereció por parte de la prensa una atención que pudiera siquiera compararse con la que hubiera suscitado una novela de televisión sobre el mismo tema. Es decir, que en este mundo cada vez más distorsionado las angustias de la ficción empiezan a ser más verosímiles y emotivas que las angustias de la realidad.

Los cosmonautas soviéticos Vladimir Lyakhov y Alexandre Alexandrov estaban a bordo de la estación orbital *Salyut* 7 desde hacía cuatro meses, pero desde hacía dos esperaban ser reemplazados. En las prácticas espaciales soviéticas esa operación de relevo es ya una rutina. Cada cierto tiempo, cuando una tripulación de dos hombres ha cumplido una misión específica en la estación orbital, una nave *soyuz* —que es como una lancha sideral— lleva otra tripulación de refresco para que la anterior regrese a disfrutar de un descanso merecido en la Tierra. La operación es compleja y requiere de un refinamiento técnico de una gran precisión, pero los soviéticos la han repetido tantas veces que tal vez no se pensaba ya demasiado en sus riesgos. Todo parece muy simple: la lancha espacial se acopla a uno de los extremos de la estación orbital y la tripulación de relevo pasa entonces por un corredor interno a la nave principal, donde inicia el cumplimiento de su misión de varios meses. Hasta el 29 de junio de 1971, este transbordo, que parecía tan elemental para los técnicos y para los aficionados a la ficción científica, no había sufrido ninguna contrariedad. Pero en esa fecha, los tres astronautas que aterrizaron sin contratiempos de regreso del *Salyut 1* fueron encontrados

Publicado originalmente el 30 de noviembre de 1983.

muertos dentro de la nave, sin que hasta el momento se conozca una explicación indudable del percance.

La gran racha de mala suerte, sin embargo, empezó en mayo pasado, cuando una tripulación de relevo del *Salyut* 7 tuvo que regresar a su base después de varias tentativas frustradas de acoplamiento. Luego, el 27 de septiembre pasado, fue la primera tragedia grande. Las instalaciones de lanzamiento de la lancha espacial fueron destruidas por una explosión en el momento del disparo, y los dos tripulantes que iban a reemplazar a Lyakhov y a Alexandrov salvaron sus vidas por puro milagro. Con la rampa de lanzamiento destruida, y la imposibilidad consiguiente de sustituirlos en una fecha inmediata, los dos tripulantes solitarios del *Salyut* 7 fueron, desde ese instante, los dos primeros náufragos en la epopeya fascinante de la conquista del espacio.

Como tantas otras, aquella tragedia no vino sola. Desperfectos que nunca se habían registrado en las estaciones orbitales soviéticas empezaron a detectarse en el *Salyut* 7, que muy pronto amenazó con convertirse en un barco cósmico al garete. Para colmo de desdichas, el aire de la cabina sufrió un envenenamiento por la fuga de un gas mortal, y los náufragos tuvieron que pasar a la lancha que los había transportado al espacio, y que continuaba acoplada a la estación, aunque no era útil para el regreso por razones técnicas muy largas de explicar. En todo caso, aquel nuevo accidente no parecía ser imprevisto, porque el aire venenoso fue purificado en pocas horas, y los náufragos pudieron regresar a su rutina. Con la perspectiva siniestra de todos los náufragos, por supuesto: el agua y los alimentos empezaban a escasear.

El hermetismo del sistema soviético —sobre todo en relación con acontecimientos que puedan interpretarse como fracasos— impidió que el mundo siguiera con la ansiedad y la emoción naturales las incidencias de aquel episodio dramático. Muchos de los pormenores, como suele ocurrir, se conocieron en Occidente por informaciones inciertas. Fue una lástima, porque al contrario de lo que la Unión Soviética podía temer, el proceso de rescate y su feliz desenlace fueron una prueba más de sus enormes avances técnicos y científicos en el dominio del espacio. A mediados de octubre, los náufragos quedaron en condiciones de resistir por varios meses más. Una nave sin tripulación logró acoplarse a la estación orbital, llevando un cargamento de aire, alimentos y otras materias indispensables para subsistir. Entre estas últimas, sin duda, no eran las menos importantes las cartas de parientes y amigos que les mandaban noticias domésticas, chismes del barrio, recortes de periódicos de esa patria planetaria que veían en el horizonte del universo, luminosa y distante, preguntándose si alguna vez volverían a sentirla bajo sus pasos.

Aun si la base de lanzamiento no hubiera estado inservible, las posibilidades del rescate eran muy escasas. En primer término, la estación orbital no sólo estaba a punto de quedarse sin combustible, sino que la mayoría de sus motores se habían ido averiando uno tras otro, como sólo hubiera podido ocurrírsele al admirable Ray Bradbury en alguno de sus delirios asombrosos. Una pregunta se imponía: ¿por qué, si había sido posible mandar una nave de abastecimiento, no podía intentarse el envío de una nave de rescate? La respuesta parecía ser —aparte de la destrucción de la base de lanzamiento— que sólo una nave tripulada podía intentar el salvamento. Ahora bien: la nave *soyuz*, que es la única capaz de acoplarse a la estación orbital, sólo podía ser llevada por dos tripulantes y sólo podía traer un náufrago de regreso a la Tierra. El otro debía quedar solo, en la soledad sin límites del universo, tal vez para siempre. ¿Cuál de los dos?

Ambos han descendido a salvo, y, según las informaciones soviéticas, han sido acogidos en tierra con toda clase de honores, pero no se dan muchas luces sobre la forma en que se resolvió el acertijo del que debía quedarse y no se quedó. Ya lo sabremos —espero— en los días por venir, si alguien se decide a contarlo. Cuántas veces en noches recientes nos había despertado la imagen tenaz de esos dos náufragos que son más de un siglo futuro que del presente, y a quienes podíamos imaginar, insomnes en su nave sin rumbo, contemplando el resplandor del planeta distante al que tal vez no volverían jamás. Eran —pensábamos— los únicos seres humanos que de algún modo podían decir que estaban viendo el mundo desde la muerte. ¿Puede concebirse una soledad más espantosa?

Joya

Daniel Arango cuenta este cuento hermosísimo que no soy capaz de mantener en secreto: un niño de unos cinco años que ha perdido a su madre entre la muchedumbre de una feria se acerca a un agente de la policía y le pregunta: «¿No ha visto usted a una señora que anda sin un niño como yo?».

LA HISTORIA VISTA DE ESPALDAS

Cuenta un amigo que en sus tiempos de conspirador juvenil se disfrazó tan bien para hacer un viaje clandestino, que había burlado retenes policiales y emboscadas de sicarios, hasta que alguien que lo conocía desde la escuela, y que no lo había reconocido de frente, lo reconoció por la espalda. Comprendí mejor esta aventura cuando vi una película en que el poeta Yevgueni Yevstushenko encarnó como actor a un anciano diseñador de aviones. Era una caracterización impecable, no sólo por el maquillaje, que era muy convincente, sino por la erosión de la voz y la incertidumbre senil. De pronto, en una escena en que el personaje se alejaba de espaldas, toda la magia de la ficción se vino abajo: por un instante ya no fue más el anciano patético que había logrado convencernos de frente, sino el amigo Yevgueni en cuerpo y alma, con sus cuarenta años apenas cumplidos y su empaque de nadador de torrentes siberianos. Cuando alguien le señaló esa única falla de su muy buena actuación, Yevstushenko dijo: «Es que nadie se conoce a sí mismo de espaldas».

Hasta hoy pensé que tenía razón. Por eso la primera sorpresa de este libro sorprendente es el dibujo de la carátula, donde el autor se ve a sí mismo de espaldas, y sin embargo el parecido es más notable que en los otros muchos dibujos en que se ha visto de frente. Lo cual revela de entrada el alto grado de su peligrosidad artística, porque nadie puede estar a salvo frente a un caricaturista que conoce tan bien a los seres humanos, que conoce inclusive el espacio reservado a su propio ángel de la guarda. Es también una prueba inmediata de la maestría de Osuna en el arte misterioso de la caricatura. Y un resumen magnífico de este libro.

El método de Osuna es la organización de las situaciones más complejas de un modo que sólo él conoce, para reducirlas luego a un símbolo único con la simplicidad y el filo sangriento de una cuchilla de afeitar. Todo un sistema de represión, de cuyo nombre seguimos

Publicado originalmente el 4 de diciembre de 1983.

acordándonos aunque no lo quisiéramos, quedó identificado para siempre con la imagen de dos caballos guasones en la caballeriza convertida en laboratorio de torturas. Al gobierno actual, desde sus orígenes, lo redujo a la complicidad de una monja escapada de un cuadro de Botero que está colgado en el palacio presidencial, y un ministro que se tiene como el más cercano a los afectos del presidente. Osuna nos ha enseñado a pensar que ésa es al mismo tiempo la complicidad y el conflicto entre la conciencia y el subconsciente de un gobierno que sabe muy bien lo que quiere, pero que no puede tanto como sabe que quiere. Es la historia vista de espaldas, con las miserias cotidianas de sus costuras, como nos ha sido servida semana tras semana durante más de veinte años con el desayuno dominical. Y con un sabor tan propio y un condimento tan variado, que ya empezamos a preguntarnos cómo serían nuestros domingos si no existiera Osuna.

Quienes lo conocemos de frente, pensamos que es un hombre que calza un alma varios números más grande que él. Para calificarlo sólo existe un adjetivo que por falta de uso está a punto de ser un arcaísmo: atildado. Todo en él es de un rigor sacramental: su atuendo metódico, su urbanidad milimétrica, su edad de niño. Uno podría creer que su sentido más útil es el de la vista, pero hablando con él se descubre que no está tan pendiente de los gestos como de los pensamientos menos pensados que se quieren esconder detrás de las palabras, y que los busca sin piedad con unos espejuelos glaciales de entomólogo, que más parecen microscopios para escuchar. Como esa materia oral se convierte en la felicidad visible de nuestros domingos, es algo tan sublime y diabólico que sin duda tiene mucho que ver con el aparato digestivo de la poesía.

Quienes sólo lo conocen por su arte dicen que Osuna no tiene corazón. Yo creo que lo tiene, y muy grande, pero dotado de una química personal que sólo asimila a los justos. Y para Osuna no hay casi nadie que lo sea en esta vida. En ese sentido es una reliquia histórica: el último cristiano puro que nos queda. Su oficio dominical es inclemente, sin una debilidad simple, sin una grieta de lástima. Aunque se le considera como el caricaturista político más lúcido y feroz que ha tenido Colombia, su ferocidad es mucho más que política, porque es sólo moral. Carece del cálculo matrero, de las pasiones efímeras, de los apetitos terrestres de los políticos. Su negocio parece ser la salvación de las almas, y su única posición legítima, en consecuencia, sólo puede ser la de los cristianos primitivos, que en el circo romano se dejaban comer por los leones cantando plegarias de amor, porque estaban tan convencidos como Osuna de que en la lógica de Dios eran ellos quienes se estaban comiendo a los leones.

EL EMBROLLO DE LA PAZ

Colombia hizo el 7 de diciembre pasado, durante dos minutos, un plebiscito en favor de la paz interna. Tal como estaba previsto, cada quien hizo una pausa a las doce del día en el sitio en que se encontraba, y cada quien hizo dentro de ella lo que le pareció más adecuado para expresar su voluntad de paz. Unos hicieron un silencio de protesta, otros elevaron plegarias a sus dioses, otros echaron las campanas al vuelo, hicieron sonar las sirenas de las fábricas, las bocinas de los automóviles. La inmensa mayoría, en los lugares más remotos del país, izaron la bandera patria y se asomaron a las ventanas agitando pañuelos blancos. Fue un estremecimiento febril e inequívoco.

Pero fue al mismo tiempo una prueba de las posibilidades y los deseos de participación directa de todo un pueblo que carece de canales propios de expresión. No tenemos mecanismos de movilización de la opinión pública distintos de la prensa, la televisión y la radio, que fueron sin duda los factores decisivos de la jornada. En el origen de la idea estuvieron los dos grandes partidos políticos, que en tiempos de elecciones son capaces de rastrear sus votos hasta en las comarcas más remotas. Se suponía que ese mismo aparato electoral sería el más apropiado para canalizar las ansias de expresión de las vastas muchedumbres nacionales que tienen votos a la hora de votar, pero que no tienen voz cuando quieren hablar. Sin embargo, este miércoles histórico el país ha hablado casi por su cuenta, con voz propia, y cada quien como pudo. Pero sin la menor posibilidad de duda: «Queremos la paz».

La ocasión es propicia para preguntarse una vez más cuáles son los factores que impiden conseguirla. Desde mucho antes de que el presidente Belisario Betancur propusiera al congreso la ley de amnistía más amplia y completa de la historia del país, ya los principales grupos armados estaban empeñados en obtenerla. El M-19 la

Publicado originalmente el 14 de diciembre de 1983.

convirtió en una bandera de lucha, hasta el punto de que logró imponerla como uno de los temas centrales de la campaña presidencial. Las Fuerzas Armadas Revolucionarias de Colombia (FARC) llegaron al extremo sin precedentes de decretar una tregua de varios meses para no entorpecer el proceso electoral y facilitar de ese modo que sus propios simpatizantes se expresaran en las urnas. De modo que el estado de ánimo de los combatientes parecía ser el mejor para sumarse a los sueños de paz de los colombianos. Era justo: estábamos a punto de dejar atrás un gobierno cuyas señas de identidad eran la represión, la tortura institucional, la violación rutinaria de los derechos humanos, y enfrentadas a ese gobierno teníamos varias organizaciones armadas que se habían desviado por los desfiladeros tenebrosos e inadmisibles del terrorismo, cuya manifestación más inhumana era el secuestro por dinero.

En estas circunstancias la ley de amnistía pareció providencial a los millones de colombianos que la creíamos indispensable para abrir una época nueva. Sin embargo, las organizaciones armadas que tanto habían clamado por la amnistía no tuvieron para ella una respuesta política. Era como si en el fondo de su alma no hubieran creído que iban a conseguirla, y cuando esto ocurrió las tomó de sorpresa y no supieron a ciencia cierta qué hacer con ella. Surgió entonces, además, otro obstáculo inesperado y muy grave: los primeros guerrilleros que se acogieron a la amnistía eran amenazados de muerte o asesinados por desconocidos. En cierto modo, para muchos guerrilleros la paz se convirtió en un riesgo terrible, mucho más peligroso que la guerra; y no fue ni mucho menos el remanso de creatividad y justicia social con que todos soñábamos. Lo más raro, en todo caso, es que en medio de la confusión y el desencanto no faltan —no faltamos— quienes siguen creyendo, de un modo empecinado y tal vez ilusorio, que la paz es posible. Y la jornada del 7 de diciembre permite pensar que no somos tan pocos como podría creerse. Es la inmensa mayoría del país la que lo cree y lo desea, y esto debe entenderse como una notificación a quienes piensan lo contrario.

Uno se pregunta, con toda razón, dónde está el nudo gordiano. Los grupos armados y el gobierno están de acuerdo por lo menos en una frase: «La amnistía no es la paz, pero es el camino para lograrla». Sobre esa base, la comisión de paz ha sostenido conversaciones constantes con los grupos armados, y en especial con las FARC: unos veinte encuentros secretos en los últimos seis meses. El M-19, que se empeñó en no conversar con nadie menos que con el presidente de la república en persona, logró su propósito, que fue además una prueba de la modestia y la disposición asombrosa del presidente de la república por conseguir la paz. En esas negociaciones difíciles, de las cuales es tan poco lo que sabe la opinión pública, se ha llegado a un acuerdo positivo: la pretensión de los movimientos armados de que el ejército se retirara de los territorios ocupados —y que era una pretensión irreal, desde luego— se ha reducido a la proposición de algo que tanto el M-19 como las FARC llaman una tregua en las hostilidades. En un país de gramáticos y leguleyos como lo es el nuestro —y tal vez para fortuna nuestra—, la sola palabra ha dado origen a una serie casi infinita de especulaciones: ¿qué se entiende por tregua? Aunque parezca mentira, esa pregunta es en la actualidad el escollo más difícil para lograr una situación que tal vez sea la misma que nuestros abuelos, en las guerras civiles, llamaban un armisticio.

Por el ministro de la Defensa, general Fernando Landazábal, se supo en el Congreso que el ejército no acepta la tregua —cualquiera sea el significado de la palabra— y que tampoco está dispuesto a conversar con los grupos armados. Otros altos oficiales opinan en privado que la ley de amnistía no ha hecho sino infundir alientos nuevos a los insurrectos, y que lo que hace falta es un consenso político para emprender una gigantesca operación militar que ponga término de una vez por todas a la subversión. Para éstos, a diferencia de lo que dice el general Landazábal en sus libros y editoriales, las condiciones políticas y sociales donde se cultivan los fermentos de la violencia son poco menos que secundarias. Sin embargo, después de treinta años de guerrillas larvadas, también la solución militar simple parece irreal.

Algunos movimientos guerrilleros, por su parte, continúan en la práctica infame de los secuestros como recurso de financiación, a pesar de que hace meses prometieron poner en libertad a sus rehenes como una prueba pública de su voluntad de paz. En síntesis, por donde quiera que se enfoque, la situación de esta guerra civil embrollada termina en un círculo vicioso. Tal vez el grito de paz que lanzó todo el país el 7 de diciembre contribuya de algún modo a romperlo. Porque hasta ahora sólo una cosa es cierta: todo el mundo dice que quiere la paz, pero nadie sabe dónde encontrarla.

VUELTA A LA SEMILLA

Al contrario de lo que han hecho tantos escritores buenos y malos en todos los tiempos, nunca he idealizado el pueblo donde nací y donde crecí hasta los ocho años. Mis recuerdos de esa época —como tantas veces lo he dicho— son los más nítidos y reales que conservo, hasta el extremo de que puedo evocar como si hubiera sido ayer no sólo la apariencia de cada una de las casas que aún se conservan, sino incluso descubrir una grieta que no existía en un muro durante mi infancia. Los árboles de los pueblos suelen durar más que los seres humanos, y siempre he tenido la impresión de que también ellos nos recuerdan, tal vez mejor que como nosotros los recordamos a ellos.

Pensaba todo esto, y mucho más, mientras recorría las calles polvorientas y ardientes de Aracataca, el pueblo donde nací y donde volví hace algunos días después de 16 años de mi última visita. Un poco trastornado por el reencuentro con tantos amigos de la niñez, aturdido por un tropel de niños entre los cuales parecía reconocerme a mí mismo cuando llegaba el circo, tenía, sin embargo, bastante serenidad para sorprenderme de que nada había cambiado en la casa del general José Rosario Durán —donde ya, por supuesto, no queda nadie de su familia ilustre—; que debajo de los camellones con que han adornado las plazas, éstas siguen siendo las mismas, con su polvo sediento y sus almendros tristes como lo fueron siempre, y que la iglesia ha sido pintada y repintada muchas veces en medio siglo, pero el cuadrante del reloj de la torre es el mismo. «Y eso no es nada», me precisó alguien: «El hombre que lo arregla sigue también siendo el mismo».

Es mucho —yo diría que demasiado— lo que se ha escrito sobre las afinidades entre Macondo y Aracataca. La verdad es que cada vez que vuelvo al pueblo de la realidad encuentro que se parece menos al de la ficción, salvo algunos elementos externos, como su calor irresistible a las dos de la tarde, su polvo blanco y ardiente y los

Publicado originalmente el 21 de diciembre de 1983.

almendros que aún se conservan en algunos rincones de las calles. Hay una similitud geográfica que es evidente, pero que no llega mucho más lejos. Para mí hay más poesía en la historia de los animes que en toda la que he tratado de dejar en mis libros. La misma palabra —animes— es un misterio que me persigue desde aquellos tiempos. El diccionario de la Real Academia dice que el anime es una planta y su resina. De igual modo define esta voz, aunque con muchas más precisiones, el excelente lexicón de colombianismos de Mario Alario di Filippo. El padre Pedro María Revollo, en sus *Costeñismos colombianos*, ni siquiera la menciona. En cambio, Sundenheim, en su *Vocabulario costeño*, publicado en 1922 y al parecer olvidado para siempre, le consagra una nota muy amplia que transcribo en la parte que más nos interesa: «El anime, entre nosotros, es una especie de duende bienhechor que auxilia a sus protegidos en lances difíciles y apurados, y de ahí que cuando se afirme de alguien que tiene animes se dé a entender que cuenta con alguna persona o fuerza misteriosa que le ha prestado su concurso». Es decir, Sundenheim los identifica con los duendes, y de modo más preciso, con los descritos por Michelet.

Los animes de Aracataca eran otra cosa: unos seres minúsculos, de no más de una pulgada, que vivían en el fondo de las tinajas. A veces se les confundía con los gusarapos, que algunos llamaban sarapicos, y que eran en realidad las larvas de los mosquitos jugueteando en el fondo del agua de beber. Pero los buenos conocedores no los confundían: los animes tenían la facultad de escapar de su refugio natural, aun si la tinaja se tapaba con buen seguro, y se divertían haciendo toda clase de travesuras en la casa. No eran más que eso: espíritus traviesos, pero benévolos, que cortaban la leche, cambiaban el color de los ojos de los niños, oxidaban las cerraduras o causaban sueños enrevesados. Sin embargo, había épocas en que se les trastornaba el humor, por razones que nunca fueron comprensibles, y les daba por apedrear la casa donde vivían. Yo los conocí en la de don Antonio Daconte, un emigrado italiano que llevó grandes novedades a Aracataca: el cine mudo, el salón de billar, las bicicletas alquiladas, los gramófonos, los primeros receptores de radio. Una noche corrió la voz por todo el pueblo de que los animes estaban apedreando la casa de don Antonio Daconte, y todo el pueblo fue a verlo. Al contrario de lo que pudiera parecer, no era un espectáculo de horror, sino una fiesta jubilosa que de todos modos no dejó un vidrio intacto. No se veía quién las tiraba, pero las piedras surgían de todas partes y tenían la virtud mágica de no tropezar con nadie, sino de dirigirse hacia sus objetivos exactos: las cosas de cristal. Mucho tiempo después de aquella noche encantada, los niños seguíamos con la costumbre de meternos en la casa de don Antonio Daconte para destapar la tinaja del comedor y ver los animes –quietos y casi transparentes– aburriéndose en el fondo del agua.

Tal vez la casa más conocida del pueblo era una esquina como tantas otras, contigua a la de mis abuelos, que todo el mundo conocía como *la casa del muerto*. En ella vivió varios años el párroco que bautizó a toda nuestra generación. Francisco C. Angarita, que era famoso por sus tremendos sermones moralizadores. Eran muchas las cosas buenas y malas que se murmuraban del padre Angarita, cuyos raptos de cólera eran temibles; pero hace apenas unos años supe que había asumido una posición muy definida y consecuente durante la huelga y la matanza de los trabajadores del banano.

Muchas veces oí decir que *la casa del muerto* se llamaba así porque allí se veía deambular en la noche el fantasma de alguien que en una sesión de espiritismo dijo llamarse Alfonso Mora. El padre Angarita contaba el cuento con un realismo que erizaba la piel. Describía al aparecido como un hombre corpulento, con las mangas de la camisa enrolladas hasta los codos, y el cabello corto y apretado, y los dientes perfectos y luminosos como los de los negros. Todas las noches, al golpe de las doce, después de recorrer la casa, desaparecía debajo del árbol de totumo que crecía en el centro del patio. Los contornos del árbol, por supuesto, habían sido excavados muchas veces en busca de un tesoro enterrado. Un día, a pleno sol, pasé a la casa vecina de la nuestra persiguiendo un conejo, y traté de alcanzarlo en el excusado, donde se había escondido. Empujé la puerta, pero en vez del conejo vi al hombre acuclillado en la letrina, con el aire de tristeza pensativa que todos tenemos en esas circunstancias. Lo reconocí de inmediato, no sólo por las mangas enrolladas hasta los codos, sino por sus hermosos dientes de negro que alumbraban en la penumbra.

Éstas y muchas otras cosas recordaba hace unos días en aquel pueblo ardiente, mientras los viejos y los nuevos amigos, y los que apenas empezaban a serlo, parecían de veras alegres de que estuviéramos otra vez juntos después de tanto tiempo. Era el mismo manantial de poesía cuyo nombre de redoblante he oído resonar en medio mundo, en casi todos los idiomas, y que, sin embargo, parece existir más en la memoria que en la realidad. Es difícil imaginar otro lugar más olvidado, más abandonado, más apartado de los caminos de Dios. ¿Cómo no sentirse con el alma torcida por un sentimiento de revuelta?

VARIACIONES

Hace unos años entré en una inmensa tienda de discos de Los Ángeles y encontré todo el ámbito ocupado por una música que no parecía de este mundo. Flotando en aquella ciénaga de belleza que me era desconocida por completo, me acerqué casi en puntillas al dependiente encargado de alimentar la música del ambiente y le pregunté con el alma en un hilo qué disco era ése, tan parecido sin duda a los que se escuchaban los domingos en el cielo. Era *La creación*, de Haydn. La revelación fue para mí un golpe de gracia, pues desde muy joven, cuando la música se me convirtió en algo tan indispensable para vivir como la comida misma, había tratado de borrar a Haydn de mi pensamiento por una razón que ninguno de mis amigos melómanos me quería perdonar: lo consideraba uno de los pocos músicos que infunden la mala suerte. El otro —que todavía no ha podido demostrarme lo contrario— es Héctor Berlioz.

Tan arraigada es esa superstición, que el viaje más terrorífico que he hecho en avión fue uno de Barcelona a Nueva York, en un jumbo cuyo programa de música tenía como plato fuerte a *Harold en Italia*, de Berlioz. Yo no conocía la pieza, por supuesto, pues la mala sombra del gran músico francés empezó a inquietarme desde mucho antes de que escuchara algo suyo. En realidad, no empezó por su música, sino por la imagen que me formé de él cuando vi por primera vez la célebre caricatura en que aparece dirigiendo una orquesta, entre cuyos instrumentos hay un cañón de guerra. La idea le vino tal vez al caricaturista por la sonoridad cataclísmica que Berlioz quiso darle a su orquesta mediante el recurso, no siempre eficaz, de aumentar el número de instrumentos, hasta el extremo de que para la ejecución de su *Réquiem* se necesitan cuatro orquestas suplementarias de instrumentos de metal. La sola visión de aquel dibujo me infundió tal terror por la música de Ber-

Publicado originalmente el 18 de enero de 1984.

lioz, que es sin duda uno de los grandes creadores de todos los tiempos, como lo atestiguan tantos tratadistas que saben muy bien lo que dicen.

Por no conocerlo me dejé llevar por la melodía fragante que encontré en los auriculares durante aquel vuelo a Nueva York, y sólo cuando concluyó me enteré por el anunciador que era *Harold en Italia*. Lo había escuchado completo, y además con un gran deleite, y a partir de aquel momento no pude seguir oyendo música, sino que permanecí pendiente de los cambios en los mínimos ruidos del avión, en sus movimientos menos pensados, y hasta repasé de memoria las instrucciones para el caso de accidentes en el océano, instrucciones que tantas y tantas veces les hemos oído a los auxiliares de vuelo sin ponerles la menor atención. Todo parecía indicar que el maleficio de Berlioz estaba conjurado, pues el cielo era diáfano hasta el infinito y la nave enorme parecía suspendida en el aire como un magnífico hotel de tierra. Sin embargo, al aproximarnos a Nueva York, el comandante anunció que las condiciones del tiempo no permitían el aterrizaje inmediato, y debíamos volar en círculos sobre la ciudad hasta que fuera posible. La verdad es que dimos vueltas durante tres horas –además de las siete que ya habíamos volado desde Barcelona– y luego aterrizamos en Boston para reabastecernos de gasolina, y volvimos a dar vueltas sobre Nueva York durante otras cuatro horas. No éramos los únicos, desde luego: por la ventanilla veíamos los otros aviones que esperaban su turno para bajar, y nos preguntábamos de qué sutil azar dependía el que no tropezáramos todos como un colosal fichero de dominó. A mí no me cabía la menor duda de que aquel contratiempo inconcebible se lo debíamos al hechizo de Berlioz, y lo único que me preguntaba mientras seguíamos girando en el cielo era si la mala sombra no sería tan intensa como para impedirnos aterrizar sanos y salvos.

La superstición de Haydn, en cambio, venía de conocerlo bastante bien. Admiraba y sigo amando su música de cámara, pero me parecía que no podía ser benéfica su afición por ciertos trucos que no tenían nada que ver con su arte. No podía soportar que en la mitad de una sinfonía ordenara un golpe de timbales que tronaba como un cañonazo sólo para despertar a la audiencia dormida, que hubiera hecho otra sinfonía para ser ejecutada con instrumentos de juguetes de niños, y que para recordarle a su príncipe la penosa situación económica de sus músicos hubiera hecho otra sinfonía en que los ejecutantes apagaban la vela de sus atriles y se retiraban uno tras otro de la escena, hasta que la orquesta quedaba exhausta y la sala en tinieblas. Esas cosas que a la edad de hoy nos parecen simples tomaduras de pelo, a las cuales tiene todo el derecho un artista del tamaño de Haydn, parecían insoportables en la juventud. Para decirlo con un término venezolano de la más alta expresividad, parecían cosas pavosas. Es decir: que por ser tan feas llevaban consigo la mala suerte, como las plumas de pavo real en los floreros, como comer mondongo en copa o hacer el amor con las medias puestas. El hecho es que el descubrimiento casual del oratorio *La creación* arrasó de raíz con mi superstición contra Haydn, entre otras cosas, dicho sea de paso, porque la belleza en sus expresiones más altas es el conjuro más eficaz contra la mala suerte.

De todos modos, estas tendencias primarias suelen manifestarse en todos los medios y oficios, y lo que más me interesa no es hablar de ellas, sino de las relaciones que los amantes de la música sostienen con los compositores consagrados. Un amigo cuya cualidad más asombrosa es que detesta a Mozart ha dicho sin que le tiemble la voz: «Mozart no existe, porque cuando es malo es mejor oír a Haydn, y cuando es bueno es mejor oír a Beethoven». Hay quienes no quieren oír hablar de Rachmaninof porque les parece un cursi —y lo peor de todo: un cursi tardío— y en cambio hay otros aficionados muy respetables que lo consideran como uno de los grandes. Entre otras razones muy justas, porque su sensibilidad está a muy pocos centímetros de los boleros tropicales, entre cuyos fanáticos nos contamos muchos de los escritores —buenos y malos— de este lado del mundo y parte del otro.

Durante muchos años, el machismo latino había repudiado a Chopin con el argumento inevitable de que la suya era música para maricas. Aparte de que no hay ninguna prueba de que los maricas tengan peor gusto que quienes no lo son, hoy no parecen ser muchos quienes se atrevan a negar que Chopin es uno de los más grandes músicos de todos los tiempos. Tanto, que se le reconoce su grandeza a pesar de la orquestación deplorable —por decir lo menos— de sus dos conciertos para piano. Beethoven, con su creatividad inagotable, hubiera sido sin duda en estos tiempos uno de los autores más solicitados para hacer música para películas en Hollywood. Sin embargo, conozco a una señora muy inteligente y seria que lo repudió para siempre cuando supo que olía tan mal que en sus conciertos había que tener muy buen estómago para ocupar la primera fila. Brahms —que para mi gusto es uno de los más grandes— me merece todavía mucho mayor respeto por haber sido pianista en un burdel de Hamburgo. Tengo un amigo, fanático de Béla Bartók, que estuvo a punto de matar a alguien cuando dijo que su primer concierto para violín –que ahora pasó a ser el número dos– era en realidad un concierto para gato y orquesta. Ernest Chausson, por su parte, suscita una ternura muy honda, no sólo por el lirismo de su música, sino por el hecho triste de que murió atropellado por una bicicleta.

¿CÓMO SE ESCRIBE UNA NOVELA?

Ésta es, sin duda, una de las preguntas que se hacen con más frecuencia a un novelista. Según sea quien la haga, uno tiene siempre una respuesta de complacencia. Más aún: es útil tratar de contestarla, porque no sólo en la variedad está el placer, como se dice, sino que también en ella están las posibilidades de encontrar la verdad. Porque una cosa es cierta: creo que quienes más se hacen a sí mismos la pregunta de cómo se escribe una novela son los propios novelistas. Y también a nosotros mismos nos damos cada vez una respuesta distinta.

Me refiero, por supuesto, a los escritores que creen en que la literatura es un arte destinada a mejorar el mundo. Los otros, los que piensan que es un arte destinada a mejorar sus cuentas de banco, tienen fórmulas para escribir que no sólo son certeras, sino que pueden resolverse con tanta precisión como si fueran fórmulas matemáticas. Los editores lo saben. Uno de ellos se divertía hace poco explicándome cómo era de fácil que su casa editorial se ganara el Premio Nacional de Literatura. En primer término, había que hacer un análisis de los miembros del jurado, de su historia personal, de su obra, de sus gustos literarios. El editor pensaba que la suma de todos esos elementos terminaría por dar un promedio del gusto general del jurado. «Para eso están las computadoras», decía. Una vez establecido cuál era la clase de libro que tenía mayores posibilidades de ser premiado, había que proceder con un método contrario al que suele utilizar la vida: en vez de buscar dónde estaba ese libro, había que investigar cuál era el escritor, bueno o malo, que estuviera mejor dotado para fabricarlo. Todo lo demás era cuestión de firmarle un contrato para que se sentara a escribir sobre medida el libro que recibiría el año siguiente el Premio Nacional de Literatura. Lo alarmante es que el editor había sometido este juego al molino de las computadoras, y éstas le habían dado una posibilidad de acierto de un ochenta y seis por ciento.

Publicado originalmente el 25 de enero de 1984.

De modo que el problema no es escribir una novela –o un cuento corto–, sino escribirla en serio, aunque después no se venda ni gane ningún premio. Ésa es la respuesta que no existe, y si alguien tiene razones para saberlo en estos días es el mismo que está escribiendo esta columna con el propósito recóndito de encontrar su propia solución al enigma. Pues he vuelto a mi estudio de México, donde hace un año justo dejé varios cuentos inconclusos y una novela empezada, y me siento como si no encontrara el cabo para desenrollar el ovillo. Con los cuentos no hubo problemas: están en el cajón de la basura. Después de leerlos con la saludable distancia de un año, me atrevo a jurar –y tal vez sería cierto– que no fui yo quien los escribió. Formaban parte de un viejo proyecto de sesenta o más cuentos sobre la vida de los latinoamericanos en Europa, y su principal defecto era el fundamental para romperlos: ni yo mismo me los creía.

No tendré la soberbia de decir que no me tembló la mano al hacerlos trizas y luego dispersar las serpentinas para impedir que fueran reconstruidos. Me tembló, y no sólo la mano, pues en esto de romper papeles tengo un recuerdo que podría parecer alentador pero que a mí me resulta deprimente. Es un recuerdo que se remonta a una noche de julio de 1955, a la víspera del viaje a Europa del enviado especial de *El Espectador*, cuando el poeta Jorge Gaitán Durán llegó a mi cuarto de Bogotá a pedirme que le dejara algo para publicar en la revista *Mito*. Yo acababa de revisar mis papeles, había puesto a buen seguro los que creía dignos de ser conservados y había roto los desahuciados. Gaitán Durán, con esa voracidad insaciable que sentía ante la literatura, y sobre todo ante la posibilidad de descubrir valores ocultos, empezó a revisar en el canasto los papeles rotos, y de pronto encontró algo que le llamó la atención. «Pero esto es muy publicable», me dijo. Yo le expliqué por qué lo había tirado: era un capítulo entero que había sacado de mi primera novela, *La hojarasca* –ya publicada en aquel momento–, y no podía tener otro destino honesto que el canasto de la basura. Gaitán Durán no estuvo de acuerdo. Le parecía que en realidad el texto hubiera sobrado dentro de la novela, pero que tenía un valor diferente por sí mismo. Más por tratar de complacerlo que por estar convencido, le autoricé para que remendara las hojas rotas con cinta pegante y publicara el capítulo como si fuera un cuento. «¿Qué títulos le ponemos?», me preguntó, usando un plural que muy pocas veces había sido tan justo como en aquel caso. «No sé», le dije. «Porque eso no era más que un monólogo de Isabel viendo llover en Macondo.» Gaitán Durán escribió en el margen superior de la primera hoja casi al mismo tiempo que yo lo decía: «Monólogo de Isabel viendo llover en Macondo». Así se recuperó de la basura uno de mis cuentos que ha recibido los mejores elogios de la crítica y, sobre todo, de los lectores. Sin embargo, esa experiencia no me sirvió para no seguir rompiendo los originales que no me parecen publicables, sino que me enseñó que es necesario romperlos de tal modo que no se puedan remendar nunca.

Romper los cuentos es algo irremediable, porque escribirlos es como vaciar concreto. En cambio, escribir una novela es como pegar ladrillos. Quiere esto decir que si un cuento no fragua en la primera tentativa es mejor no insistir. Una novela es más fácil: se vuelve a empezar. Esto es lo que ha ocurrido ahora. Ni el tono, ni el estilo, ni el carácter de los personajes eran los adecuados para la novela que había dejado a medias. Pero aquí también la explicación es una sola, ni yo mismo me la creía. Tratando de encontrar la solución volví a leer dos libros que suponía útiles. El primero fue *La educación sentimental*, de Flaubert, que no leía desde los remotos insomnios de la universidad, y sólo me sirvió ahora para eludir algunas analogías que hubieran resultado sospechosas. Pero no me resolvió el problema. El otro libro que volví a leer fue *La casa de las bellas durmientes*, de Yasunari Kawabata, que me había golpeado en el alma hace unos tres años y que sigue siendo un libro hermoso. Pero esta vez no me sirvió de nada, porque yo andaba buscando pistas sobre el comportamiento sexual de los ancianos, pero el que encontré en el libro es el de los ancianos japoneses, que al parecer es tan raro como todo lo japonés, y desde luego no tiene nada que ver con el comportamiento sexual de los ancianos caribes. Cuando conté mis preocupaciones en la mesa, uno de mis hijos —el que tiene más sentido práctico— me dijo: «Espera unos años más y lo averiguarás por tu propia experiencia». Pero el otro, que es artista, fue más concreto: «Vuelve a leer *Los sufrimientos del joven Werther*», me dijo, sin el menor rastro de burla en la voz. Lo intenté, en efecto, no sólo porque soy un padre muy obediente, sino porque de veras pensé que la famosa novela de Goethe podía serme útil. Pero la verdad es que en esta ocasión no terminé llorando en su entierro miserable, como me ocurrió la primera vez, sino que no logré pasar de la octava carta, que es aquella en que el joven atribulado le cuenta a su amigo Guillermo cómo empieza a sentirse feliz en su cabaña solitaria. En este punto me encuentro, de modo que no es raro que tenga que morderme la lengua para no preguntar a todo el que me encuentro: «Dime una cosa, hermano: ¿cómo carajo se escribe una novela?».

Auxilio

Alguna vez leí un libro, o vi una película, o alguien me contó un hecho real, con el siguiente argumento: un oficial de marina metió de contrabando a su amada en el camarote de un barco de guerra, y vivieron un amor desaforado dentro de aquel recinto opresivo, sin que nadie los descubriera durante varios años. A quien sepa quién es el autor de esta bellísima historia le ruego que me lo haga saber de urgencia, pues lo he preguntado a tantos y tantos que no lo saben, que ya empiezo a sospechar que a lo mejor se me ocurrió a mí alguna vez y ya no lo recuerdo. Gracias.

Cadaqués no es sólo uno de los pueblos más bellos de la Costa Brava —en Cataluña—, sino también uno de los mejor conservados. Esto se debe en gran parte a que la carretera que comunica con la autopista es una serpentina estrecha y retorcida: una cornisa abismal sin pavimento, donde se necesita tener el alma muy bien puesta en su almario para conducir a más de 50 kilómetros por hora. Sus casas son blancas y bajas, de acuerdo con el estilo tradicional de las aldeas de pescadores del Mediterráneo, y las casas nuevas, construidas por arquitectos de renombre, no han roto la armonía del conjunto, como ha ocurrido en la casi totalidad de los otros pueblos de esa orilla hasta la punta de Cádiz. En verano, cuando el calor parece venir de los desiertos africanos de la acera de enfrente, Cadaqués se convierte en una torre de Babel infernal, con turistas procedentes de toda Europa, que le disputan por tres meses su paraíso a los nativos y a los forasteros que tuvieron la suerte de comprar una casa a buen precio cuando todavía era posible. Sin embargo, en primavera y otoño —que es la época en que Cadaqués resulta más apetecible— un fantasma amenaza a la población: la tramontana, un viento despiadado y tenaz que, según piensan algunos nativos, lleva consigo los gérmenes de la locura.

Yo también lo creo. Hace unos 15 años yo era uno de los visitantes más entusiastas de Cadaqués. Ahora hay una autopista a la altura de las mejores de Europa, que continúa sin interrupción hasta París. Pero en aquella época la carretera a Francia era estrecha y difícil y había que contar unas cuatro horas desde Barcelona hasta la población de Rosas. Allí se toma a la derecha el ramal para Cadaqués que, por fortuna para este lugar inolvidable, sigue siendo tan primitivo y peligroso como siempre. El viaje, para mi familia, tenía un atractivo adicional. En Rosas, o en Le Perthus, ya del lado francés, nuestro lamentado amigo Juanito Durán tenía sendos restau-

Publicado originalmente el 1 de febrero de 1984.

rantes donde siempre nos sorprendía con dos especialidades que siempre me parecieron dos disparates geniales de la cocina catalana: el pollo con langosta y el conejo con caracoles. La primera vez que oí hablar de esos dos platos me parecieron conjunciones incompatibles, agua y aceite. En teoría parecían una imposibilidad metafísica. En la práctica son dos hallazgos que sólo se les podía ocurrir a inventores lunáticos, como lo son los catalanes, y desde que los gustamos por primera vez tuvimos un segundo motivo para ir a Cadaqués durante los fines de semana. Poco después surgió un tercer motivo: el cine en Perpiñán. Los españoles, aun después de que el franquismo entró en barrena, seguían viendo películas inocuas, cortadas a criterio del censor y con un recurso que sólo se les podía ocurrir a las mentes más retrógadas: los censores aprovechaban el doblaje de las películas extranjeras para convertir a los amantes en hermanos, mediante cambios en los diálogos, aunque después resultaba todo aquello más inmoral y disparatado, pues se veía a las claras que los supuestos hermanos mantenían relaciones de cama y que, a veces, tenían hijos comunes. De modo que el buen cine había que verlo en Perpiñán, en cuyas salas muchas películas se sostenían más tiempo que en París gracias a la clientela española. La apoteosis de aquellas excursiones cinematográficas, que a veces se convertían en aventuras más emocionantes que las propias películas, fue *El último tango en París*, de Bernardo Bertolucci. A las agencias de turismo de Barcelona se les ocurrió hacer programas completos a precio fijo, en el cual se incluía el valor del viaje de ida y regreso, la comida en Perpiñán y el boleto para *El último tango en París*. A veces había embotellamientos interminables en la frontera, de automovilistas ansiosos de comprobar con sus propios ojos qué era lo que Marlon Brando le hacía a la Maria Schneider con media libra de mantequilla. Las películas de Perpiñán, sumadas al pollo con langosta de Juanito Durán y a las tertulias de amigos en el bar El Marítim, de Cadaqués, daban como resultado unos fines de semana inolvidables.

Todo iba muy bien hasta que apareció la tramontana en nuestras vidas. Es un fenómeno que se presiente de pronto, sin ninguna explicación racional; uno siente que se le baja el ánimo, que se entristece sin motivo y que los amigos más amados asumen una expresión hostil. Luego empieza a escucharse un silbido que se va haciendo cada vez más agudo, más intenso, y uno empieza a cambiar de emisora en la radio, creyendo que se trata de una interferencia. Por último, el viento empieza a soplar en ráfagas espaciadas, que se van haciendo cada vez más frecuentes, hasta que llega una que se queda para siempre, sin un alivio, sin una pausa, con una intensidad y una perseverancia que tienen algo de sobrenatural. Al principio uno cree que no es más que un viento como tantos, e intenta inclusive salir a la calle para reconocerlo. Nosotros lo hicimos la primera vez, de puro inocentes, y en la primera esquina tuvimos que abrazarnos como náufragos para no ser arrastrados hasta el mar por la potencia del viento. Entonces nos dimos cuenta de que no nos quedaba más recurso que permanecer encerrados en el cuarto, con las puertas y las ventanas aseguradas por dentro, como en los ciclones del Caribe, hasta que Dios quisiera que pasara la tramontana. Y nadie tiene nunca la menor idea de cuándo Dios lo va a querer.

Al cabo de 24 horas, uno tiene la impresión de que aquel viento pavoroso no es un fenómeno meteorológico, sino un asunto personal: es algo que alguien está haciendo contra uno, y sólo contra uno. Por lo general, aquel tormento dura tres días, y uno experimenta un alivio que sólo puede compararse con una resurrección. Cuando cesa de pronto se siente demasiado el silencio, y el mar parece un remanso bajo el cielo transparente. Pero no es extraño que se repita a los pocos días, como sucedió en aquellos que tuvimos por última vez en Cadaqués. Y entonces no duró 72 horas, sino que se prolongó sin clemencia durante una semana. Cuando terminó, un portero anciano de una casa cercana a la nuestra se había colgado con una cuerda en un poste del alumbrado público, tal vez enloquecido por el delirio alucinante de la tramontana. No sin dolor, y con un sentimiento de nostalgia anticipada, salí aquella vez del pueblo con la decisión irrevocable de no volver jamás. Pensaba en García Lorca, y lo entendí en carne propia: «Aunque sepa los caminos, yo nunca llegaré a Córdoba, porque la muerte me espera entre los muros de Córdoba».

Años después de que tomé la decisión de no volver, un amigo me contó la historia de alguien que había tomado la misma decisión después de vivir la terrible experiencia de la tramontana. Sólo que su temor iba más lejos: estaba convencido de que si volvía a Cadaqués, con tramontana o sin ella, no volvería a salir con vida. Cometió el error de contarlo en una fiesta de locos en Barcelona, y a la medianoche, al calor de los duros vinos catalanes que siembran en el corazón tantas ideas desaforadas, sus amigos decidieron llevarlo a Cadaqués a la fuerza para conjurar de una vez por todas su tonta superstición. A pesar de su resistencia, lo metieron en un coche de borrachos y emprendieron a esa hora el largo viaje hacia Cadaqués. No: el coche no se precipitó en uno de los tantos abismos del último tramo, como hubiera podido ocurrir en el desenlace de un cuento malo. Lo que sucedió fue que el amigo, aterrorizado ante la proximidad de una muerte que creía segura, aprovechó un descuido de sus compañeros y se lanzó del coche en marcha. El cuerpo sólo fue rescatado al día siguiente, en una hondonada profunda que se encuentra en la última curva de la carretera.

EL ARGENTINO QUE SE HIZO QUERER DE TODOS

Fui a Praga por última vez hace unos quince años, con Carlos Fuentes y Julio Cortázar. Viajábamos en tren desde París porque los tres éramos solidarios en nuestro miedo al avión, y habíamos hablado de todo mientras atravesábamos la noche dividida de las Alemanias, sus océanos de remolacha, sus inmensas fábricas de todo, sus estragos de guerras atroces y amores desaforados. A la hora de dormir, a Carlos Fuentes se le ocurrió preguntarle a Cortázar cómo y en qué momento y por iniciativa de quién se había introducido el piano en la orquesta de jazz. La pregunta era casual y no pretendía conocer nada más que una fecha y un nombre, pero la respuesta fue una cátedra deslumbrante que se prolongó hasta el amanecer, entre enormes vasos de cerveza y salchichas con papas heladas. Cortázar, que sabía medir muy bien sus palabras, nos hizo una recomposición histórica y estética con una versación y una sencillez apenas creíbles, que culminó con las primeras luces en una apología homérica de Thelonius Monk. No sólo hablaba con una profunda voz de órgano de erres arrastradas, sino también con sus manos de huesos grandes como no recuerdo otras más expresivas. Ni Carlos Fuentes ni yo olvidaríamos jamás el asombro de aquella noche irrepetible.

Doce años después vi a Julio Cortázar enfrentado a una muchedumbre en un parque de Managua, sin más armas que su voz hermosa y un cuento suyo de los más difíciles: *La noche de Mantequilla Nápoles*. Es la historia de un boxeador en desgracia contada por él mismo en lunfardo, el dialecto de los bajos fondos de Buenos Aires, cuya comprensión nos estaría vedada por completo al resto de los mortales si no la hubiéramos vislumbrado a través de tantos tangos malevos. Sin embargo, fue ése el cuento que el propio Cortázar escogió para leerlo en una tarima frente a la muchedumbre de un vasto jardín iluminado, entre la cual había de todo, desde poetas consagrados y albañiles cesantes, hasta comandantes de la revolución

Publicado originalmente el 22 de febrero de 1984.

y sus contrarios. Fue otra experiencia deslumbrante. Aunque en rigor no era fácil seguir el sentido del relato, aun para los más entrenados en la jerga lunfarda, uno sentía y le dolían los golpes que recibía Mantequilla Nápoles en la soledad del cuadrilátero, y daban ganas de llorar por sus ilusiones y su miseria, pues Cortázar había logrado una comunicación tan entrañable con su auditorio que ya no le importaba a nadie lo que querían decir o no decir las palabras, sino que la muchedumbre sentada en la hierba parecía levitar en estado de gracia por el hechizo de una voz que no parecía de este mundo.

Estos dos recuerdos de Cortázar que tanto me afectaron me parecen también los que mejor lo definían. Eran los dos extremos de su personalidad. En privado, como en el tren de Praga, lograba seducir por su elocuencia, por su erudición viva, por su memoria milimétrica, por su humor peligroso, por todo lo que hizo de él un intelectual de los grandes en el buen sentido de otros tiempos. En público, a pesar de su reticencia a convertirse en un espectáculo, fascinaba al auditorio con una presencia ineludible que tenía algo de sobrenatural, al mismo tiempo tierna y extraña. En ambos casos fue el ser humano más impresionante que he tenido la suerte de conocer.

Desde el primer momento, a fines del otoño triste de 1956, en un café de París con nombre inglés, adonde él solía ir de vez en cuando a escribir en una mesa del rincón, como Jean-Paul Sartre lo hacía a trescientos metros de allí, en un cuaderno de escolar y con una pluma fuente de tinta legítima que manchaba los dedos. Yo había leído *Bestiario*, su primer libro de cuentos, en un hotel de lance de Barranquilla donde dormía por un peso con cincuenta centavos, entre peloteros mal pagados y putas felices, y desde la primera página me di cuenta de que aquél era un escritor como el que yo hubiera querido ser cuando fuera grande. Alguien me dijo en París que él escribía en el café Old Navy, del Boulevard Saint Germain, y allí lo esperé varias semanas, hasta que lo vi entrar como una aparición. Era el hombre más alto que se podía imaginar, con una cara de niño perverso dentro de un interminable abrigo negro que más bien parecía la sotana de un viudo, y tenía los ojos muy separados, como los de un novillo, y tan oblicuos y diáfanos que habrían podido ser los del diablo si no hubieran estado sometidos al dominio del corazón.

Años después, cuando ya éramos amigos, creí volver a verlo como lo vi aquel día, pues me parece que se recreó a sí mismo en uno de sus cuentos mejor acabados —*El otro cielo*—, en el personaje de un latinoamericano sin nombre que asistía de puro curioso a las ejecuciones en la guillotina. Como si lo hubiera hecho frente a un espejo,

Cortázar lo describió así: «Tenía una expresión distante y a la vez curiosamente fija, la cara de alguien que se ha inmovilizado en un momento de su sueño y rehúsa dar el paso que lo devolverá a la vigilia». Su personaje andaba envuelto en una hopalanda negra y larga, como el abrigo del propio Cortázar cuando lo vi por primera vez, pero el narrador no se atrevía a acercársele para preguntarle su origen, por temor a la fría cólera con que él mismo hubiera recibido una interpelación semejante. Lo raro es que yo tampoco me había atrevido a acercarme a Cortázar aquella tarde del Old Navy, y por el mismo temor. Lo vi escribir durante más de una hora, sin una pausa para pensar, sin tomar nada más que medio vaso de agua mineral, hasta que empezó a oscurecer en la calle y guardó la pluma en el bolsillo y salió con el cuaderno debajo del brazo como el escolar más alto y más flaco del mundo. En las muchas veces que nos vimos años después, lo único que había cambiado en él era la barba densa y oscura, pues hasta hace apenas dos semanas parecía cierta la leyenda de que era inmortal, porque nunca había dejado de crecer y se mantuvo siempre en la misma edad con que había nacido. Nunca me atreví a preguntarle si era verdad, como tampoco le conté que en el otoño triste de 1956 lo había visto, sin atreverme a decirle nada, en su rincón del Old Navy, y sé que dondequiera que esté ahora estará mentándome la madre por mi timidez.

Los ídolos infunden respeto, admiración, cariño y, por supuesto, grandes envidias. Cortázar inspiraba todos esos sentimientos como muy pocos escritores, pero inspiraba además otro menos frecuente: la devoción. Fue, tal vez sin proponérselo, el argentino que se hizo querer de todo el mundo. Sin embargo, me atrevo a pensar que si los muertos se mueren, Cortázar debe estarse muriendo otra vez de vergüenza por la consternación mundial que ha causado su muerte. Nadie le temía más que él, ni en la vida real ni en los libros, a los honores póstumos y a los fastos funerarios. Más aún: siempre pensé que la muerte misma le parecía indecente. En alguna parte de *La vuelta al día en ochenta mundos* un grupo de amigos no puede soportar la risa ante la evidencia de que un amigo común ha incurrido en la ridiculez de morirse. Por eso, porque lo conocí y lo quise tanto, me resisto a participar en los lamentos y elegías por Julio Cortázar. Prefiero seguir pensando en él como sin duda él lo quería, con el júbilo inmenso de que haya existido, con la alegría entrañable de haberlo conocido, y la gratitud de que nos haya dejado para el mundo una obra tal vez inconclusa pero tan bella e indestructible como su recuerdo.

LAS TRAMPAS A LA FE

El libro estaba en la primera línea de las novedades editoriales, entre memorias falsas de artistas de cine y manuales para adelgazar, con un excelente retrato a pluma en la portada y el título inequívoco en el borde superior: «Julio Cortázar: *La isla final*». Era irresistible. El escritor había muerto pocas semanas antes, las librerías estaban tapizadas de sus obras, reimpresas de urgencia para aprovechar la publicidad de la muerte, y quienes seguimos de cerca las noticias de la literatura sabíamos que había dejado dos libros sin publicar. Uno de ellos, *Los autonautas de la cosmopista*, escrito también hace año y medio, acababa de ser publicado por la editorial mexicana Nueva Imagen. El otro era un libro que la propia Carol Dunlop estaba escribiendo sobre Nicaragua cuando la sorprendió la muerte, y que Cortázar tuvo tiempo de terminar. No había ninguna sospecha de que éste hubiera dejado inédito un tercer libro, pero estaba a la vista sin ninguna duda, con su retrato inconfundible, y aquel título premonitorio que parecía una manera muy poética de llamar a la muerte: *La isla final*. Antes de comprarlo quise darle una hojeada, pero estaba envuelto en papel de plástico, y el empleado, que me vigilaba para que no pudiera robármelo, me advirtió que si rompía la envoltura tenía que comprar el libro de todos modos. Como eran las diez de la noche y no me sentía capaz de resistir los deseos de leer el último libro de Julio Cortázar, caí en la trampa con la más pura inocencia.

Sólo lo supe, por supuesto, cuando ya metido en la cama y listo para una travesía fantástica, vi que en letras muy pequeñas, casi invisibles en negro sobre azul, estaban los nombres de los compiladores. Abrí en las páginas del índice, y entonces entendí por qué el libro era el único en la mesa de novedades que estaba envuelto en un papel de plástico prohibido de romper. En efecto, era de muchos autores, muy buenos por cierto, pero sólo incluía dos artículos cono-

Publicado originalmente el 14 de marzo de 1984.

cidos de Julio Cortázar. Lo demás eran también artículos conocidos, uno de Fico, traductor al inglés, y otro de Joaquín Marcos, tan serio e interesante como todos los suyos. Los derechos de autor, por un enredo que no entendí muy bien, eran de la Oklahoma University Press desde 1978. Cuesta mucho trabajo creer que tantas trampas bien planteadas para inducir al lector a comprar un libro que no era el que parecía ser, eran una casualidad pura. Yo creo que el mismo diseño de la portada, la astucia del título en el momento actual y la envoltura inviolable fueron muy bien pensados por la editorial Ultramar, de Barcelona, para que aquella isla final pareciera ser, sin serlo, el último libro de Julio Cortázar.

Lo creo porque hay antecedentes que pasaron muy cerca de mí. Hace unos años encontré en una mesa de novedades un libro firmado por mí, cuya única falla era que yo no recordaba haberlo escrito. Se llamaba *La batalla de Nicaragua*, y la responsable de la edición era una empresa seria –Bruguera–, que además es editora de casi todos mis libros en España. Era el mismo truco: el libro incluía un estudio largo y muy bien documentado de Gregorio Selser, un ensayo del uruguayo Daniel Waksman Schinca y una denuncia de Ernesto Cardenal. Lo único mío era lo menos interesante y original dentro del conjunto: el reportaje sobre la toma del Palacio Nacional de Managua por un comando guerrillero dirigido por Edén Pastora. El texto había sido publicado como primicia mundial por la revista *Alternativa*, de Bogotá, unos tres años antes, y reproducido en varios países y en distintos idiomas. Mi agente literario había autorizado la inclusión en aquel libro, pero nadie pensó que iba a ser usado para destacar mi nombre en la portada muy por encima de los otros autores, y para anunciarlo con gran bombo en la radio y la televisión, aunque parezca mentira, como «*el último* libro de Gabriel García Márquez». Aquél parecía ser el destino de ese reportaje, pues poco antes me había visto obligado a pedir el retiro de otro libro publicado sin mi autorización por la editorial La Oveja Negra, de Bogotá, con el título de *Los sandinistas*, que no sólo se vendía por todas partes, sino que estaba incluido en el paquete de mis obras completas. Lo único mío era el manoseado reportaje del asalto al Palacio Nacional. También la sección mexicana de la Editorial Bruguera –que era la responsable del desaguisado– se comprometió a diseñar otra portada de *La batalla de Nicaragua*, que no sólo evitara cualquier confusión sino que hiciera justicia a los otros autores, colocándolos a todos con el mismo valor tipográfico y en orden alfabético. Todos aprobamos de común acuerdo la nueva presentación. Sin embargo, todavía se encuentra en las librerías de medio mundo la edición que se suponía recogida, y no recuerdo haber visto ningún ejemplar con la portada nueva.

Un incidente semejante ocurrió con *El olor de la guayaba*, la extensa entrevista que me hizo mi compadre Plinio Apuleyo Mendoza. Con la experiencia que ya tenía en estas artimañas de editores, pedí las pruebas de la portada antes del lanzamiento del libro. En efecto, sobre una foto mía a todo color y en carátula entera, decía con letras espectaculares: «Gabriel García Márquez, *El olor de la guayaba*». En la orilla inferior, de un modo casi inadvertido, decía: «Entrevista con Plinio Apuleyo Mendoza». Me pareció en primer término que aquella jerarquización astuta era una falta de respeto al estupendo y arduo trabajo de Plinio. Me pareció, en segundo término, que no se sabía al fin y al cabo quién entrevistaba a quién. Pero me pareció que más grave era la intención evidente de vender el libro como si fuera una novela mía. La portada se corrigió a tiempo en castellano, pero no fue posible en la edición francesa. El argumento del editor en Francia era muy propio de ese país: el libro estaba incluido en una colección consagrada de entrevistas con escritores, y al parecer el lector francés es demasiado inteligente para no saberlo.

Podría seguir con muchos ejemplos de libros que se compran porque ofrecen por fuera una cosa que no tienen por dentro. Uno de ellos es un excelente estudio de Robert Sklar sobre Francis Scott Fitzgerald titulado *El último Laoconte*. Lo compré y lo leí con gran placer porque es muy bueno, y ya tenía sobre él las mejores referencias. Pero viendo la portada de Barral Editores y viendo la forma en que están colocados los dos nombres y el título, me pregunto si un comprador desprevenido sabría a ciencia cierta quién es quién, de qué y sobre quién, y quién es al fin y al cabo *El último Laoconte*.

Otro ejemplo, de la mayor actualidad, es el que ha vuelto a plantear en México por estos días el conflicto ya histórico entre editores y autores. Se trata del libro titulado *Para cuando yo me ausente*, que ha sido publicado por la Editorial Grijalbo como una compilación de textos sobre Juan Rulfo, y hecha por él mismo. Juan Rulfo ha dicho en público que no es cierto que sea el padre de la criatura, que no es verdad que el título sea suyo ni haya tenido nada que ver con nada que se parezca tanto a un testimonio. Los editores, de su lado, dicen que sí, y no parece que haya ningún documento firmado para demostrar quién tiene la razón. Los escritores, por supuesto, no necesitamos de papeles para creerle a Juan Rulfo, de quien ya hemos creído y admirado tantos relatos increíbles. Lo alarmante es que mientras se calentaba la controversia, muchos lectores suyos nos precipitamos a las librerías, creyendo que se trataba de su nueva novela. No la encontramos, por desgracia. Pero en cambio encontramos un congreso mundial de editores reunido este fin de semana en Ciudad de México, que tal vez esté dispuesto a discutir en familia las trampas que se hacen a la fe de los lectores tratando de venderles libros que parecen por fuera lo que no son por dentro.

ÍNDICE ALFABÉTICO

Esta obra, publicada por
MONDADORI,
se terminó de imprimir en los talleres
de Artes Gráficas Huertas, S.A., de Madrid,
el día 21 de octubre
de 1999